AF335607

physica status solidi c

www.physica-status-solidi.com

conferences and critical reviews

Editor-in-Chief

Martin Stutzmann, Garching

Regional Editors

Martin S. Brandt, Garching
Peter Deák, Budapest
José Roberto Leite, Saõ Paulo
John I. B. Wilson, Edinburgh

Managing Editor

Stefan Hildebrandt, Berlin

Priority Programme of the Deutsche Forschungsgemeinschaft
Group III-Nitrides and Their Heterostructures:
Growth, Characterization and Applications

Guest Editors

Friedhelm Bechstedt, Bruno K. Meyer, and Martin Stutzmann

0 · 6 · 2003

WILEY-VCH

physica status solidi (c) – conferences and critical reviews

Editor-in-Chief: Martin Stutzmann

Managing Editor: Stefan Hildebrandt

Production Editors: André Danelius, Heike Höpcke, Irina Juschak

Editorial Assistance: Katharina Fröhlich, Margit Schütz

Editorial Office: physica status solidi
Bühringstr. 10, 13086 Berlin, Germany
Telephone: +49 (0) 30/47 03 13 31, Fax +49 (0) 30/47 03 13 34
e-mail: pss@wiley-vch.de

Publishers: WILEY-VCH Verlag GmbH & Co. KGaA

Postal Address: Bühringstr. 10, 13086 Berlin, Germany

Publishing Director: Alexander Grossmann

Ordering: Subscription Service, WILEY-VCH Verlag GmbH & Co. KGaA
Postfach 10 11 61, 69451 Weinheim, Germany
Telephone +49 (0) 62 01/60 64 00, Fax +49 (0) 62 01/60 61 84
e-mail: subservice@wiley-vch.de
or through a bookseller

Printing House: Druckhaus Thomas Müntzer GmbH, Bad Langensalza, Germany
Printed on chlorine- and acid free paper.

physica status solidi (c) – conferences and critical reviews is published several times per year by WILEY-VCH Verlag GmbH & Co. KGaA.

Subscription:

Volume **0** (2002/2003) of physica status solidi (c) – conferences and critical reviews is available online at Wiley InterScience (www.interscience.wiley.com). Please register for a free trial access.

Single print issues may be ordered by ISBN at www.wiley-vch.de or through your local bookseller.

Regular print and online subscriptions will be offered from 2004 in connection with subscriptions to physica status solidi (a) and (b).

ISSN 1610-1634

ISBN 3-527-40475-9

Contents

Visit our homepage on: **http://www.physica-status-solidi.com**
Full text on: **http://www.interscience.wiley.com**

Note from the Publisher: This issue of physica status solidi (c) is the first which has been produced from publication-ready manuscript files, written by the authors using the provided Word or LaTeX templates.

Review Articles

Group III-Nitrides and Their Heterostructures: Growth, Characterization and Applications

Preface 1569

Deposition and structural properties

Growth of GaN quasi-substrates by hydride vapor phase epitaxy
Wei Zhang and B. K. Meyer. (c) 1571–1582

Metalorganic chemical vapor phase epitaxy of gallium-nitride on silicon
A. Dadgar, A. Strittmatter, J. Bläsing, M. Poschenrieder, O. Contreras, P. Veit, T. Riemann,
F. Bertram, A. Reiher, A. Krtschil, A. Diez, T. Hempel, T. Finger, A. Kasic, M. Schubert,
D. Bimberg, F. A. Ponce, J. Christen, and A. Krost . . . . (c) 1583–1606

Molecular beam epitaxy of cubic III-nitrides on GaAs substrates
D. J. As, D. Schikora, and K. Lischka. . . . . (c) 1607–1626

Freestanding GaN-substrates and devices
Claudio R. Miskys, Michael K. Kelly, Oliver Ambacher, and Martin Stutzmann . . . . . (c) 1627–1650

Surfactants and antisurfactants on group-III-nitride surfaces
J. Neugebauer . . . . (c) 1651–1667

Indium distribution in epitaxially grown InGaN layers analyzed by transmission electron microscopy
D. Gerthsen, E. Hahn, B. Neubauer, V. Potin, A. Rosenauer, and M. Schowalter . . . . . (c) 1668–1683

A theoretical investigation of dislocations in cubic and hexagonal gallium nitride
A. T. Blumenau, C. J. Fall, J. Elsner, R. Jones, M. I. Heggie, and T. Frauenheim . . . . . (c) 1684–1709

Lattice dynamics in GaN and AlN probed with first- and second-order Raman spectroscopy
U. Haboeck, H. Siegle, A. Hoffmann, and C. Thomsen . . . . (c) 1710–1731

Electronic and optical properties

Electronic and vibrational properties of group-III nitrides: *Ab initio* studies
F. Bechstedt, J. Furthmüller, and J.-M. Wagner . . . . (c) 1732–1749

Phonons and free-carrier properties of binary, ternary, and quaternary group-III nitride layers measured by Infrared Spectroscopic Ellipsometry
A. Kasic, M. Schubert, J. Off, B. Kuhn, F. Scholz, S. Einfeldt, T. Böttcher, D. Hommel,
D. J. As, U. Köhler, A. Dadgar, A. Krost, Y. Saito, Y. Nanishi, M. R. Correia, S. Pereira,
V. Darakchieva, B. Monemar, H. Amano, I. Akasaki, and G. Wagner . . . . (c) 1750–1769

Mg in GaN: the structure of the acceptor and the electrical activity
    H. Alves, F. Leiter, D. Pfisterer, D. M. Hofmann, B. K. Meyer, S. Einfeld, H. Heinke,
    and D. Hommel . . . .   (c) 1770–1782

Local vibrational modes and compensation effects in Mg-doped GaN
    A. Hoffmann, A. Kaschner, and C. Thomsen . . . .   (c) 1783–1794

Optical micro-characterization of group-III-nitrides: correlation of structural, electronic and optical properties
    J. Christen, T. Riemann, F. Bertram, D. Rudloff, P. Fischer, A. Kaschner, U. Haboeck,
    A. Hoffmann, and C. Thomsen . . . .   (c) 1795–1815

Optical properties of nitride heterostructures
    Andreas Hangleiter . . . .   (c) 1816–1834

The origin of the PL photoluminescence Stokes shift in ternary group-III nitrides: field effects and
localization
    M. Strassburg, A. Hoffmann, J. Holst, J. Christen, T. Riemann, F. Bertram, and P. Fischer .   (c) 1835–1845

Devices and device issues

GaN based laser diodes – epitaxial growth and device fabrication
    T. Böttcher, S. Figge, S. Einfeldt, R. Chierchia, R. Kröger, Ch. Petter, Ch. Zellweger,
    H.-J. Bühlmann, M. Dießelberg, D. Rudloff, J. Christen, H. Heinke, P. L. Ryder,
    M. Ilegems, and D. Hommel. . . . .   (c) 1846–1859

Optical gain, gain saturation, and waveguiding in group III-nitride heterostructures
    M. Röwe, M. Vehse, P. Michler, J. Gutowski, S. Heppel, and A. Hangleiter . . . .   (c) 1860–1877

Electronics and sensors based on pyroelectric AlGaN/GaN heterostructures
    O. Ambacher, M. Eickhoff, A. Link, M. Hermann, M. Stutzmann, F. Bernardini,
    V. Fiorentini, Y. Smorchkova, J. Speck, U. Mishra, W. Schaff, V. Tilak, and
    L. F. Eastman . . . .   (c) 1878–1907

Electronics and sensors based on pyroelectric AlGaN/GaN heterostructures – Part B: Sensor
applications
    M. Eickhoff, J. Schalwig, G. Steinhoff, O. Weidemann, L. Görgens, R. Neuberger,
    M. Hermann, B. Baur, G. Müller, O. Ambacher, and M. Stutzmann . . . .   (c) 1908–1918

Influence of polarization on the properties of GaN based FET structures
    M. Neuburger, I. Daumiller, M. Kunze, M. Seyboth, T. Jenkins, J. Van Nostrand, and
    E. Kohn . . . .   (c) 1919–1939

Gallium-nitride-based devices on silicon
    A. Dadgar, M. Poschenrieder, I. Daumiller, M. Kunze, A. Strittmatter, T. Riemann,
    F. Bertram, J. Bläsing, F. Schulze, A. Reiher, A. Krtschil, O. Contreras, A. Kaluza,
    A. Modlich, M. Kamp, L. Reißmann, A. Diez, J. Christen, F.A. Ponce, D. Bimberg,
    E. Kohn, and A. Krost . . . .   (c) 1940–1949

DOI: **The fastest way to find an article online** is the *Digital Object Identifier* (DOI).
Starting with Vol. 0, No. 2 (2003), DOIs have been printed in the header of the first page of every
article. On the WWW, one can find an article for example with a DOI of 10.1002/pssc.200306190 at
http://dx.doi.org/10.1002/pssc.200306190.

Please use the DOI of the article to link from your home page to the articles in Wiley Interscience.

The DOI is a result of a cross-publisher initiative to create a system for the persistent identification
of documents on digital networks. More information is available from **www.doi.org**.

Preface

This Special Issue of *physica status solidi* (c) contains a series of articles reviewing various aspects of the epitaxy, materials science, and applications of group-III nitride films and heterostructures. All articles were written by scientists who have participated in the Priority Programme[*] 1032 *"Gruppe III-Nitride und ihre Heterostrukturen: Wachstum, materialwissenschaftliche Grundlagen und Anwendungen"*, which was funded by the German Research Foundation (Deutsche Forschungsgemeinschaft, DFG) in the six-year period from 1996 to 2002.

The purpose of the articles collected within this volume is twofold: (i) to summarize and document the main results obtained within this Priority Programme in a form which is also readily accessible for the international scientific community, and (ii) to discuss these results in the context of the current state-of-the-art in the field of the III-nitrides. Thus, most articles are organized in the form of brief reviews, which not only describe the results obtained by the authors, but also provide ample reference to related work by other groups worldwide. We hope that this format will be of particular interest to scientists looking for an authoritative and up-to-date review of both, basic and applied aspects in a field which still is developing at a very rapid pace.

For many researchers in Germany, the Priority Programme 1032 has been instrumental in helping to close the knowledge gap which existed between German university groups and those in Japan or the United States in the middle of the last decade. As a direct outcome of this Programme, which provided the necessary funds for on the average 20 research projects during the past six years, more than 1000 scientific papers have been published in conference proceedings and scientific journals, almost 200 invited talks were presented by members of the programme at topical workshops or international conferences, and 100 diploma theses as well as more than 50 PhD theses have been completed successfully. In the name of all who have benefited from this Priority Programme, we would like to thank the German Research Foundation for this important support of our work at a critical period of time. We are particularly grateful to the members of the International Peer Review Committee and to Dr. Klaus Wehrberger for their time, effort, and advice in monitoring and administrating Priority Programme 1032!

Esquinzo, August 2003

Martin Stutzmann

---

[*] The German-language term for Priority Programme is "Schwerpunktprogramm" whose abbreviation **SPP** inspired Armin Dadgar, Otto-von-Guericke-Universität Magdeburg, to assemble a collection of their bright blue GaN-on-Si LEDs (see p. 1940) for the cover illustration.

ISBLLED-2004

The 5th International Symposium on Blue Laser and Light Emitting Diodes

March 16~19, 2004
Gyeongju, Korea

The 1st Announcement

Papers are solicited on a variety of topics in rapidly advancing wide band gap semiconductor science and technologies and their applications to blue laser and light emitting diodes, including but not restricted to the following:

- Materials Growth and Characterization
- Optoelectronic Properties
- Device Design and Fabrication
- White Light Emitting Diodes
- Device Performance and Reliability
- Novel Optoelectronic Devices and Applications

Abstract Deadline: November 15, 2003

Organizing Committee
- *Honorary Chair:* Y. S. Park (Office of Naval Research, USA)
- *Honorary Co-Chair:* K. Takahashi (Teikyo Univ. of Sci. & Tech., Japan)
- *Chair:* H. J. Lee (Chonbuk National Univ., Korea)
- *Secretaries:* C.-H. Hong (Chonbuk National Univ., Korea)
  J. W. Yang (Chonbuk National Univ., Korea)

International Advisory Committee
- *Chair:* S.-K. Min (Korea Univ., Korea)
- *Co-chair:* K. H. Ploog (Paul Drude Inst., Germany)

Program Committee
- *Chairs:* E.-K. Suh (Chonbuk National Univ., Korea)
  E. Yoon (Seoul National Univ., Korea)

Sponsors
The Research Society for the Wide Band-Gap Semiconductors in cooperation with the Korean Physical Society the Office of Naval Research, the Korea Photonics Technology Institute, the Korea Science and Engineering Foundation, and the Korea Association for Photonics Industry Development.

Secretariat
Semiconductor Physics Research Center, Chonbuk National University, Chonju 561-756, Korea
Tel: +82-63-270-3927, Fax: +82-63-270-3926, E-mail: isblled@isblled2004.org

http://www.isblled2004.org

The proceedings of this conference will be published in physica status solidi.

phys. stat. sol. (c) **0**, No. 6, 1571–1582 (2003) / **DOI** 10.1002/pssc.200303136

Growth of GaN quasi-substrates
by hydride vapor phase epitaxy

Wei Zhang and **B. K. Meyer**[*]

I. Physikalisches Institut, Universität Giessen, Heinrich-Buff-Ring 16, 35392 Giessen, Germany

Received 4 March 2003, revised 16 June 2003, accepted 24 June 2003
Published online 12 August 2003

PACS 68.55.Ac, 68.55.Ln, 78.55.Cr, 78.60.Hk, 81.15.Kk

To grow high quality GaN heteroepitaxially is a difficult task due to the lattice mismatch and differences in thermal expansion coefficients between the films and the foreign substrates. A large number of structural defects form in the films, which lateron may limit the performance in the devices. Hydride vapor phase epitaxy is a high-growth-rate technique which has the potential to faciliate large area, low defect density „GaN quasi-substrates" for subsequent growth by MOCVD or MBE. We report on a flow modulated growth process which includes a nitridation of the sapphire substrate and the growth of high temperature buffer layers. The strucutral and optical properties of the films are presented and compared at different stages of the growth process.

1 Introduction

The first blue GaN LED device was brought into reality by Pankove et al. in 1971, using an insulating Zn-doped GaN layer on n-type GaN on sapphire substrate [1]. In the same year, the first lasing activity under optical excitation was observed from a GaN needle crystal by Dingle et al. [2]. However, stagnation on the work on GaN appeared until 1989. Then, with the concept of low-temperature buffer layers for the epitaxy on sapphire and the ability of p-type doping with Mg the success of the nitrides started [3–5]. Soon after, high quality multi-layer structures became available and optoelectronic devices have greatly progressed since then. The first commercial LED device appeared at the end of 1993 with an efficiency of 2.7% [6]. In the early 1996 the first nitride based laser diode was realized by Nakamura using an InGaN-based multiple-quantum-well structure [7] and by Akasaki et al. using a single-quantum-well [8], followed by Toshiba, in 1997 by Cree, and Fujitsu, and by the University of California at Santa Barbara. Shortly after that, a LD with an estimated lifetime of 10000 hours was fabricated [9].

Despite these achievements, the device performance and fundamental understanding of the material properties are still limited by the structural quality of the films. The lack of a lattice and thermally matched substrate forces to use heteroepitaxy on lattice and thermally mismatched substrates such as sapphire and SiC. Due to the differences in lattice constants and thermal expansion coefficients between the films and the foreign substrates, a large number of structural defects such as pits, surfaces cracks, and dislocations as high as 10^{10} cm^{-2} form in the films, which limit the device performance. To reduce these defects, many efforts have been made. For example, low-temperature buffer layers have been successfully introduced in the epitaxial growth of GaN [3]. The sophisticated ELOG (Epitaxial Lateral Over-Growth, or ELO) technique was also developed in the last decade. Recently, HVPE (Hydride Vapor Phase Epitaxy) has attracted much interest as it allows rapid growth of thick GaN epilayers. It is expected that this high-growth-rate technique will facilitate the growth of low-defect, thick, large-area substrates to be used for the MOVPE process. However, it is still hard to obtain high quality GaN epilay-

---

[*] Corresponding author: e-mail: bruno.k.meyer@exp1.physik.uni-giessen.de

ers – an apparent bottleneck at $\sim 10^8$ dislocations/cm^2 appears for the widely applied conventional growth processes. To overcome this problem, we therefore attempted to develop a modulated HVPE growth process. It is expected that modulating the process parameters can greatly improve the surface smoothness as well as the crystal quality. Additionally, pretreatments of the sapphire (nitridation) and growth of buffer layers, which have been used successfully in other techniques such as MOVPE and MBE, were also investigated and transferred to HVPE in this work to improve the crystal quality.

This review consists of four main parts: Section 2 reviews the history of HVPE growth of GaN and describes the home-built HVPE system.

The general aspects of the HVPE growth process are addressed in Section 3. In Section 4 the flow modulation technique is described in terms of the modulated species, i.e. HCl and/or NH$_3$. As an attempt to obtain even higher quality GaN films by taking the advantages of all the modulation techniques, an optimum modulated growth process is suggested in the last part of Section 4. Section 5 focuses on the optical properties of GaN films with respect to the growth conditions.

2 HVPE growth of GaN

2.1 History and situation

This technique is a chemical vapor phase deposition method, which is usually carried out in a hot wall reactor. The GaCl precursor is synthesized within the reactor vessel by the reaction of hydrogen chloride (HCl) with Ga metal at high temperature (750–900 °C). Afterwards the GaCl is transported to the substrate where it reacts with NH$_3$ at 900–1100 °C to form GaN. The growth process can be described by the following two-step reactions:

$$2\mathrm{Ga} + 2\mathrm{HCl} \uparrow \Leftrightarrow 2\mathrm{GaCl} \uparrow + \mathrm{H}_2 \uparrow, \text{ and} \tag{1}$$

$$\mathrm{GaCl} \uparrow + \mathrm{NH}_3 \uparrow \Leftrightarrow \mathrm{GaN} \uparrow + \mathrm{HCl} \uparrow + \mathrm{H}_2 \uparrow. \tag{2}$$

Historically, hydride vapor phase epitaxy (HVPE) and its closely related technique, halide vapor phase epitaxy, has played an important role in the development of semiconductor materials system. It was the first [10] and, until the early 1980s, the most popular method of growing epitaxial layers of GaN [11–13]. It is almost remarkable that GaN films grown by HVPE in the 1970s [12, 14] can still be considered to be of state-of-the-art quality, even by present standards.

In Table 1, we chronologically list some important progresses in the HVPE growth techniques and GaN quality improvement.

Table 1 List of important events related to HVPE growth of GaN.

Year	Important events related to HVPE growth of GaN	References
1969	HVPE was successfully used for the first time.	[10]
1972	Free carrier concentrations were strongly reduced to 2×10^{17} cm^{-3}.	[12]
1974	Study of the process kinetics	[36]
1976	Mg-doped GaN attempted	[13]
1979	HVPE process modified and GaN quality improved	[37]
1982	Nitridation of sapphire substrate	[38]
1989	Realization of p-type doping and fabrication of p-n junction	[4]
1990	Homoepitaxial growth of GaN	[39]
1992	Low-temperature buffer layer (ZnO) introduced	[23]
1995	Vertical reactor designed and used	[24]
1996	Study of the effect of reactor geometry and growth parameters on the uniformity and materials properties	[40]
2000	High-temperature buffer layer (GaN, AlN) introduced	[27, 41]

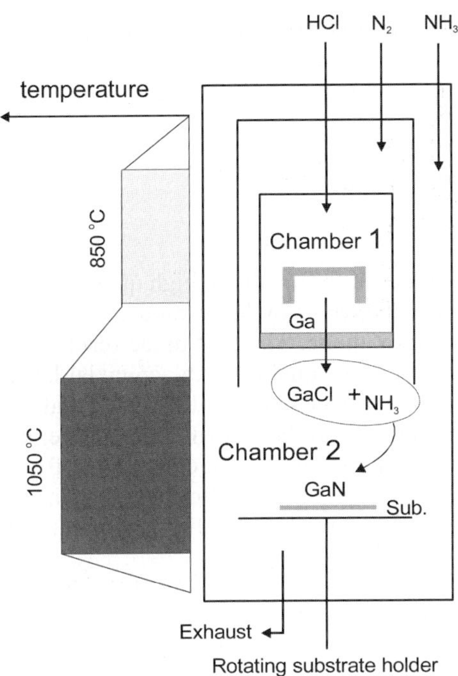

Fig. 1 (online colour at: www.interscience.wiley.com) Schematic drawing of our home-built HVPE reactor. It consists of two separate chambers: Chamber 1 is for the synthesis of the GaCl precursor, chamber 2 is for mixing the precursor GaCl and NH_3, and for GaN deposition.

2.2 Home-built HVPE system

The home-built HVPE system, as shown in Fig. 1, consists of four main parts: the gas supply and control system, the vertical hot wall HVPE reactor, the five-zone furnace, and the pressure control and exhaust system.

In the gas sources and supply system, gas sources of high purity (99.9999%) are stored in the high pressure bottles in a safe gas cabinet. At the exit of each bottle a filter is assembled to refine the gas before it is transported to the reactor. The flow rate of the gas is controlled by special mass-flow-controllers for the different gas sources, i.e., HCl, NH_3, and N_2. The switches are conveniently controlled by electromagnetic pneumatic valves which are connected to high pressure air (about 5 bar). The hot wall HVPE reactor, as shown in Fig. 1, consists of two separated chambers: Chamber 1, which is within Chamber 2, is used for the production of the GaCl precursor and Chamber 2 is for the deposition of GaN, as mentioned above. In chamber 1, a stopper is laid about 1 cm above the gallium surface. This is designed for (i) the complete reaction of HCl and Ga which is assured by a thorough mixture of HCl and Ga in the chamber and (ii) effectively preventing possible counter flow of HCl and GaCl to the cold end of the liner and condense there, which will obviously results in pollution and insufficient GaCl for GaN growth. In Chamber 2, a tube is laid coaxially out of the chamber 1 for separately transporting NH_3. Thus, both GaCl and NH_3 can mix well and be transported to the substrate uniformly. The other two tubes are symmetrically set for transporting the main carrier gas N_2. The substrate holder can be lowered and raised freely so that one can investigate the effect of the sub-nozzle distance and get an optimum distance at the deposition temperature. Due to its relative low vapor pressure at room temperature, the GaCl molecules tend to condense on unheated surfaces. This is the primary motivation for the use of a hot wall reactor and the *in-situ* synthesis of the metal chloride.

The five-zone furnace can reach temperatures as high as 1300 °C. One can obtain at least two uniform temperature zones and the length of the temperature zones can be changed as required. The difference of the temperatures between the wall and the coaxial center is less than 10 °C. The uniform temperature distribution, both in cross section and along the liner, ensures complete reaction and uniform deposition. The pressure in the reactor can be controlled in the range of 1 ~ 1000 mbar. More details on the design and equipment can be found in [15]. The optimum gas flow rates are 400–1000 sccm for NH_3, 5–20 for HCl, and for the carrier gas, N_2, a flow rate of 1000 sccm. The total pressure in the reactor, usually is at 800 mbar. Under the optimum growth conditions, growth rates up to 300–400 μm/h can be obtained.

3 Nitridation of substrate and high temperature (HT) buffer

3.1 Introduction

In the heteroepitaxy of GaN, the initial stage of growth is very important for obtaining high quality films. Generally, the epitaxial growth occurs in the two-dimensional (2D) layer-by-layer mode, the three-dimensional (3D) island mode, or the mixed (M) Stransky-Krastanov mode. The first mode results in a smooth surface, while the last two give a rough surface and lead to a low quality of the epitaxial layer. The growth mode is determined by many parameters, such as the interfacial energy of the solid and vapor phase or of the vapor phase and the substrate, which in turn depend on the growth temperature, the bond strength and bond lengths between the substrate and epilayer atoms, the impingement rate of species, surface migration rates of reactants, supersaturation of the gas phase, the size of critical nuclei, and others. Therefore, the control of these parameters and, thus, of the growth mode is the first important step to high quality films.

It is now well established that two pretreatment steps are essential to obtain good-quality MOVPE GaN layers on sapphire: a high temperature nitridation of the sapphire surface [16–18] and the deposition of a low-temperature buffer layer [3, 19, 20]. The nitridation of sapphire was found to result in the formation of a relaxed AlN layer [17, 21], which acts subsequently as a buffer layer. The mechanism of the buffer layer was studied by Akasaki et al. [5]. The essential role of a low-temperature buffer layer is both, to supply nucleation centers having the same orientations as the substrate and to promote lateral growth of the GaN film due to the decrease in the interfacial free energy between the film and substrate.

Whether these two techniques would be transferable to HVPE growth was not clear at the beginning. For nitridation of sapphire, it is expected to work. In the case of the buffer layer, the disadvantage is that the initial nucleation of the HVPE GaN on a bare sapphire substrate requires conditions of high supersaturation leading to very rapid nucleation and growth. Experimentally, several groups have attempted either *in situ* [18, 22] or *ex situ* [23–25] low-temperature buffer layer growth in HVPE. However, no clear picture has emerged. Recently, high-temperature (~ 900 °C) buffer layers have been reported to exhibit identically positive effects [26–28]. Considering these different results, and to clarify the controversial situation, it was necessary to investigate the effect of buffer layer on HVPE grown GaN systematically. The aim was to obtain the optimum growth conditions for the buffer layer.

3.2 Results and discussion

3.2.1 Nitridation of sapphire

For substrate preparation, the c-Al_2O_3 substrates were heated up to the temperature range from 1050–1200 °C in a NH_3 ambient for nitridation times from 5 to 20 min, similar to the conditions used for the MOVPE process [3]. The reason to choose high temperatures and certain times is that the slow reaction between NH_3 and sapphire surface takes place only under these conditions. The results on the linewidth (FWHM) of the symmetric (0002) reflexes of the subsequently grown GaN layers are shown in Fig. 2.

For the nitridation at different temperatures – the substrates were all heated in NH_3 for 10 min – it can be seen that layers grown on sapphire nitrided at 1100 °C have the smallest linewidth, compared to those on the substrates treated at other temperatures. Since at 1100 °C we obtained the best result, we investigated the nitridation at this temperature for various times. The best result, as shown in Fig. 2, is still the nitridation at 1100 °C for 10 min. Under such conditions, the linewidth of the rocking curve for symmetric (0002) reflection has the smallest value of 352 arcsec. Nitridation longer than 10 min finally degrades the crystal quality as also presented in [16, 17].

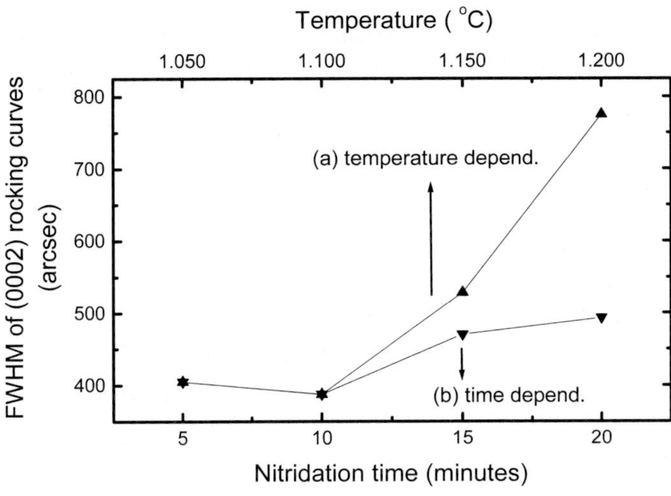

Fig. 2   GaN (0002) rocking curves for different substrate nitridation temperatures (a) and different substrate nitridation times (b). All nitridation times are 10 min in (a) and all nitridation temperatures are 1100 °C in (b).

3.2.2   GaN buffer layer

For buffer layer deposition, both GaCl and NH_3 were used as sources to form GaN with optimized flow rates. The temperatures for buffer layer growth ranged from 550 to 1100 °C. All buffer layers were grown for 10 min at different temperatures. The effect of buffer layer thickness was studied by depositing the buffers with a 2 sccm HCl flow for different times at the optimized temperature (900 °C). Additionally, buffer layers grown under optimum conditions were annealed at temperatures ranging from 1000 to 1100 °C for 5 min, as well as at 1050 °C but for various times, from 5 to 20 min. The buffer layer deposited at 900 °C had the smallest FWHM value. When the temperature was either lower or higher than 900 °C, a degradation of the crystal quality was found. It is therefore plausible that the improvement the structural quality of the HVPE grown epilayers by buffer layers occurs in a similar way as it was in the MOVPE growth [3, 5, 19]: The buffer layer supplies a nucleation layer for the growth of GaN and causes the coalescence to occur sooner.

The effect of the buffer layer was also observed with the optical differential interference contrast microscopy measurements. Figure 3 shows the morphology of HVPE GaN without (sample *a*) and with HT buffer layers (sample *b*). The morphology of sample *b* is superior to that of sample *a*. As one can see, there are almost no cracks at the surface of sample *b* while the cracks density is around 10^5 cm^{-2} in sample *a*. In addition, there are few pits at the surfaces of both layers.

In summary, for substrate nitridation in our system, the optimum process parameters are 1100 °C for the temperature and 10 min for the nitridation time, for buffer layer growth, the optimum conditions are

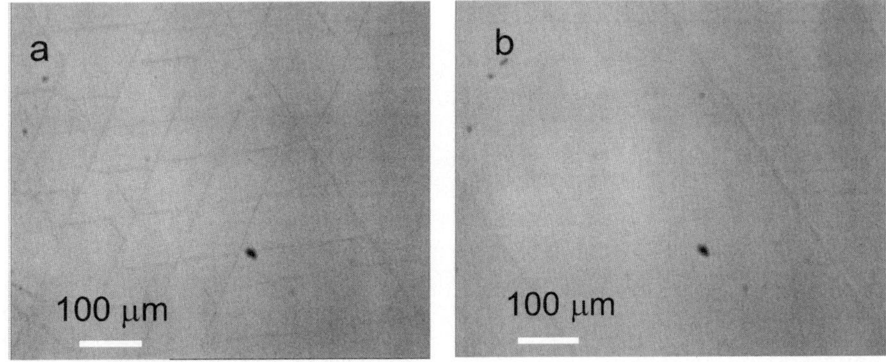

Fig. 3   Optical differential interference contrast microscopy images of GaN epilayers grown with (a) and without (b) a buffer layer.

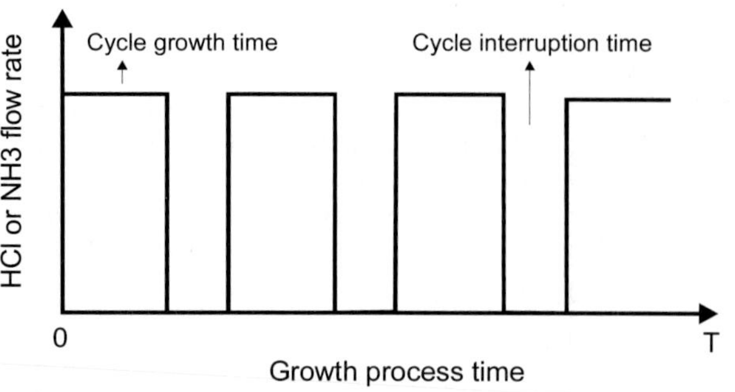

Fig. 4 Schematic view of the flow modulation growth process.

900 °C growth temperature and 10 min growth time. Annealing the buffer layers can also improve the quality of the epitaxial film when the process parameters are carefully controlled. The optimum annealing conditions are 1050 °C for 5 min.

4　Flow modulated HVPE of GaN

In addition to the nitridation of substrates and HT buffer layers, we also developed a "flow modulated growth process" for HVPE growth of GaN layers to further improve the crystalline quality of GaN. The flow modulated growth process was realized by interrupting the flow of species periodically. During growth, we periodically interrupt either the NH_3 or the HCl flow (flow of Ga source) but keep the other parameters constant. Fig. 4 shows a schematic diagram of the flow modulated growth procedure. For HCl flow modulation, a serie of growths were carried out varying intentionally the number of sequences (sub-layer) and the HCl flow ON/OFF time. For NH_3 flow modulation, the most important aspect is the periodic NH_3 flow ON/OFF (growth/interruption) time. The interruption time is on the order of minutes for HCl flow modulation and on the order of seconds for NH_3 flow modulation, respectively.

4.1　HCl flow modulation

We changed the interruption time of HCl flow in the range of 5 to 100 min while keeping a total time of 30 min for each growth run. In addition, the number of sequences was also varied while keeping the total thickness almost identical. The cross-sectional structures of GaN epilayers were investigated with chathodoluminescence (CL) and transmission electron microscopy (TEM). TEM was also used to determine the dislocation density in GaN by measuring the number of dislocations in a fixed area. Sample A has four sub layers after three times of interruptions with a thin buffer layer on the substrate [29–31]. In TEM observations (see Fig. 5) we found that there is an abrupt decrease of structural defects in two adjacent sub layers and the defects decrease greatly from the initial sublayer ($10^{10\sim12}$ cm^{-2}) above the sub-

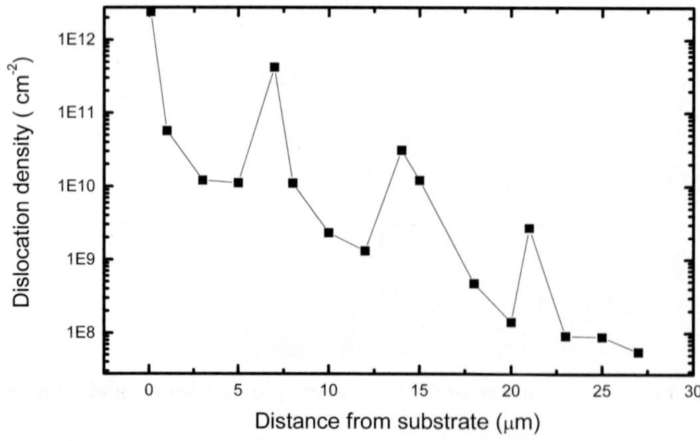

Fig. 5 Variation of the dislocation density at different thickness of GaN epilayers grown via HCl flow modulation.

strate to the sub layer at the top surface (10^7 cm^{-2}). Defects with higher density were found at the inter-faces between the neighboring sub layers.

4.2 NH$_3$ flow modulation

In addition to HCl flow modulation, NH$_3$ flow modulation growth of GaN films has also been attempted. Our work shows that, besides the reduction of dislocations and the improvement of the crystalline quality, the morphology of the films can be greatly improved via the ammonia flow modulation.

4.3 Combination of modulation techniques

It is clear that the HCl flow modulation is helpful to reduce the dislocation density, however, the surface of the sample is quite rough and there is still residual strain in the upper layers. The NH$_3$ flow modulation in addition improves the morphology of the sample. Furthermore, it can relax the strain efficiently, since there is almost no residual strain in the upper part of the film. However, the dislocation density in GaN grown via NH$_3$ flow modulation is higher than that via HCl flow modulation. Therefore, further effort is still required, since so far we can not grow GaN with both low dislocation density and smooth surface only by either HCl flow modulation or NH$_3$ flow modulation. We attempted to combine both modulations in order to take advantage of both techniques. Additionally, both the pretreatment of the substrate and the predeposition of a buffer layer were carried out prior to growth of the main epilayers.

Figure 6a shows the growth procedure including all modulation techniques. For the growth of the main epilayers, both HCl and NH$_3$ flow modulations (Fig. 6b) were used. During the growth cycle of HCl flow modulation, the NH$_3$ was pulsed so that the NH$_3$ flow modulation was also carried out simultaneously. During the HCl interruption cycle, it was found that there is no obvious influence upon the crystal structure whether the NH$_3$ flow was pulsed or not. Therefore, the NH$_3$ flow was usually kept constant during the interruption cycle of HCl flow modulation. The growth conditions applied are first the optimum growth parameters as derived in the two previous sections, i.e., the HCl flow modulation was carried out with a four-sublayer structure and 30/30 minutes for each cycle of the HCl flow ON/OFF time, while the NH$_3$ flow modulation was carried out with 15/15 seconds for each cycle of the NH$_3$ flow ON/OFF time. However, further detailed work indicated that the optimum parameters for the combination of both modulations are a little different. The highest quality samples are grown with a 14-sublayer structure on sapphire and a 5-sublayer structure on GaN templates (see Fig. 7). The NH$_3$ flow rates can also affect the sample quality. The optimum flow rates are 600 sccm for the HCl flow ON cycle and 400 sccm for the HCl flow OFF cycle, respectively.

Table 2 lists the structural parameters of three typical GaN samples A, B, and C, which were grown via the combination of all the modulated techniques. The reference samples D, E, and F are films grown directly on sapphire with and without flow modulations. An overall improvement of the crystal quality has really been achieved in the three samples A, B, and C compared to samples D, E and F. As listed in Table 2, there are also differences among the three samples A, B, and C. The samples A and B were grown on sapphire. The first sub layer of each was grown with the same thickness of 7 μm at the identical HCl flow modulation parameters. However, their other parameters are different. For example, sample A has a 5-sublayer structure while the sample B has a 14-sublayer structure. Additionally, for the NH$_3$

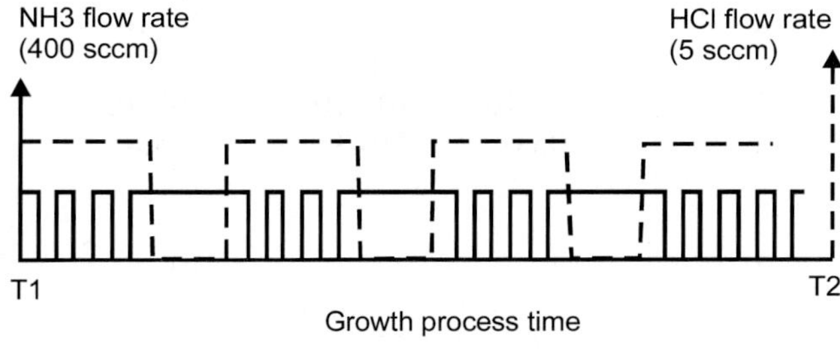

Fig. 6 Schematic flow diagram of the combination of various modulation techniques.

Sample A: 5-sublayer sequence, on sapphire

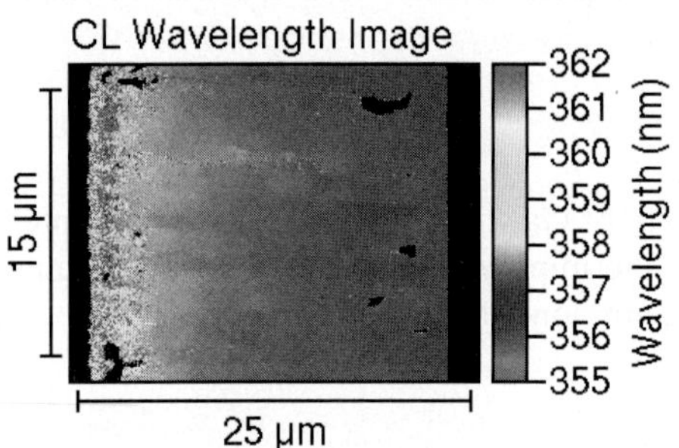

Sample B: 14-sublayer sequence, on sapphire

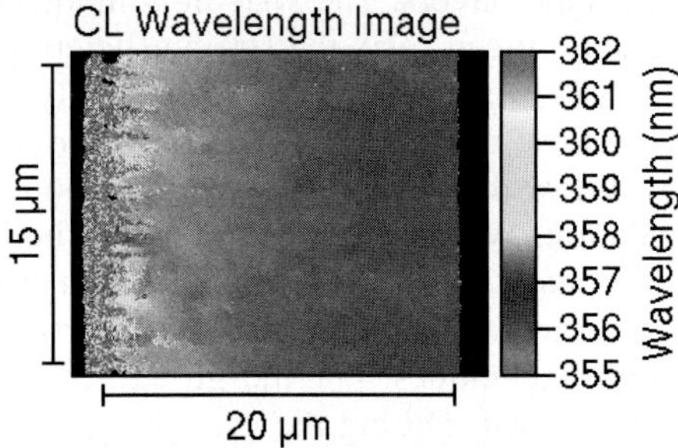

Sample C: 5-sublayer sequence, on GaN template

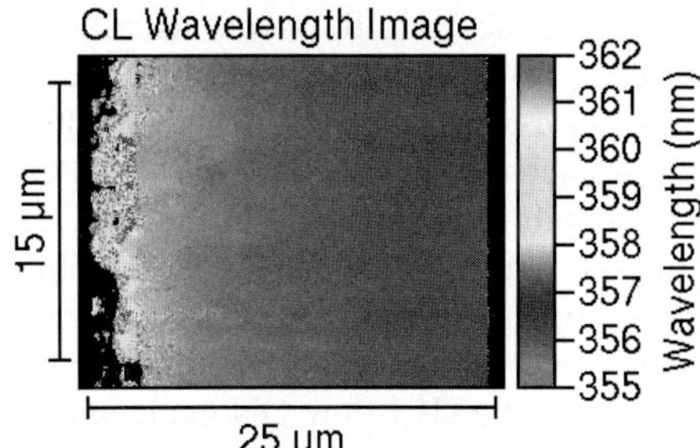

Fig. 7 (online colour at: www.interscience.wiley.com) Cross sectional views by cathodoluminescence wavelength images of GaN films grown via combined modulation techniques.

Table 2 Physical properties of GaN with different growth parameters (FM = flow modualtion).

Properties	High quality GaN epilayers			Reference samples		
	A	B	C	D	E	F
Substrate	c-Al$_2$O$_3$	c-Al$_2$O$_3$	GaN temp.	c-Al$_2$O$_3$	c-Al$_2$O$_3$	c-Al$_2$O$_3$
Growth techniques	combination of all the modulation techniques			directly	via HCl FM	via NH$_3$ FM
Number of sequences	5	14	5	1	4	1
Surface cracks (cm^{-2})	10–10^2	10–10^2	crack-free	10^5	10^2–10^3	10–10^2
Surface rough. (nm)	0.97	1.19	0.89	4.12	1.38	0.93
Dislocations (cm^{-2})	10^7	10^7	10^7	10^9–10^{10}	10^7–10^8	10^8–10^9
Rocking curve linewidth of (0002) reflection (arcsec)	281.8	263.8	176.1	405.0	310.3	334.8
PL linewidth of D^{0}X at 4.2 K (meV)	2.0	1.5	1.5	14	9	10

flow modulation growth of the sub layers, the NH_3 ON/OFF time is 15/20 and 15/15 seconds for samples A and B, respectively. Sample C was grown on GaN template under identical conditions as those of sample A. The different properties must therefore be due to the different substrate materials. The higher quality of sample C compared to samples A and B shows that a suitable substrate is always of prime importance for GaN growth whatever techniques are applied.

5 Luminescence properties

A representative PL spectrum of HVPE grown GaN recorded at $T = 4.2$ K is shown in Fig.8. The prominent line has a full width at half maximum (FWHM) of 14 meV and is located at 3.473 eV. It is assigned to the recombination of a neutral donor bound exciton, D^0X [2]. This is known to be the dominant emission in undoped GaN, because of its high n-type background doping and free exciton emission (FX) is hardly observable. In addition the zero phonon line of a donor–acceptor pair transition (DAP) appears at 3.260 eV, which is commonly seen in GaN. It is accompanied by longitudinal optical phonon (LO) replicas 92 meV, 184 meV, 276 meV, and 368 meV lower in energy, respectively. Additionally, a broad yellow band centered at 2.2 eV is visible in the luminescence (see Fig. 9). The nature of the donors and acceptors could be attributed to intrinsic defects as well as extrinsic impurities. For example, the nitrogen vacancies or the gallium interstitials should induce a shallow effective-mass-type donor-state in GaN. Extrinsic defects introducing shallow donor levels could come from the group VI elements on group V site. Oxygen would be one candidate. Among the group IV elements, the Si on Ga site is a very efficient donor. The origin of the shallow acceptors was suggested to be Mg, C, or Zn. [32–35].

By applying a combination of all modulation techniques, the best GaN films in our work could be grown. This allows to resolve more fine structures in the PL measurements, in addition to the excitonic transitions discussed above. Figure 10 shows the PL spectrum of such a GaN film. D^0X is the most intense peak with a FWHM of 1.5 meV. At higher energies, three free excitons, i.e., FX(A), FX(B), and FX(C) are resolved indicating the high crystalline quality of GaN film. The energy positions of the free excitons involving holes from the A-, B-, and C-valence bands, respectively are: FX(A) at 3.482 eV, FX(B) at 3.488 eV, and FX(C) at 3.496 eV. The energy differences of these values represent the splitting of the three valence bands of a strain-free film. In the lower energy part, the peaks are attributed to the

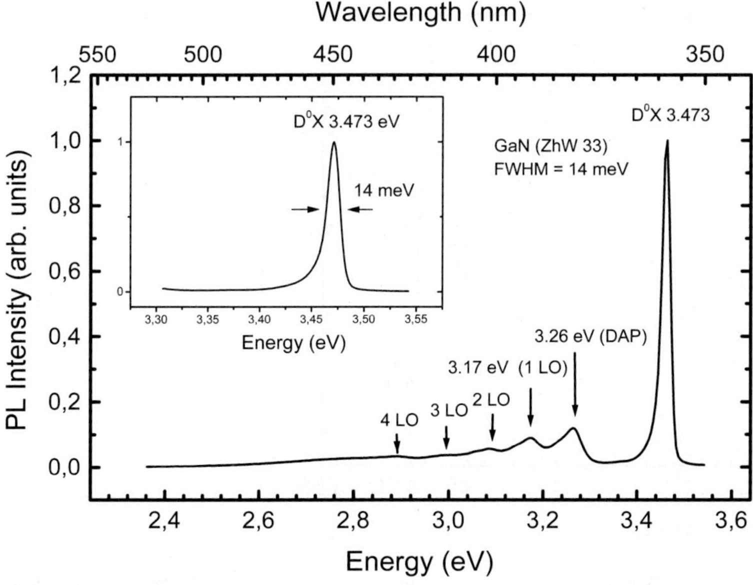

Fig. 8 Photoluminescence spectrum of GaN grown directly on sapphire ($T = 4.2$ K).

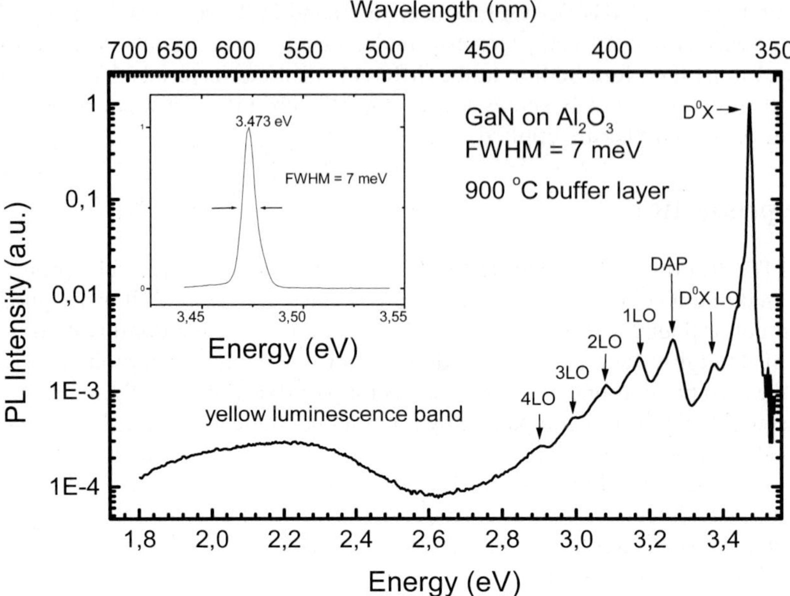

Fig. 9 Photoluminescence spectrum of GaN grown on sapphire including a buffer layer ($T = 4.2$ K).

exciton bound to an ionized donor or excitons bound at neutral acceptors (D^+X) at 3.471 eV, A_1^0X at 3.460 eV, and A_2^0X at 3.455 eV, respectively. The respective LO phonon replicas can also be seen at lower energies. Below 3.3 eV, the band acceptor transition (eA), the DAP transitions and correlated phonon replicas can be seen.

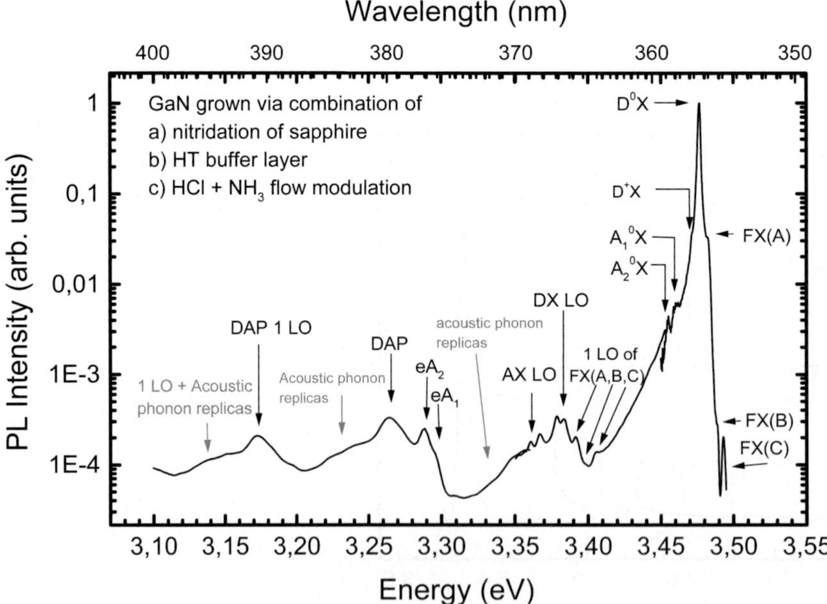

Fig. 10 (online colour at: www.interscience.wiley.com) Photoluminescence spectrum of GaN grown via combination of various modulation techniques ($T = 4.2$ K).

6 Summary

High quality GaN is very difficult to grow heteroepitaxially due to the lattice and thermal mismatch between the films and the substrates. To avoid this problem, one of the solutions is the HVPE growth of quasi-bulk, strain-free GaN, which can be used as substrates for subsequent homoepitaxy. Thick Gan layers have been grown via a modulation process in a home-built HVPE system. The growth process includes nitridation of the sapphire substrate and high-temperature buffer layers as the pre-stages, and the flow modulation growth as a central part. This concept brings the dislocation density to 10^7 cm^{-2}, the films are strain-free and have no cracks, the rocking curve linewidth of the (0002) reflection has a value of 176 arcsec and the PL linewidth of the neutral donor bound exciton at 4.2 K is 1.5 meV.

References

[1] J. I. Pankove, E. A. Miller, and J. E. Berkeyheiser, RCA Review **32**, 383 (1971).

[2] R. Dingle, K. L. Shaklee, R. F. Leheny, and R. B. Zetterstrom, Appl. Phys. Lett. **19**, 5 (1971).

[3] H. Amano, N. Sawaki, I. Akasaki, and Y. Toyoda, Appl. Phys. Lett. **48**, 353 (1986). MOCVD

[4] H. Amano, M. Kito, K. Hiramatsu, and I. Akasaki, Jpn. J. Appl. Phys. **28**, L2112 (1989).

[5] I. Akasaki, H. Amano, M. Kitoh, K. Hiramatsu, and N. Sawaki, Ext. Abstr. (175th Spring Meeting 1989) Electrochem. Soc. 673, SOA.

[6] S. Nakamura, T. Mukai, and M. Senoh, Appl. Phys. Lett. **64**, 1687 (1994).

[7] S. Nakamura, M. Senoh, S. Nagahama, N. Isawa, T. Yamada, T. Matsushita, H. Kiyoku, and Y. Sugimoto, Jpn. J. Appl. Phys. **35**, L74 (1996).

[8] I. Akasaki, S. Sota, H. Sakai, T. Tanaka, M. Koike, and H. Amano, Electron Lett. **32**, 1105 (1996).

[9] S. Nakamura, M. Senoh, S. Naghama, N. Iwasa, T. Yamada, T. Matsushita, H. Kiyoku, Y. Sugimoto, T. Kozaki, H. Umemoto, M. Sano, and K. Chocho, Proc. of the 2nd Int. Conf. on Nitride Semiconductors, 444 (October 27–31, 1997, Tokushima, Japan).

[10] H. P. Maruska and J. J. Tietjen, Appl. Phys. Lett. **15**, 327 (1969).

[11] V. S. Ban, J. Electrochem. Soc. **119**, 761 (1972).

[12] M. J. Ilegems, J. Cryst. Growth **13/14**, 360 (1972).

[13] M. Sano, and M. Aoki, Jpn. J. Appl. Phys. **15**, 1943 (1976).

[14] R. K. Crouch, W. J. Debnam, and A. L. Fripp, J. Mater. Sci. **13**, 2358 (1978).

[15] W. Zhang, GaN and its Growth by Hydride Vapor Phase Epitaxy (HVPE), Master Thesis, University Giessen, Germany (1999).

[16] S. Keller, B. P. Keller, Y. F. Wu, B. Heying, D. Kapolnek, J. S. Speck, U. K. Mishra, and S. P. DenBaars, Appl. Phys. Lett. **68**, 1525 (1996).

[17] K. Uchida, A. Watanabe, F. Yano, M. Kouguchi, T. Tanaka, and S. Minagawa, J. Appl. Phys. **79**, 3487 (1996).

[18] R. J. Molnar, W. Götz, L. T. Romano, and N. M. Johnson, J. Cryst. Growth **178** 147 (1997).

[19] S. Nakamura, Jpn. J. Appl. Phys. **30**, 1620 (1991).

[20] T. Wang, T. Shirahama, H. B. Sun, H. X. Wang, J. Bai, S. Sakai, and H. Misawa, Appl. Phys. Lett. **76**, 2220 (2000).

[21] T. D. Moustakas, T. Lei, and R. J. Molnar, Physica B **185**, 39 (1993).

[22] V. Wagner, O. Parillaud, H. J. Buhlmann, and M. Ilegems, phys. stat. sol. (a) **176**, 429 (1999).

[23] T. Detchprohm, K. Hiramatsu, H. Amano and I. Akasaki, Appl. Phys. Lett. **61**, 2688 (1992).

[24] R. J. Molnar, K. B. Nichols, P. Maki, E. R. Brown, and I. Melngailis, Mater. Res. Soc. Symp. Proc. **378**, 479 (1995).

[25] S. L. Gu, R. Zhang, J. X. Sun, L. Zhang, and T. F. Kuech, Appl. Phys. Lett. **76**, 3454 (2000).

[26] E. Valcheva, T. Paskova, S. Tungasmita, P. O. A. Persson, J. Birch, E. B. Svedberg, L. Hultman, and B. Monemar, Appl. Phys. Lett. **76**, 1860 (2000).

[27] W. Zhang, S. Roesel, P.Veit, T. Riemann, H. R. Alves, D. Meister, W. Kriegseis, D. M. Hofmann, J. Christen, B. K. Meyer, Proc. Int. Workshop on Nitride Semicond., IPAP Conf. Series **1**, 27 (2000).

[28] W. Zhang, T. Riemann, H. R. Alves, M. Heuken, P. Veit, D. Pfisterer, D. M. Hofmann, J. Blaesing, A. Krost, J. Christen, and B. K. Meyer, phys. stat. sol. (a) **188**, 453 (2001).

[29] W. Zhang, T. Riemann, H. R. Alves, M. Heuken, D. Meister, D. M. Hofmann, J. Christen, A. Krost, and B. K. Meyer, J. Cryst. Growth **234**, 616 (2002).

[30] W. Zhang, H. R. Alves, J. Blaesing, T. Riemann, M. Heuken, P. Veit, D. Pfisterer, R. Gregor, D. M. Hofmann, A. Krost, J. Christen, and B. K. Meyer, phys. stat. sol. (a) **188**, 425 (2001).

[31] W. Zhang, S. Roesel, T. Riemann, P. Veit, H. R. Alves, D. Meister, W. Kriegseis, J. Blaesing, D. M. Hofmann, A. Krost, J. Christen, and B. K. Meyer, Appl. Phys. Lett. **78**, 772 (2001).

[32] B. Monemar, Phys. Rev. B **10**, 676 (1974).

[33] S. Strite, and H. Morkoq, J. Vac. Sci. Technol. B **10**, 1237 (1992).

[34] I. Akasaki, H. Amano, M. Kito, and K. Hiramatsu, J. Lumin. **48/49**, 666 (1991).

[35] B. K. Meyer, D. Volm, A. Graber, H. C. Alt, T. Detchprohm, H. Amano, I. Akasaki, Solid State Commun. **95**, 597 (1995).

[36] A. Shintani, S. Minagawa, J. Cryst. Growth **22**, 1 (1974).

[37] Y. A. Vodakov, E. N. Mokhov, M. G. Ramm, and A. D. Roenkov, Kristall und Technik **14**, 729 (1979).

[38] B. Y. Tsaur, R. W. McClelland, J. C. C. Fan, R. P. Gale, J. P. Salerno, B. A. Vojak, and C. O. Bozler, Appl. Phys. Lett. **41**, 347 (1982).

[39] K. Naniwae, S. Itoh, H. Amano, K. Itoh, K. Hiramatsu, and I. Akasaki, J. Cryst. Growth **99**, 381 (1990).

[40] S. A. Safvi, N. R. Perkins, M. N. Horton, A. Thon, D. Zhi, and T. F. Kuech, Mater. Res. Soc. Symp. Proc. **423**, 227 (1996)

[41] X. Zhang, P. D. Dapkus, and D. H. Rich, Appl. Phys. Lett. **77**, 1496 (2000).

phys. stat. sol. (c) **0**, No. 6, 1583–1606 (2003) / **DOI** 10.1002/pssc.200303122

Metalorganic chemical vapor phase epitaxy of gallium-nitride on silicon

A. Dadgar[*, 1], **A. Strittmatter**[2], **J. Bläsing**[1], **M. Poschenrieder**[1], **O. Contreras**[3], **P. Veit**[1], **T. Riemann**[1], **F. Bertram**[1], **A. Reiher**[1], **A. Krtschil**[1], **A. Diez**[1], **T. Hempel**[1], **T. Finger**[1], *, **A. Kasic**[3], **M. Schubert**[3], **D. Bimberg**[2], **F. A. Ponce**[3], **J. Christen**[1], and **A. Krost**[1]

[1] Otto-von Guericke Universität Magdeburg, Institut für Experimentelle Physik, Fakultät für Naturwissenschaften, Postfach 4120, 39016 Magdeburg, Germany
[2] TU-Berlin, Institut für Festkörperphysik, Hardenbergstr. 36, 10623 Berlin, Germany
[3] Department of Physics and Astronomy, Arizona State University, Tempe, AZ 85287, USA
[4] Universität Leipzig, Fakultät für Physik und Geowissenschaften, Institut für Experimentelle Physik II, Linnéstr. 5, 04103 Leipzig, Germany

Received 4 March 2003, revised 22 April 2003, accepted 22 April 2003
Published online 12 August 2003

PACS 61.10.Eq, 68.37.Lp, 78.55.Cr, 78.60.Fi, 78.60.Hk, 81.05.Ea, 81.15.Gh

GaN growth on Si is very attractive for low-cost optoelectronics and high-frequency, high-power electronics. It also opens a route towards an integration with Si electronics. Early attempts to grow GaN on Si suffered from large lattice and thermal mismatch and the strong chemical reactivity of Ga and Si at elevated temperatures. The latter problem can be easily solved using gallium-free seed layers as nitrided AlAs and AlN. The key problem for device structure growth on Si is the thermal mismatch leading to cracks for layer thicknesses above 1 µm. Meanwhile, several concepts for strain engineering exist as patterning, Al(Ga)N/GaN superlattices, and low-temperature (LT) AlN interlayers which enable the growth of device- relevant GaN thicknesses. The high dislocation density in the heteropitaxial films can be reduced by several methods which are based on lateral epitaxial overgrowth using ex-situ masking or patterning and by in-situ methods as masking with monolayer thick SiN. With the latter method in combination with strain engineering by LT-AlN interlayers dislocation densities around 109 cm^{-2} can be achieved for 2.5 µm thick device structures.

1 Introduction

Due to the lack of GaN substrates GaN growth currently involves the use of heterosubstrates as sapphire and SiC. One possible alternative to these substrates which are either insulating or expensive is the usage of low-cost, large size, and thermally and electrically well conducting Si as substrate. This also opens a route towards an integration of light emitters and high speed electronics with Si technology. In this field many attempts were made in the GaAs and InP system mostly in the 80s and early 90s. But there has been little success and optical devices as lasers suffer from a short lifetime due to the high defect density originating in the large lattice mismatch of the materials and the low activation energy for the so-called dark line defects being detrimental for these devices. The only commercialized product out of these research efforts are III-V based solar cells on thin germanium substrates for space applications.

Already in 1971 by Manasevit, Erdmann, and Simpson [1] reported on the growth of AlN layers on Si(111) by metalorganic chemical vapor phase epitaxy (MOVPE). But the activities in the next 15–20 years were rather weak. With the progress in GaN growth on sapphire by the implementation of a

---

* Corresponding author: e-mail: armin.dadgar@physik.uni-magdeburg.de, Phone: +49 391 67 11384, Fax: +49 391 67 11130

low-temperature seed layer [2] and successful p-type doping [3, 4] the research activities increased significantly and with it the growth on Si substrates was also studied but still remained a rather academic work. A dramatic increase in research activities was initiated by the first LED structures grown by MBE [5, 6]. They demonstrated that p-type doping of GaN on Si and with it devices are achievable. Earning from the knowledge gained not only from growth on sapphire but also from III−V growth on Si the GaN layer quality on Si could be significantly improved and is now close to that of GaN on sapphire.

Nowadays, GaN growth on Si is motivated by two main goals: low-cost GaN-based devices and integrated high-frequency (HF) electronics and optoelectronics with Si electronics. The second point competes with hybrid integration, thus, also here the cost of device growth and manufacturing is crucial to be competitive with existing techniques. So the most important point for GaN growth on Si is always the cost thus, we focus in this article on simple techniques to overcome the three main problems: thermal mismatch causing cracks even below device-relevant layer thicknesses, a large lattice mismatch leading to a high dislocation density, and the strong chemical reactivity of Ga and Si at elevated temperatures leading to a destruction of the layers and substrate.

2 Material properties

The high lattice mismatch results in a high initial dislocation density in the GaN layer well above 10^{13} cm^{-3} but the major problem of heteroepitaxial device growth is generally the large thermal mismatch of most compound semiconductors and the substrate. Especially for Si, the large thermal mismatch hinders the growth of device-relevant layer thicknesses and cracking is a common problem.

Fortunately the high-lattice mismatch causing a high density of misfit dislocations at the substrate/Group-III-nitride interface is less severe for the nitrides than for most other III–V compound devices because of the high binding and dislocation activation energies. Commercially available LEDs on sapphire with dislocation densities above 10^9 cm^{-2} are specified with lifetimes of 100 000 hours. Even laser operation with over 10 000 hours cw is possible with dislocation densities in the 10^6 cm^{-2} range [7, 8]. In the case of GaAs and InP based lasers this would lead to device degradation within seconds, e.g. by the formation of dark line defects [9].

For all heterosubstrates the dislocation densities at the GaN / substrate interface calculated from the difference in the lattice constants (Table 1) are orders of magnitude above 10^9 cm^{-2} (2.8×10^{13} cm^{-2} (Si(111)), 1.8×10^{13} cm^{-2} (sapphire), and 1.1×10^{12} cm^{-2} (6H-SiC)), which is approximately the upper limit in threading dislocation density of commercial LEDs. To achieve initial dislocation densities suited, e.g. for commercial GaN-based laser devices ($\sim 10^6$ cm^{-2}) the lattice mismatch must be below 0.01%. In a

Table 1 Material properties of GaN, AlN, Si, SiC, and sapphire (the given thermal expansion coefficient is an averaged value and might differ significantly at very low and at high temperatures) [22–30].

Material	a (Å)	c (Å)	Thermal conductivity (W/cmK)	Thermal expansion coefficient in-plane (10^{-6}/K)	Lattice mismatch GaN/substrate (%)	Thermal mismatch GaN/substrate (%)
GaN	3.189	5.185	1.3	5.59	-	-
AlN	3.11	4.98	2.85	4.2	2.4	25
Si(111)	5.430	-	1-1.5	2.59	-16.9	54
6H-SiC	3.080	15.12	3.0-3.8	4.2	3.5	25
sapphire	4.758	12.991	0.5	7.5	16	-34

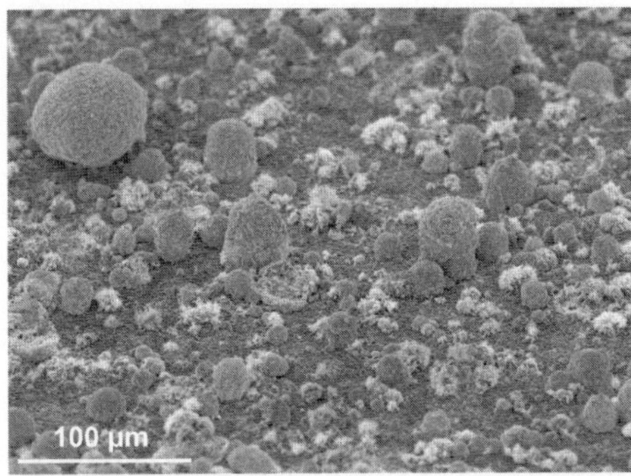

Fig. 1 Meltback etching causes a deterioration of the GaN layer and the Si substrate; SEM images of a GaN layer on an AlN seed on Si(111) (top) in plan-view and of another sample in cross section showing a rough GaN layer and deep hollows in the substrate (bottom).

typical growth process on heterosubstrates the very high initial dislocation densities in the GaN layer are reduced by dislocation annihilation the faster, the higher the dislocation density. Thus, values around 10^{10} cm^{-2} are achieved within a few hundred nanometers of GaN growth independent of the substrate used. So, the substrate dependent differences in the initially high dislocation densities are of minor importance for GaN device growth. As a consequence of the dislocation density dependent annihilation rate the only way to further reduce the dislocation density significantly is to grow very thick GaN, when growing GaN solely.

Besides growing thick layers, e.g. by HVPE, many methods are known to reduce the dislocation density and the most often applied are based on masking and lateral overgrowth commonly known as ELOG or LEO [10] and pendeo epitaxy [11, 12], all methods involving an ex-situ processing step before growth or after a first buffer layer has been grown. In-situ methods as Al(Ga)N/GaN superlattices [13] or AlN or Si_xN_y interlayers [14, 15] are also suited to reduce the dislocation density without ex-situ processing.Table 1) does, as opposite to GaN on sapphire, induce tensile stress in GaN on silicon resulting in crack formation when cooling. Additionally, due to the growth starting from islands instead of forming a continuous film, tensile stress is induced already during growth [16–18] which is enhanced for Si doped layers [19], both are general problems for the growth of thick GaN and an additional problem for the growth on Si. Besides the problems for device processing, cracks are already detrimental for device growth. Cracks which occur during growth, either of thick layers or when cooling down, e.g. to grow an InGaN layer, can partially expose the Si substrate and initiate a meltback etching reaction [20] of Ga and Si. While the exact process is not clarified yet it is likely that the alloying reaction of Si with Ga is assisted by the formation of Si_xN_y [21]. The meltback-etching process, once initiated, cannot be stopped and leads to a rough surface with deep hollows in the substrate (Fig. 1).

Si is available from high p- and n-type conductive (0.01 Ωcm) up to highly resistive material with around 10^4 Ωcm. In comparison with the intrinsic resistivity of GaN, sapphire or SiC ($>>10^9$ Ωcm) this is a rather low value. It can lead to high parasitic capacitances which are detrimental for high frequency (HF) operation of devices.

Most important to achieve a good layer quality is the Si substrate preparation to remove the native oxide usually present at the surface. It can be either removed by heating the substrate to temperatures above 900°C or by wet chemical etching with a final HF step to remove all oxides and obtain a H-terminated surface, see e.g. [31–33]. Using high temperature oxide removal for substrate preparation can cause a distorted surface due to the evaporation of deposits from the reactor walls [34].

3 Seed layer

The growth of GaN on all heterosubstrates (sapphire, SiC, Si) usually starts with a so-called seed layer which accommodates most of the lattice mismatch and helps in orienting the GaN grown on top of it. For

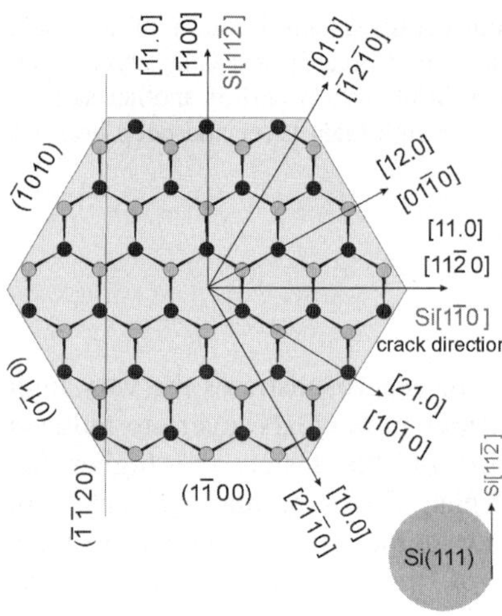

Fig. 2 Epitaxial relationship of GaN(0001) on Si(111).

a good GaN crystal orientation of the crystallites means not only the orientation along the *c*-axis, the preferred direction of growth (deviations are termed as tilt) but also the *a*-axis orientation of the crystallites parallel to each other (twist).

When GaN is grown on Si usually Si(111) is preferred which has a hexagonal arrangement of the Si atoms at the surface with a ~17% mismatch in atom positions (Table 1). Despite the large lattice mismatch the hexagonal arrangement of the Si atoms is helpful to orientate the seed layer so that the *c*-axis is oriented vertically to the substrate plane and the *a*-axis of the crystallites all parallel in-plane. In contrast to the growth on sapphire [35] where the mismatch in atom position is about the same percentage but with opposite sign no rotation of the GaN layer versus the Si-substrate is reported. Figure 2 shows the epitaxial relationship of GaN(0001) on Si(111) which is GaN[$11\bar{2}0$]∥Si[$1\bar{1}0$] [21, 36, 37]. Due to the missing six-fold symmetry of the surface atoms growth on Si(001) is very difficult and only few publications exist [38–45].

The polarity of MOVPE grown material is always reported Ga-face and no evidence for N-face polarity is given in any report [46].

In contrast to growth on the usually nitrided sapphire surface the prevention of Si_xN_y is an important issue when growth is initiated since such Si_xN_y layers are known to be amorphous [47] and a homogenous orientation of the seed layers on it not possible.

Several concepts for seed layer growth have been investigated by many groups. The simplest seed layer concept would be to adapt a well established MOVPE process on sapphire which is, e.g. the growth of a low-temperature GaN seed layer, rearrangement and formation of GaN islands from the layer during heat-up followed by a thick high temperature GaN buffer layer [2]. But due to the presence of ammonia during this process, especially when GaN islands are formed and some areas of Si are exposed to the gas ambient the Si wafer can get nitrided. As a consequence this can lead to polycrystalline GaN islands on the nitrided Si surface which are unsuited for a subsequent buffer layer growth. Also the Ga/Si reaction (meltback etching) is more likely to occur in this case when temperatures are raised as shown by Ishikawa et al. [20].

Another method involves the formation of sapphire which is the commonly used substrate for GaN growth. The advantages of Si (size, price) and sapphire with established seed layer growth procedures can be combined, e.g. by forming a sapphire pseudo-substrate on Si. Such a layer was investigated by Kobayashi and co-workers [48] who used an oxidized AlAs layer as an intermediate layer for the growth of GaN on Si. The oxidized layer is assumed to be composed of amorphous/fine grain Al_2O_3. Despite the poor quality of this seed layer single crystalline GaN(0001) could be grown on top. But an oxygen contamination of the GaN layers is indicated by cathodoluminescene measurements. They also report an ELO technique starting with an AlAs/GaAs buffer [49]. Here GaAs is selectively etched into stripes and then GaAs and AlAs oxidized whereas only GaAs forms an α-Ga_2O_3 phase separated by the amorphous/fine grain aluminum oxide. The quality of GaN laterally overgrown from the α-Ga_2O_3 is reported to be improved significantly. Despite the promising results, especially for the latter method, no devices grown on such layers were reported until now.

Besides sapphire the hexagonal modification of SiC is the only commercially relevant substrate for GaN growth. It offers the smallest lattice mismatch to GaN and especially to AlN (~1%) which is, however, still high enough to induce dislocation densities above 10^{10} cm^{-2} at the interface. SiC substrates are very expensive, not available in large diameters, and often show high defect densities. Thus it is very interesting to form a high-quality SiC buffer on Si for further GaN growth as first proposed by Takeuchi and co-workers [50]. This is usually performed by using CH_4, ethylene [51], a thin SiC layer grown with tetramethylsilane [52, 53], or by converting Si to 3C-SiC at 1360 °C using a propane/H_2 mixture [54–57] in or ex-situ the growth chamber. In this way the benefit of SiC (good lattice match, no meltback etching) and Si (price, size) can be combined in a relatively simple way. The SiC formed is always reported to be of cubic 3C-SiC type and not the commonly used 6H- or 4H-SiC. But the 3C-SiC(111) surface which is formed on Si(111) is equivalent to the 6H-SiC (0001) plane in regard of the atom arrangement. Due to the different layer sequence of cubic and hexagonal SiC stacking faults can originate at steps when growing hexagonal 2H-GaN on it. On top of the 3C-SiC layer a thin AlN seed is usually grown. One problem with converted SiC is the poor surface morphology [51, 58] which does result in a poor layer quality. This can be avoided by surface treatments prior to growth [53], special buffer layer sequences [52], or using the substrate as template for pendeo-epitaxial overgrowth [54, 55, 59–61].

AlN seed layers are applied to sapphire and SiC and differ strongly from GaN seed layers. Due to the higher Al-N bonding energy AlN seed layers hardly rearrange thus such layers are completely covering and protecting the Si substrate from nitridation. Another benefit is the smaller lattice constant inducing compressive strain on the GaN layer which is supporting the growth of thicker crack-free layers by counterbalancing the thermally induced tensile strain to some extend. AlN seed layers on Si(111) are most often reported in literature.

Most common is the growth of high temperature (HT) AlN seed layers with typical thicknesses of 20–40 nm (see, e.g. Ref. [33, 38, 62–78]).

Such AlN seed layers typically consist of grains about 20–40 nm in diameter and some dislocations originate at the grain boundaries [67].

The influence of the growth temperature of AlN seed layers on the structural quality of the subsequently grown GaN was first reported by Watanabe and co-workers [62] and more recently by Lahrèche and co-workers [33] and Zamir and co-workers [78]. Watanabe and coworkers as well as Lahrèche and coworkers obtain similar results with optimum AlN deposition temperatures around 1050 °C for different AlN thicknesses around 100–200 nm for Watanabe et al. and 20–40 nm for Lahrèche et al.. Both groups obtain polycristalline AlN layers below ~950 °C while Zamir obtains well aligned AlN seed layers above 760 °C.

In contrast to these findings are our results. We use a low-temperature AlN seed layer deposited around 720 °C. Until now such low-temperature AlN seed layers are still rarely used [79–88]. Hageman and co-workers showed that the GaN quality is strongly depending on the growth temperature of the seed layers and there is only a small window for the growth [87]. In a report on MBE grown AlN on Si [89] best results were obtained for a layer grown 25 °C below the transition temperature from a 7×7 to a 1×1 reconstructed surface [90, 91] at 830 °C. There the orientation of AlN grown below 830 °C on the 7×7 reconstructed surface is reported to yield a better oriented AlN layer than grown at higher temperatures on the 1×1 reconstructed surface. The transition temperature is, within error of the temperature reading in MOVPE, the upper limit of optimum AlN deposition temperatures given by Hageman et al.

Interestingly, Hageman et al. obtained best GaN layers (3 μm thickness, GaN(0002) and $(10\bar{1}5)$ rocking curve widths of 832 arcsec and 702 arcsec and 4 K (D^0,X) PL FWHM of 17.6 meV) with a seed layer deposition at 820 °C (Fig. 5) while we observed best results (1.3 μm GaN(0002) and $(20\bar{2}4)$ rocking curve widths of 720 arcsec and 270 arcsec) at 720 °C [79]. Below 820 °C Hageman et al. obtained only a relatively poor layer quality.

The difference of our process is the predeposition of a few monolayers of Al prior to AlN growth which is likely preventing the Si substrate from nitridation. Indeed, no amorphous Si_xN_y layer can be found at the interface by TEM measurements (Fig. 3). Such a positive effect of an Al predeposition was also observed by Chen and co-workers who applied the predeposition to HT-AlN layers [92]. They clearly show that they have a steplike AlN surface at the optimum aluminum deposition time while short deposition lead to island growth of AlN and too long deposition to a rough AlN surface. While the latter is likely due to a deterioration of the Si surface by an Al/Si alloying reaction a too short Al deposition

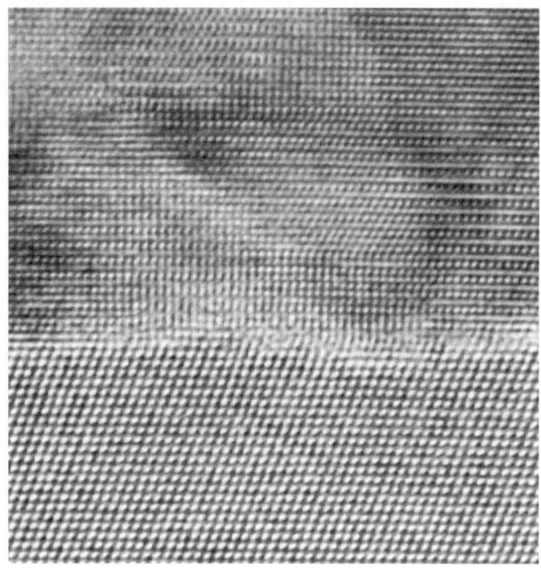

Fig. 3 High resolution TEM image of the Si/AlN interface grown with the method described in ref. 79. The interface is free of an amorphous interlayer.

might lead to a partly nitrided surface. The substrate can thus be protected by a thin covering Al layer which is transformed to AlN when ammonia is present.

One can conclude that in general, no AlN deposition method is superior in regard of crystallographic quality until now, since thin (~1 μm thick) GaN layers grown on AlN seed layers usually show best rocking curve widths around 600–700 arcsec regardless of the seed layer growth temperature. It seems, however, that most basic studies are not directly comparable, e.g. due to different AlN thicknesses, deposition temperatures, V–III ratios, reactor pressures or starting sequences (ammonia first, aluminum first or both at the same time). One reason in favor of LT-AlN is that the decomposition of GaN deposits on the reactor wall at high temperatures can strongly influence the HT-AlN seed layer growth, e.g. by nitriding or roughening the wafer surface prior to growth [34]. Another, still speculative aspect, is the higher chance to obtain low resistivity AlN due to a non-stochiometric composition. Since the Si/AlN interface is believed to have a high band-offset an AlN layer far from crystallographic and interface perfection might lower this barrier. But recent TEM results even suggest a relatively abrupt Si/AlN interface. Further investigations have to be performed to fully understand the microstruture of the AlN layer and the AlN as well as Si composition close to the interface. For the use of GaN on Si in vertically contacted devices like LEDs a low series resistance is decisive.

The problem of substrate nitridation can be completely avoided using an ammonia-free atmosphere for the seed layer growth, e.g. by using a GaAs or AlAs seed layer.

The first report on a nitrided GaAs layer on Si(111) was by Yamamoto et al. who applied it to grow InN [93]. A nitrided GaAs buffer layer was also applied by Yang et al. to grow hexagonal GaN/InGaN heterostructures on Si(111) [94]. Here the GaAs layer was deposited by MBE and subsequently nitrided in an MOVPE system.

We developed and patented a nitridation process for AlAs seed layers [95]. The seed layer growth starts with an AlAs buffer layer grown in two steps followed by a transformation to AlN [95–99]: After a low-temperature AlAs nucleation layer is grown at 450–500 °C a high-quality 50–100 nm AlAs layer is grown at 750 °C–850 °C. Subsequently, the AlAs layer is nitrided at 970 °C whereby an As–N exchange takes place resulting in an epitaxial AlN layer which serves for the growth of high-quality GaN on top of it with X-ray rocking curve widths <600 arcsec for 1.2 μm GaN and 10 K PL FWHM of 13 meV. From cross sectional TEM no formation of silicon nitride is observed at the interface (Fig. 4) and from secondary ion mass spectroscopy the Si incorporation is determined to be below the detection limit of 1×10^{16} cm^{-3} (Fig. 5). In contrast to GaN on sapphire, undoped GaN on Si is often highly insulating. Using such buffer layers on n-type Si(111) a simple LED structure with a single InGaN/GaN quantum well was realized [79].

For the fabrication of GaN-based devices on Si a low-cost seed layer should be used which enables device growth in one growth run. Both AlN and nitrided AlAs seed layers are low-cost processes. AlAs is also very interesting from several points: The AlN layer formed can be still As rich which is helpful

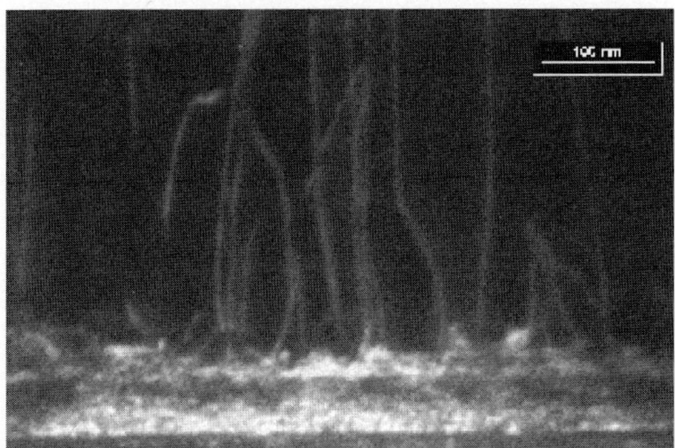

Fig. 4 TEM image of the interface region of Si (bottom), nitrided AlAs, and the GaN layer. A sharp AlN/Si interface is observed and most dislocations are anihilated in the seed layer.

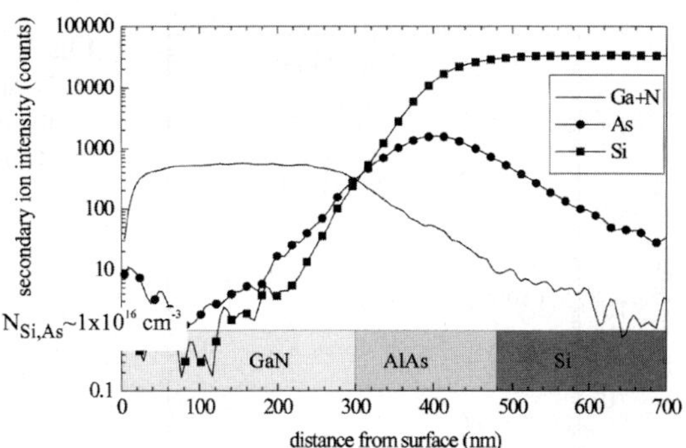

Fig. 5 SIMS measurement of a GaN/AlN(As)/Si(111) layer showing the Ga+N, As and Si signal [97]. The Si and As signals decrease below the detection limit ($<10^{16}$ cm^{-3}) after the

for substrate removal by oxidizing this layer and etching, e.g. in HF. It can also lead to low series resistance due to a lower band gap and possibly higher n-type conductivity when doped with Si.

4 Thick layers and strain engineering

For the growth of GaN-based device structures on Si by MOVPE strain engineering is the key to achieve high-quality material. Thicker layers are not only needed for light emitters but also for high-frequency devices where a good insulation of the active layer from the substrate as well as high carrier mobilities are required. For the latter, smooth surfaces, interfaces and a low defect density are essential. Another key issue is the radius of wafer curvature, which can get a serious problem in device processing, e.g. in contact lithography. The radius of wafer curvature for GaN layers on Si is often around 2 m and less [64, 80].

Because heteroepitaxial GaN growth usually starts from 3D islands the layer is always under intrinsic tensile stress. It has been reported that GaN on sapphire under standard conditions accumulates about 0.1-0.2 GPa/µm tensile stress (Fig. 13) [16, 100] which can lead to cracking of thick layers (>5 µm) during growth [101] and can be influenced by the density of initial 3D islands at the seed layer [18] (for the physical origin of intrinsic stress see, e.g. Ref 17). The large thermal mismatch adds another ~0.7 GPa/µm when cooling to room temperature, the biggest and certainly the most problematic part. Another source of tensile stress are impurities. Fu and co-workers showed that the residual stress of GaN on Si depends on the V–III ratio and can be reduced for high V–III ratios [68]. They assumed this to originate in a lower impurity incorporation. Impurities like Si and Mg used for n- and p-type doping, respectively are reported to strongly enhance the tensile stress (reduce the compressive stress at RT) of GaN and AlGaN layers on sapphire [19, 101]. Especially for Si doping, around 0.1 GPa/µm of additional tensile stress is induced for a doping concentration of 1×10^{18} cm^{-3}. This sounds negligible but for doping concentrations of 5×10^{18} cm^{-3} used in some devices ~0.5 GPa/µm are added (Fig. 13). By keeping a moderate n-type doping concentration too much stress, which leads to cracking, can be avoided. In the case of AlGaN growth the tensile stress induced by Mg doping is reported to be much smaller compared to Si doping but higher Mg concentrations are needed to achieve p-type material [101].

One general problem of substrate bowing, not only for GaN on Si, but also for GaN on SiC and sapphire, is the broadening of the X-ray rocking-curve width when a large spot is used for the measurement. On the other hand one can use the different broadening of the substrate reflection with varying spot size to determine the wafer curvature. When the GaN layer is cracked as it might be the case on Si this is preferably done by measuring from the backside of the Si substrate since then cracks are also present in the top of the Si substrate which negatively influence the measurement of the Si reflection

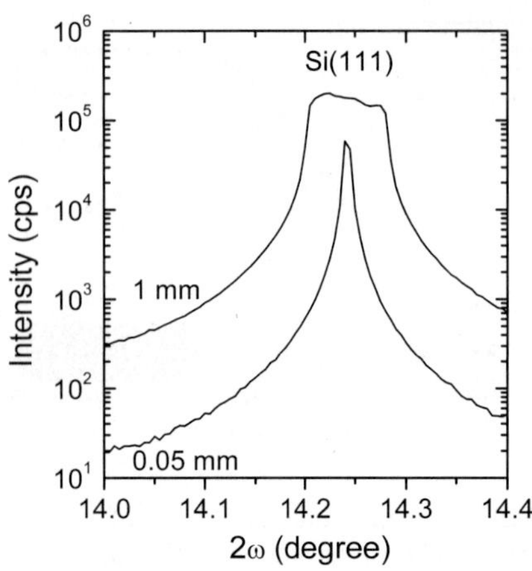

Fig. 6 Measurement of the Si(111) rocking curve of a 1.3 μm thick GaN on Si sample from ref. 80 with a primary beam slit width of 1 mm and 0.05 mm. The wafer curvature was determined to 2.9 m corresponding to an in-plane stress of 0.92 GPa.

(Fig. 6). From the wafer curvature, one can calculate [22] the basal plane stress σ_{11} and the lattice deformation σ_{11} of the GaN layer by

$$\sigma_{11} \approx \frac{E_{Si(111)} \cdot T_{Si}^2}{6 \cdot R \cdot (1 - 2 \cdot c_{13} / c_{33}) \cdot t_{GaN}} \tag{1}$$

and

$$\varepsilon_{11} = \frac{\sigma_{11}}{c_{11} + c_{12} - 2 \cdot c_{13}^2 / c_{33}} \tag{2}$$

where T_{Si} is the substrate thickness, t_{GaN} the GaN layer thickness, R the radius of the wafer bowing, $E_{Si(111)}$ Young's modulus in the (111) plane, and c_{ij} the elastic moduli of GaN. For heavily cracked layers (distance$_{cracks}$ <<100 μm) we observed that this way of determining the stress is too inaccurate because some stress might be relaxed by the cracks.

Not only cracks are likely to occur due to tensile stress but also meltback etching is initiated by exposed, not nitrided Si surfaces, e.g. when a crack occurs during growth of InGaN at lower temperatures. It has been also observed that defects on the surface, e.g. from particles, initiate cracks also when tensile stress is well below the commonly observed critical value around 0.9 GPa/μm. So, a low residual stress, a clean wafer surface, and a particle-free growth chamber are prerequisite to obtain a smooth and crack-free surface over the whole wafer.

In the following part we will describe some promising approaches for strain engineering.

One simple method to release epilayer strain and with it to solve the cracking problem, are compliant substrates which have been proposed for lattice mismatched epitaxy [102] and first applied to the nitride system with MBE [103]. The amorphous or gliding layers inserted below the surface of a substrate, e.g. by ion-implantation (SIMOX) or wafer bonding (SOI) are expected to decrease the stress simply by a slipping of the crystal on an amorphous matrix. Such substrates are used in the microelectronic industry to reduce the device capacitance and with it to increase the device speed. It is currently used for Si microelectronic devices and due to its capacitance lowering properties and high resistivity compared to standard Si substrates, it is very interesting for high-frequency GaN-based transistor applications. One limitation might be a low thermal conductivity of the amorphous SiO_2 layer. Thus it is likely less suited for high-power devices.

In SOI substrates, a thin silicon surface layer (~50 nm), which is separated by a thin (~80 nm) SiO_2 buried layer from the thick Si(001) or (111) substrate, acts as the compliant layer. For GaN growth a top Si(111) layer can be converted to 3C-SiC by carbonization [57]. MOVPE growth of GaN on such substrates resulted in high-quality material (corrected FWHM of the (0002) rocking curve of ~360 arcsec) comparable to GaN grown on 6H-SiC substrates. Another attempt to reduce the stress was tried with N^+-implanted Si(111) substrates [104].

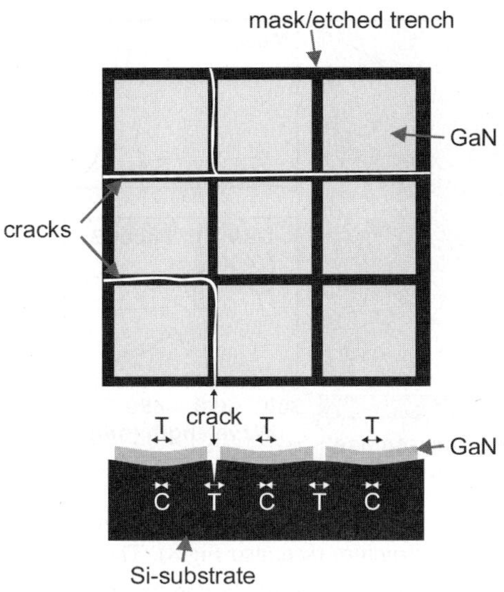

Fig. 7 Scheme of substrate masking and guiding cracks in masked areas (top). Due to the high tensile stress (bottom) cracks preferentially form in the masked regions and thus only in the Si substrate leading to a partial reduction of strain in the GaN layer for small masked regions. "C" denotes compressively strained regions, "T" tensily strained regions.

One method which we could successfully apply to grow crack-free LED structures is by patterning a silicon substrate [21, 83]. This can be done in several ways by masking or etching the substrate or a buffer layer. The idea is not to avoid cracks completely but to guide them in the masked or etched parts of the Si substrate (Fig. 7). As long as the patterned regions are smaller than the average crack distance, no cracking of the GaN layers should occur since the substrate is softer. The simplest way to apply patterning is the deposition of Si_3N_4 or SiO_2 directly on the substrate. Since such a layer is amorphous, no oriented crystalline seed layer can be grown on it. Thus, usually no or only little polycrystalline growth is observed on such a mask. The other possibility is etching trenches deep enough to avoid that the layers form a continuous film during growth. The latter was, e.g. performed by Zamir and co-workers who used etched trenches and called this technique lateral confined epitaxy. In their experiments they found that for 700 nm GaN already a field size of $14 \times 14 \ \mu m^2$ is the limit for crack-free growth [75]. We find that such thicknesses are no problem to grow crack-free on a whole wafer anyway. Thus, it is likely that a poor material quality is the reason for cracks in their experiments. They also observed an enhancement in PL intensity for smaller-sized fields and correlated this with an enhanced quality of the layer [76, 77].

Honda et al. showed that they can grow crack-free 1.5 µm thick GaN on 200x200 μm^2 fields [105]. A growth enhancement at the edges of the fields was observed [83, 105] which depends on the width of the masked region and is caused by mass transport from the masked regions. X-ray rocking curve widths of the GaN(0004) reflection of 388 arcsec and 1130 arcsec and a 77 K PL GaN band edge FWHM of 18.6 meV and 25.8 meV for a 1.5 µm thick layer on $200 \times 200 \ \mu m^2$ fields and for an unstructured layer were obtained, respectively [105].

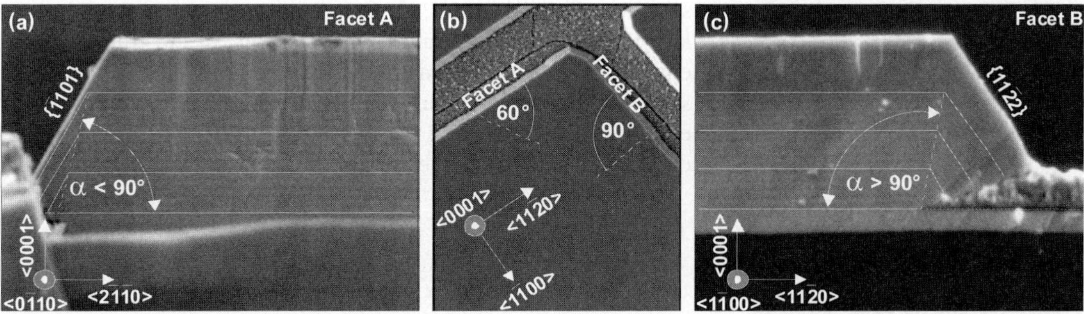

Fig. 8 Cross sectional (left and right) and top view (middle) of an LED structure with an AlGaN/GaN multilayer in the buffer grown on patterned Si(111). During growth two facets form at the sides of the unpatterned field: The growth rate of the facets differs strongly and is around 0.2 times the vertical growth rate for the $\{\bar{1}101\}$ facet and 0.8 for the $\{11\bar{2}2\}$ facet.

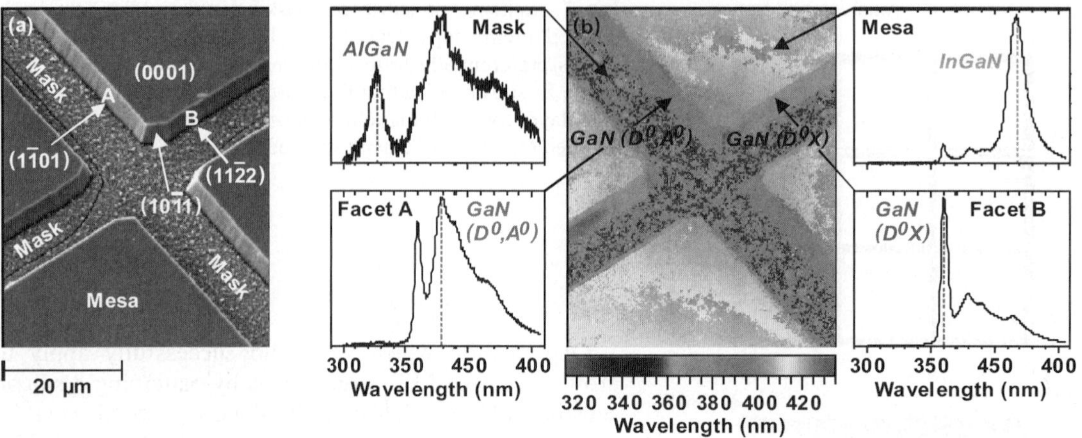

Fig. 9 (online colour at: www.interscience.wiley.com)   SEM image (left) and cathodoluminescence image (right) with some local spectra recorded at the side facets and the top of the LED structure (see also Fig.8). The luminescence of the slowly growing $\{\bar{1}101\}$ facet type is dominated by (D0,A0) luminescence likely due to an increased impurity incorporation while the luminescence of the $\{11\bar{2}2\}$ factes is dominated by (D0,X) luminescence.

By patterning we can grow crack-free device structures in fields significantly larger as described by Zamir et al. Largest areas until now are about 300x300 µm² for a 2.5 µm thick LED structure with AlGaN/GaN multilayers and 3.6 µm for 100 × 100 µm² fields (Fig. 8–10) [21, 82]. Also a reduction of the GaN (0002) X-ray rocking curve FWHM from 630" to 530" is observed for an unpatterned region of a wafer and differently sized fields around 200²–300² µm², respectively. This indicates an improvement in layer quality for GaN grown in the fields. Simultaneously, a small reduction in strain of the GaN layer is observed. Because of the high aspect ratio of these structures around 1/100 one cannot expect a relaxation and a significant influence on the strain of the layer due to the edges. Also the lateral overgrowth at the edges, which could influence our measurements since the free-hanging GaN is partially relaxed, is too small (~2 µm) to explain the data (Fig. 8). More likely strain is reduced by a modified growth as observed for AlGaN/GaN multilayers [82]. We have shown that stress relaxation occurs for the growth of such layers on unstructured substrates (Fig. 11) [82]. It can be assumed that also in the case of patterned substrates strain relaxation occurs already during growth but due to the finite size of the fields no cracks are generated in contrast to the samples reported in ref. 82. Patterning using masks might be a key to integrate GaN and Si devices on Si substrates. Using, e.g. a SiO₂ or Si₃N₄ mask to define the fields followed by selective growth one can easily integrate GaN devices with Si electronics by removing the mask and further processing the Si devices. The only disadvantage is the growth enhancement at

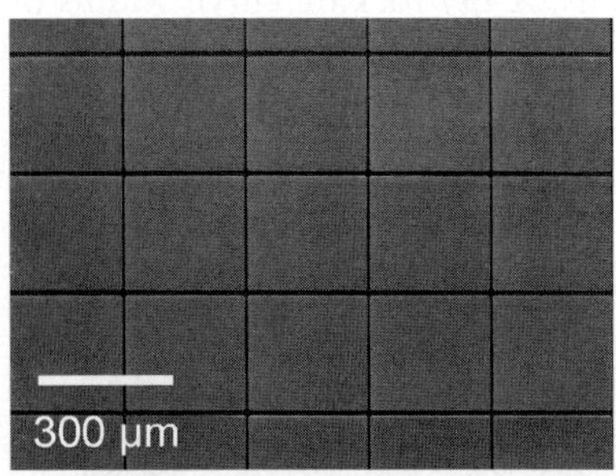

Fig. 10 Crack-free 2.5 µm thick LED structures on patterned Si(111) [67].

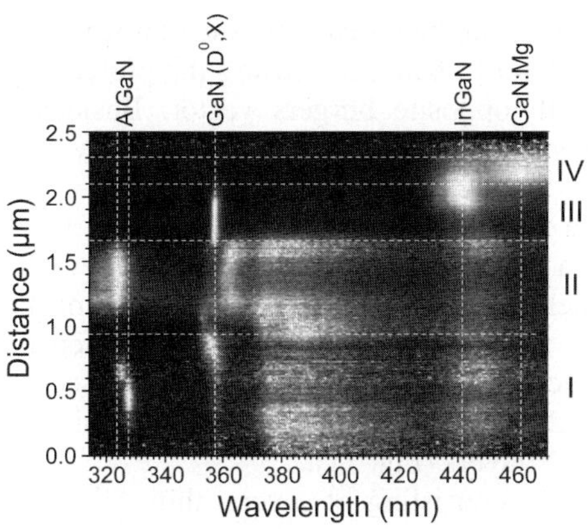

Fig. 11 Cross sectional CL-linescan (5 K, 5 keV) of a LED structure with a 15× (AlGaN:Si/GaN:Si, 50 nm/50 nm) buffer layer followed by 400 nm of GaN:Si, a threefold InGaN/GaN MQW and Mg doped 20 nm AlGaN and 200 nm GaN top layers [82]. Four regions can be distinguished: region I is the lower part of AlGaN/GaN layers with an Al concentration of ≈14% and region II the upper one with an Al concentration of ≈18%. A blue-shift of the AlGaN and GaN (D^0,X) luminescence occurs at around 0.6–0.9 µm correlated with a decrease of AlGaN luminescence likely due to a relaxation of the layer during growth. Region III comprises the n-GaN:Si, the MQW, and the AlGaN blocking layer and is dominated by InGaN MQW luminescence. In region IV, the top GaN:Mg layer, mostly Mg related luminescence is observed.

the edges of the fields due to mass transport from the masked areas to the growing surface but this does not negatively influence, e.g. LED properties. This effect can be significantly reduced when etched substrates are used which are less suited for devices to be integrated with Si electronics. The influence of the different growth facets which form during growth can be well resolved in cathodoluminescence (CL) measurements. We observe that CL from the slowly growing N-terminated {$\bar{1}$101} facet is dominated by DAP emission while the fast growing {11$\bar{2}$2} facets show predominantly (D^0,X) emission.

Another benefit is, that the cracks are propagating in the masked regions between the devices. Thus, device separation gets very easy, especially for square patterns as preferably used for LED structures (see e.g. Fig. 9 left). In Si, the primary cleavage planes are the {111} planes. Hence, for Si(111) cleavage is along the <110> directions (Fig. 3) and the wafer is naturally separated into triangles. Due to the cracks and the stress induced by the GaN layer in the substrate the separation of devices along the natural cleaving planes gets very difficult. Thus, instead of naturally triangular cleaved devices, the separation of square device structures is significantly simplified.

But patterning is also not free of problems: Due to tensile stress, cracks occur in the Si substrate during growth, especially when the temperature is lowered as required for InGaN quantum well growth. From these cracks meltback etching can be initiated. The suppression of growth on the mask for GaN layers leads to the diffusion of GaN onto the growing surface and thus enhances the growth rate at the edge of the device which can lead to problems in device processing.

Often, Al-rich buffer layers, which induce compressive stress on the subseqently grown GaN, are used for strain engineering [13, 63, 69, 70, 72]. While thick Al-rich layers are even advantageous for transistor structures, where good insulation from the substrate is required, they are negatively influencing the series resistance of vertically contacted LEDs. Additionally, such stressors can be only introduced in the lower part of the device structure and thus limit the maximum achievable thickness.

For thicker layers such stressors introduced in the lower part of the structure inevitably lead to a strong wafer bowing with convex curvature radii typically well below 1 m [13] hindering device processing. For a FET structure Marchand and co-workers [63] used a graded 800 nm thick AlGaN buffer layer on an AlN seed to induce compressive stress in the top GaN layer. In a sample with a thin 200 nm GaN cap a compressive stress in the GaN layer of 0.27 GPa at RT was introduced. But the 1 µm thick structure exhibited a high dislocation density well above 10^{10} cm^{-2}. The authors assumed that a 2-dimensional growth mode of the graded AlGaN buffer avoids that dislocations bend and annihilate. Growth of 2 µm thick GaN on Si(111) with and without a step graded AlGaN buffer on an AlN seed in an ultrahigh vacuum chemical vapor deposition reactor is reported by Kim and co-workers [72]. They observed a slight improvement of the layer quality in X-ray rocking curve FWHMs of the GaN(0002) and the asymmetric GaN (10$\bar{1}$2) reflections, a reduction of the room temperature PL FWHM, and a reduction of the surface roughness when a graded AlGaN buffer was applied. Also the crack density of the sample with the AlGaN buffer was significantly reduced.

It is already known from the cubic III–V system that dislocations can be reduced by superlattices, e.g. AlAs/GaAs. At the interfaces of such superlattices dislocations tend to bend, increasing the probability of recombination and annihilation with other dislocations with opposite burgers vector. Besides a reduction of dislocations in the nitride system by Al(Ga)N/GaN superlattices they can be applied on Si to induce a compressive stress in the layers subsequently grown.

We have shown that a stress reduction can be achieved by a 15-fold AlGaN/GaN buffer layer about 1.5 µm in total thickness which can in principle be also used as a bragg mirror for LEDs or laser devices. [82]. Here the stress of the 2.4 µm thick LED structure was likely reduced by a partial relaxation of the AlGaN/GaN layers after a thickness of approximately 0.9 µm was grown resulting in a weak tensile stress (0.35 GPa) of the top GaN layers at RT but also in cracks. A strong increase in luminescence intensity can be observed with the relaxation of the layers as well as a shift of the GaN (D^0,X) emission line and most likely also a change in Al composition of the AlGaN layers (Fig. 11).

More efficient for crack elimination is a reported buffer structure consisting of several thin AlN/GaN superlattices which induce compressive stress sufficient to grow 2.5 µm crack-free GaN on top of it with a biaxial tensile stress of 0.47 GPa at RT [13, 64–66, 74]. The AlN/GaN superlattice consisted of 3 stacks of 10 periods of 3 nm AlN and 4 nm GaN each stack separated by 200 nm GaN. The main problem with this method is the residual stress of the structure. A wafer curvature of only 1 m^{-1} was reported.

One of the most versatile methods to control strain is the introduction of LT-AlN layers which were first applied by Amano and co-workers to reduce the defect density of GaN layers [100] and to grow thick crack-free AlGaN layers on GaN on sapphire [106]. Such LT-AlN interlayers are also ideally suited to avoid cracking of GaN on Si (Fig. 12) [21, 80]. These AlN layers are usually only 10–20 nm in thickness which is sufficient to counterbalance the thermal tensile stress of approximately 0.7–1 µm thick GaN on Si while maintaining and usually even improving the layer quality. Because these layers can be repeatedly introduced in contrast to all other methods this is the only method to theoretically grow very thick GaN layers on Si free of cracks.

While such layers have been used successfully for strain engineering by many groups in recent years the mechanism leading to a compensation of tensile stress has long been unclear. We have performed a detailed study by X-ray diffraction measurements which enlightens the mechanism [86, 107]. It can be shortly described as a decoupling of the GaN layers above and below a LT-AlN layer. This decoupling is also dependent on the AlN deposition temperature which can be seen for AlN interlayers deposited at different temperatures. Here a clear dependence of stress versus growth temperature is observed. Thus, the deposition temperature is the most important parameter for a compensation of the tensile thermal stress (Fig. 13).

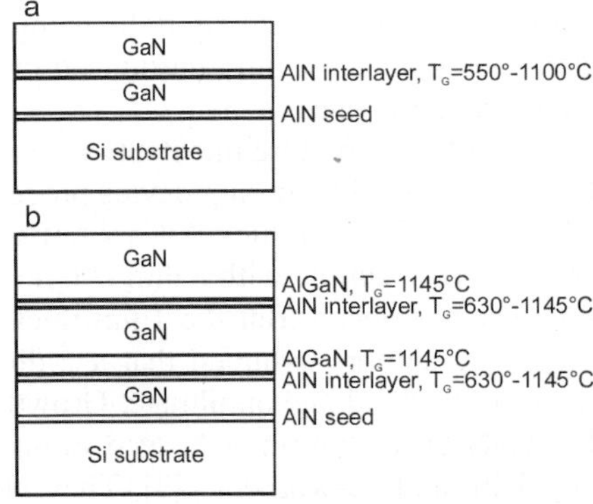

Fig. 12 Sample structure of the samples in Fig. 13(a) and Fig. 14 and 15(b).

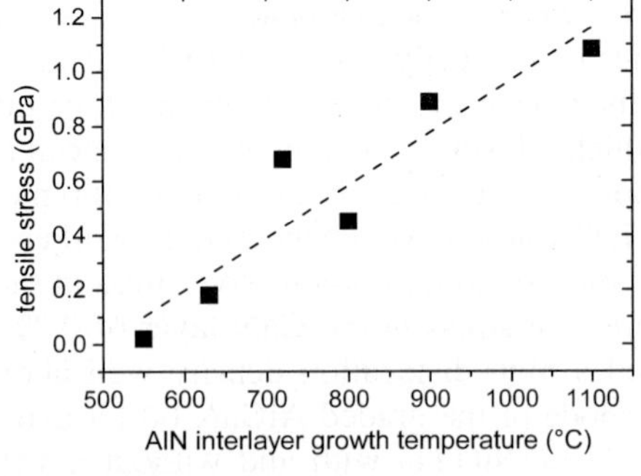

Fig. 13 Dependence of RT stress on the LT-AlN deposition temperature of a 1.2 µm thick GaN layer structure with one LT-AlN interlayer [72].

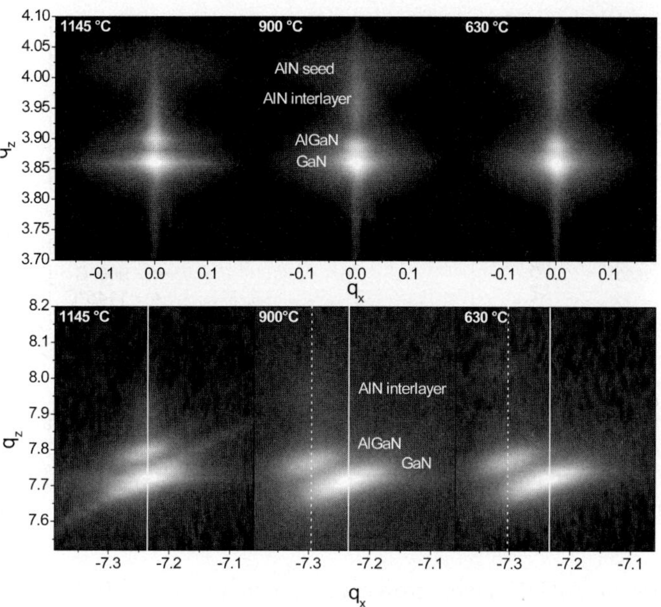

Fig. 14 X-ray reciprocal space maps of the structure b in Fig. 12 around the symmetric (0002) (top row) and the asymmetcric $(20\bar{2}4)$ reflection (bottom row) for three different AlN deposition temperatures. While the sample with a deposition temperature of 1145°C grows nearly pseudomorphic visible by the vertical alignment of the asymmetric reflection the other samples are decoupled (dotted line).

X-ray reciprocal space map measurements of a multilayer sample show (Fig. 14) that high temperature deposited AlN grows pseudomorphically on GaN. Thus, GaN or AlGaN grown on top of it has the same *a*-lattice constant as the buffer. At lower growth temperatures the coherence of the layers is destroyed and the LT-AlN layer acts more as a seed layer, however, on a high-quality substrate if compared to the Si-substrate. This incoherent growth can be also visualized in $\Theta\theta$–$2\Theta\theta$ scans of the *c*- and *a*-lattice parameter where a splitting of the GaN peak in three parts is observed corresponding to the three GaN layers separated by two AlN interlayers (Fig. 15).

The term "seed layer" or "multiple buffer layer" which is often used in literature does not match exactly our findings. By ellipsometric measurements we find that the interlayer differs in its strain state from the seed layer (Fig. 16). Because the interlayer is grown on a GaN layer the *c*-and *a*-axis orientations are maintained and even improved by the LT-AlN clearly visible in a reduction of the rocking curve widths for the GaN(0002) reflection from 720 arcsec to 600 arcsec and the asymmetric GaN $(20\bar{2}4)$ reflection from 270 arcsec to 65 arcsec for a 1.3 µm thick GaN layer with and without LT-

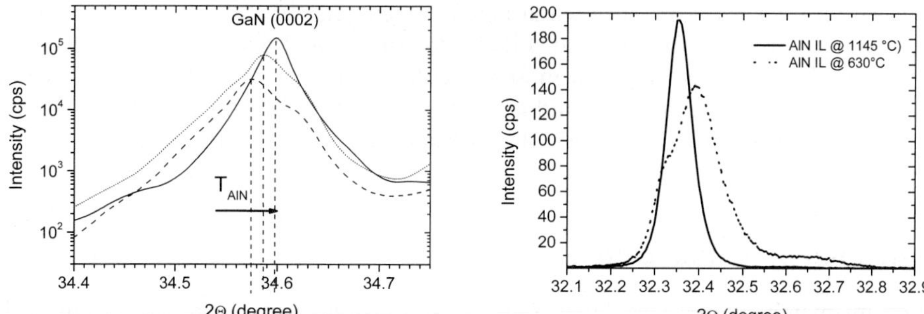

Fig. 15 Θ-2Θ scans of the GaN(0002) (left) and the GaN $(10\bar{1}0)$ (right) reflections of samples with two AlN interlayers grown at different temperatures. In the GaN (0002) scan a shift of the main GaN peak position from tensile to nearly relaxed GaN is observed for a sample with a HT-AlN interlayer (full line), with LT-AlN grown at 900°C (dotted line), and LT-AlN grown at 630°C (dashed line). The samples with LT-AlN interlayers show a splitting of the GaN(0002) reflection into three peaks corresponding to the three GaN layers decoupled by LT-AlN [86]. A similar splitting is observed for a scan of the a-lattice parameter (right)

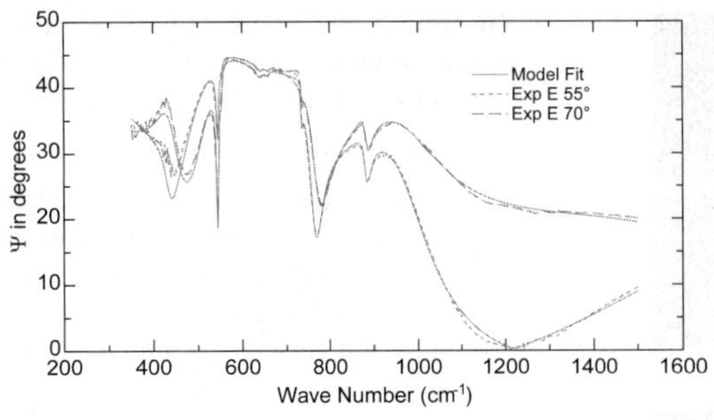

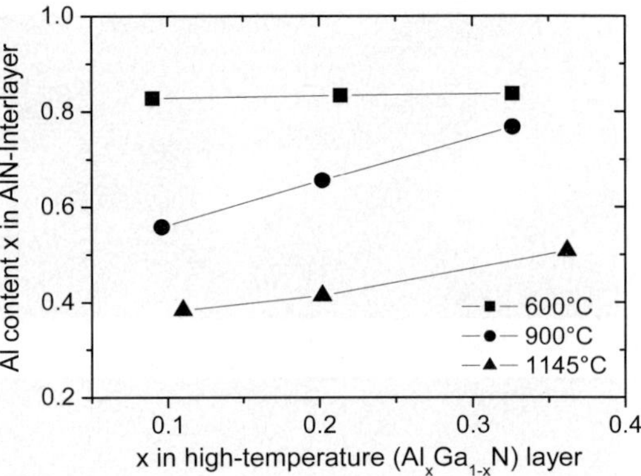

Fig. 16 (online colour at www.interscience.wiley. com) Ellipsometric measurements of a GaN/LT-AlN/GaN/AlN-seed/Si structure. The AlN E_1(TO) mode shows a slight splitting which can be attributed to differently strained AlN seed and interlayers.

Fig. 17 Ga-concentration in the AlN interlayer in dependence of the Al-content of the subsequently grown AlGaN layer and the AlN interlayer deposition temperature.

AlN interlayers, respectively [88]. The GaN or AlGaN layer grown on top of the LT-AlN layer is partially compressively stressed which counterbalances the thermal tensile stress when grown on Si. We have also observed that the Al content of the AlN interlayer is depending on the deposition temperature and the Al content of the subsequently grown layer (Fig. 17). Obviously, Ga indiffuses into the AlN interlayer and forms AlGaN. This can account partly for the temperature dependent difference in strain compensation by these interlayers.

Another important point is the influence of such interlayers on the dislocation density. One can expect that a non-pseudomorphically grown AlN film introduces new dislocations. In contrast to this assumption, Amano and co-workers reported [14] that AlN interlayers reduce the threading dislocation density. Other reports claimed that this is only the case for screw-type dislocations [108]. Our results suggest that this depends on several factors, e.g. the LT-AlN deposition temperature and thickness (Fig. 18) [109]. Evidently, a drastic reduction in threading dislocation density can be observed for some LT-AlN layers (Fig. 17 right) while it remains about the same for others (Fig. 17 left).

Many methods to reduce stress and cracks in GaN layers were demonstrated in the last years. Patterning is interesting for special device applications like LEDs where device size and field size can be easily matched or GaN devices integrated with Si electronics. The counterbalancing of tensile thermal stress by inducing compressive strain in the buffer layer is suitable for thin layers used, e.g. in FETs but it is not appropriate for thicker layers since wafer bending then gets very strong and even then very thick layers are impossible to grow by this method. Likely, the distributed decoupling of the GaN layers by multiple LT-AlN interlayers is the only and most versatile method to grow thick layers. Until now we have grown 5 µm thick crack-free layers with this method using four LT-AlN interlayers. They show X-ray rocking curve FWHMs of 370" for the symmetric (0002) reflection and 300" for the asymmetric (20$\bar{2}$4) reflecion.

5   Reduction of dislocations

The high lattice mismatch of the seed layer and the Si substrate causes an initially high dislocation density at the substrate/seed-layer interface. This high initial dislocation density above 10^{13} cm^{-2} is usually reduced to values below 10^{10} m^{-2} within a few hundred nanometers by dislocation anihilation. To further reduce the dislocation density many methods are known from the growth on sapphire and SiC. Most of these methods use ex-situ masking and/or etching and lateral overgrowth of nearly defect free GaN layers. One method also suited for a reduction in dislocation density are low-temperature AlN interlayers which have been already described in the previous section.

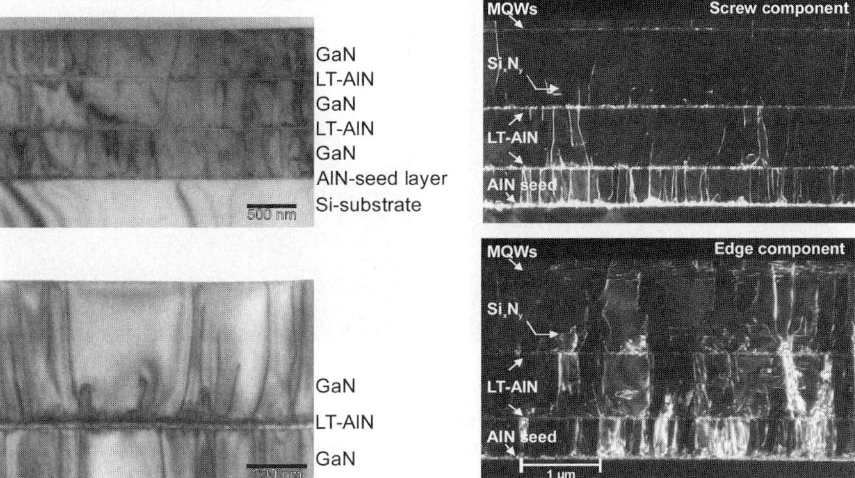

Fig. 18 Multi-beam dark field cross section TEM image of Si/AlN-seed/GaN/LT-AlN/GaN/LT-AlN/GaN buffer layer (left) and LED (right) samples [80, 85]. The dislocations are not significantly reduced at the LT-AlN interlayers for the 1.3 µm thick buffer layer sample (screw and edge: $\sim 10^{10}$ cm^{-2}) while they are significantly reduced to $\sim 10^{9}$ cm^{-2} (screw: $\sim 10^{8}$ cm^{-2}, edge: $\sim 10^{9}$ cm^{-2}) for the ~ 2.5 µm thick LED sample (right).

Epitaxial lateral overgrowth (ELOG) also named lateral epitaxial overgrowth (LEO) is a well established growth method to obtain low dislocation density, high-quality GaN. Here a GaN buffer layer is masked with stripes (e.g. SiO_2 or Si_xN_y) and overgrown. A low dislocation density is then obtained in the laterally overgrown regions except for the coalescence region.

Kung and co-workers were the first to report on MOVPE grown ELOG on Si(111) [10]. They performed growth on a 200 nm thick GaN buffer on Si and obtained GaN free of pinholes and dislocations in the overgrown region. Wing tilt, which is common for ELOG, was observed. Low dislocation densities below 10^6 cm^{-2} in the laterally overgrown regions were reported by Marchand et al. [110]. They also observed a degradation of the layers before coalescence when a too thin AlN buffer was used: This degradation is likely due to meltback etching as we could also observe in our experiments. The coalesced layers showed cracks and wing tilt.

A combination of stress-reducing buffer layers and ELOG was applied by Feltin and co-workers who used AlN/GaN superlattices in the buffer for the lateral overgrowth [64, 111]. While the layers showed a low dislocation density of 5×10^7 cm^{-2} in the overgrown region they also reported that the layers were cracked in contrast to layers on sapphire. A simpler method is the direct patterning of Si e.g. with 1 µm wide SiO_2 and 1 µm wide openings as reported by Tanaka et al. [112]. They used an AlGaN intermediate layer on the Si substrate to reduce stress and grew 2 µm GaN until full coalescence but also observe cracks. The average threading dislocation density is reported to be around 2×10^9 cm^{-2} which can be also achieved without ELOG by other methods.

A method that has been also used by NITRONEX Inc. is pendeo-epitaxy developed by the group of R.F. Davis. Here pendeo-epitaxy is applied on a Si(111) substrate covered with a 2 µm thick 3C-SiC layer [54, 55 59–61]. On this layer an AlN seed and a GaN buffer were grown in a first growth step. Then parts of the layer were masked and trenches etched into the substrate. Subsequent growth of GaN at the sidewalls of the GaN buffer layer lead to a freestanding nearly defect-free GaN layer. A tilt of the freestanding wings was observed as it is the case for most lateral growth techniques and nearly defect-free GaN was obtained. But cracking of the GaN layer was observed when the layers were coalesced. The cracking problem can be likely eliminated by using compliant SOI substrates [61] where the Si on top of the SiO_x is completely converted to SiC and the pendeo approach applied to it. This results in a low threading dislocation density, crack-free GaN layers on Si. But the technological effort to obtain

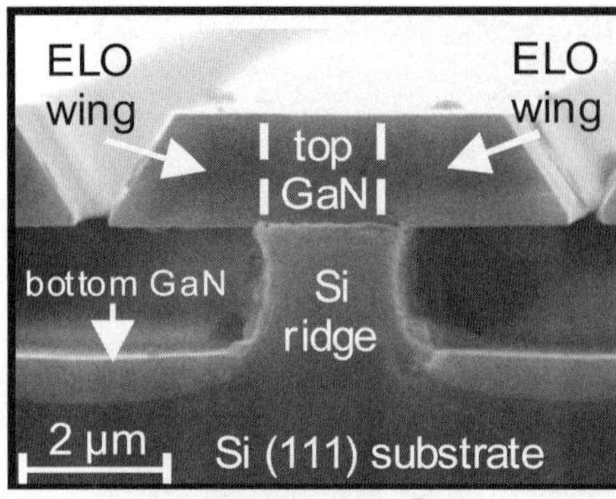

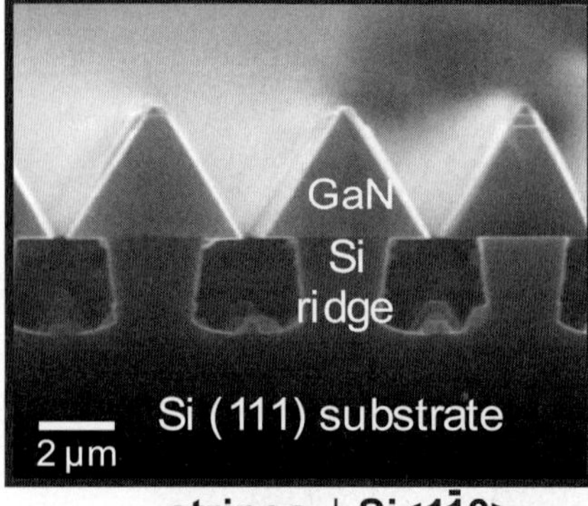

Fig. 19 SEM cross section image of GaN grown on a structured Si(111) substrate with different stripe orientations[99]. The laterally grown wings extend about 2.5 µm over the grooves (left). The thickness of the GaN layer is 0.5 µm on the bottom of the grooves and 1.4 µm on top of the ridges. for the other stripe direction the lateral growth rate is significantly lower.

such layers is high and expensive since a SOI substrate has to be overgrown at least twice (SiC formation, buffer layer growth, etching, GaN overgrowth). For light emitters vertically contacted devices are preferred which are difficult to manufacture due to the highly resistive substrate and buffer layers.

A much simpler maskless pendeo or cantilever process based on a stripe patterned Si substrate has been developed by us [99]. Here several microns deep trenches are etched into the Si(111) surface (Fig. 19, 20). On this structure an AlAs-based nitrided seed layer and GaN are grown. In the growth experiments it can be observed that the growth rate at the top of the ridges is higher than at the bottom of the trenches. By changing the growth conditions to achieve an enhanced lateral growth rate the trenches can be overgrown before the material deposited at the bottom of the trenches reaches the top GaN (Fig. 20). The resulting GaN is of high-quality in the wing regions as can be seen in AFM and cathodoluminescence measurements (Fig. 20). The integral CL-luminescence increases by two orders of magnitude in the wing region and local (D^0,X)-linewidths of around 4 meV have been measured (Fig. 21). The surface in the wing region shows regularly arranged steps and no dislocation induced holes or pits as opposite to conventionally grown GaN.

The same approach to improve GaN on Si(111) has been also used by Katona and co-workers [113]. They also observed a significant reduction in dislocation density.

In our case the GaN layers grown on stripe patterned GaN still show wing tilt in X-ray diffraction measurements. layers. Also a strong tensile stress is present leading to cracks for coalesced layers (Fig. 22). With the addition of strain engineering layers this problem might be solved soon.

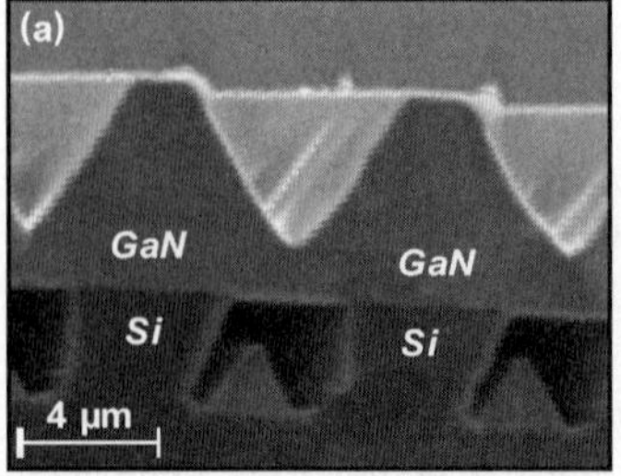

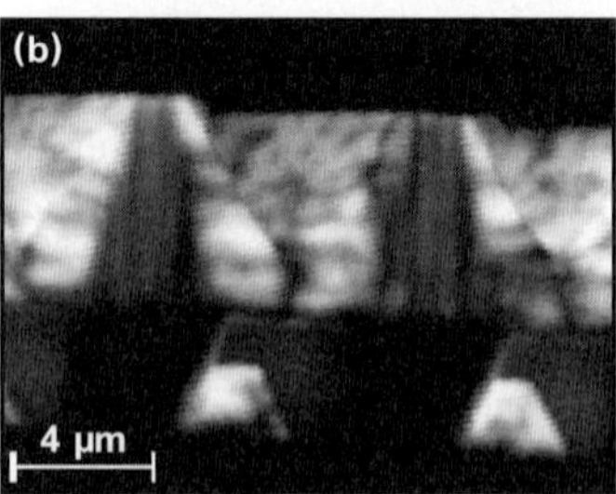

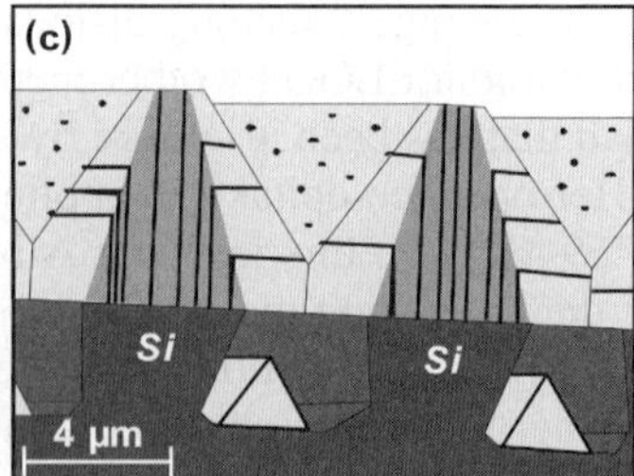

Fig. 20 SEM cross section (a), CL intensity image (b) and schematic view (c) visualising the dislocations which are still present at the surface of the GaN above the trenches. At the facet they bend during growth leading to a nearly dislocation free GaN layer above the trenches.

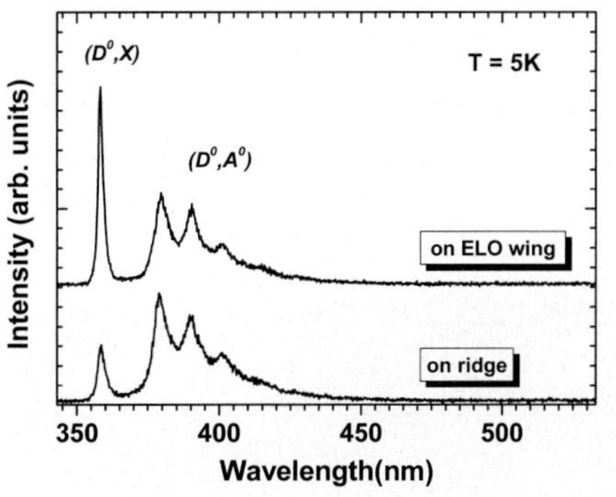

Fig. 21 cathodoluminescence recorded above the ridge and on the ELO wings.

One of the simplest methods to reduce dislocations is the usage of in-situ deposited Si_xN_y interlayers, first described by Tanaka and coworkers [114] and Lahrèche and co-workers for GaN growth on sapphire [115]. Hageman and co-workers were the first to apply it to GaN on Si [15]. The idea behind such a Si_xN_y layer in the thickness range of nominally one to a few monolayers, is to mask or bend the dislocations in-plane by masking the GaN surface and to start GaN growth from islands forming on the mask as for ELOG growth (Fig. 23, 24). The GaN growth starts from small openings in the mask due to a nonuniform coverage or or by applying a new low-temperature GaN seed layer on top of the mask (Fig. 24).

TEM investigations show that the reduction of screw type dislocations at the mask is significantly higher than for edge-type dislocations. In fact, edge-type dislocations are nearly unaffected and usually a change in the TEM images of egde components is only observed because these are mixed dislocations and the screw component has been anihilated or bent. A model for dislocation reduction has been proposed which involves pinning by the silicon impurity at surface lattice steps associated with screw dislocations. (Fig. 25)

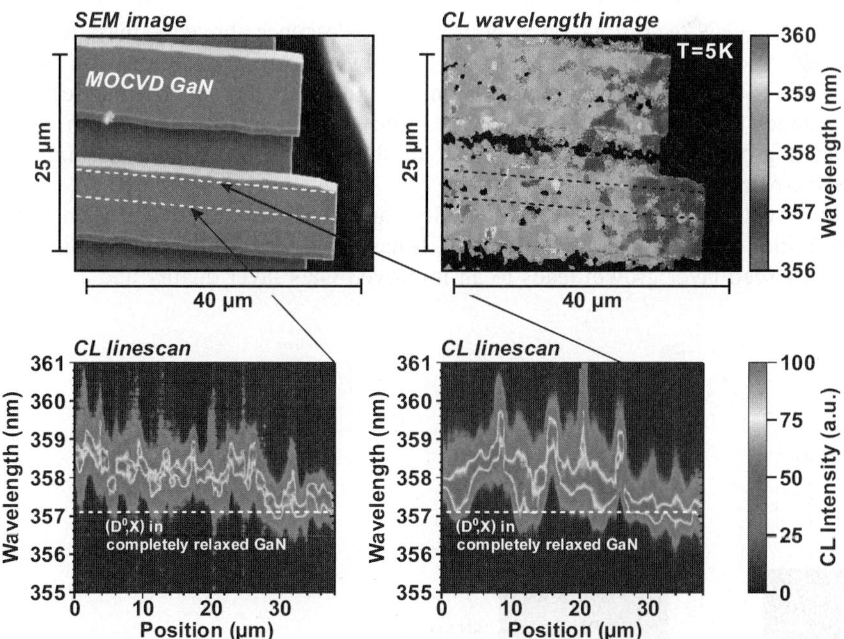

Fig. 22 (online colour at: www.interscience.wiley.com) CL top view of GaN grown on stripe patterned Si. At the edges a shift of the GaN peak emission is visible in the line scans (bottom) but also in the wavelength image (top left) due to a relaxation of the layer which is under strong tensile stress.

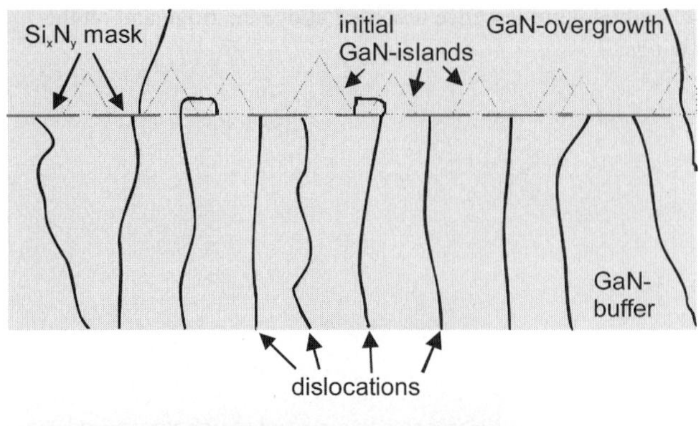

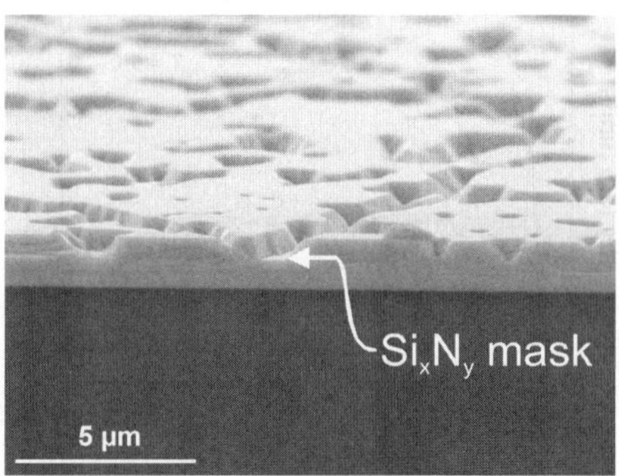

Fig. 24 SEM image of an uncoalesced GaN layer grown on a Si$_x$N$_y$ masked GaN buffer. The masking effect is visible by the regions without GaN nucleation where lateral overgrowth from the GaN islands takes place until coalescence.

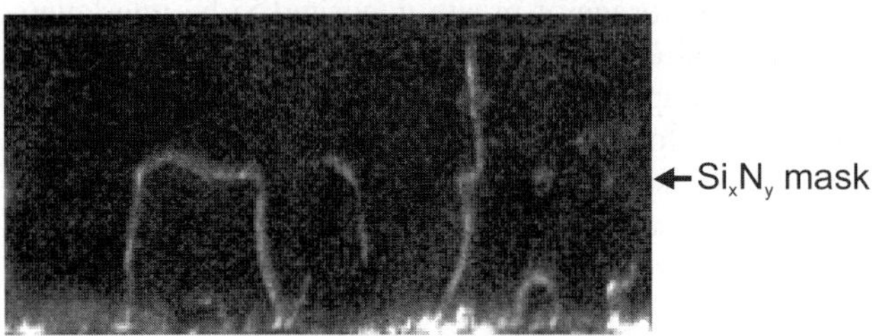

Fig. 23 Schematic view of possible dislocation termination by a Si$_x$N$_y$ in-situ mask. Dislocations are terminated directly at the mask, bend and anihilate with other dislocations at the mask, or bend and anihilate above the mask which can be forced by growing faceted islands before coalescence of the layers.

Most important for the GaN overgrown on Si$_x$N$_y$ is a fast coalescence above the mask to keep the layer thickness well below the critical value for cracking (~1 µm) which is especially difficult when the layer is doped with Si, since Si hinders the adatom diffusion and with it the lateral growth. Depending on the mask width and distance even for GaN:Si a fast coalescence can be achieved by a high ammonia flow and slightly elevated growth temperatures [84, 85]. In the case of an in-situ masking this can be influenced by the Si coverage of the surface.

A significant improvement of the GaN layer quality can be already achieved by an application of Si$_x$N$_y$ masks at the initial stages of GaN growth [116]. It has been demonstrated that a Si$_x$N$_y$ masking of sapphire prior to GaN seed layer growth leads to an improved GaN layer quality and a reduction in the

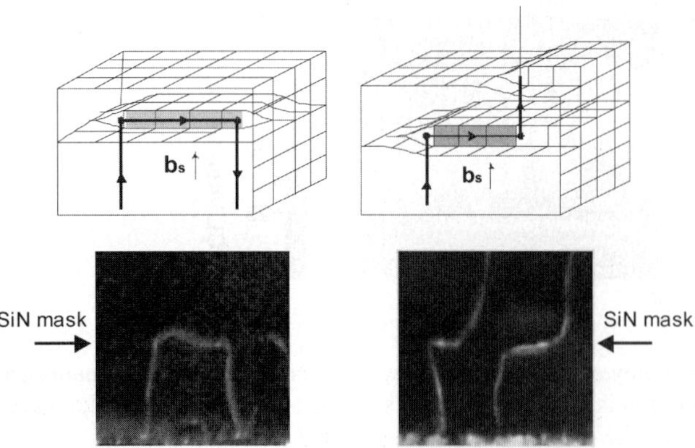

Fig. 25 Model and corresponding TEM images of screw dislocation annihilation or bending by Si$_x$N$_y$ masking

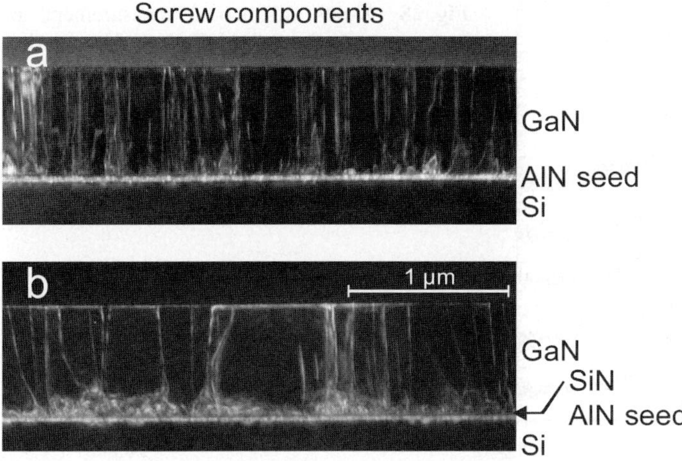

Fig. 26 TEM images of two samples without (top) and with (bottom) a SiN in-situ masking of the AlN seed layer showing a reduction in screw dislocation denisty and dislocation clustering above the seed layer.

dislocation density [117]. Unfortunately, on Si such a predeposition of Si_xN_y is not possible since Si itself rapidly nitridates which is difficult to control. But a Si_xN_y masking can be applied after the growth of, e.g. an AlN seed layer, which accommodates most of the lattice mismatch and helps in orienting the GaN layer subsequently grown. When applying such a mask a threefold reduction in the dislocation density is observed for a Si_xN_y deposition time of 90s (1–1.5 monolayers nominally) compared to a reference sample (Fig. 26) [116]. The PL intensity increases about an order of magnitude for a 750 nm thick GaN layer and the PL peak energy is shifted about 21 meV (Fig. 27). This shift in PL energy can be attributed to a decrease in tensile stress of about 0.8 GPa in agreement with X-ray diffractometry measurements. The decrease in stress can be attributed to the growth of GaN from few islands which lead to larger sized domains and thus to less boundaries likely responsible for most of the intrinsic stress during growth [17, 18] and some of the dislocations, generated at the coalescence boundary of the misoriented grains.

The last two methods, in-situ masking and cantilever epitaxy from structured substrates, are both well suited for the growth of low dislocation GaN at very low cost. While in-situ masking is less effective it is easier to combine with stress reducing interlayers and avoid cracks which are a problem for cantilever epitaxy. However, for a sub-micron structured cantilever process one can expect thin coalescence thicknesses and by this strain engineering is possible again.

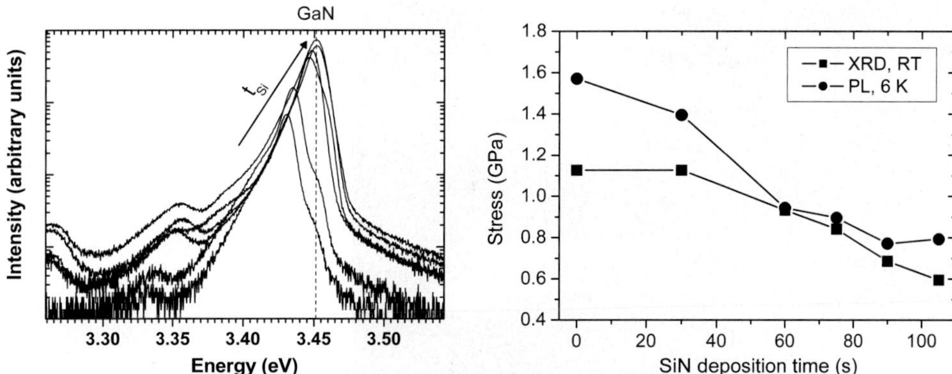

Fig. 27 PL spectrum of 750 nm thick GaN samples grown on an AlN seed layer on Si(111). On the AlN seed layer a Si_xN_y mask was deposited with different Si deposition times (t_{Si}=0-105 s; 0- ~1.5 monolayers). A strong increase in PL intensity and a shift of the PL peak energy by 21 meV can be observed. This is due to a decrease in tensile stress of ~0.8 GPa [116].

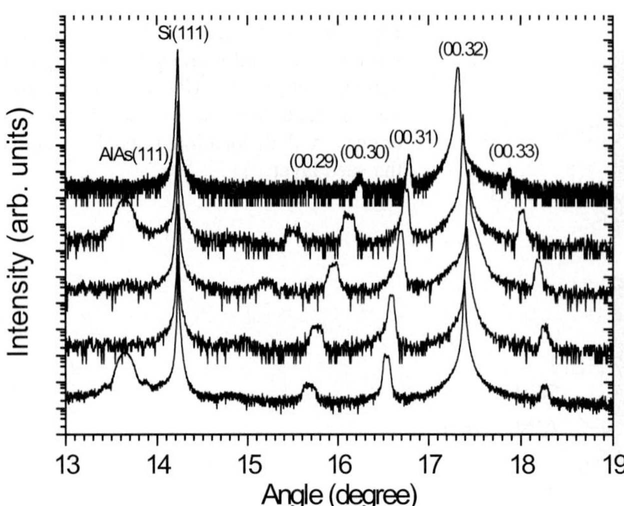

Fig. 28 X-ray diffraction measurement of spontaneously formed Al(Ga)N/GaN superlattices grown with different Al and Ga ratios.

6 Spontaneous superlattice formation

Spontaneous superlattice formation of ternary and quaternary Group-III-nitride alloys has been reported by several authors [118, 119]. Depending on the period and composition the formation of such superlattices can be also used for the growth of quantum wells. We have observed spontaneous superlattice formation for the growth of AlGaN in a wide range of compositions from low Al content around 6.5% up to 30%. When growing bulk AlGaN layers superlattice interference fringes are observed in highly-resolved XRD measurements (Fig. 28) [120]. These superlattice fringes indicate a highly ordered superlattice structure. We find for a 1.2 µm thick AlGaN layer which was grown with a TMAl/(TMAl + TMGa) ratio in the gas phase of 2.5% a superlattice structure with a period of 8.3 nm consisting of one unit cell AlN and 15 GaN unit cells. This superlattice can be also well resolved in TEM measurements (Fig. 29). Here a high interface abruptness is observed which is of high perfection. One can also observe that the high periodicity is maintained throughout the whole AlGaN layer.

The main parameters which influence the superlattice composition and period are the amount of Al, Ga, and ammonia in the gas phase while reactor pressure and temperature have a relatively low

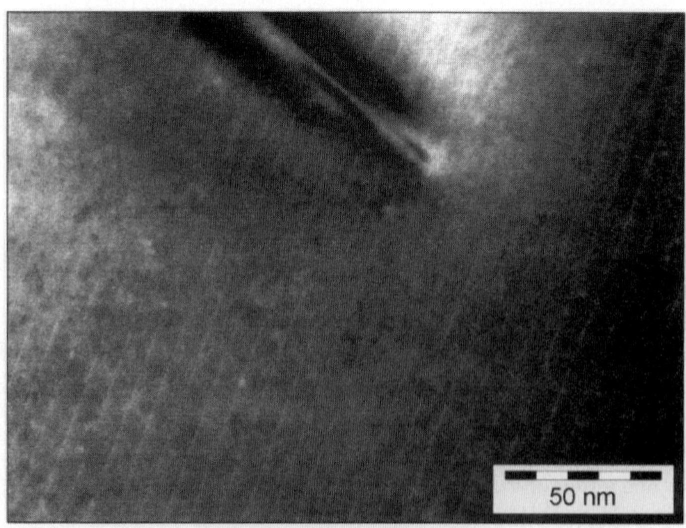

Fig. 29 TEM image of a spontaneously formed AlN/GaN superlattice.

influence. The average composition and the period of the SL depend on the TMAl/TMGa-ratio and the growth rate, respectively.

The exact mechanism leading to the formation of such superlattices is presently unknown. In our case it is likely due to a strain driven segregation of Al until it reaches a critical concentration on the surface and forms a high Al content layer. Instabilities of the MOVPE equipment can be ruled out since the AlGaN formation is stable in a wide growth regime and the growth of one superlattice period takes around 30 s which can't be explained by machine instabilities. The only difference to the growth process on Si to that of other groups is the application of an AlAs seed layer which could cause an As contamination during AlGaN growth. However, even when a clean susceptor (exchanged after AlAs growth) was used, the SL persisted.

7 Summary

In the last years several concepts for strain engineering have been presented to solve the main problem of GaN MOVPE growth on Si, cracking due to thermal mismatch. The crack-free layer thicknesses and quality achieved is presently sufficient to manufacture FET and LED devices. Electrically pumped laser devices on Si have not been reported yet and will be a question of a significant reduction in dislocation density for uncracked films. For this goal methods as ELOG, pendeo, or cantilever epitaxy must be combined with stress reducing layers to achieve uncracked films with low dislocation densities. It can be expected, that if no large homoepitaxial GaN substrates are available in the near future, Si substrates will play an important role in the production of FET and LED devices

Acknowledgements Parts of this work has been funded by the Deutsche Forschungsgenmeinschaft under contracts #Kr1239/10-1, #Ch87/4-1, #Bi284/25-2, and #Rh28/3-2. We would like to thank AIXTRON AG for the support in the early stages of this project. O. Contreras permanent address is Centro de Ciencias de la Materia Condensada, UNAM, Ensenada, Baja California, CP 22860, Mexico.

References

[1] H.M. Manasevit, F.M. Erdmann, and W.J. Simpson, J. Electrochem. Soc. **118**, 1864 (1971).

[2] H. Amano, I. Akasaki, K. Hiramatsu, N. Koide, and N. Sawaki, Thin Solid Films **163**, 415 (1988).

[3] H. Amano, M. Kito, K. Hiramatsu, and I. Akasaki, Jpn. J. Appl. Phys. **28**, L2121 (1989).

[4] S. Nakamura, N. Iwasa, M. Senoh, and T. Mukai, Jpn. J. Appl. Phys. **31**, 1258 (1992).

[5] S. Guha and N.A. Bojarczuk, Appl. Phys. Lett. **72**, 415 (1998).

[6] S. Guha and N.A. Bojarczuk, Appl. Phys. Lett. **73**, 1487 (1998).

[7] S. Nakamura, phys. stat. sol. (a) **176**, 15 (1999).

[8] S. Nagahama, N. Iwasa, M. Senoh, T. Matsushita, Y. Sugimoto, H. Kiyoku, T. Kozaki, M. Sano, H. Matsumura, H. Umemoto, K. Chocho, and T. Mukai, Jpn. J. Appl. Phys. **39**, L647 (2000).

[9] J.P. van der Ziel, R.D. Dupuis, R.A. Logan, and C.J. Pinzone, Appl. Phys. Lett. **51**, 89 (1984).

[10] P. Kung, D. Walker, M. Hamilton, J. Diaz, and M. Razeghi, Appl. Phys. Lett. **74**, 570 (1999).

[11] R.F. Davis, T. Gehrke, K.J. Linthicum, P. Rajagopal, A.M. Roskowski, T. Zheleva, E.A. Preble, C.A. Zorman, M. Mehregany, U. Schwarz, J. Schuck, and R. Grober, MRS Internet J. Nitride Semicond. Res. **6**, 14 (2001).

[12] A. Strittmatter, S. Rodt, L. Reißmann, D. Bimberg, H. Schröder, E. Obermeier, T. Riemann, J. Christen, and A. Krost, Appl. Phys. Lett. **78**, 727 (2001).

[13] E. Feltin, B. Beaumont, M. Laügt, P. de Mierry, P. Vennéguès, M. Leroux, and P. Gibart, phys. stat. sol. (a) **188**, 531 (2001).

[14] H. Amano, M. Iwaya, N. Hayashi, T. Kashima, M. Katsuragawa, T. Takeuchi, C. Wetzel, and I. Akasaki, MRS Internet J. Nitride Semicond. Res. **4S1**, G10.1 (1999).

[15] P.R. Hageman, S. Haffouz, V. Kirilyuk, A. Grzegorczyk, and P.K. Larsen, phys. stat. sol. (a) **188**, 523 (2001).

[16] S. Hearne, E. Chason, J. Han, J. A. Floro, J. Figiel, J. Hunter, H. Amano, and I. S. T. Tsong, Appl. Phys. Lett. **74**, 356 (1999).

[17] J.A. Floro, E. Chason, R.C. Cammarata, and D.J. Srolovitz, MRS Bull. **27**, 19 (2002).

[18] T. Böttcher, S. Einfeldt, S. Figge, R. Chierchia, H. Heinke, D. Hommel, and J. S. Speck, Appl. Phys. Lett. **78**, 1976 (2001).

[19] L. T. Romano, C. G. Van de Walle, J. W. Ager III, W. Götz, and R. S. Kern, J. Appl. Phys. **87**, 7745 (2000).

[20] H. Ishikawa, K. Yamamoto, T. Egawa, T. Soga, T. Jimbo, and M. Umeno, J. Cryst. Growth **189/190**, 178 (1998).

[21] A. Krost and A. Dadgar, Mater. Sci. Eng. B **93**, 77 (2002).

[22] Y. Okada and Y. Tokumaru, J. Appl. Phys. **56,** 314 (1984).

[23] see, e.g. Landolt-Börnstein, III/41A1a (Springer, Berlin/Heidelberg/New York, 2001).

[24] Properties of crystalline silicon, edited by R. Hull, EMIS datareview series No. 20 (1999).

[25] M.S. Shur, and M.A. Khan, Mater. Res. Bull. **22**, 44 (1997).

[26] G.A. Slack, R.A. Tanzilli, R.O. Pohl, and J.W. Vandersande, J. Phys. Chem. Solids **48**, 641 (1987).

[27] E.K. Sichel and J.I. Pankove, J. Phys. Chem. Solids **38**, 330 (1977).

[28] see, e.g. Landolt-Börnstein, Vol. 17 (Springer, New York, 1992).

[29] Madelung, Semiconductors Group IV and III–V compounds (Springer, Berlin, 1991).

[30] Cree Inc., product specification.

[31] M. Grundmann, A. Krost, and D. Bimberg, Appl. Phys. Lett. **58**, 284 (1991).

[32] J.W. Yang, C.J. Sun, Q. Chen, M.Z. Anwar, M.A. Khan, S.A. Nikishin, G.A. Seryogin, A.V. Osinsky, L. Chernyak, H. Temkin, C. Hu, and S. Mahajan, Appl. Phys. Lett. **69**, 3566 (1996).

[33] H. Lahrèche, P. Vennéguès, O. Totterau, M. Laügt, P. Lorenzini, M. Leroux, B. Beaumont, and P. Gibart, J. Cryst. Growth **217**, 13 (2000).

[34] I.-H. Lee, S.J. Lim, and Y. Park, J. Cryst. Growth **235**, 73 (2002).

[35] M. Sano, and M. Aoki, Jpn. J. Appl. Phys. **15**, 1943 (1976).

[36] Y. Morimoto, K. Uchiho, and S. Ushio, J. Electrochem. Soc. Solid-State Sci. Technol. **120**, 1783 (1973).

[37] T. Lei, K.F. Ludwig, Jr., and T:D. Moustakas, J. Appl. Phys. **74**, 4430 (1993).

[38] P. Kung, A. Saxler, X. Zhang, D. Walker, T.C. Wang, I. Fergusson, and M. Razeghi, Appl. Phys. Lett. **66**, 2958 (1995).

[39] L. Wang, X. Liu, Y. Zan, J. Wang, D. Wang, D.-C. Lu, and Z. Wang, Appl. Phys. Lett. **72**, 109 (1998).

[40] J.-H. Boo, S.-B. Lee, Y.-S. Kim, J.T. Park, K.-S. Yu, and Y. Kim, phys. stat. sol. (a) **176**, 711 (1999).

[41] J. Camassel, P. Vincente, N. Planes, J. Allègre, J. Pankove, and F. Namavar, phys. stat. sol. (b) **216**, 253 (1999).

[42] X .Zhang, S.-J. Chua, P. Li, K.-B. Chong, and Z.-C. Feng, Appl. Phys. Lett. **74**, 1984 (1999).

[43] X. Zhang, S.J. Chua, Z.C. Feng, J. Chen, and J. Lin, phys. stat. sol. (a) **176**, 605 (1999).

[44] J. Wan, R. Venugopal, M.R. Melloch, H.M. Liaw, and W.J. Rummel, Appl. Phys. Lett. **79**, 1459 (2001).

[45] X. Zhang, Y.-T. Hou, Z.-C. Feng, and J.L. Chen, J. Appl. Phys. **89**, 6165 (2001).

[46] L. Zhao, H. Marchand, P. Fini, S.P. DenBaars, U.K. Mishra, and J.K. Speck, MRS Internet J. Nitride Semicond. Res. **5S1**, W3.3 (2000).

[47] U. Kaiser, P.D. Brown, I. Khodos, C.J. Humphreys, H.P.D. Schenk, and W. Richter, J. Mater. Res. **14**, 2036 (1999).

[48] N.P. Kobayashi, J.T. Kobayashi, P.D. Dapkus, W.-J. Choi, and A.E. Bond, Appl. Phys. Lett. **71**, 3569 (1997).

[49] N.P. Kobayashi, J.T. Kobayashi, X. Zhang, P.D. Dapkus, and D.H. Rich, Appl. Phys. Lett. **74**, 2836 (1999).

[50] T. Takeuchi, H. Amano, K. Hiramatsu, N. Sawaki, and I. Akasaki, J. Cryst. Growth **115**, 634 (1991).

[51] H.M. Liaw, R. Doyle, P.L. Fejes, S. Zollner, A. Konkar, K.J. Linthicum, T. Gehrke, and R.F. Davis, Solid State Electron. **44**, 747 (2000).

[53] C.I. Park, J.H. Kang, K.C. Kim, E.-K. Suh, K.Y. Lim, and K.S. Nahm, J. Cryst. Growth **224**, 190 (2001); J.H. Kang, M.K. Kwon, J.I. Rho, J.W. Yang, K.Y. Lim, and K.S.Nahm, phys. stat. sol. (a) **188**, 527 (2001).

[54] R.F. Davis, T. Gehrke, K.J. Linthicum, P. Rajagopal, A.M. Roskowski, T. Zheleva, E.A. Preble, C.A. Zorman, M. Mehregany, U. Schwarz, J. Schuck, and R. Grober, MRS Internet J. Nitride Semicond. Res. **6**, 14 (2001) .

[55] R.F. Davis, T. Gehrke, K.J. Linthicum, E. Preble, P. Rajagopal, C. Ronning, C. Zorman, and M. Mehregany, J. Cryst. Growth **231**, 335 (2001).

[56] C.A. Zorman, A.J. Fleischman, A.S. Dawa, M. Mehregany, C. Jakob, S. Nishino, and P. Pirouz, J. Appl. Phys. **78**, 5136 (1995).

[57] A.J. Steckl, J. Devrajan, C. Tran, and R.A. Stall, Appl. Phys. Lett. **69**, 2264 (1996).

[58] H.M. Liaw, S.Q. Hong, K. Linthicum, R.F. Davis, and P. Fejes, MRS Symp. Proc. **572**, 219 (1999).

[59] R.F. Davis, T. Gehrke, K.J. Linthicum, T.S. Zheleva, P. Rajagopal, C.A. Zorman, and M. Mehregany, MRS Internet J. Nitride Semicond. Res. **5S1**, W2.1. (2000).

[60] T. Gehrke, K.J. Linthicum, P. Rajagopal, E.A. Preble, and R.F. Davis, MRS Internet J. Nitride Semicond. Res. **5S1**, W2.4. (2000).

[61] K.J. Linthicum, T. Gehrke, R.F. Davis, D.B. Thomson, and K.M. Tracy, US 6 255 198.

[62] A. Watanabe, T. Takeuchi, K. Hirosawa, H. Amano, K. Hiramatsu, and I. Akasaki, J. Cryst. Growth **128**, 391 (1993).

[63] H. Marchand, L. Zhao, N. Zhang, B. Moran, R. Coffie, U.K. Mishra, J.S. Speck, S.P. DenBaars, and J.A. Freitas, J. Appl. Phys. **89**, 7846 (2001).

[64] E. Feltin, B. Beaumont, M. Laügt, P. de Mierry, P. Vennéguès, M. Leroux, and P. Gibart, phys. stat. sol.. (a) **188**, 531 (2001).

[65] E. Feltin, S. Dalmasso, P. de Mierry, B. Beaumont, H. Lahrèche, A. Bouillé, H. Haas, M. Leroux, and P. Gibart, Jpn. J. Appl. Phys. **40**, L738 (2001).

[66] E. Feltin, B. Beaumont, M. Laügt, P. de Mierry, P. Vennéguès, H. Lahrèche, M. Leroux, and P. Gibart, Appl. Phys. Lett. **79**, 3230 (2001).

[67] D.M. Follstaedt, J. Han, P. Provencio, and J.G. Fleming, MRS Internet J. Nitride Semicond. Res. **4S1**, G3.72 (1999).

[68] Y. Fu, D.A. Gulino, and R. Higgins, J. Vac. Sci. Technol. A **18**, 965 (2000).

[69] H. Ishigawa, G.Y. Zhao, N. Nakada, T. Egawa, T. Soga, T. Jimbo, and M. Umeno, phys. sts. sol. (a) **176**, 599 (1999).

[70] H. Ishigawa, G.-Y. Zhao, N. Nakada, T. Egawa, T. Jimbo, and M. Umeno, Jpn. J. Appl. Phys. **38**, L492 (1999).

[71] S. Kaiser, M. Jakob, J. Zweck, W. Gebhardt, O. Ambacher, R. Dimitrov, A.T. Schremer, J.A. Smart, and J.R. Shealy, J. Vac. Sci. Technol. B **18**, 733 (2000).

[72] M.-H. Kim, Y-G. Do, H.C. Kang, D.Y. Noh, and S.-J. Park, Appl. Phys. Lett. **79**, 2713 (2001).

[73] H. Lahrèche, G. Nataf, E. Feltin, B. Beaumont, and P. Gibart, J. Cryst. Growth **231**, 329 (2001).

[74] H.P.D. Schenk, E. Feltin, M. Vaille, P. Gibart, R. Kunze, H. Schmidt, M. Weihnacht, and E. Doghèche, phys. stat. sol. (a) **188**, 537 (2001).

[75] S. Zamir, B. Meyler, and J. Salzman, Appl. Phys. Lett. **78**, 288 (2001).

[76] S. Zamir, B. Meyler, and J. Salzman, J. Cryst. Growth **230**, 341 (2001).

[77] S. Zamir, B. Meyler, J. Salzman, F. Wu, and Y. Golan, J. Appl. Phys. **91**, 1191 (2001).

[78] S. Zamir, B. Meyler, E. Zolotoyabko, and J. Salzman, J. Cryst. Growth **218**, 181 (2000).

[79] A. Dadgar, J. Christen, S. Richter, F. Bertram, A. Diez, J. Bläsing, A. Krost, A. Strittmatter, D. Bimberg, A. Alam, and M. Heuken, IPAP Conference Series **1**, 845 (2000).

[80] A. Dadgar, J. Bläsing, A. Diez, A. Alam, M. Heuken, and A. Krost, Jpn. J. Appl. Phys. **39**, L1183 (2000).

[81] A. Dadgar, M. Poschenrieder, J. Bläsing, K. Fehse, T. Riemann, A.Diez, J. Christen, and A. Krost, MRS Fall Meeting Boston (MA), **I4.7** (2001).

[82] A. Dadgar, J. Christen, T. Riemann, S. Richter, J. Bläsing, A. Diez, A. Krost, A. Alam, and M. Heuken, Appl. Phys. Lett. **78**, 2211 (2001).

[83] A. Dadgar, A. Alam, T. Riemann, J. Bläsing, A. Diez, M. Poschenrieder, M. Straßburg, M. Heuken, J. Christen, and A. Krost, phys. stat. sol. (a) **188**, 155 (2001).

[84] A. Dadgar, M. Poschenrieder, J. Bläsing, K. Fehse, A. Diez, and A. Krost, Appl. Phys. Lett. **80**, 3670 (2002).

[85] A. Dadgar, M. Poschenrieder, O. Contreras, J. Christen, K. Fehse, J. Bläsing, A. Diez, F. Schulze, T. Riemann, F.A. Ponce, and A. Krost, phys. stat. sol. (a) **192**,308 (2002).

[86] A. Reiher, J. Bläsing, A. Dadgar, A. Diez, and A. Krost, J. Cryst. Growth, in print.

[87] P.R. Hageman, S. Haffouz, V. Kirilyuk, A. Grzegorczyk, and P.K. Larsen, phys. stat. sol. (a) **188**, 523 (2001).

[88] C.A. Tran, A. Osinski, R.F. Karlicek, and I. Berishev, Appl. Phys. Lett. **75**, 1494 (1999).

[89] E.S. Hellman, D.N.E. Buchanan, and C.H. Chen, MRS Internet J. Nitride Semicond. Res. **3**, 43 (1998).

[90] F. Jona, Appl. Phys. Lett. **6**, 205 (1965).

[91] W. Telieps and E. Bauer, Surf. Sci. **162**, 163 (1985).

[92] P. Chen, R. Zhang, Z.M. Zhao, D.J. Xi, B. Shen, Z.Z. Chen, Y.G. Zhou, S.Y. Xie, W.F. Lu, and Y.D. Zheng, J. Cryst. Growth **225**, 150 (2001).

[93] A. Yamamoto, Y. Yamauchi, M.Ohkubo, A. Hashimoto, and T. Saitoh, Proceeedings of the TWN 95, Nagoya (Japan), 39 (1995).

[94] J.W. Yang, C.J. Sun, Q. Chen, M.Z. Anwar, M.A. Khan, S.A. Nikishin, G.A. Seryogin, A.V. Osinsky, L. Chernyak, H. Temkin, C.Hu, and S. Mahajan, Appl. Phys. Lett. **69**, 3566 (1996).

[95] A. Strittmatter, A. Krost, and D. Bimberg, DE 19725900.

[96] A. Strittmatter, A. Krost, J. Bläsing, and D. Bimberg, phys. stat. sol. (a) **176**, 611 (1999).

[97] A. Strittmatter, A. Krost, M. Straßburg, V. Türck, D. Bimberg, J. Bläsing, and J. Christen, Appl. Phys. Lett. **74**, 1242 (1999).

[98] A. Strittmatter, D. Bimberg, A. Krost, J. Bläsing, P. Veit, J. Cryst. Growth **221**, 293 (2000).

[99] A. Strittmatter, S. Rodt, L. Reißmann, D. Bimberg, H. Schröder, E. Obermeier, T. Riemann, J. Christen, and A. Krost, Appl. Phys. Lett. **78**, 727 (2001).

[100] H. Amano, M. Iwaya, T. Kashima, M. Katsuragawa, I. Akasaki, J. Han, S. Hearne, J.A. Floro, E. Chason, and J. Figiel, Jpn. J. Appl. Phys. **37**, L1540 (1998).

[101] S. Terao, M. Iwaya, R. Nakamura, S. Kamiyana, H. Amano, and I. Akasaki, Jpn. J. Appl. Phys. **40**, L195 (2001).

[102] C.L. Chua, W.Y. Hsu, C.H. Lin, G. Christenson, Y.H. Lo, Appl. Phys. Lett. **64**, 3640 (1994).

[103] Z. Yang, F. Guarin, I.W. Tao, W.I. Wang, and S.S. Iyer, J. Vac. Sci. Technol. B **13**, 789 (1995).

[104] E.K. Koh, Y.J. Park, E.K. Kim, C.S. Park, S.H. Lee, J.H. Lee, and S.H. Choh, J. Cryst. Growth **218**, 214 (2000).

[105] Y. Honda, Y. Kuroiwa, M. Kawaguchi, and N. Sawaki, Appl. Phys. Lett. **80**, 222 (2002).

[106] M. Iwaya, S. Terao, N. Hayashi, T. Kashima, H. Amano, and I. Akasaki, Appl. Surf. Sci **159-160**, 405 (2000).

[107] J. Bläsing, A. Reiher, A. Dadgar, A. Diez, and A. Krost, Appl. Phys. Lett. **81**, 2722 (2002).

[108] D.D. Koleske, M.E. Twigg, A.E. Wickenden, R.L. Henry, R.J. Gorman, J. A. Freitas, Jr., and M. Fatemi, Appl. Phys. Lett. **75**, 3141 (1999).

[109] A. Dadgar, M. Poschenrieder, J. Bläsing, O. Contreras, F. Bertram, T. Riemann, A. Reiher, M. Kunze, I. Daumiller, A. Krtschil, A. Diez, A. Kaluza, A. Modlich, M. Kamp, J. Christen, F.A. Ponce, E. Kohn und A. Krost, J. Cryst. Growth **248**, 556 (2003).

[110] H. Marchand, N. Zhang, L. Zhao, Y. Golan, S.Y. Rosner, G. Girolami, P.T. Fini, J.P. Ibbetson, S.P. DenBaars, J.S. Speck, and U.K. Mishra, MRS Internet J. Nitride Semicond. Res. **4**, 2 (1999).

[111] E. Feltin, B. Beaumont, P. Vennéguès, T. Riemann, J. Christen, J.P. Faurie, and P. Gibart, phys. stat. sol. (a) **188**, 733 (2001).

[112] S. Tanaka, Y. Honda, N. Sawaki, and M. Hibino, Appl. Phys. Lett. **79**, 955 (2001).

[113] T.M. Katona, M.D. Craven, P.T. Fini, J.S. Speck, and S.P. DenBaars, Appl. Phys. Lett. **79**, 2907 (2001).

[114] S. Tanaka, S.Iwai, and Y. Aoyagi, Appl. Phys. Lett. **69**, 4096 (1996).

[115] H . Lahrèche, P. Vennéguès, B. Beaumont, and P. Gibart, J. Cryst. Growth **205**, 245 (1999).

[116] A. Dadgar, M. Poschenrieder, A. Reiher, J. Bläsing, J. Christen, A. Krtschil, T. Finger, T. Hempel, A. Diez, and A. Krost, Appl. Phys. Lett. **82**, 28 (2003).

[117] T. Wang, Y. Morishima, N. Naoi, and S. Sakai, J. Cryst. Growth **213**, 188 (2000).

[118] P. Rutherana, G. DeSaint-Jores, M. Laügt, F. Omnes, and E. Bellet-Amatric, Appl. Phys. Lett. **78**, 344 (2001).

[119] N.A. El-Masry, M.K. Behbekani, S.F. LeBoeuf, M.E. Aumer, J.C: Roberts, and S.M. Bedair, Appl. Phys. Lett. **79**, 1616 (2001).

[120] A. Strittmatter, L. Reißmann, D. Bimberg, P. Veit, and A. Krost, phys. stat. sol. (b) **234**, 722 (2002).

phys. stat. sol. (c) **0**, No. 6, 1607–1626 (2003) / **DOI** 10.1002/pssc.200303133

Molecular beam epitaxy of cubic III-nitrides on GaAs substrates

D. J. As[*], **D. Schikora,** and **K. Lischka**

Department of Physics, University of Paderborn, 33098 Paderborn, Germany

Received 4 March 2003, revised 4 May 2003, accepted 26 June 2003
Published online 12 August 2003

PACS 61.10.Nz, 68.55.Ln, 72.80.Ey, 78.55.Cr, 81.15 Hi

Molecular beam epitaxy has successfully been used to grow crystalline layers of group III-nitrides (GaN, AlN and InN) with cubic (zinc-blende) structure on GaAs substrates. In this article, we discuss these efforts that, despite inherent difficulties due to the metastability of the c-III nitrides, led to substantial improvements of the structural, electrical and optical quality of these wide gap semiconductors. We review experimental work concerned with the epitaxy of c-GaN and the control of the growth process in-situ, the important issue of p- and n-type doping of c-GaN and investigations of the structural and optical properties of c-InGaN and c-AlGaN.

1 Introduction

The group III-nitrides have emerged as the most important material for blue and green light emitting diodes (LEDs) and blue laser diodes. In the last years tremendous research efforts have been taken to grow these semiconductors in high quality by various growth methods.

Most of the activity has been devoted to III-nitrides with hexagonal (wurtzite) crystal structure which is the thermodynamically stable configuration of these compounds. However, the III-nitrides can also exist in the cubic (zinc-blende) structure, which has only a slightly higher structural energy. The growth of cubic III-nitrides was strongly motivated by some advantageous properties which are all linked to their crystal structure. These are: a smaller electron and hole effective mass and an easier access to p-type doping, a higher saturated electron drift velocity and the possibility to produce laser facets by simple cleavage. However, in order to exploit their potential for practical applications, the metastable cubic III-nitrides have to be synthesized in sufficient good quality.

In this paper we summarize our activities to grow cubic III-nitrides by Molecular Beam Epitaxy (MBE) and the achievements with this material, which have been reached in the last years.

2 Molecular beam epitaxy of cubic GaN

The metastable cubic phase requires growth techniques which are operating under far from equilibrium conditions. Using MBE, which is known as nonequilibrium growth method, it has been proven by a number of researchers that c-GaN, AlGaN and InGaN of good crystalline quality can be grown [1-19].

---

[*] Corresponding author: e-mail: d.as@uni-paderborn.de, Phone: +49 5251 60 3567, Fax: +49 5251 60 3490

2.1   Substrates

This is the bodytext. The first important issue of the growth of cubic nitrides is the choice of appropriate substrates which are suitable for monocrystalline nucleation and growth. So far, 3C-SiC [1] [11] [12] [20] [21] [22], GaAs [2] [3] [5] [7] [8] [23] [13] [14] [15] [16] [19] [24] [25] [26] [27] [28] [29] Si [4] [6], GaP [18], InAs [30] and MgO [10] have been used as substrates. All these substrates were used in orientations having 4-fold symmetry, which is obviously an essential requirement for the epitaxy of metastable cubic GaN. The metric misfit among these materials varies from f ~ 20 % [GaAs/c-GaN] up to f ~ -3,5 % [3C-SiC/ c-GaN].

The most frequently used substrate material for cubic GaN is GaAs because of its commercial availability, the well- known surface preparation procedures and the possibility to generate cleaved facets. For this reason most of the growth experiments at the University of Paderborn were performed at GaAs substrates.

A common feature of the growth experiments reported so far, is the use of a low temperature nucleation stage, which was performed usually about 100-200 °C below the optimum growth temperature. The optimum growth temperatures determined on the different substrate materials are within a narrow range of about 680-740 °C. This is significantly lower than the growth temperatures reported for the MOCVD growth of hexagonal GaN, which is about 950-1100 °C. It has to be considered that in MBE the thermal decomposition of the c-GaN surface starts at temperatures above 800 °C, characterised by an initially congruent and at higher temperatures incongruent sublimation of surface atoms into the vapour phase [31].

Epitaxial growth of c-GaN at temperatures at or slightly above 800 °C can be performed under conditions near a dynamic evaporation-condensation equilibrium, where in particular displaced surface atoms exhibit a much larger probability to re-evaporate into the vapour. If the surface mobility of the species involved is rather low, as it is known from the nitrides, such a growth regime offers some advantages with respect to the crystalline perfection. However, the most frequently used GaAs-substrates do not allow substrate temperatures of about 800 °C, due to the strong loss of As and the resulting decomposition of the substrate during the growth experiment. Therefore, the optimum growth temperatures for c-GaN on GaAs substrates given in the literature [2, 3, 5, 7-9, 13-16, 19, 27, 28, 30, 32, 33] of about 680-720 °C are clearly below 800 °C and obviously represent a compromise between the c-GaN crystallinity and the substrate stability.

2.2   Nucleation stage

It is widely accepted that nucleation is the most critical step for obtaining c-GaN layers of high phase purity. Due to the lower total energy of the hexagonal phase compared to the cubic phase a structural transformation of the nucleation layer by any activation steps can be made from the cubic to the hexagonal phase, but never vice versa. Therefore it is of decisive importance to analyse the parameters which promote the formation of cubic nuclei.

Wu et al. [34] have found in MOCVD experiments that GaN nucleation layers grown at 600 °C on (0001) sapphire substrates consist mainly of cubic phase material, which can be transformed into a hexagonal phase after annealing at 950 °C. It is obvious from this result that low substrate temperatures are preferable for the formation of the cubic phase of GaN.

Kinetic studies for the nucleation of c-GaN on GaAs substrates were not reported up to now. However, from the analysis of cross sectional HRTEM images at the c-GaN/GaAs interface, a rather detailed qualitative picture of the nucleation process and the defect structure was developed [33]. Accordingly, the nucleation occurs by the formation of three-dimensional islands which are epitaxially oriented with respect to the GaAs substrate. Already in a very early pre-coalescence stage of the nucleation, the islands are plastically relaxed by the formation of edge type misfit dislocations without any climb or glide mechanism as predicted by the classical theories. Consequently, the interface between the 3D-islands and the underlying GaAs appears highly distorted.

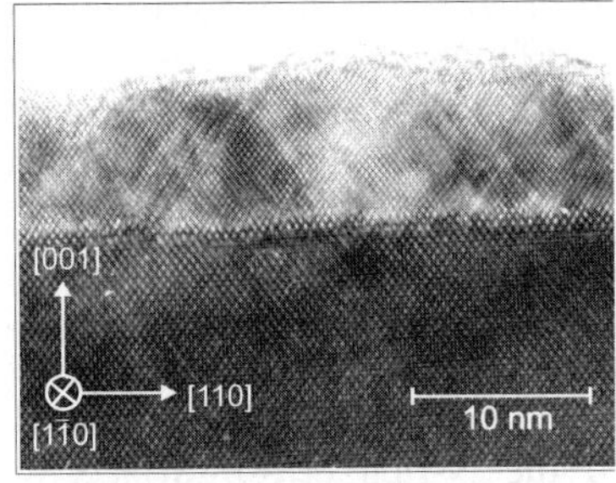

Fig. 1 Cross-sectional High Resolution Transmission Electron Microscopy (HRTEM)-image of the GaN-GaAs (001) interface taken along the [110] direction after depositing of about 20 monolayers (ML) GaN

Once formed, the isolated 3D-islands of c-GaN on GaAs grow predominantly laterally. When the coalescence is reached, planar defects like stacking faults and microtwins of high density (10^{10} cm^{-2}) are generated, which extend into the nucleation layer. In Fig. 1 a High Resolution Transmission Electron Microscopy (HRTEM) image of the c-GaN/GaAs interface region is depicted.

It shows a typical c-GaN-GaAs interface, which appears as source for long extending planar defects as stacking faults and microtwins. The c-GaN nucleation layer is nearly relaxed just above the interface, though there are many stacking fault defects at the interface. Characteristic pyramid–like structures with {111} facets, which are penetrating from the interface into the GaN layer can be observed. The relation between the parameters of the nucleation process at the one hand and structural quality of the subsequent grown c-GaN layers on the other hand is of more indirect nature. An example for the optimisation of the nucleation process with respect to the resulting structural quality c-GaN layers is given in the following.

Figure 2 illustrates the dependence of the crystalline perfection of 600 nm thick GaN layers, which were grown under constant optimum growth conditions on top of 4.5 nm thick nucleation layers. The Ga-flux during the nucleation stage, which was performed always at 600 °C, was the only parameter varied in the experiments. When the atomic N/Ga-flux ratio during the nucleation stage is of about two, the overall crystalline perfection reaches a maximum, expressed as minimum of the corresponding rocking curve halfwidth. Obviously, a slight N-excess is required during the nucleation process and promotes the formation of cubic islands.

2.3 Growth optimisation

The optimum conditions for the layer growth of c-GaN are usually determined by an experimental adjustment of mainly two parameters, the surface stoichiometry and the substrate temperature. Both parameters are interrelated, therefore an in-situ control of both substrate temperature and surface stoichiometry is highly desirable and can be achieved by monitoring the MBE growth process by Reflection High Energy Electron Diffraction (RHEED). Surface reconstructions give evidence for flat and well defined growth surfaces. It can be expected that the study of the surface reconstruction behaviour of c-GaN is one of the key issues for the understanding of the nitride growth and the improvement of the crystalline quality of the epilayers. Schikora et al. [21] first reported a surface reconstruction diagram for the growth of (001) GaN on GaAs with temperature and flux ratio depending transitions between a (1x1), (2x2) and a c(2x2) reconstruction. The surface reconstruction diagram is shown in Fig.3.

The nominal Ga-flux is related to the flux rate of atomic nitrogen, which was kept constant in the experiments. In the whole substrate temperature range, two distinctly different reconstructions were found to be stable. Under Ga-rich conditions, a c(2x2) reconstruction appears; layers grown here show n-type conductivity in Hall measurements. The (2x2) reconstruction is associated with N-rich conditions giving rise to p-type behaviour. Further increase of the N-excess leads to an unreconstructed (1x1) surface structure. In a narrow range

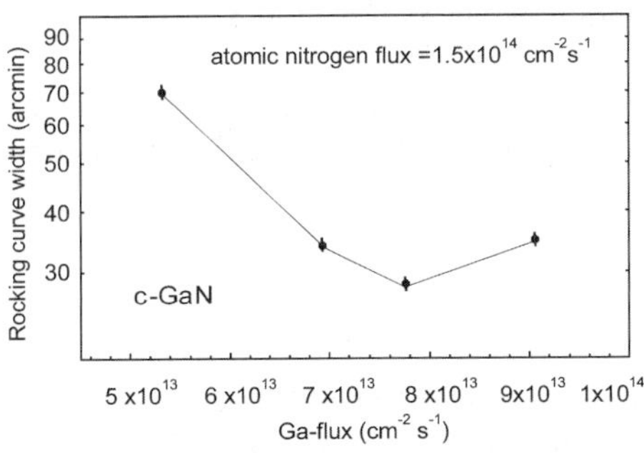

Fig. 2 Rocking curve halfwidth of the (002) reflection of 600 nm thick c-GaN layers in dependence on the Ga-flux during the nucleation stage. The atomic N-flux of about N = 1.5x 10^{14} (atoms/cm²s) was kept constant.

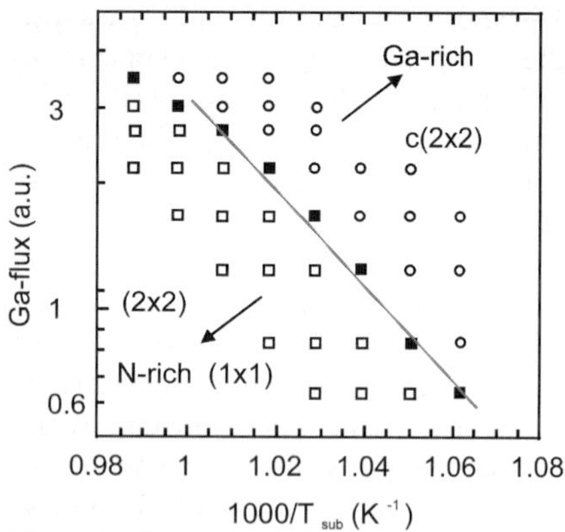

Fig. 3 (online colour at: www.interscience. wiley.com) Surface reconstruction of c-GaN(001) in dependence of the growth temperature and the Ga- lux at fixed atomic N-flow rate. The dots represent the stability range of the Ga-rich c(2x2) reconstruction, the squares show the stability range of the N-rich (2x2) reconstruction. A superposition of both reconstructions occurs in a narrow range inbetween the two surface reconstruction phases.

between the c(2x2) and (2x2) regime, both reconstructions occur simultaneously with different intensities of the reconstruction lines, indicating a nearly stoichiometric balance between atomic N species and Ga adatoms on the surface.

The surface reconstruction diagram for (001)GaN on 3C-SiC was determined by Okumura et al. [35] and Feuillet et al. [36]. They found that the intrinsic reconstructions of GaN (001) surfaces are the (1x1) and the (1x4) reconstructions. The authors exposed the N-rich (1x4) surface, as obtained during growth on 3C-SiC, to an As background pressure of about 10^{-8} Torr. Under As exposure the reconstruction changed from (1x4) to (2x2). For Ga-rich growth conditions the initial (1x1) reconstruction was transformed to a c(2x2) reconstruction under As exposure.

For an environment, where As atoms are intentionally or unintentionally introduced in the growth process, the energetic conditions are drastically changed. The chemisorption of As atoms at the surface significantly reduces the surface energy. Under these conditions a (2x2) structure with a half monolayer coverage of As dimers on a Ga-terminated surface is the most stable configuration. The total energy of this surface is much smaller than the energy of the intrinsic (1x4) reconstruction in the whole thermodynamically allowed range. This explains the experimental result that an As-contaminated (2x2) surface cannot be transformed back to a (1x4) reconstruction. In addition, the theoretical calculations have shown that structures with As in the second or deeper surface layers are energetically less favourable than structures where As stays in the top surface layer. Therefore As acts as surfactant, which reduces strongly the surface energy and which is virtually immiscible with GaN. The (2x2) and c(2x2) reconstructed surfaces obtained in As-contaminated systems are non-intrinsic surface reconstructions of the (001) GaN surface.

For the growth optimisation some important consequences can be derived [37]. For clean or As-covered c-GaN the surface energy is reduced going from N-rich conditions towards Ga-rich conditions. The stability of the (001) surface decreases and therefore the tendency of forming wurtzite clusters increases for N-rich growth conditions. The optimum growth conditions are expected under slight Ga-rich conditions where the surface energy is low.

Independent of the various types of surface reconstructions of the (001) GaN surface, the transition between the Ga-rich and N-rich existence regions is of decisive importance for the experimental control of the growth process.

2.4   Growth control

A powerful tool for the control and optimisation of epitaxial growth are RHEED oscillations which allow to control the growth rate, the layer thickness and the alloy composition with ultimate accuracy. It is generally accepted that RHEED oscillations can only be observed when the growing surface is flat and the growth occurs in a layer-by-layer growth mode. The 2-dimensional layer growth mode on the other hand requires sufficient mobility of the surface adatoms. For the nitrides Zywietz et al. [38] have investigated the surface diffusion of adatoms employing density-functional theory calculations. It was found that Ga and N adatoms exhibit a very different diffusivity. While Ga adatoms – the diffusion barrier is ~0.2 eV - are very mobile at typical growth temperatures, the diffusion of N adatoms – the diffusion

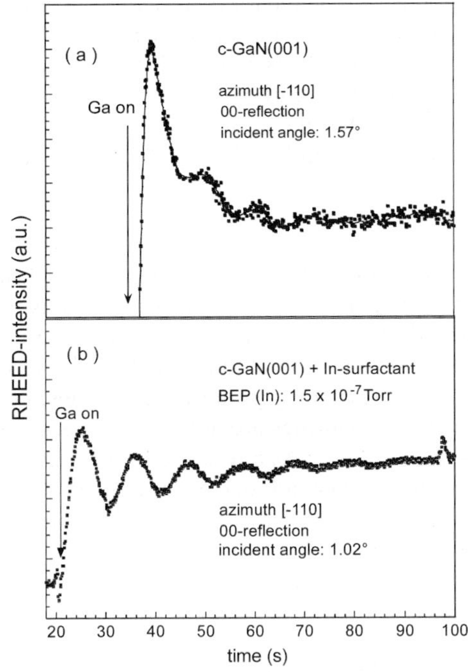

Fig. 4 RHEED-oscillations of c-GaN(001) detected at the 00-reflection in the [-110] azimuth. Curve (a) shows the growth oscillations under normal conditions, Curve (b) shows the growth oscillations, when In-adatoms are added to the surface.

barrier is ~1.5 eV - is slower by several order of magnitudes. Further calculations have shown that excess N in addition strongly reduces the mobility of Ga-adatoms, due to the formation of strong Ga-N bonds which have to be broken during the adatom migration.

Therefore RHEED oscillations during the growth of c-GaN are hardly to detect and actually could not be used up to now routinely for MBE growth control. Nevertheless, RHEED oscillations were first reported by Schikora et al. for the MBE growth of c-GaN on GaAs in 1996 [21]. The growth oscillations shown in Fig. 4a appear after growth interruption and formation of an N-rich (2x2) surface at a substrate temperature of 740 °C. The steep initial rise of the 00-reflection intensity suggests that the growth onset is accompanied by an effective surface smoothing. The oscillations are rapidly damped and entirely disappear when the 00-intensity is recovered to its steady-state level. A growth rate of about 0.08 ML/s was deduced from the oscillation period, which exactly agrees with the growth rate derived from layer thickness measurements by optical interference spectroscopy. The growth oscillations of Fig. 4b were recorded under the same growth conditions, however, the surface was exposed additionally to an In-flux of about 10^{-7} Torr beam equivalent pressure. At a substrate temperature of 740 °C one can assume that practically no In-incorporation occurs and therefore, In most probably acts as surfactant.

From this result it was concluded that the adatom mobility is significantly increased if In- adatoms are involved in the surface kinetics. This conclusion was supported recently by theoretical studies regarding surfactants in GaN-growth [39]. It was found that due to the presence of In on the surface, a second diffusion channel for the N adatoms is established, which reduces the diffusion barrier for N-adatoms by sub-surface diffusion. As consequence, the mobility of N-adatoms increases and the 2-dimensional layer-by-layer growth mode is stabilized. However, In adatoms also influence the Ga-adatom kinetics and therefore the resulting growth rate, as confirmed recently by Adelmann et al. [40]. There is no doubt that further investigations into the role of different surfactant atoms, as As, In, Bi or Sb are necessary and helpful to improve the c-GaN MBE growth.

3   Doping of cubic GaN

Controlled p- and n-type doping is crucial for advanced optoelectronic devices, like light emitting diodes (LEDs), or laser diodes (LDs) [41, 42]. The dopants are intentionally introduced to control the type and the magnitude of the electrical conductivity and are chosen for optimum device performance. The development of any semiconductor device technology requires therefore the selection and characterization of dopant impurities. The research of c-III-nitrides is currently at this stage.

Up to now, silicon and magnesium were used as standard donor and acceptor dopants in MBE. However, magnesium has several disadvantages which limit the doping efficiency of Mg in GaN to maximum hole concentrations in the upper $10^{17} cm^{-3}$ range. This is mainly ascribed to large acceptor ionization energy and compensation effects [43]. Experiments showed that Mg is volatile, requires low substrate temperatures and N-rich growth conditions [44, 45]. These conditions are disadvantageous for the epitaxy of high quality c-GaN layers, since N-rich conditions deteriorate phase purity [21].

Among possible alternative acceptor dopants, especially carbon has received a considerable interest due to its similarity to nitrogen in atomic radius and electronegativity. carbon doping of hexagonal Gallium Nitride (h–GaN) led to semi-insulating properties or to a reduction of background electron concentration [46-48]. Abernathy et al. [49] reported on p-type doping of c-GaN by carbon. However, due to the use of CCl_4 a pronounced reduction in growth rate prohibited the incorporation of higher C-concentrations and the maximum hole concentration reached was 3×10^{17} cm^{-3}.

3.1 N-type doping

Elements of the group IV on group III-site and elements of the group VI on N-site can be incorporated as donors for n-type doping of group III-nitrides. The group-VI impurities include O, S, Se and Te. Considering the atomic radius only O fits the N-atom size and is believed to be incorporated easily. The group-IV impurities include C, Si, Ge, Sn and Pb. The latter element has a large atomic radius and is therefore not expected to incorporate easily in III-Nitride semiconductors. The remaining group-IV impurities, i.e. C, Si, Ge, and Sn share different characteristics. First, all column IV impurities are amphoteric, i.e. they can occupy either the group-III or the nitrogen site of the zinc-blende lattice. As a result, group-IV impurities are donors or acceptors for cation site or anion site, respectively. Second, due to their amphoteric nature, column-IV impurities may autocompensate and the electrical activity may saturate at high impurity concentration. However, C has a size comparable to the N-atoms and is therefore expected to be mainly incorporated as an acceptor. Si, Ge and Sn are expected to replace Ga-atoms and act as a donor. These considerations show that Si and O may be the elements most appropriate for n-type doping of GaN.

Doping of MBE grown c-GaN by Si was reported by As et al. [50], Martinez-Guerrero [51] and Li et al. [52]. Elemental Si was evaporated from an effusion cell at source temperatures between 750 °C and 1200 °C.

In Fig. 5 the Si-concentration measured by SIMS (full squares) and the free electron concentration measured by Hall-effect at room temperature (full triangles) are shown as a function of the Si source temperature T_{Si}. Both the amount of incorporated Si as well as the free electron concentration follows exactly the Si-vapor pressure curve (full line in Fig.5) at $T_{Si} > 1000$°C [53]. The agreement between the values measured by Hall effect and SIMS implies that all Si atoms are incorporated at Ga sites and act as shallow donors. The maximum free electron concentration and mobility achieved so far are 5×10^{19} cm^{-3} and 75 cm^2/Vs at room temperature, respectively. The concentration and depth distribution of Si were measured by secondary ion mass spectroscopy (SIMS) using implanted calibrated standards for quantification, and an O^{2+} primary beam of 6 keV. Silicon is homogeneously distributed and no accumulation, neither at the interface nor at the surface was observed. This demonstrates the ability of Si to be incorporated in a controlled way for n-type doping of cubic GaN up to concentrations which are necessary for the fabrication of optoelectronic devices.

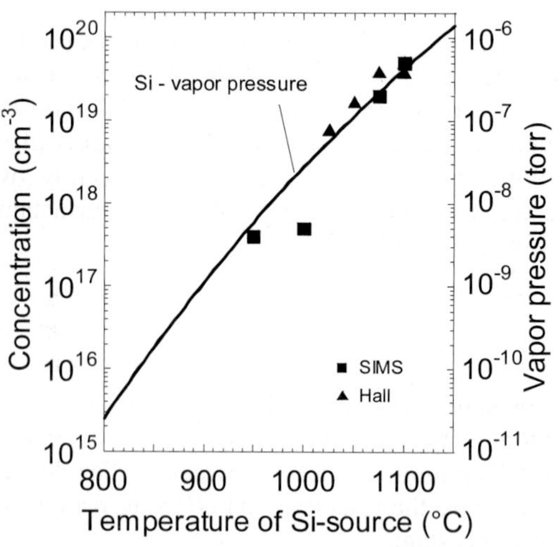

Fig. 5 Free electron concentration in Si-doped c-GaN measured by Hall-effect (full triangles) and Si concentration measured by SIMS (full squares) versus Si effusion cell temperature. The full curve represents the vapor pressure of Si. [53]

The electrical properties of Si doped c-GaN were studied as a function of the Si-flux. At Si-fluxes higher than 3×10^{10} cm^{-2}s^{-1} the samples were n-type. Temperature-dependent Hall-effect measurements further showed that these c-GaN samples were fully degenerated. However, all samples doped with a Si-flux lower than $3x10^{10}cm^{-2}s^{-1}$ were p-type with a hole concentration of about $2x10^{16}$ cm^{-3}. The hole concentration was independent of the Si-flux and was equivalent to the hole concentration of nominally undoped c-GaN epilayers. [54] [55] Thus it was concluded that a residual acceptor concentration of about $N_A= 4x10^{18}$ cm^{-3} exists in cubic GaN epilayers, and that the Si-donors have to compensate these acceptors.

From temperature-dependent Hall-effect measurements of nominally undoped c-GaN samples a thermal activation energy of about $E_A \cong 0.166$ eV and acceptor concentrations N_A of about $4x10^{18}$ cm^{-3} were estimated [55]. At room temperature the hole concentration was $p = 1x10^{16}$ cm^{-3} and the hole mobility was $\mu_p = 283$ cm^2/Vs. The acceptor concentration N_A is in excellent agreement with the value estimated from the p to n transition that occured in the Si-doping experiments.
The nature of these acceptors is attributed to states introduced by dislocations. X-ray diffraction and Rutherford Backscattering experiments revealed that the dislocation density N_{disl} in c-GaN/GaAs is about $3x10^{11}$ cm^{-2} [56, 57]. According to Weimann et al. [58] such a dislocation density results in a residual acceptor concentration of about $6x10^{18}$ cm^{-3}. The influence of the dislocation density on the electrical properties of c-GaN is reflected by the dependence of the electron mobility on the free carrier concentration. Similar to hexagonal GaN [58] the mobility first increases with carrier concentration, reaches a maximum value of about 82 cm^2/Vs at an electron concentration of $3x10^{19}$ cm^{-3} and decreases again [59]. This behaviour is characteristic for dislocation scattering and was explained by a theory recently developed for charged-dislocation-line scattering [60]. Theoretical calculations further showed that in c-GaN edge dislocations introduce acceptor-like states roughly 200 meV above the valence band maximum [61]. For such a deep acceptor level only about 1 % of the acceptors are ionized at room temperature resulting in a hole concentration of about $10^{16}cm^{-3}$. This value agrees excellently with the free hole concentrations measured in nominally undoped cubic GaN epilayers and explains in a reasonable way the p- to n-type transition observed by the Si-doping experiments. It further implies that threading edge dislocations are electrically active in cubic GaN and act as compensating acceptors.

3.2 P-type doping

Controlled p-type doping is still a challenge in the growth of group III-nitrides. In GaN the group II, Ga-site acceptors Be, Mg, Ca, Zn and Cd, together with the group IV, N-site acceptor C are potential p-type dopants. A common drawback to all these acceptors is their large activation energy of about 200 meV, which limits the fraction of ionized acceptors at room temperature to about 1 %. Therefore, acceptor concentrations in the order of 10^{20} cm^{-3} are needed for hole densities in the mid 10^{18} cm^{-3} range, as they are typical for e.g. laser diodes. However, such high acceptor concentrations severely reduce the mobility due to scattering at neutral and charged impurities. In addition, the limited solubility of the dopants or the tendency for self-compensation further prevents high doping concentrations. The great experiences accumulated with hexagonal GaN showed that Mg is the prefered p-type dopant. However, Mg is subjecte to several disadvantages like self-compensation, segregation and solubility effects, which limit the usually reported doping efficiency of Mg in hexagonal GaN to maximum hole concentrations in the upper 10^{17} cm^{-3} [43]. Doping experiments were less extensively reported in cubic GaN [62-64] and were mostly concentrated on Mg [9, 51, 65-68]. Results of Be doping were quite disappointing since no p-type conductivity could be achieved [69], however. using co-doping of Be and O, p-type conduction was demonstrated and resistivities as low as 0.02 Ωcm were measured [63]. A promising new acceptor in c-GaN is C which gave hole concentrations close to 10^{18} cm^{-3} at room temperature [70-73].

3.2.1 Mg-doping of cubic GaN

Mg doping of cubic GaN has been reported by Lin et al. [9], As et al. [66] and Martinez-Guerrero et al. [51]. Whereas Lin et al. injected Mg into the ECR source using N_2 carrier gas to enhance sublimation of Mg atoms, As et al. and Martinez-Guerrero et al. evaporated Mg from commercial effusion cells. Lin et al. obtained p-type c-GaN epilayers on GaAs (001) substrates at low growth temperatures (580°C). Hole concentrations between 8×10^{16} and 8×10^{18} cm^{-3}, and room temperature hole mobilities between 39 and 4 cm^2/Vs were obtained. Depth profiles of Mg in GaN showed a severe Mg surface segregation. Since Mg was fed through the ECR source, primarily Mg$^+$ ions were believed to be incident at the growth surface.

Using Mg effusion cells it is expected that neutral Mg is incorporated into the c-GaN epilayers. However, although the Mg-flux was varied by more than four orders of magnitude the Mg concentration varied by only one order of magnitude. The Mg doping level remains below 5×10^{18} cm^{-3} and seems to saturate at high Mg-fluxes. A similar behaviour was observed for MBE growth of hexagonal GaN [44] and GaAs [74] and was expected to be due to the high vapor pressure of Mg resulting in a high re-evaporization rate of the physisorbed Mg-atoms. It was suggested that Mg-incorporation occurs either via the presence of a saturated surface layer of Mg or via the presence of specific configurational sites for Mg-incorporation, so that the incorporation is nearly independent of the Mg arrival rate [44]. Recently, Mg incorporation as high as 1×10^{20} cm^{-3} has been reached at lower growth temperatures (T_{sub}=680°C) for cubic GaN grown on cubic SiC (001) [51]. However at Mg doping levels exceeding 10^{19} cm^{-3}, a severe degradation of the crystalline quality of the cubic GaN layers was observed [68] and unfortunately no correlation between the hole and the Mg-concentration has been found.

SIMS depth profiles of Mg in cubic GaN showed two characteristic features at the surface and at the GaN/GaAs interface and a homogeneous Mg-distribution in between [66]. At the surface, a slight increase of the Mg- signal occurred, indicating segregation of Mg. At the GaN/GaAs heterojunction an accumulation of the Mg-atoms by more than one order of magnitude was observed. This effect was attributed to an increased diffusion of Mg to the GaAs substrate, similar to the observations made in Ref. [45] for h-GaN on sapphire substrates. Due to the low operating temperature of the Mg-cell it is not very likely that transient effects of the Mg-flux just after opening of the shutter will result in this increased incorporation. However, the increased density of structural defects near the interface due to the large mismatch between substrate and epilayer may cause accumulation of Mg-atoms. A significant amount of Mg was also found in the nominally undoped GaN sample grown just after the doping experiments, due to memory effects of the growth chamber.

3.2.2   C-doping of cubic GaN

Due to the successful p-type doping of GaAs with C, it has been suggested that carbon is an acceptor for GaN. Carbon is an acceptor in GaN when substituting nitrogen [75] [76] and is the only group IV element that has a small enough atomic radius to fit the nitrogen site. Compared with other group II acceptors such as Mg, carbon is much less volatile and therefore not prone to cause memory effects in the MBE growth system. Theoretical calculations reveal that C_N is an effective mass acceptor and that the incorporation of C on the N site is preferred. However, due to the amphoteric nature of C, the concentration of C_N and C_{Ga} may be comparable. In addition, at high doping concentrations C_{Ga} and C_N have a strong tendency to form C_{Ga}^+-C_N^- nearest neighbour pairs, leading to self compensation [77, 78].

A previous study of carbon doping of c-GaN grown by metal organic molecular beam epitaxy (MOMBE) on GaAs substrates indicated that carbon created shallow acceptors and p-type conductivity [49]. However, due to the use of CCl$_4$ a pronounced reduction in growth rate prohibited the incorporation of higher C concentrations.

Carbon doping of MBE grown c-GaN on GaAs (001) substrates was achieved by e-beam evaporation of a graphite rod through adjusting the applied e-beam power [70]. The C-flux was externally calibrated by means of growing C-doped Gallium Arsenide assuming a similar sticking coefficient of C on GaAs and GaN. It followed that a maximum C-concentration of about 10^{20} cm^{-3} should be achievable in c-GaN. This was confirmed by SIMS measurements, which indeed showed a C-incorporation in c-GaN of 2×10^{20} cm^{-3}.

The room-temperature hole concentration as a function of the carbon-flux of C-doped cubic GaN epilayers is shown in Fig. 6. The hole concentration increased up to a maximum value of about 6×10^{17} cm^{-3} at a C-flux of 10^9 cm^{-2}s^{-1}. This value was about one order of magnitude higher than the value ob-

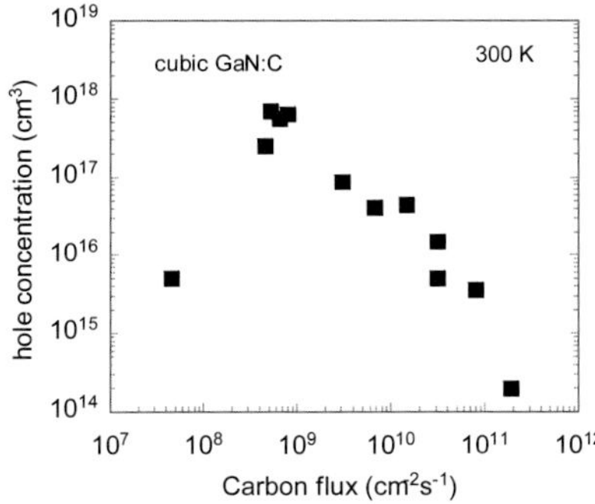

Fig. 6 (online colour at: www.interscience. wiley.com) Room temperature hole concentration of C-doped c-GaN versus flux of the Carbon evaporation source.

tained by Mg-doping of cubic GaN [62] and demonstrates the ability of C for p-type doping of cubic GaN. However, at high C-fluxes a clear reduction of the free hole concentration was observed. This indicates that at high C-concentrations additional compensating centres were generated.

The incorporation of carbon in cubic GaN has a profound influence on the PL, as can be seen in Fig.7. A photoluminescence emission band at $E = 3.08$ eV appeared in the 2 K spectra of slightly doped c-GaN layers. This transition was clearly different from the well-known line at 3.27 eV (X) and at 3.16 eV (D^0, A^0) [79]. A detailed analysis of the 3.08 eV band showed that it was a convolution of two distinct lines separated by 25 meV at low temperatures (indicated by arrows in Fig. 7). At temperatures above 100 K the low-energy contribution thermalises and only the high-energy PL emission was observed. Thus, the lower transition was attributed to a donor–acceptor recombination $(D^0, A^0)_C$ and the other to a C-related band–acceptor recombination $(e, A^0)_C$, respectively.

From temperature dependent PL-measurements $E_A = 215$ meV has been estimated for the binding energies of the shallow C related acceptor. This value is 15 meV below that of the Mg acceptors in c-GaN [66] and is also below the value known for hexagonal GaN:C [80]. The donor involved has the same ionisation energy (25 meV) as the one observed in the undoped and Mg-doped c-GaN samples [79]. Based on effective-mass-theory (EMT) an acceptor binding energy $E_A = 0.213$ eV was obtained [73] which is in excellent agreement with the experimental results and demonstrates that C indeed introduces a shallow acceptor in c-GaN.

In other cubic III-V compounds like GaAs, InP and GaP carbon is the acceptor with the smallest central cell correction. The binding energy is usually close to Mg and increases in the sequence C, Mg, Zn and Cd [80]. The same sequence is verified in cubic GaN now, since the binding energy of C is lower than that observed for the Mg acceptor in c-GaN by 15 meV [66]. Due to the higher crystal symmetry, the binding energy of the C acceptor in the cubic phase is also lower than the activation energy of C measured in hexagonal GaN [80].

Detailed analysis of the red luminescence at 2.1 eV revealed that this band was a convolution of at least three different transitions, indicated by the full and dashed arrows in the topmost spectrum of Fig. 20. The energy differences between the main peak and the two low energy peaks correspond nearly exactly to the binding energies of the omnipresent shallow acceptor ($E_A = 129$ meV) and to the binding energy of the shallow carbon acceptor ($E_A^C = 215$ meV), respectively. Therefore, the red luminescence was attributed to a superposition of three transitions involving the same deep defect level E_{Deep} located 1.185 eV below the conduction band. Since the electrical measurements on C-doped cubic GaN samples showed also a clear reduction of the free hole concentration at C-fluxes above 8×10^{10} cm⁻²s⁻¹, the deep defect acts as a deep donor [71].

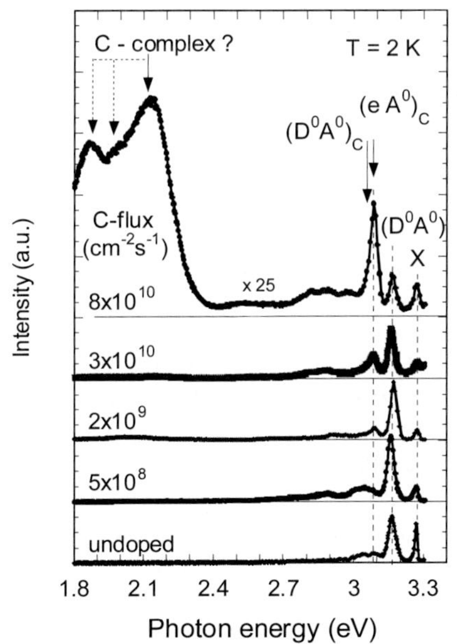

Fig. 7 Photoluminescence spectra of c-GaN layers with different Carbon doping levels. The intensity of the topmost spectrum has been multiplied by a factor of 25.

The nature of the deep compensating complex is still unclear. Kaufmann et al. [43] observed similar phenomena for Mg-doped MOVPE GaN layers and proposed self-compensation via the formation of nearest neighbour $Mg_{Ga}V_N$ pairs. However, C is placed on a N site and the formation of a comparable defect C_NV_N, which is a second nearest neighbour pair, is unlikely. For cubic GaN:C, it is more tempting to apply an analogy to C doped GaAs. In GaAs the formation of a compensating dicarbon split-interstitial center was observed at high carbon concentrations [81]. This complex involved one C atom on a As lattice site and another C atom shifted along the [111] direction [82] and acts as a double donor in GaAs. Recently, Ramos et al. [83] studied various dicarbon complexes in cubic GaN and Wright [84] suggested that a C_I-C_N complex may be responsible for self-compensation in p-type C-doped GaN grown under Ga-rich conditions.

3.3   Light emitting diodes

A GaN device grown on conducting GaAs substrates will enable easy device processing, which is compatible to the GaAs-based LED technology currently used in industry and therefore will significantly reduce production costs.

A diode fabricated by Mg and Si doping of MBE-grown c-GaN on n-type GaAs (001) has been reported by As et al. [85] .Current-voltage (I-V), capacitance-voltage (C-V), photoluminescence (PL) and electroluminescence (EL) measurements were used to characterize the c-GaN p-n homojunction at room temperature. The cubic GaN p-n homojunction structure consisted of a 760 nm thick p-type GaN:Mg on top of a 500 nm thick n-type GaN:Si. Te-doped n-type GaAs (001) with a free electron concentration of about 2×10^{18} cm^{-3} was used as substrate. The vertically structured diode was completed after evaporating electrodes on both the n-type GaAs substrate (Sn-contacts) and the p-type top layer (semitransparent two-layer Au contact).

The I-V characteristics of the diode structures was clearly rectifying [86]. Electroluminescence (EL) under dc-biased conditions was emitted through the semitransparent part of the p-type contact from the surface side of the device. Figure 8 shows the I-V characteristic of the structure measured at room temperature. The GaN n$^+$-p diode forward turn-on voltage was approximately 1.2 V, which was about the same as that measured in hexagonal GaN p-n junctions (1.2 V) [87]. At a forward bias larger than 1.2 V the current grew linearly with the applied voltage. A fit to the I-V curve yielded a series resistance of 16 Ω, which is comparable to that reported for commercial III-V nitride LEDs [88]. Because of its large band-gap (3.2 eV), the reverse bias saturation current due to minority carrier diffusion would be unmeasurably small in an ideal GaN p-n diode. However, significant reverse bias leakage was observed in c-GaN diodes, indicating the presence of trap levels within the band gap, which may act as a shunt or parallel conductance. At -5 V a dark current density of 0.25 A/cm^2 was measured. The large lattice mismatch between the GaAs substrate and c-GaN, which leads to a dislocation density in the order of 10^{11} cm^{-2} [62], may be responsible for the high dark current.

The inset of Fig. 8 shows the electroluminescence of the c-GaN n$^+$-p junction through the semitransparent part of the p-type contact. The EL-spectrum was peaked at 3.2 eV (387 nm) with an FWHM of 150 meV (20 nm) and was nearly identical to the RT PL-

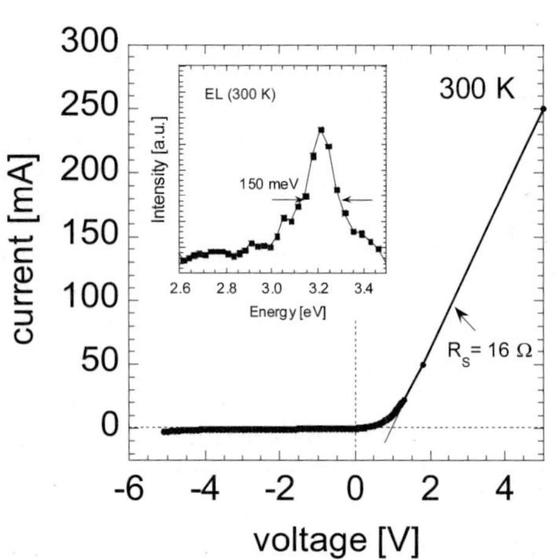

Fig. 8 The room temperature current-voltage curve of a c-GaN p-n$^+$ junction. The inset shows the room temperature electroluminescence (current = 150 mA) of this diode emitted through a semitransparent Au contact.

spectrum, suggesting that in MBE grown c-GaN LEDs the dominating recombination processes are near band-edge transitions. This is in contrast to MOCVD grown cubic LEDs where the emission originated from the blue Mg-related impurity mediated recombination [89]. Similar differences in the LED emission of MBE and MOCVD grown samples have also been reported for hexagonal GaN samples [90]. Due to the dominating near band transition the observed FWHM is also within the narrowest values ever reported for UV-LEDs. For the LEDs a linear increase of the intensity with increasing current density was measured and no shift of the EL-spectrum to lower energies was observed.

4   MBE of cubic InGaN

The growth of cubic GaN layers with excellent structural, optical and electrical properties (as demonstrated by the observation of optically stimulated emission at room temperature from cleaved c-GaN layers [91] and electroluminescence from a MBE grown c-GaN p-n junction [85]) has clearly demonstrated the potential of III-nitrides with cubic crystal structure for the realization of optoelectronic devices like light-emitting diodes and laser diodes. However, the bulk of experience accumulated with hexagonal III-nitrides shows that all working devices contain InGaN in the active region. The extensive utilization of InGaN in the active layer allows to tune the band gap energy and cover the short wavelength visible to near ultraviolet light spectrum. However, the incorporation of high amounts of In into the active region of h-InGaN-based emitters is still a serious problem. The emission efficiency of emitters still drops significantly with increasing emission wavelength. Cubic GaN has a room temperature band gap of 3.2 eV which is 0.2 eV smaller than that of hexagonal GaN and thus the availability of high-quality c-InGaN would open an attractive alternative to the hexagonal nitrides for the production of green and amber emitters.

The co-existence of a high quantum yield of photoluminescence and a high density of threading dislocations has been interpreted as being indicative for a strong localization of excitons in space. One current focus of research is to understand the mechanism of this localization. One possibility is the formation of InN quantum dots in the InGaN alloy [92-94]. The large difference in interatomic spacing between GaN and InN can give rise to a miscibility gap. Calculated phase diagrams of InN/GaN [95-97] reveal that at typical growth temperatures of 600-800 °C, a strong tendency of InGaN layers exist to separate into regions of varying In-content. Another possibility is the localization caused by natural composition fluctuations in InGaN without clustering of In-atoms [98]. Exciton recombination in In-rich clusters can also explain the observed Stokes-shift in quantum wells. However, due to the low symmetry of the wurtzite lattice and the strain in pseudomorph quantum wells, strong electric polarization fields can exist in h-InGaN quantum wells. These fields can effectively separate electrons and holes in quantum wells which may also result in a difference between the absorption and the emission wavelength (quantum-confined Stark effect).

The importance of submicrometer-scale chemical inhomogeneities of the InGaN quantum wells in addition to the effects of piezoelectric and spontaneous polarization fields has been discussed extensively [99]. However, a clear distinction between these two effects has not yet been achieved with hexagonal III-nitrides. As a consequence, it was not possible so far to gain full insight into the properties of the active region of LEDs and laser diodes based on wurtzite GaN and to unambiguously clarify the role of localized excitons in $In_XGa_{1-X}N$ for the light emission process.

Investigations of nitrides which crystallize in the metastable cubic (zincblende) configuration can help to gain a better understanding of the importance of the In-segregation and phase separation processes for the radiative recombination in light-emitting diodes and laser diodes containing $In_XGa_{1-X}N$.

Since the phase separation process is due to the large difference between the Ga-N and the In-N band lengths, which are almost identical in h- and c-III nitrides, the compositional fluctuations may be quite similar in h- and c-InGaN of equal In-content.

However, spontaneous polarization fields do not exist in the cubic nitrides and piezoelectric fields are negligible in these semiconductors due to the (001) growth directions of epitaxial layers. Therefore, the effect of In-segregation on the luminescence of $In_xGa_{1-x}N/GaN$ double heterostructures can be studied without influence of electrical fields in $c\text{-}In_xGa_{1-x}N$ layers.

4.1   Cubic InGaN-layers

Due to the large difference of the formation enthalpies of GaN and InN the growth of InGaN layers with sufficient structural perfection is much more difficult than that of GaN. The high nitrogen vapour pressure over InN results in a small In-incorporation rate at those temperatures which are usually used for GaN growth. Lower temperatures on the other hand can lead to poor crystal quality, most likely due to segregation of In onto the growth front. The difficulties of In-incorporation in InGaN are also reflected by the extremely strong temperature variation of the In sticking coefficient which has been measured in MBE experiments [100]. Reasonable In mole fractions in the layers can only be achieved at growth temperatures of less than 650 °C and a sufficiently high In/Ga flux ratio. However, the In-rich growth front does not allow to observe any surface reconstruction by RHEED which prevents the in-situ control of the growth regime. The MBE of c-InGaN layers with a well defined In-content therefore depends on the precise measurement of the In-flux and the substrate temperature.

Cubic InGaN layers with an In-content up to x = 0.4 were grown by use of rf plasma sources [101]. InGaN layers with an In-concentration of 17 % exhibited blue luminescence at temperatures up to 500 K. The broad emission band had a maximum at about 2.8 eV. Layers with even higher In-concentration (x = 0.4) emitted luminescence in the green spectral region with a maximum at about 2.45 eV [102]. The intensity of the green emission decreased by only a factor of 4 when the temperature was raised from 4 K to 400 K. Absorption and reflectivity measurements were used to determine the room temperature band gap energy of the InGaN layers yielding E_{gap} = 2.46 eV $\pm$ 0.03 eV which is close to the PL maximum. This correlation suggested that the PL stems from band edge related transitions in the InGaN layer. However, the weak temperature dependence of the intensity and the emission peak energy were rather consistent with radiative recombination between highly localized states. X-ray diffraction measurements of these layers revealed two Bragg reflexes, one due to the InGaN (x = 0.4 layer) and a weak shoulder close to the position of the Bragg reflection of c-InN indicating the existence of an In-rich phase in these layers. It seems likely that the green emission band was also due to the recombination of excitons localized in In-rich clusters.

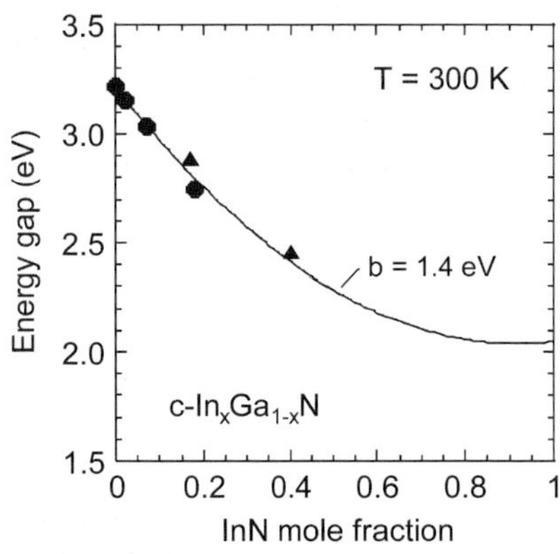

Fig. 9 The room temperature energy gap of c-In$_x$Ga$_{1-x}$N versus InN mole fraction. Full squares represent data measured by spectroscopic ellipsometry [103], full triangles are results of Ref. [102]. The full curve is a fit of the experimental data using a bowing factor b= 1.4 eV.

Reliable data of the gap energy of c-GaN were obtained from spectroscopic ellipsometry measurements. R. Goldhahn et al. [103] used this method to determine the complex refractive index of c-InGaN layers in the energy range from 1.5 to 4.0 eV. The c-InGaN layers were grown by plasma-assisted MBE at a substrate temperature between 610 and 680 °C on top of a c-GaN/GaAs hybrid substrate. The InGaN layers had a thickness of 100 to 200 nm. Their alloy composition was obtained from the lattice parameter measured by high resolution x-ray diffraction and by Rutherford backscattering (RBS) [56]. The InGaN layers were fully relaxed as inferred from reciprocal space maps of the symmetric (002) and the asymmetric (-1-1 3) Bragg reflexes and also from RHEED reflex measurements during growth.

Figure 9 shows the room temperature gap energy of five c-InGaN layers. The circles are the result of spectrometric ellipsometry, the triangles are the results of transmission measurements [101, 102]. The experimental data were fitted by a parabolic expression of the gap energy versus In-content (full curve) using a bowing parameter b = 1.4 eV.

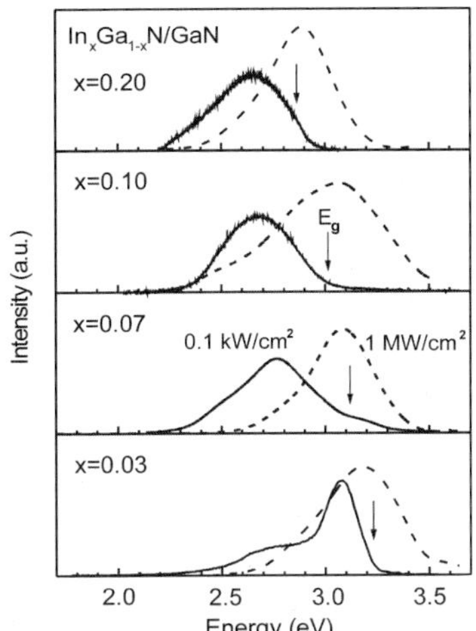

Fig. 10 2K photoluminescence spectra of c-In$_x$Ga$_{1-x}$N films (full curves) and the high-excitation edge emission (dashed curves). The gap energies from spectroscopic ellipsometry and reflectivity measurements are indicated by arrows.

Low temperature photoluminescence (PL) spectra of the c-InGaN/GaN heterostructures are shown in Fig. 10 as full curves [104]. Luminescence was exited by the 325 nm radiation of a continuous-wave (cw) HeCd laser. The full width at half maximum (FWHM) of the InGaN layer emission increases from 200 meV to 450 meV with increasing x. The energy of the emission peak shifts to lower energies with increasing In-content and is about 100-200 meV lower than the gap energy which was obtained from spectrally resolved ellipsometry.

The dashed curves in Fig. 10 are the spectra of light emitted from a cleaving edge of the heterostructure. In this case, the InGaN layer surface was excited by the 340 nm light of a pulsed laser with an intensity of 1 MW/cm^2. The peak energy of the edge emission is shifted with respect to the luminescence peak close to the gap energy, indicating that localized states contribute to the PL at weak excitation. The localized states are filled by carriers with increasing excitation intensity and at 1 MW/cm^2 all low-energy radiative recombination processes are saturated and the emission observed at high excitation is the near band edge luminescence from the InGaN layers. This suggestion is corroborated by time-resolved measurements where a non-exponential decay of the PL was observed [104]. The kinetics of the recombination was described by the model of stretched exponentials which was introduced to describe the radiative decay of strongly disordered systems with localized recombination centres. It was applied successfully to h-InGaN [105], II-VI quantum dots [106], and porous Si [107].

Localization of excess electron-hole pairs has also a strong influence on the optical gain in c-InGaN layers. With increasing excitation (i) the crossover between gain and absorption is shifted to higher energies, indicating the filling of the localized states, (ii) the gain structures broaden and (iii) the maximum gain shifts to higher energies. This behaviour is a further proof that the optical amplification is due to localized states, because the gain of an electron-hole plasma is expected to shift to lower energies with increasing density [108]. Optical gain was observed up to 500 nm indicating that c-InGaN may have some advantages for device applications in the green spectral range.

The correlation of structural and optical properties was further investigated by cathodoluminescence (CL) microscopy of c-InGaN layers with different In-content. The CL spectra of all c-InGaN layers exhibited broad emission bands, but their peak position changed according to the local composition. In samples with higher average In-content (more than 10%) the fluctuations of the composition were strongly enhanced [109].

To monitor the influence of the of In-content on the optical amplification, gain spectra of samples with varying In-content were measured at a fixed excitation density of 7 MW/cm^2. The largest gain was observed for samples with an In-content of about 7 %, while for In-contents above 20 % the optical amplification was fully suppressed.

Evidence for In-rich phases in c-In$_x$Ga$_{1-x}$N layers was first obtained from investigations of their vibrational properties [110] using micro Raman spectroscopy.

Room temperature micro Raman spectra of 100-300nm thick c-In$_x$Ga$_{1-x}$N layers are depicted in Fig. 11. The layers had an In-content of x = 0.07, 0.19 and 0.31 respectively. The composition of the layers was obtained by high resolution x-ray diffraction. Beside the TO- and LO-peaks of the layers (indicated by arrows in Fig. 27), the layer with x = 0.31 showed an extra Raman peak at about 625 cm^{-1}. This peak was ascribed to the LO-phonon mode of In-rich inclusions in the layer most likely formed by a phase separation process during growth. The existence of an In-rich phase was corroborated by x-ray diffraction which showed that the c-InGaN layers contained strained crystalline cubic In-rich inclusions.

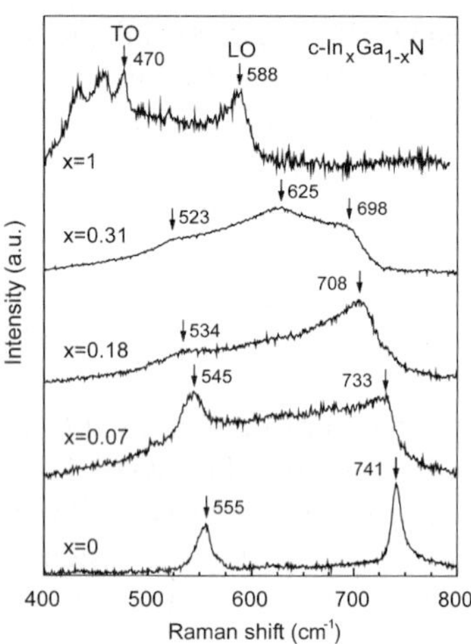

Fig. 11 Room temperature micro Raman spectra (excitation energy E_L = 2.6 eV) of GaN, c-In$_x$Ga$_{1-x}$N layers with x = 0.07, 0.18, 0.33 and of c-InN.

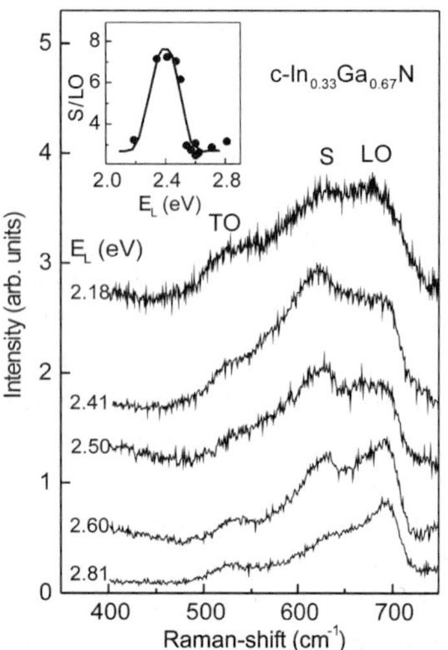

Fig. 12 Room temperature micro Raman spectra of c-In$_x$Ga$_{1-x}$N (x = 0.33). The energy of the exciting laser radiation E_L is indicated for each spectrum. The inset shows the ratio of the integrated intensity of the S- and the LO-peak as a function of the laser energy E_L.

Electronic excitations of the In-rich phase in c-In$_x$Ga$_{1-x}$N layers were studied by resonant Raman scattering [111]. Raman spectra of c-In$_x$Ga$_{1-x}$N (x = 0.33) which were recorded at different excitation laser energies E_L are shown in Fig. 12. In the spectrum measured at E_L = 2.81 eV, the LO- and the weak TO-phonon peak of the layer can be distinguished. By lowering the excitation energy a peak at about 630 cm^{-1} (indicated by S) becomes evident and is further enhanced at even lower excitation energies. In the range of E_L between 2.5 eV and 2.34 eV, the S-peak becomes the strongest line in the spectrum. The integrated intensity of the S-peak as a function of excitation energy E_L is shown in the inset of Fig. 12. The plot shows a clear resonance with a maximum at about 2.4 eV. Assuming that the maximum corresponds to an outgoing resonance process and taking into account the phonon energy of In$_x$Ga$_{1-x}$N (E_{LO} = 78 meV) the energy of the resonant electronic excitation was calculated to be 2.32 eV. This value was in excellent agreement with the peak energy of the photoluminescence (PL) of the layer [112].

It was supposed in Ref. [112] that the observed photoluminescence is due to radiative recombination of electron-hole pairs in In-rich regions of the layers. The composition of these regions was obtained by high resolution x-ray diffraction. By comparing the gap energy of the In-rich phase and the measured optical transition energy, it was found that the transition energy exceeds the gap energy by more than 100 meV. It has been concluded that this is due to confinement effects of quantum dot-like structures formed by phase separation in the c-In$_x$Ga$_{1-x}$N layers. From the confinement energy the size of these structures has been estimated to be about 3 nm using an effective mass model. These experiments demonstrated that the emission from c-In$_x$Ga$_{1-x}$N layers is associated with In-rich structures of nm-size.

The composition of these regions was obtained by high resolution x-ray diffraction. By comparing the gap energy of the In-rich phase and the measured optical transition energy, it was found that the transition energy exceeds the gap energy. It has been concluded in Ref. [112] that this is due to confinement effects of quantum dot-like structures formed by phase separation in the c-In$_x$Ga$_{1-x}$N layers. From the confinement energy the size of these structures has been estimated to be about 3 nm using an effective mass model. These experimental results are considered to proof *directly* that the emission from c-In$_x$Ga$_{1-x}$N layers is associated with In-rich structures of nm-size.

4.2   Cubic-GaN/InGaN/GaN double heterostructures and quantum wells

In the following, recent experimental investigations of c-GaN/InGaN/GaN double heterostructures (DH) and quantum wells (QW) are summarized. All investigations demonstrate the strong influence of chemical inhomogeneities and In-segregation.

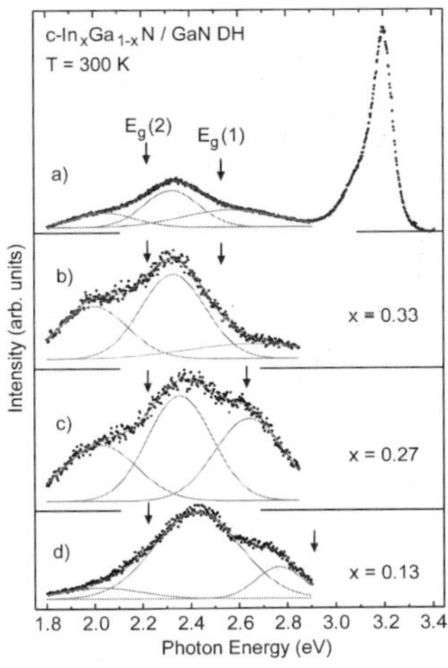

Fig. 13 (online colour at: www.interscience.wiley.com) Room-temperature PL spectra of c-InGaN/ GaN double hetero-structures with different In-content. Spectrum a) was excited above the gap energy of the GaN barrier, spectra b) to d) were excited below the GaN gap energy. The corresponding gap energies of the InGaN(1) and InGaN(2) phase are indicated.

The DHs on GaAs(100) substrates [113] consisted of a 300 nm thick GaN buffer layer grown under stoichiometric conditions at a substrate temperature of 720 °C and a growth rate of about 40 nm/h. Then 30 nm thick c-InGaN layers were deposited at a temperature of 600 °C. During the InGaN growth the Ga-flux was reduced by about 20 %. The variation of the In mol fraction in different samples was achieved by varying the In-flux between 0.5×10^{13} and 6×10^{13} atoms/s^{-1} cm^{-2}. A 30 nm GaN cap layer was finally grown at 600 °C.

The lattice parameters and the strain in the DH-structures were measured by high resolution x-ray diffraction (reciprocal space maps, RSM). Three maxima of the scattered x-ray intensity, one from c-GaN, one which was associated with the InGaN layer (InGaN(1)) and a weak reflex from an In-rich phase In-GaN(2) were obtained.

Raman spectra of these DHs were recorded in backscattering geometry using laser light with an energy of 2.4 eV for excitation. The c-GaN phonon frequencies and two other peaks, corresponding to the LO-phonon mode of the InGaN(1) layer and to LO-phonons propagating in the In-rich InGaN(2) phase were clearly identified. The fact that the LO-phonon frequencies of c-InGaN vary linearly with x allowed to determine the InGaN alloy composition [110] revealing that the Raman results are in good agreement with the x-ray data [113].

Room-temperature PL spectra of three different c-In$_x$Ga$_{1-x}$N/GaN DHs are depicted in Fig. 13. Curve a) is a PL spectrum which was excited by HeCd laser radiation (E$_L$ = 3.51 eV) above the bandgap of c-GaN. The emission at 3.2 eV is due to exciton recombination in the c-GaN barrier [79]. The emission bands at lower energy are also present in spectrum b) where E$_L$ was well below the gap energy of GaN, indicating that this emission stems from c-InGaN. A Gaussian curve fit of the long wavelength part of the spectrum yields peak energies of 2.35 eV and 1.9 eV, respectively.

The PL emission from two other DHs with InGaN layers with a lower In-concentration (curves c, d) are also plotted in Fig. 13. It is a common feature of all these PL spectra that (i) the energy of the main emission peak is only slightly varying, and (ii) the gap energy of the In-rich InGaN(2) phase is below the energy of the dominating emission band. The latter may be inferred from the c-InGaN gap energies also shown in Fig. 13. They were obtained from the relation between Eg and the composition of relaxed c-InGaN layers [103] and the calculated gap shift due to biaxial compressive strain.

The weak PL at about 2.05 eV is only observed with samples grown at relatively high In-fluxes and may be due to radiative recombination in InN micro-crystals. Low temperature PL of c-InN was recently measured at about 2.1eV [114].

The experimental data have been explained by Husberg et al. [113] assuming that the observed luminescence at about 2.4 eV is due to the radiative recombination of excitons localized at In-rich, strained quantum-dot like structures which are embedded in the InGaN(1) layers. Since no spontaneous polarization- or piezoinduced electric fields exist in c-III-nitrides, the difference between the InGaN(2) gap and the luminescence peak energy is equal to the localization energy of the excitons E$_{loc}$ which is decreasing with increasing In-content of InGaN(1). In order to explain the variation of E$_{loc}$ (i) a decrease of the localization energy due to the decreasing confining potential V$_0$ (which is equal to the difference of the band gap energies of the InGaN phases), and (ii) a variation of the size of the QDs were considered. In an effective mass model calculation of the ground state transition energy, a three-dimensional quantum confinement structure in the form of a cube of dimension L was assumed, yielding that the QD structures have an average size of about 10 nm and the shift of the PL-peak energy is mainly due to a decrease of the QD-size with decreasing In-content of the layers. The influence of strain on the energy gap of InGaN alloys has been recently calculated more accurately [115]. If these values are used to obtain the average dot size, it is found to be about 3 nm smaller. More sophisticated ab-initio calculations of the InGaN QD energy gave an even smaller average dot size [116].

Quantum dot like-structures which were observed in relaxed c-InGaN layers and DHs [112, 113] seem to be absent in fully strained c-InGaN/GaN QWs. This conclusion is in good agreement with recent investigations of strained and relaxed c-InGaN/GaN DHs grown on GaAs substrates which revealed that the formation of QD-like structures in InGaN is suppressed by strain in these layers [117].

4.3   Indium segregation and quantum dot formation

Experiments demonstrated the importance of exciton localization in c-InGaN layers, DHs and QWs due to nano-structural disorder as a result of composition modulation. Details of the formation process of these In-rich nanostructures are not clear yet. Theoretical models of the In-segregation process in InGaN layers have focused on phase separation in the bulk (spinodal alloy decomposition). Calculated phase diagrams of bulk InGaN [95] [96] [97] revealed large miscibility gaps at those temperatures which are applied for the epitaxial growth of InGaN layers. However, it is questionable if the assumption of thermodynamic equilibrium, which is made in these calculations, is valid for MBE growth. Bulk spinodal decomposition may also be hindered by the relatively slow self-diffusion in InGaN. Using experimental values of the In-diffusion coefficient in h-InGaN [118], it is calculated that the In-atoms in our InGaN layers will diffuse less than 1 nm during the growth of a complete DH structure. This is obviously not sufficient to explain the formation of spatial fluctuations of the chemical composition on a nm-scale.

In-rich clusters may also be formed by a separation of the alloy constituents at the surface during growth [119]. Since the In-adatom mobility is quite high In-rich clusters can aggregate in time intervals comparable to the time needed for the growth of one monolayer, and are buried then in the bulk of the layer because of the deposition of material. The spontaneous formation of In-clusters at the (0001) surface of h-InGaN by surface-induced lateral composition modulations has recently be observed by tunneling microscopy [120].

Annealing experiments were performed to gain insight into the formation process of the In-rich phase in c-In$_x$Ga$_{1-x}$N/GaN DHs [121]. Samples were annealed in vacuum for 10 hours. The annealing temperature was between 450 °C and 750 °C and was raised by 50 °C for each annealing step. Subsequent to each annealing step PL and X-ray diffraction was measured. These annealing parameters were chosen in order to establish thermodynamic equilibrium at a temperature well above the growth temperature. This and the annealing time, which was ten times longer than the growth time of the InGaN layers, yield an In-diffusion length of several nanometer. A significant modification of the In-rich QDs can be expected if diffusion would not be hindered by other processes.

Figure 14 shows the room temperature PL spectra of an as grown c-InGaN/GaN DH ($x = 0.33$) and the PL of the same sample after 10 h annealing steps at different temperatures. All spectra were measured at room temperature with an excitation laser energy of $E_L = 3.42$ eV and are normalized to the maximum intensity of the c-GaN peak at 3.2 eV. The dominating

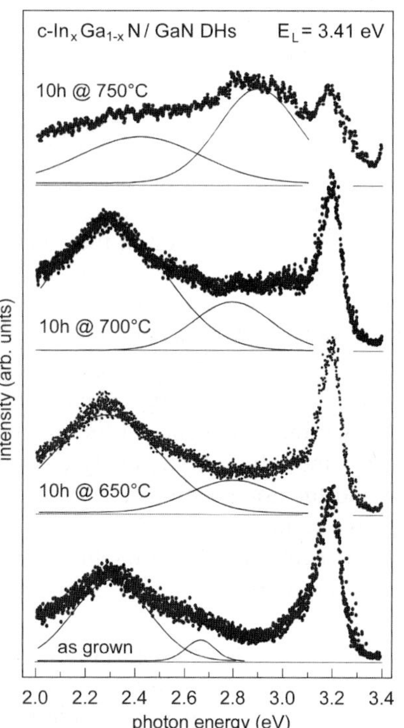

Fig. 14 The room-temperature PL of an as grown c-InGaN/GaN double heterostructure and the PL of the same sample after annealing for 10 hours in vacuum at temperatures between 650 and 750°C. The spectra are fitted by two emission lines at about 2.3 eV and 2.8 eV, respectively.

2.3 eV luminescence of the as grown sample has been identified to be an emission from QD-like In-rich structures [112]. This emission is also found in the spectra of the DH annealed at temperatures up to 700 °C. However, already after the 700 °C annealing step a second PL peak appears at about 2.8 eV. This line dominates the spectrum after a final 750 °C/10 h annealing step while the 2.3 eV emission is broadened and its intensity is significantly reduced. Assuming that the 2.8 eV PL is an emission from a homogeneous c-InGaN layer, we can use the measured dependence of the gap energy vs. In-content [103] to estimate the composition of the layer yielding $x = 0.2$.

X-ray diffraction measurements of the as grown sample revealed two Bragg reflections, a strong one from InGaN with $x = 0.33$ and a weaker from an $x = 0.56$ phase. These Bragg reflections remained almost unchanged in position and intensity after annealing up to 700 °C, however after the 750 °C annealing step the intensity of the In-rich phase Bragg peak was below the detection limit of our apparatus.

These results together with Raman scattering experiments [121] suggest that the In-rich phase in c-InGaN layers is thermodynamically stable up to temperatures of about 700 °C. However, at about 750 °C diffusion of In out of the In-rich clusters into the InGaN-layers seems to be effective. The fact that we also observe a reduction of the In-content of the layers itself from $x = 0.33$ to $x = 0.2$ may be an evidence for an additional diffusion of In-atoms into the GaN barrier layers. Extrapolating the In-diffusion coefficient to the c-InGaN growth temperature of 600 °C, we obtain a diffusion length of In during growth below 0.01 nm. However, at 750 °C and 10 hours annealing time, the diffusion length is in the order of 10 nm, which can explain our experimental findings.

Due to the low diffusibility of the In-atoms at growth temperature it seems unlikely that the In-rich QDs are formed by a process including bulk diffusion. It may rather be assumed that spontaneous formation of nanometer-size compositional inhomogeneities at the surfaces is the main mechanism. Due to the high mobility of In-adatoms, In-rich clusters can aggregate in time intervals comparable to the time needed for the growth of one monolayer. During growth these In-conglomerates are buried then in the bulk of the layer forming the observed In-rich QD-like structures. Further interchange of the In- and Ga-atoms is then suppressed due to the low In-bulk diffusion coefficient.

5   Conclusions

In this article we reviewed recent progress in the MBE of cubic III-nitrides. Inspite of the inherent difficulties connected with the growth of compounds with metastable crystal structure, quite impressive progress has been achieved within the last years. This was in some part due to the possibility of in-situ control of the MBE process. From a technological point of view, GaAs seems to be the best choice for the substrate, since it is cheap and available in large quantities. However, the large lattice misfit generates high dislocation densities in the c-III nitride layers and the small thermal stability of GaAs can cause problems in epitaxial growth at elevated substrate temperatures. Since spontaneous polarization and piezoelectric fields do not exist in c-III nitrides, double heterostructures and quantum wells were considered as being optimally suited to clarify the origin of some peculiar properties of the luminescence from InGaN/GaN quantum wells. Experiments with MBE grown c-III nitrides have demonstrated the influence of quantum dot-like structures on photoluminescence emission from quantum structures. Doping is an inevitable prerequisite for any device application. Like in their hexagonal counterparts, high n-type conductivity in c-GaN can be achieved by doping with Si. For the MBE of c-GaN with p-type conduction doping by carbon seems to be an interesting new alternative to Mg-doping. Up to now, no commercial device application based on c-MBE-grown group III nitrides has been reported. However, the observation of optical stimulated emission at room temperature from cleaved c-GaN layers and electroluminescence from c-GaN p-n junction has clearly demonstrated the potential of the c-III nitrides for future realisation of optoelectronic devices.

Acknowledgements   The authors acknowledge the support of this work by "Deutsche Forschungsgemeinschaft" (DFG). We like to thank DAAD and PROBRAL for a long-termed and successful collaboration with the group of J.R. Leite at Sao Paulo University, Brazil.

References

[1]   M. J. Paisley, Z. Sitar, J. B. Posthill and R. F. Davis, J. Vac. Sci. Technol. A **7**, 701 (1989).
[2]   S. Strite, J. Ruan, Z. Li, N. Manning, A. Salvador, H. Chen, D. J. Smith, W. J. Choyke and H. Morkoç, J. Vac. Sci. Technol. B **9**, 1924 (1991).
[3]   H. Okumura, S. Misawa and S. Yoshida, Appl. Phys. Lett. **59**, 1059 (1991).
[4]   T. Lei, M. Fanciulli, R. J. Molnar, T. D. Moustakas, R. J. Graham and J. Scanlon, Appl. Phys. Lett. **59**, 944 (1991).
[5]   S. Fujieda and Y. Matsumoto, Jpn. J. Appl. Phys. **30**, L 1665 (1991).
[6]   T. Lei, T. D. Moustakas, R. J. Graham, Y. He and S. J. Berkowitz, J. Appl. Phys. **71**, 4933 (1992).
[7]   S. Yoshida, H. Okumura, S. Misawa and E. Sakuma, Surf. Sci. **267**, 50 (1992).
[8]   S. Miyoshi, K. Onabe, N. Ohkouchi, H. Yaguchi, R. Ito, S. Fukatsu and Y. Shiraki, J. Cryst. Growth **124**, 439 (1992).
[9]   M. E. Lin, G. Xue, G. L. Zhou, J. E. Greene and H. Morkoç, Appl. Phys. Lett. **63**, 932 (1993).
[10] R. C. Powell, N. E. Lee, Y. W. Kim and J. E. Greene, J. Appl. Phys. **73**, 189 (1993).
[11] H. Liu, A. C. Frenkel, J. G. Kim and R. M. Park, J. Appl. Phys. **74**, 6124 (1993).
[12] H. Okumura, S. Misawa, T. Okahisa and S. Yoshida, J. Crystal Growth **136**, 361 (1994).
[13] A. Kikuchi, H. Hoshi and K. Kishino, Jpn. J. Appl. Phys. **33**, 688 (1994).
[14] M. Nagahara, S. Miyoshi, H. Yaguchi, K. Onabe, Y. Shiraki and R. lto, Jpn. J. Appl. Phys. **33**, 694 (1994).
[15] Z. Q. He, X. M. Ding, X. Y. Hou and X. Wang, Appl. Phys. Lett. **64**, 315 (1994).
[16] J. N. Kuznia, J. W. Yang, Q. C. Chen, S. Krishnankutty and M. A. Khan, Appl. Phys. Lett. **65**, 2407 (1994).
[17] J. G. Kim, A. C. Frenkel, H. Lie and R. M. Park, Appl. Phys. Lett. **65**, 91 (1994).
[18] T. S. Cheng, L. C. Jenkins, S. E. Hooper, C. T. Foxon, J. W. Orton and D. E. Lacklison, Appl. Phys. Lett. **66**, 1509 (1995).
[19] O. Brandt, H. Yang, B. Jenichen, Y. Suzuki, L. Däweritz and K. H. Ploog, Phys. Rev. B **52**, R2253 (1995).
[20] J. G. Kim, A. C. Frenkel, H. Liu and R. M. Park, Appl. Phys. Lett. **65**, 91 (1994).
[21] D. Schikora, M. Hankeln, D. J. As, K. Lischka, T. Litz, A. Waag, T. Buhrow and F. Henneberger, Phys. Rev. B **54**, R8381 (1996).
[22] E. Martinez-Guerrero, C. Adelmann, F. Chabuel, J. Simon, N. T. Pelekanos, G. Mula, B. Daudin, G. Feuillet and H. Mariette, Appl. Phys. Lett. **77**, 809 (2000).
[23] M. E. Lin, G. Xue, G. L. Zhou, J. E. Greene and H. Morkoç, Appl. Phys. Lett. **63**, 932 (1993).
[24] G. Mula, C. Adelmann, S. Moehl, J. Oullier and B. Daudin, Phys. Rev. B **64**, 195406 (2001).
[25] C. Adelmann, R. Langer, G. Feuillet and B. Daudin, Appl. Phys. Lett. **75**, 3518 (1999).
[26] H. Okumura, H. Hamaguchi, G. Feuillet, Y. Ishida and S. Yoshida, Appl. Phys. Lett. **72**, 3056 (1998).
[27] B. Yang, O. Brandt, B. Jenichen, J. Müllhäuser and K. H. Ploog, J. Appl. Phys. **82**, 1918 (1997).
[28] H. Yang, O. Brandt, M. Wassermeier, J. Behrend, H. P. Schönherr and K. H. Ploog, Appl. Phys. Lett. **68**, 244 (1996).
[29] A. Trampert, O. Brandt, H. Yang and K. H. Ploog, Appl. Phys Lett. **70**, 583 (1997).
[30] A. P. Lima, T. Frey, U. Köhler, C. Wang, D. J. As, K. Lischka and D. Schikora, J. Crystal Growth **197**, 31 (1999).
[31] Z. A. Munir and A. W. Searcy, J.Chem.Phys. **42**, 4223 (1965).
[32] S. Shokhovets, R. Goldhahn, V. Cimalla, T. S. Cheng and C. T. Foxon, J. Appl. Phys. **84**, 1561 (1998).
[33] A. Trampert, O. Brandt, H. Yang and K. H. Ploog, Appl. Phys. Lett. **70**, 583 (1997).
[34] X. H. Wu, D. Kapolnek, E. J. Tarsa, B. Heyening, S. Keller, B. P. Keller, U. K. Mishra, S. P. DenBars and J. S. Speck, Appl.Phys.Lett **68**, 1371 (1996).
[35] H. Okumura, K. Ohta, G. Feuillet, K. Balakrishnan, S. Chichibu, H. Hamaguchi, P. Hacke and S. Yoshida, J. Crystal Growth **178**, 113 (1997).
[36] G. Feuillet, H. Hamaguchi, K. Ohta, P. Hacke, H. Okumura and S. Yoshida, Appl. Phys. Lett. **70**, 1025 (1997).
[37] J. Neugebauer, Z. Zywietz, M. Scheffler, J. E. Northrup and C. G. V. d. Walle, Phys.Rev. Lett. **80**, 3097 (1998).
[38] T. Zywietz, J. Neugebauer and M. Scheffler, Appl.Phys.Lett **73**, 4 (1998).
[39] J. Neugebauer, private communication
[40] C. Adelmann, R. Langer, E. Martinez-Guerrero, H. Mariette, G. Feuillet and B. Daudin, J. Appl. Phys. **86**, 4322 (1999).
[41] H. Morkoç, Nitride Semiconductors and Devices, Springer, Berlin 1999
[42] E. F. Schubert, Doping of III-V Semiconductors, University Press, Cambridge 1993
[43] U. Kaufmann, P. Schlotter, H. Obloh, K. Köhler and M. Maier, Phys. Rev. B **62**, 10867 (2000).

[44] S. Guha, N. A. Bojarczuk and F. Cardone, Appl. Phys. Lett. **71**, 1685 (1997).
[45] B. Y. Ber, Y. A. Kudriavtsev, A. V. Merkulov, S. V. Novikov, D. E. Lacklison, J. W. Orton, T. S. Cheng and C. T. Foxon, Semicond. Sci. Technol. **13**, 71 (1998).
[46] S. Shimizu and S. Sonoda, Proc. Int. Workshop on Nitride Semiconductors, IPAP Conf. Series **1**, 740 (2001).
[47] R. Zhang and T. F. Kuech, Appl. Phys. Lett. **72**, 1611 (1998).
[48] U. Birkle, M. Fehrer, V. Kirchner, S. Einfeldt, D. Hommel, S. Strauf, P. Michler and J. Gutowski, MRS Internet J. Nitride Semicond. Res. **4S1**, G5.6 (1999).
[49] C. R. Abernathy, J. D. MacKenzie, S. J. Pearton and W. S. Hobson, Appl. Phys. Lett. **66**, 1969 (1995).
[50] D. J. As, R. Richter, J. Busch, B. Schöttker, M. Lübbers, J. Mimkes, D. Schikora and K. Lischka, MRS Internet J. Nitride Semicond. Res. **5S1**, W3.81 (2000).
[51] E. Martinez-Guerrero, B. Daudin, G. Feuillet, H. Mariette, Y. Genuist, S. Fanget, A. Philippe, C. Dubois, C. Bru-Chevallier, G. Guillot, P. Aboughe Nze, T. Chassagne, Y. Monteil, H. Gamez-Cuatzin and J. Tardy, Mat. Sci. & Eng. **B82**, 59 (2001).
[52] Z. Q. Li, H. Chen, H. F. Liu, L. Wan, M. H. Zhang, Q. Huang, J. M. Zhou, N. Yang, K. Tao, Y. J. Han and Y. Luo, Appl. Phys. Lett. **76**, 3765 (2000).
[53] J. L. Souchiere and V. T. Binh, Surface Science **168**, 52 (1986).
[54] D. J. As, D. Schikora, A. Greiner, M. Lübbers, J. Mimkes and K. Lischka, Phys.Rev. **B54**, R11118 (1996).
[55] J. R. L. Fernandez, V. A. Chitta, E. Abramof, A. Ferreira da Silva, J. R. Leite, A. Tabata, D. J. As, T. Frey, D. Schikora and K. Lischka, MRS Symp. Proc. Vol. **595**, W3.40 (2000).
[56] J. Portmann, C. Haug, R. Brenn, T. Frey, B. Schöttker and D. J. As, Nuclear Instruments and Methods in Physics Research B **115**, 489 (1999).
[57] R. Brenn, D. J. Jamieson, A. Cimmino, K. K. Lee, T. Frey, D. J. As and S. Prawer, Nuclear Instruments and Methods B **161-163**, 435 (2000).
[58] N. G. Weimann, L. F. Eastman, D. Doppalapudi, H. M. Ng and T. D. Moustakas, J. Appl. Phys. **83**, 3656 (1998).
[59] D. J. As, Defect and Diffusion Forum **206-207**, 87 (2002).
[60] D. C. Look and J. R. Sizelove, Phys. Rev. Lett. **82**, 1237 (1999).
[61] A. T. Blumenau, J. Elsner, R. Jones, M. I. Heggie, S. Öberg, T. Frauenheim and P. R. Briddon, J. Phys.: Condens. Matter **12**, 10223 (2000).
[62] D. J. As and K. Lischka, phys. stat. sol. (a) **176**, 475 (1999).
[63] O. Brandt, H. Yang, H. Kostial and K. H. Ploog, Appl. Phys. Lett. **69**, 2707 (1996).
[64] K. H. Ploog and O. Brandt, J. Vac. Sci. Technol. A **16**, 1609 (1998).
[65] D. J. As, phys. stat. sol. (b) **210**, 445 (1998).
[66] D. J. As, T. Simonsmeier, B. Schöttker, T. Frey and D. Schikora, Appl. Phys. Lett. **73**, 1835 (1998).
[67] D. J. As, T. Simonsmeier, J. Busch, B. Schöttker, M. Lübbers, J. Mimkes, D. Schikora and K. Lischka, MRS Internet J. Nitride Semicond. Res. **4S1**, G.3.24 (1999).
[68] E. Martinez-Guerrero, E. Bellet-Amalric, L. Martinet, G. Feuillet, B. Daudin, H. Mariette, P. Holliger, C. Dubois, P. Aboughe Nze, T. Chassagne, G. Ferro and Y. Monteil, J. Appl. Phys. **91**, 4983 (2002).
[69] O. Brandt, in: Group III nitride semiconductor compounds - physics and applications, B. Gil, Clarendon Press, Oxford 1998 (p.417)
[70] D. J. As and U. Köhler, J. Phys.: Condens. Matter **13**, 8923 (2001).
[71] D. J. As, U. Köhler, M. Lübbers, J. Mimkes and K. Lischka, phys. stat. sol. (a) **188**, 699 (2001).
[72] M. Schubert, A. Kasic, S. Einfeldt, D. Hommel, U. Köhler, D. J. As, J. Off, B. Kuhn, F. Scholz, J. A. Woollam and C. M. Herzinger, SPIE - Int. Soc. Opt. Eng. Proc. **4449**, 58 (2001).
[73] D. J. As, U. Köhler and K. Lischka, Mat. Res. Soc. Symp. Proc. **693**, 17 (2002).
[74] P. D. Kirchner, J. M. Woodall, J. L. Freeouf, D. J. Wolford and G. D. Petit, J. Vac. Sci. Technol. **19**, 604 (1981).
[75] J. Neugebauer and C. G. v. d. Walle, J. Appl. Phys. **85**, 3003 (1999).
[76] M. Marques, L. E. Ramos, L. M. R. Scolfaro, L. K. Teles and J. R. Leite, in Proc. 25th Intern. Conf. Phys. Semicond. (ICPS 2000), Osaka 1411 (2000).
[77] P. Boguslawski, E. L. Briggs and J. Bernholc, Appl. Phys. Lett. **69**, 233 (1996).
[78] P. Boguslawski and J. Bernholc, Phys. Rev. B **56**, 9496 (1997).
[79] D. J. As, F. Schmilgus, C. Wang, B. Schöttker, D. Schikora and K. Lischka, Appl. Phys. Lett. **70**, 1311 (1997).
[80] S. Fischer, C. Wetzel, E. E. Haller and B. K. Meyer, Appl. Phys. Lett. **67**, 1298 (1995).
[81] B.-H. Cheong and K. J. Chang, Phys. Rev. B **49**, 17436 (1994).
[82] J. Wagner, R. C. Newman, B. R. Davidson, S. P. Westwater, T. J. Bullough, T. B. Joyce, C. D. Latham, R. Jones and S. Öberg, Phys. Rev. Lett. **78**, 74 (1997).

[83] L. E. Ramos, J. Furthmüller, L. M. R. Scolfaro, J. R. Leite and F. Bechstedt, phys. stat. sol. (b) **234**, 864 (2002).
[84] A. F. Wright, J. Appl. Phys. **92**, 2575 (2002).
[85] D. J. As, A. Richter, J. Busch, M. Lübbers, J. Mimkes and K. Lischka, Appl. Phys. Lett. **76**, 13 (2000).
[86] D. J. As, A. Richter, J. Busch, M. Lübbers, J. Mimkes and K. Lischka, phys. stat. sol. (a) **180**, 369 (2000).
[87] P. Kozodoy, J. P. Ibbetson, H. Marchard, P. T. Fini, S. Keller, J. S. Speck, S. P. DenBaars and U. K. Mishra, Appl. Phys. Lett. **73**, 975 (1998).
[88] S. Nakamura, T. Mukai and M. Senoh, Appl. Phys. Lett. **64**, 1687 (1994).
[89] H. Yang, L. H. Zheng, J. B. Li, C. Wang, D. P. Xu, Y. T. Wang, X. W. Hu and P. D. Han, Appl. Phys. Lett. **74**, 2498 (1999).
[90] N. Grandjean, J. Massies, P. Lorenzini and M. Leroux, Electron. Lett. **33**, 2156 (1997).
[91] J. Holst, A. Hoffmann, I. Broser, B. Schöttker, D. J. As, D. Schikora and K. Lischka, Appl. Phys. Lett. **74**, 1966 (1999).
[92] S. Chichibu, T. Azuhata, T. Sota and S. Nakamura, Appl. Phys. Lett. **70**, 2822 (1997).
[93] N. A. El-Masry, E. L. Piner, S. X. Liu and S. M. Bedair, Appl. Phys. Lett. **72**, 40 (1998).
[94] M. D. McCluskey, L. T. Romano, B. S. Krusor, D. P. Bour, N. M. Johnson and S. Brennan, Appl. Phys. Lett. **72**, 1730 (1998).
[95] I. Ho and G. B. Stringfellow, Appl. Phys. Lett. **69**, 2701 (1996).
[96] T. Saito and Y. Arakawa, Phys. Rev. B **60**, 1701 (1999).
[97] L. K. Teles, J. Furthmüller, L. M. R. Scolfaro, J. R. Leite and F. Bechstedt, Phys. Rev. B **62**, 2475 (2000).
[98] L. W. Wang and A. Zunger, J. Chem. Phys. **100**, 2394 (1994).
[99] S. F. Chichibu, A. C. Abare, M. P. Mack, M. S. Minsky, T. Deguchi, D. Cohen, P. Kozodoy, S. B. Fleischer, S. Keller, J. S. Speck, J. E. Bowers, E. Hu, U. K. Mishra, L. A. Coldren, S. P. DenBaars, K. Wada, T. Sota and S. Nakamura, Mat. Science and Engineering **B59**, 298 (1999).
[100] T. Frey, PhD thesis, University of Paderborn (2000).
[101] J. R. Müllhäuser, B. Jenichen, M. Wassermeier, O. Brandt and K. H. Ploog, Appl. Phys. Lett. **71**, 909 (1997).
[102] J. Müllhäuser, O. Brandt, A. Trampert, B. Jenichen and K. H. Ploog, Appl. Phys. Lett. **73**, 1230 (1998).
[103] R. Goldhahn, J. Scheiner, S. Shokhovets, T. Frey, U. Köhler, D. J. As and K. Lischka, Appl. Phys. Lett. **76**, 291 (2000).
[104] J. Holst, A. Hoffmann, D. Rudloff, F. Bertram, T. Riemann, J. Christen, T. Frey, D. J. As, D. Schikora and K. Lischka, Appl. Phys. Lett. **76**, 2832 (2000).
[105] M. Pophristic, F. H. Long, C. Tran, I. Ferguson and J. R.F. Karlicek, Appl. Phys. Lett. **73**, 3550 (1998).
[106] X. Chen, B. Henderson and K. P. O'Donnell, Appl. Phys. Lett. **60**, 2672 (1992).
[107] L. Pavesi and M. Cheschin, Phys. Rev. B **48**, 17625 (1993).
[108] J. Holst, L. Eckey, A. Hoffmann, I. Broser, B. Schöttker, D. J. As, D. Schikora and K. Lischka, Appl. Phys. Lett. **72**, 1439 (1998).
[109] J. Holst, A. Hoffmann, I. Broser, D. Rudloff, F. Bertram, T. Riemann, J. Christen, T. Frey, D. J. As, D. Schikora and K. Lischka, phys. stat. sol. (b) **216**, 471 (1999).
[110] A. Tabata, J. R. Leite, A. P. Lima, E. Silveira, V. Lemos, T. Frey, D. J. As, D. Schikora and K. Lischka, Appl. Phys. Lett. **75**, 1095 (1999).
[111] E. Silveira, A. Tabata, J. R. Leite, R. Trentin, V. Lemos, T. Frey, D. J. As, D. Schikora and K. Lischka, Appl. Phys. Lett. **75**, 3602 (1999).
[112] V. Lemos, E. Silveira, J. R. Leite, A. Tabata, R. Trentin, L. M. R. Scolfaro, T. Frey, D. J. As, D. Schikora and K. Lischka, Phys. Rev. Lett. **84**, 3666 (2000).
[113] O. Husberg, A. Khartchenko, D. J. As, H. Vogelsang, T. Frey, D. Schikora, K. Lischka, O. C. Noriega, A. Tabata and J. R. Leite, Appl. Phys. Lett. **79**, 1243 (2001).
[114] T. Yodo, H. Yona, H. Ando, D. Nosei and Y. Harada, Appl.Phys.Lett. **80**, 968 (2002).
[115] L. K. Teles, J. Furthmüller, L. M. R. Scolfaro, J. R. Leite and F. Bechstedt, Phys. Rev. B **63**, 085204 (2001).
[116] P. R. C. Kent and A. Zunger, Applied Physics Letters **79**, 1977 (2001).
[117] A. Tabata, L. K. Teles, L. M. R. Scolfaro, J. R. Leite, A. Khartchenko, T. Frey, D. J. As, D. Schikora, K. Lischka, J. Furthmüller and F. Bechstedt, Appl. Phys. Lett. **80**, 769 (2002).
[118] C.-C. Chuo, C.-M. Lee and J.-I. Chyi, Appl. Phys Lett. **78**, 314 (2001).
[119] F. Leonard and R. C. Desai, Phys. Rev. B **57**, 4805 (1998).
[120] H. Chan, R. M. Feenstra, J. E. Northrup, T. Zywietz, J. Neugebauer and D. W. Greve, Phys. Rev. Lett. **85**, 1902 (2000).
[121] O. Husberg, A. Khartchenko, D. J. As, K. Lischka, E. Silveira, O. C. Noriega, J. R. L. Fernandez and J. R. Leite, phys. stat. sol. (c) **0**, No. 1, 293 (2002).

phys. stat. sol. (c) **0**, No. 6, 1627–1650 (2003) / **DOI** 10.1002/pssc.200303140

Freestanding GaN-substrates and devices

Claudio R. Miskys, Michael K. Kelly [*], **Oliver Ambacher** [**], and **Martin Stutzmann**[***]

Walter Schottky Institut, Technische Universität München,
Am Coulombwall 3, 85748 Garching, Germany

Received 27 June 2003, accepted 29 July 2003
Published online 28 August 2003

PACS 68.37.Ps, 78.55.Cr, 81.15.-z, 81.15.Kk, 81.65.Cf, 85.60.Jb

Various physical aspects and potential applications of the laser-induced separation of GaN epilayers from their sapphire substrate are reviewed. The effect of short laser pulses on the thermal decomposition of GaN and possible applications of the laser-induced dissociation of GaN for fast etching of this material is discussed. Particular emphasis is placed on the defect-free delamination of large area GaN films with thicknesses ranging from 3 to 300 µm from sapphire substrates. The use of the resulting freestanding GaN films in device technology and homoepitaxy of III-nitrides are outlined. Specific examples are the flip-chip bonding of freestanding InGaN/GaN LEDs to a silicon submount and the production of pseudosubstrates for the homoepitaxy of high quality GaN epilayers.

1   Introduction

Compared to well established semiconductor materials systems such as Si/Ge or GaAs/AlGaAs, some of the most challenging aspects of GaN-based thin-film technology come from the fact that the most common substrate materials used today (sapphire and SiC) have very different properties than the device layer itself. In contrast to the majority of modern semiconductors, present commercial GaN–based devices are exclusively prepared by heteroepitaxy onto foreign substrate materials. Surprisingly, the ensuing dislocation densities in the range of 10^7 to 10^8 cm^{-2} did not prevent the commercial success of III-nitrides for light-emitting devices in the visible to ultraviolet spectral range. However, it is now widely accepted that the ultimately possible performance of GaN-based devices can only be reached by the use of homoepitaxy onto bulk GaN or AlN substrates. Unfortunately, despite many years of intense research, the high equilibrium pressures and temperatures of the III-nitrides so far have rendered the growth of bulk substrates for homoepitaxy with production grade sizes and quality impossible.

An alternative approach, which has been persued since a couple of years, is the use of thick GaN-layers grown by hydride vapour phase epitaxy (HVPE) on sapphire. These thick layers are subsequently removed from the sapphire substrate by a suitable process and then can serve as homosubstrates for device fabrication. The purpose of this review article is to summarize the present state of research concerning such freestanding GaN homosubstrates, with special emphasis on laser-based lift-off processes for the separation of the thick GaN-layers from their sapphire substrates. The review is organized as follows: after a brief resumé of the various substrates used for heteroepitaxy so far, we will discuss in detail the physical and technological aspects of the laser lift-off (LLO) processing for the production of freestanding GaN substrates and devices. In the final section of this article, the surface preparation of laser-separated HVPE-layers and typical results obtained by subsequent homoepitaxy will be addressed. For a

---

[*] Permanent address: Agilent Technologies, 71034 Böblingen, Germany.
[**] Permanent address: Institut für Festkörperelektronik, TU Ilmenau, 98693 Ilmenau, Germany.
[***] Corresponding author: e-mail: mst@wsi.tu-muenchen.de

more detailed discussion of the growth of III-nitrides on heterosubstrates such as sapphire, GaAs, or Si the reader is refered to other articles in this volume.

2   Substrates for III-nitride heteroepitaxy: a brief survey

In the case of the standard vapor phase epitaxy methods for GaN growth (HVPE, MOCVD), the high growth temperatures (>1000 °C) and the high concentration of ammonia and hydrogen considerably reduce the choice of possible substrates. The deposition of one or even multiple low-temperature buffer layers in the first step of device fabrication is essential to allow the use of foreign substrates such as sapphire, SiC, Si, GaAs or GaP. Even for molecular beam epitaxy (MBE) with growth temperatures which are about 250 °C lower than those for HVPE and MOCVD growth, the stability of the substrate surface at 800 °C and under the influence of nitrogen radicals is a critical issue. For device production on an industrial scale, the substrate has to fulfill further criteria such as minimum size (2"), atomically flat surfaces, and availability in large quantities at an acceptable price. Sapphire was and still is the most common substrate for the deposition of GaN-based light-emitting diodes (LEDs), because of its reasonably low cost and wide availability, and despite the fact that it has a large lattice constant and thermal expansion coefficient mismatch with respect to GaN.

However, high dislocation densities caused by this mismatch are detrimental to the performance of more demanding device structures than LEDs, such as laser diodes operating at high power densities [1]. Therefore, the lattice constant mismatch has been viewed as the most important criterion for determining the suitability of a material as a substrate for GaN growth. In practice, other material's issues including crystal structure, composition, chemical reactivity and surface termination, or thermal and electrical properties also have a strong influence on the final epitaxial layer. In the end, the substrate determines the crystal orientation, polarity, surface morphology, strain, and the defect concentration of the heteroepitaxial III-nitride films and, thus, is directly related to an optimal device performance. By now, many different substrate materials have been tried out GaN epitaxy, including metals, oxides, nitrides, and semiconductors. A still incomplete list of the more commonly used substrate materials, their crystal structure, lattice constants and thermal expansion coefficients is given in Table 1. Among many other issues, the influence of the substrate on the crystal polarity and macroscopic polarization of the group III-nitrides is particularly important. E.g., the chemical reactivity and the optimum growth conditions required for high quality epitaxy generally depend on the polarity of the crystal, which in turn is determined by the surface termination of the substrate. The magnitude and sign (tensile or compressive) of the strain incorporated into the heteroepitaxial film depends on many details of the nucleation and growth process, and its optimization still is more a "black art" rather than a well understood phenomenon. The densities of misfit and threading dislocations in GaN films deposited on the substrates in Table 1 are typically between 10^8 and 10^{10} cm^{-2}. For comparison, the dislocation densities in GaAs or silicon homoepitaxy are as low as 10^2 to 10^4 cm^{-2} or almost zero [2]. In addition, inversion domain boundaries and stacking faults caused by the heteroepitaxy give rise to non-radiative recombination centers in the band gap and reduce the minority carrier lifetimes [48][49]. Threading dislocations form an undesirable path for enhanced impurity diffusion, causing an inhomogeneous impurity distribution throughout the heteroepitaxial films and giving rise to an overall degradation of electronic and optical properties [50].

Therefore, many efforts have been made to improve the crystal quality of the heteroepitaxial layers, mainly in the specific case of GaN on sapphire. Successful approaches include an appropriate surface preparation of the substrate such as nitridation, deposition of low-temperature (LT) AlN or GaN buffer layers, multiple LT buffer layers [51], epitaxial lateral overgrowth (ELOG) [52], PENDEO-epitaxy [53], and other techniques [54-56]. Despite of all these efforts, heteroepitaxial GaN layers with dislocation densities below 10^7 cm^{-2} cannot be obtained in a practical manner.

3 Freestanding GaN-substrates and devices by laser lift-off

As a viable alternative to direct heteroepitaxy of device relevant layer sequences on heterosubstrates, the use of bulk crystals or bulk-like "pseudo-substrates" consisting of thick (several 100 μm) GaN layers

Table 1 Some characteristic structural properties of substrate materials for GaN heteroepitaxy.

Material	Structure	Lattice constants (Å)			Thermal expansion coefficient (10^{-6} K^{-1})	References
		(a)	(b)	(c)		
w-GaN	wurtzite	3.1885		5.185	5.45	[3]
zb-GaN	zincblende	4.511				[4]
r-GaN	rocksalt	4.22				[5]
w-AlN	wurtzite	3.1106		4.9795	2.2	[6]
zb-AlN	zincblende	4.38				[7][8]
r-AlN	rocksalt	4.04				[9]
ZnO	wurtzite	3.2496		5.2065	2.9	[10][11]
ß-SiC	3C-zincblende	4.3596			3.9	[12][13]
SiC	4H-wurtzite	3.073		10.053		[12]
SiC	6H-wurtzite	3.0806		15.1173	4.46	[12][13]
BP	zincblende	4.538				[14][15]
GaAs	zincblende	5.6533			6.0	[16]
GaP	zincblende	5.4309			4.65	[17][18]
Si	diamond	5.4310			3.59	[19]
Al$_2$O$_3$	rhombohedral	4.758		12.982	7.50	[20][21]
MgAl$_2$O$_4$	spinel	8.083			7.45	[22][23][24]
MgO	rocksalt	4.216			10.5	[7][25]
LiGaO$_2$	orthorhombic	5.4063	6.3786	5.0129	9.0	[26]
Γ-LiAlO$_2$		5.169		6.267	15	[27][28][29]
NdGaO$_3$	orthorhombic	5.428	5.498	7.71	11.9	[30][31][32]
ScAlMgO$_4$	wurtzite	3.236		25.15		[33][34]
Ca$_8$La$_2$(PO$_4$)$_6$O$_2$	apatite	9.446		6.922		[35]
MoS$_2$						[36][37]
LaAlO$_3$	rhombohedral	5.364		13.11		[38]
(Mn,Zn)Fe$_2$O$_4$	spinel	8.5				[39]
Ca$_8$La$_2$(PO$_4$)$_6$O$_2$	apatite	9.446		6.922		[35]
Hf	hcp	3.18		5.19		[40][41]
Zr	hcp	3.18		5.19		[41]
ZrN	rocksalt	4.5776				[42]
Sc	hcp	3.309		5.4		[43]
ScN	rocksalt	4.502				[43][44]
NbN	rocksalt	4.389				[45]
TiN	rocksalt	4.241				[46][47]

grown with a high deposition rate on sapphire substrates by HVPE has attracted considerable attention in recent years. There are various motivations for such an approach. First, the capability to produce material with a superior quality, e.g. a largely reduced density of dislocations and much narrower photoluminescence linewidths compared to the conventional heteroepitaxial material on sapphire or SiC has been demonstrated by different groups [57, 58]. Secondly, as far as the use of such bulk or bulk-like substrates for homoepitaxy is concerned, two-dimensional growth can be achieved without the need for additional steps such as surface nitridation or the growth of additional nucleation or buffer layers. Moreover, in the case of freestanding GaN-substrates, laser cavities can be obtained by simple facet cleaving [59, 60], and the GaN substrate can be rendered electrically conductive, allowing for a significant simplification in device processing and circuit integration. In particular the latter two issues require a complete removal of the sapphire substrate after HVPE growth of the thick GaN pseudo-substrate.

Recent progress in the HVPE growth of GaN has demonstrated that critical problems such as obtaining large-area crack-free wafers of up to 2" diameter and 300 μm thickness can indeed be overcome [61, 62]. However, the surface roughness of such thick HVPE layers often is in the range of several microns, which is detrimental for any device application. Therefore, surface polishing after the HVPE growth is required to prepare epi-ready GaN pseudo-substrates. The necessary polishing procedures are hampered by the mechanical brittleness of sapphire and the significant bowing of the GaN/Al$_2$O$_3$ bilayers, which is caused by the differences in thermal expansion coefficients between GaN and sapphire and leads to a sometimes critical accumulation of thermal stress during the cool down from the high growth temperature (≈ 1050 °C) to room temperature. Again, a complete removal of the Al$_2$O$_3$ substrate also would remove most of these problems. Finally, as far as devices are concerned, Nakamura et al. have reported a significant improvement in the lifetime of laser diodes and cleaved facets after polishing the sapphire substrate away [59, 63].

All of the issues mentioned above make it highly desirable to develop a process which enables a fast, reliable, and high-yield detachment of thick HVPE-grown GaN layers from their sapphire heterosubstrates. In practice, such a process will involve a sacrificial layer somewhere between the substrate and the GaN-layer, which is removed by a specific chemical or thermal treatment. One possibility would be to deposit a sacrificial layer which makes use of selective chemical etching, e.g. of AlN versus GaN in KOH. However, in this case the detachment of large area GaN pseudo-substrates would be slowed down by chemical diffusion and would impose additional boundary conditions for the deposition of buffer layers. Just recently, an epitaxial procedure was developed to cause GaN films to self-detach from its sapphire substrate during cool down from growth temperature due to the accumulation of thermal stress and the lattice mismatch between GaN and sapphire [64]. In this case, a rather complex pretreatment of the wafer including lithography is needed, which limits the relevance of such a processing step for future mass production schemes. A third approach is the complete removal of the substrate by etching or polishing. This, however, is quite time-consuming, especially in the case of sapphire or SiC substrates.

A process which is much more flexible and significantly faster than the above mentioned methods is the laser-induced delamination of III-nitrides from a transparent substrate such as sapphire, or laser lift-off (LLO) for short. In this process, the separation of a GaN layer from the substrate is achieved by irradiation of the substrate-film-interface through the substrate with high power laser pulses at a wavelength which is transmitted by the substrate, but is strongly absorbed in the GaN layer. In the case of GaN on sapphire, the most suitable laser systems in this context are the third harmonic of a Nd:YAG laser [65, 66] or excimer lasers [67]. The absorption of such high intensity laser pulses causes a rapid thermal decomposition of the irradiated GaN interfacial layer into metallic Ga and gaseous N$_2$ (Fig. 1). The intrinsic advantages of such a LLO process compared the other possibilities mentioned above are:

- Sapphire as a readily available, well developed, and reasonably expensive substrate material can be used. It would be even possible to recycle the separated sapphire substrates after LLO for further deposition runs.
- The LLO process does not require any specific sacrificial layers in the growth sequence. Even better, the usual low-temperature and, thus, low-quality AlN or GaN buffer layers are removed together with the substrate in the same process.
- The LLO method is quite fast and can be scaled up easily. Using standard Nd:YAG or excimer laser systems with pulse repetition rate ot 1000 Hz, a 2" wafer in principle could be lifted-off in a few seconds.
- Since LLO does not require any direct mechanical or chemical contact with the wafer, the lift-off process could also occur directly after deposition of the III-nitride film in the deposition reactor before cool-down to room temperature. This would just require a suitable optical window and an external laser scanning arrangement, but at the same time would avoid the build-up of critical thermal strain,

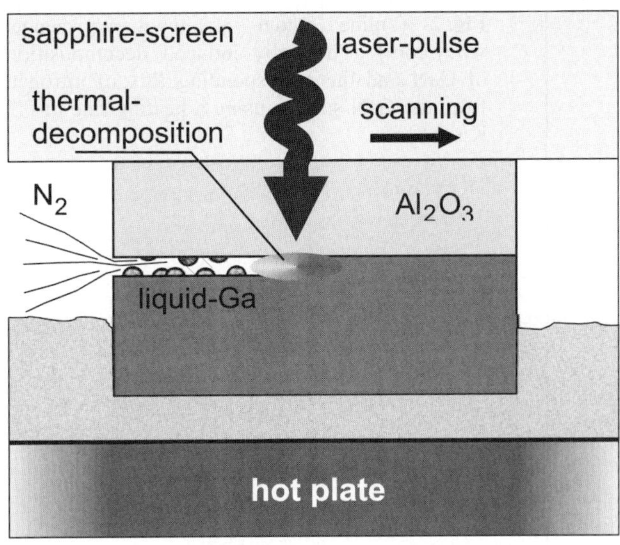

Fig. 1 (online colour at: www.interscience. wiley.com) Schematic view of the laser lift-off (LLO) process. High intensity laser pulses enter the sample via the sapphire substrate and thermally decompose a thin GaN-layer at the substrate interface. The shock waves resulting from the explosive production of nitrogen gas during each laser pulse are damped by placing the GaN-sample into sapphire powder. A hot plate can be used to raise the substrate temperature during the process, in order to relieve some of the accumulated thermal strain.

which occurs in thick GaN films on heterosubstrates during the cool down from the deposition temperature of 1100 °C to room temperature because of the large mismatch between the thermal lattice expansion coefficients of the two materials.

3.1 Laser-induced decomposition of GaN

At sufficiiently high temperatures, the stability of GaN is limited by the decomposition of the crystal into nitrogen gas and liquid gallium: $2\,GaN(s) \rightarrow N_2(g) + 2\,Ga(l)$. The flux of nitrogen molecules, $\Phi(N_2)$, leaving the crystal surface in vacuum shows an exponential increase with temperature above 830 °C, which can be parametrized as:

$$\Phi(N_2) = 1.2 \times 10^{31} \ cm^{-2}s^{-1} \ \exp(-3.9 \ eV/\,kT) \,. \tag{1}$$

The resulting thermal decomposition rate of GaN as a function of temperature is shown in Fig. 2 for a linear temperature ramp of 0.3 K/s [68]. The decomposition rate reaches approximately one monolayer per second at a temperature of 930 °C. Thus, GaN can be removed very efficiently via thermal decomposition by methods which enable a controlled local heating of the sample to temperatures above 900 °C.

As mentioned above, one possibility to locally decompose GaN is by absorption of intense light with photon energies above the bandgap of GaN (3.42 eV), e.g the 355 nm (3.49 eV) third harmonic of a Nd:YAG pulsed laser [69, 70]. The properties of neodymium-doped yttrium aluminum garnet (Nd:YAG) are the most widely studied and best understood of all solid-state laser media. The active medium is Nd^{3+}, which is optically pumped by flash lamps, generating pulses at 1064 nm. The resulting pulse length is shorter than 10 ns, the peak of the optical power is tens of megawatts, and the pulse power is very stable, with intensity variation from pulse to pulse of less than 3%. The high peak power of the pulses permits efficient frequency conversion to 532 nm (2nd harmonic) and 355 nm (3rd harmonic) in non-linear potassium dideuterium phosphate crystals (KD*P).

In Fig. 3, the effects of a single shot of a Nd:YAG-laser with energy densities in the range between 180 and 410 mJ/cm^2 on the macroscopic optical appearance of thin GaN layers on sapphire are shown. Below a critical absorbed pulse intensity of approximately 250 mJ/cm^2, no visible alteration of the GaN-layer occurs. For higher intensities, however, the absorbed photon energy leads to local heating of the layer above the critical sublimation temperature of about 900 °C, causing the destruction of the GaN.

The remaining metallic Ga is responsible for the dark areas appearing after the laser tretment in the optical pictures. As can be seen from Fig. 3, small lateral inhomogeneities of the laser beam have a dras-

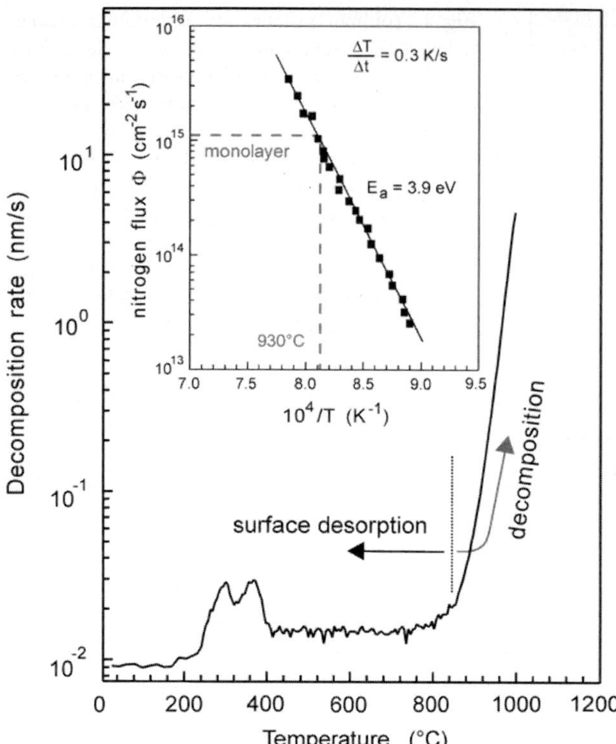

Fig. 2 (online colour at: www.interscience. wiley.com) Thermally induced decomposition of GaN and the corresponding flux of nitrogen from the GaN surface using a heating rate of 0.3 K/s [68].

tic effect on the macroscopic quality of the transformation process. In this respect, the better coherence properties of Nd:YAG lasers compared to Excimer lasers are somewhat problematic, since it is more difficult to obtain flat intensity profiles over macroscopic areas due to interference pattern formation. On the other hand, the shorter pulse length of Nd:YAG lasers (6-10 ns) compared to Excimer lasers (40 ns) can be of some advantage in the LLO-processing of thin GaN-based device structures, where thermal and mechanical strain is much more critical than for thick HVPE-grown quasi-substrates.

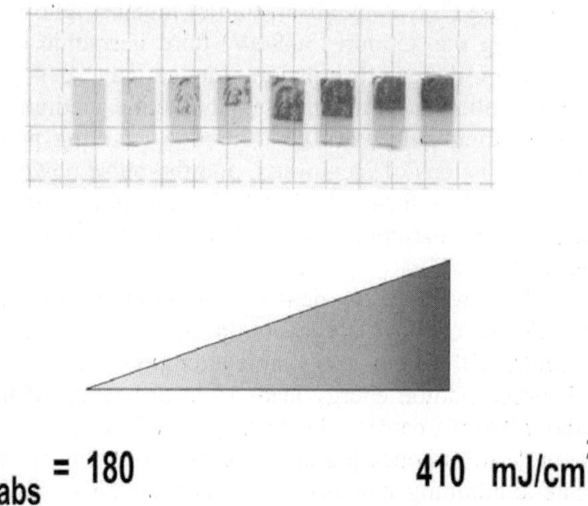

Fig. 3 (online colour at: www.interscience. wiley.com) Effect of a single shot from a Nd:YAG-laser on the optical appearance of a GaN layer on sapphire. For absorbed energy densities below 200 mJ/cm^2, no visible effects can be observed. For energy densities above the sublimation threshold of approx. 250 mJ/cm^2, the entire surface of the GaN epilayer is transformed into metallic Ga, giving rise to the dark colour in the upper half of the specimen pictures. Close to the sublimation threshold, inhomogeneities of the intensity profile of the laser beam are clearly visible in the irradiation patterns.

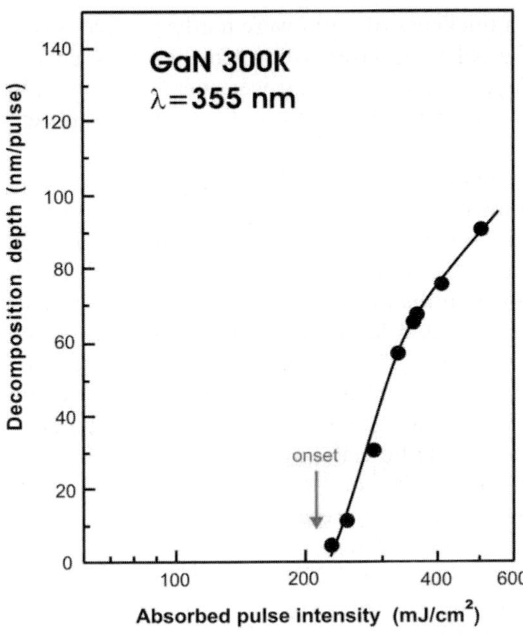

Fig. 4 (online colour at: www.interscience.wiley.com) Experimentally determined decomposition depth caused by a single pulse of a Nd:YAG laser in GaN at room temperature as a function of the pulse intensity absorbed in the thin GaN layer on sapphire. The change in slope for pulse intensities exceeding 350 mJ/cm^2 is probably due to absorption or reflection of the laser light by the metallic Ga layer formed during the process.

The extinction depth of the 355 nm Nd:YAG laser line in undoped GaN at room temperature is about 100 nm, corresponding to an absorption coefficient of 10^5 cm^{-1}. Given the sublimation threshold of 250 mJ/cm^2 determined above, and pulse intensities up to 100J/cm^2 which can be obtained by beam focussing for typical laboratory ND:YAG laser systems, GaN layers up to a thickness of 500 nm can in principle be removed by a single laser shot.

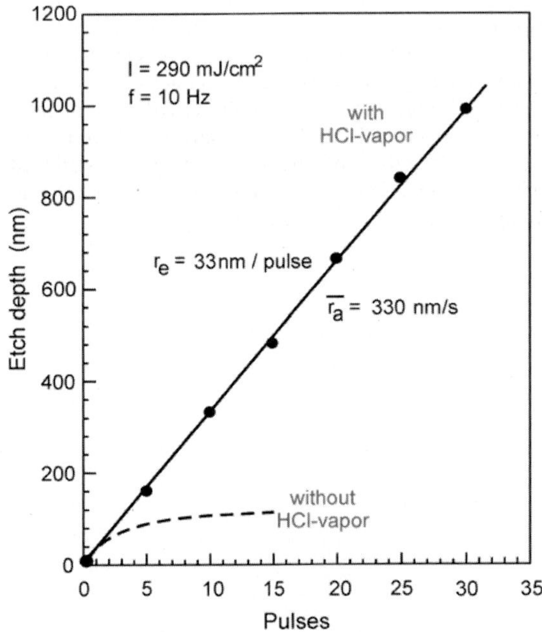

Fig. 5 (online colour at: www.interscience.wiley.com) Dependence of the laser-induced etch depth of GaN on the number of laser pulses with and without the presence of HCl-vapor. The laser treatment was performed at 300 K with an absorbed pulse intensity of 290 mJ/cm^2 and a pulse repetition rate of 10 Hz. From the slope of the straight line an average etch rate of 330 nm/s can be deduced.

For a more quantitative investigation, GaN layers with a thickness of 1 μm were partly covered with a reflective coating und then submitted to single pulses of a Nd:YAG laser with intensities between 100 and 600 mJ/cm^2. After laser exposure, the metallic Ga layer was removed by wet chemical etching in HCl, and the effective depth of the removed material was determined by depth profiling. The results are summarized in Fig. 4.

The obtained decomposition depth for a single pulse increases monotonically from 0 at the threshold intensity of approx. 250 mJ/cm^2 to almost 100 nm for an intensity of 600 mJ/cm^2. Above 320 mJ/cm^2 a slight saturation sets in, probably due to the self absorption or the high reflectivity of the metallic Ga layer formed on the surface of the irradiated GaN. This effect can be reduced by performing the laser irradiation in an HCl atmosphere, which causes a fast removal of the metallic Ga in the form GaCl$_3$. This chemical removal of the Ga is particularly important for repeated laser treatments of GaN, e.g. with longer sequences of pulses. An example is shown in Fig. 5, where GaN was irradiated with a pulse repetition rate of 10 Hz in the presence of HCl. As could be expected, a linear dependence of the etch depth on the number of pulses was obtained, with a removal of 33 nm GaN per pulse or, equivalently, an average etch rate of 330 nm/s. Without HCl, the accumulation of metallic Ga at the surface causes a saturation of the etch depth versus the number of pulses at aproximately 100 nm.

As shown in Fig. 6, this laser-induced etch rate is much higher than presently employed ion-assisted etch methods, such as electron cyclotron resonance plasma etching (ECR), chemically assisted ion beam etching (CAIBE), high-density magnetron reactive ion etching (RIE-MIE), or inductively coupled plasma etching (ICP). For more details and further references see [71, 72, 73].

Despite of the advantage of having fast etching rates, light diffraction at mask edges and borders will be detrimental for the application of laser-induced etching in the processing and isolation of small scale devices (e.g. mesa-structures). However, a high resolution laser structuring of GaN down to a dimension of 100 nm can be achieved with holographic patterning without the use of masks and therefore avoiding the effect of light scattering at edges and borders [69]. Another issue is the amount of surface damage caused by the various etch methods. In the case of laser-etching, little systematic experience exists so far, but first structural investigations indicate that crystal damage caused by the laser treatment is restricted to a surface layer of about 100 nm [74–78], which can easily be removed by conventional etching methods.

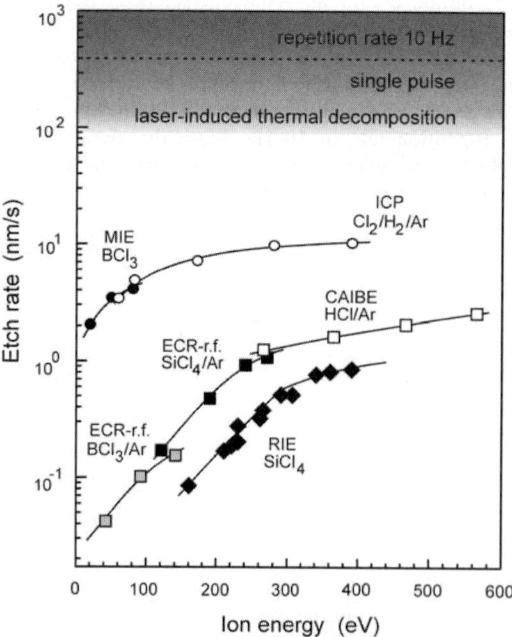

Fig. 6 (online colour at: www.interscience.wiley.com) Comparison between the maximal etch rates for GaN achieved by laser-induced etching and conventional reactive ion etching methods (ICP: inductively coupled plasma, MIE: magnetron ion etching, CAIBE: chemically assisted ion beam etching, ECR: electron cyclotron resonance, RIE: reactive ion etching). Using laser-induced thermal decomposition of GaN, much higher etch rates in the range of 1 μm/s can be realized.

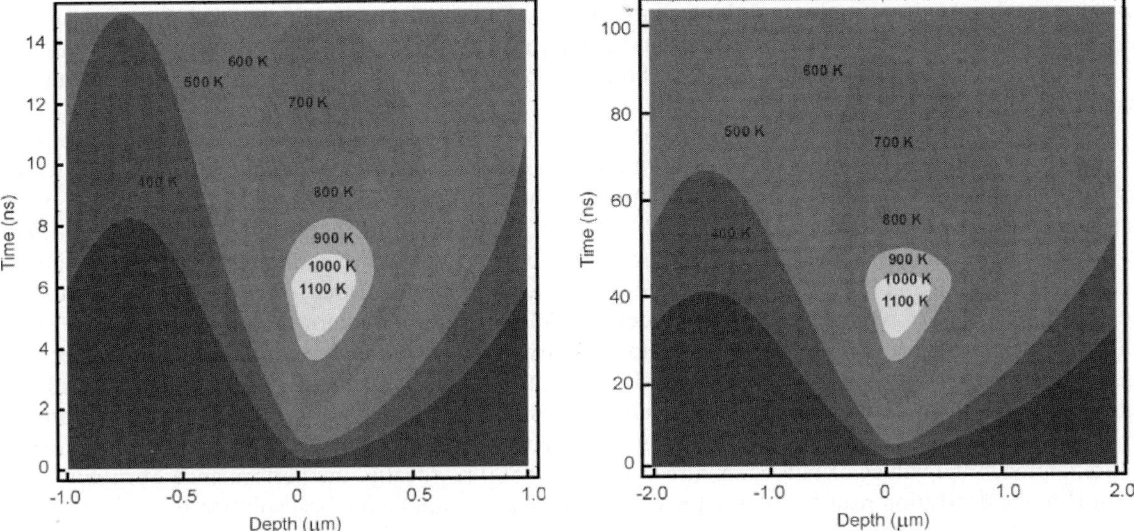

Fig. 7 (online colour at: www.interscience.wiley.com) Temporal and spatial variation of the temperature at the sapphire/GaN interface (depth zero) during and after a Nd:YAG laser pulse (left figure, $\lambda = 355$ nm, $\tau = 6$ ns, $I = 300$ mJ/cm^2) and a KrF excimer laser pulse (right figure, $\lambda = 248$ nm, $\tau = 38$ ns, $I = 600$ mJ/cm^2). The intensities of the two pulses were chosen such as to obtain the same maximum temperature of 1100 K (sublimation temperature, cf. Fig. 2).

A more detailed understanding of the laser lift-off process requires a quantitative analysis of the temporal and spatial temperature profiles at the substrate/III-nitride interface during and after the short laser pulse. As an example, we discuss here briefly the comparison between the two main laser systems used so far for LLO-processing of GaN: (i) the third harmonic of the Nd:YAG laser at $\lambda = 355$ nm with a pulse length of $\tau = 6$ ns, and (ii) the KrF excimer laser with $\lambda = 248$ nm and $\tau = 38$ ns. Because of the much longer pulse duration in the case of the KrF laser, a higher pulse energy of typically 600 mJ/cm^2 is necessary to heat the GaN above the sublimation threshold, whereas pulse energies of 300 mJ/cm^2 are sufficient in the case of the Nd:YAG laser [69, 74]. Since optical thermalization and relaxation processes in GaN occur on a much faster timescale of ps or a few ns, the temperature profiles generated by the different laser pulses can be calculated quite easily in a one-dimensional model based on the temporal shape of the laser pulse, the known absorption coefficients of GaN, and the thermal properties (heat capacitance and conductivity) of GaN and sapphire [79–83]. Fig. 7 shows a comparison of the resulting temperature profiles for the two above cases. Both profiles look qualitatively quite similar, but the thermal load on the sappire/GaN heterosystem is significantly reduced in the case of the Nd:YAG laser compared to KrF irradiation. Because of the shorter pulse length in the former case, high temperatures above 700 K persist only for about 14 ns and are restricted to the first 500 nm next to the interface, whereas in the case of the KrF pulse, the same temperature is present during almost 100 ns and extends over a range of 1 µm into the GaN layer. Thus, Nd:YAG laser systems will have a significant advantage especially for the LLO-processing of thin, temperature sensitive device structures such as InGaN-based light emitting diodes or laser diodes (see Section 3. 4).

3.2   Laser-induced lift-off of thick HVPE GaN wafers

A successful separation of thick HVPE-GaN layers from their sapphire substrates depends on several factors, including film thickness and thickness homogeneity across the wafer, growth procedure, and the resulting stresses and stress gradients caused by the large thermal expansion coefficient mismatch. During the past years, we have processed HVPE-GaN wafers prepared by different growth methods, but mainly thick GaN multilayer samples deposited with *in-situ* growth stops and HVPE-GaN grown on MOCVD-GaN templates structured with the ELOG technique. Due to the potential commercial relevance of GaN pseudosubstrates and the technological difficulty of growing thick, crack-free GaN films, it was not possible to obtain relevant information about further growth details. Nevertheless, some useful correlations between the structural and morphological properties of the GaN samples and the results of their delamination could be obtained.

For samples with thicknesses ranging from 70 to 300 μm (i. e. comparable to the sapphire substrate thickness of 330 μm), the differences of the thermal expansion coefficients result in substantial wafer bowing. Despite of this macroscopic bowing, most of the wafers exhibited very little or even no near-surface cracking across the whole wafer area. However, the bowing of samples with thicker GaN films (> 100 μm) conflicted with the laser lift-off procedure, because the strain inhomogeneity at the boundary of already released and thus relaxed areas and still attached portions of the wafer often gave rise to extensive fracturing. To reduce the bowing-induced fracturing, we used homogeneous heating of the sample during the laser lift-off. Already at temperatures around 600 °C, the strain in the sample and, consequently, the bowing was visibly reduced and hardly noticeable through simple optical inspection of the GaN surface. This is consistent with temperature-dependent X-ray diffraction measurements performed by Leszczynski et al. [84], who have shown that the lattice constants of a 2 μm thick GaN film on sapphire approach the values of bulk GaN already at about 800 K. Ideally, the performance of *in-situ* laser lift-off in the deposition reactor close to growth temperature would be the best solution to avoid all bowing or cracking problems caused by the thermal expansion mismatch between GaN and the sapphire substrate.

As examplified by Fig. 8, extended areas could be routinely released without sample damage by performing the laser treatment at temperatures above 600 °C, but below 830 °C (decomposition temperature of GaN). As shown in Fig. 1, the wafer is positioned in a steel holder filled with a thin layer of sapphire powder. A transparent sapphire window on top avoids unwanted movements of the wafer during laser scanning. The role of the sapphire powder with small grain size is to allow a free changing of the bowing radius with temperature during the heat-up period. The powder is also important as a damping system during the laser processing as well as after the delamination during the cool-down time to support the free-standing GaN film.

After removal of any residual GaN deposit on the backside of the substrate, which sometimes occurs in HVPE growth and would interfere with the laser lift-off, the laser treatment was performed in a computer-controlled scanning system which allowed a specific choice of scanning speed, step size, overlap between subsequent laser shots, and trajectory across the wafer. Even for the laboratory-size laser system used in our studies, the laser scanning of a 2" wafer itself takes only 3 min for a step size of 2 mm. A spiral scan trajectory towards the center of the wafer was found to be best suited for the lift-off of samples with circular shape, because of the symmetrical stress distribution between the free and still attached parts of the wafer. This is particularly important for the laser processing of GaN wafers with strong thickness inhomogeneity.

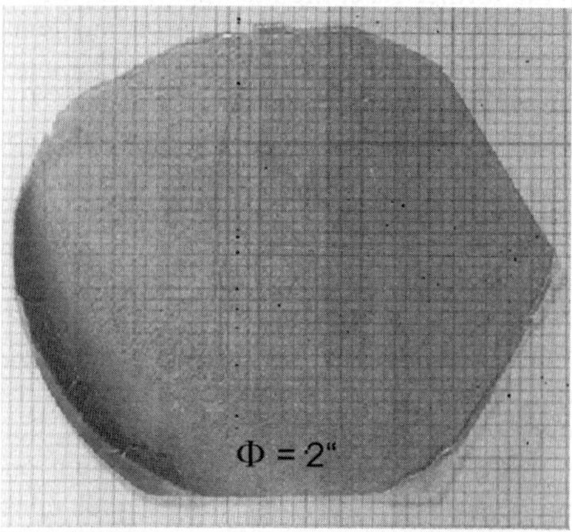

Fig. 8   Photograph of a 275 μm thick free-standing GaN film, after removal from the 2" sapphire substrate [65]. The missing pieces at the wafer border actually had broken off right after the HVPE growth.

3.3 Laser-induced lift-off of thin GaN films

A much more challenging problem than the lift-off of thick HVPE films as described in the previous section is the delamination of GaN films in the thickness range from a few micrometers up to about 60 µm. These films are mechanically too fragile to be separated from their substrate in the same way as described above. It is necessary to attach a support layer to the GaN layer in order to avoid large area fracturing after laser lift-off. The choice of a proper material which can be used for this support of thin freestanding GaN films is of crucial importance. Such a material should have an elastic modulus comparable to that of GaN and should be easily removable. On the other hand, it should be flexible enough to accommodate changes in the bowing radius during liftoff. Finally, it would be desirable to have a support material which can stand high temperatures up to 650 °C. Unfortunately, to the best of our knowledge, such a material is not yet commercially available. We thereforeby performed a series of experiments using different types of commercial adhesives and glues like a two-component epoxy resin, and silicone thermoresistant elastomer. Sticking GaN samples to a "blue tape" used in semiconductor production for fixing samples during sawing is also a possible approach. All delamination processes described below were performed at room temperature, due to the relatively low temperature resistance of the applied support materials.

As an example, completely crack-free GaN films were delaminated by fixing them to two-component epoxy resin prior to laser scanning. A glass carrier was used to support the GaN samples, which were pressed onto a thin layer of the transparent resin with a thickness of less than 0.5 mm. The laser liftoff was performed 5 min after curing the glue at room temperature. Figure 9 shows the delaminated GaN layer and the sapphire substrate, which is completely free at the end of the process. The strong adhesion of the used two-component epoxy resin is advantageous for a perfectly crack-free GaN liftoff, but on the other hand it is difficult to separate from the GaN film after sapphire removal due to its high chemical stability. Nevertheless, the epoxy can be dissolved with dichloromethane or in boiling water after several minutes.

An alternative procedure for the laser lift-off of free-standing full size 2" GaN membranes makes use of a silicone elastomer during laser processing and a thermoplastic adhesive to support the film after sapphire removal. A schematic step-by-step view of this process is shown in Fig. 10. The free side of the HVPE-GaN film on sapphire is covered with the thermoresistant (up to 300 °C) silicone elastomer forming a ~3 mm thick support layer. The film is then fixed to a metallic plate with doublesided adhesive tape to avoid moving of the flexible silicone elastomer. After laser scanning at room temperature, the sapphire substrate is removed. Then, the free side of the delaminated GaN film is protected with a ~3 mm thick transparent thermoplastic adhesive, which allows a safe peel-off of the silicone elastomer (Fig. 10 d)).

A 60 µm thick GaN membrane with a diameter of 2" is shown in Fig. 11. The GaN membrane is still protected by the thermoplastic adhesive, which can be easily dissolved in acetone without critical mechanical stress. The flexibility of the silicone elastomer allows the accommodation of the GaN film bowing. After fixing the membrane with the much more viscous thermoplastic adhesive, the residual bowing

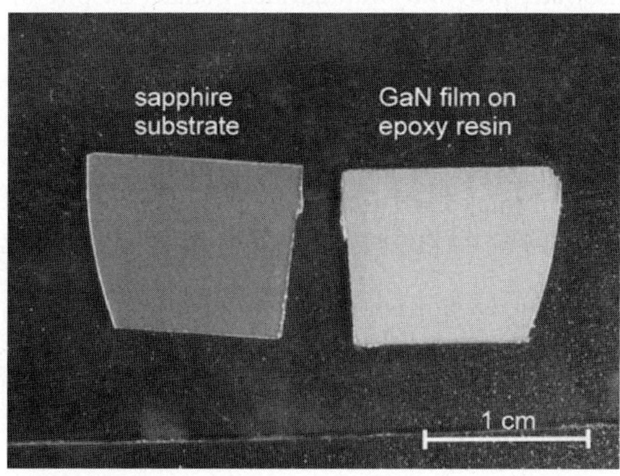

Fig. 9 (online colour at: www.interscience.wiley.com) Image of a HVPE-GaN sample with an area of about 1 cm^2 and a thickness of 30 µm fixed with epoxy resin to a glass holder after laser lift-off. The delaminated sapphire substrate is shown on the left.

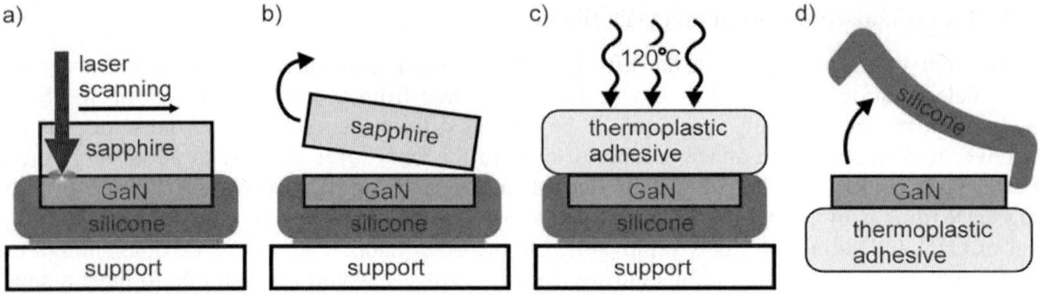

Fig. 10 (online colour at: www.interscience.wiley.com)　Process sequence for the laser lift-off of 2" GaN membranes. a) Laser lift-off of the GaN film covered with silicone elastomer and mounted onto a support; b) sapphire removal after laser scanning; c) deposition of a ~3 mm thick layer of thermoplastic adhesive at 120 °C; d) peel-off of the silicone elastomer. Subsequently, the GaN film can be easily isolated by dissolving the thermoplastic adhesive in an acetone bath.

of the GaN film and some cleavage lines can be observed through specular light reflection (left side of Fig. 11). Despite of some cracks appearing over the GaN membrane, the procedure described above is fully reproducible and, thus, applicable to the delamination of large HVPE-GaN films with a thickness of some tens of micrometers. A different approach to produce completely crack-free thin and large area GaN-based heterostructure wafers for device production is based on the wafer bonding technique as discussed in the following section.

3.4　Delamination of thin wafer-bonded GaN heterostructures and devices

The monolithic integration of two dissimilar materials onto one platform with wafer-bonding and laser lift-off is a convenient alternative to heteroepitaxial growth when this is not possible or too problematic. In the case of group III-nitrides, the integration is motivated by combining the superior optoelectronic properties of GaN-based heterostructures with receptor substrates which possess much better thermal and electrical properties than the original sapphire substrate. For devices processed on sapphire, all electrical contacts must be made from the top side. This complicates contact and packaging schemes and results in higher operation voltages as well as a spreading-resistance handicap [67]. Most importantly, the poor thermal conductivity of sapphire is detrimental for high-power device performance of LEDs, laser diodes, and high-power transistors. Hence the transfer of GaN-devices to other substrates with better thermal properties would be of enormous benefit. In addition to the cheap and readily available Si, GaAs, Ge, Mo, and copper are receptor wafers with certain advantages.

A major issue in wafer bonding is the choice of the optimal bonding layer. The bonding material has to adhere to both the epitaxial film and the receptor substrate, providing full surface contact despite of sub-micron surface asperities. The temperature necessary to achieve bonding should be low (< 200 °C) to avoid thermal stress, and without leaving behind any phase with a melting point below the bonding temperature. Moreover, it should have low electrical resistance to allow easy contact formation with the bonded devices. For the specific case of the III-nitride material system, Wong et al. have used a Pd-In metal bonding to connect GaN to Si [85], whereas Jasinski et al. have succeeded in producing GaN/GaAs bond-structures by direct wafer fusion [86]. 2 μm thick GaN films grown on sapphire were fused to a GaAs substrates in nitrogen ambient by applying uniaxial pressure of 2 MPa at bonding temperatures of 550 and 750 °C.

Individual devices on a GaN- wafer can be scribed or isolated via mesa-etching prior to wafer bonding, thus enabling the production of separated devices after the laser lift-off. A possible technological process flow diagram for the production of efficient nitride based LEDs using a metal bonding and laser lift-off procedure is illustrated in Fig. 12. In this way, the individualized devices on the sapphire wafer can be transferred to a receptor substrate like Si, GaAs or Cu.

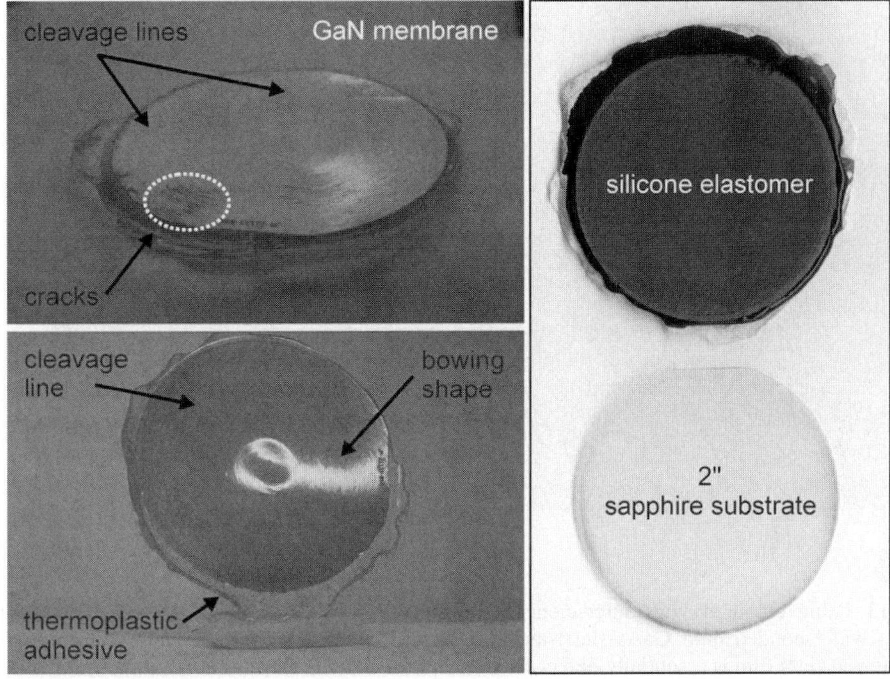

Fig. 11 (online colour at: www.interscience.wiley.com)   The left hand side shows the HVPE-GaN membrane with a diameter of 2" and thickness of 60 μm attached to thermoplastic adhesive after the laser liftoff. The specular light reflection due to the bowing of the film and some cleavage lines are indicated. The silicon elastomer used as a support during laser processing can be completely peeled off, as shown on the right hand side. The separated sapphire substrate is also shown.

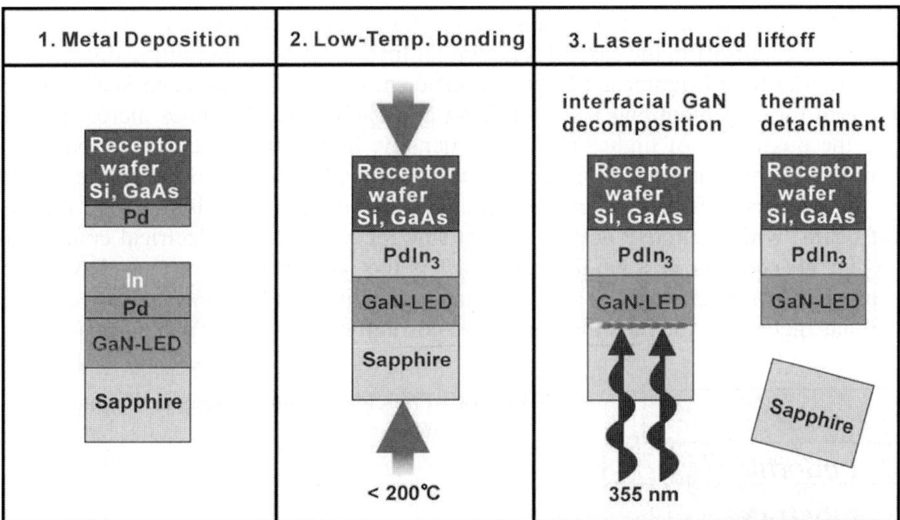

Fig. 12 (online colour at: www.interscience.wiley.com)   Process flow for bonding and transfer of a GaN-based LED from sapphire onto a receptor wafer using PdIn$_3$ as the bonding material.

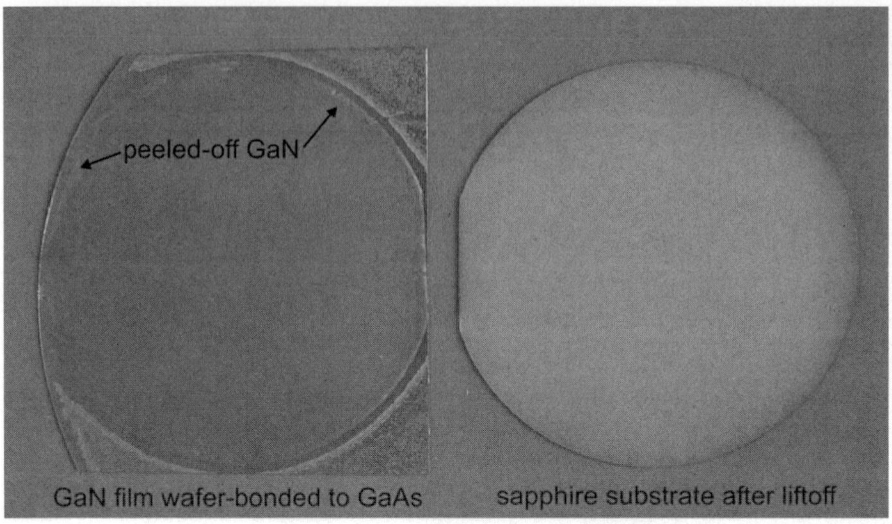

Fig. 13 (online colour at: www.interscience.wiley.com)   Delaminated 3 μm thick 2" GaN film (not structured) wafer-bonded onto GaAs (left) and the corresponding GaN-free sapphire substrate (right). The transferred GaN film is essentially defect free, except for some peeled-off areas at the wafer rim.

The GaN-based heterostructures are typically thinner than 5 μm, so that practically no bowing of the wafers occurs, and laser lift-off can be performed at room temperature. The energy density of the laser shots should be kept as close as possible to the threshold value necessary for GaN decomposition. Any additional energy will lead to unwanted thermal and mechanical stress which can favour film spalling and peel-off after sapphire removal.

Figure 13 shows a successfully delaminated GaN film wafer-bonded to GaAs, where some peeling–off occurred only at the very border of the wafer. The isolation of the devices prior to wafer-bonding can prevent the propagation of cracks and spalling, thus conserving the integrity of the device structures during laser delamination. A yield as high as 90% and above can be reproducibly achieved for 2" wafers. In order to illustrate the potential of the LLO processing for commercial device production, we now have a closer look at the laser lift-off in the case of blue/violet InGaN/GaN light emitting diodes.

For the realization of high-brightness LEDs with efficient current injection and heat sinking, the flip-chip technique used for chip mounting is a promising approach which becomes increasingly important. Advantages are the possibility of higher packaging density, higher frequency operation due to shorter contacts, a minimum of interconnects ("direct die attach"), and a low height of the whole assembly, as no bonding wires are necessary. Basically, in the flip-chip technology the die is mounted onto a substrate (printed circuit board) with the active device contact side face down. The electrical connection is made simultaneously for all contacts in a single step. One of the contact partners must have "bumps", i.e. raised areas composed of electrically conductive material. Different galvanic and mechanical processes with or without melting of the metallic bumps are currently available for this purpose, but one of the

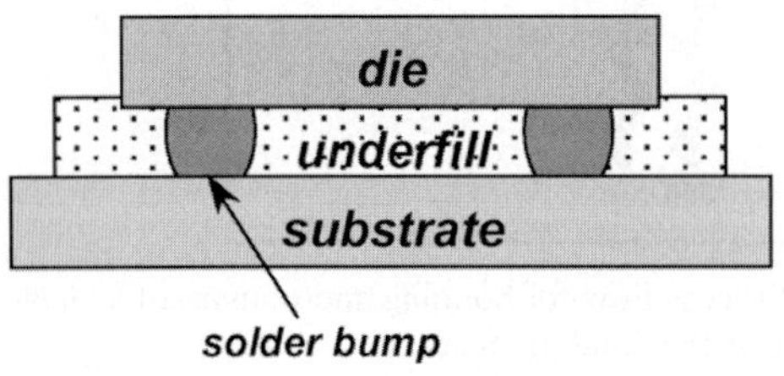

Fig. 14   Flip-chip bonding of a device die onto a substrate with solder bumps.

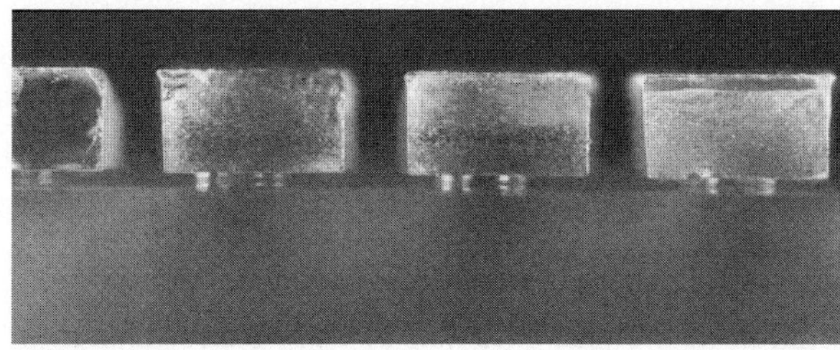

Fig. 15 (online colour at: www.interscience. wiley.com) Side view of individualized In-GaN/GaN LEDs flip-chip bonded to a Si-submount.

most widely used processes is traditional soldering technology modified for direct chip mounting. In this case, metallic alloy solders for the bumping and connection process are used. A schematic view is given in Fig. 14.

In the specific example described here, InGaN/GaN LED chips provided by the IAF Fraunhofer Institut in Freiburg, Germany were flip-chip bonded to metallized Si-submounts by ball bonding with a wire bonding machine. First, Au-balls were positioned on the interconnecting metallization of the Si-submount, consisting of a Ti/Au/Ti plating base covered by an electroplated Au-layer. Subsequently, the LED chips were mounted upside-down by thermal compression welding of the bump/LED contact interface at a temperature of ~200 °C. In Fig. 15, a series of isolated LEDs is shown, where a sapphire stripe with the devices was sawed away from the wafer and the individualized LEDs were finally flip-chip bonded to the Si-submount. Entire device stripes can be connected to the base substrate in a similar way.

The laser-induced delamination method was applied directly to the flip-chip contacted devices grown on sapphire. As the size of the devices is much smaller than the laser-shot area, a single pulse is sufficient to remove the sapphire substrate, which is beneficial to maintain the structural integrity and the performance of the device even after the laser treatment. As in the delamination of GaN wafers described above, an important aspect for a successful laser processing of the fragile GaN devices is to to minimize the structural and thermal stress during the short laser / target interaction. The use of a suitable underfill is indispensable to absorb the mechanical impact caused by the high power laser pulse on the thin film supported only by the contact bumps. Wafer stripes of size 0.5×25 mm^2 containing a series of LEDs in line as well as individually sawed LEDs like the ones shown in Fig. 15 with a device area of 250×250 µm^2 and a thickness of 3.8 µm were prepared with an optimized underfill for the laser lift-off. After removal of the sapphire substrate, the residual Ga on the exposed backside was etched away by

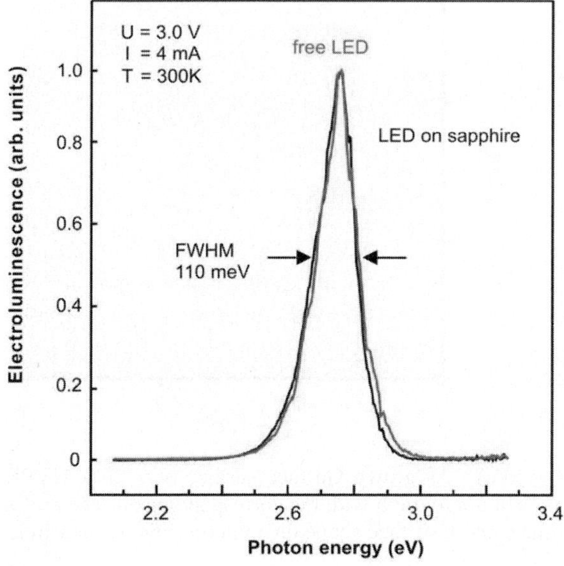

Fig. 16 (online colour at: www.interscience.wiley. com) Electroluminescence spectra of an InGaN/GaN-LED flip-chip bonded to a silicon submount before and after removal of the sapphire substrate by laser lift-off. The thickness of the freestanding, flip-chip bonded device structures is 3.8 µm.

contact with HCl vapour for 10 sec. After removing the protective adhesive with acetone, the perform-ance of the freestanding flip chip bonded LEDs was compared to that prior to the LLO process. As dem-onstrated by the electroluminescence spectra in Fig. 16, no degradation of device performance due to the laser processing was observed.

4 GaN homoepitaxy on freestanding substrates

As already mentioned in the introduction, the exploitation of the real potential of GaN-based devices is still hindered by the lack of a suitable bulk III-nitride substrate (GaN and/or AlN). Here, the use of free-standing GaN pseudosubstrates obtained e.g. by laser lift-off from thick HVPE-GaN grown on sapphire [87, 88] may provide a real alternative to the still difficult and time-consuming equilibrium growth of bulk substrates. Although homoepitaxy of GaN on such bulk substrates has resulted in the best material quality so far [2, 89], the available substrate sizes are still incompatible with a commercially viable de-vice production. In this section, we will therefore review recent results obtained on homoepitaxial GaN layers grown on freestanding pseudosubstrates produced by laser lift-off.

4.1 Surface conditioning of freestanding substrates

The surface morphology of the thick HVPE-GaN films used in this work was characterized by optical microscopy and surface profiling. Complementary, the microscopic structure of the surface was deter-mined by atomic force microscopy, which shows a smooth surface morphology with a rms roughness of 0.4 nm. Dislocation densities between 8×10^6 and 5×10^7 cm^{-2} can be estimated from the pin-hole con-centration at the surface (black spots in the AFM image in Fig. 17) , which agrees well with the values published by Vaudo et al. [90]. However, on a larger scale, the Ga-face oriented growth surface of HVPE-GaN exhibits a much rougher morphology, consisting of hexagonal pyramids with diameters of several 10 µm and a height of typically 1 µm, as shown by the optical micrograph on the left hand side of Fig. 17. Some samples also exhibit characteristic large surface voids on the as grown side of the sam-ple shown in Fig 18. These large volume crystal defects are mostly localized at the borders of the 2" HVPE-GaN wafers, and are probably caused by a deviation from the optimal growth conditions together with the fast growth rates of 100 µm/h and more. They are obviously detrimental for homoepitaxial

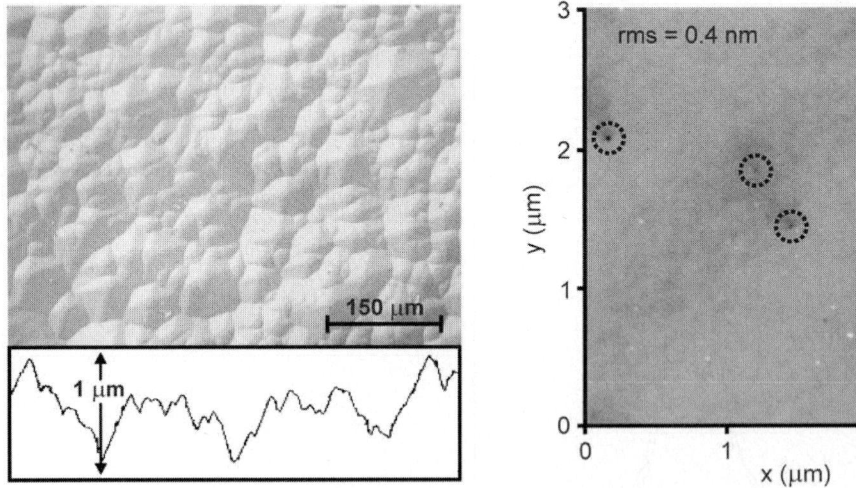

Fig. 17 (online colour at: www.interscience.wiley.com) As grown Ga-face surface of a thick HVPE-GaN film on sapphire. On the left hand side, an optical micrograph with the corresponding surface profile is shown. The AFM picture on the right presents the smooth surface shape on a microscopic scale. Circles indicate the sites of dislocations.

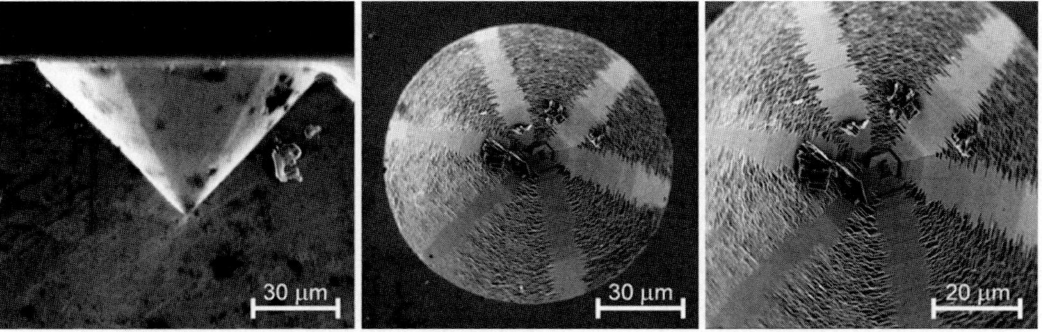

Fig. 18 Typical large defects found at the Ga-face surface of thick HVPE-GaN samples. The voids can reach a diameter of 100 µm and depth of 50 µm, as shown by the side view on the left. From the top view it is possible to recognize the hexagonal symmetry of the voids, despite of their almost perfect circular shape.

overgrowth and have to be removed by lapping and polishing. On the substrate side of the thick HVPE GaN layer, the laser-induced delamination gives rise to a characteristic surface morphology consisting of small grains with a size of approximately 2 µm, and a similar rms roughness.

In order to produce free-standing GaN pseudo-substrates with adequate surface characteristics for further epitaxial overgrowth, a lapping and fine-polishing procedure was developed for the delaminated GaN films [91]. First a planarization step was introduced to get rid of residual sample bowing after laser liftoff by using a diamond lapping pad with a grain size of 40 µm. Afterwards a polishing sequence under controlled pressure of about $2\,kg/cm^2$ was performed with diamond grain sizes of 15, 7, 3 and 0.25 µm for approximately 10 minutes per step. The last step for the preparation of the Ga-face surface was a long mechanical polishing with a special fine polishing solution (grain size of 0.04 µm). For the N-face surface , a chemical polishing step with a 1:10 KOH : H_2O solution was applied to obtain the final surface smoothness. The rms roughness achieved in this way, as measured by AFM, is of the order of 1 nm for N-face and 2.2 nm for Ga-face surfaces. An additional chemo-mechanical polishing step to remove subsurface damage caused by the mechanical polishing was found to be inefficient for Ga-face surfaces. For this purpose, chemically assisted ion beam etching (CAIBE) or reactive ion etching (RIE) were also investigated by other groups [73, 92]. As confirmed by photoluminescence measurements, an optical quality similar to the unpolished original GaN surface after the HVPE growth could be recovered by this treatment.

4.2 Structural and electronic properties of homoepitaxial GaN layers on freestanding GaN pseudosubstrates

After optimization of the surface conditioning of freestanding GaN pseudosubstrates as described in the last section, homoepitaxial overgrowth on such substrates was performed by a conventional MOCVD process [93]. Prior to the growth, the HVPE-GaN pseudosubstrate was annealed at 800 °C under a flow of 2 slm of nitrogen and hydrogen (N_2 / H_2 = 1) for 10 min. Then NH_3 was added to the transport gases and the temperature raised to 1050 °C within a time period of 5 min. A 2 µm thick epitaxial GaN film was grown on the Ga-face side of the substrate, with a V/III ratio of 40,000 and a growth rate of 600 nm/h. As demonstrated by AFM investigations, epitaxial layers with an outstanding surface morphology can be obtained (Fig. 19).

In the cross section of the AFM micrograph shown on the left hand side of Fig. 19 it is possible to distinguish bilayer steps separated by terraces with a width of about 200 nm. Compared to the reference sample on the right hand side of Fig. 19 (an *in-situ* overgrown MOCVD-GaN sapphire based template) the estimated dislocation density is reduced by almost 3 orders of magnitude for the homoepitaxial layer. The estimated dislocation density of $2 \times 10^{-7}\,cm^{-2}$ was obtained based on a series of AFM micrographs and is in the range determined for the original HVPE-GaN substrate.

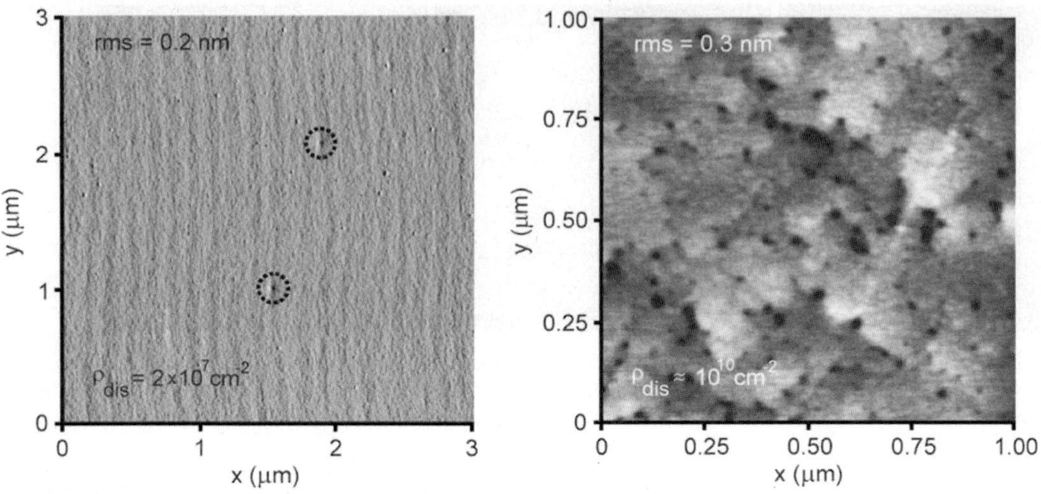

Fig. 19 (online colour at: www.interscience.wiley.com)   AFM micrograph of homoepitaxial GaN grown by MOCVD on the Ga–face surface of a freestanding, polished HVPE-GaN substrate (left). On the right hand side, a layer grown on a MOCVD-GaN template substrate under identical conditions is shown for comparison. Circles mark the site of isolated dislocations.

High resolution X-ray diffraction (HRXRD) measurements were performed to check the structural quality, strain and mismatch between the GaN pseudosubstrate and the homoepitaxial layer. In Fig. 20, 2θ–ω scans of the (0002), (0004) and (0006) reflexes are compared, showing a full width at half maximum (FHWM) as low as 17 arcsec. From the high dispersion (0006) diffraction peak and reciprocal space

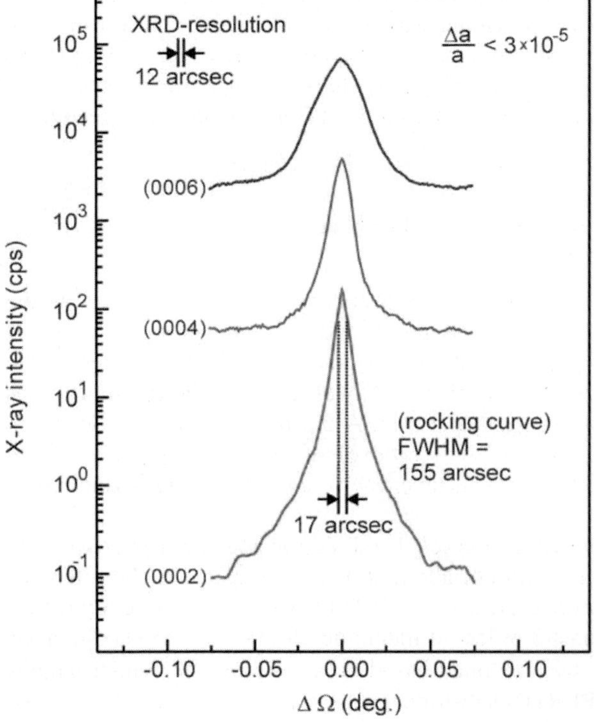

Fig. 20 (online colour at: www.interscience. wiley.com)   High resolution X-ray diffraction of GaN grown on the Ga-face of a 300 µm thick HVPE-GaN substrate. The full width at half maximum of the *2θ–ω* scan is only 17 arcsec for the (0002) reflex. The resolution limit of the XRD-system is 12 arcsec.

maps of the (205) reflex, the maximum lattice mismatch between substrate and the homoepitaxial GaN film can be estimated to $\Delta a/a = 3 \times 10^{-5}$. The FWHM of rocking curves is limited by a small residual curvature of the free-standing GaN substrate, caused by remaining biaxial compressive strain.

To evaluate the optical and electronic quality of the homoepitaxial GaN layers, photoluminescence (PL) measurements were performed at liquid helium temperature. The 333.6 nm line of an Argon ion laser was used for excitation. The photoluminescence was dispersed using a triple-grating spectrometer with a focal length of 800 mm and was detected by a lN_2 cooled CCD camera. With this setup, a spectral resolution of better than 1 cm^{-1} or 120 µeV can be obtained. The observed PL transitions in Fig. 21 are typical for a nominally undoped GaN films with a small density of residual shallow impurities. The high optical quality of the GaN film is confirmed by the presence of the well resolved free-excitonic lines X_A, X_B, and the excited state $n=2$ of X_A at 3.480, 3.486, and 3.500 eV, respectively, and the neutral donor bound-exciton transition ($D_2^0 - X$) at 3.474 eV with a FWHM of 550 µeV [94]. Other PL lines due to the neutral acceptor bound-exciton ($A^0 - X$) at 3.469 eV, the neutral donor bound-exciton ($D_1^0 - X$) at 3.478 eV, and the transition involving the excited *2s*-like state of the donor ($D_2^0 - X)_{n=2}$ at 3.453 eV are also seen [95].

On the low energy side of the spectra in Fig. 21, small bands emerge at 3.384 eV and 3.391 eV in the PL spectra of the GaN substrate, which are attributed to phonon replicas of the free-exciton line and the acceptor bound-exciton, respectively. A strong band centered at 3.403 eV (FWHM = 19 meV) in the homoepitaxial GaN film is assigned to overlapping LO-phonon replicas associated with the free and bound exciton transitions.

A comparison of the results presented above with homoepitaxial films grown by MOCVD and MBE on high pressure/high temperature synthesized bulk GaN substrates reveals similar linewidths, but slightly different energy positions of the donor bound-exciton transitions [96, 97]. This shift in energy is most likely due to slight differences in residual strain. In contrast to the films grown on HVPE-GaN substrates, the ten times larger lattice mismatch of $\Delta a/a = 2 \times 10^{-4}$ reported for homoepitaxial GaN films on high pressure /high temperature grown bulk GaN substrates is due to the significantly higher residual carrier concentration of the order of 10^{19} cm^{-3} in the latter substrates [98].

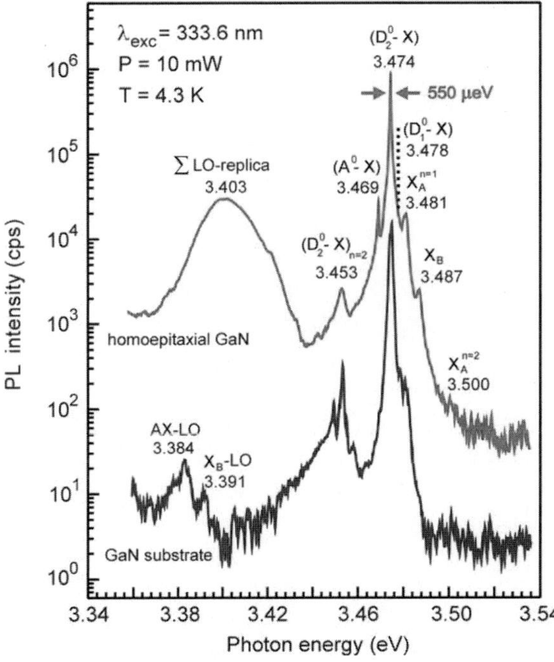

Fig. 21 (online colour at: www.interscience.wiley.com) Low temperature PL-spectra of a homoepitaxial GaN film on a free-standing HVPE-GaN substrate. The corresponding PL spectra of the substrate prior to homoepitaxial overgrowth are shown by the lower spectrum.

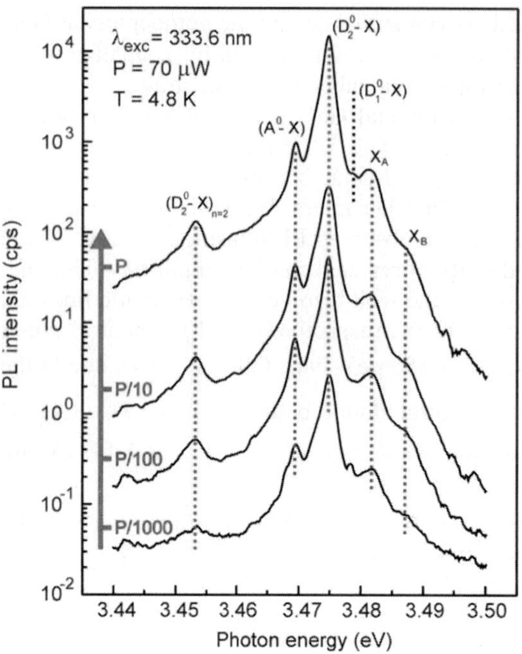

Fig. 22 (online colour at: www.interscience.wiley.com) Dependence of the photoluminescence spectra of homoepitaxial GaN films grown on freestanding HVPE GaN substrates on the excitation power.

Figure 22 shows the PL spectra of homoepitaxial GaN on a freestanding GaN substrate for various excitation intensities. The excitonic character of the observed transitions is confirmed by the absence of an energy shift with increasing excitation intensity. The integrated intensities of each transition increases superlinearly with the excitation power. According to Schmidt et al. [99], such a dependence of the luminescence intensity I on excitation power is ecpected for excitonic transitions.

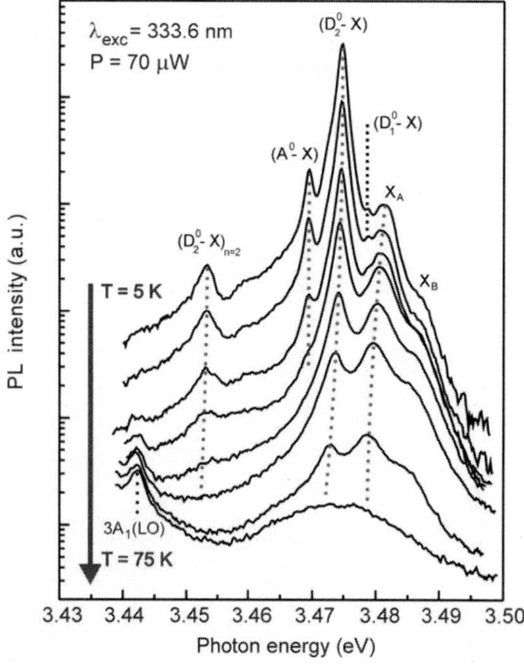

Fig. 23 (online colour at: www.interscience.wiley.com) Temperature dependence of the photoluminescence for a homoepitaxial GaN film on a freestanding HVPE substrate. See text for further details.

Finally, Fig. 23 summarizes the temperature dependence of the various transitions in Figs. 21 and 22. All PL-lines show a significant change with increasing temperature in the range from 5 to 55 K. The higher energy peaks labeled X_A and X_B broaden and become relatively stronger, eventually becoming the dominant Pl lines above 50 K. This is consistent with their assignment to free-exciton lines. Pronounced quenching due to thermal dissociation is observed for the bound-exciton recombinations ($D_1^0 - X$), ($D_2^0 - X$), ($D_2^0 - X)_{n=2}$, and ($A^0 - X$). The shift in the exciton peak positions is in agreement with the band-gap shrinkage of wurtzite GaN [100], except for a small energy offset due to residual strain.

4   Summary and conclusions

The present state-of-the-art concerning the laser-induced lift-off of both thin and thick GaN films and heterostructures from sapphire substrates has been reviewed. The physical background of the laser-induced thermal decomposition of GaN by short intense pulses of KrF excimer lasers and the third harmonic of Nd:YAG lasers is discussed, and potential applications for rapid etching of GaN have been outlined. Of particular interest for future applications is the possibility to produce freestanding GaN films by laser lift-off of heteroepitaxial layers from transparent sapphire substrates. Specific applications include the production of freestanding GaN pseudosubstrates starting from thick HVPE grown epilayers, as well as the delamination of thin GaN device heterostructures for the purpose of wafer bonding onto foreign substrates or for flip-chip bonding in device technology. Using optimized processes, the defect free lift-off of entire 2" wafers can be achieved for GaN film thicknesses ranging from 3 to 300 µm. Separation of thin device heterostructures from their sapphire substrates opens up new possibilities for the formation of electrical contacts, the extraction of photons, and for thermal management. Thick GaN films grown by HVPE can be removed from their sapphire substrates to obtain freestanding GaN pseudosubstrates for the homoepitaxial growth of high quality epilayers. The required processing steps, including surface preparation after laser lift-off, have been described and the structural and optoelectronic properties of homoepitaxial GaN layers deposited on freestanding pseudosubstrates have been investigated. All results suggest that the laser lift-off method may become a major technique in future III-nitride device technology.

Acknowledgements   The financial support of our work by the Deutsche Forschungsgemeinschaft in the framework of Sonderforschungsbereich 348 and the Schwerpunktsprogramm Gruppe III-Nitride is gratefully acknowledged. Special thanks are due to Bob Vaudo (ATMI), Volker Härle (OSRAM Opto), and Joachim Wagner (Fraunhofer Institut Freiburg) for providing many of the samples used in this study. The authors are also grateful to Ana Cros and Gema Martinez-Criado for their help in understanding what really goes on in GaN.

References

[1] S. Nakamura, Semicond. Sci. Technol. **14** (1999) R27.
[2] F.A. Ponce, in: B. Gil (Ed.), Group III Nitride Semiconductor Compounds, Oxford University Press, Oxford, 1998, p. 123.
[3] R.R. Reeber, K. Wang, Mater. Res. Soc. Symp. **622** (2000) T6.35.1.
[4] D.P. Xu, Y.T. Wang, H. Yang, S.F. Li, D.G. Zhao, Y. Fu, S.M. Zhang, R.H. Wu, Q.J. Jia, W.L. Zheng, and X.M. Jiang, J. Appl. Phys. **88** (2000) 3762.
[5] P. E. Van Camp, V. E. Van Doren, and J. T. Devreese, Solid State Commun. **82** (1992) 23.
[6] M. Tanaka, S. Nakahata, K. Sogabe, H. Nakata, and M. Tobioka, Jpn. J. Appl. Phys. **36** (1997) L1062.
[7] S. Okubo, N. Shibata, T. Saito, and Y. Ikuhara, J. Cryst. Growth **189/190** (1998) 452.
[8] H. Okumura, H. Hamaguchi, T. Koizumi, K. Balakrishnan, Y. Ishida, M. Arita, S. Chichibu, H. Nakanishi, T. Nagatomo, and S. Yoshida, J. Cryst. Growth **189/190** (1998) 390.
[9] V. Pankov, M. Evstigneev, and R. H. Prince, Appl. Phys. Lett. **80** (2002) 4142.
[10] E.S. Hellman, D.N.E. Buchanan, D. Wiesmann, and I. Brener, MRS Internet J. Nitride Semicond. Res. **1** (1996) 16.
[11] F. Hamdani, M. Yeadon, D.J. Smith, H. Tang, W. Kim, A. Salvador, A.E. Botchkarev, J.M. Gibson, A.Y. Polyakov, M. Skowronski, and H. Morkoç, J. Appl. Phys. **83** (1998) 983.

[12] G.L. Harris, in: G.L. Harris (Ed.), Properties of Silicon Carbide, Inspec/IEE, 1995, p. 4.

[13] Kh.S. Bagdasarov, E.R. Dobrovinskaya, V.V. Pishchik, M.M. Chernik, Yu. Yu, A.S. Gershun, and I.F. Zvyagintseva, Sov. Phys. Crystallogr. **18** (1973) 242.

[14] S. Nishimura and K. Terashima, Appl. Surf. Sci. **159/160** (2000) 288.

[15] T. Izumiya, A. Hatano, and Y. Ohba, Inst. Phys. Conf. Ser. **129** (1993) 157.

[16] J. C. Brice, in: M.R. Brozel, G.E. Stillman (Eds.), Properties of Gallium Arsenide, 3rd ed., Inspec/IEE, 1996, p. 8.

[17] Y. Xin, P.D. Brown, R.E. Dunin-Borkowski, C.J. Humphreys, T.S. Cheng, and C.T. Foxon, J. Cryst. Growth **171** (1997) 321.

[18] V.Yu. Davydov, I.N. Goncharuk, A.N. Smirnov, R.V. Zolotareva, A.V. Subashiev, T.S. Cheng, and C.T. Foxon, J. Cryst. Growth **189/190** (1998) 430.

[19] Y. Okada, in: Robert Hull (Ed.), Properties of Crystalline Silicon, IEE, London, 1999.

[20] H. Morkoç, S. Strite, G. B. Gao, M. E. Lin, B. Sverdlov, and M. Burns, J. Apply. Phys. **76** (1994) 1363

[21] L.M. Belyaev, Rubby and Sapphire, Amerind Publishing Co., New Delhi (translated from Russian, RUBIN I SAPFIR, Nauka Publishers, Moscow, 1974), 1980.

[22] M.T. Duffy, C.C. Wang, G.D. O'Clock, S.H. McFarlane, and P.J. Zanzhcchi, J. Electron. Mater. **2** (1973) 359.

[23] S. Nakamura, M. Senoh, S. Nagahama, N. Iwasa, T. Yamada, T. Matsushita, H. Kiyoku, and Y. Sugimoto, Appl. Phys. Lett. **68** (1996) 2105.

[24] H. Ohsato, T. Kato, S. Koketsu, R.D. Saxena, and T. Okuda, J. Cryst. Growth **189/190** (1998) 202.

[25] J.R. Heffelfinger, D.L. Medlin, and K.F. McCarty, J. Mater. Res. **13** (1998) 1414.

[26] C.J. Rawn and J. Chaudhuri, J. Cryst. Growth **225** (2001) 214.

[27] E.S. Hellman, Z. Liliental-Weber, and D.N.E. Buchanan, MRS Internet J. Nitride Semicond. Res. **2** (1997) 30.

[28] C. Klemenz and H.J. Scheel, J. Cryst. Growth **211** (2000) 62.

[29] P. Waltereit, O. Brandt, M. Ramsteiner, R. Uecker, P. Reiche, and K.H. Ploog, J. Cryst. Growth **218** (2000) 143.

[30] H. Okazaki, A. Arakawa, T. Asahi, O. Oda, and K. Aiki, Solid-State Electron. **41** (1997) 263.

[31] C. Fechtmann, V. Kirchner, S. Einfeldt, H. Heinke, D. Hommel, T. Lukasiewicz, Z. Luczynski, and J. Baranowski, Mater. Res. Soc. Symp. Proc. **482** (1997) 295.

[32] V.V. Mamutin, S.V. Sorokin, V.N. Jmerik, T.V. Shubina, V.V. Ratnikov, S.V. Ivanov, P.S. Kop'ev, M. Karlsteen, U. Södervall, and M. Willander, J. Cryst. Growth **201/202** (1999) 346.

[33] E.S. Hellman, C.D. Brandle, L.F. Schneemeyer, D. Wiesmann, I. Brener, T. Siegrist, G.W. Berkstresser, D.N.E. Buchanan, and E.H. Hartford, Jr., Mater. Res. Soc. Symp. Proc. **395** (1996) 51.

[34] E.S. Hellman, C.D. Brandle, L.F. Schneemeyer, D. Wiesmann, I. Brener, T. Siegrist, G.W. Berkstresser, D.N.E. Buchanan, and E.H. Hartford, MRS Internet J. Nitride Semicond. Res. **1** (1996) 1.

[35] A. Yoshikawa, V.V. Kochurikhin, N. Futagawa, K. Shimamura, and T. Fukuda, J. Cryst. Growth **204** (1999) 302.

[36] A. Yamada, K.P. Ho, T. Akaogi, T. Maruyama, and K. Akimoto, J. Cryst. Growth **201/202** (1999) 332.

[37] A. Yamada, K.P. Ho, T. Maruyama, and K. Akimoto, Appl. Phys. A **69** (1999) 89.

[38] J.J. Lee, Y.S. Park, C.S. Yang, H.S. Kim, K.H. Kim, K.Y. Yang, T.W. Kang, S.H. Park, and J.Y. Lee, J. Cryst. Growth **213** (2000) 33.

[39] J. Ohta, H. Fujioka, H. Takahashi, and M. Oshima, phys. stat. sol. (a) **188** (2001) 487.

[40] R. Beresford, K.S. Stevens, C. Briant, R. Rai, and D.C. Paine, Mater. Res. Soc. Symp. Proc. **395** (1996) 55.

[41] R. Beresford, D.C. Paine, and C.L. Briant, J. Cryst. Growth **178** (1997) 189.

[42] S.D. Wolter, B.P. Luther, S.E. Mohney, R.F. Karlicek, Jr., and R.S. Kern, Electrochem. Solid-State Lett. **2** (1999) 151.

[43] R. Kaplan, S.M. Prokes, S.C. Binari, and G. Kelner, Appl. Phys. Lett. **68** (1996) 3248.

[44] F. Perjeru, X. Bai, M.I. Ortiz-Libreros, R. Higgins, and M.E. Kordesch, Appl. Surf. Sci. **175/176** (2001) 490.

[45] M.I. Kotelyanski, I.M. Kotelyanski, V.B. Kravchenko, Technol. Phys. Lett. 26 (2000) 163.

[46] S. Ruvimov, Z. Liliental-Weber, J. Washburn, K.J. Duxstad, E.E. Haller, Z.-F. Fan, S.N. Mohammad, W. Kim, A.E. Botchkarev, and H. Morkoç, Mater. Res. Soc. Symp. Proc. **423** (1996) 201.

[47] P. Ruterana, G. Nouet, Th. Kehagias, Ph. Komninou, Th. Karakostas, M.A. di Forte Poisson, and F. Huet, Mater. Res. Soc. Symp. Proc. **595** (2000) W11.75.

[48] L.T. Romano, in: J.H. Edgar, S.S. Strite, I. Akasaki, and H. Amano (Eds.), Properties, Processing and Applications of Gallium Nitride and Related Semiconductors, INSPEC, The Institution of Electrical Engineers, Stevenage, UK, 1999, p. 209.

[49] J.E. Northrup, L.T. Romano, in: J.H. Edgar, S.S. Strite, I. Akasaki, and H. Amano (Eds.), Properties, Processing and Applications of Gallium Nitride and Related Semiconductors, INSPEC, The Institution of Electrical Engineers, Stevenage, UK, 1999, p. 213.

[50] T. Suski, J. Jun, M. Leszczyski, H. Teisseyre, S. Strite, A. Rockett, A. Pelzmann, M. Kamp, and K.J. Ebeling, J. Appl. Phys. **84** (1998) 1155.

[51] H. Amano, M. Iwaya, T. Kashima, M. Katsuragawa, I. Akasaki, J. Han, S. Hearne, J.A. Floro, E. Chason, and J. Figiel, Jpn. J. Appl. Phys. **37** (1998) L1540.

[52] B. Beaumont, Ph. Vennegures, P. Gibart, Phys. Stat. Sol. (b) 227 (2001) 1.

[53] T.S. Zheleva, S.A. Smith, D.B. Thomson, K.J. Linthicum, P. Rajagopal, and R.F. Davis, J. Electron. Mater. **28** (1999) L5.

[54] H. Lahreche, P. Vennegues, B. Beaumont, and P. Gibart, J. Cryst. Growth **205** (1999) 245.

[55] C.I.H. Ashby, C.C. Mitchell, J. Han, N.A. Missert, P.P. Provencio, D.M. Follstaedt, G.M. Peake, and L. Griego, Appl. Phys. Lett. **77** (2000) 3233.

[56] T. Detchprohm, M. Yano, S. Sano, R. Nakamura, S. Mochizuki, T. Nakamura, H. Amano, and I. Akasaki, Jpn. J. Appl. Phys. **40** (2001) L16.

[57] M. Mayer, A. Pelzmann, M. Kamp, K. J. Ebeling, H. Teisseyre, G. Nowak, M. Leszczynski, I. Grzegory, S. Porowsky, and G. Karczewski, Jpn. J. Appl. Phys. **36** (1997) L1634.

[58] F. A. Ponce, D. P. Bour, W. Götz, N. M. Johnson, H. I. Helava, I. Grzegory, J. Jun, and S. Porowsky, Appl. Phys. Lett. **68** (1996) 917.

[59] S. Nakamura, M. Senoh, S. Nagahama, N. Iwasa, T. Yamada, T. Matsushita, H. Kiyoku, Y. Sugimoto, T. Kozaki, H. Umemoto, M. Sano, and K. Chocho, Appl. Phys. Lett. **73** (1998) 832.

[60] M. Kuramoto, C. Sasaoka, Y. Hisanaga, A. Kimura, A. A. Yamaguchi, H. Sunakawa, N. Kuroda, M. Nido, A. Usui, and M. Mizuta, Jpn. J. Appl. Phys. **38** (1999) L184.

[61] R.P. Vaudo, V.M. Phanse, X. Wu, Y. Golan, and J.S. Speck, 2nd Int. Conf. Nitride Semiconductors, Tokushima, 1997.

[62] S.S. Park, I.-W. Park, and S.H. Choh, Jpn. J. Appl. Phys. **39** (2000) L1141.

[63] S. Nakamura, M. Senoh, S. Nagahama, N. Iwasa, T. Yamada, T. Matsushita, H. Kiyoku, Y. Sugimoto, T. Kozaki, H. Umemoto, M. Sano, and K. Chocho, Appl. Phys. Lett. **72** (1998) 2014.

[64] K. Tomita, T. Kachi, S. Nagai, A. Kojima, S. Yamasaki, and M. Koike, phys. stat. sol. (a) **194** (2002) 563.

[65] M.K. Kelly, R.P. Vaudo, V.M. Phanse, L. Görgens, O. Ambacher, and M. Stutzmann, Jpn. J. Appl. Phys. **38** (1999) 217.

[66] M. Kelly, O. Ambacher, M. Stutzmann, M. Brandt, R. Dimitrov, and R. Handschuh, United States Patent 6,559,075 B1 (2003)

[67] W.S. Wong, T. Sands, N.W. Cheung, M. Kneissl, D.P. Bour, P. Mei, L.T. Romano, and N.M. Johnson, Appl. Phys. Lett. **75** (1999) 1360.

[68] O. Ambacher, M.S. Brandt, R. Dimitrov, T. Metzger, M. Stutzmann, R.A. Fischer, A. Miehr, A. Bergmaier, and G. Dollinger, J. Vac. Sci. Technol. B **14** (1996) 3532.

[69] M.K. Kelly, O. Ambacher, B. Dahlheimer, G. Groos, R. Dimitrov, H. Angerer, and M. Stutzmann, Appl. Phys. Lett. **69** (1996) 1749.

[70] M.K. Kelly, O. Ambacher, R. Dimitrov, R. Handschuh, and M. Stutzmann, phys. stat. sol. (a) **159** (1997) R3.

[71] O. Ambacher, M.K. Kelly, C.R. Miskys, L. Höppel, C. Nebel, and M. Stutzmann, Mat. Res. Soc. Symp. **617** (2000) J1.7.1.

[72] O. Ambacher, J. Phys. D: Appl. Phys. **31** (1998) 2653.

[73] S.J. Pearton, J.C. Zolper, R.J. Shul, and F. Ren, J. Appl. Phys. **86** (1999) 1.

[74] W.S. Wong, T. Sands, and N.W. Cheung, Appl. Phys. Lett. **72** (1998) 599.

[75] E.A. Stach, M. Kelsch, W.S. Wong, E.C. Nelson, T. Sands, and N.W.Cheung, Mat. Res. Soc. Symp. **617** (2000) J3.5.1

[76] E.A. Stach, M. Kelsch, E.C. Nelson W.S. Wong, , T. Sands, and N.W.Cheung, Appl. Phys. Lett. **77** (2000) 1819.

[77] W.S. Wong, Y. Cho, E.R. Weber, T. Sands, K.M. Yu, J. Krüger, A.B. Wengrow, and N. W. Cheung, Appl. Phys. Lett. **75** (1999) 1887.

[78] Hyeon-Soo Kim, M.D. Dawson, and Geun-Young Yeom, J. Korean Phys. Soc. **40** (2002) 567.

[79] D.A. Ditmars, S. Ishihara, S.S. Chang, G. Berstein, and E.D. West, J. Res. Natl. Bur. Stand. (USA) **87** (1982) 159.

[80] Union Carbide Corporation-Crystal Products, Properties of Sapphire Manual.

[81] X.L. Chen, Y.C. Lan, J.K. Liang, X.R. Chen, Y.P. Xu, T. Xu, P.Z. Jiang, and K.Q. Lu, Chin. Phys. Lett. **16** (1999) 107.

[82] R.P. Vaudo, G.R. Brandes, J.S. Flynn, X. Xu, M.F. Chriss, C.S. Christos, D.M. Keogh, and F.D. Tamweber, Int. Workshop on Nitride Semiconductors, Nagoya, Japan, 2000.

[83] I. Barin, O. Knacke, and O. Kubaschewski, Thermochemical Properties of Inorganic Substances, Springer, 1977, Berlin.

[84] M. Leszczynski, T. Suski, H. Teisseyre, P. Perlin, I. Grzegory, J. Jun, S. Porowski, and T.D. Moustakas, J. Appl. Phys. **76** (1994) 4909.
[85] W.S. Wong, T. Sands, N.W. Cheung, M. Kneissl, D.P. Bour, P. Mei, L.T. Romano, and N.M. Johnson, Appl. Phys. Lett. **77** (2000) 2822.
[86] J. Jasinski, Z. Liliental-Weber, S. Estrada, and E. Hu, Appl. Phys. Lett. **81** (2002) 3152.
[87] Y. Golan, X.H. Wu, J.S. Speck, R.P. Vaudo, and V.M. Phanse, Appl. Phys. Lett. **73** (1998) 3090.
[88] S.T. Kim, Y.J. Lee, D.C. Moon, C.H. Hong, and T.K. Yoo, J. Crystal Growth **194** (1998) 37.
[89] H. Teisseyre, P. Perlin, T. Suski, I. Grzegory, S. Porowski, J. Jun, A. Pietraszko, and T.D. Moustakas, J. Appl. Phys. **76** (1994) 2429.
[90] R.P. Vaudo, V.M. Phanse, X. Wu, Y. Golan, and J.S. Speck, 2nd Int. Conf. Nitride Semiconductors, Tokushima, 1997.
[91] C.R. Miskys, M.K. Kelly, O. Ambacher, and M. Stutzmann, phys. stat. sol. (a) **176** (1999) 443.
[92] M. Schauler, F. Eberhard, C. Kirchner, V. Schwegler, A. Pelzmann, M. Kamp, K.J. Ebeling, F. Bertram, T. Riemann, J. Christen, P. Prystawko, M. Leszczynski, I. Grzegory, and S. Porowsky, Appl. Phys. Lett. **74** (1999) 1123.
[93] C.R. Miskys, M.K. Kelly, O. Ambacher, G. Martínez-Criado, and M. Stutzmann, Appl. Phys. Lett. **77** (2000) 1858.
[94] B.J. Skromme, J. Jayapalan, R.P. Vaudo, and V.M. Phanse, Appl. Phys. Lett. **74** (1999) 2354.
[95] G. Martínez-Criado, C.R. Miskys, A. Cros, O. Ambacher, A. Cantarero, and M. Stutzmann, J. Appl. Phys. **90** (2001) 5627.
[96] T. Taliercio, M. Gallart, P. Lefebvre, A. Morel, B. Gil, J. Allègre, N. Grandjean, J. Massies, I. Grzegory, and S. Porowski, Solid State Commun. **117** (2001) 445.
[97] Z.X. Liu, K.P. Korona, K. Syassen, J. Kuhl, K. Pakula, J.M. Baranowski, I. Grzegory, and S. Porowski, Solid State Commun. **108**, (1998) 433.
[98] I. Grzegory and S. Porowski, Thin Solid Films **367** (2000) 281.
[99] T. Schmidt, K. Lischka, and W. Zulehner, Phys. Rev. B **45** (1992) 8989.
[100] F. Calle, F.J. Sánchez, J.M.G. Tijero, M. A. Sánchez-García, E. Calleja, and R. Beresford, Semicond. Sci. Technol. 12 (1997) 1396.

phys. stat. sol. (c) **0**, No. 6, 1651–1667 (2003) / **DOI** 10.1002/pssc.200303132

Surfactants and antisurfactants on group-III-nitride surfaces

J. Neugebauer*[1]

[1] Fritz-Haber-Institut der Max-Planck-Gesellschaft, Faradayweg 4-6, D-14195 Berlin (Dahlem), Germany

Received 15 May 2003, accepted 18 June 2003
Published online 28 August 2003

PACS 61.72.Ji, 68.35.Bs, 82.70.Uv

We have studied the effect of various dopants/impurities on structure and stability of group-III-nitride surfaces employing first-principles density-functional theory. For the example of H, Si, and In it is discussed whether and under what conditions impurities can affect the growth morphology and may act as surfactant or antisurfactant. Based on these results surface structures and growth conditions have been identified which allow to overcome intrinsic (thermodynamic) materials properties and limitations. Examples which will be discussed concern the enhancement of the doping efficiency, the formation of novel (tailor-made) alloys, and the self-organized growth of nano-structures.

1   Introduction

In order to grow group-III nitride devices it is crucial to have a high control over the surface morphology: Growth conditions have to be identified where e.g. two-dimensional growth (to realize sharp heterointerfaces) or spontaneous formation (self-organization) of nanostructures occurs. Experimentally, it is well known that the surface morphology sensitively depends on the specific growth conditions such as the ratio between III/V elements (stoichiometry), the growth method (e.g. molecular-beam epitaxy (MBE), metal-organic chemical vapor deposition (MOCVD), hydride vapor-phase epitaxy (HVPE)) temperature, or the presence of impurities. While the first parameters are intrinsic to the materials system and allow only a rather limited control over the surface morphology the use of impurities gives much wider flexibility.

Experimentally, impurities are well known to affect the growth morphology. A particularly rich behavior of effects is found for the example of In on GaN surfaces. The presence of small amounts of indium leads to smoother surfaces (surfactant effect)[1, 2, 3] while higher indium concentrations lead to the formation of nanostructures[4]. When going to even higher concentrations formation of extended defects (inverted pyramid defects) occurs[5, 6]. Impurities may also be used to stabilize metastable bulk phases (e.g. arsenic to stabilize cubic GaN)[4] or to realize phases not present in nature (ordered InGaN or AlGaN alloys)[7, 8, 9]. Even doping may strongly affect the surface morphology. For example, silicon which is the donor impurity of choice, can severely affect the growth morphology: At high concentrations (above 5×10^{19} cm^{-3}) and MOCVD growth the GaN (0001) surface becomes unstable against surface roughening effectively limiting the doping level[10].

In order to improve group-III nitride growth or to technologically employ the above described phenomena it is crucial to understand the origin of these effects. While until recently such phenomena could be described only qualitatively using simple models and fitted parameters recent advances in parameter free (*ab initio*) total energy calculations allowed in the last couple of years to address these issues. In the present paper we will give a brief overview about the underlying concepts, and the results achieved with this methodology.

---

* Corresponding author: e-mail: neugebauer@fhi-berlin.mpg.de, Phone: +49 30 8413 4826, Fax: +49 30 8413 4701

The paper is organized as follows. In Sec. 2 a brief overview about the applied methods will be given. Sec. 3 briefly summarizes our present knowledge regarding structure and properties of bare GaN surfaces. Based on these results applications of this method for a variety of technologically important examples will be discussed in the following sections. Specifically, section 4 covers the effect hydrogen (which is highly abundant in MOCVD or HVP GaN growth) has on GaN surfaces. Section 5 addresses Si (important for n-type doping of GaN) and Sec. 6 the complexity of structures and growth phenomena In exhibits on GaN surfaces. The results/discussions are summarized in Sec. 7.

2   Methods

In order to describe the effect impurities have on growth and growth morphology two principal scenarios exist: If the surface is close or in equilibrium with its environment (which is given by temperature and partial pressure of all species present in the growth chamber) an equilibrium thermodynamic approach can be chosen. In this case the stable surface can be found by calculating the Gibbs-free energy as function of temperature and partial pressures for all possible surface structures and identifying the one which minimizes this energy.

The second scenario occurs if activation barriers exist which fully or partly block the system in going into the thermodynamic ground states: In this case the observed surface would not be the energetically most favorable one (i.e. the one which minimizes the Gibbs free energy) but a metastable surface structure which is driven by kinetics. To theoretically address the latter case is significantly more complex than the thermodynamic case and consists of three steps: (i) Construct an initial surface structure (a good guess is typically the equilibrium surface structure), (ii) calculate the kinetic parameters (i.e. identify the path of adatoms and their energetic barriers on the surface), and (iii) calculate how the kinetic mechanisms modify the surface structure. In principle, the whole procedure has to be done self-consistently, i.e., until an initial structure is found which is *not* modified by the kinetic processes. This is then the steady-state kinetically stabilized surface.

In the following the key concepts of the two approaches will be described. It will be further discussed how to combine these approaches with modern *ab initio* (i.e. parameter free) total energy calculations and how this combination allows to obtain a very detailed insight into various aspects of nitride growth.

2.1   Fundamentals

Under realistic conditions adatoms attach and detach from the surface, i.e., from a thermodynamic point of view the system we want to describe is an open system. The key quantity to describe such a system is the partition function:

$$Z(V,T) = \sum_i e^{-E_i/k_B T} \quad .$$

(1)

Here, V is the volume of the system, T its temperature, and k_B the Boltzmann constant. E_i is the total energy of a configuration i where the index i runs over all possible atomic configurations. A configuration is given by a set of atoms $\{Z_I^{\mathrm{nuc}}, \vec{R}_I\}$ (with Z_I^{nuc} the nucleus charge determining the chemical species of the atom, $\vec{R}_I$ its position and the index I running over all atoms within the set). The number of atoms within such a set is variable; however, according to Eq. (1) the volume must be kept fixed.

It is interesting to note that the latter condition poses a a formal problem: The concept of a volume is a *mesoscopic* property and for finite atomic systems no unique definition of a volume can be given. In principle, only for infinite crystals the volume per unit cell is well defined. Fortunately, for the systems considered here – surfaces at growth conditions characteristic for MBE or MOCVD – this issue poses not a real problem. The idea is to split the growth system in two parts (the substrate with surface and the gas phase) and treat both phases separately. The important thing to note is that the two parts have extremely different compressebilities: The gas phase at characteristic growth pressures (upper boundary ≈ 1 atm) has a compressibility which is orders of magnitude larger than that of the solid phase (bulk+surface). Thus, the exact definition of how the volume is defined for the surface system (i.e. whether one assumes that the surface boundary is given by a plane going through the topmost surface layer or whether this layer is shifted by a few Å outward) will be small. A simple estimate shows that even for the highest relevant pressure (≈ 1 atm) an outward shift of the surface gives rise to an energy change which is 4(!) orders smaller than characteristic surface energies. Thus, when calculating the partition function for a surface it is save to neglect the condition of constant volume. The volume (or we will later use the pressure) enters only by enforcing thermodynamic equilibrium between surface and gas phase.

Using modern parameter free total energy calculations (see Subsec. 2.3) the energy E_i needed in Eq. (1) can be in principle calculated for all possible atomic configurations with sufficiently high accuracy. The probably easiest approach to sample all possible configurations would be molecular dynamics, i.e., providing initial coordinates for the atoms and velocities (the kinetic energy of the atoms determines the temperature of the system) and solving the Newton equations for the atoms. Then, the system maps after a sufficiently long time all possible configurations (i.e. the entire phase space). For surface growth studies this approach has two major disadvantages: First, since activation barriers on surfaces (order of eV) are typically much higher than the thermal energy (≈ 0.1 eV) such simulations would be way too time consuming even on the fastest computers. A second problem with this approach is that it allows only to describe systems with fixed particle numbers, i.e., open systems as surfaces are not straightforward to describe.

The approach to overcome the above limitations is to split the complete phase space into smaller regions each of which can be easily calculated. Since the barriers are much larger than the thermal energy the mapping of the total phase space can be restricted to regions close to the global and local minima. In the following, the corresponding minima are labeled by λ (this index runs over all minima corresponding to inequivalent configurations) and σ_λ (which runs for a given λ over all symmetrically equivalent configurations). Then, the sum over phase space in Eq. (1) can be written as:

$$Z(V,T) = \sum_\lambda \sum_{\sigma_\lambda} \sum_\mu e^{-E_{\lambda\sigma_\lambda}(\mu)/k_B T} \quad . \tag{2}$$

Here, μ runs over the local region of phase space around the minima. In a good approximation (except for very light elements such as hydrogen and for not too high temperatures) this region can be described by expanding the potential energy surface up to the second order (harmonic approximation).

Based on the partition function (which is fully determined by the *ab initio* calculated potential energy surface) all thermodynamic variables can be determined. The *Helmholtz free energy* is given by

$$F(V,t) = -k_B T \ln\{Z(V,T)\} \quad . \tag{3}$$

Also, for a given pressure p and a fixed number of particles the Gibbs free energy

$$G(p,T) = F(T) + pV \tag{4}$$

can be calculated. The latter quantity is the relevant energy for an open system: To find the thermodynamically stable phase/configuration it has to be minimized at constant pressure and temperature.

An immediate implication from the Gibbs free energy being minimum is that an open system in thermodynamic equilibrium is stable against an infinitesimal increase or decrease in the number of particles. This condition is typically formulated in terms of chemical potentials:

$$\mu_A := \left(\frac{\partial G(A^n)}{\partial n} \right)_{T,p} = 0 \tag{5}$$

Here, μ_A is the chemical potential of atoms of species A, n the number of atoms of species A, and $G(A^n)$ the Gibbs free energy of a system consisting of n atoms A. Thermodynamic equilibrium means then that the chemical potential μ_A must be the same in the bulk (where it determines the vacancy/impurity concentration), on the surface (where it determines the surface structure and stoichiometry) and in the gas phase above the surface (which is given by the specific growth conditions such as e.g. temperature and partial pressure).

2.2 Chemical potentials

To be more specific let us consider MOCVD growth of GaN. Then, three relevant species exist (Ga, N, and H). The abundance of each of these species is described by its corresponding chemical potential (μ_{Ga}, μ_N, and μ_H). It is important to note that desirable growth conditions and the choice of these potentials are closely connected. For example, to ensure that GaN is thermodynamically stable the following condition applies:

$$\mu_{GaN(bulk)} = \mu_{Ga} + \mu_N \quad . \tag{6}$$

$\mu_{GaN(bulk)}$ is the chemical potential of bulk GaN which can be directly calculated using *ab initio* methods. An immediate consequence of Eq. (6) is that the Ga and N chemical potential cannot be chosen independently but only one of them. Throughout the paper we will choose the N-chemical potential μ_N as independent variable.

The choice of the chemical potentials is further limited by the condition of preventing the formation of undesirable phases (e.g., precipitates in the hosts, droplets on the surface, or formation of molecules). For the case of hydrogen two possibilities of undesired phases exists: The formation of hydrogen molecules or of ammonia (NH_3). To prevent the formation of these phases one has to chose conditions where they are thermodynamically unstable. For example, to prevent the formation of H_2

$$\mu_H < \mu_{H(H_2 molecule)} \tag{7}$$

with $\mu_{H(H_2 molecule)}$ the chemical potential of a hydrogen atom in an H_2 molecule at temperature $T = 0\,K$. Similarly, to prevent the formation of ammonia the following condition has to be fulfilled:

$$\mu_H < \mu_{H(NH_3 molecule)} \quad . \tag{8}$$

By choosing specific values for the chemical potentials different chemical environments (growth conditions) can be simulated. For example, extreme Ga-rich conditions would be expressed by $\mu_{Ga} = \mu_{Ga(bulk)}$. For gaseous species the chemical potential can be expressed as function of temperature and pressure, i.e. by quantities which are experimentally in principle directly accessible. It should be noted, however, that during crystal growth the chemical potential of the gas phase will be higher than the chemical potential on the surface: It is the difference/gradient in the chemical potential that drives crystal growth. Only in specially designed experiments where steady-state equilibrium is realized the chemical potential in the gas phase and on the surface will be identical (see e.g. [11]).

For hydrogen, this dependence is given by[12, 13]

$$2\mu_H = E_{H_2} + k_B T \left[\ln \left(\frac{pV_Q}{k_B T} - \ln Z_{rot} - \ln Z_{vib} \right) \right] \tag{9}$$

where E_{H_2} is the energy of an H_2 molecule at $T = 0\,K$, k_B is the Boltzmann constant, T is the temperature, and p is the pressure. $V_Q = (h^2/2\pi mkT)^{3/2}$ is the quantum volume and Z_{rot} and Z_{vib} are the rotational and vibrational partition functions.

Once the chemical potentials are specified the Gibbs free energy to form a surface can be calculated. To be more specific let us consider bare and hydrogen terminated GaN surfaces. Choosing the ideal, bare 1×1 surface as reference, the Gibbs free energy of formation is[13]:

$$\Delta G^f = E_{tot}[GaN(0001)] - E_{tot}[GaN(0001), ideal] + \Delta F_{vib} - n_{Ga}\mu_{Ga} - n_N\mu_N - n_H\mu_H \quad . \tag{10}$$

$E_{\text{tot}}[\text{GaN}(0001)]$ is the calculated total energy for the surface under study, and $E_{\text{tot}}[\text{GaN}(0001), \text{ideal}]$ is the total energy of the reference system. $n_{\text{Ga,N,H}}$ is the number of Ga, N, H atoms. ΔF_{vib} includes the vibrational contributions and is discussed in [12, 13]. The formalism described above allows to identify the equilibrium surface structure by identifying the structure which minimizes the Gibbs free energy for a given set of chemical potentials (for the example discussed here μ_{H} and μ_{Ga}). It should be noted that the method is completely general and can be applied to study surfaces under clean conditions (i.e. in the absence of any impurities) and in the presence of foreign chemical species.

2.3 Total energy calculations

As pointed out above the key input in the above formalism are accurate total energies. The method of choice to perform such calculations is density-functional theory using either the local-density approximation[14, 15] or the generalized gradient approximation (GGA)[16]. Most of the theoretical studies discussed in the following were based on soft *ab initio* pseudopotentials[17], a plane wave basis set (with an energy cutoff up to 70 Ry). The surfaces have been modeled by a repeated slab geometry which consists of 4 up to 10 layers of GaN. The slabs are separated by a vacuum region of at least 7 Å. Details of the method can be found elsewhere[18, 13].

3 Bare GaN surfaces

Before discussing the effect impurities/dopants have on nitride surfaces first a brief summary of the properties of bare III-nitride surfaces will be given. Here and in the following the focus will be on GaN and the wurtzite phase: This phase is the thermodynamic stable phase and it is the relevant phase used in all group-III nitride based commercial devices. Experimental and theoretical studies on surfaces of cubic GaN, which can be epitaxially stabilized on cubic substrates, can be found e.g. in Refs. [19, 20, 21].

3.1 Relevant surface orientations

The relevant surface orientations for bulk and epitaxial growth of wurtzite GaN are the polar (0001) and ($000\bar{1}$) surfaces. While epitaxial growth along both orientations is possible by a suitable treatment of the substrate and even both orientations on the same substrate can be grown in a controlled manner[22] various experiments reported growth on (0001) superior to growth on ($000\bar{1}$). A surface orientation often observed in connection with nano-structural defects such as e.g. as side facets of inverted pyramids [6, 5, 23, 24] or of epitaxial layer overgrowth (ELOG) structures is the polar ($1\bar{1}01$) surface. A surface orientation with potentially great promises for the future is the neutral ($1\bar{1}00$) surface: In contrast to all the polar surfaces discussed before growing along this surface avoids the buildup of macroscopic electric fields due to spontaneous polarization or piezo-electricity.

Figure 1 shows the ideal (bulk-truncated) structure of the (0001), ($000\bar{1}$), and ($1\bar{1}01$) surfaces. For the (0001) surface this structure consists of Ga in the top-surface layer where each surface Ga atom is three-fold coordinated to the N atoms in the second surface layer. First-principles calculations showed this structure to be highly unstable. The relevant surface reconstructions have been identified by combining STM and first principles studies[25]: For N-rich and modest Ga-rich conditions surface reconstructions with Ga and/or N adatoms are formed. At extreme Ga-rich conditions a surface structure consisting of two Ga monolayers is found to be most stable [26]. This structure – which is commonly called Ga-bilayer structure – has been confirmed by various experimental studies [26, 27, 28].

The ideal bulk-truncated structure of GaN ($000\bar{1}$) consists of nitrogen in the topmost layer with each N forming three bonds to the underlying Ga layer. Similar like for the (0001) surface the ideal bulk truncated structure is highly unstable: First principles calculations in combination with STM revealed that under N-rich and modest Ga-rich conditions a Ga monolayer is most stable [25]. An interesting feature of this structure is that each Ga atom in the surface layer forms only a single bond to the underlying N layer -

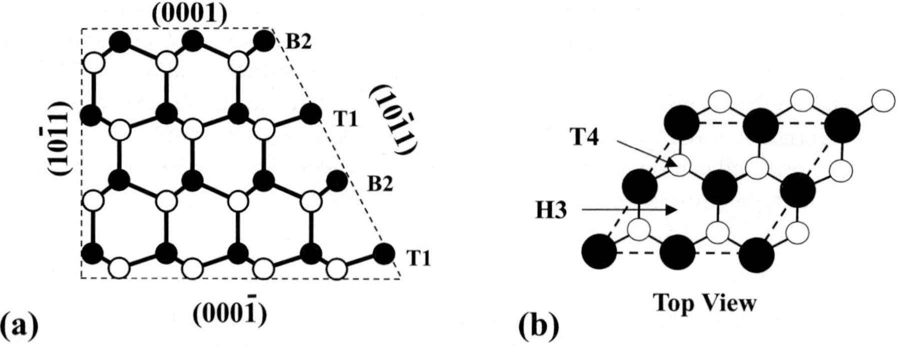

Fig. 1   Schematic atomic geometry of the relevant surface orientations of wurtzite GaN ((a)–side view) and of the (0001) orientation ((b)–top view). Black balls mark Ga atoms, white balls nitrogen atoms. $T1$, $B2$, $H3$, and $T4$ label different surface sites ($T1$: singly coordinated top position; $B2$: doubly coordinated bridge position; $H3$: three-fold coordinated hollow site, and $T4$: hollow position above N atom).

before the discovery of this surface such a structure had been considered as energetically highly unfavorable [4]. Going towards more Ga-rich conditions varios surface reconstructions are formed by adding additional Ga atoms on the adlayer [26].

Experimentally, no information regarding the $(1\,\bar{1}01)$ atomic surface structure are available so far. *Ab initio* studies identified two energetically favorable structures [29, 30]. One, which is stable under very Ga-rich conditions, is shown in Fig. 1. The structure consists of a Ga adlayer consisting of two inequivalent Ga atoms per surface unit cell: One Ga atom is singly coordinated (i.e. it forms a single Ga-N bond; this site is called $T1$) the other is doubly coordinated (and called $B2$ site). Under more N-rich conditions half of the Ga atoms are removed. The remaining Ga adatoms are three-fold coordinated.

One of the first theoretical studies on GaN surfaces had been performed for the non-polar (1-100) surface[31]. For N-rich conditions the surface consists of tilted Ga-N dimers. Going towards more Ga-rich conditions the N is replaced by Ga and the surface consists of Ga-Ga dimers. Very recent experimental and theoretical studies showed that the latter structure is unstable and that the surface consists of a compressed Ga adlayer [32].

3.2   Trends and discussion

As obvious from the above description a general feature of GaN surfaces is the presence of single or multiple Ga layers except for very N-rich conditions. This is a unique property of group-III nitrides and has not been found for traditional III-V or II-VI compound semiconductors. The reason for this unusual behavior has been analyzed in detail in Ref. [33]: It is due to (i) the small lattice constant of GaN – compared to that of As or P based III-V semiconductors it is smaller by 20 % – and (ii) the large difference in the chemical potential of its constituents (the bonding energy of a nitrogen atom in a N_2 molecule is roughly 2 eV larger than that of a Ga atom in Ga bulk). Due to the small lattice constant the spacing between two Ga atoms in bulk GaN is close to that of bulk GaN. For the surface this means that even in an ideal adlayer (i.e. an adlayer where the Ga atoms have the same spacing as the Ga atoms in bulk GaN) strong metallic bonds between the Ga atoms are formed. The presence of these bonds strongly stabilizes Ga adlayer structures. An additional stabilization is realized by mechanism (ii): The large difference in chemical potentials makes it energetically much more favorable to add Ga atoms than N atoms.

The large stability of Ga-rich surface structures (with one up to several monolayers) and the large energy it needs to build up more N-rich surfaces (due to the large difference in chemical potentials) have important consequences on the properties and growth of these surfaces. Here only properties relevant to identify/discuss surfactants and antisurfactants will be discussed.

One important feature is the *high surface energy* compared to those of As or P based III-V semiconductors. As shown in Table 1 the GaN surface energy for extreme Ga-rich conditions is similar for all studied surfaces and between $110\ldots125\,\text{meV/Å}^2$. In contrast, for GaAs and related compounds it is roughly a factor of two smaller.

Table 1: GaN surface energies for various orientations in the Ga-rich limit. As comparison also the surface energy of GaAs (111) is given. After Refs. [30, 34].

Material	Surface	Energy (meV/Å^2)	Surface structure
GaN	$(10\bar{1}0)$	110	Ga terminated
GaN	$(10\bar{1}0)$	118	stoichiometric
GaN	(001)	125	1×4 Ga tetramer
GaN	$\frac{1}{2}\left[(0001) + (000\bar{1})\right]$	110	Ga-bilayer+1×1 Ga adlayer
GaN	$\frac{1}{2}\left[(10\bar{1}1) + (10\bar{1}\bar{1})\right]$	125	1×1 Ga adlayer+1×1 Ga adlayer
GaAs	(111)	125	2×2 Ga vacancy

The large surface energy indicates that GaN surfaces are chemically highly reactive: The chemisorption energy of adatoms and molecules is expected to be much stronger than on surfaces with low surface energy. An important consequence is that GaN surfaces efficiently trap impurities and/or dissociate molecules (such as e.g. oxygen). This effect leads to an accumulation of impurities on the surface. As has been pointed out in Sec. 2 for growth conditions away from thermodynamic equilibrium this would lead to enhanced incorporation of impurities compared to the thermodynamic solubility limit and thus to unintentional high background impurity levels. The high surface energy has also important consequences for growth. For example, the growth surface can become thermodynamically unstable against formation of facets and/or surface roughening[21].

In order to control these effects, which are often detrimental to device performance, it is crucial to identify conditions which modify the surface energy. An efficient method to do this is to introduce an additional chemical species to the system which adsorbs on the surface and modifies its structure and energetics. However, to be actually useful for practical applications such a species must obey the following criteria:

1. It must form a continuous thin film on the surface (i.e. it must wet the surface),

2. the chemical bonds between the additional species and the substrate should be weak enough to allow an exchange with the original atoms (otherwise further growth would be prevented), and

3. the solubility of the species in the host material must be negligible.

If a species obeys the above conditions and leads to smoother surfaces it is called surfactant (surface active species), if it leads to rougher surfaces antisurfactant. Surfactants are potentially interesting when growing two-dimensional structures such as e.g. sharp interfaces or quantum wells while antisurfactants are potentially interesting to form nanostructures such as quantum dots. It is important to note that the addition of surfactants and antisurfactants often occurs unintentionally: For example, as will be shown later, dopants (e.g. Si), species used to grow alloys (e.g. In) or which are omnipresent in certain growth modes (e.g. H) can significantly modify surface structure and energetics.

The formalism described in Sec. 2 allows to directly study how and under what growth conditions impurities affect surface structure and energetics. In the following we will present the results for H, Si

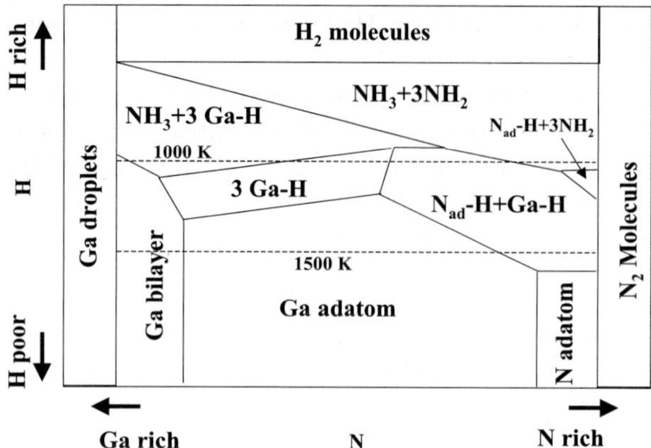

Fig. 2 Surface phase diagram of possible equilibrium surface reconstructions of bare and H covered GaN (0001) surfaces as function of the H and N chemical potential. The dashed lines show the position of the H chemical potential at a pressure of 200 Torr and temperatures of 1000 and 1500 K. After Ref. [13].

and In on GaN surfaces and discuss how these results can be used to get a detailed insight into growth and doping aspects.

4　H on GaN surfaces

The effect hydrogen has on the atomic structure and energetics of the GaN (0001) surface has been studied by combining *ab initio* density-functional theory with thermodynamic concepts in Ref. [13]. The calculated surface phase diagram (Fig. 2) shows the thermodynamically most stable surface structures for a given set of hydrogen and nitrogen chemical potentials. Under hydrogen-rich and nitrogen-rich conditions the GaN surface is mainly covered by NH_3 and NH_2 molecules. Going towards more Ga-rich conditions N is removed from the surface leaving Ga dangling bonds which are passivated by hydrogen. Going towards less hydrogen-rich conditions some intermediate structures are formed and eventually the bare GaN (0001) surfaces (N adatom, Ga adatom, Ga bilayer) become stable.

From the phase diagram and the structures found to be stable a number of conclusions can be deduced. First, the fact that at high hydrogen chemical potentials (i.e. at hydrogen-rich conditions) surface structures which are completely covered by hydrogen are more stable than the bare GaN surfaces implies that the formation of a hydrogen film is exothermic: At these conditions hydrogen exhibits a wetting behavior. Second, previous density function calculations showed that hydrogen – even at optimum conditions (high temperatures and presence of acceptors) – has a maximum bulk concentration of $\approx 10^{20}\,\mathrm{cm}^{-3}$[35, 36]. Since the number of Ga and N atoms in GaN is $\approx 10^{23}\,\mathrm{cm}^{-3}$ the number of hydrogen atoms in GaN is orders of magnitude smaller than the number of host atoms. As can be deduced from the phase diagram (Fig. 2) at the surface a very different scenario occurs: Under hydrogen-rich conditions structures with a coverage of several monolayers (ML) (up to three ML) of hydrogen are found to be stable. This means, that *the concentration of hydrogen on GaN surfaces may exceed the corresponding bulk concentration by orders of magnitude*. We can thus conclude that hydrogen obeys the two criteria discussed in Sec. 3.2 (it wets the surface and has a comparatively low bulk solubility).

In order to understand whether and when the conditions necessary to stabilize these surface structures are realized under realistic growth conditions the explicit temperature and pressure dependence can be calculated[12, 13]. The corresponding hydrogen chemical potential for two examples ($T = 1000$ and $1500\,\mathrm{K}$ at a pressure of 200 Torr) are shown in Fig. 2. As can be seen at the lower temperature of $1000\,\mathrm{K}$ (which is characteristic for gas source MBE or for MOCVD InGaN growth) surfaces are always

H-terminated except for extrem Ga-rich conditions. At 1500 K, however, a complete different scenario is observed: Except for very N-rich conditions H adsorption on GaN surfaces is energetically unfavorable, i.e., the bare GaN surfaces are observed.

Based on the above discussion a tentative explanation of the role of hydrogen on the GaN (0001) surface can be given. At a typical MOCVD growth temperature of 1300 K surface structures are found where hydrogen passivates N and Ga dangling bonds. This effect energetically stabilizes the surface. At the same time the thermodynamically allowed region where N adatoms are stable on the surface becomes significantly larger - in contrast to the hydrogen-free case where N adatoms are stable only under extreme N-rich conditions the presence of H significantly stabilizes N adatoms at the surface making them stable even at modest Ga-rich conditions. This process significantly enhances the number of available N atoms for growth and is thus crucial. The phase diagram reveals also another mechanism important for growth. The fact that at 1500 K the GaN (0001) surface becomes essentially free of hydrogen implies that at typical growth temperatures (which are only by 200 K lower) the hydrogen is only weakly bound to the surface. Thus, at these conditions hydrogen can be easily exchanged with the host atoms, i.e., while the hydrogen layer is important to stabilize the surface and N adatoms it does not prevent further growth.

All the above features highlight the important role of hydrogen in MOCVD growth. However, it should be noted that many questions still remain to be addressed: For example, to perform quantitative growth simulations calculations regarding adatom kinetics on hydrogenated GaN surfaces and the effect of hydrogen on other surface orientations (to understand whether H stabilizes/destabilizes certain orientations and thus acts as selective surfactant) are needed.

5   Si on GaN surfaces

As a next example of how impurities modify surfaces we will consider the case of dopants. The focus will be on silicon since (i) it is the donor impurity of choice to make GaN n-type conductive and (ii) the surface morphology is strongly dependent on growth conditions and techniques. MOCVD studies by Tanaka *et al.* [37] reported that Si concentrations in the range of $10^{18} \ldots 10^{19}$ cm^{-3} modify growth on GaN (0001) from step flow to a three-dimensional mode, suggesting that Si may act as anti-surfactant. At sufficiently high Si concentrations even the formation of quantum dots has been observed[37]. On the other hand Si doping on GaN surfaces using MBE has been found to lead to smooth surfaces, indicating that Si does not adversely effect growth and that it may even act as surfactant[38].

In order to identify why Si has such a dramatically different effect on GaN surfaces first-principles calculations have been performed[38]. Before discussing the surface results let us first focus on the incorporation/energetics of Si in bulk GaN. The results of this discussion will be used later to show that not only impurities may have a strong effect on the surface morphology but that also *surfaces may have a significant effect on doping efficiency and solubility*.

Using the formalism as described in Sec. 2 the Si concentration in bulk GaN has been calculated as function of the Si and N chemical potential[39]. Figure 3 shows the results for a temperature of 1300 K (i.e., for a temperature characteristic for MOCVD growth). The corresponding phase diagram marks also the region where the incorporation of Si on the Ga substitutional site becomes thermodynamically unstable against the formation of Si_3N_4. As can be seen, the formation of this phase limits the equilibrium concentration of Si in GaN (i.e. the Si solubility) to $\approx 5 \times 10^{19}$ cm^{-3} at N-rich conditions. It is interesting to note that the maximum solubility is not reached under nitrogen-rich conditions (where the removal of Ga atoms is easiest) but under Ga-rich conditions. The reason is that also a competing process exists: Nitrogen-rich conditions promote the formation of Si_3N_4 and as can be seen in Fig. 3 this mechanism dominates.

An important conclusion which can be drawn from the above analysis is that N-rich conditions (which are characteristic for MOCVD growth; see Sec. 4) have a lower Si solubility than Ga-rich conditions (which are characteristic for MBE growth). However, even for MBE conditions (i.e., extreme Ga-rich, temperature $T = 900$ K) the Si solubility is limited to $\approx 10^{19}$ cm^{-3}. The question is therefore whether

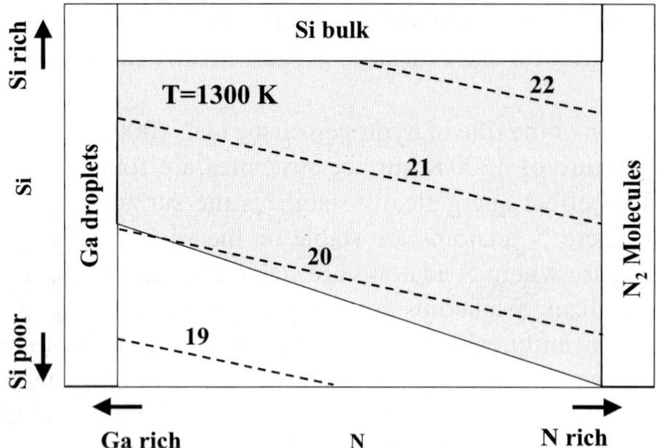

Fig. 3 Equilibrium concentration of Si donors in bulk GaN as function of the nitrogen (μ_N) and (μ_Si) chemical potentials at $T = 1300\,\mathrm{K}$. The solid lines are the phase boundaries for the thermodynamically most stable phases. The numbers at the contour lines give the $\log_{10}$ Si concentration. After Ref. [39].

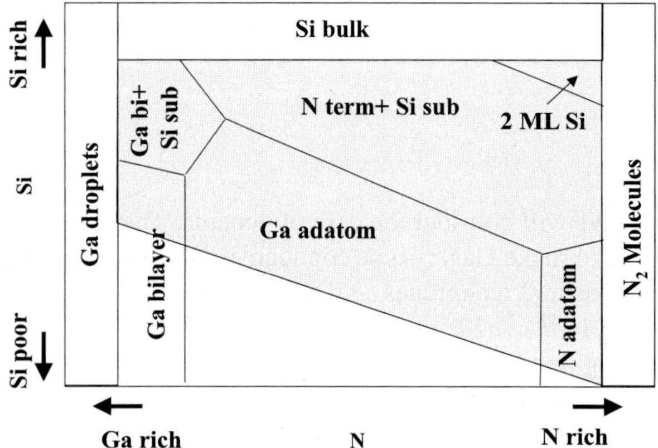

Fig. 4 Surface phase diagram of possible equilibrium surface reconstructions of bare and Si covered GaN (0001) surfaces as function of the N and Si chemical potential. The grey area marks the region where surfaces are thermodynamically unstable against the formation of Si_3N_4. The corresponding geometries are shown in Fig. 5. After Ref. [38].

kinetic surface effects (as discussed in Sec. 2) can be used to control/enhance the doping efficiency. To address this question and also to resolve the puzzling behavior how Si affects GaN surfaces a combined experimental (scanning-tunneling microscopy) and experimental study (density-functional theory) had been performed[38]. The results of this study are briefly summarized below.

Starting from the bare GaN surface structures (i.e. the Ga and N adatom and the Ga-bilayer structure; see Sec. 3) and systematically adding Si atoms and/or substituting Ga and N atoms by Si atoms equilibrium geometry and formation energy for a large number of Si induced surface reconstructions have been investigated. Based on these results the corresponding surface phase diagram as function of the Si and N chemical potential could be derived (Fig. 4). The relevant surface geometries are shown in Fig. 5. Let us first focus on Si-rich conditions. For Ga-rich conditions the phase diagram shows that the most favorable structure is almost identical to the Ga-bilayer structure: The only difference is that a Ga atom in the third surface layer has been replaced by a Si atom. Going towards more N-rich conditions a surface structure

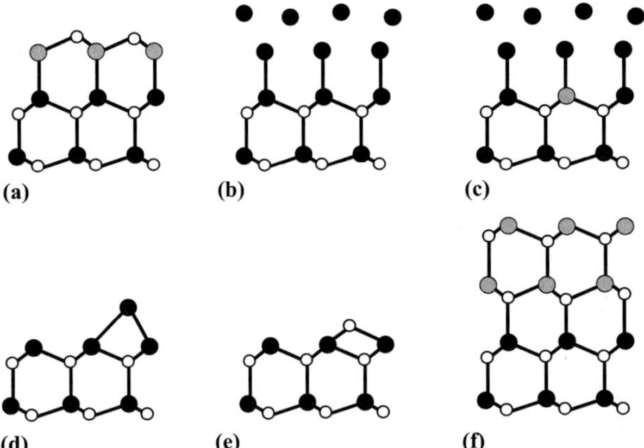

Fig. 5 Schematic geometry (side view) of the surface structures listed in the Si on GaN surface phase diagram (Fig. 4). White balls mark N atoms, black Ga, and gray Si atoms. The abbreviations used in Fig. 4 correspond to the structures as follows: (a) N terminated surface with Si subsurface ("N term+Si sub"), (b) Ga bilayer, (c) Ga bilayer with Si subsurface ("Ga bi + Si sub"), (d) Ga adatom, (e) N adatom, and (f) two Si-N layers on top of GaN ("2 ML Si"). After Ref. [38].

is formed which consists of a N-Si adlayer. For Si-poor conditions the bare GaN surface structures are reproduced.

An interesting result emerging from the phase diagram is that all Si induced surface structures (i.e., structures containing Si in the first few surface layers) are thermodynamically unstable against the formation of Si_3N_4. However, as has been pointed out in Ref. [38] the kinetics (activation barriers) to form such precipitates may be very different for the two classes of Si induced structures. For more N-rich conditions surface reconstructions with high Si concentration in the surface layer are found. A comparison with the Si donor concentration in bulk for identical growth conditions (i.e. for identical chemical potentials) (Fig. 3) clearly reveals that the surface Si concentration is orders of magnitude larger. Therefore, surface segregation might be an important issue. For more Ga-rich conditions a fundamentally different behavior is observed – Si prefers subsurface configurations rather than surface sites. Under these conditions surface segregation does not occur and Si can be easily incorporated in GaN bulk *allowing to overcome the thermodynamic solubility limit*. Ga-rich conditions are, therefore, the optimum regime to incorporate Si in GaN.

The above discussion also gives insight into the puzzling difference between MBE and MOCVD when doping GaN with Si. As pointed out above under more N-rich conditions (as characteristic for MOCVD or gas-source MBE growth where H stabilizes N on the surface; see also Sec. 4) Si segregates to the surface where it can form Si_3N_4 islands. In fact, an analysis of the surfaces stable under these conditions (Fig. 5a) revealed the Si-N layer characteristic for these surfaces to be under large tensile strain: Without being bonded to the GaN surface the lateral lattice constant of this layer would be smaller by $\approx 20\,\%$. This surface is thus highly unstable against the formation of Si_3N_4 and can thus be considered as a precursor state for Si_3N_4 precipitation. Since it has been observed that Si_3N_4 islands/precipitates chemically passivate GaN surfaces and block growth,[37] GaN growth occurs only in regions of the surface not covered by Si_3N_4 – three-dimensional growth results. Under these conditions Si acts as anti-surfactant and thus can be used to grow quantum dots[37] or to significantly reduce the dislocation density by causing dislocations to bend[37, 40].

Under Ga-rich conditions (which are characteristic for MBE growth), the Si-induced surfaces are essentially free of Si in the top surface layer and topologically very similar to the corresponding bare GaN surface structure (Ga bilayer). This result immediately explains why experimentally under these condition

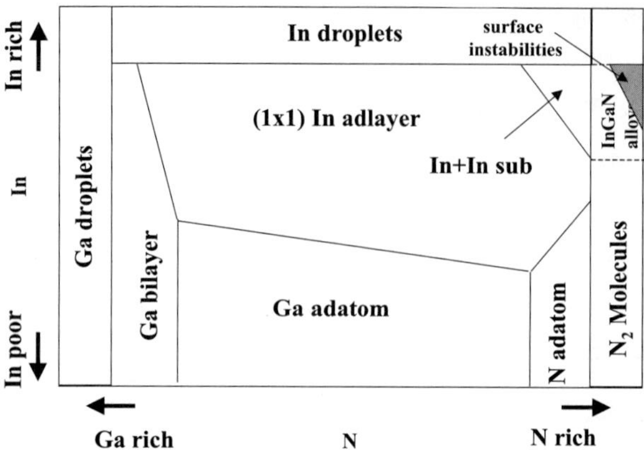

Fig. 6 Surface phase diagram of possible equilibrium surface reconstructions of bare and In covered GaN(0001) surfaces as function of the In and N chemical potential. The corresponding geometries are shown in Fig. 7. The gray area indicates the region where the (0001) orientation becomes unstable against facet and micro-facet formation (see text).

no change in the growth mode occurs, i.e., why Si doping in MBE (even at high concentrations) has no adverse effect on the surface morphology.

6 In on GaN surfaces

As a final example how impurities modify surface morphology and growth we will now focus on In on GaN surfaces. This choice is motivated by the fact that (i) the formation of In$_{1-x}$Ga$_x$N alloys is one of the technologically key steps in making III-nitride optoelectronic devices (it is used as active layer in light emitting and laser diodes) and (ii) depending on In concentration and growth conditions it shows a large variety of phenomena. As will be discussed in the following it can act as surfactant, as selective surfactant, it can be used to form nanostructures on the surface or to overcome thermodynamic limitations.

6.1　Surface phase diagram

The effect In has on the energetics and atomic structure of GaN surfaces has been extensively studied both theoretically and experimentally[41, 42, 2, 43, 44, 45, 4]. Specifically, based on density-functional theory calculations the energetically most favorable structures for a given set of chemical potentials have been calculated[4, 44, 45]. The results are summarized in the surface phase diagram shown in Fig. 6. The corresponding structures are given in Fig. 7. In addition to results for the (0001) orientation also results for the (1$\bar{1}$01) surface have been included.

Under In-poor conditions the phase diagram reproduces the bare GaN (0001) surfaces. For high In concentrations a variety of structures are found. At extreme Ga-rich conditions In is thermodynamically unstable: Under these conditions the Ga-bilayer structure is energetically most favorable. Going to slightly more N-rich conditions a structure consisting of an In-adlayer becomes most stable (see Fig. 7a). The In atoms in the adlayer sit on-top of the underlying Ga atoms, i.e., each In atom forms a single bond to a Ga atom. This structure is free of any In-N bonds implying that under such conditions In just floats on-top of the surface[46]. Going to even more N-rich conditions additional In is incorporated and substitutes Ga in the second layer (Fig. 7b). Under these conditions In-N bonds are formed which act as precursor for the formation of InGaN alloys[44]. Increasing even further the N-chemical potential increases the number of In atoms in the second layer[44]. Since an In-N bond is ≈10% larger than a Ga-N bond the incorporation of In atoms in the second layer (which is accompanied with the formation of In-N bonds) induces a substantial

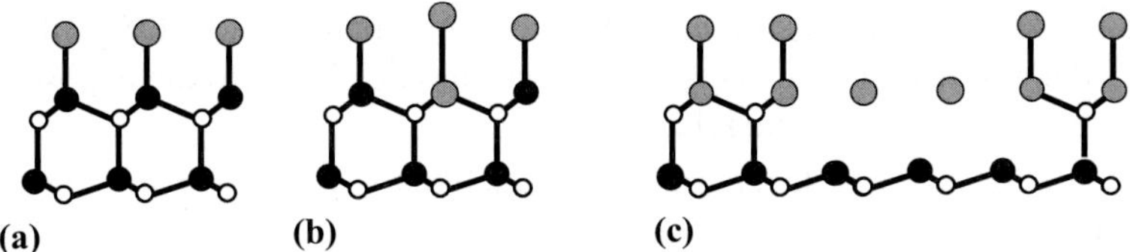

Fig. 7   Schematic geometry (side view) of the surface structures listed in the surface phase diagram (Fig. 6). White balls mark N atoms, black Ga, and gray In atoms. The abbreviations used in Fig. 6 correspond to the structures as follows: (a) unreconstructed In adlayer, (b) subsurface incorporation of In atoms ("In+in sub") and (c) example of a surface instability (micro-facet). After Ref. [4].

lateral strain in the surface which increases with increasing In concentrations. An interesting consequence of this behavior is that when going towards more and more N-rich conditions the In incorporation (and thus the surface strain) increase. An interesting consequence of this behavior is that after exceeding a critical In concentration the (0001) growth surface becomes unstable against the formation of In-covered $(1\,\bar{1}01)$ GaN surfaces[30, 29, 4].

6.2   Low In concentrations: Surfactant behavior

The presence of these surface structures has direct consequences on surface morphology and bulk properties. Let us first focus on the In-adlayer structure (Fig. 7a) which is stable over a wide range of chemical potentials. Under conditions where this surface is realized the solubility of In in bulk GaN is negligible. Thus, In will float on top of the surface and not be incorporated – this structure is thus an ideal candidate for a surfactant/antisurfactant. Indeed, a recent combined theoretical/experimental study showed this structure to largely affect the kinetics of adatoms.

To discuss the effect this surface structure has on adatom diffusion let us briefly summarize the behavior of Ga and N adatoms on the bare GaN surfaces. DFT calculations showed that the diffusion barriers for Ga adatoms on GaN are generally lower than for N adatoms – the lateral surface diffusion is thus limited by the diffusion barrier of N[47, 33]. The reason for the very different behavior of the two species is directly related to the fact that equilibrium GaN surfaces are Ga-stabilized. Thus, for Ga adatoms the adsorbate-surface interaction is predominantly realized by Ga-Ga bonds. The fact, that Ga bulk melts already slightly above room temperature ($T_{\mathrm{melt}} = 30°$) implies Ga-Ga bonds to be weak: Ga adatoms behave almost like a liquid film on the surface. N adatoms, however, form strong Ga–N bonds resulting in significantly higher diffusion barriers (on GaN (0001) $\approx 1.3\,\mathrm{eV}$). The rather high diffusion barrier of N adatoms limits the lateral surface transport and causes three-dimensional growth as indeed observed under more N-rich conditions in experiment[48].

As has been pointed out recently the presence of an In adlayer on the surface quantitatively affects the diffusion of N adatoms: Lateral transport of N adatoms on this surface is *not realized by diffusion on the top surface layer but below the In adlayer*[46]. This mechanism reduces the N diffusion barrier dramatically (by more than a factor of two) and opens an efficient and hitherto unexpected diffusion channel for lateral surface transport. The significantly enhanced adatom diffusivity on this surface favors step-flow growth over three-dimensional growth and thus allows to achieve a smooth surface morphology even at rather low growth temperatures as characteristic for MBE.

The identification of this mechanism allows the interpretation of various growth studies. As has been pointed out above the surfactant behavior of In is related to the presence of an In adlayer structure (which enables subsurface diffusion). As can be seen from the phase diagram (Fig. 6) this structure is stable over a large range of possible chemical potentials except for extreme Ga-rich conditions where the Ga-bilayer structure becomes energetically most favorable. Thus, In is expected to behave as a surfactant only for growth conditions where no Ga-bilayer is formed. This is consistent with recent experiments which clearly showed that under extreme Ga-rich conditions In has no effect on the surface morphology[3]. Only when going to growth conditions where in the absence of In a rough growth morphology is found the exposure of In leads to a dramatic improvement in surface morphology[1, 2, 3].

Subsurface diffusion is also expected for the Ga-bilayer structure since like the In-adlayer structure it has a thin metallic film (consisting of three layers of Ga) on the surface[46]. Again, this is consistent with experimental studies finding extreme Ga-rich conditions (i.e., conditions where the Ga-bilayer structure is stable; see Fig. 6) to lead to smooth surface morphology[49]. Therefore, Ga can be considered as auto-surfactant[28].

6.3   Medium In concentrations: Alloy fluctuations

The surface structures found for In- and N-rich conditions (i.e. in the upper-right corner of the phase diagram) are relevant for the formation and composition of InGaN alloys. As can be seen in the phase diagram the deposition of the GaN surface with In is relatively easy – except for extreme Ga-rich conditions In is always stable on the surface. However, the actual incorporation, i.e., the substitution of Ga atoms by In and the formation of In-N bonds, occurs only in a rather small region of the phase diagram (only under very N and In-rich conditions where there is a substantial deficiency of Ga atoms). This behavior is consistent with experimental observations: In is well know to segregate to the surface and In incorporation is only possible under rather N-rich conditions[4].

The strong surface segregation has important implications on the surface morphology. As has been shown in Ref. [30, 29] the relative surface energy between the $(1\bar{1}01)$ and (0001) surfaces significantly changes with In content: Increasing the In content reduces the surface energy of the $(1\bar{1}01)$ relative to the (0001), i.e., the (0001) surface becomes less and less stable against the formation of $(1\bar{1}01)$ facets. Based on the fact that the strong reduction in surface energy is only observed for $(1\bar{1}01)$ and not for (0001) In has been called *selective surfactant* [29]. It should be noted that this is restricted to conditions where In is actually incorporated in bulk GaN, i.e., under conditions where In-N bond are formed.

The origin for the selective behavior of In has been given in Ref. [29]. A detailed analysis of the different surface structures showed that the coordination (number of nearest neighbor bonds) of surface cations is very different for the two orientations. While on the (0001) surface the In atoms forming In-N bonds are three-fold coordinated on $(1\bar{1}01)$ In and Ga atoms prefer low coordinated sites with only one or two N neighbors. Since In-N bonds are significantly weaker than Ga-N bonds a general tendency is that In atoms prefer low coordinated sites while Ga atoms prefer high coordinated site. Thus, when In atoms segregate from the bulk to the surface the energy gain to go to the $(1\bar{1}01)$ surface (where low coordinated sites are available) is significantly larger than by going to triply coordinated sites on GaN(0001)[30].

The selective behavior of In has important consequences on the growth morphology and the bulk properties of InGaN alloys. Depending on the In concentration qualitatively different effects arise. Already at rather modest bulk In concentrations of only 1-2% the (0001) surface as shown in Fig. 7b becomes energetically unstable against the formation of vacancy islands[4]. These islands have a diameter of 10–20 Å and their depth is up to 2.5 Å. The walls around the vacancy islands can be considered as $(1\bar{1}01)$ micro-facets (see Fig. 7c) explaining the stability of these structures under In-rich conditions. The formation of these islands locally relieves the large compressive strain induced by the In.

A consequence is that in the close vicinity around the vacancy islands significantly higher In concentrations can be incorporated then on the regular (0001) surface (as shown in Fig. 7b). A second consequence is that the In distribution on the surface becomes inhomogeneous with a length scale given by the distance between the vacancy islands (≈ 50 Å). Fluctuations of this length scale have indeed been observed in experiments [50] and have been shown to dramatically enhance the luminescence efficiency[51]. These results indicate that the properties of InGaN-films (e.g. alloy composition and fluctuations) are surface controlled – surface engineering thus might thus be used to control alloy properties on a nanometer scale.

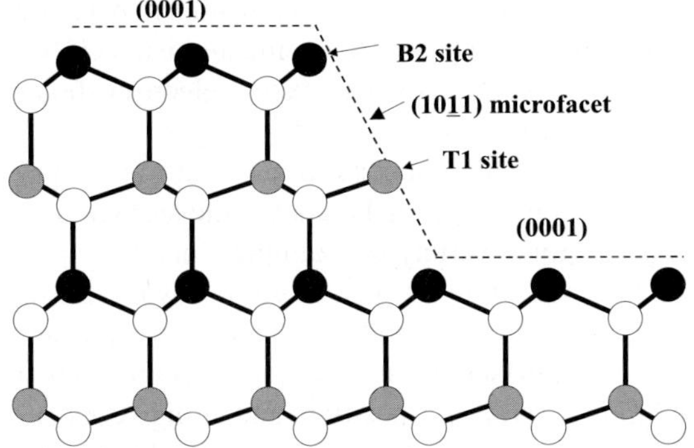

Fig. 8 Schematic representation of a step of height c on a GaN (0001) surface. At such a step the cations become attached at either $B2$ (doubly coordinated) or $T1$ (singly coordinated) sites. The Ga binds preferentially in $B2$ sites while the In binds in $T1$ sites (see text). This leads to the formation of alternately Ga-rich and In-rich (0001) planes as the steps flow across the surface. Black disks mark Ga atoms, grey disks In atoms, and white disks N atoms. After Ref. [29].

6.4   High In concentrations: Defect formation and chemical ordering

The selective surfactant behavior of In on GaN surfaces has also important consequences when growing pseudomorphic InGaN films with higher In concentrations. Under these conditions the In stabilization of the $(1\bar{1}01)$ surface becomes so dominant that defects of a much larger size than the vacancy islands are formed. These defects have the form of an inverted hexagonal pyramid with the six sidewalls formed by $(1\bar{1}01)$ facets and are called inverted hexagonal pyramids, hexagonal pinholes, or as V-defects[6, 5, 24]. The peak of the pyramid ends in a dislocation. The energetic driving force to form such a defect over a dislocation is the reduction in energy achieved by removing strained material in the region near the dislocation core. A detailed energy balance which includes the strain energy of the dislocation and the *ab initio* calculated surface energies showed that pinholes with very large diameters of $\approx 100\,\text{nm}$ may be formed[30].

The existence of low coordinated surface sites on the $(1\bar{1}01)$ orientation (see Fig. 1a) has also another important implication on the properties of InGaN alloys. As has been pointed out in Ref. [29] the existence of two low coordinated and inequivalent sites on the same surface may result in chemical ordering. This is schematically shown in Fig. 8: In incorporation will mainly occur on the lowest (singly) coordinated site while Ga prefers the doubly coordinated site. Consequently, if steps sweep across the surface, Ga incorporation will occur mainly on the Ga planes while In incorporation will occur mainly on the In planes. Chemically ordered InGaN alloys along (0001) have been indeed observed experimentally[7]. It should be noted that the mechanism causing ordering is solely due to kinetics. Calculations performed on ordered structures such as $(0001)\,(1+1)\,\text{In}_{0.5}\text{Ga}_{0.5}$ superlattices showed them to be unstable against the formation of random alloys[52, 29].

7   Conclusion

A key issue when fabricating/growing semiconductor devices is that one has not only to deal with host atoms but also with foreign chemical species. As has been discussed in detail these species may be introduced by the specific growth techniques (e.g. hydrogen in MOCVD growth), when doping the semiconductor (e.g. Si in GaN) or when growing alloys (e.g. In in InGaN alloys). A general feature found for all impurity species studied here was a strong tendency towards surface segregation: In thermodynamic equilibrium the calculated impurity concentration on the surface is typically orders of magnitude larger than in bulk. The origin of the large discrepancy between bulk and surface concentration is the mismatch between impurity and substituted host atoms with respect to atomic radius (causing strain), chemical mismatch (reducing the bond strength), or electronic mismatch (if the impurity is not isovalent excess electrons/holes are introduced which increase the electrostatic energy of the system).

An important consequence of the large ratio between surface and bulk concentration is that to incorporate even modest amounts of foreign species (e.g to dope the material or to make alloys) the surface may be completely covered by it. As has been shown in this paper for a variety of examples this may have dramatic effects on the surface structure, on surface properties and often on the growth behavior. Some impurities show a very complex behavior depending on the specific growth conditions and concentrations. For example, In may act at extreme low bulk concentrations as surfactant, at medium concentrations it causes the formation of nanostructured surface defects and thus alloy decomposition, and at high concentrations it drives the formation of extended defects (with a size of ≈ 100 nm) or it can be used to kinetically stabilize chemically ordered alloys. The last example demonstrates how crucial a deeper understanding of surface related effects is to overcome intrinsic materials properties such as thermodynamic limits of dopant or alloy solubility, surface roughening (e.g. to prevent droplet formation) or to use them as a powerful tool to enable/control self-organized formation of nano-structures. The application of these processes will be a first step in what could be called *surface engineering*, i.e., the active design/processing of surfaces on an atomic level to optimize nitride-materials or to design/realize completely new structures with tailor-made properties. As has been demonstrated here for a variety of examples *ab initio* based theoretical methods/simulations are nowadays a mature and accurate tool to help in this quest.

Acknowledgements   This work has been done in close collaboration with John Northrup and Chris Van de Walle (both Palo Alto Research Center), Huajie Chen and Randy Feenstra (both Carnegie Mellon University), and Liverios Lymperakis, Andreia da Rosa, Tosja Zywietz and Matthias Scheffler (all Fritz-Haber-Institut). Stimulating discussions with M. Albrecht and H. Strunk (Uni-Erlangen), O. Ambacher (TU-Ilmenau), D. Hommel (Uni-Bremen), H. Riechert (Infineon), and M. Stutzmann (TU-München) are greatly acknowledged. The work was supported by the Deutsche Forschungsgemeinschaft (Schwerpunktprojekt "Gruppe-III-Nitride").

References

[1] F. Widmann *et al.*, Appl. Phys. Lett. **73**, 2642 (1998).
[2] C. Adelmann, R. Langer, G. Feuillet, and B. Daudin, Appl. Phys. Lett. **75**, 3518 (1999).
[3] C. Kruse, S. Einfeldt, T. Böttcher, and D. Hommel, Appl. Phys. Lett. **79**, 3425 (2001).
[4] H. Chen *et al.*, Phys. Rev. Lett. **85**, 1902 (2000).
[5] Z. Liliental-Weber, Y. Chen, S. Ruvimov, and J. Washburn, Phys. Rev. Lett. **79**, 2835 (1997).
[6] L. T. Romano *et al.*, Appl. Phys. Lett. **73**, 1757 (1998).
[7] P. Ruterana *et al.*, Appl. Phys. Lett. **72**, 1742 (1998).
[8] E. Iliopoulos *et al.*, Mat. Sci. & Eng. B **87**, 227 (2001).
[9] M. Benamara *et al.*, Appl. Phys. Lett. **82**, 547 (2003).
[10] L. T. Romano *et al.*, J. Appl. Phys. **87**, 7745 (2000).
[11] A. Munkholm *et al.*, Phys. Rev. Lett. **83**, 741 (1999).
[12] J. Northrup, R. D. Felice, and J. Neugebauer, Phys. Rev. B **56**, R4325 (1997).
[13] C. G. Van de Walle and J. Neugebauer, Phys. Rev. Lett. **88**, 066103 (2002).
[14] P. Hohenberg and W. Kohn, Phys. Rev. **136**, B864 (1964).
[15] W. Kohn and L. J. Sham, Phys. Rev. **140**, A1133 (1965).
[16] J. P. Perdew, K. Burke, and Y. Wang, Phys. Rev. B **54**, 16533 (1996).
[17] N. Troullier and J. L. Martins, Phys. Rev. B **43**, 1993 (1991).
[18] M. Bockstedte, A. Kley, J. Neugebauer, and M. Scheffler, Comp. Phys. Comm. **107**, 187 (1997).
[19] G. Feuillet *et al.*, Appl. Phys. Lett. **70**, 1025 (1997).
[20] M. Wassermeier *et al.*, Surf. Sci. **385**, 178 (1997), sTM pictures of cubic GaN(001).
[21] J. Neugebauer *et al.*, Phys. Rev. Lett. **80**, 3097 (1998).
[22] M. Stutzmann *et al.*, Phys. Stat. Sol. B **228**, 505 (2001).
[23] Y. Chen *et al.*, Appl. Phys. Lett. **72**, 710 (1998).
[24] X. Wu *et al.*, Appl. Phys. Lett. **72**, 692 (1998).
[25] A. R. Smith *et al.*, Phys. Rev. Lett. **79**, 3934 (1997).
[26] J. E. Northrup, J. Neugebauer, R. M. Feenstra, and A. R. Smith, Phys. Rev. B **61**, 9932 (2000).
[27] R. M. Feenstra *et al.*, Surf. Rev. & Let. **7**, 601 (2000).
[28] C. Adelmann *et al.*, J. Appl. Phys. **91**, 9638 (2002).
[29] J. E. Northrup, L. Romano, and J. Neugebauer, Appl. Phys. Lett. **74**, 2319 (1999).
[30] J. E. Northrup and J. Neugebauer, Phys. Rev. B **60**, R8473 (1999).
[31] J. E. Northrup and J. Neugebauer, Phys. Rev. B **53**, R10477 (1996).
[32] C. D. Lee *et al.*, Appl. Phys. Lett. **82**, 1793 (2003).
[33] J. Neugebauer, Phys. Stat. Sol. B **227**, 93 (2001).
[34] A. Kley, Ph.D. thesis, TU-Berlin, 1997.
[35] J. Neugebauer and C. G. Van de Walle, Appl. Phys. Lett. **68**, 1829 (1996).
[36] S. M. Myers *et al.*, J. Appl. Phys. **89**, 3195 (2001).
[37] S. Tanaka, M. Takeuchi, and Y. Aoyagi, Jpn. J. Appl. Phys. 2 **39**, L831 (2000).
[38] A. L. Rosa, J. N. J. E. Northrup, C. D. Lee, and R. M. Feenstra, Appl. Phys. Lett. **80**, 2008 (2002).
[39] J. Neugebauer and C. G. Van de Walle, in *Festkörperprobleme / Advances in Solid state Physics*, edited by R. Helbig (Vieweg, Braunschweig/Wiesbaden, 1996), Vol. 35, pp. 25–44.
[40] O. Contreras *et al.*, Appl. Phys. Lett. **81**, 4712 (2002).
[41] T. Bottcher *et al.*, Appl. Phys. Lett. **73**, 3232 (1998).
[42] R. Averbeck and H. Riechert, Phys. Stat. Sol. (a) **176**, 301 (1999).
[43] D. F. Storm and C. A. B. Daudin, Appl. Phys. Lett. **79**, 1614 (2001).
[44] H. J. Chen *et al.*, JVST B **18**, 2284 (2000).
[45] H. J. Chen *et al.*, MRS Internet JNSR **6**, 11 (2001).
[46] J. Neugebauer *et al.*, Phys. Rev. Lett. **90**, 056101 (2003).
[47] T. Zywietz, J. Neugebauer, and M. Scheffler, Appl. Phys. Lett. **73**, 487 (1998).
[48] E. J. Tarsa *et al.*, J. Appl. Phys. **82**, 5472 (1997).
[49] B. Heying *et al.*, J. Appl. Phys. **88**, 1855 (2000).
[50] Y. Narukawa and et al., Appl. Phys. Lett. **70**, 981 (1997).
[51] S. Nakamura, T. Mukai, and M. Senoh, Appl. Phys. Lett. **72**, 2014 (1998).
[52] F. Grosse and J. Neugebauer, Phys. Rev. B **63**, 085207 (2001).

phys. stat. sol (c) **0**, No. 6 1668–1683 (2003) / **DOI** 10.1002/pssc.200303129

Indium distribution in epitaxially grown InGaN layers analyzed by transmission electron microscopy

D. Gerthsen[*], **E. Hahn**, **B. Neubauer**, **V. Potin**, **A. Rosenauer**, and **M. Schowalter**

Laboratorium für Elektronenmikroskopie, Universität Karlsruhe (TH), 76128 Karlsruhe, Germany

Received 4 March 2003, accepted 23 April 2003
Published online 28 August 2003

PACS 64.75 +g, 68.37.Lp, 68.55.Nq, 81.05 Ea, 81.15Gh, 81.15.Hi

An overview is given about microstructure and composition analyses of InGaN quantum wells embedded in Ga(Al)N barriers to study the mechanisms which determine the In distribution in epitaxially grown InGaN layers. The applied technique is transmission electron microscopy (TEM). The main prerequisite for this work was the development of a technique based on high-resolution lattice fringe images that allows quantitative chemical analyses of InGaN on an atomic scale. A large variety of samples was investigated that were produced by molecular beam epitaxy (MBE) and metal-organic vapor phase epitaxy (MOVPE). The effect of the deposition temperature, growth rate, strain and high-temperature annealing treatments on the average In concentration and In distribution was studied to assess the influence of phase separation, In surface segregation and In desorption. Composition fluctuations in InGaN are always observed on two different lateral scales independent of the growth technique and particular set of growth parameters but the strength of the composition fluctuations can be influenced by the details of the growth.

1 Introduction

The investigation of the structural properties and composition – in particular composition fluctuations - of InGaN contained in group-III nitride heterostructures has been the goal of a very large number of studies because the defects and In distribution strongly affect the performance of light-emitting and electronic devices. For example, Chichibu et al. [1] demonstrated that localized excitons play an important role in the lasing process in InGaN-based quantum well (QW) structures. The inhomogeneous In distribution is attributed primarily to the large miscibility gap at typical growth temperatures between 600 and 800 °C that is deduced from calculated phase diagrams by Ho and Stringfellow [2] or Ito [3]. However, the situation for epitaxially grown strained-layer structures may be considerably more complex. Deviations from the compositions suggested by the phase diagram are expected due to kinetic limitations and the misfit strain in InGaN QWs. Surface segregation of indium was shown to occur during molecular beam epitaxy (MBE) growth by Chen et al. [4]. A redistribution of the group-III element occurs at the surface, if a transition from two-dimensional to three-dimensional growth (Stranski–Krastanow (SK) growth mode) occurs as observed under selected growth conditions for MBE by Damilano et al. [5] and metal-organic vapor phase epitaxy (MOVPE) by Tachibana et al. [6].

Transmission electron microscopy (TEM) is particularly adequate to study microstructure and composition on a sub µm and nm scale. Qualitative indications of an inhomogeneous In distribution in InGaN are already visible in conventional TEM images [1]. Z-contrast imaging in a scanning transmission electron microscope performed by Lakner et al. [7] yielded further evidence for an inhomogeneous In distribution and asymmetrical composition profiles along the growth direction. Energy dispersive X-ray analyses by Narukawa et al. [8] showed distinct composition variations in an InGaN QW within a region

---

[*] Corresponding author: e-mail: gerthsen@lem.uni-karlsruhe.de, Phone: +49 721 608 3200, Fax: +49 721 608 3721

of only a few nanometers. Kisielowski et al. [9] presented first composition analyses based on high-resolution (HR) TEM images, which also demonstrated composition fluctuations on a scale of only a few nanometers.

Based on these early observations it appeared to be attractive to develop a technique, which allows quantitative composition analyses on an atomic, i.e. sub-nanometer, scale. This implies that it must rely on HRTEM images. The method should be easy to use and insensitive towards changes of the imaging conditions in order to be able to evaluate a reasonable number of images to obtain statistically relevant data. This review article contains the present state of our composition analysis technique, which will be outlined in Section 2. The technique has been improved over the years and has reached a state where quantitative analyses with a high accuracy are possible. Readers not particularly interested in the details of the analysis method may skip a large part of this section, which was included for TEM specialists. In Section 3, typical examples for In distributions in InGaN will be presented. Although a large number of MBE and MOCVD samples was analyzed within a wide range of growth parameters, composition fluctuations are always present in InGaN. In particular, In-rich agglomerates with sizes of only a few nm are a characteristic feature, which are often suggested to act optically as quantum dots. The mechanisms which are expected to have an effect on the In distribution are discussed in Section 4.

2 Quantitative analysis of the composition of InGaN on an atomic scale by high-resolution transmission electron microscopy

Composition fluctuations in InGaN have been recognized already a number of years ago to determine significantly the properties of devices fabricated on the basis of InGaN/GaN heterostructures. It is therefore of considerable interest to study quantitatively the In distribution of InGaN and the spatial dimension of compositional fluctuations. With typical QW thicknesses in the order of a few nanometers, an analysis technique is required that provides on the one hand high spatial resolution and reasonable accuracy with respect to the obtained concentration values. On the other hand, it must be possible to obtain statistically relevant data, which is particularly important for a material with a pronounced compositional inhomogeneity. For this reason, we developed a technique based on high-resolution (HR) TEM lattice fringe images, which fulfill the above requirements. The first development stage of the method applied to InGaN was published by Gerthsen et al. [10]. It is implemented in the software package DALI developed by Rosenauer et al. [11] that is also used for the analysis of numerous zincblende-type semiconductors [12, 22]. The technique is now in an improved state with respect to its accuracy after the optimization of the diffraction conditions, which is presented in the following.

The composition evaluation is based on a rather simple principle by assuming that the lattice parameter of a ternary compound, e.g. InGaN with a lattice parameter d_{InGaN}, is linearly dependent on the In concentration x_{In}.

$$d_{\mathrm{InGaN}} = d_{\mathrm{GaN}} + x_{\mathrm{In}} \left(d_{\mathrm{InN}} - d_{\mathrm{GaN}} \right) \tag{1}$$

InGaN is a system that is well suited for composition analyses on the basis of Eq. (1) because the lattice parameter mismatch between InN and GaN is 10 % (c lattice parameters: $c_{\mathrm{InN}} = 0.5760$ nm, $c_{\mathrm{GaN}} = 0.5185$ nm). Calculations by Mattila and Zunger [13] for zincblende InGaN showed that only a small downward bowing of the lattice parameter from the linear interpolation between the binary endpoints could exist. Experimental evidence for a large deviation of the lattice parameter behavior from Eq. (1) has not been presented so far. In materials with wurtzite structure, the distances between the (0002) basal planes can be measured if the growth proceeds along the [0001] direction. Composition analyses based on the measurement of local lattice parameters (LLPs) were first presented by Bierwolf et al. [14] who determined LLPs between local intensity maximum positions in zone-axis HRTEM images. Assuming that the distance between the intensity maximum positions are a fingerprint of the LLP, the composition can be determined on an atomic scale in the image plane. The local compositions are measured with respect to a reference region with known composition contained in the same image, which can be e.g. a buffer layer consisting of a binary compound. Despite the high lateral resolution, it has to be kept in mind, that the composition is averaged through the sample thickness, typically between a few and 30 nm for HRTEM samples.

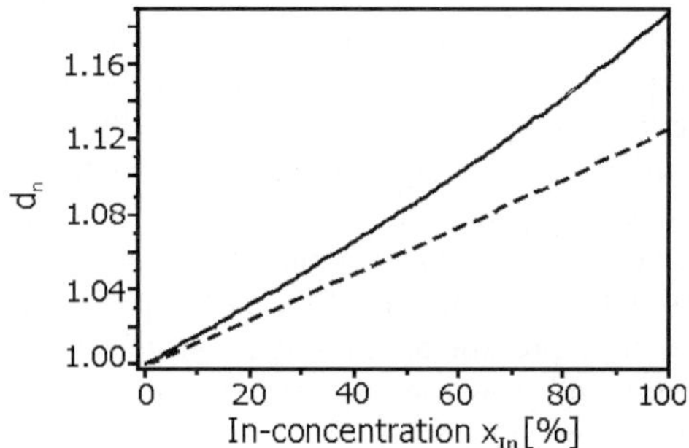

Fig. 1 Influence of the relaxation state on the normalized lattice parameters d_n plotted as a function of the In concentration. Biaxial strain state for a "thick" sample (solid line) and monoaxial strain state for a "thin" sample (dashed line).

In strained-layer InGaN/GaN heterostructures the tetragonal distortion of the InGaN lattice has to be taken into account, if In concentrations are determined from the measured LLPs. Two limiting cases can be distinguished. A biaxial strain state exits for "thick" bulk-like TEM sample thicknesses in contrast to a complete relaxation of the strain along the electron-beam direction in very "thin" samples which results in a monoaxial strain state. The strain induced by the lattice parameter mismatch can be calculated analytically using elasticity theory. Figure 1 shows the influence of the relaxation state on the normalized LLP d_n, i.e. the ratio between the InGaN and the GaN c-lattice parameters, plotted as a function of the In concentration. Since the local sample thickness cannot be easily determined at a high precision, an average In concentration from the "thick" and "thin" sample approximation is always given in our data.

Another inherent error source, that has to be addressed, is the (in)accuracy of the elastic constants that are required to calculate the tetragonal strain. This – by the way – also applies to all X-ray diffraction data, which relies on the precise knowledge of the elastic constants. There are large discrepancies between different theoretical and experimental data sets. We use calculated values for GaN and InN obtained by Wright [15], which agree well with experimental data for GaN by Polian et al. [16] and Takagi et al. [17]. The elastic constants for InGaN are linearly interpolated between the binary endpoints.

It is well known that the image pattern of zone-axis HRTEM images is sensitive towards changes of the local sample thickness t and objective lens defocus Δf. It has been also shown that HRTEM images of materials with wurtzite structure are particularly affected by a slight local misorientation of the sample with respect to the exact the zone-axis orientation [18]. Changes of the image pattern due to the factors mentioned above will induce artifacts if the LLPs are evaluated from an image of a sample region where t, Δf or the orientation of the sample changes. Only a relatively small change of imaging conditions is acceptable for the reliable evaluation zone-axis HRTEM images, which can be determined by HRTEM image simulations as demonstrated e.g. by Ruterana et al. [19]. The number of images that fulfill the required conditions is usually further reduced by local sample misorientations which cannot be controlled on a scale of a few 10 nm. Therefore, it appeared reasonable to look for different imaging conditions that minimize artifacts due to the influence of t and/or Δf changes in order to be able to obtain a statistically relevant number of images that can be evaluated for each sample.

Some general considerations can be used as a guide to optimize the imaging conditions. To verify the validity of the considerations, image simulations will be presented in the following. The simulations were carried out assuming a 100 GaN/5 nm $In_{0.2}Ga_{0.8}N$/100 nm GaN QW structure. An abrupt compositional transition was assumed between the GaN and the $In_{0.2}Ga_{0.8}N$ QW. Finite element model (FEM) simulations were carried out as a function of the sample thickness to obtain a realistic strain distribution due to the lattice parameter mismatch and the elastic relaxation of the thin HRTEM sample along the electron-beam direction. An atomic model was generated on the basis of the FEM calculations, which was subdivided into slices with a thickness corresponding to a (0002)-plane distance. The exit wave function was calculated using the multislice formalism [20]. Finally, the image was simulated using the parameters of the Philips CM200 FEG/ST transmission electron microscope which is characterized by a spherical aberration coefficient $C_s = 1.2$ mm and a defocus spread of 7 nm. The images were evaluated with exactly the same routine as the experimental images. Further details of the simulation procedure will be presented elsewhere [21].

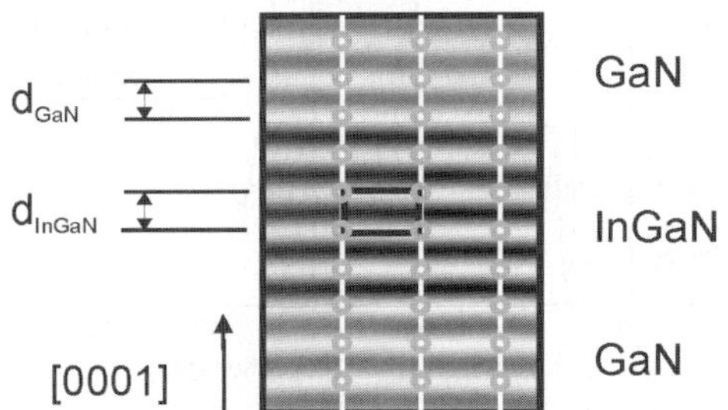

Fig. 2 (0002) lattice fringe image of a GaN/InGaN/GaN QW structure. Line scans are used to determine the intensity maximum positions of the fringes. The size of an image unit cell is indicated by the black rectangle.

Instead of using a zone-axis HRTEM image, where typically at least nine beams are used to form an image along the $[11\bar{2}0]$ or $[1\bar{1}00]$-zone axes, two-beam conditions simplify considerably the image formation process. The interference of only two beams leads to a fringe image. The apparent loss of resolution along the (0002) fringes compared to a conventional zone-axis image does not present a problem because only the distances between the (0002) fringes need to be measured. This can be achieved by carrying out closely spaced line scans. An example for a lattice fringe image is displayed in Fig. 2 where a thin InGaN QW is embedded in GaN. The distances between the (0002) planes d_{0002} are measured by detecting the intensity maximum positions of the bright fringes by line scans along the [0001] direction (white lines in Fig. 2). Technical details on the noise reduction procedure and the determination of the intensity maximum positions are outlined in [12, 22 AR Habilschrift]. The normalized LLPs d_n given by d_{InGaN}/d_{GaN} (d_{InGaN}: local (0002) fringe distance in InGaN, d_{GaN}: fringe distance in GaN typically averaged over a larger reference region) can be evaluated for image unit cells with the size of approximately $d_{0002} \times d_{0002}$. An example for an image unit cell is indicated by the black rectangle in Fig. 2.

Figure 3 shows the normalized LLPs as a function of the distance in growth direction that were obtained from simulated images on the basis of our model structure. The simulated images were evaluated in exactly the same way as the experimental images. Figure 3a was obtained for a $[11\bar{2}0]$-zone axis assuming 9 beams to contribute to the image. The sample thickness was kept constant at $t = 15$ nm while Δf was varied between + 50 and –200 nm ("+" denotes overfocus, "–" underfocus values). The real position of the QW is indicated by the two lines at 0 and 5 nm. The input profile is reasonably reproduced only at $\Delta f = -200$ nm. Large deviations occur at most other Δf values. It has to be noted that the oscillations at distances around –7.5 nm result from the nonperiodicity of the structure at the supercell boundary, which induces simulation artifacts. These simulations show that great attention has to be paid to the optimum choice of Δf, which depends in addition on the sample thickness. Fig. 3b shows results for a two-beam excitation condition with strongly excited (0000) and (0002) beams. Choosing the same sample thickness and Δf values, the profiles can be reasonably evaluated for the two-beam condition over a wide range of Δf values. However, another artifact of the image formation becomes obvious. The evaluated profile is shifted to the right-hand side as Δf is changed from –200 to +50 nm. The "delocalization" of the image is induced by the spherical aberration of the objective lens which imposes a phase shift on the electron wave function that depends on Δf and the transmitted spatial frequency. If the image is dominated by only one spatial frequency – as for two-beam conditions – the profile is only shifted without severe modifications of the profile shape. The delocalization can be minimized for one particular spatial frequency, e.g. the (0002) fringe distance, by choosing an appropriate Δf value, the Lichte defocus [23]. Delocalization contributes significantly to the strong deviations of the evaluated zone-axis images from the input profile because the fringes of each contributing spatial frequency experience a different amount of phase shift.

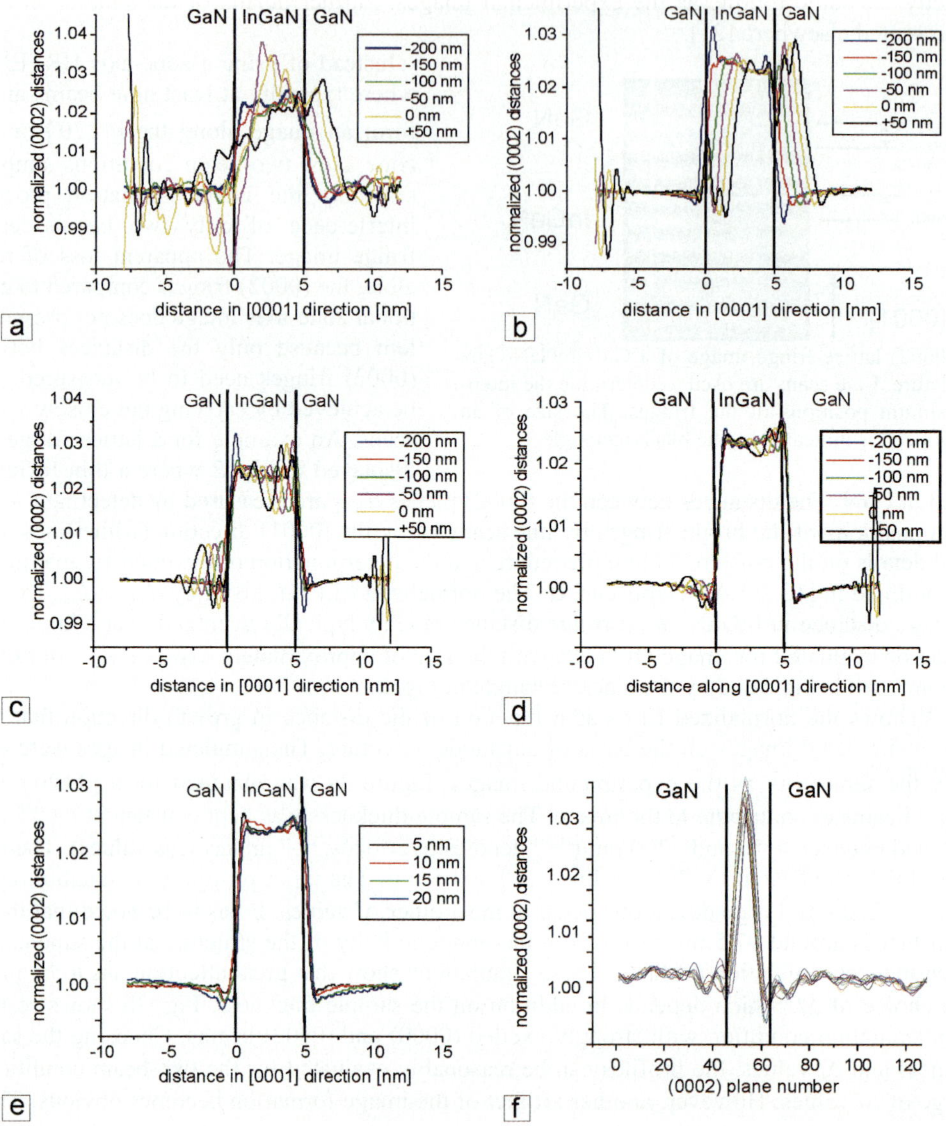

Fig. 3 Evaluation of the normalized lattice parameter d_n on the basis of simulated (a) – (e) and experimental (f) lattice fringe images of a GaN/InGaN/GaN QW structure plotted as a function of the distance in growth direction using different excitation and imaging conditions (see text for further details).

It is easy to suppress the shift of the profiles if the (0002) beam is centered on the optic axis where the phase shift by the objective lens is always zero independent of Δf. The results of the evaluation of the simulated images is shown in Fig. 3c where again two-beam conditions were chosen but with the (0002) beam aligned on the optic axis. The profiles are clearly not shifted anymore but show some oscillations, Fresnel fringes, which are induced by the abrupt step of the mean inner potential at the interface between the GaN and the InGaN. These fringes would be less pronounced if a gradual chemical transition is present at the interface.

A further optimization of the imaging conditions can be achieved if (0004) two-beam conditions are chosen, the (0002) beam is again centered on the optic axis and the image is formed by only selecting the (0000) and (0002) beams for the image formation. The difference compared to the conditions used in Fig. 3c consists in the fact that an excitation error exists for the (0002) beam. The result is plotted in Fig. 3d, which demonstrates that delocalization is again negligible. In addition, the oscillations of d_n are strongly reduced. Values for d_n below 1 in the vicinity of the interfaces are not an artifact of the simulation. They result from the compression of the GaN lattice by the InGaN QW. A reliable measurement of the input profile is achieved under these conditions in a wide range of Δf values. The thickness dependence was tested by choosing specific defocus values, here −150 nm, and varying the sample thickness between 5 and 20 nm. The profiles shown in Fig. 3e display the negligible influence of the sample thickness variation.

The optimum imaging condition was chosen to take a series of six experimental images at different Δf values in the same sample region. The defocus interval between the images was approximately 50 nm. The evaluation of the experimental images shown in Fig. 3f yields almost the same profile for all images.

3 Dependence of composition fluctuations on growth technique and growth parameters

The following section focuses on the In distribution of samples grown by MOVPE and MBE under various growth conditions. The goal of the comparison of a large number of samples was the assessment of the origin of composition fluctuations and in particular to explore possibilities how these fluctuations can be controlled or – at least – influenced by the growth process. This section will be subdivided into subsections presenting results for MOVPE as well as MBE-grown samples in Section 3.1 and Section 3.2. The analysis of the redistribution of the group-III elements in a sample that was subjected to an annealing treatment will be shown in Section 3.3. In Section 3.4, the In distribution in a thick InGaN layer will be presented. Since several differently grown samples are discussed, the growth conditions will be always given in context with the results and not in a separate subsection on the growth procedures. Finally conclusions will be drawn regarding the effects that contribute to the In distribution.

3.1 MOVPE-grown samples

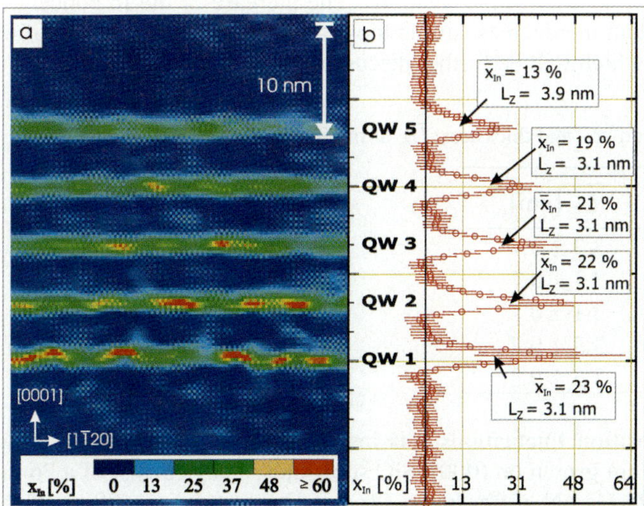

Fig. 4 shows the results of the analysis of an InGaN/GaN QW structure, which contains 5 In-GaN layers separated by 5 nm GaN spacers. The structure was grown on a SiC(0001) substrate on Si-doped 380 nm AlGaN and 75 nm GaN buffer layers. It was capped by 75 nm GaN and 240 nm AlGaN doped with Mg. The growth temperature of the In-GaN/GaN QW structure was 700 °C.

In the color-coded map Fig. 4a, we observe a strongly inhomogeneous In distribution in all QWs. Particularly striking are the small In-rich clusters with lateral extensions below 4 nm, which are present at an ex-

Fig. 4 a) Color-coded map of the In distribution of an InGaN/GaN multiple QW structure and b) averaged In concentration along the [11$\bar{2}$0] direction plotted as a function of the (0002)-plane. number.

tremely high density. It is only an effect of the scale of the color coding that the composition fluctuations in the upper QWs appear to be less pronounced than in the lower ones because the average In concentrations decrease in the upper QWs. It has to be noted that the evaluated composition of the clusters is only an apparent In concentration. The real In concentration inside the clusters will be higher because they are embedded in an InGaN QW with a lower In concentration. The measured In concentrations are an averaged value determined by the composition of the cluster and the embedding matrix along the electron-beam direction for HRTEM sample thicknesses which typically range between 5 and 20 nm.

The In concentrations averaged over the width of the image along the $[11\bar{2}0]$ direction are plotted in Fig. 4b as a function of the (0002) plane number in growth direction. The presence of strong composition fluctuations in the QWs can be clearly distinguished from apparent composition fluctuations in the GaN induced by image noise because the errors bars, representing the standard deviation of the compositions obtained from the individual image unit cells along one particular (0002) plane, are larger in the QWs than in the GaN. In this particular region of the sample, the average In concentration $\bar{x}_{In}$ decreases from the bottom QW (23 %) to the top QW (13 %) although the growth conditions were not changed. The QW thickness L_z was almost constant. The determination of L_z is by no means straightforward for a QW with gradual chemical transitions at the interfaces. It was measured here by the determination of the distance between the zero transitions of the composition profile. Alternatively, the full width at half maximum (FWHM) of the profile could have been chosen.

More than 50 images were evaluated for this sample. An overall average In concentration $\bar{x}$ was determined by averaging the mean values $\bar{x}_{In}$ of the individual images. The results for $\bar{x}$ and the averaged thickness values $\bar{L}_z$ are given in Table 1 for each of the five QWs. The given errors $\Delta\bar{x}$ and $\Delta\bar{L}_z$ correspond to the standard deviations of the locally average values. They can be interpreted in terms of composition fluctuations on a larger scale, which is determined by the width of the individual images, which varies between 20 and 50 nm. The considerable variation of the locally averaged In concentrations with respect to the overall average concentration is visualized in Fig. 4b where $\bar{x}_{In}$ is between 13 and 23 % which deviates significantly from the overall average values between 9 and 13 %. We do not want to speculate here on the decrease of $\bar{x}$ despite unaltered growth conditions. It indicates, however, that the precise control of the composition is extraordinary difficult in this system.

None of the composition profiles shows abrupt chemical transitions. The increase of the In concentration at the lower interfaces is often steeper than the decrease at the top interfaces. Asymmetrical profiles are a characteristic feature of In segregation which will be further discussed in Section 4.

Table 1 Overall average In concentration $\bar{x}$ and thickness $\bar{L}_z$ of the five quantum wells.

	$\bar{x}$ [%]	$\bar{L}_z$ [nm]
QW 1	13 ± 6	3.3 ± 0.7
QW 2	12 ± 6	3.1 ± 0.7
QW 3	11 ± 5	3.1 ± 0.7
QW 4	10 ± 4	3.3 ± 0.7
QW 5	9 ± 4	3.3 ± 0.8

With respect to the strength of the composition fluctuations it is instructive to study the effect of an InGaN growth rate variation. The samples were grown on (0001)SiC substrates. They consist of a 76 nm GaN nucleation layer, a 600 nm Si-doped $Al_{0.1}Ga_{0.9}N$ buffer layer. The cap layer containing 78 nm GaN, 240 nm $Al_{0.1}Ga_{0.9}N$ and 38 nm GaN was undoped. The InGaN QW between the buffer and cap layer were grown with growth rates of 2.5 and for the second sample 7.5 nm/min. Table 2 gives the overall average In concentration $\bar{x}$, thickness $\bar{L}_z$ and in addition the averaged In concentration in the clusters $\bar{x}_{\text{In,cluster}}$.

Table 2 Dependence of the overall average In concentration $\bar{x}$, QW thickness $\bar{L}_z$ and averaged In concentration in the clusters $\bar{x}_{\text{In,cluster}}$ on the growth rate

	$\bar{x}$ [%]	$\bar{L}_z$ [nm]	$\bar{x}_{\text{In,cluster}}$ [%]
2.5 nm/min	16 ± 6	4.2 ± 1.0	46 ± 24
7.5 nm/min	12 ± 2	3.9 ± 0.3	27 ± 8

The decrease of the standard deviations for $\bar{x}$ and $\bar{x}_{\text{In,cluster}}$ is a clear indication for a more homogeneous In distribution in the sample grown at the higher rate.

Another parameter of interest is the growth temperature, which is expected to influence the In distribution and average In concentration. For this purpose, two series of GaN/InGaN/GaN-heterostructures were investigated which were grown on $Al_2O_3(0001)$ substrates by MOVPE. In both series only the growth temperature T_g of the InGaN QW was varied between 860 and 800 °C (series 1) and between 770 and 690 °C (series 2) leaving all other parameters unchanged. In series 1, the nominal QW thickness was 3.5 nm. The QW is embedded in two 7 nm thick InGaN layers containing 2 % indium, a Si-doped GaN buffer layer with a thickness of 2 μm and an undoped 10 nm GaN cap layer. In the second series, the 5 nm thick InGaN QW was grown on a 1.5 μm thick Si-doped GaN buffer layer with an undoped 75 nm thick GaN layer on top.

Table 3 Dependence of the overall average In concentration $\bar{x}$, QW thickness $\bar{L}_z$ and averaged In concentration in the clusters $\bar{x}_{\text{In,cluster}}$ on the growth temperature for series 1 and series 2.

Series 1	860 °C	840 °C	820 °C	800 °C
$\bar{x}$ [%]	9 ± 4	12 ± 8	17 ± 6	17 ± 10
$\bar{L}_z$ [nm]	2.8 ± 0.6	4.1 ± 1.4	3.1 ± 0.3	3.8 ± 0.5
$\bar{x}_{\text{In,cluster}}$ [%]	28 ± 15	48 ± 32	47 ± 30	56 ± 35
Series 2	770 °C	710 °C	690 °C	
$\bar{x}$ [%]	8 ± 3	8.5 ± 3	8 ± 3	
$\bar{L}_z$ [nm]	7.0 ± 2.0	7.8 ± 1.4	11.0 ± 2.0	
$\bar{x}_{\text{In,cluster}}$ [%]	65 ± 18	-	59 ± 16	

The change of the T_g between 800 and 860 °C in the first series leads to a strong reduction of $\bar{x}$ from 17 to 9 % while a constant $\bar{x}$ around 8 % is found for the second series in the range of lower temperatures. The measured QW thickness in the first series is close to the nominal value in contrast to the second series where the QW thickness increases with decreasing temperature. In-rich clusters are observed in all samples. The In concentration of the clusters $\bar{x}_{\text{In,cluster}}$ is correlated with $\bar{x}$ but the standard deviation is rather large because only approximately ten images were analyzed. The sizes of the clusters are below 5 nm and do not depend on T_g.

3.2 MBE-grown samples

Since fundamental differences between the growth processes in MOVPE and MBE exist, it was of considerable interest to study the In distribution of MBE-grown samples. The presented results were obtained from a representative sample that was grown on an Al_2O_3(0001) substrate. After the deposition of a low-temperature GaN nucleation layer (25 nm, 500 °C) and a thick GaN buffer layer (800 °C, 4 µm), the InGaN QW was grown at 530 °C. Finally, the structure was capped at 800 °C with 30 nm GaN. The InGaN was grown under N-rich conditions and the temperature was ramped up to 800 °C after the In-GaN deposition during a growth interruption.

Figure 5 depicts the In distribution of this sample which shows composition fluctuations as the MOVPE samples. The sizes of the In-rich clusters (red regions) are extremely small between 1 and 3 nm.

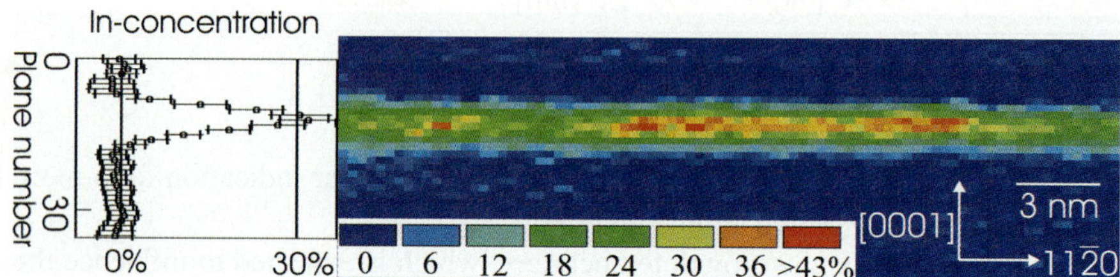

Fig. 5 Color-coded map of the local In concentration of a MBE-grown InGaN/GaN/GaN QW structure (right-hand side) and averaged In concentration along the [11$\bar{2}$0] direction plotted as a function of the (0002)-plane number (left-hand side).

The maximum concentrations in the clusters reach (42 ± 6) %. These values correspond approximately to twice the average In concentration of (21 ± 3) %.

The local In concentrations averaged along the [11$\bar{2}$0] direction are plotted as a function of the distance along the growth direction given (0002)-plane numbers (left-hand side of Fig. 5). The profile along the [0001] growth direction shows a different type of asymmetry compared to the MOVPE samples, because the In concentration increases gradually at the lower interface and decreases more abruptly at the upper interface. This is also visualized by the positions of the In-rich clusters, which tend to be located in the middle of the QW in the MBE-grown sample. Analyses have been carried out for several ten areas, which yields $\bar{x} = (17.4 \pm 5.1)$ %, $\bar{L}_z = (3.1 \pm 0.7)$ nm for the average QW thickness and (1.7 ± 0.4) ML for the In content (expressed in units of InN monolayers).

3.3 Modification of the In distribution by an annealing treatment

Since the change of growth conditions did not significantly affect the size of the In-rich clusters, a sample was subjected to a postgrowth annealing treatment to find out whether an increase of the size of the In-rich clusters could be achieved. The investigated heterostructure was grown by MOVPE on a SiC(0001) substrate. It consists of a 600 nm $Al_{0,09}Ga_{0,91}N$ buffer layer deposited at a substrate temperature of 980 °C. The InGaN QW grown at 670 °C with a nominal thickness of 1.5 nm was embedded in two ~77 nm thick GaN cladding layers. The structure was capped at 980 °C with 240 nm $Al_{0,09}Ga_{0,91}N$ and 38 nm GaN. The buffer and the lower cladding layers are Si doped. A high-temperature furnace was used for the annealing treatment for one hour at 980 °C under a nitrogen atmosphere.

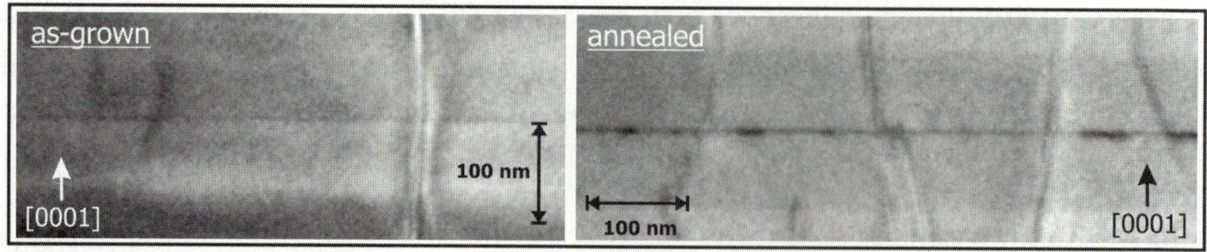

Fig. 6 Overview cross-section TEM images of the InGaN QW region before (left-hand side) and after the annealing treatment (right-hand side).

Fig. 6 shows overview cross-section TEM images of the sample before and after the annealing treatment. At this low magnification, the as-grown InGaN QW (Fig. 6 left-hand side) is characterized by a homogeneous width and morphology. After the annealing treatment (Fig. 6 right-hand side), width fluctuation can be recognized along the QW, and the varying image intensity is indicative of composition and strain fluctuations.

The DALI evaluations show more details about the In distribution, which are displayed in Fig. 7 as grey-scale coded maps. On this high-resolution scale, the small In-rich clusters are found as usual for the as-grown sample (Fig. 7a), which are not visible in the overview image Fig. 6 (left-hand side). The extension and composition of the In-rich agglomerates are more clearly displayed in Fig. 7c below, where the maximum In concentrations in the QW are plotted as a function of the distance along the $[11\bar{2}0]$ direction. The size of the In-rich agglomerates is – as usual – in the order of a few nm and the In concentration in the clusters reaches almost 60 %. The composition profile along the [0001] direction of the as-grown QW is plotted in Fig. 7e. Gradual transitions of the In concentration at the interfaces towards the GaN cladding layers can be recognized. The measured total QW thickness (4.2 ± 1.0) nm indicates in this case a significant broadening compared to the nominal QW thickness of 1.5 nm marked by the two lines in Fig. 7e. The overall average In concentration was determined to be (16 ± 6) %.

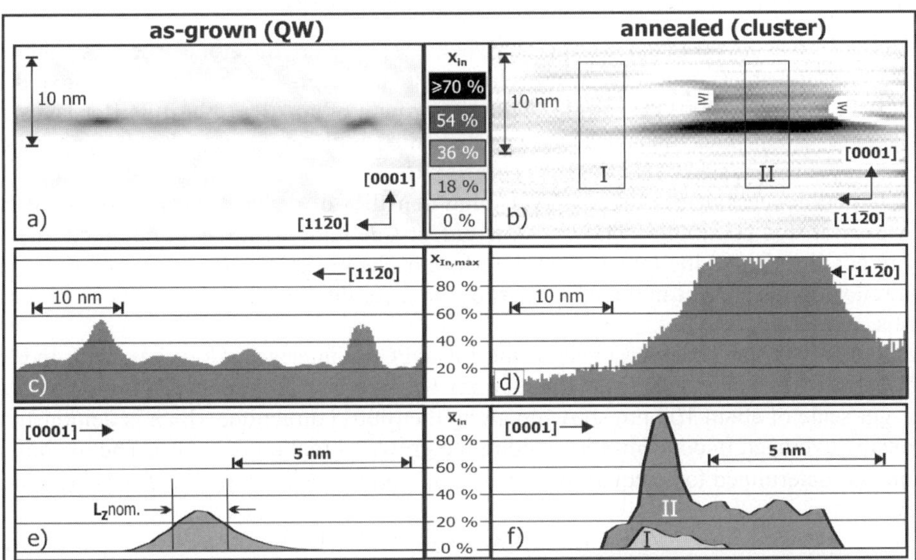

Fig. 7 Examples for the In-distribution in the InGaN QW before (as-grown) and after the annealing treatment with a), b) the grey-scale coded local In-concentrations, c), d) the maximum In-concentration as a function of the distance along the $[11\bar{2}0]$ direction and e), f) the In-concentration profiles in the [0001]-growth direction averaged along the $[11\bar{2}0]$ direction.

The In distribution changes significantly after the annealing treatment and can be classified into two different regions.

Region 1: The annealed QW between the dark regions in Fig. 6 (right-hand side) was only slightly modified by the annealing treatment. Compared to the as-grown QW, $\bar{x}$ decreases to (15 ± 4) % whereas the average QW thickness increases to (4.7 ± 2) nm. The In-rich clusters (occurrence, size, and maximum In concentrations) do not appear to be strongly affected by the annealing treatment.

Region 2: The quantitative analyses of the dark regions in the overview image Fig. 6 (right-hand side) with lateral sizes between 10 and 100 nm and a density in the order of 10^{11} cm^{-2} yield very high In concentrations as shown in Fig. 7b,d,f. The In distribution along the [0001] direction in the In-rich cluster (area II in Fig. 7b) and in the vicinity of the cluster (area I in Fig. 7b) was analyzed in more detail and plotted in Fig. 7f. Whereas a strong In depletion is observed in area I next to the cluster with an average In concentration $\overline{x}_{In} = 9\,\%$, the In concentration inside the cluster (area II) increases up to 100 %. Remarkable is the broadened plateau of the In-concentration profile in area II with $\overline{x}_{In} \sim 30\,\%$ in the upper region of the cluster. The HRTEM images of these clusters exhibit a high density of stacking faults, especially in large ones. Therefore, only few quantitative composition analyses were possible, because the analysis technique is based on the measurement of the LLPs, which are distorted near the stacking faults. However, the analyses of smaller clusters (as presented in Fig. 7b), where stacking faults only occur at the edges of the cluster, point out that the dark regions in Fig. 6 (right-hand side) contain indeed very high In concentrations.

The results demonstrate a significant redistribution of the group-III elements in the InGaN QW, which leads the formation of In-rich clusters with sizes between 10 and 100 nm. Therefore, annealing treatments are a possibility to adjust the size of the clusters if suitable parameters are chosen. More details about the results of this study and the correlation of the with the photoluminescence are presented in [24].

3.4 In distribution in a thick InGaN layer grown by MOVPE

A thick InGaN layer was investigated, to find out whether significant differences between the In distributions in thin QWs and thick, partially strain-relaxed InGaN layers occur. The investigated layer was grown by MOVPE on a SiC(0001) substrate. After the growth of an AlN (20nm) buffer layer and a GaN (780 nm) layer, a 100nm thick InGaN layer was deposited with an average In concentration of 16 % according to X-ray diffractometry (XRD).

Due to the large thickness of the InGaN layer, high densities of dislocations and stacking faults were present. However, it was possible to analyse some defect-free regions. It has to be noted that misfit dislocations were not observed at the InGaN/GaN interface of the investigated regions but it can be assumed that a partial relaxation is present in the upper part of the layer due to the high density of dislocations and stacking faults.

In-rich agglomerates were observed close to the InGaN/GaN interface. However, with increasing distances from the interface, the small-scale fluctuations become less pronounced. In-composition fluctuations on a larger scale of about 100 nm show up along the [0001] direction, which is demonstrated by the plot of the locally averaged In concentrations along the growth direction (Fig. 8b). The overall average In concentration was determined to be equal to $16 \pm 6\,\%$, in good agreement with the XRD measurements.

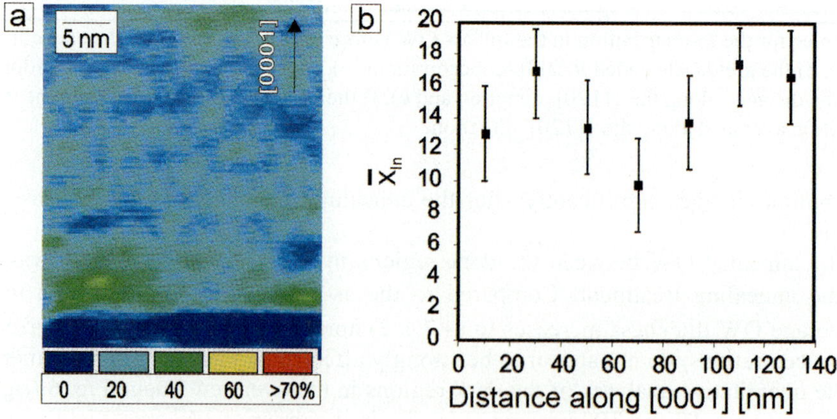

Fig. 8 a) Color-coded map of the local In concentration of a thick InGaN layer on a SiC(0001) substrate and b) locally averaged In concentrations along the [0001] direction showing pronounced large-scale composition fluctuations.

4 Discussion

The experimental results can be summarized by the following points.

1. Pronounced composition fluctuations are always present in InGaN independent of the growth technique (MBE or MOVPE) and particular set of growth parameters. Composition fluctuations are even observed at small average In concentrations below 5 % as shown by Schowalter et al. [25].

2. Composition fluctuations occur on two typical scales:
   - *small In-rich clusters with lateral extensions of a few nm and high In concentrations*
   - *weaker composition fluctuations on a scale of a some 10 to 100 nm, which are "superimposed" on the In-rich clusters*

3. The size of the In-rich clusters could not be significantly modified by the variation of the growth parameters. However, the size of the clusters increases by an appropriate postgrowth high-temperature annealing treatment (Section 3.3).

4. The strength of the composition fluctuations can be influenced by the InGaN growth rate (Section 3.1) indicating that kinetics plays a role in their formation.

5. Large discrepancies between the nominal average In concentration and QW thickness are frequently observed. This was in particular demonstrated by a set of samples grown under identical growth conditions where only the growth duration was varied [26]. In these samples, the average In concentration changed between 6 and 15 % as the growth duration was increased from 1 to 10 min.

In the following, we will discuss the effects, which have been considered to act during epitaxial growth.

It is on the first sight tempting to invoke the SK growth mode to explain the origin of the In-rich clusters. It is, however, quite obvious, that an SK transition cannot be responsible for the In-rich clusters for the following reasons. The clusters are already observed at low average In concentrations where the strain in the epilayer is small due to lattice-parameter misfits below 1 % [26]. The clusters are located inside the QW rather than sitting on top of a wetting layer, which would be expected if three-dimensional islands were generated on a two-dimensional wetting layer. There are only few reports in the literature where convincing evidence for the formation of islands – significantly larger in size than the clusters - by SK growth was presented. Damilano et al. [5] managed to grow small – approximately 10 nm – islands by MBE. They demonstrated that the critical thickness for the transition from the two to the three-dimensional growth mode is determined by the NH_3 flux. The formation of islands with a similar size was reported for MOVPE in a very narrow window of growth conditions by Tachibana al. [6]. Larger islands with lateral dimensions of a few 10 nm and a small aspect ratio were observed under selected MOVPE growth conditions by Kobayashi et al. [27].

As far as the origin of the In-rich clusters is concerned, calculations by Teles et al. [28] for cubic $In_{0.2}Ga_{0.8}N$ are interesting to consider. They suggest that random alloy fluctuations, i.e. local random deviations from the average In concentration, could induce In-rich clusters with a rather large size of a few nanometers. It is unclear, however, whether this effect would apply analogously to wurtzite material for average In concentrations between 5 and 25 %.

The most plausible explanation for the composition inhomogeneities is still provided by the thermodynamic instability of InGaN [2,3]. It seems to be impossible to suppress the formation of small and large-scale composition fluctuations although the strength of the composition fluctuations can be influenced by the growth kinetics as demonstrated by the variation of the InGaN growth rate (Section 3.1, Table 2). Only one single sample did not contain In-rich clusters, which was an *uncapped* InGaN layer with an average thickness of 3 nm and an average In concentration of 20 % grown by MOVPE [29]. A temperature dependence of the clusters size was not observed for the MOVPE samples in the temperature range between 690 and 860 °C. A conventional TEM study by Westmeyer and Mahajan [30] comes to the same conclusion regarding the temperature dependence of the periodicity of composition fluctuations, which they determined to be around 3 nm. We find a slight tendency towards smaller cluster sizes in

MBE samples (Section 3.3), which were grown at low temperatures around 530 °C. The effect of phase separation is clearly visible after a postgrowth high-temperature annealing treatment, which yields a significant redistribution of the group-III elements and changes of the cluster sizes and cluster compositions (see Section 3.4). Qualitatively similar results were obtained by McCluskey et al. [33] after a high-temperature treatment. The analysis of a thick InGaN layer (Section 3.4) provides some evidence that strain enhances the small-scale composition fluctuations, because the In concentration of the clusters decrease with increasing distance from the GaN/InGaN interface. It is adequate to note that large-scale fluctuations show also up in cathodoluminescence (CL) studies of InGaN performed e.g. by Chichibu et al. [36], which show pronounced fluctuations of the CL peak energy on a 100 nm scale.

In-rich clusters with sizes that are independent on the growth conditions are not compatible with the formation of precipitates by a "nucleation and growth" mechanism (see e.g. Haasen [31]) where the sizes of precipitates strongly depend on the diffusion coefficient, i.e. the temperature. Although it would not be reasonable to assume that bulk phase separation is the relevant process, phase separation at the growth surface should also be dependent on the surface diffusivities of the group-III elements. Spinodal decomposition appears to be a more adequate process, because a narrow range of spatial periodicities for the composition fluctuations are expected and the amplitude of the composition fluctuations increases with time. The latter point would be compatible with the growth rate dependence of the strength of the composition fluctuations (Section 3.1, Table 2). Phase separation on the basis of spinodal decomposition in InGaN was theoretically studied by Okamura et al. [32]. Their results indicate that In-rich clusters will indeed be formed. In the very early stages of decomposition, only the In concentration but not the size of the clusters should be affected by the surface diffusion length, which would be qualitatively consistent with our experimental results.

There remain, however, two problems with the picture presented so far. Spinodal decomposition is not expected to occur at low average In concentrations because the spinodal points for the relevant temperatures are predicted at In concentrations of approximately 20 and 80 % [2]. Two completely different length scales – we observe small and large-scale fluctuations – are also incompatible with the theory of spinodal decomposition according to which one spatial periodicity should dominate [31]. The presence of a second characteristic length scale is in particular visualized by the In distribution along the [0001] direction in the thick MOVPE layer (Fig. 8b) which shows a periodicity close to 100 nm. A solution to the latter problem might be provided by taking strain into account, which could be sufficient even at low average In concentrations to induce strained-induced composition fluctuations. Strain-induced large-scale composition fluctuations in strained SiGe epilayers with thickness undulations were considered by Christiansen [34], which demonstrated that the concentration of the Ge atoms is higher in regions with lower strain at the top of the thickness undulations.

A phase diagram was calculated for InGaN strained on GaN by Karpov [35], which shows, however, a significant decrease of the miscibility gap, which does not agree with the experimental observations. Other theoretical predictions regarding spinodal decomposition in thin strained InGaN epilayers do not exist according to our knowledge.

The situation during the InGaN growth is further complicated by In surface segregation and In desorption that influence the In distribution along the growth direction. Detailed studies of these processes during MBE growth are already available [37–40] which show, that the In incorporation strongly depends on the V/III-flux ratio, the ratio of the metal fluxes, the polarity of the substrate surface and the substrate temperature.

It is a characteristic feature of In segregation that the indium is only partially incorporated during the InGaN growth which leads to a gradual increase of the In concentration along the growth direction and the formation of an In-rich floating layer on the growth surface. An In-rich floating layer was shown to exist during MBE growth of InGaN by Chen et al. [38] with an In content that is influenced by the V/III-flux ratio. After the termination of the InGaN deposition, the indium contained in the floating layer is only gradually consumed during the GaN cap layer growth, which results in an asymmetrical In-concentration profile, significant In concentrations in the cap layer and, as a consequence, to a broadening of the InGaN QW compared to the intended layer thickness. The broadening will be dependent on the amount of indium in the floating layer after the termination of the InGaN deposition. Using the phenomenological model of Muraki and al. [41] to describe the profile, a temperature-dependent segregation coefficient R can be determined, which yields the fraction of indium atoms that segregate into the floating layer from the atomic layer below. MOVPE-grown samples show frequently an asymmetrical In-concentration profile as shown in Fig. 4b. A detailed analysis of a similarly grown sample yielded $R = 0.69$ for a growth temperature of 720 °C, while a stronger segregation with $R = 0.85 \pm 0.05$ at 530 °C was obtained for the MBE sample (Fig. 5) by Potin et al. [42]. It has to be noted, that the In-concentration profile of the MBE sample (Fig. 5, left-hand side) does not exhibit the characteristic asymmetrical shape, i.e. the exponential decay of the In concentration at the upper interface. The more abrupt compositional transition can be explained in this case by indium desorption from the floating layer. This was achieved by raising the temperature from 530 to 800 °C immediately after the InGaN deposition, which is a suitable method of reducing In segregation during the GaN cap layer growth and an uncontrolled broadening of the InGaN QW.

A further indication of In desorption is obtained from the measured average In concentrations for the first sample series with growth temperatures between 800 and 860 °C (Table 3) where a significant loss of indium occurs above 820 °C. In desorption was also measured directly in an uncapped MOVPE sample grown at 780 °C by Potin et al. [29] which shows an indium depletion in the uppermost four atomic layers of a 3 nm thick InGaN QW. This sample was the only one ever analyzed without In-rich clusters. Although the link between the lack of clusters and desorption is unclear, it is suggestive to postulate a connection between the two effects. The capped sample that was grown under exactly the same conditions as the uncapped one, contained again the typical In-rich clusters.

It is interesting to note that the highest overall average In concentrations, that we observed in all investigated samples, did not exceed 30 % despite extensively varied growth conditions. Higher average In concentrations appear to be extremely difficult to achieve although this would be very desirable with respect to shifting the emission wavelength of laser diodes into the blue/green spectral range.

Despite the different processes that occur at the growth surface during MBE and MOVPE, the same phenomena (phase separation, In segregation and In desorption) appear do govern the InGaN growth. Due to the strong dependence of the In incorporation and final QW width on the details of the growth procedure, large discrepancies between the nominal and real InGaN QW properties are often observed. It cannot be emphasized enough that many contradicting results in the literature may be related to the discrepancy between the real and the nominal InGaN QW data. In particular, measuring the composition of thin InGaN QW by XRD is not sufficient because the QW widths can be completely different compared to the nominal value. This may be due to In segregation but also due to growth rates that are different in the early stages of growth compared to the growth rates that are deduced from thick InGaN calibration layers [26]. This problem has been realized in a still relatively small number of other studies, e.g. Waltereit et al. [39] and Pereira et al. [43].

5 Summary

The In distribution of InGaN quantum wells was analyzed on an atomic scale by a technique based on high-resolution TEM lattice-fringe images. By optimizing the diffraction conditions, the technique is easy to use and robust against changes of the objective lens defocus and TEM sample thickness, which is a prerequisite to gain statistically relevant data by evaluating a sufficiently large number of images.

To gain a comprehensive overview about the effects that govern the In distribution, MBE and MOVPE samples were investigated that were grown in a wide range of deposition conditions. It is common to all samples that composition inhomogeneities occur on two different scales. In-rich clusters with lateral sizes below 5 nm and weaker composition fluctuations on a scale of approximately 100 nm are observed. The size of the clusters does not show a pronounced dependence on the growth technique and growth conditions, but the amplitude of the composition fluctuations can be influenced by the kinetics of the growth. The gradual increase of the amplitude of the composition fluctuations with time is typical for spinodal decomposition. However, decomposition at very low average In concentrations and two different length scales are inconsistent with this mechanism. There are indications, that strain plays a role, but detailed theoretical studies are not available yet.

The shapes of the In-concentration profiles along the growth direction demonstrate that In segregation takes place with a segregation efficiency that is higher in MBE than in MOVPE growth. In desorption becomes dominant above growth temperatures of approximately 800 °C. In segregation and In desorption contribute to significant deviations between the nominal and experimentally measured quantum well width and average In concentration which are frequently observed.

Acknowledgements Many collaborators have contributed to the results presented in this review article. We are grateful to the crystal growth groups who have provided a large number of interesting samples: Prof. Dr. M. Heuken (Aixtron Company), Prof. Dr. D. Bimberg and Dr. A. Strittmatter (Institute of Solid State Physics, Technical University Berlin), Dr. F. Scholz (IV. Physics Institute, University of Stuttgart) and Dr. N. Grandjean (Centre de Recherche sur l′ Hétéro-Epitaxie et ses Applications CRHEA/CNRS, Valbonne) for the MBE material. The financial support of the German Research Foundation (Deutsche Forschungsgemeinschaft, DFG) during six years of this project and the German Department for Education and Research (Bundesministerium für Bildung and Forschung, BMBF) is gratefully acknowledged.

References

[1] S. Chichibu, T. Sota, K. Wada, and S. Nakamura, J. Vac. Sci. Technol. B **16**, 2204 (1998).
[2] I-Hsui Ho and G. B. Stringfellow, Appl. Phys. Lett. **69**, 2701 (1996).
[3] T. Ito, Jpn. J. Appl. Phys. **36**, L1065 (1997).
[4] H. Chen, R. M. Feenstra, J. E. Northrup, T. Zywietz, and J. Neugebauer, Phys. Rev. Lett. **85**, 1902 (2000).
[5] B. Damilano, N. Grandjean, S. Dalmasso, and J. Massies, Appl. Phys. Lett. **75**, 3751 (1999).
[6] K. Tachibana, T. Someya, and Y. Arakawa, Appl. Phys. Lett. **74**, 383 (1999).
[7] H. Lakner, Q. Liu, G. Brockt, A. Radefeld, A. Meinert, and F. Scholz, Mater. Sci. Eng. B **51**, 44 (1998).
[8] Y. Narukawa, Y. Kawakami, M. Funato, S. Fujita, S. Fujita, and S. Nakamura, Appl. Phys. Lett. **70**, 981 (1997).
[9] C. Kisielowski, Z. Liliental-Weber, and S. Nakamura, Jpn. J. Appl. Phys. **36**, 6932 (1997).
[10] D. Gerthsen, E. Hahn, B. Neubauer, A. Rosenauer, O. Schön, M. Heuken, and A. Rizzi, phys. stat. sol. (a) **177**, 145 (2000).
[11] A. Rosenauer, S. Kaiser, T. Reisinger, J. Zweck, W. Gebhardt, and D. Gerthsen, Optik **102**, 63 (1996).
[12] A. Rosenauer and D. Gerthsen, Adv. Imag. Electron Phys. **107**, 121 (1999).
[13] T. Mattila and A. Zunger, J. Appl. Phys. **85**, 160 (1999)).
[14] R. Bierwolf, M. Hohenstein, F. Philipp, O. Brandt, G. E. Crook, and K. Ploog, Ultramicroscopy **49**, 273 (1993).
[15] A. F. Whright, J. Appl. Phys. **82**, 2833 (1997).
[16] A. Polian, M. Grimsditch, and I. Grzegory, J. Appl. Phys. **79**, 3343 (1996).
[17] Y. Takagi, M. Ahart, T. Azuhata, T. Sota, K. Suzuki, and S. Nakamura, Physica B **219/220**, 547 (1996).
[18] P. Vermaut and P. Ruterana, G. Nouet, Philos. Mag. A **76**, 1215 (1997).
[19] P. Ruterana, S. Kret, A. Vivet, G. Maciejewski, and P. Dluzewski, J. Appl. Phys. **91**, 8979 (2002).
[20] A. F. Moodie, Z. Naturf. **27a**, 437 (1972).
[21] A. Rosenauer, V. Potin, and D. Gerthsen, in preparation.
[22] A. Rosenauer, Transmission electron microscopy of semiconductor nanostructures: an analysis of composition and strain state, Springer Tracts in Modern Physics, Vol. 182 (Springer Verlag, Berlin/Heidelber/New York, 2003), in press.
[23] H. Lichte, Ultramicroscopy **38**, 13 (1991).
[24] E. Hahn, A. Rosenauer, D. Gerthsen, B. Kuhn, and F. Scholz, phys. stat. sol. (b) **234**, 738 (2002).
[25] M. Schowalter, B. Neubauer, A. Rosenauer, D. Gerthsen, O. Schön, and M. Heuken, Mater. Res. Soc. Online Symp. Proc., spring meeting 2001, Symposium E, paper E 3.6. http://www.mrs.org/publications/epubs/proceedings/spring2001/.
[26] D. Gerthsen, B. Neubauer, A. Rosenauer, T. Stephan, H. Kalt, O. Schön, and M. Heuken, Appl. Phys. Lett. **79**, 2552 (2001).
[27] Y. Kobayashi, V. Perez-Solarzano, J. Off, B. Kuhn, H. Gräfeldinger, H. Schweizer, and F. Scholz, J. Cryst. Growth **243**, 103 (2002).
[28] L. K. Teles, J. Furthmüller, L. M. R. Scolfaro, J. R. Leite, and F. Bechstedt, Phys. Rev. B **62**, 2475 (2000).
[29] V. Potin, A. Rosenauer, D. Gerthsen, B. Kuhn, and F. Scholz, phys. stat. sol. (b) **234**, 947 (2002).
[30] A. N. Westmeyer and S. Mahajan, Appl. Phys. Lett. **79**, 2710 (2001).
[31] P. Haasen, Physical Metallurgy, 3rd edition, Chapter 9 (Cambridge University Press), and references therein.

[32] T. Okamura and Y. Akagi, J. Cryst. Growth **223**, 43 (2001).

[33] M. D. McCluskey, L. T. Romano, B. S. Krusor, D. P. Bour, and N. M. Johnson, Appl. Phys. Lett. **72**, 1730 (1998).

[34] S. H. Christiansen, The interaction square in heteroepitaxial growth: strain-topography-defects-composition, Series Mikrostrukturelle Materialforschung, Vol. 4, Chapter 7, edited by H.P. Strunk (xxx, xxx, 1997).

[35] S. Yu. Karpov, MRS Internet J. Nitride Semicond. Res. **3**, 16 (1998).

[36] S. Chichibu, K Wada, and S. Nakamura, Appl. Phys. Lett. **71**, 2346 (1998).

[37] M. L. O´Steen, F. Fedler, and R. J. Hauenstein, Appl. Phys. Lett. **75**, 2280 (1999).

[38] H. Chen, R. M. Feenstra, J. Northrup, J. Neugebauer, and D. W. Grewe, MRS Internet J. Nitride Semicond. Res. **6**, 11 (2001).

[39] P. Waltereit, O. Brandt, K. H. Ploog, M. A. Tagliente, and L. Tapfer, phys. stat. sol. (b) **228**, 49 (2001).

[40] A. Dussaigne, B. Damiliano, N. Grandjean, and J. Massies, J. Cryst. Growth, 251, 471 (2003).

[41] K. Muraki, S. Fukatsu, and Y. Shiraki, Appl. Phys. Lett. **61**, 557 (1992).

[42] V. Potin, A. Hahn, A. Rosenauer, D. Gerthsen, B. Kuhn, F. Scholz, A. Dussaigne, B. Damilano, and N. Grandjean, submitted to J. Cryst. Growth.

[43] S. Pereira, M.R. Correira, T. Monteiro, E. Pereira, M.R. Soares, and E. Alves, J. Cryst. Growth **230**, 448 (2001).

phys. stat. sol. (c) **0**, No. 6, 1684–1709 (2003) / **DOI** 10.1002/pssc.200303126

A theoretical investigation of dislocations in cubic and hexagonal gallium nitride

A. T. Blumenau[*,1,2], **C. J. Fall**[2], **J. Elsner**[1], **R. Jones**[2], **M. I. Heggie**[3], and **T. Frauenheim**[1]

[1] Department of Physics, Faculty of Science, University of Paderborn, 33098 Paderborn, Germany
[2] School of Physics, University of Exeter, Exeter EX4 4QL, United Kingdom
[3] CPES, University of Sussex, Falmer, Brighton BN1 9QJ, United Kingdom

Received 4 March 2003, revised 2 May 2003, accepted 7 May 2003
Published online 28 August 2003

PACS 61.72.Lk, 71.15.Mb, 71.55.Eq, 79.20.Uv

In this article we review our theoretical work on dislocations in GaN. The methods applied are two distinct approximations to density functional theory: Density functional based tight-binding total energy calculations allow the prediction of low energy core structures and energies of extended defects embedded in larger regions of perfect material. However, whenever less approximate electronic structure calculations are required they are obtained in a localised basis pseudopotential approach.

1 Introduction

Wide band-gap semiconductors play an important role in today's optoelectronics. In particular hexagonal gallium nitride (GaN), with a gap of 3.4 eV, became the basic material for laser devices emitting blue light [1, 2]. Also the slightly less stable cubic polytype of GaN seems a promising material for laser applications, but device technology is still at a very early stage. However, first light-emitting diodes based on cubic GaN have been constructed [3, 4]. Dislocations, emerging in high densities in most GaN growth processes, can induce electronic states deep in the band-gap. These might alter the optical and electrical performance. Also point defects or defect complexes are likely to be trapped in the dislocation stress field. The resulting electrostatic field can result in carrier scattering and hence affect the carrier mobilities [5]. Especially in laser devices "parasitic" components in the emission spectrum, resulting from dislocations or point defects trapped at dislocations, are undesirable. A good example is the often observed yellow luminescence in hexagonal GaN.

1.1 Classes of dislocations in GaN

The two classes of dislocations treated in this work are basal plane dislocations and threading dislocations. Dislocations in the pyramidal plane of hexagonal GaN are out of the scope of this work.

Basal plane dislocations in GaN are dislocations of the main slip system of the crystal, in hexagonal material the $\{0001\} \langle 11\bar{2}0 \rangle$ system and in cubic material the $\{111\} \langle 110 \rangle$ system[1]. In those slip systems dislocations inhabit a $\{0001\}$ or $\{111\}$ glide plane and possess $\langle 11\bar{2}0 \rangle$ or respective $\langle 110 \rangle$ type line directions and Burgers vectors.

---

* Corresponding author: e-mail: blumenau@phys.upb.de

[1] Here the term "basal plane" is used for the basal plane in hexagonal material and the $\{111\}$ planes in cubic material alike. Concerning next neighbour atoms both types of planes are equivalent.

During crystal growth, unevenness of the substrate surface or the collision of growth islands leads to the nucleation of dislocations with line direction parallel to the direction of growth. These dislocations thread the epilayer and are very common in hexagonal GaN [6]. The two basic types are the threading screw (ts) and the *threading edge dislocation* (te), both with a line direction of $\ell = [0001]$ and Burgers vectors $\boldsymbol{b}_{\mathrm{ts}} = [0001]$ and $\boldsymbol{b}_{\mathrm{te}} = \frac{1}{3}[1\bar{2}10]$ respectively, assuming minimum Burgers vectors.

In our work we model basal plane dislocations and threading dislocations in GaN, investigating their stability, their atomic and electronic structure and to some extent their interaction with point defects. Since the electronic structure of threading dislocations still remains controversial, we further calculate electron energy-loss spectra for these dislocations, which can be compared with experiments.

Further types of dislocations are misfit dislocations, which compensate the lattice mismatch between the substrate and the GaN layer during growth, and grain boundary dislocations. However, as these structures are very similar (or even equivalent) in structure to basal plane or threading dislocations [7], they are not modelled explicitly in this work.

1.2 Outline

Section 2 gives a brief description of the methods used – including an introduction into the simulation of electron energy-loss as applied in this work. Section 3 presents an overview over the three basic atomistic disclocation models – the cluster, the supercell and the hybrid approach. Finally, our results on basal plane and threading dislocations in GaN are reported in Sections 4 and 5, respectively.

2 Computational approaches

In this work we use a density functional based tight-binding approach (DFTB) to find the low energy dislocation core structures. Whenever more reliable electronic structure calculations of those structures are required, these are modelled within a localised basis pseudopotential scheme. In the following both methods will be discussed briefly.

2.1 Density functional based tight-binding

In DFTB the electronic wavefunctions are approximated by a linear combination of atomic orbitals involving a minimal basis set of s and p-valence orbitals – in the case of GaN further the Ga-3d orbitals are included. The two centre Hamiltonian and overlap matrix elements result from atom-centred valence electron orbitals and the atomic potentials from single-atom calculations in density functional theory (DFT). Exchange and correlation contributions in the total energy as well as the ionic core–core repulsion are taken into account via a repulsive pair potential. The latter is obtained by comparison with DFT calculations. Thus the DFTB method can be seen as an approximate density functional scheme [8, 9]. In this approach all DFT integrals (Hamiltonian and overlap matrix) as well as the repulsive potential can be calculated in advance and are tabulated over the interatomic distance of two atoms. This considerably reduces the computational effort, allowing the treatment of larger models compared to less approximate DFT-based methods.

For some systems with strong ionic bonding character and strong charge transfer the approximations of standard DFTB fail: For these systems the mere superposition of electron densities of neutral free atoms is not sufficient to describe the electron density of the modelled system. Considerable charge transfer between atoms of different electronegativity then requires an explicit treatment of long-range coulomb interactions. To some extent the resulting effects are included in the self-consistent charge extension of the DFTB method (SCC-DFTB). Here the charge transfer is approximated by spherical atom-centred charge fluctuations. These fluctuations explicitly occur in the respective total energy expressions up to the second order terms. For further details see [9].

When modelling extended defects, and especially dislocations, medium and long range stress/strain effects play an important role. For example the elastic strain energy contained in a cylinder of radius R around a dislocation diverges with $\ln(R/\text{Å})$ and dislocation–dislocation interaction forces drop with only R^{-1} if R is the distance between the two dislocations [10]. Therefore, modelling dislocations often requires large supercells or clusters to give a good representation of the surrounding bulk material or, in case of a supercell approach, to minimise dislocation–dislocation interaction with the periodic images in neighbouring cells. Here the simplicity of the SCC-DFTB method proves to be a major advantage over less approximate and computationally more demanding methods: On today's workstations the SCC-DFTB method as described above allows a structural relaxation of systems containing more than 700 atoms.

The DFTB method has been applied to model dislocation core structures and energies in various semiconductor materials [11, 12, 13] and was found to give results in good agreement with DFT pseudopotential calculations (AIMPRO [14, 15]). Also, excellent agreement was found with other DFT calculations [16] carried out on the 90° glide partial dislocation in diamond [12].

2.2   A localised basis pseudopotential approach

Although the DFTB method sufficiently describes dislocation structures and energetics, results regarding the electronic structure are only of limited reliability. This stems both from the minimal basis approach as well as from the rather crude representation of the charge density in tight-binding. Therefore, to give a more precise description of the electronic structure, a less approximate DFT pseudopotential method in local density approximation (LDA) is applied. In this method, as implemented in the AIMPRO [14, 15] code, we use pseudopotentials [17] to describe the ion cores. In the specific case of GaN a non-linear core correction is included to account for the Ga-3d electrons. Test calculations are also performed with the Ga-3d electrons included as valence electrons. The charge density is expanded in plane waves up to a cutoff of 300 Ry. Wavefunctions, however, are described using localized s, p, and d atom-centred Gaussian orbitals that are optimised for bulk systems. Structural optimisation is performed with a dppp basis on each atom – i.e. one spd function with a small Gaussian exponent, and three sp functions with increasing Gaussian exponents. Tests performed with other basis sets have shown only small changes in the atomic and band structures. A grid of Monkhorst–Pack (MP) k-points is used to perform the Brillouin zone (BZ) integrations [18]. The bulk lattice parameters, computed with 24 reduced k-points, are given in Table 1 and the bulk band structures are shown in Fig. 1 [19]. A comparison between our band structures and previous calculations shows only small differences [24, 25]. For example, we obtain valence band widths of 1.13, 2.58, and 7.56 eV with the Ga-3d states treated as valence states (against band widths of 0.98, 2.65, and 7.26 eV respectively in a previous linear-combination-of-atomic-orbitals calculation [24]). The computed GaN band-gap is then found to be 1.92 eV, compared with 2.04 eV previously [25]. For AlN, we obtain valence band widths of 3.01 and 6.18 eV (against 2.89 and 6.16 eV previously [24]).

Table 1   Theoretical bulk lattice constants (a_0, c_0) and bulk modulus (B) for hexagonal GaN and AlN. The value with an asterisk is obtained with the Ga-3d states treated as valence bands.

	GaN		AlN	
	AIMPRO	Exp.	AIMPRO	Exp.
a_0 [a.u.]	5.98	6.03 [20]	5.79	5.88 [21]
c_0/a_0	1.626	1.626 [20]	1.62	1.601 [21]
B [GPa]	208, 216*	237±31 [22]	213	207.9±6.3 [23]

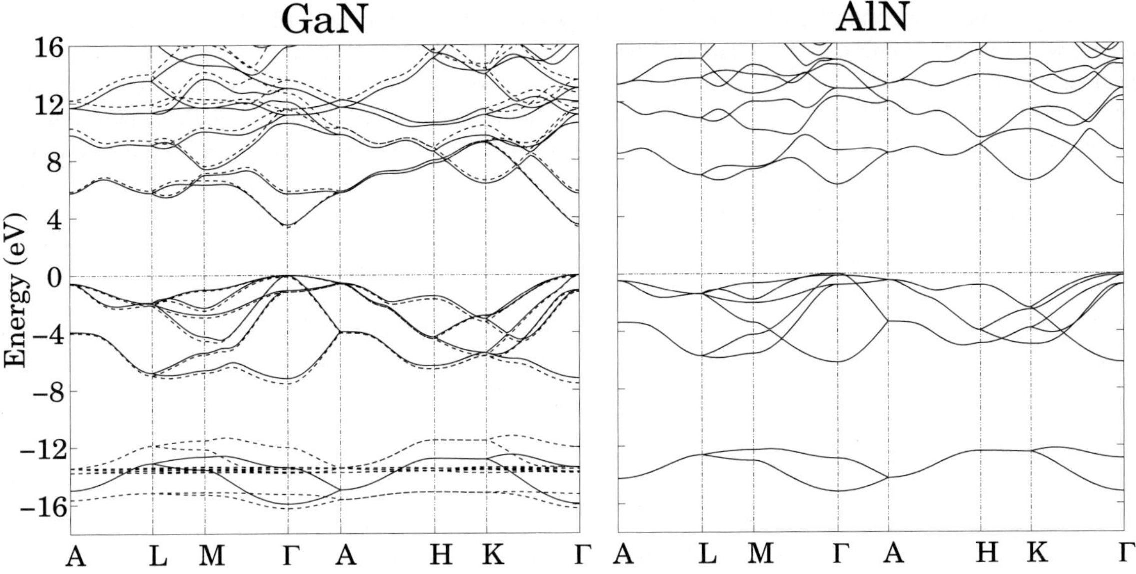

Fig. 1 Calculated band structures of hexagonal GaN and AlN (pseudopotential approach). The energies are referred to the top of the valence band in each case. Left: The solid lines give the band structure of GaN with the Ga-3d electrons treated as core electrons, whereas the dashed lines represent a calculation in which the Ga-3d electrons are treated as valence electrons. The GaN conduction bands have been shifted by 1.58 and 1.25 eV when the Ga-3d electrons are treated as valence or respectively core electrons to bring the computed gaps in line with the experimental values. Right: For AlN, the conduction bands are shifted by 1.5 eV.

Pseudopotential methods require greater computing resources than DFTB and thus in this work AIM-PRO is usually applied on the low energy structures found with the DFTB method. These structures are then further structurally relaxed and their electronic structure ist determined.

2.3 The modelling of electron energy-loss spectra

Electron energy-loss spectroscopy (EELS) is one of the few experimental techniques which, if combined with transmission electron microscopy (TEM), allow to directly probe the gap states of an extended defect [2]. In EELS primary high-energy electrons pass through a thin sample and some are scattered inelastically and lose a fraction of their energy. The energy is lost as electronic transitions between the valence band and conduction band states of the solid are induced, resulting in so called *low-loss* spectra. The interaction can be that of single valence electron excitation events or can also encompass collective modes of the solid, called plasmons. If the transitions occur between tightly bound atomic core states and the conduction band one speaks of *core-excitation* EELS or *energy-loss near edge structure* (ELNES). The interaction of high-energy electrons passing through a solid can be described as an interaction with the field of polarisation which is caused by the electric field of the solid's electrons [26, 27]. In this description the signal obtained is representative of $-\mathrm{Im}(\varepsilon(E))^{-1}$ (also known as the loss function), where $\varepsilon(E) = \varepsilon_1(E) + i\varepsilon_2(E)$ is the macroscopic dielectric function [28, 29]. In the single particle approximation the imaginary part of the dielectric function can be written as [30, 31]

$$\varepsilon_2(\boldsymbol{q}, \omega) = \frac{4\pi e^2}{\Omega q^2} \sum_{n,j} |\langle \Psi^n | e^{i\boldsymbol{q}\cdot\boldsymbol{r}} | \Psi^j \rangle|^2 \, \delta(E^n - E^j - \hbar\omega). \tag{1}$$

---

[2] Typically EEL spectrometer are combined with scanning electron microscopes (STEM) and allow high spatial resolutions of the order of atomic distances.

Here $\boldsymbol{q}$ gives the momentum transfer, $\omega = E/\hbar$, Ω is the unit cell volume, and $|\Psi^n\rangle$ are single particle states, with respective energies E^n. The real part of the dielectric function can be obtained easily via a Kramers–Kronig transformation. Equation (1) basically represents a joint density of states, weighted by matrix elements $\langle\Psi^n|e^{i\boldsymbol{q}\cdot\boldsymbol{r}}|\Psi^j\rangle$. Electronic structure calculations can be used to calculate these matrix elements. In this work EEL spectra are shown which were obtained based on pseudopotential calculations (AIMPRO). The next paragraphs will give some details on how the necessary matrix elements can be approximated in a simple approach for low-loss and core-excitation EELS, respectively.

2.4　Low-loss EELS

As mentioned above, low-loss EELS involves excitations from occupied valence band states to empty conduction band states. In the long-wavelength dipole approximation ($q \to 0$), the matrix elements in Eq. (1) can be simplified further using $e^{i\boldsymbol{q}\cdot\boldsymbol{r}} \approx 1 + i\boldsymbol{q}\cdot\boldsymbol{r}$

$$\varepsilon_2^\ell(E) = \frac{4\pi e^2}{\Omega} \sum_{c,v} \int_{BZ} |\langle\Psi_{\boldsymbol{k}}^c|r^\ell|\Psi_{\boldsymbol{k}}^v\rangle|^2\, \delta(E_{\boldsymbol{k}}^c - E_{\boldsymbol{k}}^v - E)\, \mathrm{d}^3\boldsymbol{k}. \tag{2}$$

Here c and v denote conduction and valence bands respectively and $\boldsymbol{k}$ gives the Brillouin zone vector – excitations with $\Delta\boldsymbol{k} \neq 0$ are excluded in this simplified approach. As the dielectric function depends on the polarisation of the exciting wave through $\boldsymbol{q}$, it has to be calculated for different orientations ℓ of $\boldsymbol{r}$ of the electron beam. The integral over $\boldsymbol{k}$ covers the whole Brillouin zone. As LDA underestimates the band-gap, a scissors operator is applied to the band structure, shifting the conduction bands upwards to match the experimental gap. Defect states in the gap are shifted proportionally to their distance from the valence band maximum. Traditionally, EEL spectra are presented with a Lorentzian broadening scheme, which has the advantage of allowing Kramers–Kronig transformations to be computed analytically [32]. However, Lorentzian functions have long tails which tend to mask spectral features at low energy in the band-gap region. Instead, we use here a new polynomial broadening scheme. The advantage of this broadening function is that is has both a finite width and an analytical Kramers–Kronig transform. In Eq. (2), the $\delta(x)$ function is replaced by a normalised polynomial function

$$f(x) = \begin{cases} \frac{15}{16\beta^5}(x^4 - 2\beta^2 x^2 + \beta^4), & \text{for } -\beta \leq x \leq +\beta \\ 0, & \text{otherwise} \end{cases}, \tag{3}$$

which has a full width at half maximum of $\Delta x = 2\beta\sqrt{1 - 2^{-1/2}}$. In the following, we set $\Delta x = 0.8$ eV.

2.5　Core-excitation EELS

K-edge core-excitation EEL spectra in GaN result from electronic transitions between N-1s core electrons and empty gap or conduction band states. Neglecting final-state interactions between the excited electron and the core hole as a first approximation, and in view of the strong localization of the initial state, the core EEL spectra are related to the p-like projected local density of states on the excited N atoms. The measured core-loss EEL intensity $I(E)$ is hence given by

$$I(E) \approx \sum_{\boldsymbol{R}} \sum_c \int \mathrm{d}^3\boldsymbol{k}|\langle\Psi_{\boldsymbol{k}}^c|p_{\boldsymbol{R}}\rangle|^2\delta(E_{\boldsymbol{k}}^c - E_{\boldsymbol{R}}^i - E), \tag{4}$$

where the sum over $\boldsymbol{R}$ includes all N atoms that contribute to the measured signal, $p_{\boldsymbol{R}}$ represents a p-like state centred at $\boldsymbol{R}$, and $E_{\boldsymbol{R}}^i$ is the energy of the core level at $\boldsymbol{R}$. In the vicinity of a defect, such as a dislocation, changes in the atomic bonding will induce local potential variations that shift the core-level energies. We therefore decompose $E_{\boldsymbol{R}}^i$ into $E_{\boldsymbol{R}}^i = \overline{E} + \Delta E_{\boldsymbol{R}}$, where $\overline{E}$ is an average value of the core energy level, taken over a suitable volume, and $\Delta E_{\boldsymbol{R}}$ is the energy difference between this average and

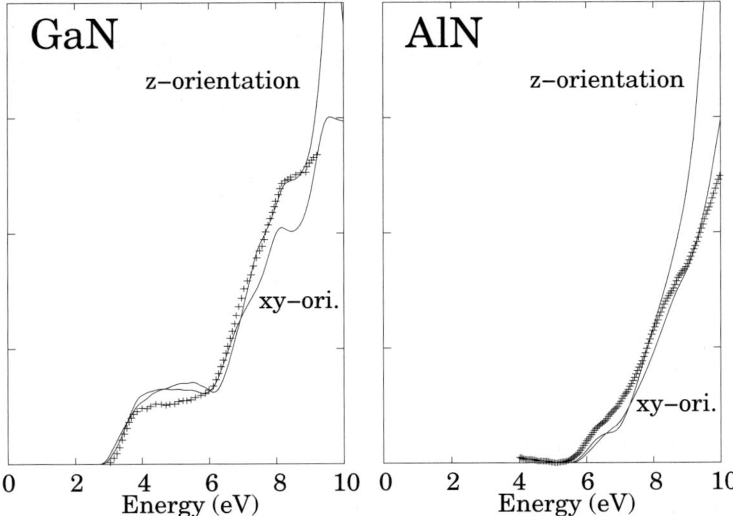

Fig. 2 Theoretical (lines) and experimental[35] (crosses) low-loss EEL spectra of bulk hexagonal GaN (left) and AlN (right). The theoretical curves are given for x, y orientations of the electron beam, perpendicular to c (solid line), and for a z orientation parallel to c (dashed line), in arbitrary units. Experimental spectra are taken with a $[2\bar{1}\bar{1}0]$ incident beam.

the core energy on the N atom of interest. We take the total potential differences – including the Hartree, exchange-correlation, and local pseudopotential terms – to represent the shifts of the N core levels. We display core-level EEL spectra with a Gaussian broadening of 0.8 eV. Excitonic effects are neglected here, but were found previously in bulk silicon to affect core-excitation peak positions by up to 1 eV [33]. Very recently, excitonic effects in GaN core excitations have been included using the $Z + 1$ approximation [34].

The BZ integrations in Eqs. (2) and (4) are performed in bulk material using 1024 MP k-points and a similar k-point density in supercells containing dislocations. A test calculation for bulk material with 2000 MP k-points yielded very similar results.

2.6 Bulk low-loss EEL spectra

Figure 2 shows the theoretical low-loss EEL spectra of bulk GaN and AlN, together with corresponding experimental data [35], acquired in a cross-sectional geometry with an electron beam in the x, y (basal) plane, perpendicular to the c-axis. Computed EEL spectra for x and y orientations are identical because of the 3-fold symmetry of the hexagonal crystal. A good qualitative agreement between the peaks and shoulders of the GaN x, y computed and measured EEL signals is found. This close correspondence gives us confidence in our approach and parameters, and allows us to extend our study to examine the effect of dislocations on the EEL spectra. The EEL spectrum of bulk AlN provides a supplementary check that our theoretical approach reproduces experimental spectra accurately.

The effect of the Ga-3d states on the GaN EEL spectra has been investigated in bulk test calculations by including them as valence states in the pseudopotential. Only small changes in the resulting EEL spectra were found in the 0–10 eV range.

Experimental EEL spectra acquired on nitride thin films may be influenced by surface plasmons or surface states. However, no quantitative experimental or theoretical data on the importance of these effects are known yet, and they are thus not considered further here.

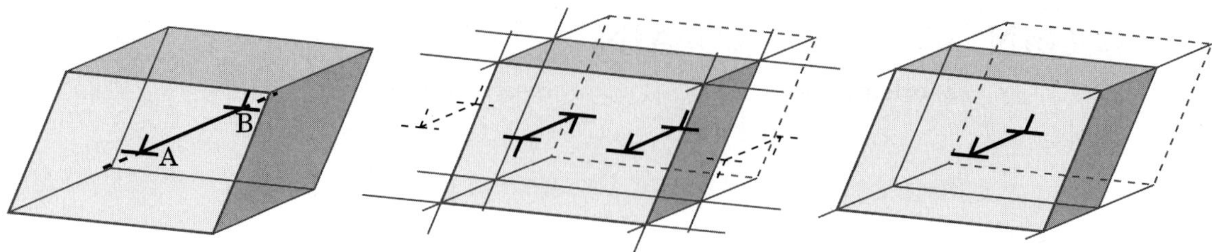

Fig. 3 (online colour at: www.interscience.wiley.com) Schematic sketches of distinct atomistic dislocation models. Left: A cluster containing a single dislocation. A and B denote the points of dislocation–surface intersection. Centre: A dislocation dipole in a periodic supercell. Right: A dislocation in a supercell-cluster hybrid model, periodic along the dislocation line and cluster-like perpendicular to the line.

3   General remarks on dislocation modelling

Generally there are two main approaches used to model defects in semiconductors atomistically: The *cluster* and the *supercell* approach. For the case of dislocations this means that either an atom cluster containing a single dislocation is considered or a supercell containing a dislocation multipole. The long range elastic effects are treated differently in each case but neither approach treats these effects rigorously, as will be discussed below.

3.1   The cluster approach

Here a single straight dislocation is inserted into an atomic cluster. As a result, the dislocation line intersects with the surface of the cluster. Figure 3 (left) depicts a cluster, whose surfaces are cut as crystallographic planes. For small clusters it is advisable to impose the displacements expected from elasticity theory for a dislocation in an infinite crystal at the surface of the finite cluster. In other words one then has to embed the cluster in a continuous medium, which represents the surrounding crystal environment. However, if the cluster is large enough, the difference with a freely relaxed surface is negligible [36].

In DFT-based methods where the electronic energy is explicitly included, the cluster surfaces are usually terminated with hydrogen atoms to avoid electronic gap states associated with dangling bonds or surface reconstructions, which might induce additional strain in the cluster core.

The main advantage of the cluster model is the possibility to study a *single and isolated* dislocation, but its principal disadvantage stems from the part of the dislocation line which intersects the surface. The dislocation core structure will be distorted near the surface, where it loses its periodicity. Thus, the cluster has to be considerably extended in the direction of the dislocation line. Experience from modelling dislocations in Si shows that clusters of around 700 atoms with approximately 6 repeat distances along the dislocation line are necessary, as applied in [11].

3.2   The supercell approach

Here a dislocation dipole or multipole is placed in a supercell. The periodic boundary conditions require an overall Burgers vector of $b_\mathrm{tot} = 0$. Thus, the most simple configuration is a dipole with two dislocations of opposite Burgers vector. Figure 3 (middle) depicts an example supercell containing a dipole, indicating its periodically repeated images in one dimension as an example. In the case of a dipole, the volume per single dislocation is only half that of the cell. Since DFT calculations are nowadays limited to a few hundred atoms, usually the two dislocations are separated by a relatively small distance. A result of this is a considerable dislocation–dislocation interaction within and across the cell boundaries to neighbouring cells and in compound semiconductors also a considerable charge transfer might occur. The elastic interaction can be described by a procedure similar to an Ewald sum, but the resulting expressions are not easy to handle [37]. Blase et al. [16] perform this summation by taking advantage of an analytical solution found for tilt boundaries. However, the same authors conclude that the core reconstruction of a dislocation depends on the stress state. This implies that the stress of the selected supercell geometry influences the core structure and hence possibly the core energy.

Seeing the advantages and disadvantages of both the cluster and the supercell, it is a rather straightforward idea to combine the advantages of both in a *supercell-cluster hybrid* approach – minimising the problems associated with either of the preceding.

3.3   The supercell-cluster hybrid approach

Here the dislocation is placed in a model which is periodic along the dislocation line, however, it is non-periodic with a hydrogen-terminated surface perpendicular to the line direction [38, 39]. This allows to maintain the 'natural' dislocation periodicity as a line defect and at the same time avoid the interaction between dislocations in different cells. The single dislocation in the hybrid model as in Fig. 3 (right) is surrounded by twice the amount of bulk crystal compared to the pure supercell containing a dipole – assuming both models are of the same volume and neglecting surface effects. In other words, in the hybrid model it is by far easier to give a good representation of the surrounding bulk crystal. The latter in combination with the capability of modelling a single and isolated dislocation (avoiding dislocation–dislocation interaction) makes the hybrid model the ideal choice to describe line defects.

The hybrid approach – just as the pure cluster approach – is however not well suitable for DFT methods based exclusively on plane wave basis sets, since these methods have considerable problems dealing with the large amount of vacuum surrounding the model: The then necessary high energy-cutoff makes the cluster and hybrid models computationally very demanding. Thus whenever plane wave methods are applied to describe dislocations, the pure supercell is still a very popular model.

In Section 5 the supercell and the hybrid approach will be compared in more detail with threading dislocations in hexagonal GaN as an example.

4   Basal plane dislocations

Dislocations in the basal plane are common defects in hexagonal GaN. They often occur dissociated in the vicinity of the surface interface [7, 40, 41]. As demonstrated by Albrecht et al. [42], plastically deforming a GaN crystal induces dislocations in the basal and in the pyramidal plane. In the basal plane $60°$ and screw dislocations were observed, however no indication of dissociation into Shockley partials was found. Cathodoluminescence experiments revealed a 2.9 eV radiative recombination at $60°$ dislocations, probably a donor–acceptor pair transition, whereas screw dislocations act mainly as non-radiative recombination centres [42]. No correlation between basal plane dislocations in GaN and the yellow luminescence could be found experimentally.

There are only very few experimental results on the same type of dislocation in cubic GaN. However, there are reports of $60°$ dislocations, twins and stacking faults seen by TEM in cubic GaN grown on GaAs (001) [43]. The $60°$ dislocations emerge from the GaN/GaAs interface. So far, experimentally not much is known concerning their core structures. As et al. [44] find from Hall measurements on Si doped cubic GaN, that the electron mobility increases with increasing Si-doping for electron concentrations smaller 3×10^{19} cm^{-3}. For higher concentrations the mobility decreases again. This behaviour is characteristic for electron trapping at dislocations suggesting that dislocations are active [5]. Temperature dependent Hall measurements on undoped material showed it to be p-type ($p \sim 10^{16}$ cm^{-3}) with an acceptor level at ~ 166 meV above the valence band maximum [44]. This indicates that many acceptors are neutral at room temperature and leads to an acceptor concentration of $\sim 10^{18}$ cm^{-3}, which is comparable with the density of dislocation core atoms. One problem is that the electrical properties might be related to interfaces between cubic and hexagonal phases.

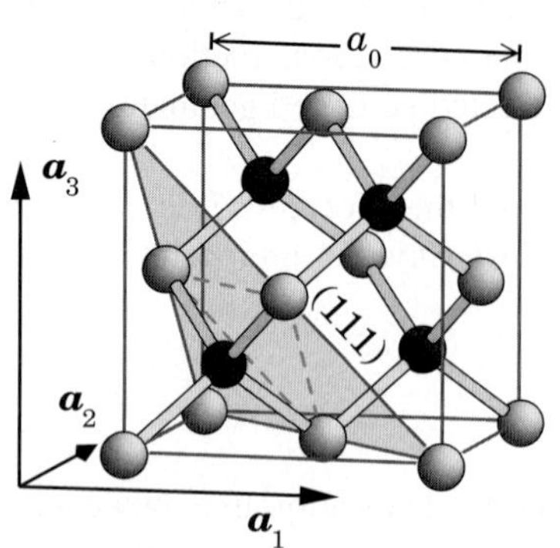
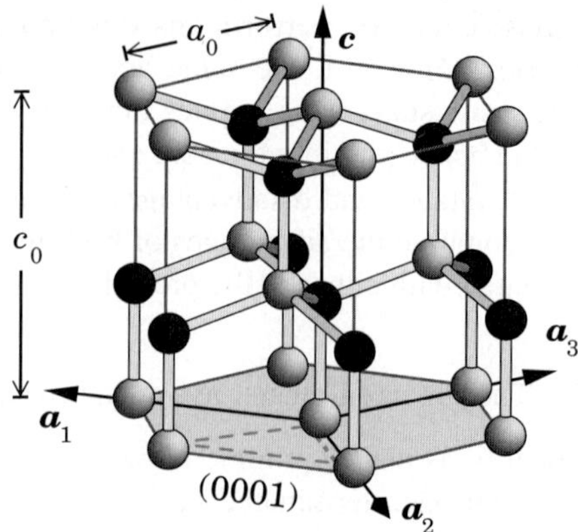

Fig. 4 (online colour at: www.interscience.wiley.com) Unit cells of GaN. Ga atoms are drawn in dark grey and N atoms in light grey. Left: The cubic structure, also known as the zincblende structure. It can be constructed as a face-centred cubic lattice with a two-atom basis. The lattice constant a_0 is given as the edge of the cubic cell. The three perpendicular basis vectors are labelled a_1 to a_3. Right: The hexagonal 2H structure, also known as the wurtzite structure. It can be constructed as a hexagonal closed-packed lattice with a two-atom basis. The two lattice constants are given as one edge of the basal hexagon (a_0) and the height of the unit cell (c_0). The basis vectors are labelled a_1 to a_3 and c. The a_i lie in one plane and form angles of $120°$. a_3 is conventionally introduced for reasons of symmetry in the description. In both unit cells a plane with the highest density of lattice sites is drawn: (111) and (0001) for cubic and hexagonal symmetry respectively. The minimum lattice translations within these planes are shown as dashed lines.

4.1   Burgers vectors and line directions

With the elastic energy of a dislocation proportional to the squared Burgers vector $|\boldsymbol{b}|^2$ [10], low energy dislocations are those with minimum Burgers vector. Since a Burgers vector is always given as a lattice translation, dislocations are preferentially formed on planes containing the minimum lattice translations. These planes are the preferred slip planes. A slip plane contains both the Burgers vector and the line direction of the generated dislocations. Hence the plane of crystal slip is identical with the glide plane of the dislocations involved in the process.

GaN occurs in two distinct crystal structures, the cubic (or zincblende) structure and the slightly more stable hexagonal 2H (or wurtzite) structure[3]. As can be seen in Fig. 4, in both structures atoms are fourfold tetrahedrally bonded. In the cubic structure the planes of highest lattice site density are the $\{111\}$ planes. Fig. 4 (left) shows one specific plane of that family. These planes contain the minimum lattice translations and are thus preferred slip planes. The slip system is defined by the plane and the direction of slip, conventionally written as $\{111\}\langle110\rangle$. For each plane we have three possible directions. Specifically in the (111) plane the minimum lattice translations (Burgers vectors) are $\frac{1}{2}[1\bar{1}0]$, $\frac{1}{2}[\bar{1}01]$, and $\frac{1}{2}[01\bar{1}]$. They are drawn as dashed lines in Fig. 4 (left). Whilst in the cubic system there is a whole family of planes with the highest lattice site density, in the hexagonal structure there is only one, the (0001) or *basal plane*, as shown in Fig. 4 (right). This plane is basically equivalent to the $\{111\}$ planes in cubic material. The slip system is of the $\{0001\}\langle11\bar{2}0\rangle$ type and the minimum lattice translations are given as $\frac{1}{3}[11\bar{2}0]$, $\frac{1}{3}[1\bar{2}10]$, and $\frac{1}{3}[\bar{2}110]$. They are drawn as dashed lines in Fig. 4 (right).

Since the line direction, at least locally, is also given by lattice translations, the afore mentioned directions yield two basic dislocation types with minimum Burgers vector:

---

[3] The underlying lattices are the face-centred (closed-packed) cubic (fcc) and the hexagonal closed-packed (hcp) lattice – both with a two atom basis.

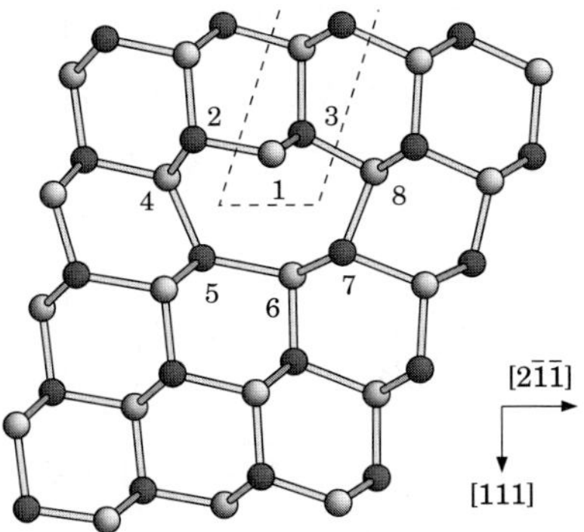
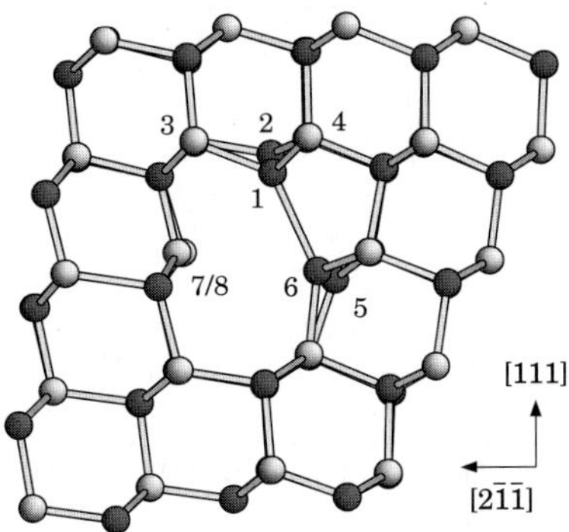

Fig. 5   Cross section view (along $[01\bar{1}]$) of the relaxed core of the two different types of 60° shuffle dislocations in cubic GaN. Both dislocations lie along $[01\bar{1}]$, perpendicular to the projection plane. Ga atoms are drawn in dark grey, N atoms in light grey. Left: The N core dislocation. The dashed line indicates the extra half-plane. Right The Ga core dislocation. This is a double-period structure with a bond between the two Ga atoms 1 and 6 but no bond between 2 and 5.

- The screw dislocation with line direction and Burgers vector parallel, e.g. in cubic material $\ell = [1\,\bar{1}0]$ and $b_s = \frac{1}{2}[1\bar{1}0]$ or in hexagonal material $\ell = [11\bar{2}0]$ and $b_s = \frac{1}{3}[11\bar{2}0]$.

- The 60° dislocation, where line direction and Burgers vector form an angle of 60°, e.g. in cubic material $\ell = [1\bar{1}0]$ and $b_{60} = \frac{1}{2}[0\bar{1}1]$ or in hexagonal material $\ell = [11\bar{2}0]$ and $b_{60} = \frac{1}{3}[2\bar{1}\bar{1}0]$.

4.2   The perfect 60° dislocation

As the 60° dislocation has partially edge character, its structure involves an extra half-plane of material inserted into the crystal. If the half-plane terminates between two closely-separated {111} or {0001} planes we obtain the so called glide structure, if it terminates between widely separated planes we obtain the shuffle structure [10]. For the shuffle structure the extra half-plane is indicated in Fig. 5 (left). The transition between shuffle and glide is a climb process and thus involves the trapping or release of interstitial atoms or vacancies. As both types can either terminate on a line of Ga or a line of N atoms, overall there are four different types.

We investigated the two shuffle structures in hybrid models containing 412 atoms using the DFTB method [38]. The structures are geometrically optimised in a conjugate gradient algorithm until all forces are well below 5×10^{-3} eV/Å. Figure 5 (left) shows the projection of the relaxed 60° shuffle dislocation with N core. The {111} planes appear slightly bent, as the surface of the hybrid model was allowed to relax freely. With respect to the perfect lattice the bonds between the core atom 1 and its next Ga neighbours 2/3 are 7% compressed while those between 4, 5, 6, 7, and 8 are expanded 7%. Although the three-fold coordinated N core atoms (1) relax along $-[111]$ they remain sp^3 like, with a dangling bond each. These dangling bonds, as shown in Fig. 6, give rise to a band (a) above the valence band maximum. Further we find states (b) just above the Ga-3d band. For an uncharged dislocation line, band (a) is approximately half filled and lies ~ 0.2 eV above the valence band maximum.

Test calculations have further shown, that the exact position of the band is influenced by the strain environment of the surrounding lattice: Allowing a relaxation with torsion of the dislocation core shifts the band ~ 0.1 eV towards the valence band maximum compared to relaxation where torsion is prevented by constraints. All numbers given in this paper are calculated without allowing torsion.

The relaxed Ga core 60° shuffle dislocation is shown in Fig. 5 (right). Unlike for the N core type, the Ga–N bonds between columns (5/6) and (7/8) (corresponding to 5 and 6 for the N core type dislocation) are broken, allowing those columns to relax further outwards. Every second three-fold coordinated Ga atom

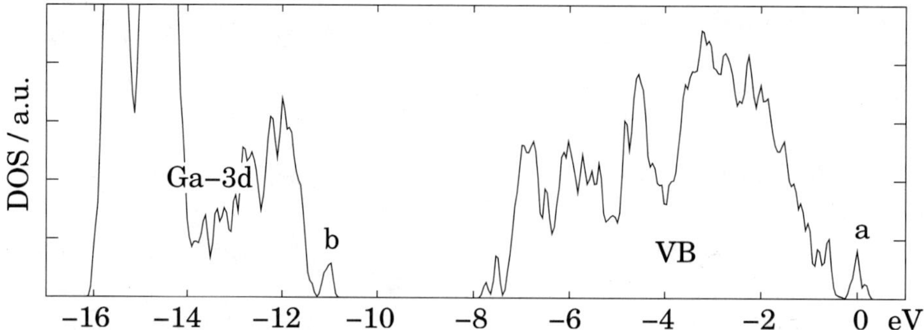

Fig. 6 Calculated density of states (DOS) of the 60° shuffle dislocations with N core (cubic GaN, DFTB method). The dislocation induces states (b) in the gap between the low lying states of the Ga-3d electrons and the valence band as well as states (a) above the valence band maximum. If we scale the band-gap (the conduction band minimum is not shown here) to fit the experimental value, we find (a) to be 0.2 eV above the valence band maximum. 0 eV refers to the fermi level for an uncharged dislocation line. Thus the dislocation induced band (a) is then half filled.

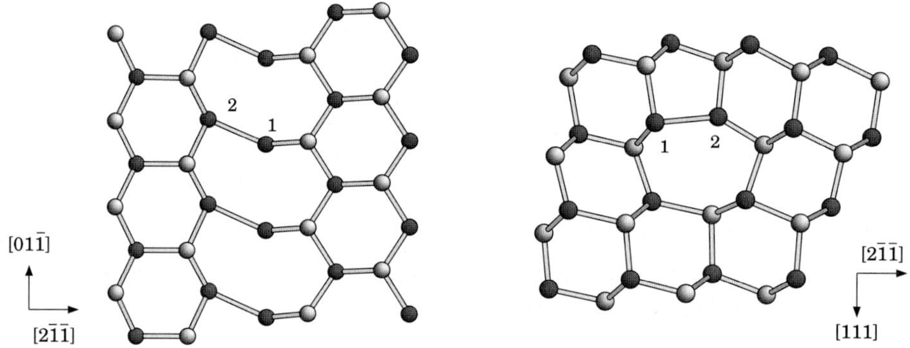

Fig. 7 Relaxed core of the 60° glide dislocation with Ga core (cubic GaN, DFTB method). Left: View projected onto the {111} glide plane. The dislocation lies vertically along [01$\bar{1}$]. Right: Cross sectional view along [01$\bar{1}$].

along the dislocation line is distorted inwards, forming a double period structure with a bond between the two Ga atoms 1 and 6 (2.5 Å), but no bond between 2 and 5. This effect goes alongside with a charge modulation of ±0.2 electrons along the dislocation line. The corresponding density of states (not shown) for the uncharged dislocation indicates an empty band around mid-gap and a partially filled band $\sim$ 0.3 eV above the valence band maximum.

Test calculations on the same structures in hexagonal GaN gave almost identical results. In both cases, the N and the Ga terminated 60° shuffle structure, the band induced close to the valence band maximum can be expected to act as an acceptor in p-type material. The position and character of this band appears to agree reasonably well with Hall experiments in cubic material [44] and CL studies in hexagonal material [42].

Removing the column of three-fold coordinated N atoms (1) from the 60° shuffle dislocation with N core (Fig. 5 (left)) leads to a glide dislocation with a Ga core. Figure 7 shows its corresponding relaxed structure. We find reconstructed Ga–Ga bonds with a length of 2.37 Å between atoms 1 and 2. For the neutral charge state, the reconstructed core structure gives rise to a filled band 0.3 eV above the valence band maximum and a half-filled band 1.6 eV above the valence band maximum and thus the dislocation will be positively charged in p-type material. Even under N-rich growth conditions we find the formation

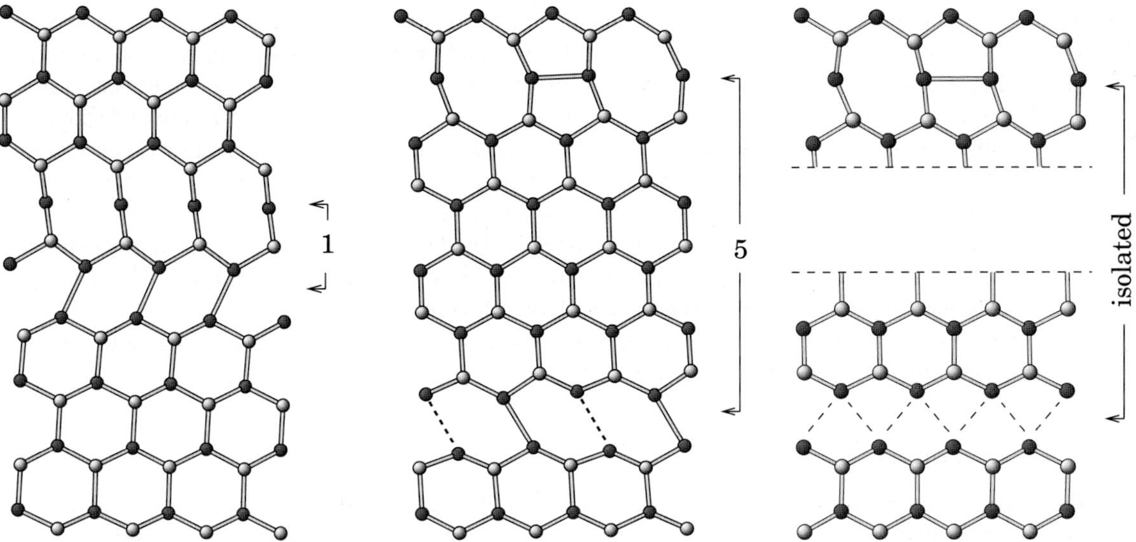

Fig. 8 The dissociated 60° glide dislocation (Ga core) in the glide plane. The dissociation is shown for three different distances between the partial dislocations. The arrows show the location of the partial cores, the 90° partial being on the bottom and the 30° partial on the top, separated by an intrinsic stacking fault. As discussed in the text, bond reconstruction and geometry of the core depends on the distance between the two partials. Left: Separation of 1 spacing of ~ 2.7 Å between the cores. Centre: Separation of 5 spacings of ~ 2.7 Å. Right: Isolated partials with no interaction.

energy of the 60° glide dislocation (Ga core) to be at least 0.3 eV/Å lower than that of the 60° shuffle dislocation (N core). Thus the Ga core glide type is more stable than the corresponding N core shuffle type.

First calculations on the N core glide structure, not shown, but similar in structure to the Ga core glide structure, gave occupied bands in the gap close to the valence band. These bands reach at least 0.5 eV into the gap. However, as determining the position of the valence band edge turned out to be difficult, further investigations are needed.

4.3 The dissociation of the 60° Ga core dislocation

In many tetrahedrally bonded semiconductors the 60° glide dislocation dissociates into two Shockley partial dislocations lying in the glide plane, and separated by an intrinsic stacking fault [10]. The dissociation reaction in cubic material is given by

$$\frac{1}{2}[1\bar{1}0] \longrightarrow \frac{1}{6}[1\bar{2}1] + \frac{1}{6}[2\bar{1}\bar{1}], \tag{5}$$

where the term on the right hand side corresponds to a 30° partial and the second to a 90° partial dislocation[4].

We modelled the two partials with the DFTB method in a 520-atom supercell-cluster hybrid. We were able to compare dislocations with different separations between the two partial dislocations. Figure 8 shows the relaxed Ga core structures for three different separations. The structures for the isolated partials

---

[4] Burgers vectors of perfect dislocations are given as lattice translations. However, since a partial dislocation is bordered by a stacking fault in the glide plane, the lattice offset between faulted and unfaulted region, $\frac{1}{6}\langle 112\rangle$ and $\frac{1}{3}\langle 1\bar{1}00\rangle$ for cubic and hexagonal material, respectively, has to be subtracted and hence allows smaller Burgers vectors, which are *not* given as lattice translations.

without any interaction were obtained from additional models containing only one partial dislocation each. One can see from Fig. 8 that the core structures of the 30° and the 90° partials strongly depend on their separation. For the smallest separation (1 spacing), the 90° partial (bottom) shows a single period bond reconstruction whereas the 30° partial (top) remains unreconstructed. For a larger separation of 5 spacings, we find a double period bond reconstruction for the 90° partial, where every second pair of Ga core atoms along the dislocation line is weakend. For the same separation also the core of the 30° partial shows a double period reconstruction. Here pairs of Ga atoms are bound along [01 $\bar{1}$] (bond length 2.46 Å). Finally, for very large separation we obtain quasi isolated partials. The 90° partial core remains unreconstructed but the 30° partial shows a similar double period structure as for 5 spacings separation. However, the reconstruction bond length of 2.81 Å is considerably larger than for 5 spacings.

The change in core structures with increasing separation leads to a corresponding change in the (bond) charge at the Ga core atoms. For details see Ref. [38].

4.4 Energetics

Whether dissociation really occurs can only be decided through the energy of the process. Isotropic elasticity theory gives two energy contributions: The reduction of the strain field and the corresponding elastic energies with increasing distance between the partials (90°–30° partial repulsion) and as a compensating effect the increase in stacking fault energy with increasing distance [10]. In addition to this, one has to consider the core energies of the undissociated dislocation and the two partials – contributions which cannot be obtained in elasticity theory only. Comparing the total energies of our models for different distances, which were all of the same stoichiometry, gives actually an increase in energy for the first three dissociation steps of almost 0.5 eV/Å [38]. The main contribution for small distance seems to be the difference in the core formation energies at the very first dissociation step. Also, for small distances, the electrostatic interaction between the two partials might play a role, since the core charge is then not fully screened. From the fourth step onwards the energy decreases again. However, due to computational restrictions, dissociation distances beyond 5 spacings could not be modelled yet. Therefore, at this point it cannot be decided if dissociation is overall energetically favourable. One would guess though, that the observed increase is just a short range barrier and for large distances the energy drops below the energy of the undissociated dislocation.

4.5 Basal plane dislocations – conclusions

We have modelled the core structures of perfect and dissociated basal plane 60° dislocations in GaN with the DFTB method. Test calculations with a pseudopotential approach (AIMPRO) in smaller models gave similar results. The 60° dislocations are electrically active and especially the undissociated 60° dislocation with a nitrogen terminated core can induce acceptor states roughly 0.2 eV above the valence band maximum. Thus these dislocations are good candidates for the ~0.166 eV acceptors observed experimentally in cubic GaN [44]. Our results are further in accordance with recent CL studies in hexagonal material [42].

We predict a considerable barrier of 0.4–0.5 eV/Å to the dissociation of the 60° dislocation. This barrier is several elementary dissociation steps wide. As such wide barriers are not easily overcome thermally, this might explain why no dissociation of basal plane 60° dislocations could be observed in indented and originally dislocation free GaN crystals [42].

5 Threading dislocations

As mentioned in the introduction, the second main class of dislocations in GaN are the threading dislocations – occuring in hexagonal GaN – with line directions along [0001]. Device quality hexagonal GaN can be grown using metalorganic chemical vapor phase deposition (MOCVD) on (0001) sapphire substrates.

The lattice mismatch results in extremely high dislocation densities near the interface but isolated threading dislocations of edge, screw or mixed type, with densities of up to 10^9 cm^{-2}, persist beyond 0.5μm above the interface region [6, 45, 46]. Hence they cross the active region of the devices. Unlike in AlN, where light emitting devices fail already at densities greater than 10^4 cm^{-2}, highly efficient GaN-based light emitting diodes (LEDs) have been reported to contain up to 10^{10} cm^{-2} [47, 48, 49]. However, in particular in GaN- and InGaN-based LEDs and laser diodes threading dislocations appear to form the leakage current pathway. Therefore there is a considerable effort to grow dislocation lean material. Epitaxial lateral overgrowth (ELOG) has been applied successfully to reduce the number of threading dislocations and produce long-lifetime devices [50, 51].

Despite years of investigation, the atomic and electronic structures of threading dislocations is still a current topic of interest and remains controversial [52]: Full-core structures have been seen for edge, screw, and mixed dislocations using Z-contrast imaging [53]. But also nanopipes associated with open-core threading screw dislocations have been imaged by transmission electron microscopy (TEM) [54, 55]. The electrical and structural properties, of screw dislocations have been found to depend sensitively on the growth conditions [56]. Full-core structures are expected to lead to gap states, which, if filled, would lead to dislocation line charging. Scanning capacitance spectroscopy results have suggested that GaN threading dislocations are negatively charged [57], as have electron holography measurements [58]. Contradicting the latter, in ballistic electron emission microscopy studies it was concluded that threading dislocations do not have a fixed negative charge, and are, if anything, positively charged at the surface [59].

Our initial theoretical investigations of GaN edge dislocations suggested that dislocation core reconstructions lead to a lack of deep gap states [60]. Subsequent calculations, however, have predicted deep levels in the top half of the band-gap associated with GaN edge dislocations, leading in particular to charge accumulation at dislocations [61, 62, 63]. Theoretical work has also shown that under Ga-rich growth conditions Ga-filled screw dislocations are more likely to be formed than full-core structures [39], which is compatible with experimental observations.

5.1 The threading edge dislocation

We modelled the pure threading edge dislocation, with a line direction $[0001]$ and a Burgers vector of $\frac{1}{3}[1\bar{2}10]$, using the DFTB method and AIMPRO. Figure 9 shows the relaxed geometry of a neutral threading edge dislocation in GaN using the pure supercell approach with a dislocation dipole (left) and the supercell-cluster hybrid approach (right) as described in Section 3. The coordinates displayed were aquired using the pseudopotential method (AIMPRO). As one can see, the large Burgers vector perpendicular to the line direction leads to substantial strain, resulting in the $(10\bar{1}0)$ planes being bent considerably in the hybrid model. This artificial bending does not occur in the pure supercell model as here the planes are constrained due to the boundary conditions. Furthermore, in the central row of three-fold atoms (A in Fig. 9 (left)) the Ga and N atoms are seperated along $[10\bar{1}0]$ by 0.21 and 0.31 Å for the supercell and hybrid model respectively.

As shown in Fig. 9, in both approaches localised states in the upper half of the gap are found. The DFTB method yields qualitatively the same (not shown)[5]. The isosurfaces given in Fig. 10 show the central peaks of the wavefunctions – the wavefunction tails extend into the strained region. The central part is mainly composed of Ga-sp orbitals. As can already be guessed from Fig. 10, the wavefunction associated with the lower gap state mixes orbitals related to Ga atoms in the columns labelled A, D, and F, whereas for the higher gap state mainly A, E, B, and H contribute.

Figure 11 shows the band structure of a negatively charged full-core edge dislocation, with 2 electrons per repeat distance (c_0) on the core of the dislocation. This is the charge state observed by electron holography [58]. The addition of electrons to the dislocation is stabilised by a change in atomic structure that

---

[5] The results found here are in variance with earlier ones, obtained in smaller cluster models. Those clusters gave rise to band-gaps larger than the experimental one. Empty dislocation bands with a broad dispersion can then be confused with conduction band states.

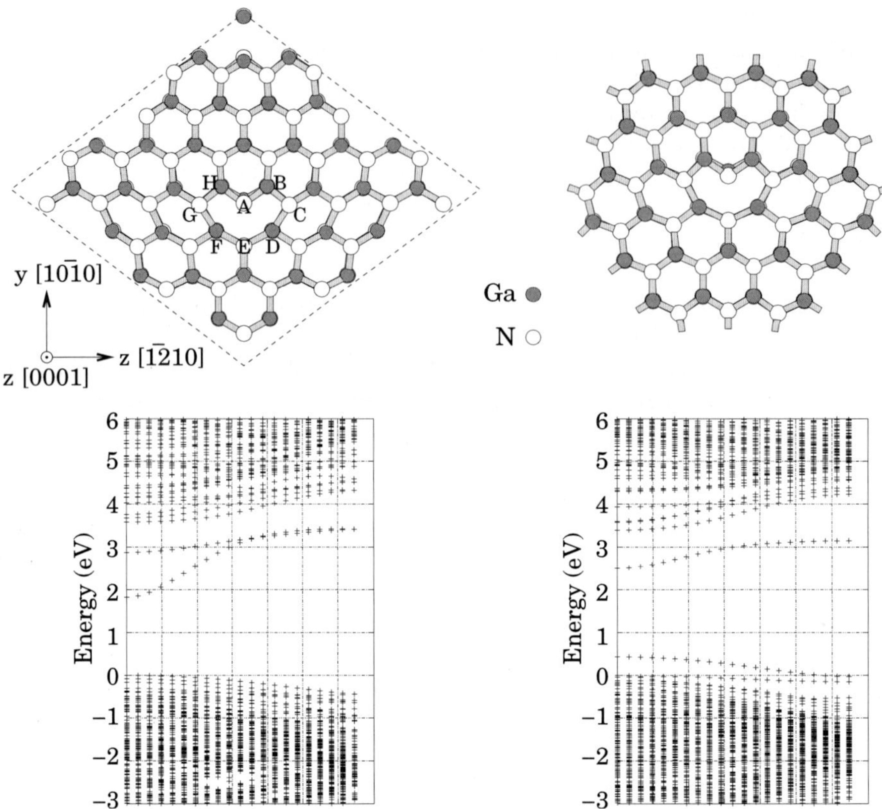

Fig. 9 GaN threading edge dislocation. The two complementary approaches are shown: Upper panels: A supercell containing two edge dislocations (left) and a hybrid model containing a single edge dislocation surrounded by vacuum (right). In the latter the surface bonds are saturated by fractional hydrogen atoms (not shown). Both structures have been fully relaxed (AIMPRO) in the neutral charge state. In the core the atomic columns are indexed with capital letters. Lower panels: The corresponding projected band structures in the [0001] direction. The reference energy has been set at the top of the filled valence states. In the neutral charge state all levels in the top half of the gap are empty, and respectively filled in the bottom half.

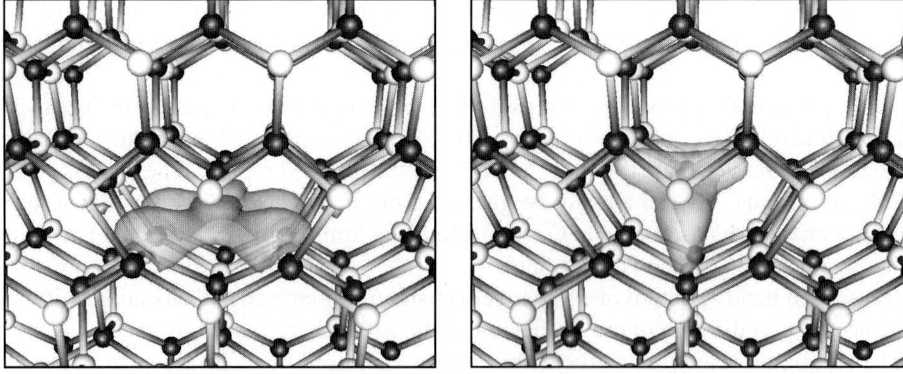

Fig. 10 Isosurface of the wave functions localized on the GaN edge dislocation, computed at Γ in the supercell shown in Fig. 9 (left). The left panel shows the lower gap state and the right one the upper.

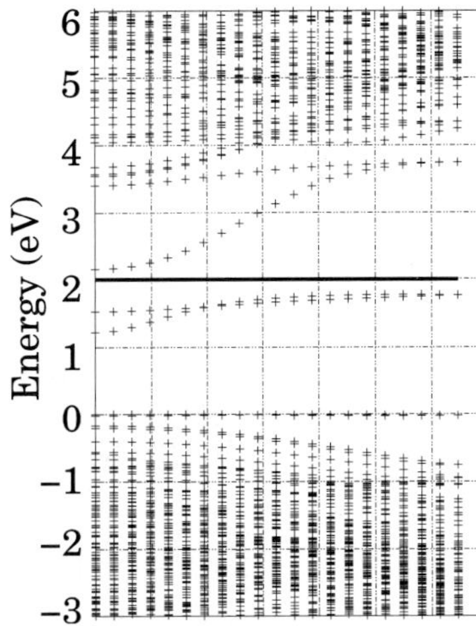

Fig. 11 Band structure of an edge dislocation charged with 2 electrons/c_0, calculated in the supercell shown in Fig. 9 (left). All atoms were relaxed in the charged state. The solid horizontal line separates filled and empty states (Fermi level). As explained in the text, the splitting of the two dislocation related bands at the zone centre is an artificial effect of the supercell approach.

lowers the localized gap states of the neutral dislocation, pushing them deeper into the gap. The change in structure has been reported previously by Wright and Grossner [61], and consists of a repulsive outward movement of the N atoms in columns A, E, F, and D and an attractive inward movement of the Ga atoms in these columns in the 8-fold ring at the dislocation core.

Summarising, these results conclusively show that neutral edge dislocations, expected to be present in p-type material, and also the negatively charged edge dislocation possess localised levels in the gap.

5.2   Comparison of the supercell and the hybrid approach:

Even though in terms of atomic and electronic structure both the pure supercell and the hybrid approach yield qualitatively the same results, some differences can be seen. In the supercell approach the strain induced by the dislocations is over-estimated with respect to an isolated dislocation in an otherwise perfect infinite crystal because of the boundary conditions of the computationally-restricted supercell size. Conversely, in the hybrid approach, because the hybrid surface is free to relax entirely, the dislocation strain is underestimated, and creates a bending of the $(10\bar{1}0)$ planes parallel to the Burgers vector. This distortion gives rise to an artificially raised filled level $\sim$0.4 eV above the valence band (Fig. 9 (right)). As the supercell contains a dislocation dipole with only a small separation of the two dislocations, a 1 eV splitting of the dislocation related band located at around 2.5 eV can be observed at the Brillouin zone centre (Fig. 9 (left)). Comparison with the analogous band in the hybrid approach (Fig. 9 (right)) shows that this artificial splitting approximately preserves the centre of mass. The same effect can be observed for the charged edge dislocation (Fig. 11).

5.3   Alternative core structures

We further investigated alternative cores structures of the edge dislocation, the empty core (the central row of atoms removed, A in Fig. 9 (left)) and N- and Ga-vacancy structures. These structures might occur, depending on the exact growth conditions and doping. The alternative cores were modelled with DFTB in 624 atom supercells of the same shape as in Fig. 9 (left), however period-doubled ($2\,c_0$) and with larger diameter. This minimises the dislocation–dislocation interaction and allows to simulate cross sectional high resolution transmission electron microscopy (HRTEM) images of edge dislocation surrounded by a considerable amount of bulk GaN. These simulated images were compared with defocus series from HRTEM experiments on unintentionally n-doped GaN grown in hydride vapour phase epitaxy (HVPE) and MOCVD under N-rich growth conditions (Remmele et al. [64]). This allowed the preliminary identification of a Ga-vacancy core structure, where all Ga atoms from the central row of atoms have been removed.

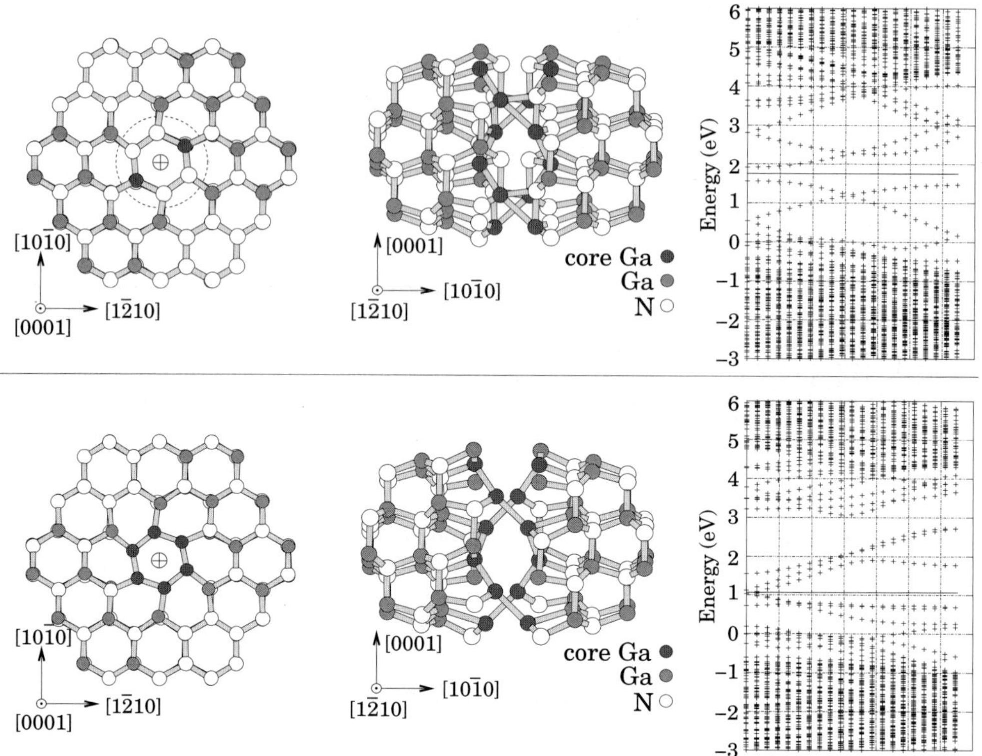

Fig. 12 Neutral [0001] threading screw dislocation. The upper panels show the full-core structure and the lower panels show the Ga-filled core structure. The N atoms are shown in white and Ga atoms in grey. Core Ga atoms are shaded darker grey to highlight the double helix of Ga bonds. Left panel: The relaxed atomic structures projected in the (0001) plane. The cores of the dislocations are indicated by a cross. In the full-core structure a dashed circle encompasses the core region which, if removed, leaves the lowest energy open core screw dislocation (see text for further details). Centre panel: The same structures projected into the (1$\bar{2}$10) plane. The supercell is repeated twice in the [0001] direction. Right panel: The corresponding projected band structures. The reference energy has been set at the estimated position of the bulk band edge. The solid horizontal lines separate filled and empty states in undoped material.

However it could not be excluded that the core structure fluctuates along the dislocation line and both Ga-vacancy and open-core structures are present in the samples investigated. Furthermore, Ruterana et al. [65][7] observed HRTEM contrast patterns in hexagonal GaN grown by molecular beam epitaxy (MBE), which can be explained by a 5/7 atom ring configuration – nothing else than an open core edge dislocation as discussed above.

5.4 The threading screw dislocation

In this work we consider three structures of the neutral screw dislocation: The full-core structure, the Ga-core structure [39], and the open-core structure. The relaxed atomic structure of a full-core threading screw dislocation is shown in Fig. 12. In the full-core model, within the central hexagon, the Ga atoms in adjacent [0001] columns are separated by 2.56 Å, creating additional loose bonds between them, shown highlighted in the top panel of Fig. 12. In the Ga-core model, the central Ga atoms are separated from neighboring Ga atoms by 2.46 Å. The corresponding band structures are also displayed, and show filled and empty states throughout the gap. The deep states are created by the dangling bonds of the three-fold coordinated N and Ga core atoms. Further analysis at the full-core structure reveals that the lower bands near the valence band are localised on N core atoms, whereas the bands close to the conduction band are of mixed Ga and N character.

Summarising, both the full-core and the Ga-core screw dislocation are electrically active. At least full-core screws have been observed experimentally in high resolution Z-contrast imaging [53]. Further studies using various forms of electron microscopy and atomic force microscopy in combination with photoluminescence experiments have confirmed that these dislocations act as non-radiative centres [66, 67].

Removing the hexagonal core of the threading screw dislocation – as indicated by a dashed circle in Fig. 12 – yields a core with a narrow opening of about 7.2 Å. The atoms on the walls of this opening adopt threefold coordination similar to those found on a $(10\bar{1}0)$ surface. Thus Ga (N) atoms develop sp^2 (p^3) hybridisation which lower the surface energy and clear the central region of the gap of deep states. The remaining states near the band edges are induced by the distortion arising from the strain field. This was verified by comparison with the undistorted $(10\bar{1}0)$ surface which shows no such states. Therefore screw dislocations with larger openings can be expected to be completely free of gap states. DFTB supercell calculations on the full-core and an open-core gave a 0.3 eV/Å lower formation energy for the open core structure described above. The energy required to form the surface at the wall is apparently compensated by the energy gained by removing the highly strained core region. As a further opening gave a higher line energy, we conclude that the equilibrium diameter is approximately 7.2 Å. Openings of the screw dislocation – three atomic rows wide – have indeed been confirmed experimentally [55], however, they seem not to occur very frequently. Recently the stability of threading screw dislocations has been discussed in depth, finding nitrogen lean core structures to be more stable than full or empty core – even under nitrogen rich growth conditions [68]. The latter results are consistent with experimental evidence [55, 69].

5.5 The simulation of EEL spectra

As explained in Section 2.3, spatially-resolved EELS allows to probe the electronic structure of extended defects. When taken on a dislocation core, EEL spectra can in principle yield valuable information about the electronic structure of the dislocation, both within and above the band-gap. In comparison with calculated EEL spectra, this information could prove crucial to help resolve whether any electrical activity is associated with undecorated dislocations. Energy-loss spectra induced by excitations of N core electrons conveniently allow unoccupied conduction and gap states to be probed. In particular, nitrogen K-edge spectra collected near a pure GaN edge dislocation have shown an increase in absorption just above the band-gap compared to bulk regions [53]. Low-loss EEL spectroscopy (Section 2.3) is another sensitive technique that can be performed in a scanning TEM that leads directly to detailed information about gap states. Cross-sectional low-loss EEL experiments in GaN have shown differences in the onset of the EEL spectrum on and off threading dislocations [35].

Figure 13 shows the obtained energy-loss spectra for edge dislocations in the supercell approach. In view of the gap states created by the edge dislocation, additional absorption in the EEL spectrum is found below the bulk band edge. Absorption with the electron beam in the basal plane is seen to lead to absorption at lower energies than with the beam oriented along the [0001] axis. Supplementary absorption is also observed in the 5–7 eV range. Test calculations using the hybrid model gave a stronger rise of the spectra above 6 eV, although the changes in absorption compared with bulk are similar to those found in Fig. 13 (left)[6]. These results illustrate the effect of boundary conditions when calculating the electronic properties of dislocations. The computed EEL spectrum for a negatively charged edge dislocation is shown in Fig.

---

[6] As mentioned in Section 5.1, the dipole approach leads to an artificial splitting of dislocation induced bands near the zone centre. As the splitting preserves the centre of mass, this results in a broadening of peaks in the spectra without considerable effect on the peak positions. However since also those transitions away from $k = 0$ are included in the EELS simulation, the broadening is much less pronounced than one would expect from the rather wide splitting at $k = 0$.

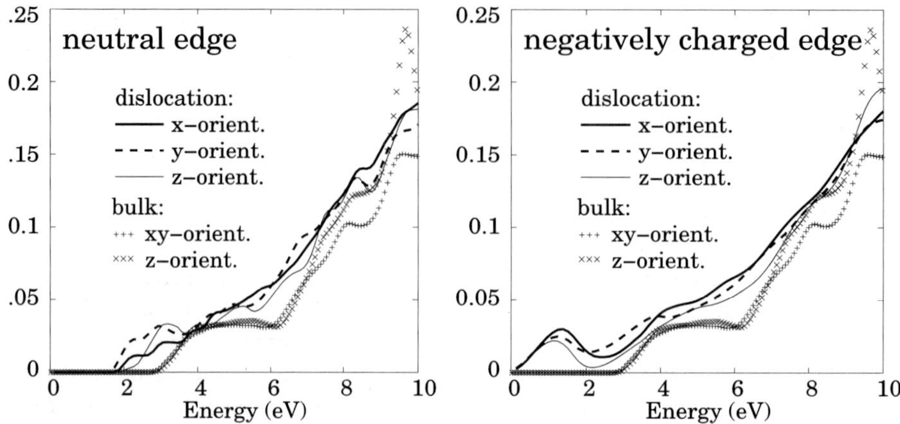

Fig. 13 Comparison between the computed EEL spectra of a region containing a GaN edge dislocation (lines) and bulk GaN (symbols). The results are shown for the supercell model shown in Fig. 9 (left). Results are given for an electron beam oriented along x, y, and z, using the coordinate system defined in Fig. 9. Left: The neutral threading edge dislocation. Right: The negatively charged threading edge dislocation. The dislocation carries a charge of two electrons per c_0.

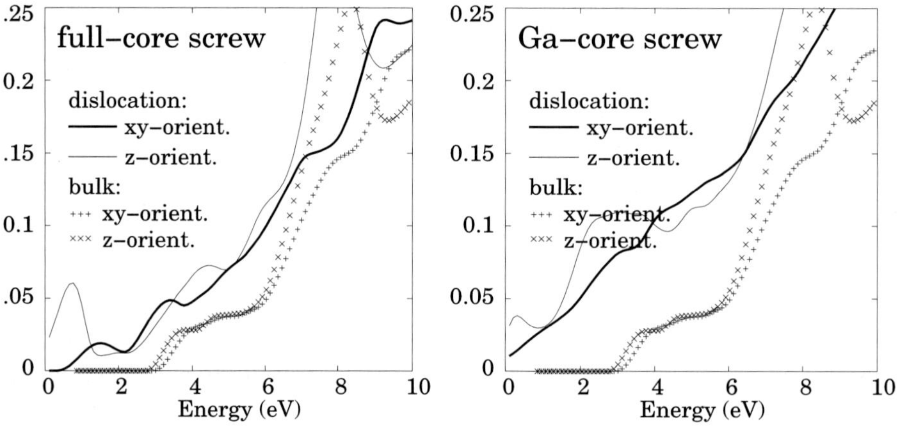

Fig. 14 Comparison between the computed EEL spectra of a region containing a neutral screw dislocation (lines) and bulk GaN (symbols). Results are given for an electron beam oriented along x, y, and z, using the coordinate system defined in Fig. 9. Because of the threefold symmetry of the screw dislocation, EEL spectra for x and y orientations are identical. Left: The full-core screw dislocation. Right: The Ga-core screw dislocation.

13 (right). The absorption in the band-gap for all three orientations of the electron beam is shifted to lower energies compared to the neutral charge state, and consists of a broad peak centred around 1 eV.

Figure 14 gives the theoretical energy-loss spectra for the two neutral screw dislocation structures shown in Fig. 12. Strong absorption within the gap region is found and clear differences between x, y- (perpendicular to c) and z-oriented (parallel to c) beams are noted. The Ga-core structure appears to differ much more than the full-core structure from the bulk spectrum. This stems from the metallic-like bonding in the centre of the Ga-core dislocation. Indeed, the EEL spectrum approaches a linear behavior at low energy losses – particularly for x, y orientations – as predicted from the dielectric function of a free-electron

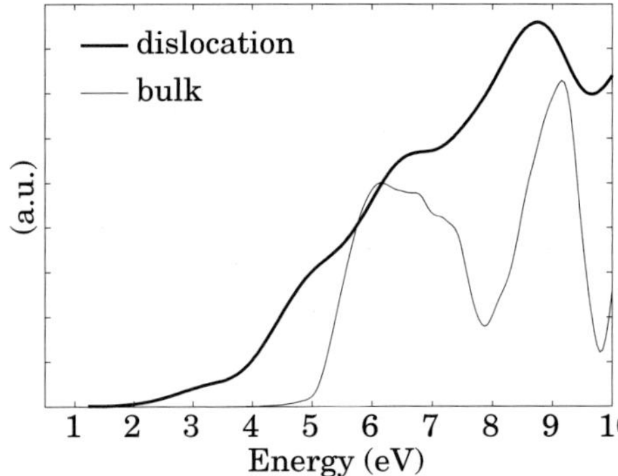

Fig. 15 Theoretical nitrogen K-edge core-level spectrum for an edge dislocation (thick line), in neutral or p-type GaN, compared to a similar bulk spectrum (thin line). The energy scale is referenced to the top of the valence band.

jellium metal [28]. These results demostrate how sensitively the EEL spectrum depends on the bonding within the dislocation core – and how useful low-loss EELS might be to identify certain core structures.

As discussed in Section 2.3, in a similar approach also core-EEL spectra can be simulated. At a dislocation the strain induces, via a piezoelectric effect, an electrical field. This field then affects the energies of the electron core-levels and the shift can be observed in the core-EEL spectra. We investigated this effect for the N-core levels near a neutral threading edge dislocation (Fig. 9 (left)). Atoms in the compressive region near the dislocation (in particular columns A, B and H) appear with shallower core levels than those in the tensile region (in particular columns D, E and F). Differences in potential are as large as 1 eV, a significant fraction of the band-gap[7]. These potential shifts broaden the core EEL spectra as they directly modify the energies of the N-core states. The final empty states, however, are more delocalised and hence average over the potential fluctuations. Figure 15 shows the theoretical N K-edge EEL spectrum for a dislocation compared with one taken in a bulk region. The bulk core EEL spectrum displays a strong absorption only above 5 eV (relative to the valence band maximum), a value significantly larger than the band-gap. Indeed, states near the conduction band minimum are formed mainly of Ga orbitals, and hence have only a small overlap with the initial N-core state. This effect may cause difficulties in identifying all gap states from the sole observation of N-related core EEL spectra. Near the dislocation, substantial absorption is predicted within the bulk band-gap. We also note that the peaks found in the bulk spectrum are smeared out near the dislocation because of the core-level shifts.

For charged dislocations, we expect an even stronger smearing of the peaks in view of the supplementary potential changes at N sites. Therefore, core excitation EEL spectra should be interpreted with caution, as absorption seen below the conduction band edge does not necessarily imply the presence of an empty gap state at that energy in the band-gap. We also note that the potential shifts found near uncharged dislocations are not expected to significantly influence low-loss EEL spectra as both the initial and the final states are delocalized and therefore weakly susceptible to short-range potential fluctuations.

5.6   Interaction with oxygen

The most commonly observed parasitic defect emission in n-type hexagonal GaN grown on sapphire is the yellow luminescence (YL), centred around 2.2 to 2.3 eV [71]. Most models for the YL in GaN assume a transition between a shallow donor and a deep acceptor [72, 73]. Cathodoluminescence (CL) studies have shown that the YL is spatially non-uniform but correlated with extended defects, in particular with low angle grain boundaries containing dislocations [74]. The most likely acceptor is an intrinsic defect like a Ga vacancy, V_{Ga}, probably complexed with an impurity. Indeed, positron anihilation experiments found the concentration of V_{Ga}, or of vacancy complexes, to be related to the intensity of the YL [75]. There is experimental evidence that oxygen acts as a donor in bulk GaN [76] and total energy calculations show

---

[7] Using the piezoelectric parameters of Park and Chuang [70] combined with the stress field as obtained in linear elasticity theory gives shifts of the electrostatic potential of the same order of magnitude.

that O substitutionally replaces N [77]. Theory has therefore suggested that V_{Ga} forms defect complexes with oxygen which might be involved with the YL [77, 78].

5.7 Oxygen trapped at edge dislocations

The V_{Ga}–$(O_N)_2$ and V_{Ga}–(O_N) complexes in GaN act as single and double acceptors, respectively. If these defects are trapped in the strain field of a dislocation, then we would expect an intense YL localised at the dislocation. To test the trapping, we placed several defects (V_{Ga}^{3-}, O_N^+, V_{Ga}–$(O_N)^{2-}$, V_{Ga}–$(O_N)_2^-$, V_{Ga}–$(O_N)_3$) far away from the dislocation in a period-doubled 624 atom supercell containing an edge dipole. The structure was relaxed with the DFTB method. Their energies were compared with the same defects sited within the dislocation core. For V_{Ga}^{3-} and O_N^+ we found the formation energy at the dislocation to be approximately 1.6 eV lower than in bulk GaN. An even more dramatical drop in formation energy is found for the V_{Ga}–$(O_N)_n$ complexes: Here the neighbouring pair of threefold coordinated Ga and N atoms is removed from the core and an O atom is inserted into the N site. The oxygen atom then lies in bridge site between two Ga atoms in a normal bonding configuration – the formation energies are lowered by 2 to 3 eV (see earlier work [79] for more details). Hence, if oxygen is mobile, the dislocation core will spontaneously oxidise. The resulting high density of electrically active donor acceptor pairs decorating the dislocation will possibly give rise to a localised broad band YL. Similar mechanisms can be imagined at other extended defects, however, unless these extended defects have similar bonding configurations as the threading edge dislocation, the binding energies with V_{Ga}–$(O_N)_n$ complexes will probably be considerably lower.

5.8 The formation of nanopipes

Nanopipes in hexagonal GaN thread along the $[0001]$ axis and appear with hexagonal cross sections. They are enclosed by $\{10\bar{1}0\}$ type walls. Neither our DFTB and pseudopotential calculations [60] nor Frank's theorem [80] support the idea of nanopipes being simply empty core screw dislocations of large diameter. Experimentally it was found, that both the diameter and densities of nanotubes increase with the presence of impurities like O, Mg, In, and Si [55]. It was argued that these impurities decorate the $\{10\bar{1}0\}$ walls of the nanopipes inhibiting overgrowth. As oxygen is the main source of unintentional doping, we investigated the low energy sites of O substituting N on $\{10\bar{1}0\}$ type surfaces [81]. We find that the energy of a neutral O_N defect is 0.8 eV lower at the relaxed $\{10\bar{1}0\}$ surface. This shows there is a tendency for oxygen to segregate to the surface. The additional electron of the added oxygen occupies a state near the valence band maximum. Therefore this high energy defect would attract acceptors resulting in a neutral complex. Similar to the situation at the edge dislocation, Ga vacancies are possible acceptors. With V_{Ga} being a triple acceptor, neutral V_{Ga}–$(O_N)_3$ complexes are likely to form at the surface. Our calculations show, that V_{Ga}–$(O_N)_3$ is more stable at a $\{10\bar{1}0\}$ type surface than in bulk GaN by 2.2 eV. Two O neighbours of the surface vacancy lie below the surface and each is bonded to three Ga atoms, but the surface O is bonded to only two subsurface Ga atoms in a normal oxygen bridge site. The defect is electrically inactive with the the O atoms passivating the vacancy in the same way as VH_4 in Si. For details see [81].

The question then arises as to the influence of the defect on the growth of the material. Growth over the defect must proceed by adding a Ga atom to the vacant site but this leaves three electrons in shallow levels near the conduction band resulting in a very high energy. Hence the defect will stabilise the surface and thus inhibit further growth. Assuming that during growth of the epilayers, either nanopipes with very large radii are formed wich gradually shrink when their surfaces grow out, or if there is a rapid drift of oxygen to a pre-existing nanopipe, then V_{Ga}–$(O_N)_3$ complexes can lead to stabilised nanopipes. In either scenario the concentration of oxygen at the walls of the nanopipe increases. The maximum concentration of this defect would be reached if half of the first layer N atoms and all of the second layer N atoms were replaced by O, preventing further growth. It is however likely that far less than the maximum concentration is sufficient to stabilise the surface and prevent further shrinkage of the nanopipe. Provided oxygen could diffuse to the surface fast enough, the diameter and density of the nanopipe holes would be related to the initial oxygen concentration in the bulk. This model requires that the walls of the nanopipe are coated with oxygen although the initial stages of the formation of the nanopipe are obscure.

5.9 Threading dislocations – conclusions

Since the atomic and electronic structure of threading dislocations in hexagonal GaN is still discussed controversially, with contradicting results from different ab initio calculations and experiments, we re-investigated the threading edge and the threading screw dislocation in supercell and hybrid models. In both approaches the neutral edge dislocation is found with empty states in the upper half of the gap. The negatively charged edge dislocation as well as neutral screw dislocations with a full-core and a Ga-filled core induce even more gap states.

Based on our pseudopotential calculations we computed low-loss and core-EELS spectra of threading dislocations in GaN to compare them with experimental studies of high energy resolution. Experimentally dislocations are found to perturb the EEL spectrum eliminating the plateau seen in bulk GaN spectra [82]. However, so far there is no indication of deep mid-gap states on dislocations in the obtained spectra. There is some evidence for gap states close to the conduction band but no dramatic difference between p and n-type regions as expected from theory. It may be that the doping is ineffective or that impurities or point defects affect the experimental spectra actually removing states in the gap. However, calculations incorporating hydrogen into the dislocation core in GaN did not find this to occur. One must bear in mind though that there is a tradeoff in low-loss EEL measurements between deep levels on the one hand, which create strong changes in the EEL spectra, but are very localised and hard to pinpoint spatially, and shallow levels on the other hand, which lead to smaller changes in absorption, but are easier to probe spatially. Therefore, to obtain representative spectra, precisely positioning the electron probe on the core of the dislocation is of importance but unfortunately not aquired easily.

We found the stress field and core bonding configuration of threading edge dislocations likely to trap gallium vacancies and oxygen as well as their complexes. The gallium vacancy and oxygen related defect complexes are electrically active and suggest that they increase the intensity of the yellow luminescence near threading edge dislocations consistent with cathodoluminescence studies [74].

Oxygen tends to segregate to the $\{10\bar{1}0\}$ surface and forms stable and chemically inert V_{Ga}–$(O_N)_3$ defects. In a scenario for nanopipe growth, these defects increase in concentration when the internal surfaces of the nanopipe grow out. When a critical concentration of the order of a monolayer is reached, further growth is prevented. The model leads to nanopipes with $\{10\bar{1}0\}$ walls coated with GaO and supports the suggestions of Lilienthal-Weber et al. [55] that nanopipes are linked to the presence of impurities.

6 Summary

We have presented our theoretical work on dislocations in cubic and hexagonal gallium nitride. Two different methods were applied to model dislocations: A densitiy functional theory based tight-binding scheme (DFTB) and a pseudopotential method (AIMPRO). The former, being more approximate and computationally less demanding, allowed a convenient embedding of a dislocation core into a rather large region of bulk GaN, giving good estimates for structures and energies, whereas the strong point of the latter method was the precise modelling of the associated electronic structures. This combination of methods allowed to cover various aspects and effects of dislocations in gallium nitride, including core reconstructions and energies, dissociation energetics, the interaction with point defects, the electronic structure and based on the latter also the simulation of electron energy-loss spectra obtained at dislocation cores.

The predicted gap levels at basal plane dislocations were in accordance with Hall [44] and cathodoluminescence measurements [42] and a calculated barrier to disscociation in the basal plane might explain why this dissociation of $60°$ dislocations could not be observed in recent indentation experiments [42].

Our re-investigation of the threading dislocations lead to more conclusive statements in the ongoing debate concerning their electronic structure: The threading edge dislocation is found to possess states in the upper half of the band-gap, which contradicts earlier predictions obtained in less precise calculations.

The trapping mechanism found for Ga vacancy–oxygen complexes at threading dislocations and further at $\{10\bar{1}0\}$ type surfaces can explain both the formation of nanopipes and the spatial correlation of the observed yellow luminescence with dislocations.

Acknowledgements We thank T. Remmele, M. Albrecht, D. As, and K. Lischka for collaboration and discussions. This work was supported by the Deutsche Forschungsgemeinschaft (Schwerpunktprogramm Gruppe III-Nitride, Grant No. FR889/10, Germany) and the Engineering and Physical Sciences Research Council (Grant No. GR/M53523, UK).

References

[1] S. Nakamura and G. Fasol, The Blue Laser Diode (Springer, Berlin, 1997).

[2] S. Nakamura, M. Senoh, S. Nagahama, N. I. T. Yamada, T. Matsushita, H. Kiyohu, and Y. Sguimoto, Jpn. J. Appl. Phys. **35**, L74 (1996).

[3] H. Yang, L. X. Zheng, J. B. Li, X. J. Wang, and X. W. Hu, Appl. Phys. Lett. **74**, 2498 (1999).

[4] D. J. As, A. Richter, J. Busch, M. Lübbers, J. Mimkes, and K. Lischka, Appl. Phys. Lett. **76**, 13 (2000).

[5] D. C. Look and J. R. Sizelove, Phys. Rev. Lett. **82**, 1237 (1999).

[6] X. J. Ning, F. R. Chien, and P. Pirouz, J. Mater. Res. **11**, 580 (1996).

[7] P. Ruterana and G. Nouet, phys. stat. sol. (b) **227**, 177 (2001).

[8] D. Porezag, T. Frauenheim, T. Köhler, G. Seifert, and R. Kaschner, Phys. Rev. B **51**, 12947 (1995).

[9] T. Frauenheim, G. Seifert, M. Elstner, Z. Hajnal, G. Jungnickel, D. Porezag, S. Suhai, and R. Scholz, phys. stat. sol. (b) **217**, 41 (2000).

[10] J. P. Hirth and J. Lothe, Theory of Dislocations, Vol. 2 (Wiley, New York, 1982).

[11] A. T. Blumenau, R. Jones, S. Öberg, P. R. Briddon, and T. Frauenheim, Phys. Rev. Lett. **87**, 187404 (2001).

[12] A. T. Blumenau, M. I. Heggie, C. J. Fall, R. Jones, and T. Frauenheim, Phys. Rev. B **65**, 205205 (2002).

[13] C. J. Fall, A. T. Blumenau, R. Jones, P. R. Briddon, T. Frauenheim, A. Gutiérrez-Sosa, U. Bangert, A. E. Mora, J. W. Steeds, and J. E. Butler, Phys. Rev. B **65**, 205206 (2002).

[14] R. Jones and P. R. Briddon, in: Identification of Defects in Semiconductors, edited by M. Stavola, Semiconductors and Semimetals, Vol. 51a, Chap. 6 (Academic Press, Bristol, Boston, MA, 1998).

[15] P. Briddon and R. Jones, phys. stat. sol. (b) **217**, 131 (2000).

[16] X. Blase, K. Lin, A. Canning, S. G. Louie, and D. C. Chrzan, Phys. Rev. Lett. **84**, 5780 (2000).

[17] G. B. Bachelet, D. R. Hamann, and M. Schlüter, Phys. Rev. B **26**, 4199 (1982).

[18] H. J. Monkhorst and J. D. Pack, Phys. Rev. B **13**, 5188 (1976).

[19] C. J. Fall, R. Jones, P. R. Briddon, A. T. Blumenau, T. Frauenheim, and M. I. Heggie, Phys. Rev. B **65**, 245304 (2002).

[20] M. Leszczynski, H. Teisseyre, T. Suski, I. Grzegory, M. Bockowski, K. Pakula, J. M. Baranowski, C. T. Foxon, and T. S. Cheng, Appl. Phys. Lett. **69**, 73 (1996).

[21] J. Singh, Physics of semiconductors and their heterostructures (McGraw-Hill, New York, 1993).

[22] M. Ueno, M. Yoshida, A. Onodera, O. Shimomura, and K. Takemura, Phys. Rev. B **49**, 14 (1994).

[23] M. Ueno, A. Onodera, O. Shimomura, and K. Takemura, Phys. Rev. B **45**, 10123 (1992).

[24] Y.-N. Xu and W. Y. Ching, Phys. Rev. B **48**, 4335 (1993).

[25] A. F. Wright and J. S. Nelson, Phys. Rev. B **50**, 2159 (1994).

[26] J. Geiger, Elektronen und Festkörper (Vieweg, Braunschweig, 1968).

[27] P. Schattschneider, Fundamentals in inelastic scattering (Springer, Wien, New York, 1986).

[28] D. Nozières and D. Pines, Phys. Rev. **113**, 1254 (1959).

[29] P. M. Platzman and P. A. Wolff, Waves and interactions in solid state plasmas, Solid State Phys., Vol. 13 (Academic Press, New York, London, 1972).

[30] H. Ehrenreich and M. H. Cohen, Phys. Rev. **115**, 786 (1959).

[31] G. F. Bassani and G. P. Parravicini, Electronic states and optical transitions in solids, International series of monographs in the science of the solid state, edited by R. A. Ballinger, Vol. 8 (Pergamon Press, Oxford, 1975).

[32] L. X. Benedict, T. Wethkamp, K. Wilmers, C. Cobet, N. Esser, E. L. Shirley, W. Richter, and M. Cardona, Solid State Commun. **112**, 129 (1999).

[33] R. Buczko, G. Duscher, S. J. Pennycook, and S. T. Pantelides, Solid State Commun. **112**, 129 (1999).

[34] I. Arslan and N. D. Browning, Phys. Rev. B **65**, 075310 (2002).

[35] A. Gutiérrez-Sosa, U. Bangert, A. J. Harvey, W. R. Flavell, W. R. Jacobs, I. Moermann, and A. Rizzi, Inst. Phys. Conf. Ser. **169**, 255 (2001).

[36] M. Heggie and M. Nylén, Philos. Mag. B **50**, 543 (1984).

[37] T. A. Arias and J. D. Joannopoulos, Phys. Rev. Lett. **73**, 680 (1994).

[38] A. T. Blumenau, J. Elsner, R. Jones, M. I. Heggie, S. Öberg, T. Frauenheim, and P. R. Briddon, J. Phys., Condens. Matter **12**, 10223 (2000).

[39] J. E. Northrup, Appl. Phys. Lett. **78**, 2288 (2001).

[40] Z. Lilienthal-Weber, C. Kisielowski, S. Ruvimov, Y. Chen, and J. Washburn, J. Electron. Mater. **25**, 1545 (1995).

[41] N. E. Lee, R. C. Powell, Y. W. Kim, and J. E. Greene, J. Vac. Sci. Technol. A **13**, 2293 (1995).

[42] M. Albrecht, H. P. Strunk, J. L. Weyher, I. Grzegory, S. Porowski, and T. Wosinski, J. Appl. Phys. **92**, 2000 (2002).

[43] S. Kaiser and D. J. As (1999), TEM experiments, private communication.

[44] D. J. As, A. Richter, J. Busch, B. Schöttker, M. Lübbers, J. Mimkes, D. Schikora, K. Lischka, W. Kriegseis, W. Burkhardt, et al., MRS Internet J. Nitride Semicond. Res. **5S1**, W3.81 (2000).

[45] X. H. Wu, L. M. Brown, D. Kapolnek, S. Keller, B. Keller, S. P. DenBaars, and S. J. Speck, J. Appl. Phys. **80**, 3228 (1996).

[46] F. A. Ponce, D. Cherns, W. T. Young, and J. W. Steeds, Appl. Phys. Lett. **69**, 770 (1996).

[47] S. D. Lester, F. A. Ponce, M. G. Cranford, and D. A. Steigerwald, Appl. Phys. Lett. **66**, 1249 (1996).

[48] L. Sugiura, J. Appl. Phys. **81**, 1633 (1997).

[49] D. Cherns, J. Phys., Condens. Matter **12**, 10205 (2000).

[50] S. Nakamura, M. Senoh, S. Nagahama, N. Iwasa, T. Yamada, T. Matsushita, H. Kiyoku, Y. Sugimoto, T. Kozaki, H. Umemoto, et al., J. Cryst. Growth **190**, 820 (1998).

[51] S. Nakamura, M. Senoh, S. Nagahama, N. Iwasa, T. Matushita, and T. Mukai, MRS Internet J. Nitride Semicond. Res. **4**, G1.1 (1999).

[52] S. C. Jain, M. Willander, J. Narayan, and R. Van Overstaeten, J. Appl. Phys. **87**, 965 (2000).

[53] Y. Xin, S. J. Pennycook, N. Browning, P. D. Nellist, S. Sivananthan, F. Omnès, B. Beaumont, J. P. Faurie, and P. Gibart, Appl. Phys. Lett. **72**, 2680 (1998).

[54] P. Vennéguès, B. Beaumont, M. Vaille, and P. Gibart, Appl. Phys. Lett. **70**, 2434 (1997).

[55] Z. Lilienthal-Weber, Y. Chen, S. Ruvimov, and J. Washburn, Phys. Rev. Lett. **79**, 2835 (1997).

[56] J. W. P. Hsu, M. J. Manfra, S. N. G. Chu, C. H. Chen, L. Pfeiffer, and R. J. Molnar, Appl. Phys. Lett. **78**, 3980 (2001).

[57] D. M. Schaadt, E. J. Miller, E. T. Yu, and J. M. Redwing, Appl. Phys. Lett. **78**, 88 (2001).

[58] D. Cherns and C. G. Jiao, Phys. Rev. Lett. **87**, 205504 (2001).

[59] H.-J. Im, Y. Ding, J. P. Pelz, B. Heying, and J. Speck, Phys. Rev. Lett. **87**, 106802 (2001).

[60] J. Elsner, R. Jones, P. K. Sitch, V. D. Porezag, M. Elstner, T. Frauenheim, M. I. Heggie, S. Öberg, and P. R. Briddon, Phys. Rev. Lett. **79**, 3672 (1997).

[61] A. F. Wright and U. Grossner, Appl. Phys. Lett. **73**, 2751 (1998).

[62] K. Leung, A. F. Wright, and E. Stechel, Appl. Phys. Lett. **74**, 2495 (1999).

[63] S. M. Lee, M. A. Belkhir, X. Y. Zhu, Y. H. Lee, Y. G. Hwang, and T. Frauenheim, Phys. Rev. B **61**, 16033 (2000).

[64] T. Remmele, M. Albrecht, H. P. Strunk, A. T. Blumenau, M. I. Heggie, J. Elsner, T. Frauenheim, H. P. D. Schenk, and P. Gibart, in: Microscopy of Semiconducting Materials 2001, Inst. Phys. Conf. Ser. Vol. 169 (IOP, Bristol, UK, 2001), p. 323.

[65] P. Ruterana, V. Potin, and G. Nouet, Mater. Res. Soc. Symp. Proc. **482**, 435 (1998).

[66] T. Hino, S. Tomiya, T. Miyajima, K. Yanashima, S. Hashimoto, and M. Ikeda, Appl. Phys. Lett. **76**, 3421 (2000).

[67] S. J. Rosner, E. C. Carr, M. J. Ludowise, G. Girlami, and H. I. Erikson, Appl. Phys. Lett. **70**, 420 (1997).

[68] J. E. Northrup, Phys. Rev. B **66**, 045204 (2002).

[69] D. Cherns, W. T. Young, J. W. Steeds, F. A. Ponce, and S. Nakamura, J. Cryst. Growth **178**, 201 (1997).

[70] S.-H. Park and S.-L. Chuang, Phys. Rev. B **59**, 4725 (1999).

[71] C. V. Reddy, K. Balakrishhan, H. Okumura, and S. Yoshida, Appl. Phys. Lett. **73**, 244 (1998).

[72] T. Ogino and M. Aoki, Jpn. J. Appl. Phys. **19**, 2395 (1980).

[73] M. Godlewski, Mater. Sci. Forum **258/63**, 1194 (1997).

[74] F. A. Ponce, D. B. Bour, W. Gotz, and P. J. Wright, Appl. Phys. Lett. **68**, 57 (1996).

[75] K. Saarinen, T. Laine, S. Kuisma, J. Nissilä, P. Hautojärvi, L. Dobrzynski, J. M. Baranowski, K. Pakula, R. Stepniewski, M. Wojdak, et al., Phys. Rev. Lett. **79**, 3030 (1997).

[76] C. Wetzel, T. Suski, J. W. Ager, E. R. Weber, E. E. Haller, S. Fischer, B. K. Meyer, R. J. Molnar, and P. Perlin, Phys. Rev. Lett. **78**, 3923 (1997).

[77] J. Neugebauer and C. G. Van de Walle, Festkörperprobleme **35**, 25 (1996).

[78] T. Mattila and R. M. Nieminen, Phys. Rev. B **55**, 9571 (1997).

[79] J. Elsner, R. Jones, M. Haugk, T. Frauenheim, M. I. Heggie, S. Öberg, and P. R. Briddon, Phys. Rev. B **58**, 12571 (1998).

[80] F. C. Frank, Acta Crys. **4**, 497 (1951).

[81] J. Elsner, R. Jones, M. Haugk, R. Gutierrez, T. Frauenheim, M. I. Heggie, S. Öberg, and P. R. Briddon, Appl. Phys. Lett. **73**, 3530 (1998).

[82] A. Gutiérrez-Sosa, U. Bangert, A. J. Harvey, C. J. Fall, R. Jones, P. Briddon, and M. I. Heggie, Phys. Rev. B **66**, 035302 (2002).

phys. stat. sol. (c) **0**, No. 6, 1710–1731 (2003) / **DOI** 10.1002/pssc.200303130

Lattice dynamics in GaN and AlN probed with first- and second-order Raman spectroscopy

U. Haboeck[*], H. Siegle, A. Hoffmann, and **C. Thomsen**

Technische Universität Berlin, Institut für Festkörperphysik, Sekr. PN 5-4, Hardenbergstr. 36, D-10623 Berlin, Germany

Received 4 March 2003, revised 15 May 2003, accepted 6 June 2003
Published online 28 August 2003

PACS 63.20.Dj, 78.20.-e, 78.30.Fs

We present a selection of our contributions to basic research on the lattice dynamical properties of group-III nitrides and their alloys. We used first-order Raman scattering to determine the zone-center phonons and their dependence on structural attributes such as stress, chemical composition, impurities, and doping. Results on the angular dispersion of the polar modes, strain distribution, coupled LO-phonon plasmon modes, multi-mode behavior in $Al_xGa_{1-x}N$, and the quantitative determination of the phase purity of cubic and hexagonal GaN are shown. Second-order Raman-scattering experiments on GaN and AlN provide information on the vibrational states throughout the entire Brillouin zone. Based on a comparison of experimental data and calculated phonon-dispersion curves we assigned the observed structures to particular phonon branches and points in the Brillouin zone. We also discuss the behavior of the optical modes under large hydrostatic pressure.

1 Introduction During the last decade group-III nitrides have been an object of increasing research because of their interesting physical properties such as an adjustable band gap (1.9 eV to 6.2 eV) by varying the composition of the alloys [1] and application as basic materials for optoelectronic devices working in the blue and ultraviolet spectral region [2–4]. In particular GaN and AlN can withstand high temperatures and have large piezoelectric constants which makes them suitable for applications in high-frequency devices, sensors and low-dimensional structures [5, 6].

Although basic information about group-III nitrides e.g. crystal- and band structure have been available for a decade several details still remained unclear. In the following review of our work we present results of first- and second-order Raman investigations on various GaN, AlN, and $Al_xGa_{1-x}N$ samples and point out some remarkable details of the structural and optical properties of these materials.

2 Experimental
2.1 Samples The samples investigated were grown by Hydride Vapor Phase Epitaxy (HVPE), Metal-Organic Vapor Phase Epitaxy (MOVPE), and Molecular Beam Epitaxy (MBE). GaN samples with thicknesses ranging between 200 and 400 µm were grown direct or with a buffer layer on (0001) sapphire substrate using HVPE [7–9]. Their free carrier concentrations (n-type) varied between 10^{17} and 10^{20} cm^{-3}. The MOVPE samples, predominantly also n-conducting, were grown with GaN or AlN buffer layers in order to decrease the background carrier concentration [10, 11]; they are only a few µm thick. The cubic GaN samples were exclusively grown with MBE on (001) GaAs substrates [12, 13]. Their thickness runs from a few nm to 1.8 µm and their free carrier concentrations (n-type) from 10^{18} to 10^{20} cm^{-3}. The investigated $Al_xGa_{1-x}N$ layers with $0 < x < 1$ were grown with plasma-assisted MBE on (0001) sapphire and are about 1 µm thick [14, 15]. The AlN samples were fabricated by a direct reaction

---

[*] Corresponding author: e-mail: haboeck@physik.tu-berlin.de, Phone: +49 30 314 24440, Fax: +49 30 314 22064

of aluminum vapor with nitrogen at high temperatures (1900 °C) [16, 17]. These whiskers are some 10 µm thick and up to 1 cm long. Although they are undoped they have background carrier concentrations ranging from 10^{17} up to 10^{20} cm^{-3}.

2.2 Characterization technique The Raman scattering experiments were carried out using a Dilor XY800 triple-grating spectrometer and a Dilor LABRAM single-grating spectrometer with a super-notch filter. Both systems have a charge-coupled device (CCD) as detector and are equipped with confocal optics. The samples were excited either parallel (in-plane) or perpendicular (on-plane) to the surface using several lines of an Ar$^+$–Kr$^+$ mixed-gas laser and the 632.8 nm line of a He–Ne laser. The micro optics focused the laser on a point spot of about 1 µm diameter and detected the scattered light in back-scattering geometry, which corresponds to an $x(...)\underline{x}$ configuration for in-plane excitation and a $z(...)\underline{z}$ configuration for on-plane excitation. Spectra have been taken at room temperature and at 4.2 K using an Oxford microscope cryostat. With this setup we obtained a spatial resolution of about 1 µm and a spectral resolution better than 1 cm^{-1} for the triple-grating and about 2 cm^{-1} for the single-grating spectrometer.

3 Results and discussion

3.1 First-order Raman scattering We start with the first-order Raman spectra of GaN and AlN that are the basics for further discussion. Under equilibrium growth conditions they both crystallize in the hexagonal wurtzite structure [18] and belong to the point group C_{6v} (6mm) with four atoms per unit cell. Near $k = 0$ group theory predicts the following eight modes: $2A_1 + 2B + 2E_1 + 2E_2$ of which one A_1 and one E_1 correspond to acoustic phonons. The B modes are silent. Because there is no inversion symmetry phonons can be infrared-active as well as Raman-active (i.e., they are polar modes). The Raman-active modes permitted in backscattering geometries[1] of hexagonal GaN and AlN are listed in Table 1.

The corresponding first-order Raman spectra of hexagonal (α-) GaN and AlN are shown in Fig. 1. The spectra were taken at room temperature with an excitation energy of 632.8 nm in backscattering geometry. The samples c-axis is parallel to the z-direction. The x- and y-axes are perpendicular to each other and the z-axis but arbitrary in the z-plane. Both samples may be considered unstrained because of their bulk-like properties (thick GaN layer, AlN whisker). The Raman modes and scattering geometries are indicated. Forbidden modes which are weakly visible in some spectra originate from non-perfect polarization and scattering geometry or low crystal quality leading to a relaxation of the selection rules. The E_1(LO) mode is actually not allowed in backscattering geometry but visible in the spectrum taken in $x(zz)\underline{x}$ configuration. It likely originates from scattering processes permitted via deformation potential and particularly the Fröhlich interaction. The latter is represented by an electron-phonon interaction Hamiltonian H_{ep}. The matrix elements of H_{ep} between an initial and final electronic state taking into account the conservation of $\boldsymbol{q}$ lead to two terms. One so-called intraband term is inversely proportional to $|q|$ and thus singular at $|q| = 0$. The other describes interband scattering. Reference [19] shows in detail how both terms can contribute to a 'forbidden' LO-scattering.

The phonon frequencies found in our investigations are listed in Table 2. They are in good agreement with other published values [20–24].

Table 1 First-order Raman modes in wurtzite crystals for different backscattering geometries.

Configuration	Permitted Raman Modes
$z(xx)\underline{z}$	E_2(low, high), A_1(LO)
$z(xy)\underline{z}$	E_2(low, high)
$x(zz)\underline{x}$	A_1(TO), E_1(LO) (see text)
$x(yy)\underline{x}$	A_1(TO), E_2(low, high)
$x(yz)\underline{x}$	E_1(TO)

---

[1] Forward scattering, induced by back reflections due to the fact, that the sample is transparent to the laser beam, also occurs.

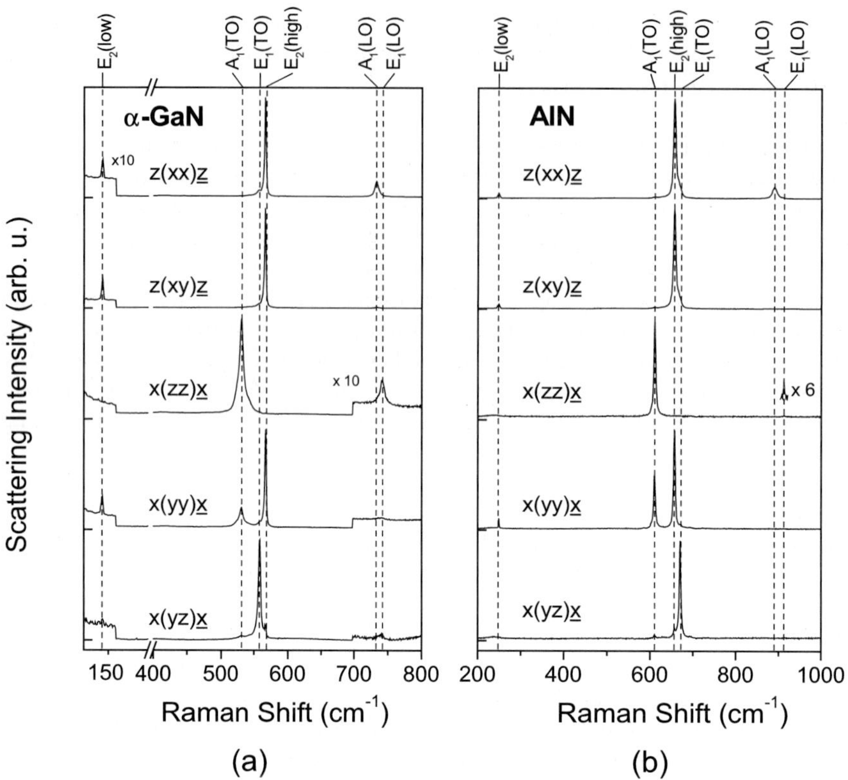

Fig. 1 Room temperature first-order Raman spectra of hexagonal a) GaN and b) AlN in different scattering geometries normalized to the maximum intensity.

3.2 Angular dispersion of the polar Raman modes Since AlN and GaN usually crystallize in the uniaxial hexagonal wurtzite structure the orientation of the c-axis, which coincides with the optical axis, plays an important role when measuring phonon modes in Raman scattering experiments. A phonon polarized along the optical axis is necessarily an A_1 phonon in this structure, one with a polarization in the plane perpendicular to the optical axis is an E_1 phonon. If the phonon propagation direction is along the crystal axes only purely longitudinal or transverse phonons of well-defined symmetry character are observed in Raman scattering. When the propagation direction is not along one of these axes the frequencies of the polar A_1 and E_1 modes depend on the angular dispersion with respect to the c-axis and a mixing of the modes, so-called quasi-modes, occurs. Their energy level is located between the high-symmetry modes [25, 26]. If one knows about the angular dispersion the orientation of the c-axis can be determined.

In uniaxial crystals it is necessary to consider simultaneously two independent forces: the long-range electrostatic forces responsible for the longitudinal-transverse splitting, and the short-range interatomic forces, which reflect the anisotropy of the force constants. For an approximate characterization of the quasi modes wurtzite crystals are divided in two categories [26]: the A_1–E_1 splitting is larger than the LO-TO splitting, i.e., the short-range interatomic forces preponderate the electrostatic forces and the

Table 2 Frequencies of the first-order Raman modes in hexagonal GaN and AlN at room temperature.

	E$_2$(low)	A$_1$(TO)	E$_1$(TO)	E$_2$(high)	A$_1$(LO)	E$_1$(LO)
GaN	145	533	560	567	735	742
AlN	249	610	669	656	891	912

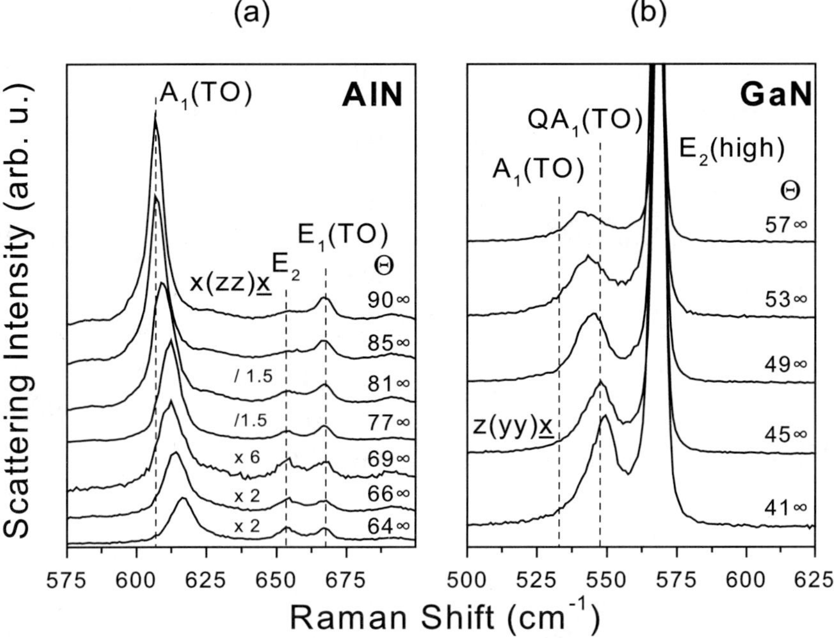

Fig. 2 Series of Raman spectra for different directions of phonon propagation Θ relative to the c-axis of a) AlN and b) hexagonal GaN. The spectra were taken at room temperature with an excitation energy of 632.8 nm.

quasi modes are a mixture between transverse and longitudinal modes, but still keep their A_1- or E_1-symmetry. On the other hand the LO$-$TO splitting can be much larger than the A_1-E_1 splitting, i.e., the long-range electrostatic forces dominate over the anisotropy of the interatomic forces. The quasi modes then are pure transverse or longitudinal but their symmetry is a mixture of A_1 and E_1. The latter is the category to which GaN and AlN belong. The energies of the angular-depend LO and TO modes are then given by the equations

$$\omega\left(\Theta\right)^2_{\mathrm{LO}} = \omega^2_{A_1(\mathrm{LO})}\cos^2\left(\Theta\right) + \omega^2_{E_1(\mathrm{LO})}\sin^2\left(\Theta\right), \tag{1}$$

$$\omega\left(\Theta\right)^2_{\mathrm{TO}} = \omega^2_{A_1(\mathrm{TO})}\sin^2\left(\Theta\right) + \omega^2_{E_1(\mathrm{TO})}\cos^2\left(\Theta\right), \tag{2}$$

in which $\omega(\Theta)$ denotes the frequency of the quasi mode, ω_{E1} and ω_{A1} the frequency of the pure E_1 and A_1 modes, respectively. The LO$-$TO as well as the A_1-E_1 splitting are listed in Table 3.

In order to compare theoretical calculations (DFT-LDA) such as described in Ref. [27] with experimental data we performed Raman spectroscopy on the AlN whisker crystals described in 2.1 and on a 200 µm thick hexagonal GaN layer grown by HVPE. The spectra plotted in Fig. 2 were taken at room temperature either in backscattering geometry corresponding to $x(...)\underline{x}$ configuration and in right angle scattering geometry, i.e., $z(...)\underline{x}$ configuration.

The samples' c-axes were tilted with respect to the incident laser beam. From the angle of the incident laser beam and the scattered light relative to the crystal surface we determined the direction of phonon

Table 3 Comparison between A_1-E_1-splitting and LO$-$TO-splitting of GaN and AlN.

	$A_1(\mathrm{LO})-A_1(\mathrm{TO})$	$E_1(\mathrm{LO})-E_1(\mathrm{TO})$	$\gg$	$E_1(\mathrm{LO})-A_1(\mathrm{LO})$	$E_1(\mathrm{TO})-A_1(\mathrm{TO})$
GaN	202 cm⁻¹	182 cm⁻¹	$\gg$	7 cm⁻¹	27 cm⁻¹
AlN	281 cm⁻¹	243 cm⁻¹	$\gg$	21 cm⁻¹	59 cm⁻¹

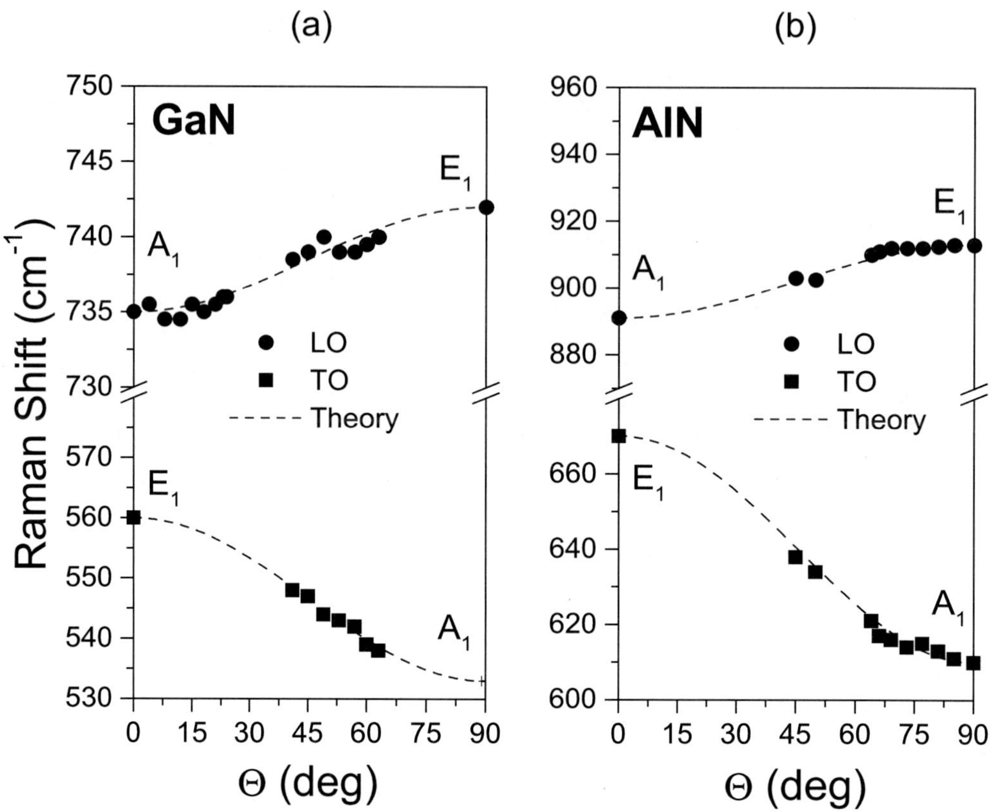

Fig. 3   Angular dispersion of the LO and TO modes in hexagonal a) GaN and b) AlN for various directions of phonon propagation Θ relative to the c-axis. The dashed lines represent calculations according to equations (1) and (2), the symbols the experimental data.

propagation relative to the optical axis [27]. In the Raman spectra Fig. 2 the shift of the A$_1$(TO) and the quasi-TO mode, respectively, caused by the intermixture of the E$_1$ phonon, towards higher energies is clearly visible. Despite the E$_2$ and the E$_1$(TO) modes are forbidden in $x(zz)x$ configuration they are weakly detectable because non-perfect crystal structure and polarization alignment (Fig. 2a). They both show no angular dispersion. We point out that the E$_1$(TO) phonon does not shift although it is also a polar one. The reason is the double degeneracy of this mode. The crystal has been moved around the y-axis, i.e., the one E$_1$(TO) which is polarized in y-direction remains unmixed because there is no change with respect to the c-axis. The second E$_1$(TO) polarized in x-direction mixes with the A$_1$(TO) and thus generate the appearance of the quasi mode in the spectra.

Fitting the experimental data with a Lorentzian line shape and plotting it as a function of the phonon propagation direction Θ leads to the angular dispersion shown in Fig. 3. The symbols represent the measured energies, the dashed lines the calculated distribution. Because the size and geometry of the samples not all angles could be measured but the experimental data confirm the theoretical prediction quite well: in both GaN and AlN the long range electrostatic forces dominate over the anisotropy of the interatomic forces.

3.3 Impact of biaxial stress on the vibronic properties   Most of our investigated GaN samples were grown on sapphire. The large mismatch of lattice constant (13.5%) and the different thermal expansion coefficients between hexagonal GaN and sapphire result in a strongly inhomogeneous strain distribution. GaN is biaxially compressively stressed on the sapphire substrate and via the deformation potentials the phonon frequencies change. In hexagonal GaN not all phonon modes are suited for the determination of

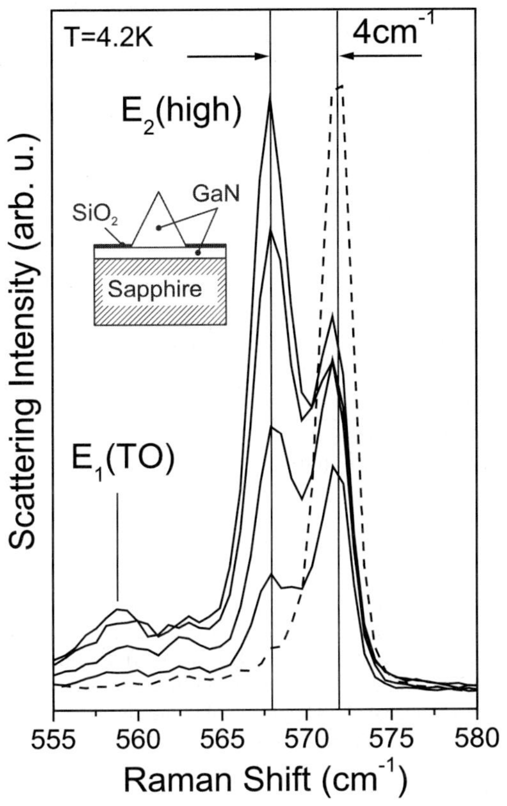

Fig. 4 Raman spectra of a 10 µm high GaN pyramid grown on sapphire substrate (the sample structure is depicted in the inset). The dashed line shows the spectrum of the strongly compressive strained GaN buffer layer. The solid lines represent spectra taken in different depths of the pyramid.

Fig. 5 (a) Depth profile of a 25 µm thick GaN layer grown on GaN substrate visualizing the shift of the E$_2$(high) mode dependent on the distance from the substrate-sample-interface (b)

the strain distribution. As mentioned in 3.2 the polar A$_1$- and E$_1$-modes show an angular dispersion and thus their frequencies depend on the scattering geometry. The E$_2$(low) mode is non-polar and in principle suitable but has a very small pressure coefficient [22]. Hence, the most suited and commonly used line is the E$_2$(high). The phonon shift in case of biaxial (0001)-plane oriented stress can be described by the relation [22]

$$\Delta\omega = K \, \sigma_{xx = yy} , \tag{3}$$

where $\sigma_{xx = yy}$ is the component of the stress tensor and K the pressure coefficient. The values of K are still a matter of discussion and vary in different publications [28–30]. An overview and discussion is given in the recent work of Wagner and Bechstedt [31].

Figure 4 displays Raman spectra of a 10 µm high GaN pyramid. The structure of the selectively epitaxially grown sample is depicted in the inset, the growth details are given elsewhere [32, 33]. The spectra were recorded spatially resolved with an excitation energy of 514.5 nm at 4.2 K in different sample depths including the top and the GaN-sapphire interface. The latter is the higher-energy dashed line and visualizes the strong compressive stress of about 1 GPa (using the value $K = 4.2$ cm^{-1}/GPa [30, 34]) in this region of the sample. The unstressed E$_2$(high) frequency is taken to be 567 cm^{-1} at room temperature and 568 cm^{-1} at 4.2 K, respectively, as determined with measurements on thick 'quasi-bulk' samples. Because of the non-perfect scattering geometry a residual E$_1$(TO) peak is also detectable. A remarkable result is the gradual decrease in intensity of the E$_2$ peak from the pyramid together with a corresponding

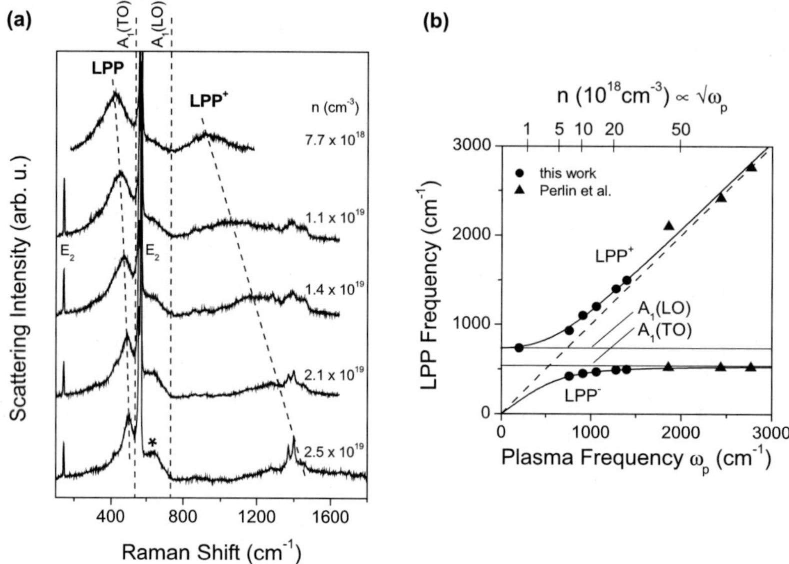

Fig. 6   a) Raman spectra of MOVPE-grown GaN/sapphire layers having different free-carrier concentrations. The spectra were taken at room temperature in $z(...)\bar{z}$ configuration with z parallel to the samples c-axis. Besides the permitted E_2 modes the coupled LO phonon plasmon modes (LPP modes) are observable whereas the also allowed A_1(LO) phonon disappears. The other structures in the high-energy range originate from second-order scattering and the ruby luminescence of the sapphire substrate. For the peak marked with an asterisk around 650 cm^{-1} see text. b): Frequency-dependence of the LPP modes on the plasma frequency and the carrier concentration.

increasing in intensity from the peak near the interface without any shift from lower to higher energies, i.e., most of the pyramid is not stressed. The reason is most likely that the 'freestanding' GaN of the pyramid is able to expand undisturbed [35].

Usually a step by step relaxation of GaN layers with increasing thickness is observed. Figure 5 represent the result of a spatially resolved line scan at room temperature over a 25 µm thick GaN layer grown on a GaN template. The stress at the substrate-sample interface is about 0.4 GPa and thus not as strong as in the above mentioned pyramid which was grown on sapphire. However, there is also a relaxation from the interface towards the surface; in this case we see the expected gradual shift of the E_2(high) mode.

3.4 Coupled LO-phonon plasmon modes (LPP modes)   In polar material such as GaN and AlN the macroscopic electric field of the longitudinal-optical phonons can couple to the field of the collective excitation of the free carriers, the so-called plasmons. This results in coupled LO phonon plasmon modes (LPP) [36–39] with one low-energy branch denominated LPP⁻ and a high-energy one called LPP⁺. The frequencies of this modes depend strongly on the free carrier concentration n. The dielectric function is composed of a plasmon and a phonon contribution:

$$\varepsilon(\omega) = \varepsilon_\infty\left(1 - \frac{\omega_p}{\omega(\omega + i\gamma)}\right) + \frac{\omega_{LO}^2 - \omega_{TO}^2}{\omega_{TO}^2 - \omega^2 - i\Gamma\omega}, \tag{4}$$

where ω_{LO} and ω_{TO} are longitudinal and transverse optical phonon frequencies, respectively, and ε_∞ is the high-frequency dielectric constant. Γ and γ are phonon and electron damping constants. ω_p is the uncoupled plasma frequency, defined as

$$\omega_p^2 = \frac{ne^2}{m^*\varepsilon\varepsilon_0}, \tag{5}$$

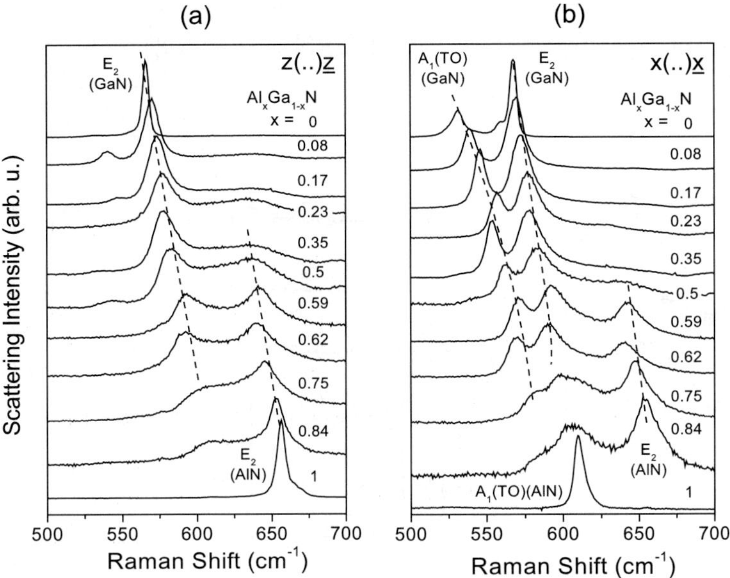

Fig. 7 Raman spectra of an Al$_x$Ga$_{1-x}$N series with varying Al content ($0 < x < 1$). The spectra are taken at room temperature in (a) on-plane as well as (b) in-plane configuration and are normalized to their maximum intensity. The dashed lines are guide to the eye emphasizing the two-mode-behavior of the E$_2$(high) and the A$_1$(TO), respectively.

where n is the free carrier concentration and m^* the effective mass of conduction electrons. Neglecting the damping terms, the two LPP frequencies are the roots of the equation:

$$(\omega_{LPP}^{\pm})^2 = \frac{1}{2}\left\{\omega_{LO}^2 + \omega_p^2 \pm \left[\left(\omega_{LO}^2 + \omega_p^2\right)^2 - 4\omega_p^2\omega_{TO}^2\right]^{\frac{1}{2}}\right\}. \tag{6}$$

This equation yields the typical non-linear curves plotted in Fig. 6b. In some wide-gap semiconductors such as GaP [40], CdS [41], and SiC [42] LPP-modes could not be observed because of the strong plasmon damping. Some authors also reported on LPP-modes in GaN [39, 43–46] with controversial interpretations. Kozawa et al. [44] only detected the LPP$^+$ mode in a small doping range up to 10^{18}cm^{-3} and deduced an overdamping of the plasmons. In contrast Perlin et al. [39] observed both LPP branches and pointed out that they found no evidence of overdamping. Their measurements were performed on highly doped samples (triangles in Fig. 6b).

With our work [47] we added experimental data in the lower frequency range and furthermore found the dependence of the LPP$^-$ mode on the free-carrier concentration n as visible in the Raman spectra of Fig. 6a. The LPP$^-$ exhibits a shift starting at 420 cm^{-1} up to 500 cm^{-1} and the LPP$^+$ from 930 cm^{-1} to 1500 cm^{-1}, respectively. The values are plotted in Fig. 6b (circles) together with those of Ref. [39] and the calculated lineshape according to relation (6). We used $m^* = 0.2$ m$_0$ [48], $\varepsilon_\infty = 5.76$ [39], and the frequencies of the A$_1$(TO) and A$_1$(LO) phonons given in Table 2. Experimental and theoretical data are in good agreement and evidence only a low damping of the plasmon. From the energy position of the LPP-modes we calculated the free-carrier density and assigned them to the corresponding spectra of Fig. 6a. The broader structure at 650 cm^{-1} marked with an asterisk is also an indication of high carrier density. In Fig. 6a the increasing intensity according to increasing carrier concentration is evident. Demangeot et al. [46] attributed this mode to the q-dependent electron charge density fluctuation mechanism and concluded a continuous scattering in the TO-LO frequency range activated by the relaxation of q conserving mirrors the density of LO phonon states renormalized by the interaction with carriers. Knowing about the relation between LPP modes and free carrier concentration Raman spectroscopy can be used as a tool for sample characterization, in particular in inhomogeneous microstructures.

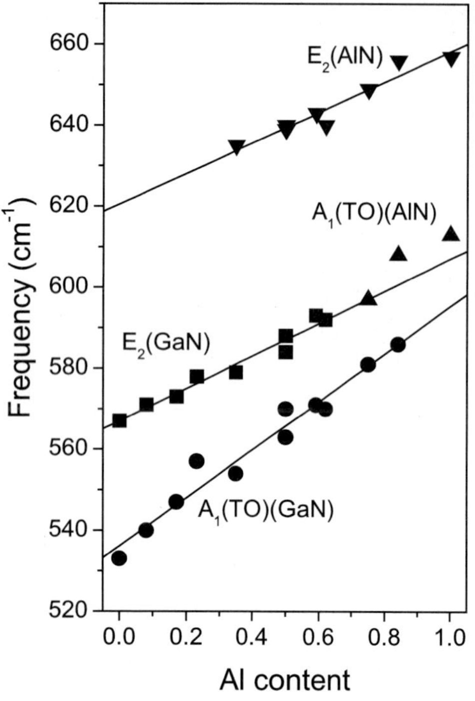

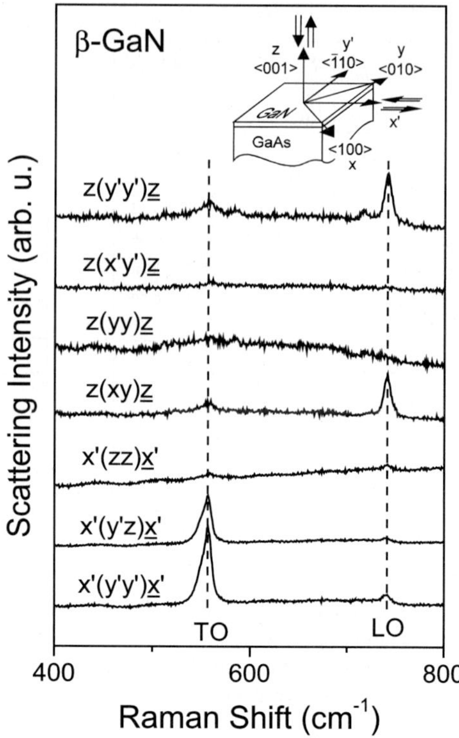

Fig. 8 Frequencies of phonon modes depending on the Al content of Al$_x$Ga$_{1-x}$N. The circles and rectangles represent GaN-like modes, the triangles those of AlN.

Fig. 9 Room temperature Raman spectra taken from a cubic GaN sample in various configurations. The inset shows a sketch of the corresponding scattering geometries.

3.5 Al$_x$Ga$_{1-x}$N: dependence of the phonon energies on the Al-content The phonon frequencies of alloys depend on their stoichiometry in a characteristic way [49–51]: phonons exhibit one- or multi-mode behavior. In the case of one-mode behavior the number of modes does not change with stoichiometry and their frequencies depend approximately linearly on the composition of the alloy. If separate modes of both possible pairs of adjacent atoms exist this is called two-mode behavior, reported e.g. for Al$_x$Ga$_{1-x}$As [52]. Raman measurements on Al$_x$Ga$_{1-x}$N samples with a maximum Al-content of $x = 0.15$ have been performed by Hayashi et al. [53]. They did not find local modes and deduced a one-mode behavior for Al$_x$Ga$_{1-x}$N. Cros et al. [54] did similar work but over the whole range of Al-contents $0 < x < 1$. They observed a one-mode behavior of the A$_1$(LO) phonon and a two-mode behavior of the E$_2$(high). We contributed to this with the missing data for the A$_1$(TO) phonon. Our results confirmed those of Cros et al. and revealed also a two-mode behavior of the A$_1$(TO) phonon.

Figure 7 displays Raman spectra of a series of Al$_x$Ga$_{1-x}$N samples with $0 < x < 1$ taken at room temperature with an excitation energy of 632.8 nm in (a) on-plane and (b) in-plane configuration. The dashed lines indicate the behavior of the E$_2$(high)- and A$_1$(TO)-mode as guide to the eye. Comparing both scattering geometries allows the classification of the mode symmetry, i.e., A$_1$ or E$_2$ symmetry. The necessity of such a treatment is visualized in the spectra in Fig. 7b with Al-contents of 0.75 and 0.84 where the AlN-A$_1$(TO) mode overlap with the GaN-like E$_2$(high). Only a comparison of the different scattering geometries permits the unambiguous identification of the A$_1$(TO) character of this mode. The typical broadening of the lines during the transition to a local mode is also discernible. From linear fits of the measured data we obtained

$$\text{A}_1\text{(TO)-GaN:} \quad \omega(\text{Al}_x\text{Ga}_{1-x}\text{N}) = 536 \pm 2 \text{ cm}^{-1} + (59 \pm 4 \text{ cm}^{-1}) \cdot x_{\text{Al}}, \tag{7.1}$$

$$E_2\text{(high)-GaN:}\quad \omega(Al_xGa_{1-x}N) = 567 \pm 1\ \text{cm}^{-1} + (40 \pm 3\ \text{cm}^{-1}) \cdot x_{Al}, \tag{7.2}$$

$$E_2\text{(high)-AlN:}\quad \omega(Al_xGa_{1-x}N) = 621 \pm 3\ \text{cm}^{-1} + (38 \pm 4\ \text{cm}^{-1}) \cdot x_{Al}, \tag{7.3}$$

which gives us a rough estimate for the further characterization of samples with Raman spectroscopy. The advantage of this method over X-ray diffraction measurements is that it can be done spatially resolved and thus detect inhomogeneities or cluster formation. The experimental values together with the linear fits are displayed in Fig. 8.

The absolute error of the samples Al-content was declared to be maximally 5 % by Angerer et al. [15].

3.6 Determination of the hexagonal minority phase in cubic GaN   Although GaN crystallizes under equilibrium conditions in the hexagonal wurtzite structure improved growth conditions enable the fabrication of the metastable cubic modification. The aim of high carrier mobility in the more easily doped cubic GaN compared to the hexagonal phase [1, 55] has been one of the reasons to focus on this subject. A major difficulty in the growth of cubic (β-) GaN is the polytypism of this material. During the growth process subdomains of hexagonal phases may be built in via a stacking fault mechanism [56]. X-ray measurements showed that cubic domains exist in predominately hexagonal GaN and vice versa [57, 58] whereas the syntactic growth is more dominant in the cubic phase. The standard θ–2θ X-ray scans frequently used to demonstrate the crystal quality are unable to determine a possible minority phase present in a sample. If the c-axis, e.g., in hexagonal subdomains is tilted by a few degrees with respect to the cubic majority phase only, Bragg peaks originating from the majority phase become observable. Therefore, θ–2θ scans are less suitable for obtaining information about a possible minority-phase content.

Regardless of the particular advantages that one of the two crystal structures of GaN may have over the other, it is important to unambiguously distinguish the two phases when comparing experimental results. Raman spectroscopy is able to verify the purity of GaN layers with respect to polytype-domain formation [59]. The different polarization scattering selection rules allow us to distinguish between the two phases by selecting appropriate scattering geometries. The selection rules for backscattering configurations used in our work is listed in Table 4:

There are two principally different possibilities for the orientation of hexagonal subdomains in cubic GaN. In the first case, the c-axis [0001] of the hexagonal phase of GaN is tilted only a few degrees with respect to the c-axis of the otherwise cubic material. In the second case the [0001] axis is parallel to a cubic $\langle 111 \rangle$ direction, i.e., the space diagonal of the cubic unit cell. In the first case standard θ–2θ X-ray scans are unable to detect hexagonal material. In contrast to the coherent process of X-rays, where the scattering intensity is obtained by the first adding and then squaring, the Raman intensity is the sum of

Table 4   Raman selection rules for backscattering configurations[1] used in this work (the c-axis of the cubic GaN is parallel to the z-direction). The selection rules are only weakly relaxed if the angle between the hexagonal [0001] axis and k is small.

Surface	Incident Polarization	Scattered Polarization	Porto Notation	Allowed Modes	
				Cubic	Hexagonal
[001]	[$\underline{1}$10]	[$\underline{1}$10]	$z(y'y')\underline{z}$	LO	E_2, A_1(LO)
	[110]	[$\underline{1}$10]	$z(x'y')\underline{z}$	–	E_2
	[010]	[010]	$z(yy)\underline{z}$	–	E_2, A_1(LO)
	[100]	[010]	$z(xy)\underline{z}$	LO	E_2
[110]	[001]	[001]	$x'(zz)\underline{x}'$	–	A_1(TO)
	[$\underline{1}$10]	[001]	$x'(y'z)\underline{x}'$	TO	E_1(TO)
	[$\underline{1}$10]	[$\underline{1}$10]	$x'(y'y')\underline{x}'$	TO	A_1(TO), E_2

---

[1] Forward scattering, induced by back reflections due to the fact, that the sample is transparent to the laser beam, also occurs.

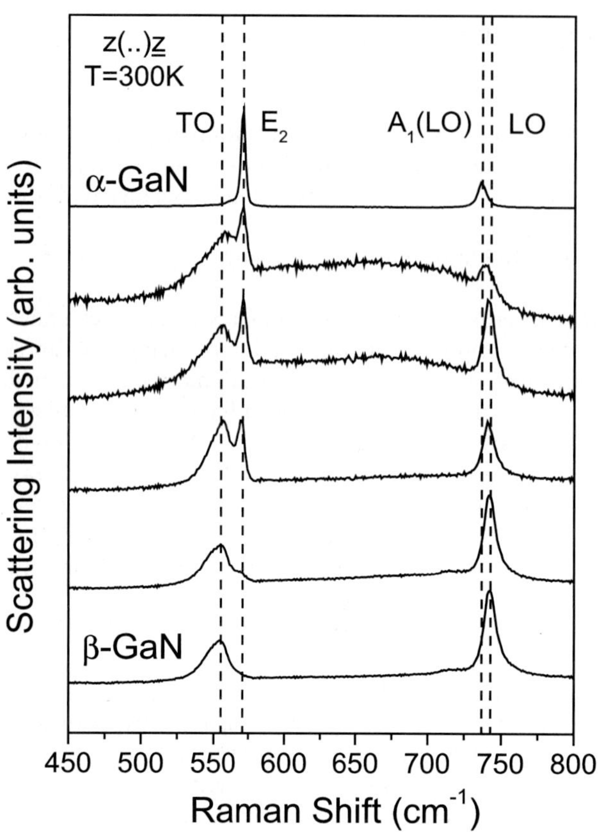

Fig. 10 Series of GaN samples with top down increasing cubicity. The dashed lines show the position of the hexagonal and cubic phonon modes. For comparable amounts of both phases a broad disorder-activated band between TO and LO mode can be detected.

the squared amplitudes [19, 60]. Therefore, small deviations from principal angles do not affect the Raman scattering intensity significantly, and the minority-phase orientation can be considered as approximately parallel to the cubic phase.

Cubic GaN has zincblende structure and belongs to the point group $T_d = 43$ m and has thus close to $k = 0$ a doubly degenerate TO and a single LO phonon with a higher frequency. The wurtzite material may be considered having a slightly distorted zincblende structure in nearest neighbor distance; consequently the energy difference between the Raman modes of both phases are not very large and make it difficult to determine the minority-phase content. The frequencies of the hexagonal phase are given in Table 2. Those of the cubic phase obtained by the Raman spectra displayed in Fig. 9 are $\omega_{(TO)} = 555$ cm^{-1} and $\omega_{(LO)} = 740$ cm^{-1}.

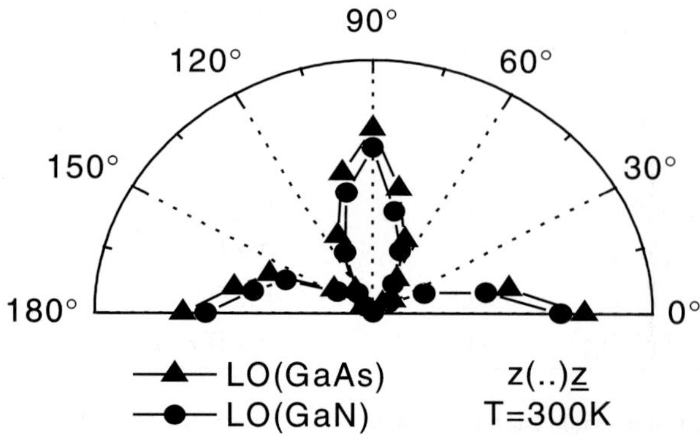

Fig. 11 Polarization dependence of the parallel polarized GaN and GaAs LO-phonon intensity as the sample is rotated. An angle of 45° corresponds to $z(yy)\underline{z}$.

These values agree well with the results of other workers [61, 62]. As can be seen from Tab. 4 it is possible to distinguish the two phases of GaN using the $z(...)\underline{z}$ configuration because of the different selection rules of the cubic LO and the hexagonal $A_1(LO)$ mode in spite of their closeness in frequency. While in $z(y'y')\underline{z}$ configuration both modes are observable, in the $z(yy)\underline{z}$ (rotated by 45° in the polarization plane) only the hexagonal $A_1(LO)$ mode is allowed. To compare the lines originating from cubic or hexagonal GaN we investigated a series of samples with different degrees of cubicity. Figure 10 shows the corresponding Raman spectra.

Even the slightly different phonon frequencies (vide the shift of the $A_1(LO)$ towards the LO frequency) and particularly the E_2-mode, which does not appear in cubic material, make it possible to estimate the phase purity of the samples. This becomes more difficult the less hexagonal material is built in. To determine the phase purity of GaN samples using the $z(...)\underline{z}$ configurations one has to be sure of the epitaxial growth of the investigated layer. For a cubic GaN layer oriented only in growth direction but not perpendicular to it there is obviously no selection rule involving x- and y-polarizations. In order to estimate the cubicity quantitatively we fitted a Lorentzian to the allowed polarization $(y'y')$ and compared it to the spectrum in yy-polarization. For the sample which spectra are shown in Fig. 9 we found a cubicity of better than 98% (having corrected for the different thicknesses and relative Raman coefficients). An unambiguous method for the quantitative determination of the cubicity is obtained for Raman excitation in plane of the substrate. By comparing the $x'(zz)\underline{x}'$ and $x'(y'y')\underline{x}'$ configuration the hexagonal phase content can easily be determined. The $E_1(TO)$ phonon which appears nearly at the same frequency as the cubic TO mode is forbidden in the $x'(y'y')\underline{x}'$ configuration. Therefore comparing the intensity ratio as described above of the cubic TO and the hexagonal E_2(high) mode, which are furthermore separated more than 10 cm^{-1} from each other, allows the determination of the sample's cubicity.

In the second case the [0001] axis is parallel to a cubic $\langle 111 \rangle$ direction, i.e., the space diagonal of the cubic unit cell. Even then it is difficult to detect the hexagonal minority phase with $\theta–2\theta$ X-ray scans whereas in Raman measurements an E_2-signal is available in all configurations mentioned so far. In order to obtain quantitative information we measured the polarization dependence of the phonon modes of a cubic sample. Figure 11 shows polar plots of the intensity of the LO phonon mode of GaN compared to the LO mode of GaAs. The incident and scattered polarizations were kept fixed and parallel polarized, while the sample was rotated in total by 180° in the plane defined by the polarizations. The GaN mode exhibits the same polarization dependence as the GaAs mode indicating an epitaxial growth. If both phases coexisted in the sample, the different polarization dependences would overlap and the distinct d-wave like minima of the scattering intensity would disappear with increasing hexagonal-phase content.

3.7 Zone-boundary phonons in GaN and AlN   In contrast to first-order Raman measurements second-order experiments provide information on the vibrational states throughout the entire Brillouin zone, which is important in considering, for example, the electronic conduction or nonradiative relaxation processes of electrons. An earlier publication about second-order Raman scattering on GaN was done by Murugkar et al. [63]. They investigated only hexagonal GaN for frequencies higher than 1280 cm^{-1}. Our measurements on hexagonal and cubic GaN covered the spectral region from 200 cm^{-1} until 1550 cm^{-1} and thus contain the acoustic as well as the optical overtone and combination part. We used all necessary scattering geometries to obtain selection rules and performed a calculation of the dispersion curves [64, 65]. In order to derive the selection rules for second-order Raman scattering we carried out a group-theoretical analysis. Using these selection rules in addition to the calculated dispersion curves we assigned the observed peaks of GaN and AlN to particular phonon branches and determined the points in the Brillouin zone from which the scattering originates.

To calculate the phonon dispersion of hexagonal GaN we used a modified valence-force model of Kane [66] which is detailed in Ref. [65]. By adjusting our model to results of *ab initio* calculations [67, 68] we obtained corresponding results. Figure 12 displays the phonon dispersion of hexagonal GaN and AlN.

The phonon dispersion is related to the phonon density of states. The latter is reflected by the second-order intensity since phonons with wave vectors throughout the whole Brillouin zone are involved. The various structural characteristics of the second-order spectra are due to variations in the density of pho-

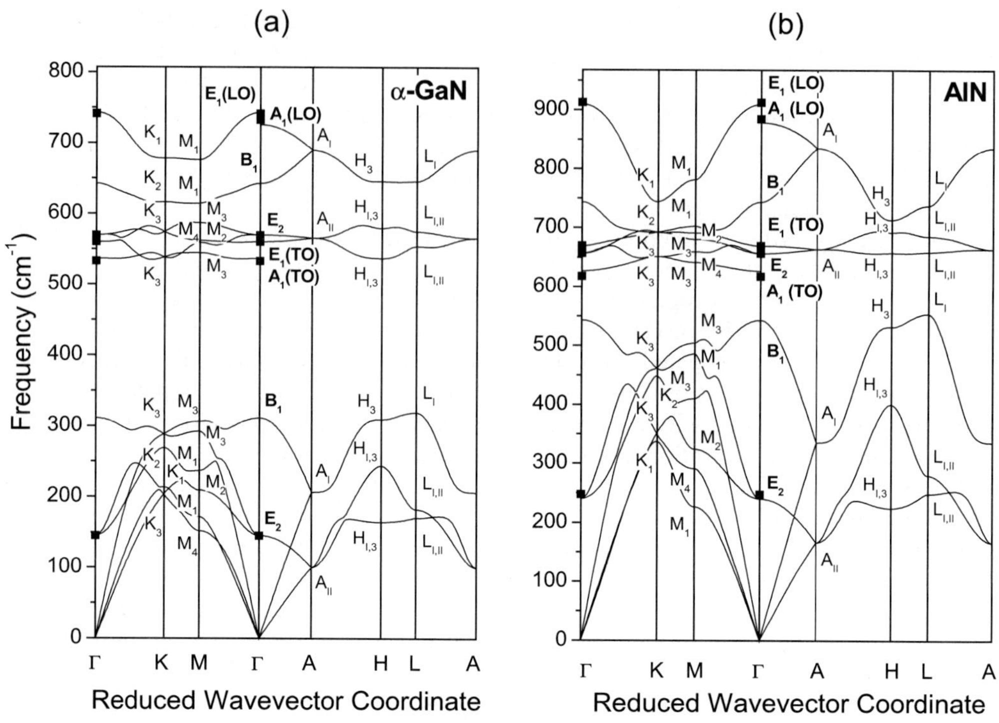

Fig. 12   Calculated phonon-dispersion curves for a) hexagonal GaN and b) AlN.

non states and selection rules for various scattering processes (combinations and overtones). From the results of our group-theoretical analysis the following conclusions can be drawn (for details vide Ref. [65]):

Overtones always contain the representation A_1 while combinations of phonon states belonging to different irreducible representations never contain A_1.

More overtones or combinations become allowed with decreasing symmetry of the point in the Brillouin zone considered.

A consequence of the different packing sequence of hexagonal GaN (ABAB) compared to the cubic material (ABCABC) is halving the Brillouin-zone dimension in the cubic (111) direction, which is followed by a folding of the phonon branches at the new zone boundary. Frequencies which previously occurred at the symmetry point L in the zinc-blende Brillouin zone now occur at the center of the wurtzite Brillouin zone. Hence, the different phonon branches from the Γ point, e.g. the optical $\Gamma_6 = E_2$ and the acoustic $\Gamma_3 = B$ run together at the A point. Therefore, we use $A_I = A_1 + A_3$ and $A_{II} = A_5 + A_6$. The same applies to the points H and L.

It is important to point out that, apart from the selection rules, the density of two-phonon states determines the intensity of the second-order Raman signal [19]. Figure 13 shows an overview of a room temperature Raman spectrum of a hexagonal and a cubic GaN layer which exhibit a rich structure due to second-order scattering processes beside those known to be first-order originated. The samples were excited either perpendicular or parallel to the sample surface depending on the scattering geometry using the 632.8 nm line of a He−Ne laser and the 514.53 nm line of an Ar^+ laser. The scattered light was detected in backscattering geometry which corresponds to either an $x(zz)\underline{x}$ (A_1), an $x(yz)\underline{x}$ (E_1), or a $z(xy)\underline{z}$ (E_2) configuration.

The total spectrum may be divided into three parts: the low-frequency region (200−650 cm^{-1}) is dominated by acoustic overtones; optical and acoustic combination bands appear below and above the first-order optical scattering (650−1000 cm^{-1}); and, finally, the second-order optical bands in the high-frequency range (1000−1500 cm^{-1}). The corresponding spectra taken from the hexagonal GaN sample in

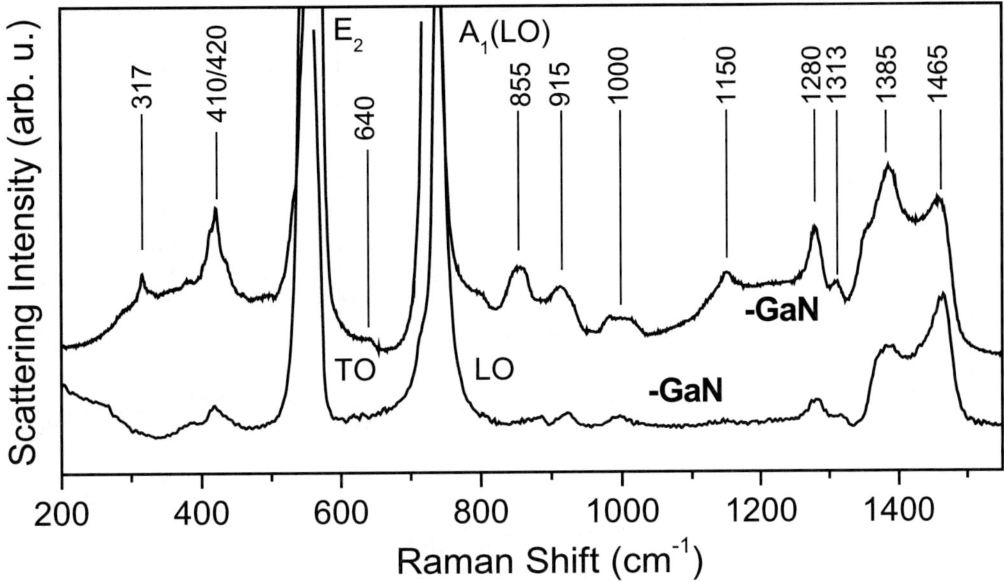

Fig. 13 Second-order Raman spectrum of hexagonal (α-, upper curve) and cubic (β-, bottom curve) GaN grown on (0001) sapphire and (001) GaAs, respectively. The spectra were recorded at room temperature in backscattering geometry, corresponding to $z(...)\underline{z}$.

A_1, E_1, and E_2 scattering geometry are shown in Fig. 14. In the following the strongest experimentally observed structures along with our calculated phonon-dispersion curves (Fig. 12) will be discussed and compared to the modes observed in cubic GaN. Recently, inelastic x-ray scattering has become an interesting technique for studying the phonon dispersion throughout the Brillouin zone [69]. When we compare our calculated dispersion curves with these experiments we find an, in general, excellent agreement. Differences arise in the splitting of B_1 and LO-derived branches which in our model is larger by a factor of about 2 compared to the one in Ref [69]. For details see Fig. 12 and Ref. [69] where the experimental work is compared to *ab initio* calculations and the results of [65, 70, 71].

The low-frequency region displayed in Fig. 14a exhibits two strong Raman bands, one located at 317 cm^{-1} and a doublet at 410/420 cm^{-1}. Whereas the 317 cm^{-1} and the 410 cm^{-1} modes unambiguously have A_1 symmetry the 420 cm^{-1} line also shows E_2 symmetry (see inset of Fig. 14a). From the lineshape and it's low frequency we assign the 317 cm^{-1} mode to an overtone process of acoustic phonons. As can be seen in the phonon dispersion (Fig. 12) the energy as well as the observed symmetry fits well with the flat branch at the H point in the Brillouin zone. This is emphasized by the fact that this mode was not observed in cubic GaN. The doublet at 410/420 cm^{-1} is also due to acoustic phonons. Comparing the phonon-dispersion curves (Fig. 12) it can be attributed to an overtone of transverse-acoustic phonons either at the symmetry point A, K, or M. According to the selection rules the part of the doublet having pure A_1 symmetry, i.e., the 410 cm^{-1} line, is assigned to originate from the *A* or *K* point whereas the 420 cm^{-1} mode having A_1 and E_2 symmetry belongs to the M point. The last mode of the low-frequency region lying between the first-order modes at around 640 cm^{-1} can be attributed to an overtone process of the highest acoustic-phonon branch either at the zone boundary (L point) or from the B mode at the Γ point. The middle-frequency region is dominated by three modes which we attribute to combinations between acoustic and optical phonons considering the phonon-dispersion and the fact that most of them are found in the spectra of the cubic sample, too. Overtones can be excluded because of the gap between the acoustic- and optical-phonon branches which extends in the overtone spectrum from approximately 650–1050 cm^{-1}. The optical combination and overtone region is shown in Fig. 14c. Most of the modes are strongest in A_1 symmetry. The structure located at 1150 cm^{-1} is probably a consequence of the flat dispersion curve of GaN in the energy region between the A_1(TO) = 533 cm^{-1} and E_2 = 569 cm^{-1}. From

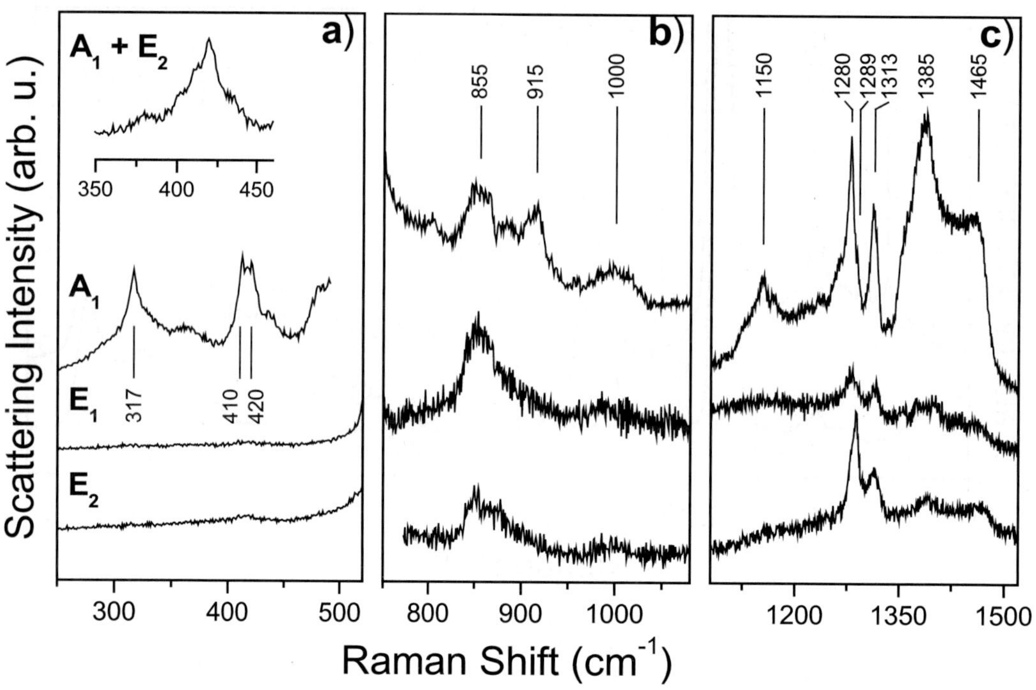

Fig. 14 Second-order Raman spectra of hexagonal GaN taken at room temperature for various scattering geometries corresponding to symmetry A$_1$, E$_1$, and E$_2$ in the a) low-, b) middle-, and c) high-frequency region.

Table 5 Frequencies and symmetries of the strongest modes found in the first- and second-order Raman spectra of hexagonal GaN and their assignments.

Frequency (cm^{-1})	Symmetry	Process	Point in the BZ
317	A$_1$	acoustic overtone	H
410	A$_1$	acoustic overtone	A, K
420	A$_1$, E$_2$	acoustic overtone	M
533	A$_1$	first-order process	Γ = A$_1$(TO)
560	E$_1$	first-order process	Γ = E$_1$(TO)
567	E$_2$	first-order process	Γ = E$_2$(high)
640	A$_1$	overtone	Γ = [B]^2, L
735	A$_1$	first-order process	Γ = A$_1$(LO)
742	E$_1$	first-order process	Γ = E$_1$(LO)
855	A$_1$, E$_1$, E$_2$	acoustic-optical comb.	
915	A$_1$	acoustic-optical comb.	
1000	A$_1$, (E$_2$)	acoustic-optical comb.	
1150	A$_1$	optical overtone	
1280	A$_1$, (E$_1$)	optical combination	
1289	E$_2$	optical combination	
1313	A$_1$, (E$_1$, E$_2$)	optical combination	
1385	A$_1$	optical combination	A, K
1465	A$_1$	optical overtone	Γ = [A$_1$, LO]^2
cutoff 1495	A$_1$, E$_2$	optical overtone	Γ = [E$_1$, LO]^2

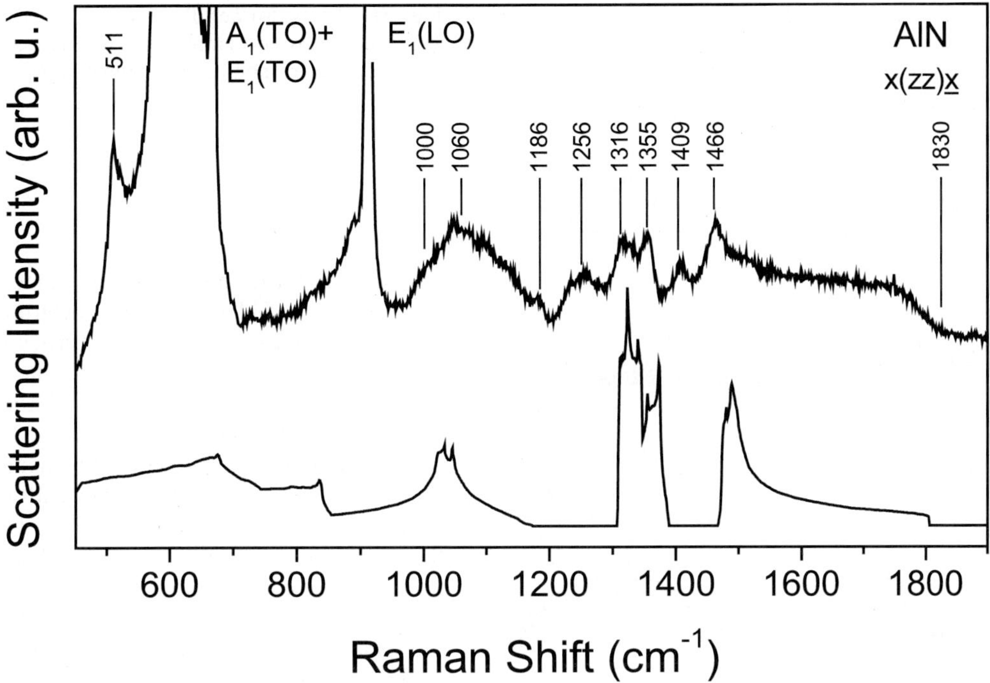

Fig. 15 Second-order Raman spectrum of hexagonal AlN taken at room temperature in backscattering geometry, corresponding to $x(...)\underline{x}$ (upper curve) and calculated overtone phonon density (bottom curve) of cubic AlN [67].

the observed A$_1$ symmetry we consider an optical overtone most likely originating from the H point. The narrow peaks at 1280 and 1313 cm^{-1} are most intensive in A$_1$ symmetry but they also appear in the E$_1$ spectrum. In the E$_2$ spectrum the 1313 cm^{-1} mode is likewise visible but instead of the 1280 cm^{-1} peak a mode appears at 1289 cm^{-1}. Despite the fact that combinations of zone-center modes are possible from energy considerations we assign them to combination processes at the zone boundary because of the observed symmetry and their appearance in the cubic spectrum, too. The highest-energy distribution with a cutoff at 1495 cm^{-1} is attributed to an overtone of the zone-center E$_1$(LO) mode according to the observed symmetry. The upper maximum of the camelback structure at around 1465 cm^{-1} corresponds to an overtone of the zone-center A$_1$(LO) mode. The lower camelback at 1385 cm^{-1} arises from the highest phonon branches which become flat at the Brillouin boundary. Hence, they originate either from the A or K point. The frequencies of the observed first- and second-order features together with their symmetry and their assignments are listed in Table 5.

The last item of this paragraph about the results of second-order scattering will be on AlN. Figure 15 exhibits a room temperature Raman spectrum of hexagonal AlN (upper curve) together with a calculated density of two-phonon states [67] of cubic AlN (bottom curve).

The gap between the acoustic and the optical phonon branches of AlN amounts only to about 60 cm^{-1} and is thus much smaller than that of GaN (see also Fig. 12) because of the smaller mass difference between Al and N compared to Ga and N. A classification of frequency regions with absolute acoustic or optical character as done above for GaN is therefore not possible. According to Fig. 12 acoustic overtones can occur until approx. 1100 cm^{-1}. Optical combinations and overtones are expected from 1220 cm^{-1} on. Acoustic-optical combinations can appear in a broad overlap region. Figure 16 shows the full Raman spectra of a hexagonal AlN sample in the relevant A$_1$, E$_1$ and E$_2$ configuration. In the following we discuss the strongest structures in comparison with the calculated phonon dispersion (Fig. 12) and the density of two-phonon states (Fig. 15) and assign them according to the selection rules of second-order scattering.

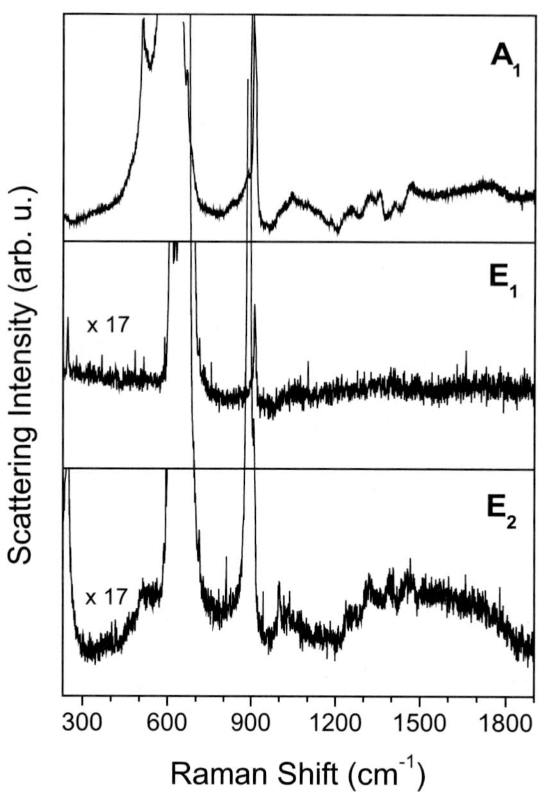

Fig. 16 Second-order Raman spectra of hexagonal AlN taken at room temperature for various scattering geometries corresponding to symmetry A_1, E_1, and E_2, respectively.

Below the first-order Raman modes a sharp line appears at 511 cm^{-1} in the spectrum with A_1 symmetry whereas only a broad structure is visible in E_2 symmetry. In the latter the structure seems to be split into two with maxima at 511 and 553 cm^{-1}. Whether the line at 553 cm^{-1} has even A_1 symmetry or not cannot be deduced from the spectra because of its overlap with the strong peak of the first-order A_1(TO) mode. Comparing the calculated phonon dispersion (Fig. 12) one finds that the energies possible due to the symmetry differ from the experimental data. Comparing the spectra with those of GaN we find a similarity to the lineshape of the structure at 317 cm^{-1} and thus assign the line at 511 cm^{-1} in the spectrum of AlN to the flat running acoustic phonon branch of the H point. This is supported by the fact that it is not visible in the calculated cubic overtone density. Due to the above mentioned difficulty to determinate the symmetry of the shoulder at 553 cm^{-1} an unambiguous assignment is not possible. It might result from a combination, overtone or difference process from the M point.

Table 6 Frequencies and symmetries of the strongest modes found in the first- and second-order Raman spectra of hexagonal AlN and their assignments.

Frequency (cm^{-1})	Symmetry	Process	Point in the BZ
511	A_1, (E_2)	acoustic overtone	H
553	E_2, (A_1)		M
610	A_1	first-order process	$\Gamma = A_1$(TO)
656	E_2	first-order process	$\Gamma = E_2$(high)
669	E_1	first-order process	$\Gamma = E_1$(TO)
891	A_1	first-order process	$\Gamma = A_1$(LO)
912	E_1	first-order process	$\Gamma = E_1$(LO)
1000	E_2, (A_1)	acoustic overtone	M
1060	A_1, E_2	acoustic overtone, comb.	M
1186	A_1, E_2	acoustic-optical comb.	M
1256	A_1, E_2		M
1316	A_1, E_2	optical overtone	$\Gamma = [E_2]^2$
1355	A_1	optical overtone	K, H
1393	E_2	optical combination	M
1409	A_1	optical overtone	K, H
1466	A_1	optical overtone	Γ, K
cutoff 1830	A_1, E_2	optical overtone	$\Gamma = [E_1]^2$

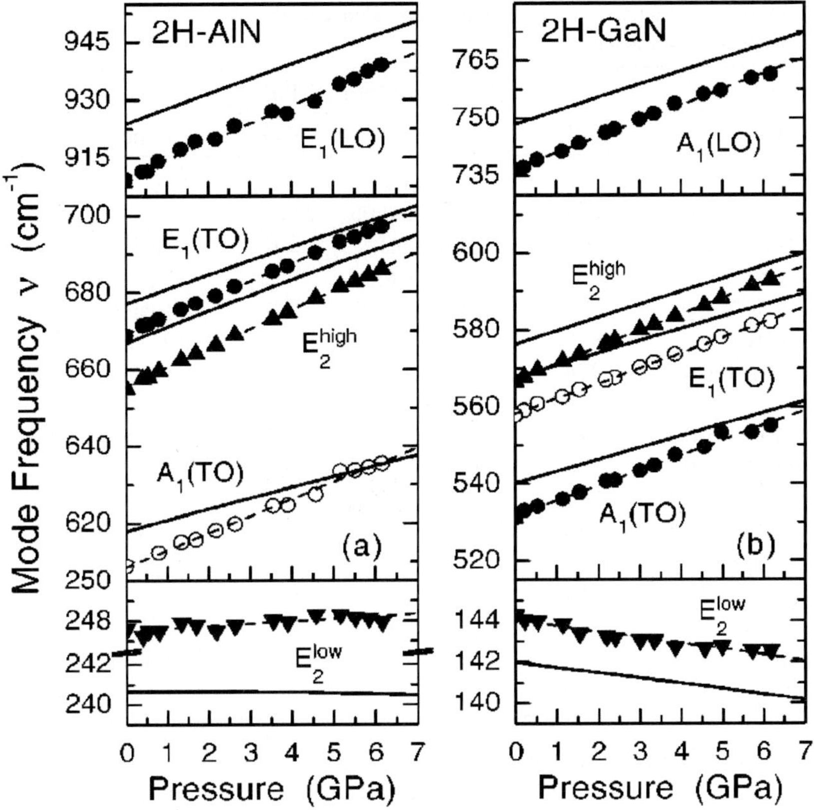

Fig. 17 Measured zone center optical phonon frequencies (symbols) versus hydrostatic pressure for *2H*-AlN (left panels) and 2H-GaN (right panels). Solid lines represent the results of *ab initio* calculations. Dashed lines are linear fits to the experimental data.

Beyond the first-order phonons a broad structure appears at around 1060 cm^{-1} having A_1 and E_2 symmetry. Additionally a sharp peak at 1000 cm^{-1} is detected in the E_2 symmetry. Comparing this with the calculated overtone density leads to the assignment to originate from the M point. The energy of the peak at 1186 cm^{-1} is located in the gap between the acoustic and optical branches and must thus stem from an acoustic optical combination process. Because of the observed A_1 and E_2 symmetry it originates most likely from the M point. Considering the calculated phonon dispersion the structure at around 1256 cm^{-1} can be explained as an acoustic-optical combination ($M_1 \times M_1$ with $\omega_{calc} = 1267$ cm^{-1}) as well as an optical overtone ($[M_3]^2$ with $\omega_{calc} = 1276$ cm^{-1}). Both processes fulfill the selection rules. Davydov et al. suggested an optical overtone from the K or M point [70]. The doublet at 1316 and 1355 cm^{-1} is assigned to optical overtones; the first one at 1316 cm^{-1} having A_1 and E_2 symmetry coincides with the double E_2(high) energy of the first-order scattering and is hence its overtone. The structure at about 1400 cm^{-1} is composed of one line at 1409 cm^{-1} with pure A_1 symmetry and a second line at 1393 cm^{-1} with defined E_2 symmetry. The latter we assign because of its symmetry to a $M_1 \times M_2$ process ($\omega_{calc} = 1394$ cm^{-1}) whereas the lines at 1409 and 1355 cm^{-1} do not allow an unambiguous classification. Overtones originating from the K or H point are possible. The last well resolved peak at 1466 cm^{-1} visible in A_1 symmetry we consider to be an optical K_1 or B_1 overtone. Hence, we obtain the frequency of the silent B_1 [Γ] mode to be 733 cm^{-1} what fits to the *ab-initio* calculations of Gorczyca et al [72]. Davydov et al. found this structure in A_1 and E_2 symmetry and assigned it to an optical combination process from the M point [70]. The cutoff at around 1830 cm^{-1} is attributed analogous to GaN to an overtone of the zone-center E_1(LO) mode according to the observed symmetry. All phonon frequencies together with their symmetries and assignments are listed in Table 6.

Table 7 Fitting parameters used for the pressure dependence of the phonon frequencies in GaN and AlN and resulting Grüneisen parameters γ.

mode	ω (cm^{-1})		$\partial\omega/\partial P$ (cm^{-1}/GPa)		γ	
	Expt.	Calc.	Expt.	Calc.	Exper.	Calc.
			2H-AlN			
E_2(low)	247.5 ± 0.5	241	0.12 ± 0.05	-0.03	0.10 ± 0.05	-0.02
A_1(TO)	608.5 ± 0.5	618	4.4 ± 0.1	3.0	1.51 ± 0.05	1.02
E_2(high)	655.5 ± 0.1	667	4.99 ± 0.03	4.2	1.58 ± 0.01	1.34
E_1(TO)	669.3 ± 0.1	677	4.55 ± 0.03	3.8	1.41 ± 0.01	1.18
A_1(LO)	891	898		3.5		0.82
E_1(LO)	910.1 ± 0.4	924	4.6 ± 0.1	4.0	1.06 ± 0.03	0.91
			2H-GaN			
E_2(low)	144.1 ± 0.2	142	-0.3 ± 0.1	-0.24	-0.4 ± 0.1	-0.35
A_1(TO)	531.7 ± 0.3	540	3.9 ± 0.1	3.1	1.47 ± 0.04	1.21
E_1(TO)	558.2 ± 0.1	568	3.94 ± 0.03	3.3	1.41 ± 0.01	1.19
E_2(high)	567.0 ± 0.1	576	4.24 ± 0.03	3.6	1.50 ± 0.01	1.28
A_1(LO)	736.5 ± 0.2	748	4.4 ± 0.1	3.5	1.20 ± 0.05	0.98
E_1(LO)	742 ± 0.2	757		3.6		0.99
			3C-GaN			
TO	553 ± 2	560	4.0 ± 0.2	3.2	1.4 ± 0.1	1.19
LO	743 ± 2	750	4.5 ± 0.2	3.7	1.20 ± 0.05	1.02

3.8 Lattice vibrations under hydrostatic pressure Aside from work on the optical transitions [73] and the local vibrational modes of Mg [74] under pressure and for different temperatures, we also studied the effect of pressure on the optical phonon frequencies in GaN and AlN. This work which contains experiments and *ab initio* work focuses on the possible softening of the E_2(low) mode under pressure and on the effective charge as defined by the splitting of TO and LO modes. The experiments were performed on hexagonal GaN grown on SiC and on AlN whiskers. Thessample were pressurized in a dia-

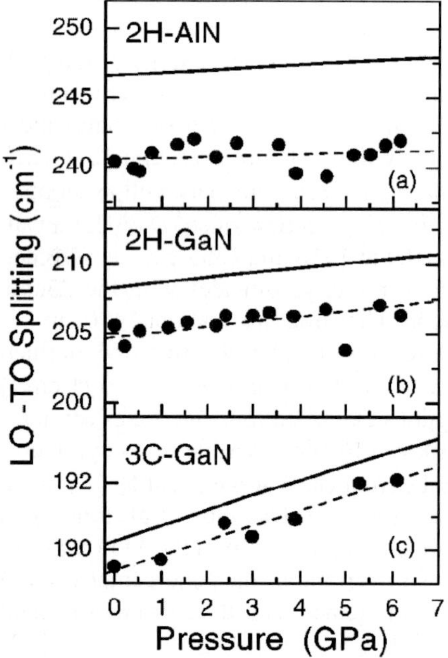

Fig. 18 Experimental LO-TO splittings (symbols) as a function of pressure for a) the 2H-AlN E_1 modes, b) the 2H-GaN A_1 modes, and c) 3C-GaN. Solid lines represent the results of *ab initio* calculations. Dashed lines are linear fits to the experimental data.

Table 8 Pressure coefficients of the LO$-$TO splitting $\Delta\omega$ and transverse effective charge e_T^* (in units of the elementary charge e).

material	mode	$\Delta\omega\ (\mathrm{cm}^{-1})$		$\partial(\Delta\omega)/\partial P\ (\mathrm{cm}^{-1}/\mathrm{GPa})$		$e_T^*\ (e)$		$\partial e_T^*/\partial P\ (10^{-3}e/\mathrm{GPa})$	
		Expt.	Calc.	Expt.	Calc.	Expt.	Calc.	Expt.	Calc.
2H-AlN	E_1	240.8 ± 0.5	247	0.10 ± 0.12	0.21	2.63 ± 0.05	2.54	0.15 ± 0.70	-1.0
2H-GaN	A_1	205.3 ± 0.2	208	0.4 ± 0.1	0.39	2.73 ± 0.03	2.74	-1.3 ± 07	-2.0
3C-GaN	(cub.)	189 ± 1	190	0.45 ± 0.05	0.50	2.65 ± 0.03	2.65	0.0 ± 0.5	-1.9

mond-anvil cell; the ruby method was used to determine the pressure. The experimental results were complemented by *ab initio* density functional calculations using a plane-wave method. Details of the calculation are given in [68, 75], those of the experiment in [76].

The pressure dependence of the various zone center modes in GaN and AlN are given in Fig. 17 together with the results of the calculation. The overall agreement is quite good, in particular the shapes and the curious behavior of the E_2(low) mode which has a small positive (zero-theory) in AlN and even a negative slope in GaN. The evaluation of the data leads to the pressure derivatives $\partial\omega/\partial P$ and Grüneisen parameters

$$\gamma = -\partial\ln\omega/\partial\ln V = (B_0/\omega_0)\partial\omega/\partial P , \tag{8}$$

where $B_0 = 200$ GPa was taken as a bulk modulus for GaN [30] and 208 GPa for AlN [77]. Our results are summarized in Table 7. Note that the small negative E_2(low) pressure slopes are reproduced also in the calculations. They were interpreted as being due to a critical balance between central and non-central forces.

The TO$-$LO splitting has been reported somewhat controversially in the literature. Perlin et al. found that the separation decreased with increasing pressure [78] but had combined data taken on different samples. Our results, shown in Fig. 18, display an increase in splitting and hence effective charge, and they were recorded on the same sample making them somewhat more reliable.

The pressure slopes of hexagonal GaN and AlN as well as of cubic GaN are summarized in Table 8 together with the derived theoretical values. They, too, show a reasonable agreement. Different from, e.g. GaAs, the effective charge changes only little under hydrostatic pressure, which we believe is a consequence of the strong directional covalent bonding of the nitrides; they behave similar to SiC or diamond but less pronounced.

4 Conclusion   In summary we presented a review of our work on the lattice dynamics of group-III nitrides. With first-order Raman spectroscopy we determined the zone-center phonons of GaN and AlN and their dependence on structural properties. Our investigations on the angular dispersion of the polar A_1 and E_1 modes confirmed the theory, that in case of larger TO$-$LO splitting compared to the A_1-E_1 splitting the angular dependent quasi modes are still pure transverse or longitudinal with mixed symmetry. We showed details of the strain distribution of different GaN samples and added experimental data about coupled LO phonon plasmon modes (LPP modes) in the lower frequency range, i.e. the dependence on carrier densities between 10^{18} and 10^{19} cm^{-3}. Studies on Al$_x$Ga$_{1-x}$N samples covering the whole composition range $0 < x < 1$ certified the two-mode behavior of the E_2(high) phonon and revealed the same with regard to the A_1(TO) mode. We analyzed the incorporation of the hexagonal minority phase in cubic GaN samples and presented methods for its quantitative determination. With second-order Raman spectroscopy we measured the frequencies of zone-boundary phonons of hexagonal GaN and AlN as well as cubic GaN in various scattering geometries corresponding to different symmetries. We compared the experimental data with a calculated phonon dispersion and, considering the selection rules obtained from a group-theoretical analysis, discussed the observed phonon modes. Assignments for hexagonal GaN and AlN were given and the behavior under hydrostatic pressure discussed.

Acknowledgements The work presented here was done in part in cooperation with F. Bechstedt and K. Syassen. It was furthermore partly supported by the Deutsche Forschungsgemeinschaft. L. Eckey, L. Filippidis, G. Kaczmarczyk, A. Kaschner, A. R. Goñi, and A. P. Litvinchuk promoted this project with their dedicated work. We are very grateful to several growth teams for providing specimens: T. Detchprohm and K. Hiramatsu (Nagoya University), H. Amano and I. Akasaki (Meijo University), D. Schikora, D. J. As and K. Lischka (Universität Paderborn), M. Stutzmann (TU München), O. Briot (Université de Toulouse), D. K. Wickenden (Hopkins University), M. Hommel (Universität Bremen), and J. Pastrňák (former Czechoslovak Academy of Sciences, Prague).

References

[1] S. Strite and H. Morkoç, J. Vac. Sci. Technol. B **10(4)**, 1237 (1992).
[2] S. Nakamura, T. Mukai, and M. Senoh, Jpn. J. Appl. Phys. Part 2, **30**, L1998 (1991).
[3] S. Nakamura, M. Senoh, S. Nagahama, N. Iwasa, T. Yamada, T. Matsushita, Y. Sugimoto, and H. Kiyoku, Appl. Phys. Lett. **69**, 4056 (1996).
[4] T. N. Oder, J. Li, J. Y. Lin, and H. X. Jiang, Appl. Phys. Lett. **77**, 791 (2000).
[5] H. Morkoç, S. Strite, G. B. Gao, M. E. Lin, B. Sverdlov, and M. Burns, J. Appl. Phys. **76**, 1363 (1994).
[6] C. M. Lueng, H. L. W. Chan, C. Surya, and C. L. Choy, J. Appl. Phys. **88**, 5360 (2000).
[7] K. Naniwae, S. Itoh, H. Amano, K. Itoh, K. Hiramatsu, and I. Akasaki, J. Cryst. Growth **99**, 381 (1990).
[8] T. Detchprohm, K. Hiramatsu, H. Amano, and I. Akasaki, Appl. Phys. Lett. **61**, 2688 (1992).
[9] K. Hiramatsu, T. Detchprohm, and I. Akasaki, Jpn. J. Appl. Phys. **32**, 1528 (1993).
[10] H. Amano, N. Sawaki, I. Akasaki, and Y. Toyoda, Appl. Phys. Lett. **48**, 353 (1986).
[11] I. Akasaki, H. Amano, N. Koide, M. Kotaki, and K. Manabe, Physica B **185**, 428 (1993).
[12] D. J. As, D. Schikora, A. Greiner, M. Lübbers, J. Mimkes, and K. Lischka, Phys. Rev. B **54**, R11118 (1996).
[13] D. Schikora, M. Hankeln, D. J. As, K. Lischka, T. Litz, A. Waag, T. Buhrow, and F. Henneberger, Phys. Rev. B **54**, R8381 (1996).
[14] H. Angerer, O. Ambacher, R. Dimitrov, T. Metzger, W. Rieger, and M. Stutzmann, MIJ-NSR **1**, Article 15 (1996).
[15] H. Angerer, D. Brunner, F. Freudenberg, O. Ambacher, M. Stutzmann, R. Höpler, T. Metzger, E. Born, G. Dollinger, A. Bergmaier, S. Karsch, and H. J. Körner, Appl. Phys. Lett. **71**, 1504 (1997).
[16] J. Pastrňák and L. Roskovcová, phys. stat. sol. **7**, 331 (1964).
[17] J. Pastrňák and L. Roskovcová, phys. stat. sol. **26**, 591 (1968).
[18] P. Lawaetz, Phys. Rev. B **5**, 4039 (1972).
[19] M. Cardona, in: Light Scattering in Solids II, edited by M. Cardona and G. Güntherodt, Topics Appl. Phys. **50**, Springer, Berlin, Heidelberg 1982 (p. 19ff).
[20] D. D. Manchon, Jr., A. S. Barker, Jr., P. J. Dean, and R. B. Zetterstrom, Solid State Commun. **8**, 1227 (1970).
[21] A. Cingolani, M. Ferrara, M. Lugara, and G. Scamarcio, Solid State Commun. **58**, 823 (1986).
[22] P. Perlin, C. Jauberthie-Carillon, J. P. Itie, A. San Miguel, I. Grzegory, and A. Polain, Phys. Rev. B **45**, 83 (1992).
[23] T. Azuhata, T. Sota, K. Suzuki, and S. Nakamura, J. Phys. C **7**, L129 (1995).
[24] L. E. McNeil, M. Grimsditch, and R. H. French, J. Am. Ceram. Soc. **76**, 1132 (1993).
[25] R. Loudon, Adv. Phys. **13**, 423 (1964).
[26] C. A. Arguello, D. L. Rousseau, and S. P. S. Porto, Phys. Rev. **181**, 1351 (1969).
[27] L. Filippidis, H. Siegle, A. Hoffmann, C. Thomsen, K. Karch, and F. Bechstedt, phys. stat. sol. (b) **198**, 621 (1996).
[28] T. Kozawa, T. Kachi, H. Kano, H. Nagase, N. Koide, and K. Manabe, J. Appl. Phys. **77**, 4389 (1995).
[29] F. Demangeot, J. Frandon, M. A. Renucci, O. Briot, B. Gil, and R. L. Aulombard, Solid State Commun. **100**, 207 (1996).
[30] C. Kisielowski, J. Krüger, S. Ruvimov, T. Suski, J. W. Ager III, E. Jones, Z. Lilienthal-Weber, M. Rubin, E. R. Weber, M. D. Bremser, and R. F. Davis, Phys. Rev. B **54**, 17745 (1996).
[31] J.-M. Wagner and F. Bechstedt, Phys. Rev. B **66**, 115202 (2002).
[32] K. Hiramatsu, S. Kitamura, and N. Sawaki, Mater Res. Soc. Symp. Proc. **395**, 267 (1996).
[33] S. Kitamura, K. Hiramatsu, and N. Sawaki, Jpn. J. Appl. Phys. **34**, L1184 (1995).
[34] H. Siegle, Dissertation, Technische Universität Berlin, p. 96 (1998).
[35] A. Hoffmann, H. Siegle, A. Kaschner, L. Eckey, C. Thomsen, J. Christen, F. Bertram, M. Schmidt, K. Hiramatsu, S. Kitamura, and N. Sawaki, J. Cryst. Growth **189/190**, 630 (1998).
[36] B. B. Varga, Phys. Rev. **137**, A1896 (1965).
[37] A. Mooradian and G. B. Wright, Phys. Rev. Lett. **16**, 999 (1966).

[38] E. Burstein, A. Pinczuk, and S. Iwasa, Phys. Rev. **157**, 611 (1967).
[39] P. Perlin, J. Camassel, W. Knap, T. Taliercio, J. C. Chervin, T. Suski, I. Grzegory, and S. Porowski, Appl. Phys. Lett. **67**, 2524 (1995).
[40] D. T. Hon and W. L. Faust, Appl. Phys. **1**, 241 (1973).
[41] J. F. Scott, T. C. Damen, J. Ruvalds, and A. Zawadowski, Phys. Rev. B **3**, 1295 (1971).
[42] M. V. Klein, B. N. Ganguly, and P. J. Colwell, Phys. Rev. B **6**, 2380 (1972).
[43] R. Ruppin and J. Nahum, J. Phys. Chem. Solids **75**, 1311 (1974).
[44] T. Kozawa, T. Kachi, H. Kano, Y. Taga, M. Hashimoto, N. Koide, and K. Manabe, J. Appl. Phys. **75**, 1098 (1994).
[45] F. A. Ponce, J. W. Steeds, C. D. Dyer, and G. D. Pitt, Appl. Phys. Lett. **69**, 2650 (1996).
[46] F. Demangeot, J. Frandon, M. A. Renucci, C. Meny, O. Briot, and R. L. Aulombard, J. Appl. Phys. **82**, 1305 (1997).
[47] H. Siegle, A. Hoffmann, L. Eckey, C. Thomsen, J. Christen, F. Bertram, D. Schmidt, D. Rudloff, and K. Hiramatsu, Appl. Phys. Lett. **71**, 2490 (1997).
[48] M. Drechsler, D. M. Hofmann, B. K. Meyer, T. Detchprohm, H. Amano, and I. Akasaki, Jpn. J. Appl. Phys. **34**, L1178 (1995).
[49] R. Besermann, C. Hirlimann, M. Balkanski, and J. Chevalier, Solid State Commun. **20**, 485 (1976).
[50] B. Jusserand and S. Slempkes, Solid State Commun. **49**, 95 (1984).
[51] S. Emura, S. Gonda, Y. Matsui, and H. Hayashi, Phys. Rev. B **38**, 3280 (1988).
[52] X. Wang and X. Zhang, Solid State Commun. **59**, 869 (1986).
[53] K. Hayashi, K. Itoh, N. Sawaki, and I. Akasaki, Solid State Commun. **77**, 115 (1991).
[54] A. Cros, H. Angerer, O. Ambacher, R. Höpler, T. Metzger, and M. Stutzmann, Solid State Commun. **104**, 35 (1997).
[55] K. Karch, F. Bechstedt, and T. Pletl, Phys. Rev. B **56**, 3560 (1997).
[56] S. Strite, M. E. Lin, and H. Morkoç, Thin Solid Films **231**, 197 (1993).
[57] T. Lei, K. F. Ludwig, Jr., and T. D. Moustakas, J. Appl. Phys. **74**, 4430 (1993).
[58] O. Brandt, H. Yang, B. Jenichen, Y. Suzuki, L. Däweritz, and K. H. Ploog, Phys. Rev. B **52**, R2253 (1995).
[59] H. Siegle, L. Eckey, A. Hoffmann, C. Thomsen, B. K. Meyer, D. Schikora, M. Hankeln, and K. Lischka, Solid State Commun. 96, 943 (1995).
[60] C. Thomsen, R. Wegerer, H.-U. Habermeier, and M. Cardona, Solid State Commun. **83**, 199 (1992).
[61] S. Miyoshi, K. Onabe, N. Ohkouchi, H. Yaguchi, R. Ito, S. Fukatsu, and Y. Shiraki, J. Cryst. Growth **124**, 439 (1992).
[62] M. Giehler, M. Ramsteiner, O. Brandt, H. Yang, and K. H. Ploog, Appl. Phys. Lett. **67**, 733 (1995).
[63] S. Murugkar, R. Merlin, A. Botchkarev, A. Salvador, and H. Morkoç, J. Appl. Phys. **77**, 6042 (1995).
[64] H. Siegle, L. Filippidis, G. Kaczmarczyk, A. P. Litvinchuk, A. Hoffmann, and C. Thomsen, in: Proceedings of the 23rd International Conference on Physics of Semiconductors (ICPS 23), Berlin, Germany, World Scientific 1996 (p. 533 ff).
[65] H. Siegle, G. Kaczmarczyk, L. Filippidis, A. P. Litvinchuk, A. Hoffmann, and C. Thomsen, Phys. Rev. B **55**, 7000 (1997).
[66] E. O. Kane, Phys. Rev. B **31**, 7865 (1985).
[67] K. Karch and F. Bechstedt, Phys. Rev. B **56**, 7404 (1997).
[68] K. Karch, J.-M. Wagner, and F. Bechstedt, Phys. Rev. B **57**, 7043 (1998).
[69] T. Ruf, J. Serrano, M. Cardona, P. Pavone, M. Pabst, M. Krisch, M. D'Astuto, T. Suski, I. Grzegory, and M. Leszczynski, Phys. Rev. Lett. **86**, 906 (2001).
[70] V. Yu. Davydov, Yu. E. Kitaev, I. N. Goncharuk, A. N. Smirnov, J. Graul, O. Semchinova, D. Uffmann, M. B. Smirnov, A. P. Mirgorodsky, and R. A. Evarestov, Phys. Rev. B **58**, 12899 (1998).
[71] J. M. Zhang, T. Ruf, M. Cardona, O. Ambacher, M. Stutzmann, J.-M. Wagner, and F. Bechstedt, Phys. Rev. B **56**, 14399 (1997).
[72] I. Gorczyca, N. E. Christensen, E. L. Peltzer y Blancá, and C. O. Rodriguez, Phys. Rev. B **51**, 11936 (1995).
[73] Z. X. Liu, A. R. Goñi, K. Syassen, H. Siegle, C. Thomsen, B. Schöttker, D. J. As, and D. Schikora, J. Appl. Phys. **86**, 929 (1999).
[74] G. Kaczmarczyk, A. Kaschner, A. Hoffmann, and C. Thomsen, Appl. Phys. Lett. **78**, 198 (2001).
[75] J.-M. Wagner and F. Bechstedt, Phys. Rev. B **62**, 4526 (2000).
[76] A. R. Goñi, H. Siegle, K. Syassen, C. Thomsen, and J.-M. Wagner, Phys. Rev. B **64**, 035205 (2001).
[77] M. Ueno, A. Onodera, O. Shimomura, and K. Takemura, Phys. Rev. B **45**, 10123 (1992).
[78] P. Perlin, T. Suski, J. W. Ager III, G. Conti, A. Polian, N. E. Christensen, I. Gorczyca, I. Grzegory, E. R. Weber, and E. E. Haller, Phys. Rev. B **60**, 1480 (1999).

phys. stat. sol. (c) **0**, No. 6, 1732–1749 (2003) / **DOI** 10.1002/pssc.200303131

Electronic and vibrational properties of group-III nitrides: *Ab initio* studies

F. Bechstedt[*], **J. Furthmüller**, and **J.-M. Wagner**

Institut für Festkörpertheorie und Theoretische Optik, Friedrich-Schiller-Universität, Max-Wien-Platz 1, D-07743 Jena, Germany

Received 4 March 2003, revised 5 June 2003, accepted 12 June 2003
Published online 28 August 2003

PACS 61.66.Fn, 63.20.Dj, 63.22.+m, 71.20.Nr, 77.70.+a

We present *first-principles* calculations based on pseudopotentials and the density functional theory. Structural, electronic and vibrational properties are investigated. The influence of strain and alloy effects is also discussed. Examples for novel exciting results obtained from such calculations are the energy gap of InN, the spontaneous polarization fields, the phonon modes of short-period superlattices, the deformation potentials of the Raman modes, and the gap bowing parameter of InGaN in the entire composition range.

1 Introduction

The modern solid state physics and materials science try to understand the properties of solids starting from basic notions and to predict the properties of novel materials. For that reason these theoretical approaches are called *first-principles* or *ab initio* methods. They really allow predictions without taking any parameter from the experiment. Most of these techniques are based on the density functional theory (DFT) [1] in the local density approximation (LDA) [2]. The full interaction of the electrons including exchange and correlation is taken into account. The form of the interaction functional is taken from exact results obtained for the homogeneous electron gas [3, 4]. However, being a ground-state theory the DFT-LDA method cannot describe excited states correctly, in particular not electronic excitation energies. The band structures derived from the single-particle Kohn-Sham eigenvalues [2] cannot directly be compared with measured transition energies [5]. Quasiparticle corrections calculated at least in the so-called GW approximation [6, 7] of the exchange-correlation self-energy have to be included to account for the excitation aspect. On the other hand, theoretical approaches based on DFT and LDA allow accurate calculations of the changes of the total energy with atomic displacements. The density-functional perturbation theory (DFPT) [8] gives the opportunity to compute the dynamical matrix and, hence, the lattice vibrations of a crystal without the knowledge of experimental input parameters apart from atomic numbers and masses.

The group-III nitrides AlN, GaN, and InN, their alloys, and their low-dimensional structures represent outstanding examples of the applicability of the *ab initio* theories mentioned above. One reason is related to their extreme properties, such as high ionicity, very short bond lengths, large energy gap (except for InN), low compressibility, high thermal conductivity, and high melting temperature of the nitride subgroup of the III-V compounds. These properties also make the growth of high-quality single crystals and the epitaxy of perfect layers complicated. As a consequence only insufficient data basis sets existed or exist for several

---

[*] Corresponding author: e-mail: bech@ifto.physik.uni-jena.de, Phone: +49 3641 947150, Fax: +49 3641 947150

properties. For instance, authors are frequently using elastic coefficients that have been calculated for zinc-blende and wurtzite AlN, GaN, and InN [9] but not measured. Another example is the determination of electronic properties of wurtzite InN. The fundamental energy gap of about 1.9 eV has been extracted from room-temperature optical absorption data of thin films prepared by reactive radio-frequency sputtering [10] (and references therein). The polycrystalline films consist of oriented needle-like wurtzite crystallites with the c-axis parallel to the growth direction and diameters in the range of $10 - 30$ nm. Meanwhile, better samples of InN have been grown by means of molecular beam epitaxy (MBE). Despite the large number of free carriers photoluminescence measurements of the fundamental gap were possible [11]. They indicate a gap below 1 eV in agreement with theoretical predictions. Calculations are thus helpful tools to clarify the properties of intrinsic InN crystals.

The theoretical description of the properties of crystals, heterostructures, and mixed crystals consisting of group-III nitrides is a challenge for several reasons. First of all, these compounds contain a first-row element, nitrogen. Due to the lack of p electrons the N core is extremely small. Consequently, an expansion of the one-electron eigenfunctions requires many plane waves in order to represent correctly the wave functions in the near-core region. Moreover, semicore d shells occur in the compounds GaN and InN. In GaN the semicore $3d$ shell of Ga is not inert as in most Ga-based materials and affects both the electronic and structural properties [12]. The situation is similar for the In $4d$ levels in InN [13]. The shallow Ga $3d$ and In $4d$ levels are resonant with the N $2s$ states. This makes a hybridization of these states and contributions of the shallow d electrons to the chemical bonding very likely in GaN and InN. The Ga $3d$ and In $4d$ states are highly localized in real space. Consequently, methods based on normconserving pseudopotentials cannot readily be applied, because they are difficult to converge in a plane-wave expansion of the eigenfunctions.

Another problem of the theoretical treatment of the group-III nitrides concerns the extreme differences between the cations and the nitrogen anion, in particular for GaN and InN. The extremely different atomic sizes and electronegativities cause considerable charge transfers between cations and anions and, hence, strongly ionic bonds. Altogether, these properties of the free atoms already indicate the zinc-blende–wurtzite polytypism of the group-III nitrides [14].

As a consequence of the extreme atomic properties of the constituents, the III-V nitride semiconductors AlN, GaN, and InN have electronic and optical properties that make them attractive candidates for fabricating novel electronic and in particular optoelectronic devices [15]. These devices usually consist of layered structures. Heterostructures involving wurtzite layers are influenced by the phenomenon of spontaneous polarization [16], which is not present in layered (unstrained) systems of cubic crystals. The accompanying built-in electric fields, stemming from polarization charges at the heterointerfaces, have a major impact on the electronic and vibrational properties of single and multiple quantum wells or superlattices. The misfit between the lattice constants of the nitrides can induce considerable biaxial strains in the layers. The strain affects the layer properties directly or via the induced piezoelectric polarization fields in addition to the spontaneous ones. In order to tailor further the electrical and optical properties of the layered systems, ternary or quaternary alloys of the group-III nitrides are applied. Alloys such as $In_xGa_{1-x}N$ and $Al_xGa_{1-x}N$ allow a variation of the fundamental energy gaps and effective masses in a wide parameter range. The alloying also drastically influences the spontaneous and piezoelectric polarization fields.

Consequently, in this article important properties of group-III nitride crystals, their layered structures, and their mixed crystals are studied on the basis of results of theoretical considerations and calculations within the framework of the *ab initio* DFT approach. Three aspects of the nitride systems are discussed in detail: strain influence, alloys, and interplay of different wurtzite layers resulting in spontaneous polarization fields. After a brief description of the computational methods, the results for basic properties of electrons and phonons in bulk nitride crystals and superlattices are summarized.

2   Computational approaches

The majority of results presented in this article is based on two versions of parameter-free calculations within the framework of DFT-LDA [1, 2]. In the first version the semicore Ga $3d$ and In $4d$ electrons are explicitly treated as valence electrons [17, 18]. Their interaction with the atomic cores is treated by non-normconserving *ab initio* Vanderbilt pseudopotentials. They allow a substantial softening also for first-row elements with the lack of core p electrons as well as for the attraction of shallow d electrons [19]. As a consequence of the optimization the plane-wave expansion of the one-electron eigenfunctions can be restricted by a kinetic-energy cutoff of 14.4 (AlN), 16.2 (GaN), or 15.5 Ry (InN). The cutoffs have been tested carefully. In the case of alloys the cutoff is substantially increased to 22 Ry to account for possible shorter bond lengths. The many-body electron-electron interaction is described by the Ceperley-Alder functional [3] as parametrized by Perdew and Zunger [4]. The **k**-space integrals are approximated by sums over special points. For instance, 10 or 12 mesh points of the Chadi-Cohen type [20] are used within the irreducible part of the Brillouin zone (BZ) of the zinc-blende or wurtzite structure. In the case of superlattices and alloys more dense meshes of special points of the Monkhorst-Pack type [21] are used. The calculations employ the conjugate-gradient method to minimize the total energy with respect to the atomic coordinates. Explicitly, the Vienna Ab-initio Simulation Package [22, 23] is used. Self-energy corrections in the GW approximation [5, 6, 7] and, hence, quasiparticle band structures are computed using a novel numerical approach [24].

In the second version [25, 26] the valence electron-ion interaction is treated by soft *ab initio* normconserving pseudopotentials for Ga, Al, and N which are generated according to the scheme of Troullier and Martins [27]. The shallow Ga $3d$ electrons are frozen into the core. Their effect on the other electrons is taken into account [28] via nonlinear core corrections [29]. Nevertheless, the small N core requires relatively high cutoff energies for the plane-wave expansion of the eigenfunctions up to 80 Ry for the various polytypes and structures based on GaN and AlN. The second version can be easily combined with the self-consistent DFPT [8]. It allows the calculation of dielectric and lattice-dynamical properties. The dynamical matrix follows from the self-consistent determination of the linear electron-density response to a change in the ionic potential due to the atoms being displaced according to a phonon mode. The high-frequency dielectric tensor and the Born effective charge tensor can be directly calculated for each inequivalent atom in the unit cell. Their combination gives the nonanalytic part of the dynamical matrix of the considered polar crystals.

3   Electronic states: fundamental gap of InN

Band structures of AlN, GaN, and InN calculated using DFT-LDA are presented for the two polytypes, $3C$ (zinc blende) and $2H$ (wurtzite), in Fig. 1. They are calculated using the structural parameters also obtained within the DFT-LDA. Quasiparticle corrections [5, 6, 7] are not taken into account. In the average the fundamental gaps decrease along the series AlN, GaN, and InN as a consequence of the decreasing covalent bonding and total bond energies. In the InN case the direct gap at Γ is even negative. The sequence of the Γ_{15v}^{3C} and Γ_{1c}^{3C} (Γ_{6v}^{2H} and Γ_{1c}^{2H}) levels is inverted in the zinc-blende (wurtzite) case. This is mainly due to the overestimation of the pd repulsion. The negative energy gaps are also observed in several other DFT-LDA calculations [13, 31]. The band structures for GaN and InN also show the flat bands related to the shallow Ga $3d$ and In $4d$ core states. Within the DFT-LDA approach they lie in the same energetical region as the lowest valence bands governed by N $2s$ states. However, in excitation spectra the d-bands are below the N $2s$ bands [32]. The energetical overlap of s- and d-states gives rise to a hybridization and an unphysical splitting of the lowest valence bands. In wurtzite crystals the conduction-band minima are located at the Γ point in the BZ. In the zinc-blende case the band minima of GaN and InN are still at Γ. All these crystals are therefore direct semiconductors. However, cubic AlN is an indirect semiconductor with the conduction-band minimum situated at the X point in the fcc BZ. Experimentally, there is a controversy concerning the direct or indirect character of cubic AlN. The energetical position and the decrease of

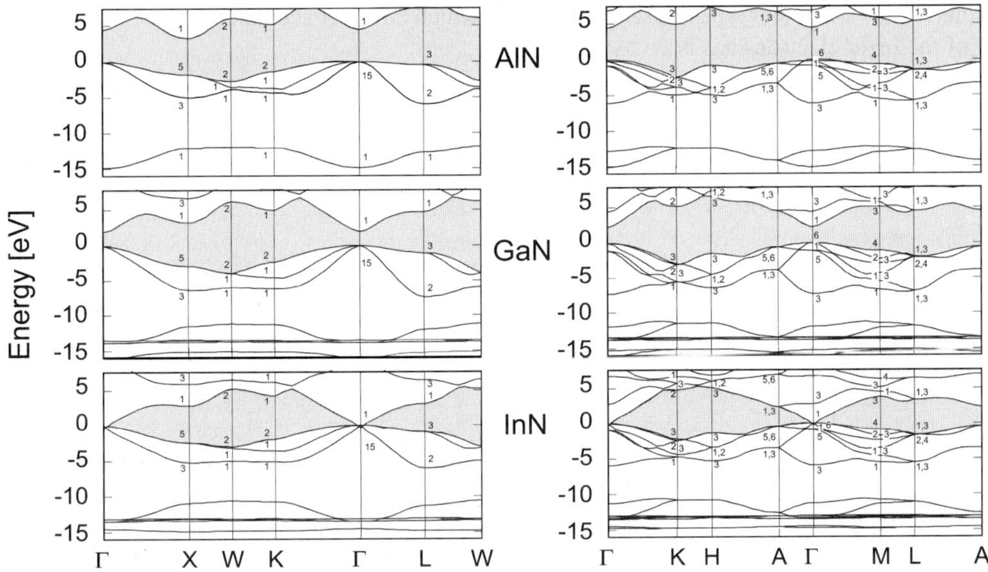

Fig. 1 Energy band structures of group-III nitrides crystallized in zinc-blende (left panels) or wurtzite (right panels) structure. The energies are calculated within DFT-LDA. The numbers indicate the irreducible representations at the respective symmetry points. The Γ_{15} (zinc blende) and Γ_6 or Γ_1 (wurtzite) valence-band maxima are set to zero. The fundamental gap is indicated by the shaded region. After Ref. [30].

emission intensity is believed to imply the indirect band structure of cubic $Al_xGa_{1-x}N$ in the mid-range of composition x [33] in agreement with the theoretical findings. However, recent cathodoluminescence studies of cubic AlN and $Al_xGa_{1-x}N$ epilayers do not show a significant decrease of the luminescence signal [34]. Hence, a direct character is assumed contrary to the theoretical predictions with 1.17 eV difference in the energies of the conduction-band minima at X and Γ.

Quasiparticle band structures have been calculated for AlN and GaN [35]. They show the characteristic widening of all energy gaps between empty and occupied bands. The quasiparticle corrections vary between 1.2 – 1.9 eV for AlN and 1.2 – 1.5 eV for GaN [35]. They give rise to fundamental quasiparticle gaps of 4.7 eV (3C-AlN), 5.8 eV (2H-AlN), 3.1 eV (3C-GaN), and 3.5 eV (2H-GaN) [35]. These values are close to experimental gaps, 6.3 eV (2H-AlN [36]), 3.4 eV (3C-GaN [37]), and 3.5 eV (2H-GaN [38]).

The electronic-structure theory seems to fail completely in the case of InN. The experimental value for the optical gap of 2H-InN accepted until 2001 amounts to about 1.9 eV [10]. This is in sharp contrast to the estimated quasiparticle gap with quasiparticle corrections of about 0.8 – 1.0 eV. Therefore careful studies have been done in addition for InN.

The structural optimizations have been performed including the In $4d$ electrons for both the zinc-blende and wurtzite polytype. This gives theoretical lattice constants $c = 5.688$ Å and $a = 3.523$ Å for wurtzite which are only slightly smaller than the measured ones [10, 11]. The corresponding zinc-blende value is $a_0 = 4.967$ Å. The resulting band structures are also presented in Fig. 1. They show the small overlap of conduction and valence bands near the Γ point as a consequence of the overestimation of the pd repulsion. The VBM is strongly p-like. The In $4d$ bands appear within the s-like valence bands. The pd repulsion shifts the VBM towards higher energies. In the wurtzite case, the energetical distance of the $4d$ bands to the VBM amounts about 13.5 eV. This energy is considerably smaller than the measured values 14.9 eV [39] or 16.7 eV [40].

In order to make an estimate for the effect of pd repulsion we have repeated the electronic-structure calculations for InN with In $4d$ electrons frozen into the core instead of treating the $4d$ electrons as valence electrons ("d_{val}"). Another type of pseudopotentials which however account for self-interaction corrections (SIC) of the In $4d$ electrons has been used.

The band structure in Fig. 2 demonstrates the influence of two effects. Since the In $4d$ electrons are

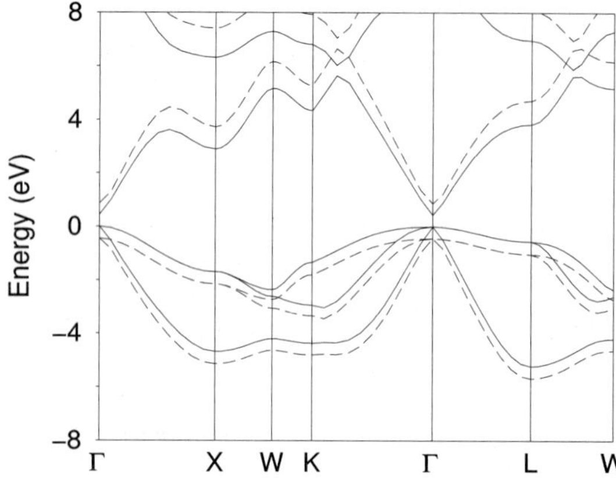

Fig. 2 DFT-LDA (solid lines) and quasiparticle (dashed lines) band structures calculated with self-interaction-corrected (SIC) pseudopotentials for zinc-blende InN. Only the region of the fundamental gap is plotted.

not explicitly taken into account, the p-like SIC bands are not anymore pushed towards higher energies. As a consequence the Γ_{15} valence-band maximum is below the Γ_1 conduction band. There is a small but positive gap. The quasiparticle correction shifts the occupied and empty bands away from each other. As a consequence all optical transition energies between valence- and conduction-band states are increased. The gaps are opened.

The results for the fundamental gaps are listed in Table 1. While the QP corrections are nearly the same, the gap of the zinc-blende modification is by about 0.2 eV smaller than that of the wurtzite structure. The QP shifts are nearly independent of the crystal structure. The small variations are mainly a consequence of the used electronic dielectric constants computed within the independent-particle approximation, $\varepsilon_\infty = 7.92$ (zinc blende) and $\varepsilon_\infty = 7.16$ (wurtzite, on average). The resulting shifts of about 0.9 eV are similar to the values 0.80 eV [41] and 1.03 eV [42] estimated by means of the semiempirical Bechstedt-Del Sole formula [43].

Table 1 DFT-LDA and QP gaps of InN. The QP corrections of the energy gaps are also given. All values in eV.

Polytype	LDA (d_{val})	LDA (SIC)	QP correction	QP (d_{val})	QP (SIC)
zinc blende	−0.36	0.43	0.87	0.52	1.31
wurtzite	−0.19	0.58	0.93	0.74	1.50

The resulting QP gaps 0.52 (d_{val}) or 1.31 (SIC) eV for zinc blende [0.74 (d_{val}) or 1.50 (SIC) eV for wurtzite] differ by 0.79 (0.76) eV as a consequence of the overestimation (d_{val}) or non-inclusion (SIC) of the pd repulsion. The true percentage contribution of the pd repulsion may be estimated from the true position of the In $4d$ levels 14.9 eV below the VBM in wurtzite InN [39]. Using perturbation-theory arguments and the DFT-LDA eigenvalue difference of 13.5 eV, a fraction of 13.5/14.9 of the total repulsion should be included. This estimate gives a fundamental energy gap $E_g = 0.59$ eV ($E_g = 0.81$ eV) for the zinc-blende (wurtzite) polytype. Gap values of about 0.8 eV have been indeed observed recently for wurtzite InN [11]. However, one has to mention an inaccuracy in the determination of the true pd repulsion. Using a different value of 16.7 eV [40] for the energetical distance In $4d$-VBM, slightly larger gaps $E_g = 0.67$ (0.89) eV are predicted.

4 Spontaneous polarization fields

As materials with partially ionic bonds the group-III nitrides exhibit the piezoelectric effect. Piezoelectric polarization fields are induced for internal displacements of the group-III atoms relative to the nitrogen atoms in an elementary cell. In the zinc-blende case (T_d^2) this happens only for shear strains. For wurtzite crystals with lower symmetry (C_{6v}^4) piezoelectric polarization fields can be also induced by diagonal strains. However, even in the equilibrium the lower-symmetry wurtzite polytypes are expected to possess already an intrinsic spontaneous polarization field, since the four tetrahedral bonds around one atom are not equivalent (cf. Fig. 3), so that wurtzite crystals represent pyroelectrics.

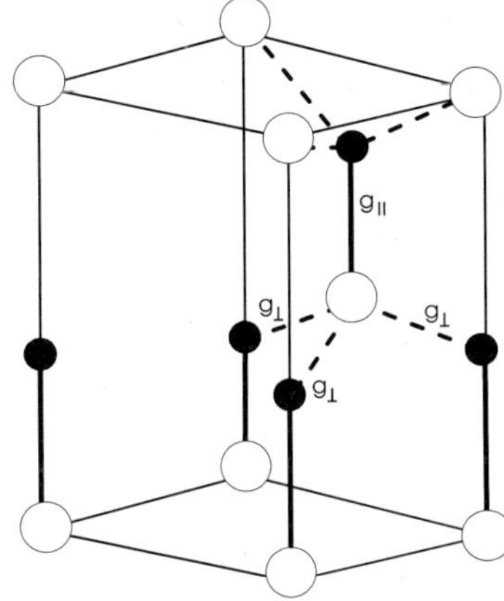

Fig. 3 The unit cell of a wurtzite crystal. The eight bond dipoles are characterized by the bond ionicities $g_\parallel$ and $g_\perp$ as well as the tetrahedron vectors $\boldsymbol{\tau}_i$ ($i = 1, 2, 3, 4$) and the corresponding negative vectors rotated by 60° around the c-axis.

In the bulk case the rearrangement of surface charges nullifies a spatially uniform spontaneous polarization. The effect of the spontaneous polarization fields can therefore only be observed in layered structures such as heterostructures consisting of wurtzite crystals of different compounds. In general, in heterostructures or inhomogeneous alloy layers variations in the composition are expected to create nonvanishing polarization fields. Indeed, the influence of such intrinsic polarization fields has been observed via the quantum-confined Stark effect in quantum-well structures based mainly on the $Al_xGa_{1-x}N/GaN$ system. However, the same effect should also occur in systems with variations of the crystal structure, e.g. in heterocrystalline structures consisting of nitride layers of both the wurtzite and zinc-blende polytype.

The macroscopic polarization in an electronic system is related to the charge density by its definition as electric dipole moment per unit volume. Fortunately, in the *ab initio* theories one has direct access to the charge density. Unfortunately, for an infinite crystal without center of inversion symmetry and using periodic boundary conditions in the computations, the polarization is an ill-defined quantity. In explicit calculations one has to eliminate truncation and surface effects. In the literature such calculations have therefore been performed within the Berry phase approach [44] or a superlattice treatment [18].

The polarization field has only a nonvanishing component parallel to the c-axis of the wurtzite structure (see Fig. 3), which is indicated by the z-axis in the following. In the superlattice approximation the macroscopic termination-dependent longitudinal polarization P_{Lz} in the presence of a nonvanishing depolarization field [45, 46], i.e., the total polarization in the wurtzite layers in the absence of external charges and a piezoelectric polarization, can be derived from *first principles* as the difference of the polarizations in a cubic ($3C$) reference material and in the wurtzite ($2H$) structure. For symmetry reasons the polarization field should vanish in a real cubic crystal. However, the accompanying electric field is nonzero in the cubic layers leading to a saw-tooth potential in a superlattice arrangement of zinc-blende and wurtzite layers (see Fig. 4). The macroscopic polarization P_{Lz} is related to homogeneous electric fields in each polytype region and, hence, to the gradient of the spatially averaged total one-electron potential by

$$P_{Lz} = -\varepsilon_0 \left[E_z^{2H} - E_z^{3C} \right] = \varepsilon_0 \left[\frac{d}{dz} \bar{V}_{\text{tot}}(z) \big|_{2H} - \frac{d}{dz} \bar{V}_{\text{tot}}(z) \big|_{3C} \right]. \tag{1}$$

The total single-particle potential occurs in the Kohn-Sham equation of the DFT-LDA. It is represented in Fig. 4 for a $(3C)_6(2H)_6$ superlattice containing 30 [111]- and [0001]-oriented III-N bilayers. The lattice constant of the cubic system is adapted to the in-plane constant of the wurtzite structure to avoid piezoelectric fields. The potential averaged over the xy-plane perpendicular to the c-axis shows the typical atomic oscillations. The additional spatial average in z-direction then gives the mentioned macroscopic saw-tooth potential (cf. Fig. 4) related to the macroscopic electric fields.

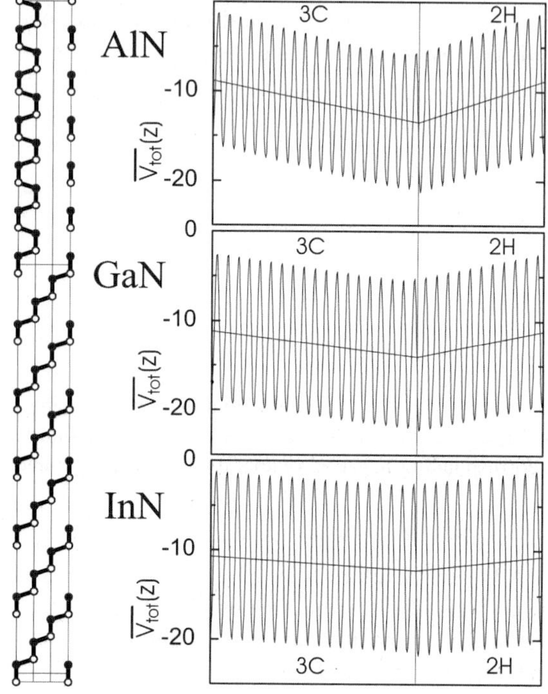

Fig. 4 Structure of a $(3C)_6(2H)_6$ superlattice cell and averaged total one-electron potentials. The vertical lines separate cubic and hexagonal stacking regions in AlN, GaN, and InN. After Ref. [18].

The polarization field strengths P_{Lz} derived from expression (1) are listed in Table 2. In order to determine the macroscopic zero-field or transverse polarization P_{Tz} [45, 46], the polarization induced in

Table 2 Calculated spontaneous polarization fields (in units of 10^{-3} C/m^3) of wurtzite group-III nitrides.

Polarization	Approach	AlN	GaN	InN
P_{Lz}	superlattice [18]	-26	-14	-6
P_{Tz}	superlattice [18]	-120	-74	-50
P_{Tz}	Berry phase [44]	-81	-29	-32
P_{Tz}	model [18]	-51	-50	-22

the electronic system by the electric fields in the hexagonal and cubic regions have to be taken into account. This effect is proportional to the electronic susceptibility. It holds

$$P_{Tz} = \bar{\varepsilon}_\infty P_{Lz} \qquad (2)$$

with an effective electronic dielectric constant of the considered superlattice in z-direction

$$\bar{\varepsilon}_\infty = \frac{d_{3C}\varepsilon_\infty + d_{2H}\varepsilon_\infty^{\|}}{d_{3C} + d_{2H}}, \qquad (3)$$

where the thicknesses, d_{2H} and d_{3C}, as well as the components of the dielectric tensor parallel to the c-axis, $\varepsilon_\infty^{\|}$ and ε_∞, appear. From the *ab initio* calculations we take $\bar{\varepsilon}_\infty = 4.52$ (AlN) [25], 5.41 (GaN)

[26], and 8.49 (InN) [45]. The values resulting for P_{Tz} are also given in Table 2 and compared with results of the Berry-phase approach [44]. The Berry-phase approach has been demonstrated to predict accurately polarization-related quantities in semiconductors from first principles [47]. Taking into account the sensitivity of the polarization with respect to the internal parameter u, excellent agreement between Berry-phase and supercell calculations can be stated. Particularly this holds for the variation from AlN to GaN. The agreement of Berry-phase predictions (supplemented by nonlinearity in the alloys [48]) with experiment as evaluated through device simulation has also been reported [49].

Interpreting the charge-asymmetry coefficients $g_{||}$ and $g_{\perp}$ [50] as the ionicities of the bonds parallel to the c-axis or with an angle of about $70°$ to the c-axis, one has to deal with bond dipoles $\mathbf{p}_{||/\perp} = eg_{||/\perp}\boldsymbol{\tau}_{||/\perp}$, where $\boldsymbol{\tau}_{||/\perp}$ denote vectors of the deformed tetrahedra in the wurtzite structure (cf. Fig. 3). The summation over the eight bonds per $2H$ unit cell with the volume Ω_0 leads to the expression

$$P_{Tz} \approx -\frac{2ec}{\Omega_0}\left[4g_{\perp}\left(\frac{3}{8}-u\right)+(g_{\perp}-g_{||})\,u\right]. \tag{4}$$

There are two contributions to the spontaneous polarization field parallel to the c-axis. The first contribution $\sim \left(\frac{3}{8}-u\right)$ is directly related to the nonideality of the tetrahedra in the wurtzite structure. The second term is proportional to the difference of the ionicities of the two inequivalent bonds in the structure. The values $g_{||}$ and $g_{\perp}$ are almost identical with those g of the zinc-blende structure. For AlN it holds $g_{\perp} = g(1-0.0025)$ and $g_{||} = g(1+0.0259)$ [25]. Despite of the small variations with the bond direction, the second term is not negligible, in particular for nitrides with geometry parameters u and c/a close to the ideal values. Results using the geometry parameters [17, 18], the g-values from Ref. [50], and the model described above [Eq. (4)] are also given in Table 2.

The two *ab initio* calculations and the estimate in the framework of the bond-orbital model give rise to the same order of magnitude of the intrinsic polarization. Also the only experimental value $P_{Tz} = 0.055$ C/m^2 available for AlN [51] is in the range of the calculated ones. Physically relevant are only the polarization differences at the interfaces. They are in reasonable agreement for the combination AlN-GaN but not for GaN-InN. The variation of the absolute numbers between the two *first-principles* calculations is mainly a consequence of the extreme accuracy of the structural optimization that is required for polarization calculations. Small changes in the geometry parameters may induce big variations in the polarization fields.

5 Lattice vibrations

The lattice-vibrational properties of group-III nitrides have been calculated using two types of *ab initio* approaches [18, 25, 26, 52]. For the cubic polytypes the complete dispersion relations and the resulting one-phonon density of states are presented in Fig. 5. Zone-center frequencies are summarized in Table 3 for the wurtzite nitrides [52]. The overall agreement of the *first-principles* calculations with Raman observations is excellent. In the case of the acoustic modes the maximum deviations amount to 4 cm^{-1}. The deviations in the region of the optical branches are slightly larger, mainly due to accuracy requirements in calculation of the LO-TO splittings and sample-quality problems.

Figure 5 clearly shows the dependence of the frequency gap between the acoustical and optical phonon branches for the nitrides AlN, GaN, and InN. This dependence can be quite easily explained within the classical two-atomic linear chain model where the frequency gap is given by zone-boundary vibrations with the nitrogen mass (upper limit) or the group-III atom mass (lower limit): the larger the mass difference between the two atoms the larger the frequency gap is. In the case of AlN there is only a small gap between the acoustical and optical phonon branches. This is due to the fact that the Al mass is only slightly larger than the nitrogen mass. The gap is opened with increasing ratio of cation and anion masses. Additionally the optical branches shift towards smaller frequencies. Other interesting features visible in the density of states are the remarkable changes of the dispersion of the optical branches. There is a strong reduction of the LO dispersion from the stronger bonded AlN to GaN and InN. Also the dispersion of the TO phonon

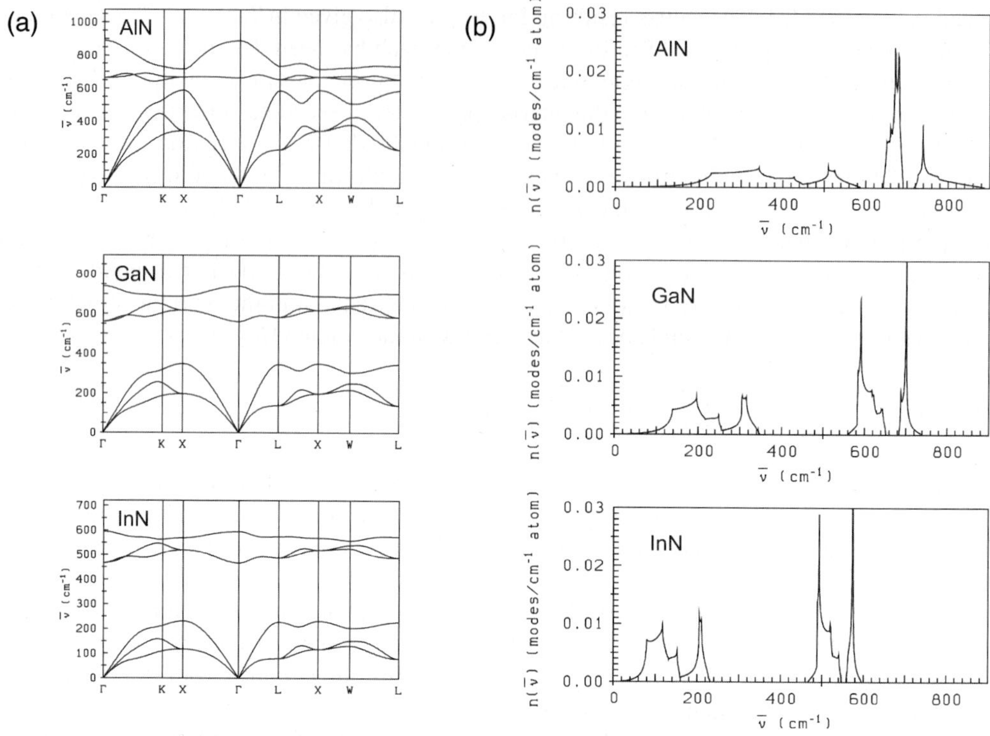

Fig. 5   Calculated dispersion of phonon branches (a) and corresponding density of states (b) of the zinc-blende nitrides. After Ref. [18].

branches shows a clear chemical trend accompanied by the inversion of the band dispersion. The strongest contributions to the density of the TO phonons arise from the Brillouin zone center. As a consequence of the nearly linear increase of the ratios $\omega_{TO}(\Gamma)/\omega_{TO}(X)$ and $\omega_{TO}(\Gamma)/\omega_{TO}(L)$ with decreasing cation mass, the frequencies of the zone-center TO phonons of GaN and InN are smaller than those of the zone-boundary phonons. The TO phonon branch of AlN is nearly dispersionless, at least along the ΓX directions. Hence, the strongest contributions to the TO phonon density of states occur at the upper (lower) boundary for AlN (GaN and InN). The comparison of the phonon densities of states of AlN, GaN, and InN with measured ones [53, 54] indicates that the main features are fairly well reproduced. This holds for both the peak positions and the peak intensities.

The mechanical, optical, and electrical properties of the nitrides can be remarkably influenced by their arrangement in layered structures. Interesting examples are short-period superlattices (SLs). For GaN/AlN SLs, one has a quite different physical situation compared to the case of GaAs/AlAs SLs: first, the layers are mutually strained due to different lattice constants, and there is an internal relaxation of the atomic layers at the interfaces. Second, the optical phonon branches partially overlap (cf. Fig. 5), and, third, the common anion is the lightest atom. The question arises if the phonons are also sensitive to the layered structure and to the accompanying electric fields. The characterization of the layered structures by means of Raman and IR spectroscopy therefore requires the knowledge of the special vibrational properties of the nitride SLs and their dependence on strain as well as on the arrangement and thickness of the layers. Furthermore, the knowledge of the SL phonons is important for the optimization of intersubband optical devices [55].

The zone-center phonon frequencies of a $(AlN)_2(GaN)_2(001)$ superlattice made by strained cubic nitrides are plotted in Fig. 6 versus the propagation angle Θ between the [001] growth direction of the

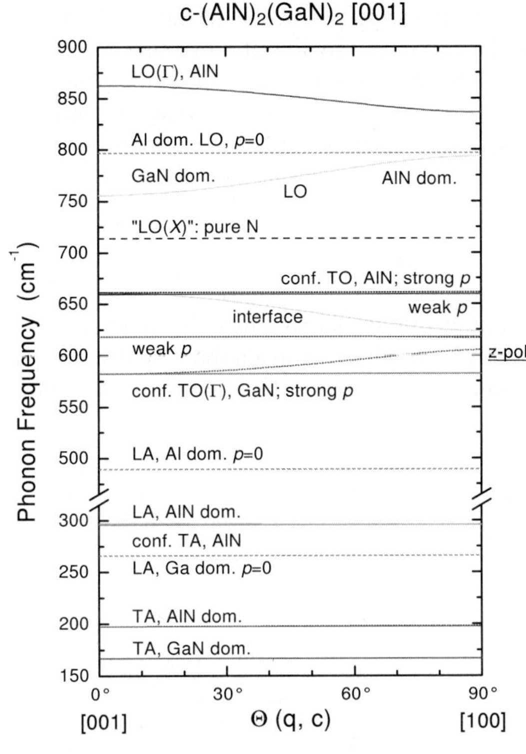

Fig. 6 Angular dispersion of zone-center phonon modes of a cubic 2×2 superlattice. Θ refers to the angle between the *c*-axis of the superlattice and the phonon propagation direction **q**. The dominant (dom.) material character and the strength of the accompanying dynamical polarization are indicated. An average in-plane lattice constant has been assumed. From [56].

superlattice and the direction of the phonon wave vector **q**. In general, concerning the angular dispersion, all transversal phonons split into a dispersionless transversal and an angle-dependent mode with a mixed polarization, with the exception of the strongly IR-active mode which exhibits angular dispersion despite its TO nature (marked "z-pol." in Fig. 6). All other modes showing angular dispersion can be identified with interface modes of the macroscopic theory. The angular dispersion of the interface mode in the TO-gap region is nearly independent of strain. For the strongly IR-active LO(Γ) [TO(Γ)] mode in Fig. 6, the angular variation amounts to 39 cm^{-1} (34 cm^{-1}) for the AlN in-plane lattice constant, and it reduces to 14 cm^{-1} (12 cm^{-1}) for the GaN in-plane lattice constant. Furthermore, the uppermost LO mode is special with respect to the fact that it is the only one that is determined completely by the long-range electric field. All other LO branches are either 'intrinsic' (i.e., they are eigenmodes of the SL that are completely determined by the analytical part of the dynamical matrix and are not influenced by the long-range macroscopic electric field), which here holds for the mode with vanishing dipole field, or at least one of the limiting frequencies is intrinsic, which here are the ones for in-plane propagation. Therefore, the uppermost LO mode must be related to a TO mode in a similar way as in the case of bulk crystals where the degeneracy of modes of the same symmetry is lifted by the macroscopic electric field, giving rise to the well-known LO-TO splitting. Then, the angular dispersion of this LO mode is determined by the one of the corresponding TO mode and of the strength of the LO-TO splitting.

6   Strain effects

Lattice and thermal misfits can cause large biaxial stresses in the epitaxial layers. For instance, a compressive stress results for hexagonal GaN grown on sapphire whereas a tensile one is observed for GaN on 6*H*-SiC. The situation is more complex in the case of heterostructures or superlattices, e.g. based on the AlN/GaN material combination (cf. Fig. 6), for which a mutual influence of the different material layers may occur.

In hexagonal crystals biaxial strains (or stresses) in the xy-plane perpendicular to the c-axis belong to the same class of perturbations as uniaxial ones along the c-axis and hydrostatic pressure. All these deformations conserve the C_{6v}^4 space-group symmetry. The influence of a biaxial strain in the plane perpendicular to the c-axis (i.e., the z-axis) of the wurtzite structure can be modelled by fixing the in-plane lattice constant a at a value different from the equilibrium one, a_{eq}. Then, the diagonal components of the strain tensor are

$$\epsilon_{xx} = \epsilon_{yy} = (a - a_{\mathrm{eq}})/a_{\mathrm{eq}},$$
$$\epsilon_{zz} = (c - c_{\mathrm{eq}})/c_{\mathrm{eq}}. \tag{5}$$

The actual lattice constant c follows from the boundary conditions for the stress tensor. Within the framework of *ab initio* calculations it is determined self-consistently by minimizing again the total energy with respect to c and u but for a fixed value a. This procedure simultaneously gives the internal strain, defined as $(u - u_{\mathrm{eq}})/u_{\mathrm{eq}}$.

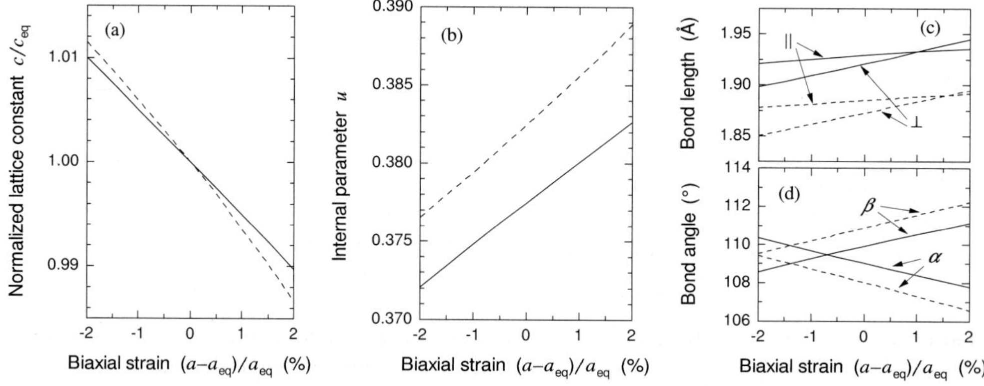

Fig. 7　Parameters of the atomic geometry of the wurtzite structure versus biaxial strain. GaN: solid line, AlN: dashed line. The lattice constant c (a), the internal parameter u (b), the bond lengths (c), and the bond angles (d) are plotted. The bonds along the c-axis (forming angles of about $70°$ with the c-axis) are labeled by $\parallel$ ($\perp$). The label α (β) denotes the angle between the c-axis and one of the three bonds nearly perpendicular to it (between two bonds nearly perpendicular to the c-axis). After Ref. [52].

Results of total energy optimizations using normconserving pseudopotentials for structural parameters of GaN and AlN [52] are represented in Fig. 7 versus given biaxial strain. As a consequence of the Poisson effect one observes a more or less linear decrease of the lattice constant c with increasing tensile biaxial strain. The behaviour of GaN and AlN is rather similar. Surprisingly the effect of a biaxial strain on the internal-cell parameter u and, hence, on the internal strain $(u - u_{\mathrm{eq}})/u_{\mathrm{eq}}$ is completely different. The u parameter increases with rising tensile biaxial strain.

The rather strong internal-strain effects dominate the variations of the bond lengths and bond angles (Figs. 7c and d). The length uc of the bonds parallel to the c-axis increases only slightly as a consequence of opposite variations of u and c. The length $\sqrt{\frac{1}{3}a^2 + (\frac{1}{2} - u)^2 c^2}$ of the bonds nonparallel to the c-axis increases much stronger with a tensile biaxial strain. Both strain and internal strain give rise to a stretching of these bonds. The two different types of bond angles, α and β, behave differently versus biaxial strain. There is a tendency for the reduction of the deformation of the bonding tetrahedra towards compressive biaxial strain. The angles approach the value $109.47°$ of the ideal tetrahedra. In the opposite direction, i.e., for tensile biaxial strain, the bonding zig-zag chains in $(11\bar{2}0)$ planes parallel to the c-axis are flattened. The bonding tetrahedra are compressed along the c-axis by shrinking the vertical distance $(\frac{1}{2} - u)\,c$ within an X-N bilayer (X = Ga, Al) towards a planar structure. The angle α tends towards $90°$, whereas the angle

β tends towards $120°$. This implies a tendency for dehybridization from ideal sp^3 hybrids towards sp^2 and p_z orbitals.

Table 3 Phonon deformation potentials $a(j)$ and $b(j)$ (in cm^{-1}) for GaN and AlN. Additionally, the resulting isotropic and shear deformation potentials, $K^{\mathrm{iso}}(j) = 2a(j) + b(j)$ and $K^{\mathrm{sh}}(j) = 2[a(j) - b(j)]$, are given. The mode frequencies $\omega(j)$ (in cm^{-1}) calculated in the strain-free case are also listed. From [52].

	E_2^{low}	B_1^{low}	$A_1(\mathrm{TO})$	$E_1(\mathrm{TO})$	E_2^{high}	B_1^{high}	$A_1(\mathrm{LO})$	$E_1(\mathrm{LO})$
GaN								
$\omega(j)$	142	337	540	568	576	713	748	757
$a(j)$	75	-334	-640	-717	-742	-661	-664	-775
$b(j)$	4	-275	-695	-591	-715	-941	-881	-703
$K^{\mathrm{iso}}(j)$	154	-943	-1975	-2025	-2199	-2263	-2209	-2253
$K^{\mathrm{sh}}(j)$	142	-118	110	-252	-54	560	434	-144
AlN								
$\omega(j)$	241	552	618	677	667	738	898	924
$a(j)$	149	-580	-776	-835	-881	-601	-739	-867
$b(j)$	-223	-197	-394	-744	-906	-757	-737	-808
$K^{\mathrm{iso}}(j)$	75	-1357	-1946	-2414	-2668	-1959	-2215	-2542
$K^{\mathrm{sh}}(j)$	744	-766	-764	-182	50	312	-4	-118

Since phonons are straightforward signatures of the chemical bonds, Raman scattering spectroscopy is widely used to measure stress and strain fields in epitaxial semiconductor layers. It represents a rather sensitive, local, nondestructive, and fast technique for the detection of these elastic fields. Their knowledge is particularly important for wurtzite crystals like 2H-GaN and -AlN. Their optical properties and the properties of strained epitaxial layers based on GaN, AlN, and their alloys are remarkably influenced by spontaneous and piezoelectric polarization fields. The Raman detection of the underlying stress and strain fields, however, needs the knowledge of the precise values of the phonon deformation potentials which relate the shifts of the vibrational frequencies linearly to the strain or stress present in the crystal. In the range of the validity of Hooke's law, the magnitude of the frequency shift of each phonon mode $j = A_1$ (LO and TO), E_1 (LO and TO), and E_2 (high and low) is determined by the two deformation potential constants $a(j)$ and $b(j)$ according to

$$\Delta\omega(j) = 2a(j)\epsilon_{xx} + b(j)\epsilon_{zz} \tag{6}$$

for a given strain field characterized by the diagonal components $\epsilon_{xx} = \epsilon_{yy}$ and ϵ_{zz}.

The calculated phonon deformation potentials [52, 57] are also listed in Table 3. The agreement with measured values is excellent, in particular if the strain is correctly accounted for and reliable elastic constants are used [52]. The situation is more complicated for the lower E_2 mode where the strain influence is weak and the error bars become large.

Any strain that conserves the hexagonal C_{6v}^4 symmetry can be decomposed into an isotropic and a pure shear component,

$$\epsilon = \epsilon^{\mathrm{iso}}\mathrm{diag}(1, 1, 1) + \epsilon^{\mathrm{sh}}\mathrm{diag}(1, 1, -2). \tag{7}$$

This decomposition is unique only up to an arbitrary factor that determines the magnitude of the shear deformation parameter ϵ^{sh}. Since the relative volume change is given by $\Delta V/V_{\mathrm{eq}} = 3\epsilon^{\mathrm{iso}}$, for the isotropic deformation parameter one has

$$\epsilon^{\mathrm{iso}} = \epsilon_{xx}(2 + R)/3, \tag{8}$$

where $R = \epsilon_{zz}/\epsilon_{xx}$. For the decomposition chosen here, this leads to

$$\epsilon^{\mathrm{sh}} = \epsilon_{xx}(1 - R)/3. \tag{9}$$

Correspondingly, any linear phonon frequency shift can be related to the isotropic and shear components of the strain tensor. According to Eq. (7), one has

$$(\Delta\omega)^{\mathrm{iso}} = (2a + b)\epsilon^{\mathrm{iso}} \quad \text{and} \quad (\Delta\omega)^{\mathrm{sh}} = 2(a - b)\epsilon^{\mathrm{sh}}. \tag{10}$$

The resulting isotropic and shear deformation potentials are also listed in Table 3. Whereas the isotropic deformation potentials are rather similar for GaN and AlN, the shear deformation potentials show significant differences.

The dependence of the zone-center phonon frequencies on biaxial strain is represented in Fig. 8 for AlN and GaN. It is similar to that in the case of a negative hydrostatic pressure [58]. Apart from the lower E_2

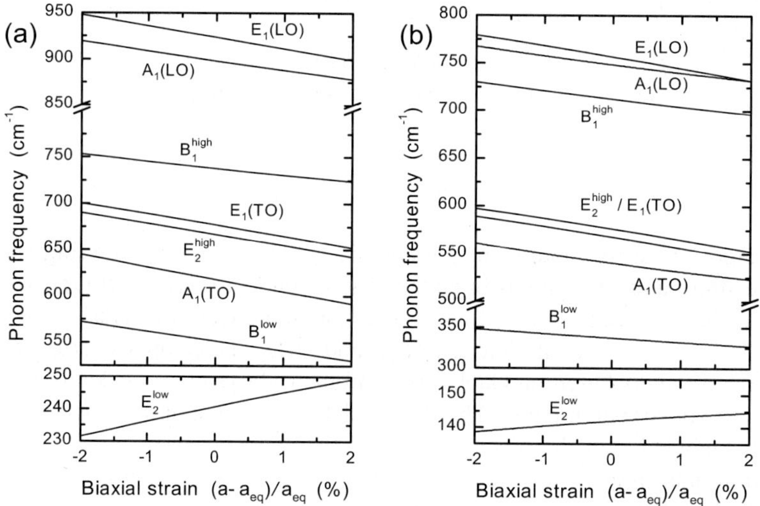

Fig. 8   Zone-center phonon frequencies versus biaxial strain for AlN (a) and GaN (b). After Ref. [52].

mode, all other modes are softened with increasing biaxial strain. In contrast to the case of pure internal strain, this happens for all mode symmetries. The frequency decrease versus biaxial strain in Fig. 8 is rather similar for the majority of mode frequencies. The opposite behaviour of the lower E_2 modes is correlated with their mode softening in the presence of hydrostatic pressure [58]. The mode softening of the lower E_2 modes in Fig. 8 may be understood in detail from the changes of the total-energy contributions, the band-structure energy, the Hartree term counted twice in the band-structure energy, the difference of the exchange-correlation (XC) energy and the XC-potential, and the electrostatic Ewald term [59].

7 Alloying effects

In order to investigate the stability and the properties of unstrained and strained pseudobinary $X_x Y_{1-x} N$ alloys (X,Y = In, Ga, Al), one has to calculate the Helmholtz free energy $F(x, T)$ as a function of the molar fraction x and the absolute temperature. Since wurtzite and zinc-blende mixed crystals represent lattices of nearly tetrahedrally coordinated atoms, their thermodynamical behaviour should be similar. Moreover, an applied biaxial strain, at least in a plane perpendicular to a cubic axis, can be easily simulated. For that reason, the calculations discussed in the following are restricted to cubic mixed crystals. The N atoms occupy one fcc sublattice, whereas the X and Y atoms are randomly distributed over the other fcc sublattice.

The mixing free energy of a random alloy, $\Delta F(x,T)$, can be described within a special cluster expansion approach [60], the so-called generalized quasichemical approximation (GQCA) [61]. In this approximation the considered mixed crystal is divided into an ensemble of clusters, each of which is taken to be independent – statistically and energetically – of the surrounding atomic configuration. As a random alloy the material is assumed to be spatially homogeneous everywhere on a macroscopic (or at least mesoscopic) length scale. According to the Connolly-Williams method [62], the configurationally averaged and hence composition-dependent (and in general also temperature-dependent) value is given by

$$P(x,T) = \sum_{j=0}^{J} x_j P_j \tag{11}$$

for a property P of interest of the pseudobinary alloy with mainly compositional disorder. The considered property is different for each cluster and depends on the cluster class index j according to P_j. Each cluster class j is realized with a probability x_j.

Within the GQCA the clusters with the energy E_j used in the calculation of the internal energy should have a reasonable size. The smallest clusters that consider local correlation are 16-bond clusters with a central N atom and four alloying atoms bound to the environment by 12 second-neighbour N atoms [63]. In general, the cluster energies E_j do not depend linearly on j and, hence, give rise to an asymmetric dependence of the mixing internal energy and the mixing enthalpy. They cannot be described anymore by an expression such as $\Omega x(1-x)$ for the mixing enthalpy, as in the limit of the ideal solution model. Rather, a composition-dependent interaction parameter Ω has to be considered, influencing the miscibility gap.

Mixing free energies resulting for different growth temperatures together with the accompanying temperature-composition phase diagram have been calculated for InGaN, AlGaN [63], BGaN, BAlN [64], and InAlN [65, 66]. For In$_x$Ga$_{1-x}$N thermodynamic results are summarized in Fig. 9. The shapes of the ΔF

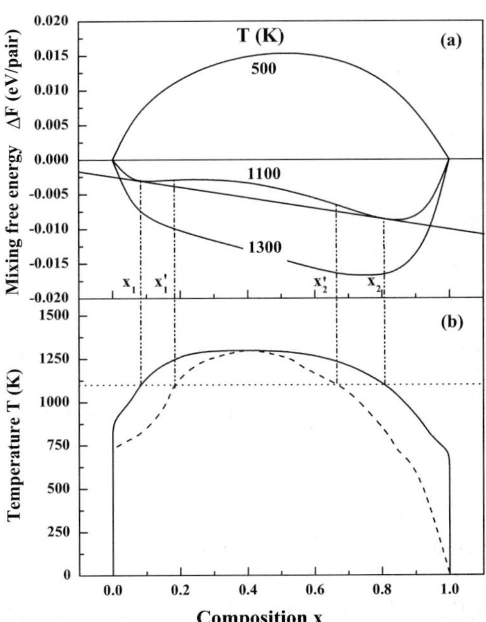

Fig. 9 (a) Mixing free energy $\Delta F(x,T)$ of In$_x$Ga$_{1-x}$N alloys versus composition. In the 1100 K case also the common tangent is plotted. (b) $T-x$ phase diagram of In$_x$Ga$_{1-x}$N; solid line: binodal curve, dashed line: spinodal curve. After Ref. [67].

curves versus the In molar fraction are asymmetric. The behaviour of the excess free energy below the critical temperature at $T = 1295$ K in Fig. 9 indicates that there is a miscibility gap for x in the interval $x_1 < x < x_2$. Here x_1 and x_2 are the points at which the common tangent line touches the ΔF curve. There are two other special composition values, x'_1 and x'_2, called the spinodal points (the inflection points of ΔF). For $x_1 < x < x_1'$ or $x_2' < x < x_2$, the alloy is metastable against local decomposition, whereas for $x_1' < x < x_2'$, the mixed crystal is inherently unstable. It may decompose into homogeneous random alloys with the averaged compositions x_1 and x_2 on a length scale larger than the cluster sizes, i.e., larger than 1 nm.

The influence of a biaxial strain on all alloy properties is remarkable. An inhomogenous strain distribution is assumed in the alloy on a microscopic length scale in agreement with the assumption of pseudomorphic growth. When the strong biaxial strain effects are introduced, the mixing free energy ΔF becomes negative in the entire range of the alloy composition and exhibits a pronounced single minimum. In comparison to Fig. 9 the $T - x$ phase diagram undergoes a dramatic change leading to an almost zero critical temperature. Spinodal decomposition taking place in unstrained samples is predicted to be fully suppressed in the samples under biaxial strain $\epsilon_{||}$, at least in the extreme cases of about $\left| \epsilon_{||} \right| = 0.1x$. However, also for smaller in-plane strains of about $\left| \epsilon_{||} \right| = 0.01x$ remarkable changes happen [63]. The critical temperature is lowered and, simultaneously, the miscibility gap as well as the composition region of spontaneous decomposition are reduced. There is a clear tendency for suppression of the phase separation in $In_xGa_{1-x}N$ due to elastic strain. There are also experimental indications in this direction [68].

The total-energy minimizations directly give the structural parameters as the lattice constants, the bond lengths, as well as the second nearest-neighbour (2NN) distances for each cluster type. The calculated average Ga-N, Al-N, and In-N bond lengths versus x are compared in Fig. 10 with Extended X-Ray Absorption Fine Structure (EXAFS) measurements [69, 70]. In agreement with experiment two different

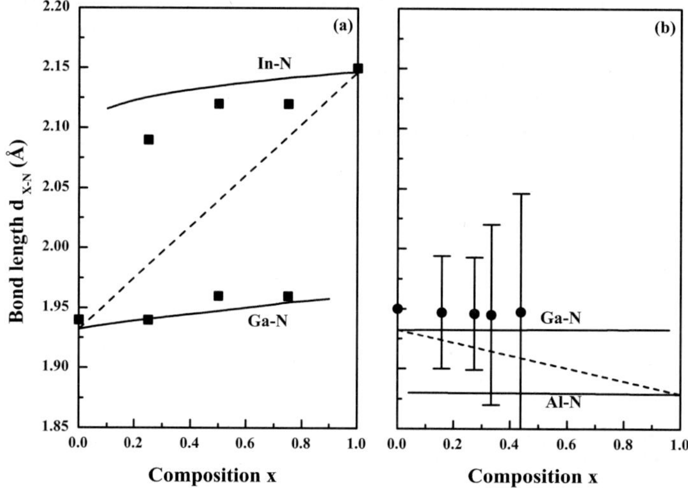

Fig. 10 Averaged bond lengths (solid lines) versus composition in $In_xGa_{1-x}N$ (a) and $Al_xGa_{1-x}N$ (b). The dashed line corresponds to the fictitious common bond length according to Vegard's rule. The filled squares and dots describe measured values [69, 70]. A growth temperature $T = 950$ K is assumed. From Ref. [63].

bond lengths d_{X-N} and d_{Ga-N} occur in each ternary $X_xGa_{1-x}N$ alloy. Apart from the small overbinding effect due to the used DFT-LDA and the fact that a possible strain in the samples influences the measured lengths, there is quantitative agreement with the experimental results. The same reasonable agreement can be stated with recent calculations performed within the Valence Force Field method, using extremely large clusters [71]. Figure 10 makes obvious that Vegard's rule fails for the bond lengths. The situation is similar for the 2NN distances [63]. These distances also do not follow Vegard's rule. The magnitude of the 2NN distances is more related to the covalent radii of the contributing atoms, although there is an influence of the surrounding cation distribution. Taking into account the overbinding tendency in DFT-LDA, again qualitative and quantitative agreement with experimental [69, 70] and other theoretical [71] predictions can be stated, in particular if the uncertainties due to the undetermined strain state of the samples, the length fluctuations, and the restrictions of the cluster-expansion method are taken into consideration.

The alloying also influences the electronic properties of the pseudobinary or ternary group-III nitrides. In particular, there is a huge, in general nonlinear composition dependence of the fundamental energy gap. Experimental data for the energy gap of hexagonal $In_xGa_{1-x}N$ alloys are summarized in Fig. 11. They

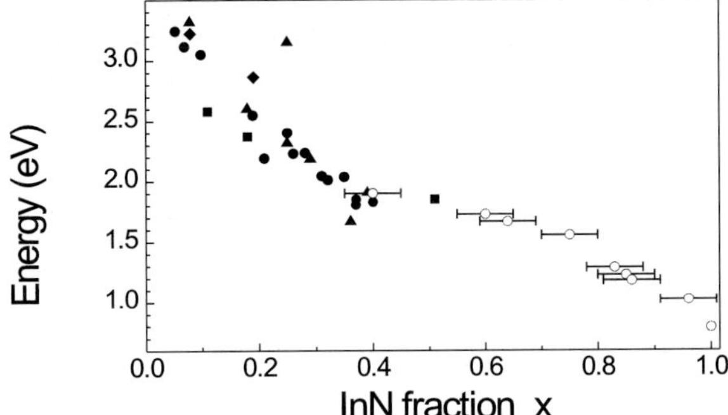

Fig. 11 Selected experimental data for the energy gap of $In_xGa_{1-x}N$ alloys. Filled circles and triangles: [72], filled squares: [73], filled diamonds: [74], and open circles: [75]. From Ref. [76].

are mainly derived using photoluminescence (PL). While many data exist for In molar fractions $x < 0.4$, the majority of gaps for alloys with $0.5 < x < 1.0$ has been measured recently on MBE samples [75]. The measurements of these values are influenced by a variety of effects, such as spinodal decomposition, alloy fluctuations, strain, chosen gaps of GaN and InN, and measurement technique. Nevertheless, Fig. 11 indicates a remarkable dependence of the gap on the composition and, hence, a considerable bowing. Using a conventional picture of the x-dependence of the gap,

$$E_g(x) = (1 - x)E_g(\text{GaN}) + xE_g(\text{InN}) - bx(1 - x),\qquad(12)$$

an average bowing parameter of about $b = 2.5$ eV follows [75]. In the $In_xGa_{1-x}N$ system the bowing is however not constant but itself composition-dependent. One finds roughly $b(x) = (1 - x)[11.4 - 19.4x]$ eV, i.e., $b(0.25) = 4.9$, $b(0.5) = 0.9$, and $b(0.75) = -0.8$ eV [63]. As a consequence, the so-called bowing parameter depends on the composition interval considered.

Not only the procedure to derive the band-gap bowing parameter from experimental data contains uncertainties, but also the determination of the theoretical value depends on many details of the calculation, the alloy description, the cluster size, and the definition of the alloy gap [63, 77, 78]. Considering large 64-atom clusters and the cubic polytype we indeed found a composition-dependent bowing, however, the absolute values are smaller than discussed above [77]. Within the interval $0 \leq x \leq 0.25$ the calculated values vary between 2.5 and 1.5 eV. For higher In molar fractions the b parameter decreases further to $b(0.5) = 1.4$ eV and $b(0.75) = 1.3$ eV. The main contribution to the bowing is due to a structural effect, the composition-induced disorder in the bond lengths [77]. This holds for the configurationally averaged gap. Studying other gaps, e.g. $E_g(x) - \Delta E_g(x)$ (which may be measured more or less in PL) with $\Delta E_g(x)$ as the rms deviations, larger bowing parameter are extracted [63].

8   Conclusions

In this article some of the ideas have been presented how properties of group-III nitrides can be predicted from *first-principles* calculations. They are based on the density functional theory in its local version, pseudopotentials, and a plane-wave expansion of the eigenfunctions. We have seen that the modern theory of solids can be successfully applied to the study of the energetical, structural, and dynamical properties of the group-III nitrides including their polytypic structures, strained layers, and alloys. After inclusion of many-body effects, as for instance the quasiparticle self-energy shifts, even electronic excitation energies can be predicted. There is a high degree of agreement between different calculations as well as between theory and experiment.

Acknowledgements We thank M. Ferhat, U. Grossner, K. Karch, G. Portisch, L.E. Ramos, and L.K. Teles for collaboration. This work was supported by the Deutsche Forschungsgemeinschaft (Schwerpunktprogramm Gruppe-III-Nitride, Grant No. Be 1346/8-5). Part of the calculations was performed at the John von Neumann Institute of Computing in Jülich.

References

[1] P. Hohenberg and W. Kohn, Phys. Rev. **136**, B864 (1964).
[2] W. Kohn and L.J. Sham, Phys. Rev. **140**, A1133 (1965).
[3] D.M. Ceperley and B.J. Alder, Phys. Rev. Lett. **45**, 566 (1980).
[4] J.P. Perdew and A. Zunger, Phys. Rev. B**23**, 5048 (1981).
[5] F. Bechstedt, Adv. Solid State Phys. **32**, 161 (1992).
[6] L. Hedin and S. Lundqvist, Solid State Phys. **23**, 1 (1969).
[7] M.S. Hybertsen and S.G. Louie, Phys. Rev. B**34**, 5390 (1986).
[8] P. Gianozzi, S. de Gironcoli, P. Pavone and S. Baroni, Phys. Rev. B**43**, 7231 (1991).
[9] A.F. Wright, J. Appl. Phys. **82**, 2833 (1997).
[10] T.L. Tansley and C.P. Foley, J. Appl. Phys. **59**, 3241 (1986).
[11] V.Yu. Davydov, A.A. Klochikhin, R.P. Seisyan, V.V. Emtsev, S.V. Ivanov, F. Bechstedt, J. Furthmüller, H. Harima, V. Mudryi, J. Aderhold, O. Semchinova, and J. Fraul, phys. stat. sol. (b) **229**, R1 (2002).
[12] A.F. Wright and J.S. Nelson, Phys. Rev. B**50**, 2159 (1994).
[13] A.F. Wright and J.S. Nelson, Phys. Rev. B**51**, 7866 (1995).
[14] C.-Y. Yeh, Z.W. Lu, S. Froyen, and A. Zunger, Phys. Rev. B**46**, 10085 (1992).
[15] S. Nakamura and G. Fasol, *The blue laser diode* (Springer, Heidelberg 1997).
[16] F. Bernardini and V. Fiorentini, Appl. Surf. Sci. **166**, 23 (2000).
[17] U. Grossner, J. Furthmüller, and F. Bechstedt, Phys. Rev. B**58**, R1722 (1998).
[18] F. Bechstedt, U. Grossner, and J. Furthmüller, Phys. Rev. B**62**, 8003 (2000).
[19] J. Furthmüller, P. Käckell, F. Bechstedt, and G. Kresse, Phys. Rev. B**61**, 4576 (2000).
[20] D.J. Chadi and M.L. Cohen, Phys. Rev. B**8**, 5747 (1973).
[21] H.J. Monkhorst and J.D. Pack, Phys. Rev. B**13**, 5188 (1976).
[22] G. Kresse and J. Furthmüller, Phys. Rev. B **54**, 1169 (1996).
[23] G. Kresse and J. Furthmüller, Computat. Materials Science **6**, 15 (1996).
[24] J. Furthmüller, G. Cappellini, H.-C. Weissker, and F. Bechstedt, Phys. Rev. B**66**, 165328 (2002).
[25] K. Karch and F. Bechstedt, Phys. Rev. B**56**, 7404 (1997).
[26] K. Karch, J.-M. Wagner, and F. Bechstedt, Phys. Rev. B**57**, 7043 (1998).
[27] N. Troullier and J.L. Martins, Phys. Rev. B**43**, 1993 (1991).
[28] K. Karch, F. Bechstedt, and T. Pletl, Phys. Rev. B**56**, 3560 (1997).
[29] S.G. Louie, S. Froyen, and M.L. Cohen, Phys. Rev. B**26**, 1738 (1982).
[30] U. Grossner, J. Furthmüller, and F. Bechstedt, IPAP Conf. Series **1**, 56 (2000).
[31] C. Stampfl and C.G. Van de Walle, Phys. Rev. B**59**, 5521 (1999).
[32] W.R.L. Lambrecht, B. Segall, S. Strite *et al.*, Phys. Rev. B**50**, 14155 (1994).
[33] A. Nakadeira and H. Tanaka, Jpn. J. Appl. Phys. **37**, 1449 (1998).
[34] T. Koizumi, H. Okumura, K. Balakrishnan *et al.*, J. Cryst. Growth **201/202**, 341 (1999).
[35] A. Rubio, J.L. Corkill, M.L. Cohen, E.L. Shirley, and S.G. Louie, Phys. Rev. B**48**, 11810 (1993).
[36] P.B. Perry and R.F. Rutz, Appl. Phys. Lett. **33**, 319 (1978).
[37] H. Hong, D. Pavlidis, S.W. Brown, and S.C. Rand, J. Appl. Phys. **77**, 1705 (1995).
[38] B. Monemar, Phys. Rev. B**10**, 676 (1974).
[39] Y.X. Guo, M. Nishio, H. Ogawa, A. Wakahara, and A. Yoshida, Phys. Rev. B**58**, 15304 (1998).
[40] Y. Bu, L. Ma, and M.C. Lin, J. Vac. Sci. Technol. A**11**, 2931 (1993).
[41] U. Grossner, J. Furthmüller, and F. Bechstedt, phys. stat. sol. (b) **216**, 675 (1999).
[42] C. Persson, R. Ahuja, A. Ferreira da Silva, and B. Johansson, J. Phys.: Condensed Matter **13**, 8945 (2001).
[43] F. Bechstedt and R. Del Sole, Phys. Rev. B**38**, 7710 (1988).

[44] F. Bernardini, V. Fiorentini, and D. Vanderbilt, Phys. Rev. B**56**, R10024 (1997).

[45] F. Bernardini, V. Fiorentini, and D. Vanderbilt, Phys. Rev. Lett. **79**, 3958 (1997).

[46] M. Posternak, A. Baldereschi, A. Catellani, and R. Resta, Phys. Rev. Lett. **64**, 1777 (1990).

[47] F. Bernardini, V. Fiorentini, and D. Vanderbilt, Phys. Rev. B**63**, 193201 (2001).

[48] A. Zoroddu, F. Bernardini, P. Ruggerone, and V. Fiorentini, Phys. Rev. B**64**, 045208 (2001).

[49] V. Fiorentini, F. Bernardini, and O. Ambacher, Appl. Phys. Lett. **80**, 1204 (2001).

[50] A. Garcia and M.L. Cohen, Phys. Rev. B**47**, 4215 (1993).

[51] M. Albrecht, S. Christiansen, T. Remmele *et al.*, Inst. Phys. Conf. Ser. No. **164**, 361 (1999).

[52] J.-M. Wagner and F. Bechstedt, Phys. Rev. B**66**, 115202 (2002).

[53] V.Yu. Davydov, V.V. Emtsev, I.N. Goncharuk *et al.*, Appl. Phys. Lett. **75**, 3297 (1999).

[54] V.Yu. Davydov, Yu.E. Kitaev, I.N. Goncharuk et al., Phys. Rev. B**58**, 12899 (1998).

[55] S. Tomić, V. Milanović, and Z. Ikonić, IEEE J. Quant. Electronics **37**, 1337 (2001);
       K. Kishino, A. Kikuchi, H. Kanazawa, and T. Tachibana, Appl. Phys. Lett. **81**, 1234 (2002).

[56] J.-M. Wagner, J. Gleize, and F. Bechstedt, IPAP Conf. Series **1**, 669 (2000).

[57] J.-M. Wagner and F. Bechstedt, Appl. Phys. Lett. **77**, 346 (2000).

[58] J.-M. Wagner and F. Bechstedt, Phys. Rev. B**62**, 4526 (2000).

[59] J.-M. Wagner and F. Bechstedt, phys. stat. sol. (b) **235**, 464 (2003).

[60] A. Zunger, *Statics and dynamics of alloy phase transformations* (Plenum Press, New York 1994), p. 361.

[61] A. Sher, M. Van Schilfgaarde, A.-B. Chen, and W. Chen, Phys. Rev. B**36**, 4279 (1987).

[62] J.W.D. Conolly and A.R. Williams, Phys. Rev. B**27**, 5169 (1983).

[63] L.K. Teles, J. Furthmüller, L.M.R. Scolfaro, J.R. Leite, and F. Bechstedt, Phys. Rev. B**62**, 2475 (2000).

[64] L.K. Teles, J. Furthmüller, L.M.R. Scolfaro, A. Tabata, J.R. Leite, F. Bechstedt, T. Frey, D.J. As, and K. Lischka,
       Physica E**13**, 1086 (2002).

[65] M. Ferhat and F. Bechstedt, Phys. Rev. B**65**, 075213 (2002).

[66] L.K. Teles, L.M.R. Scolfaro, J.R. Leite, J. Furthmüller, and F. Bechstedt, J. Appl. Phys. **92**, 7109 (2002).

[67] L.K. Teles, L.M.R. Scolfaro, J. Furthmüller, J.R. Leite, and F. Bechstedt, IPAP Conf. Series **1**, 689 (2000).

[68] A. Tabata, L.K. Teles, L.M.R. Scolfaro, J.R. Leite, T. Frey, A. Karchenko, D.J. As, D. Schikora, K. Lischka, J.
       Furthmüller, and F. Bechstedt, Appl. Phys. Lett. **80**, 769 (2002).

[69] N.J. Jeffs, A.V. Blant, T.S. Cheng *et al.*, MRS Proc. **512**, 519 (1998).

[70] K.E. Miyano, J.C. Woicik, L.H. Robins, C.E. Bouldin, and D.K. Wickenden, Appl. Phys. Lett. **70**, 2108 (1997).

[71] T. Mattila and A. Zunger, J. Appl. Phys. **85**, 160 (1999).

[72] K.P. O'Donnell, MRS Proc. **595**, W11.26.1 (2000).

[73] M. Kim, J.-K. Cho, I.-H. Lee, and S.-J. Park, phys. stat. sol. (a) **176**, 269 (1999).

[74] W. Shan, B.D. Little, J.J. Song, Z.C. Feng, M. Schurman, and R.A. Stall, Appl. Phys. Lett. **69**, 3315 (1996).

[75] V.Yu. Davydov, A.A. Klochikhin, V.V. Emtsev, S.V. Ivanov, V.V. Vekshin, F. Bechstedt, J. Furthmüller, H.
       Harima, A.V. Mudryi, A. Hashimoto, Y. Yamamoto, J. Aderhold, J. Graul, and E.E. Haller, phys. stat. sol. (b)
       230, R4 (2002); unpublished.

[76] F. Bechstedt, J. Furthmüller, M. Ferhat, L.K. Teles, L.M.R. Scolfaro, J.R. Leite, V.Yu. Davydov, O. Ambacher,
       and R. Goldhahn, phys. stat. sol. (a) **195**, 628 (2003).

[77] M. Ferhat, J. Furthmüller, and F. Bechstedt, Appl. Phys. Lett. **80**, 1394 (2002).

[78] L.K. Teles, J. Furthmüller, L.M.R. Scolfaro, J.R. Leite, and F. Bechstedt, Phys. Rev. B**63**, 085204 (2001).

phys. stat. sol. (c) **0**, No. 6, 1750–1769 (2003) / **DOI** 10.1002/pssc.200303135

Phonons and free-carrier properties of binary, ternary, and quaternary group-III nitride layers measured by Infrared Spectroscopic Ellipsometry

A. Kasic[*1,2], M. Schubert[1], J. Off[**3], B. Kuhn[***3], F. Scholz[#3], S. Einfeldt[4], T. Böttcher[4], D. Hommel[4], D. J. As[5], U. Köhler[5], A. Dadgar[6], A. Krost[6], Y. Saito[7], Y. Nanishi[7], M. R. Correia[8], S. Pereira[8], V. Darakchieva[2], B. Monemar[2], H. Amano[9], I. Akasaki[9], and G. Wagner[10]

[1] Institut für Experimentelle Physik II, Universität Leipzig, Linnéstraße 5, 04103 Leipzig, Germany

[2] Department of Physics and Measurement Technology, Linköping University, 581 83 Linköping, Sweden

[3] 4. Physikalisches Institut, Universität Stuttgart, Pfaffenwaldring 57, 70569 Stuttgart, Germany

[4] Institut für Festkörperphysik, Universität Bremen, Kufsteiner Straße NW 1, 28359 Bremen, Germany

[5] Fachbereich Physik, Universität Paderborn, Warburger Straße 100, 33095 Paderborn, Germany

[6] Institut für Experimentelle Physik, Otto-von Guericke Universität Magdeburg, Universitätsplatz 2, 39016 Magdeburg, Germany

[7] Faculty of Science and Engineering, Ritsumeikan University, 1-1-1 Noji-Higashi, Kusatsu, Shiga 525-8577, Japan

[8] Departamento de Física, Universidade de Aveiro, 3810-193 Aveiro, Portugal

[9] High-Tech Research Center and Department of Materials Science and Engineering, Meijo University, 1-501 Shiogamaguchi, Tempaku-ku, Nagoya 468-8502, Japan

[10] Institut für Nichtklassische Chemie, Universität Leipzig, Permoserstraße 15, 04318 Leipzig, Germany

Received 1 March 2003, revised 6 June 2003, accepted 12 June 2003
Published online 28 August 2003

PACS 63.20.Dj, 78.30.Fs, 78.66.Fd, 78.67.Pt

This work reviews recent ellipsometric investigations of the infrared dielectric functions of binary, ternary, and quaternary group-III nitride films. Spectroscopic Ellipsometry in the mid-infrared range is employed for the first time to determine phonon and free-carrier properties of individual group-III nitride heterostructure components, including layers of some ten nanometer thickness. Assuming the effective carrier mass, the free-carrier concentration and mobility parameters can be quantified upon model analysis of the infrared dielectric function. In combination with Hall-effect measurements, the effective carrier masses for wurtzite n- and p-type GaN and n-type InN are obtained. The mode behavior of both the $E_1(TO)$ and $A_1(LO)$ phonons are determined for ternary compounds. For strain-sensitive phonon modes, the composition and strain dependences of the phonon frequencies are differentiated and quantified. Information on the crystal quality and compositional homogeneity of the films can be extracted from the phonon mode broadening parameters. A comprehensive IR dielectric function database of group-III nitride materials has been established and can be used for the analysis of complex thin-film heterostructures designed for optoelectronic device applications. Information on concentration and mobility of free carriers, thickness, alloy composition, average strain state, and crystal quality of individual sample constituents can be derived.

---

[*] Corresponding author: e-mail: aleka@ifm.liu.se, Phone: +46 13 284 411.

[**] Now with: OSRAM Opto Semiconductors, Wernerwerkstr. 2, 93049 Regensburg, Germany.

[***] Now with: Robert Bosch GmbH, Tübinger Str. 123, 72762 Reutlingen, Germany.

[#] Now with: Abteilung Optoelektronik, Universität Ulm, 89069 Ulm, Germany.

1 Introduction

The driving force for the recent technological development of the wurtzite (α) and zincblende (β) group-III nitrides is the accessibility of materials with large direct band gaps, which span the range from ~0.8 eV (α-InN) [see, e.g., 1, 2] over 3.4 eV (α-GaN) to ~6.2 eV (α-AlN), covering the whole visible region and extending well into the UV spectral range. By varying the alloy composition, the band gaps of the ternary compounds can – in principle – be tuned to any value within this range. Accordingly, some of the ternary group-III nitride alloys are currently used as active materials for light emitting devices. In particular, appropriately tailored optoelectronic devices based on group-III nitrides operate in the violet, blue, and green spectral range, which was previously difficult to accomplish.

Except for α-GaN, the research in the field of group-III nitrides has been focused mainly on device performance, and basic materials research has not been exhaustive. Many fundamental physical properties, particularly of the ternary and quaternary compounds, are still unknown. On the other hand, detailed knowledge of these properties is mandatory for appropriate design of heterostructure devices and could be used for the characterization of complex multi-layered device structures. Metrology and control of individual layer properties in complex heterostructures, such as free-carrier concentration and mobility, strain, and alloy composition, represent still a challenge.

Optical tools are nondestructive and noninvasive, and are beneficial for both research and production environments. Spectroscopic ellipsometry (SE) is known as an excellent technique for the precise determination of the complex-valued dielectric functions of semiconductor materials. Inaccuracies due to the Kramers-Kronig analysis of reflection or transmission data in order to obtain the complex dielectric function spectra are avoided. While SE from the near-infrared (near-IR) to the ultraviolet (UV) spectral range has become a standard investigation technique in many thin-film research and technology sectors, only little effort has been spent to extend the abilities of SE for the exploration of the long-wavelength dielectric function properties of semiconductor compounds and their heterostructures. In 1990, Röseler published a pioneering book merging the viable FTIR technique with the SE approach [3], but it was not until the end of the last decade when ellipsometers operating in the mid-IR spectral range became commercially available.

The present report briefly reviews the recent comprehensive work on the exploration of the mid-IR dielectric functions (~350–6000 cm^{-1}) of binary, ternary, and quaternary group-III nitride compounds using infrared spectroscopic ellipsometry (IRSE). The analysis of the dielectric function leads to the determination of fundamental material quantities, such as phonon mode parameters and free-carrier properties. Lattice and free-carrier contributions to the mid-IR dielectric functions of the group-III nitride materials are differentiated and quantified precisely upon model lineshape analyses. Generally, IRSE is highly sensitive to the dielectric functions of all components in complex heterostructures, even to those of films with some ten nanometer thickness only. For ternary and quaternary compounds, the mode behavior of IR-active phonons is determined. Effects of strain and composition on certain phonon mode frequencies can be separated from each other. Layer inhomogeneities such as the existence of thin carrier-depleted surface films and composition gradients along the growth direction are detected. Information on the crystal quality and compositional homogeneity of the films are extracted from transverse optical phonon mode broadening parameters. Longitudinal optical phonon-plasmon coupling effects are quantified and allow the determination of the free-carrier concentration and mobility in individual sample layers, provided the effective carrier mass is known. Otherwise, if the free-carrier concentration is, e.g., taken from electrical Hall-effect measurements, the effective carrier mass can be obtained. The requirement of performing electrical contacting for Hall-effect measurement in order to obtain the carrier concentration value as input parameter for the effective mass determination by IRSE is dispensed with by the recently demonstrated magnetooptical generalized ellipsometry extension to group-III-V compound materials [4]. The application to group-III nitrides is proposed in this review.

From the IRSE analyses of various group-III nitride compounds, a comprehensive IR dielectric function database is established. This database is used for precise, nondestructive, and contactless IR optical characterization of complex device heterostructures by IRSE. Examples are demonstrated for light-emitting device structures, where thickness, alloy composition, crystal quality, free-carrier concentration and mobility of individual sample constituents are determined and the *p-n* junction is localized.

2 Theory

The wurtzite structure of the binary group-III nitrides belongs to the space group C_{6v}^{4} with four atoms (two formula units) per unit cell leading to 12 phonon branches, 9 optical and 3 acoustic. The optical phonon modes belong to the following irreducible representation at the Γ point of the Brillouin zone:

$$\Gamma_{opt} = 1A_1 + 2B_1 + 1E_1 + 2E_2 \tag{1}$$

The branches of E_1 and E_2 symmetry are twofold degenerate. Both A_1 and E_1 modes are Raman and IR-active. The two non-polar E_2 modes are Raman-active only, and the B_1 modes are neither IR- nor Raman-active (silent modes). Both A_1 and E_1 modes are polar, that is, they split into transverse optical (TO) and longitudinal optical (LO) components with different frequencies due to the macroscopic electric fields associated with the atomic displacement of the longitudinal phonons. The comparatively large mode polarity observed for the group-III nitride compounds is caused by the large ionicities of the bonds between group-III and N atoms. The short-range interatomic forces cause the mode anisotropy, that is the A_1 and E_1 modes possess different frequencies. Because in the case of group-III nitride compounds the electrostatic forces dominate the anisotropy in the short-range forces, the TO-LO splitting is larger than the A_1-E_1 splitting. For the lattice vibrations with A_1 and E_1 symmetry, the atoms move parallel and perpendicular to the hexagonal c axis, respectively.

The mode behavior of long-wavelength optical phonons in a ternary randomly mixed crystal $A_yB_{1-y}C$ is conventionally categorized into two main classes, referred to as one- and two-mode behavior. Crystals exhibiting a one-mode behavior have only one set of frequencies that each shows a continuous and nearly linear dependence on the composition y, from the AC component to the BC component, while the strength of each mode remains approximately constant. On the other hand, in the two-mode class, two energetically well separated sets of phonons are observed for a given composition y, each set corresponding to one of the two components. The two sets of phonons occur at frequencies close to those of the end components, i.e., for both branches the composition dependence of the phonon frequency deviates strongly from the linear interpolation between the respective phonon frequencies of the binary end components. One set of frequencies corresponds to the AC component and one set to the BC component. Also, the oscillator strength of each phonon mode branch is approximately proportional to the mole fraction of the component it represents.

Considering transverse and longitudinal phonons in the two-mode case, strictly speaking, each TO lattice resonance is accompanied by its LO mode. However, the spectral strength of one of the LO modes may vanish, i.e., Re$\{\varepsilon\}$ remains less than zero at the LO mode position ($\varepsilon = $ Re$\{\varepsilon\} + $ Im$\{\varepsilon\} = \varepsilon_1 + i\cdot\varepsilon_2$ denotes the dielectric function). In such a case, which can be interpreted as coupling of both LO modes by their macroscopic electric fields, the LO phonon displays an effective one-mode behavior.

Many attempts have been made to account for the phonon mode behavior of mixed crystals [e.g., 5]. Chang and Mitra [6] applied a modified random element isodisplacement (MREI) model in order to derive a criterion to determine whether a given mixed crystal $A_yB_{1-y}C$ exhibits a one- or two-mode behavior for its optical phonons. Applying their theory to the ternary group-III nitride alloys, a one-mode behavior is predicted for all TO and LO phonons. However, the experimental findings, including those of the present work, confirm this prediction only in part. Recent more sophisticated MREI model calculations were performed to account for the mode behavior in all hexagonal and cubic ternary group-III nitride alloys [7]. However, there is – to our best knowledge – no general criterion which would allow immediate prediction of the phonon behavior of ternary and quaternary group-III-V alloys. Furthermore, deviations from the idealized one- and two-mode behavior scheme are observed for some mixed crystals, e.g., by the occurrence of disorder-induced phonon mode bands [7] or the loss of well defined phonon symmetry [8].

Besides composition and atomic ordering in multinary alloys, strain due to non-lattice-matched heteroepitaxial growth influences the lattice mode behavior in group-III nitride films. Since the group-III nitride crystals retain their space-group symmetry in the presence of biaxial strain normal to the c axis when growth is performed along [0001], there is only a shift of the phonon frequencies but no splitting of the modes. In the range of the validity of Hooke's law, the magnitude of the frequency shift $\Delta\omega_\lambda$ for each phonon mode λ is determined by the two deformation potential constants a_λ and b_λ [9]:

$$\Delta\omega_\lambda = 2a_\lambda\varepsilon_{xx} + b_\lambda\varepsilon_{zz}. \tag{2}$$

The in-plane (ε_{xx}, ε_{yy}) and normal (ε_{zz}) strain components are given by $\varepsilon_{xx} = \varepsilon_{yy} = (a-a_0)/a_0$ and $\varepsilon_{zz} = (c-c_0)/c_0$, respectively, where a_0 (a) and c_0 (c) denote the in-plane and normal lattice constants of the unstrained (strained) crystal, respectively.

In order to determine the deformation potential constants from an observed mode frequency shift, one more quantity – the bulk Grüneisen parameter – needs to be known. This parameter, which is related to the elastic constants and phonon deformation potential constants [9], characterizes the phonon mode frequency shift at hydrostatic compression and is commonly determined by high-pressure experiments on single bulk crystals.

For biaxially strained layers (normal stress component $\sigma_{zz} = 0$), the frequency shift $\Delta\omega_\lambda$ is

$$\Delta\omega_\lambda = 2\left(a_\lambda - C_{13}/C_{33}\, b_\lambda\right)\varepsilon_{xx}. \tag{3}$$

In the case of a ternary $A_yB_{1-y}C$ layer, phonon mode shifts with respect to the unstrained binary end components can be induced – in addition to the film in-plane strain – also by the alloy composition y. It is often necessary to include a bowing term, $\alpha_1 y(1-y)$, for the alloy-induced mode shift in order to describe the experimental findings appropriately:

$$\Delta\omega_\lambda = \Delta\omega_\lambda(y) + \Delta\omega_\lambda(\varepsilon_{xx}) = \alpha_{0,\lambda}y + \alpha_{1,\lambda}y(1-y) + \beta_\lambda\varepsilon_{xx}. \tag{4}$$

3 Method

Ellipsometry in reflection mode determines the complex ratio ρ of the reflection coefficients R_p and R_s for light polarized parallel (p) and perpendicular (s) to the plane of incidence (see Fig. 1) [10]. The result of an ellipsometry measurement is usually presented by the real-valued ellipsometric parameters Ψ and Δ, where $\tan\Psi$ is defined as the absolute value of ρ, and Δ denotes the relative phase change of the p and s components of the electric field vector upon interaction with the sample:

$$\rho \equiv \left(\frac{B_p}{A_p}\right)\Big/\left(\frac{B_s}{A_s}\right) = \frac{R_p}{R_s} = \tan\Psi\,\exp(i\Delta). \tag{5}$$

A_p, A_s, B_p, and B_s denote the complex amplitudes of the p and s components of the electric field vector before (A) and after (B) reflection, respectively. The parameters Ψ and Δ depend on the photon energy, the layer sequence within the sample, the dielectric function ε and thickness d of each layer, the dielectric function of the substrate material, and the angle of incidence Φ.

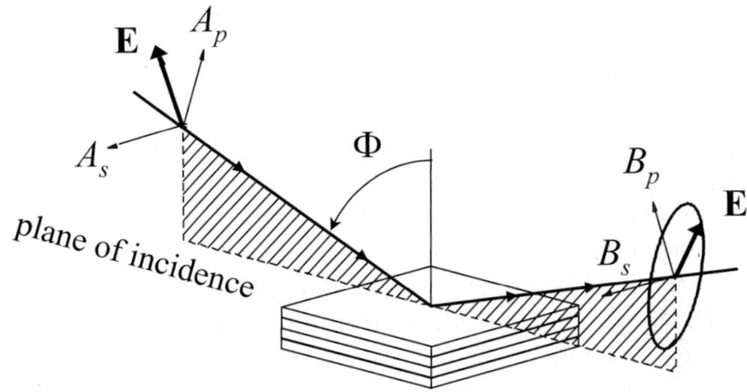

Fig. 1 Geometry of an ellipsometric experiment, showing the plane of incidence (hatched) and the angle of incidence Φ.

In the general case of an optically anisotropic crystal with arbitrary axes orientation with respect to the laboratory coordinate system, the generalized ellipsometry approach is required [11, 12]. The ellipsometric parameters then depend on the polarization state of the incident light beam and are described by three ratios of the polarized-light reflection coefficients instead of the single ratio given by Eq. (5). All wurtzite group-III nitride films studied in the present work were grown with their c axes parallel to the sample normal and to the substrate hexagonal c axis. For such high-symmetric c axes orientations, cross-conversion between p- and s-polarized light is suppressed and – as shown in Ref. [11] – a single complex ratio as in Eq. (5) is still sufficient for a complete description.

4 Experimental

A commercially available Fourier-transform-based variable angle-of-incidence spectroscopic ellipsometer was used (IR-VASE®, J. A. Woollam Co., Inc., Lincoln, USA). The ellipsometer setup consisted of a FTIR spectrometer source, a rotating polarizer, a high-precision goniometer stage and a rotating compensator. A 25 W silicon carbide globar served as the light source. Upon focusing, the IR probe beam spot size at the sample surface was approximately 8×5 mm^2, i.e., the probed sample area was macroscopic. Intensity spectra were obtained from a DTGS detector at a number of settings for the compensator and polarizer azimuth angles. From this information, the ellipsometric Ψ and Δ spectra were determined. Details of the data acquisition procedure and system alignment, as well as treatment of systematic and random error sources are given in Ref. [3]. For all samples discussed in the present work, the mid-IR ellipsometric spectra were recorded between 300 and 6000 cm^{-1} with a spectral resolution of 2 cm^{-1}. The spectra were taken at room temperature and multiple angles of incidence Φ between 35° and 72°.

In contrast to pure intensity reflectometry experiments, ellipsometry measurements are not or only little influenced by intensity fluctuations and the spectral intensity distribution of the light source. Hence, as long as parts of the IR beam reach the detector, the IRSE data are not influenced by atmospheric absorption. In general, ellipsometry is superior to reflectometry because two parameters, instead of one, are determined independently at each wavelength and angle of incidence. Both real and imaginary parts of the complex dielectric functions of materials can be measured without the necessity to perform further Kramers-Kronig analysis. The two parameters also place tighter constraints to model calculations than the single rather unsecured intensity parameter. Therefore, determination of the optical properties from very thin layers can be achieved by ellipsometry [13].

In principle, the dielectric function and thickness of all sample layers can be determined from an ellipsometry experiment by adjusting calculated data to measured data using parameterized model dielectric functions (MDF). The standard model for analyzing SE data consists of a sequence of parallel layers with perfectly abrupt interfaces and spatially homogeneous dielectric functions, bound between a semi-infinite substrate and the ambient. Then, a regression analysis (Levenberg-Marquardt fitting algorithm) is performed, where model parameters are varied until calculated and experimental data match as closely as possible. Details of the MDF´s used in the SE data analyses are given in Refs. [14–18].

The A_1(LO) phonon mode frequency can be located precisely through the well-known Berreman-polariton effect in thin polar dielectric films [19]. The IR radiation excites surface polaritons of transverse magnetic character at the boundary of two media whose dielectric functions fulfill certain conditions [3, 20, 21]. The incident wave is guided along the thin-film interfaces near wave numbers where the index of refraction approaches unity. In this spectral region, the ellipsometry data provide high sensitivity to the A_1(LO) phonon mode frequency and broadening parameter. The E_1(LO) mode frequency follows from the polarity of the E_1(TO) mode, which is resonantly excited by the IR probe beam and thus gives rise to a distinct spectral feature in the mid-IR range, if mode damping is not too large. Since the IRSE sensitivity to the E_1(LO) phonon mode parameters follows from the line shape analysis in a small spectral range, these parameters can only be obtained precisely, if the film is sufficiently thick and strong spectral features caused by other sample constituents are absent in the frequency range near the E_1(LO) phonon. Due to the geometry of an IRSE experiment, the ellipsometry data from c-plane oriented films are insensitive to the A_1(TO) mode parameters. (In the case of r-plane, a-plane, or m-plane oriented samples, the A_1(TO) phonon mode parameters are accessible by generalized IRSE, see, e. g., Ref. [22].) For one free-carrier species, two out of the three free-carrier parameters N (concentration), μ^{opt} (mobility), and m^* (effective mass) can be extracted from the IRSE data, since the actual determinable quantities are N/m^* and $N\mu^{opt}$. A detailed discussion about parameter accessibility through the IRSE approach is given in Refs. [14] and [17] for the group-III nitrides.

5 Results and Discussion

5.1 GaN

Figures 2–4 present the mid-IR ellipsometric Ψ and Δ spectra of three highly resistive α-GaN films (samples A–C) grown on different substrate materials. The ~890 nm thick GaN epilayer of sample A was directly deposited on c-plane sapphire substrate by plasma-assisted MBE. A ~110 nm thick AlN buffer layer was deposited onto c-plane 6H-SiC substrate in order to improve the crystal quality of the subsequently grown ~1070 nm thick GaN film of sample B. Sample C consists of two, ~470 and ~1290 nm thick GaN films, which are separated by a ~20 nm thick low-temperature AlN interlayer and grown on n-type (111) Si. The group-III nitride films of samples B and C were grown by MOCVD. For comparison, the ellipsometric spectra of the respective substrates without group-III nitride films are included in Figs. 2–4 as well. For all three samples, excellent agreement between experimental (dashed lines) and calculated best-fit (solid lines) data was obtained in the entire spectral range.

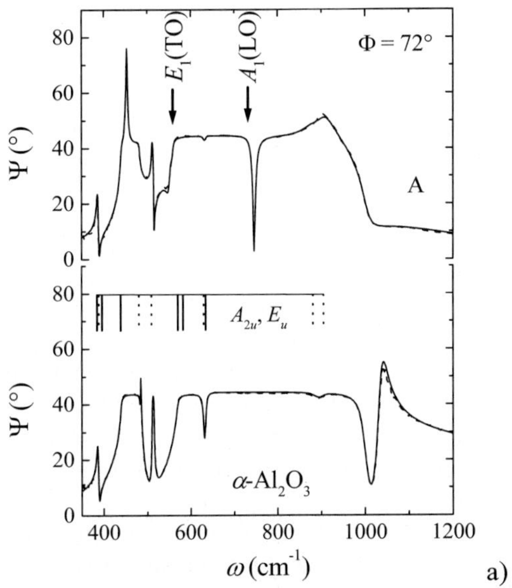
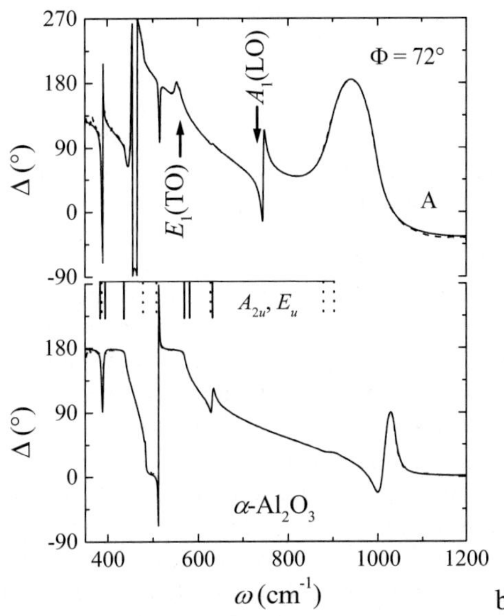

Fig. 2 Measured (dashed lines) and calculated best-fit (solid lines) Ψ and Δ spectra of a ~890 nm thick, nominally undoped α-GaN film grown on c-plane sapphire substrate (upper panels) and of a bare sapphire substrate (lower panels). The positions of the E_1(TO) and A_1(LO) phonons of the GaN film, as obtained by IRSE, are marked by arrows. The IR-active sapphire phonon modes of A_{2u} and E_u symmetry are indicated by vertical markers (TO, solid; LO, dotted).

In the case of sample A, the measured IR dielectric response is strongly affected by distinct spectral features due to the c-plane sapphire substrate possessing six IR-active phonon mode bands of A_{2u} and E_u symmetry [18], respectively, which in part overlap with the GaN phonon bands. However, a precise data regression analysis renders the assignment of the measured spectral features to phonon modes of individual layers within heterostructure samples unique. In the ellipsometric spectra the Berreman surface polariton effect near the A_1(LO) phonon mode and the E_1(TO) lattice vibration of the GaN films cause resonance-like features, which are marked by arrows in Figs. 2–4.

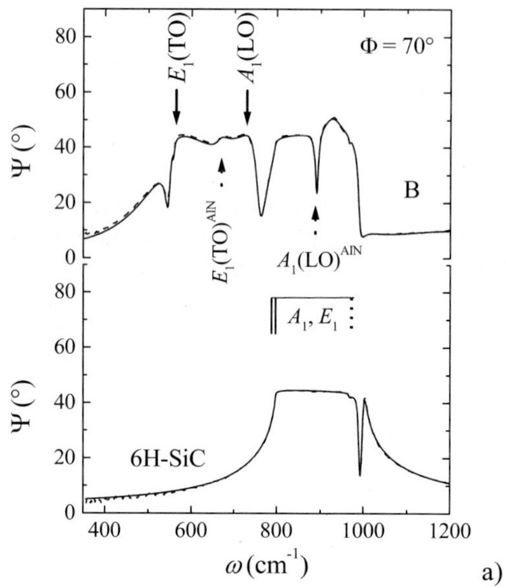
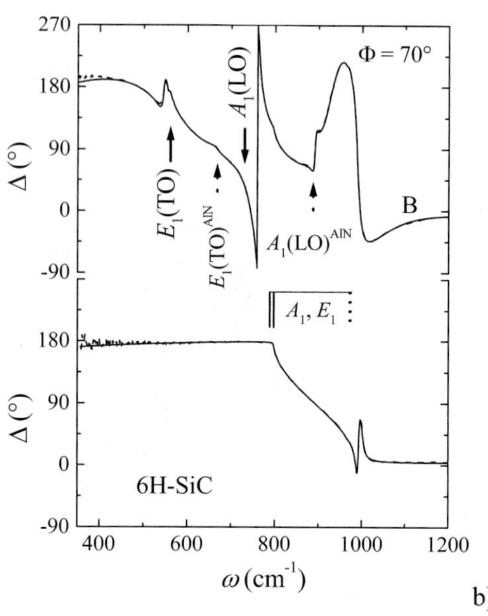

Fig. 3   Measured (dashed lines) and calculated best-fit (solid lines) Ψ and Δ spectra of a ~1070 nm thick, nominally undoped α-GaN film grown on c-plane 6H-SiC substrate (upper panels) and of a bare SiC substrate (lower panels). The positions of the E_1(TO) and A_1(LO) phonons of the GaN film and the AlN buffer layer, as obtained by IRSE, are marked by solid and dotted arrows, respectively. The SiC phonon modes of A_1 and E_1 symmetry are indicated by vertical markers (TO, solid; LO, dotted).

The GaN phonon mode frequencies determined by IRSE are consistent with those obtained previously by other methods within the usual variation range [23–25]. Furthermore, the IRSE results agree well with those obtained from Raman scattering experiments performed on the same samples [26]. From the measured spectral position of the E_1(TO) phonon mode the average strain state of the GaN epilayer can be derived. Employing phonon deformation potential constants, elastic constants, and the phonon mode frequency for totally relaxed material determined previously [9, 27, 28], the average in-plane GaN film stress σ_{xx} is obtained as –0.63, –1.11, and +0.73 GPa for samples A, B, and C, respectively. The E_1(TO) resonance broadening parameter, which can usually be determined with high precision by IRSE, is a measure of the degree of strain homogeneity and crystal perfection for binary compounds.

In order to demonstrate the influence of free carriers on the IR dielectric response, we consider a series of n-type GaN layers where the Si doping concentration was varied over approximately two orders of magnitude (samples D–I). The ~1.6 μm thick GaN layers were grown by MOCVD on c-plane sapphire substrate. Figure 5a displays the IRSE Ψ spectra of these samples. The drastic changes of the ellipsometric spectra are primarily caused by different free-electron concentrations ranging from 2×10^{17} to 1×10^{19} cm^{-3}. For high N_e values, the free carriers effectively screen the polar phonon modes of sapphire and GaN. Hence, the spectral features due to the E_1(TO) and A_1(LO) phonons appear strongly damped. In particular, the large dip in Ψ observed for the low-conductive films near the uncoupled A_1(LO) phonon (~730 cm^{-1}) almost vanishes for high N values due to strong LO-phonon plasmon coupling. However, a small dip still remains (marked by asterisks in Fig. 5a) originating from a thin carrier-depleted layer at the sample surface, where the A_1(LO) phonon mode is not screened by free carriers.

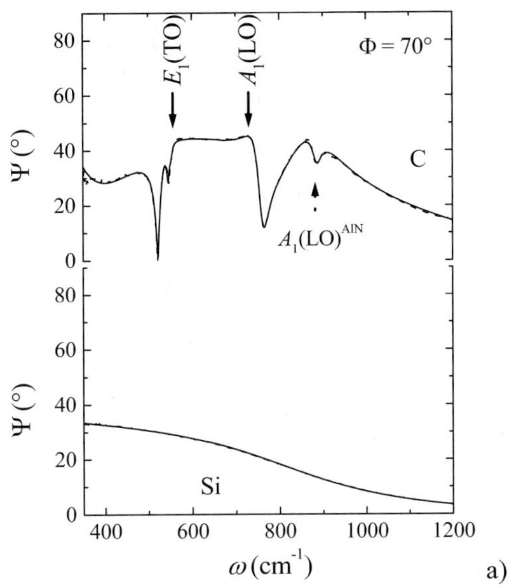
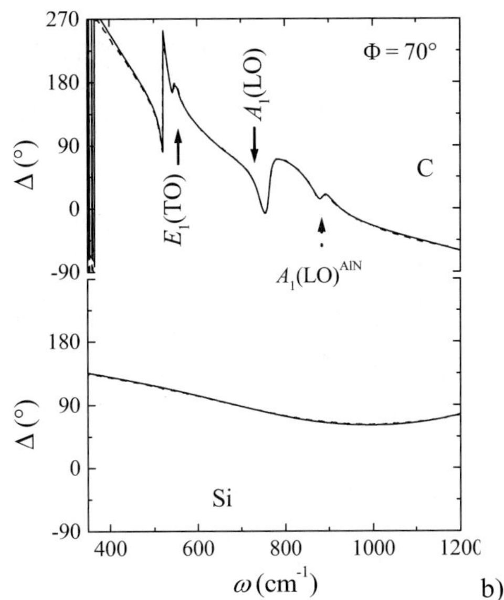

Fig. 4 Measured (dashed lines) and calculated best-fit (solid lines) Ψ and Δ spectra of a ~1760 nm thick, nominally undoped α-GaN film grown on n-type (111) Si substrate (upper panels) and of a bare Si substrate (lower panels). The positions of the E_1(TO) and A_1(LO) phonons of the GaN film and the AlN buffer layer, as obtained by IRSE, are marked by solid and dotted arrows, respectively.

Figure 5b shows the best-fit $\varepsilon_{2,\perp}$ and $\mathrm{Im}\{-\varepsilon_\parallel^{-1}\}$ spectra ($\perp$ and $\parallel$ denote components parallel and perpendicular to the c axis) of samples D–I illustrating the well-known LPP coupling effects: While the E_1(TO) lattice resonance is not significantly influenced by free carriers, the peak in $\mathrm{Im}\{-\varepsilon_\parallel^{-1}\}$ related to the LPP$^+$ mode becomes broadened and strongly blue-shifted with increasing N_e. For the GaN layers with the highest free-electron concentrations additionally a fairly broadened lower maximum in $\mathrm{Im}\{-\varepsilon_\parallel^{-1}\}$ occurs, which is subject to a high-frequency shift with increasing N_e. The lower (upper) peak in $\mathrm{Im}\{-\varepsilon_\parallel^{-1}\}$ is attributed to the LPP$^-$ (LPP$^+$) coupled mode of A_1 symmetry. This LPP coupling behavior has been observed by Raman scattering experiments for n-type GaN in Refs. [29] and [30].
Following the same approach as used by Barker and Ilegems [31] and Perlin *et al.* [32] for n-type GaN, the free-carrier concentration values derived from Hall-effect measurements were used as input parameters for the IRSE data analysis, and the effective carrier masses for p-type and n-type GaN film with $N_\mathrm{e} = 1.0\times10^{19}$ cm^{-3} and $N_\mathrm{h} = 8\times10^{17}$ cm^{-3}, respectively, were determined [14]. The effective hole mass was obtained as $m_\mathrm{h}^*/m_0 = 1.40 \pm 0.33$. According to our result, the anisotropy of the effective hole mass parameters is not larger than the uncertainty of m_h^*, i.e., the anisotropy should be less than ~25 %. The effective electron mass values were determined as $m_{\mathrm{e},\parallel}^*/m_0 = 0.228 \pm 0.008$ and $m_{\mathrm{e},\perp}^*/m_0 = 0.237 \pm 0.006$, i.e., no significant anisotropy was found for m_e^*. These m_e^* values agree well with those reported by Perlin *et al.* employing IR-reflectometry [32].

The analysis of free-carrier properties by IRSE is limited by the fact, that the IR dielectric function provides access to the plasma frequency and broadening parameters, i.e., two of the three free-carrier parameters (concentration, mobility, effective mass) can be extracted from the dielectric function only. Hence, one free-carrier parameter needs to be known from a different experiment and is used as input parameter in the IRSE data analysis. All three free-carrier parameters can be obtained simultaneously by employing IR magnetooptical generalized ellipsometry (IR-MOGE), a novel approach, which is currently being developed [4]. In an applied magnetic field, the IR dielectric response becomes a complex-valued tensor, which – in principle – contains sufficient information to differentiate between the three free-carrier parameters and to determine the sign of the majority carriers. In the near future, IR-MOGE

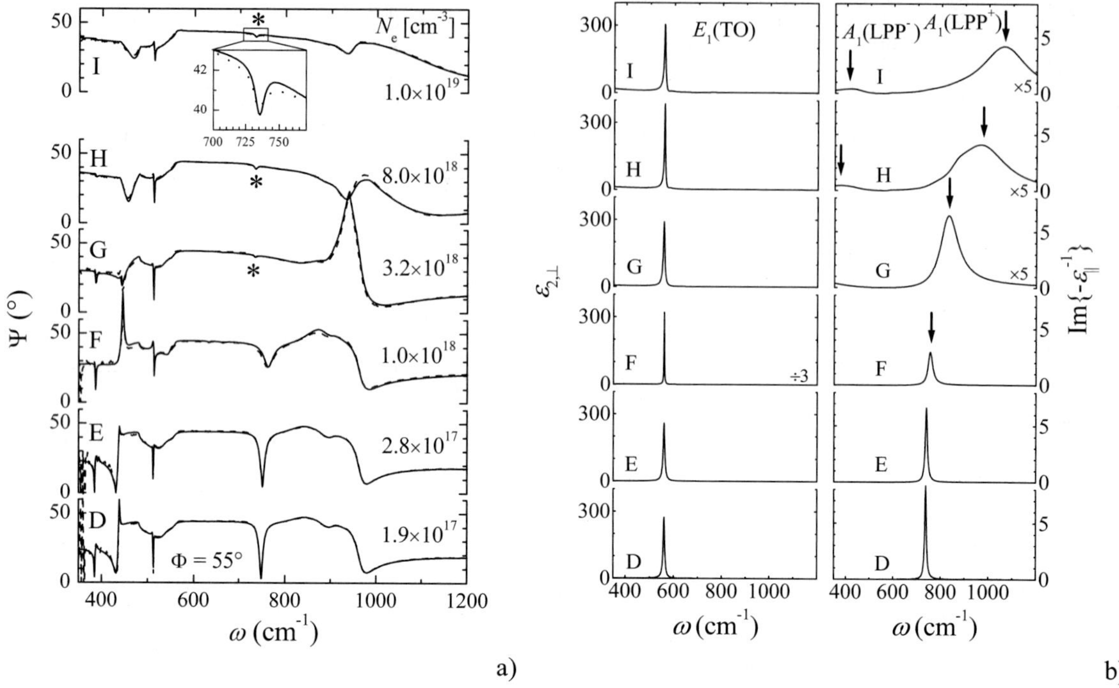

Fig. 5 a) Measured (dashed lines) and best-fit calculated (solid lines) Ψ spectra of ~1.6 μm thick, differently high Si-doped n-type GaN films (samples D–I) deposited on c-plane sapphire substrates. The free-electron concentration N_e in the GaN films varies over approximately two orders of magnitude and drastically influences the line shape of the ellipsometric spectra. For sample I, the inset enlarges the small spectral feature (marked by asterisks), which is related to the unscreened GaN A_1(LO) phonon mode existing in a thin carrier-depleted surface layer. b) $\varepsilon_{2,\perp}$ and Im$\{-\varepsilon_\parallel^{-1}\}$ spectra for samples D–I, derived from the best-fit IRSE model calculation. While the peak corresponding to the E_1(TO) resonance (left column) remains unaffected by free carriers, the spectral features due to the LO-like excitations with A_1 symmetry (right column) exhibit the characteristic LPP coupling behavior. The asymmetric line shape of the upper LPP branch in the Im$\{-\varepsilon_\parallel^{-1}\}$ spectra of samples G–I is caused by additional low-polar modes observed within the GaN reststrahlen range.

will become a useful tool for the determination of valence and conduction band effective mass parameters of group-III nitrides independent of input parameters.

In contrast to the well established frequencies of the lattice phonon modes in pure undisturbed GaN, only little is known about disorder-activated IR-active vibrational modes in the spectral region between 350 and 1500 cm^{-1}. There, the strong IR absorption bands of the commonly used substrate material sapphire dominate the IR absorption and transmission spectra. IRSE overcomes this problem by precisely separating the dielectric response of highly Si-doped n-type GaN films from that of the substrate. As a result, in the IRSE spectra of these films three low-polar IR-active modes are observed, which occur within the GaN reststrahlen range [15].

5.2 InN

Using the free-electron concentration value determined by Hall-effect measurements ($N_e^{\text{Hall}} = 2.8 \times 10^{19}$ cm^{-3}) as well as the plasma frequency and the high frequency dielectric constant both obtained from IRSE, the isotropically averaged effective electron mass of InN is derived as $m_e^* = (0.141 \pm 0.014)m_0$ by IRSE. The uncertainty of m_e^* results mainly from the uncertainty of the free-electron concentration value. Previous experimental effective mass values for InN were reported by Tyagai *et al.* [33], Inushima *et al.* [34], and Wu *et al.* [35]. The E_1(TO) phonon resonance frequency was found at 477.1 ($\pm$ 0.6) cm^{-1}. The occurrence of a strong LO-like feature in the Raman scattering spectra near the suspected mode frequency of the unscreened A_1(LO) phonon despite the high free-electron concentration cannot arise from a surface depletion layer, which has often been regarded to be responsible for this phenomenon [34]. The IRSE analysis rather suggested that wave vector non-conservation, possibly arising from elastic scattering by impurities in the InN film, leads to large wave vectors of the LO-like excitation. Hence, the observed excitation was attributed to the lower branch of the A_1-LPP mode and constitutes a lower limit (590 cm^{-1}) for the unscreened A_1(LO) phonon mode frequency in InN. Further details are presented in Ref. [16].

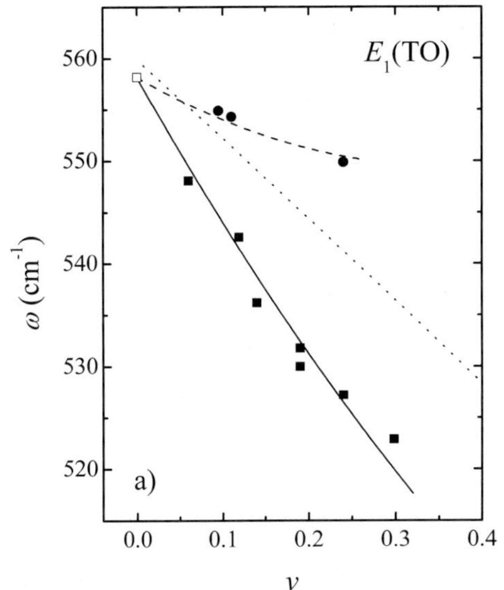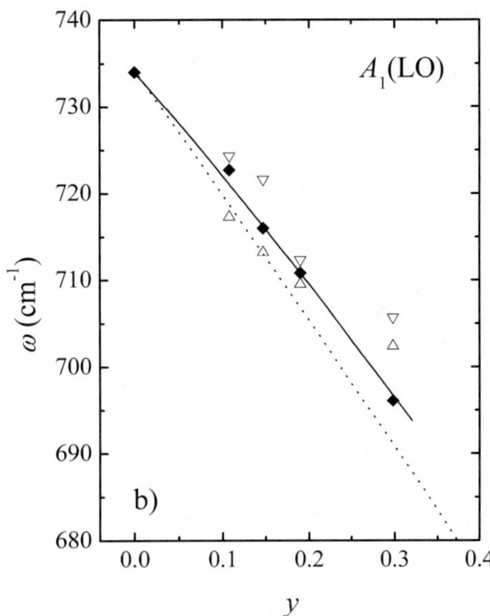

Fig. 6   a) Composition dependence of the In$_y$Ga$_{1-y}$N E_1(TO) phonon mode frequency. Squares and solid line, fully relaxed films; circles and dashed line, pseudomorphically strained films on GaN; open square, Ref. [28]; dotted line, theory, Ref. [7]. b) Composition dependence of the In$_y$Ga$_{1-y}$N A_1(LO) phonon mode frequency. Diamonds and solid line, differently strained films. The triangles (inverted triangles) refer to μ-Raman scattering results obtained for the same samples employing the excitation energy 2.41 eV (2.71 eV). Dotted line, theory, Ref. [7].

5.3   In$_y$Ga$_{1-y}$N

According to our IRSE investigations on differently strained In$_y$Ga$_{1-y}$N films, both the In$_y$Ga$_{1-y}$N E_1(TO) and A_1(LO) phonons exhibit a one-mode behavior for the composition range studied here ($0 \leq y \leq 0.30$). In Fig. 6a, the E_1(TO) phonon mode frequencies obtained by IRSE are depicted vs. y. Squares indicate fully relaxed ternary layers and circles indicate pseudomorphically strained films with respect to the underlaying GaN buffer layers. As expected from consideration of the phonon frequencies observed for the binary end components, the In$_y$Ga$_{1-y}$N E_1(TO) phonon is shifted toward lower frequencies with increasing y. In addition, the E_1(TO) phonon mode frequency is obviously affected by the film strain. Both the composition and strain effects on the E_1(TO) phonon mode frequency can be assessed – according to Eq. (4) – by

$$\omega[E_1(\text{TO})] = 558.2 \text{ cm}^{-1} - 81.1 \text{ cm}^{-1} \cdot y + \alpha_1 y(1-y) + \beta\varepsilon_{xx}. \tag{6}$$

Here, the E_1(TO) phonon mode frequencies for unstrained GaN and InN are adopted from Ref. [28] and according to the results of section 5.2, respectively. The strain-induced E_1(TO) mode shift is assumed to depend linearly on the film in-plain strain ε_{xx}, while a bowing term, $\alpha_1 y(1-y)$, is included for the y-dependent mode shift. A least-squares fit of the above equation to the experimental data yields $\alpha_1 =$ $-67\,(\pm 6)$ cm^{-1} for fully relaxed material (solid line in Fig. 6a). This bowing value deviates from results of MREI calculations [7] (dotted line in Fig. 6a), which predict a small positive E_1(TO) phonon mode bowing parameter. For the pseudomorphically strained In$_y$Ga$_{1-y}$N films, where ε_{xx} depends linearly on y, the strain effect on the E_1(TO) phonon mode frequency is best reflected by $\beta = -887\,(\pm 25)$ cm^{-1} (dashed line in Fig. 6a).

The observation of the E_1(TO) phonon mode of In$_y$Ga$_{1-y}$N has been reported only once before: IR-reflectivity measurements under near-normal incidence suggested a one-mode behavior for the E_1(TO) phonon ($0 \leq y \leq 0.86$) with an almost linear composition dependence [36]. Due to the polycrystallinity of these films, however, the data scattered considerably, and thus no reliable bowing value for $\omega[E_1$(TO)$]$ could be derived from the data in Ref. [36]. It should be pointed out that none of the Raman scattering studies on In$_y$Ga$_{1-y}$N reported so far have revealed any features attributed to In$_y$Ga$_{1-y}$N TO phonon modes.

As depicted in Fig. 6b by diamonds, the A_1(LO) phonon mode frequencies determined for differently strained films ($-4\times10^{-3} \leq \varepsilon_{xx} \leq 0$) shift almost linearly as a function of the In content, which is indicative of a one-mode behavior. Neglecting strain effects, the composition dependence of the measured A_1(LO) phonon mode frequency values can be well described by

$$\omega[A_1(\text{LO})] = 734\ \text{cm}^{-1} - 144\ \text{cm}^{-1}\cdot y + \alpha_1 y(1-y). \tag{7}$$

for $0 \leq y \leq 0.30$ with a fairly small positive phonon mode bowing parameter of $\alpha_1 = 27\,(\pm 3)$ cm^{-1} (solid line in Fig. 6b). This value is consistent with the bowing parameter of 43 cm^{-1} reported by Wieser et al. for In contents up to 0.33 [37]. A nearly linear composition dependence of the In$_y$Ga$_{1-y}$N A_1(LO) phonon mode frequency was also predicted by MREI calculations [7, 38]. The composition dependence obtained by Grille et al. [7] is shown in Fig. 6b by the dotted line and is in reasonable agreement with our data. Apparently, at least for compressive in-plane film strains $|\varepsilon_{xx}| \leq 4\times10^{-3}$ the influence of strain on the A_1(LO) phonon mode frequency is negligible. A small stress influence on the A_1(LO) mode frequency was already obtained for α-GaN by means of Raman scattering at 77 K [39].

Strong electron-phonon Fröhlich interaction permits the detection of the A_1(LO) phonon mode by near-resonant μ-Raman scattering measurements as well. However, the position of the Raman peak related to the A_1(LO) mode depends considerably on the used excitation energy [triangles in Fig. 6b]. This effect can be explained in terms of local compositional inhomogeneities in the ternary films, if Fröhlich-type scattering processes dominate [40].

The E_1(TO) and A_1(LO) mode behavior as well as the α_1 and β values derived from IRSE investigations of the three ternary hexagonal group-III nitride compounds Al$_x$Ga$_{1-x}$N ($0 \leq x \leq 0.21$), In$_y$Ga$_{1-y}$N ($0 \leq y \leq 0.30$), and Al$_{1-y}$In$_y$N ($0.11 \leq y \leq 0.21$) [41, 42] are summarized in Table 1.

Table 1   Mode behavior, α_1, and β parameters [according to Eq. (4)] determined for the E_1(TO) and A_1(LO) phonons of the three hexagonal ternary group-III nitride compounds.

	E_1(TO)			A_1(LO)		
	mode behavior	α_1 [cm^{-1}]	β [cm^{-1}]	mode behavior	α_1 [cm^{-1}]	β [cm^{-1}]
Al$_x$Ga$_{1-x}$N ($0 \leq x \leq 0.21$)	2			1		[a]
In$_y$Ga$_{1-y}$N ($0 \leq y \leq 0.30$)	1	-67 (6)	-887 (25)	1	27 (3)	~ 0
Al$_{1-y}$In$_y$N ($0.11 \leq y \leq 0.21$)	1	-25 (1)	-987 (25)	1		

[a]Neglecting possible strain effects, the composition dependence of the Al$_x$Ga$_{1-x}$N A_1(LO) phonon frequency can be well described by $\omega/\text{cm}^{-1} = 736.5\,(\pm 0.3) + 215\,(\pm 8)x - 243\,(\pm 32)x^2$ for $x \leq 0.21$.

5.4 Cubic $Al_xGa_{1-x}N$

The infrared optical properties and electronic interband transitions of cubic $Al_xGa_{1-x}N$ films ($0 \leq x \leq 0.20$) are studied and presented in detail in Ref. [17].

5.5 $Al_{1-y}In_yN$

As shown by IRSE studies in Ref. [41], the $E_1(TO)$ phonon in α-$Al_{1-y}In_yN$ ($0.11 \leq y \leq 0.21$) reveals a distinct one-mode behavior in contrast to theoretical predictions based on MREI calculations [7]. Figure 7 presents Ψ spectra of $50 - 320$ nm thick partially tensile strained $Al_{1-y}In_yN$ films in the range of the $E_1(TO)$ and $A_1(LO)$ phonon modes, which each gives rise to a distinct spectral dip. The damping of the $A_1(LO)$ phonon mode increases rapidly with increasing In content (and increasing film thickness), but even for the film with $y = 0.157$ the spectral dip related to the $A_1(LO)$ phonon mode can be well resolved. As for the lattice vibrations with E_1 symmetry, one TO-LO phonon band with A_1 symmetry is sufficient to describe the experimental data. Thus, we conclude a one-mode behavior of the $Al_{1-y}In_yN$ $A_1(LO)$ phonon mode in accordance with the theoretical results of Ref. [7].

The $A_1(LO)$ phonon mode frequency values determined for the partially strained $Al_{1-y}In_yN$ films (circles in Fig. 8) are fairly consistent with the composition dependence of this mode calculated for strain-free material in Ref. [7] (dotted line in Fig. 8). Because of the few available samples, a possible influence of strain on the $A_1(LO)$ mode frequency cannot be assessed at this point. Nevertheless, the (small) devia-

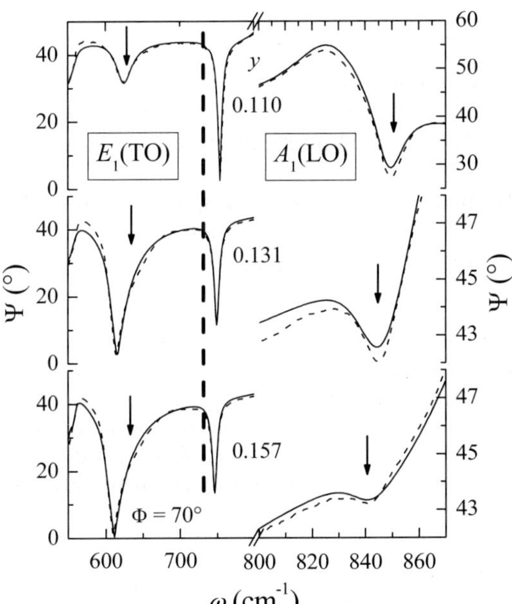
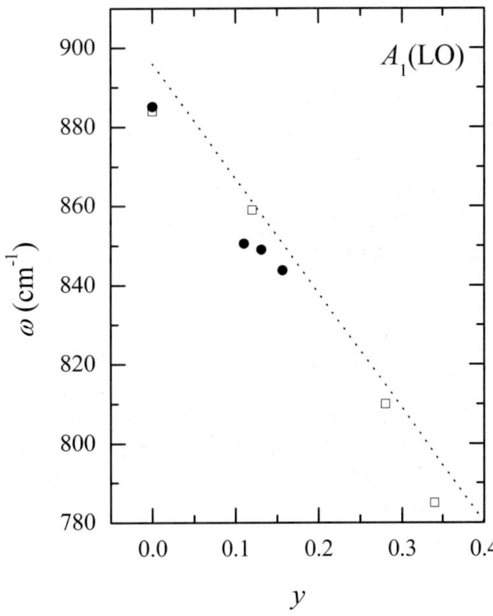

Fig. 7 Measured (dashed lines) and calculated best-fit (solid lines) Ψ spectra of $Al_{1-y}In_yN$/GaN/AlN/Al_2O_3 heterostructures near the $Al_{1-y}In_yN$ $E_1(TO)$ and $A_1(LO)$ phonon modes, both causing well-resolved spectral features. The respective phonon mode frequencies determined by IRSE are marked by arrows. The vertical dashed line indicates the $A_1(LO)$ phonon mode position for the GaN buffer layers. Note the different scales for the ordinates and abscissa below and above 800 cm^{-1}.

Fig. 8 Composition dependence of the $Al_{1-y}In_yN$ $A_1(LO)$ phonon mode frequency. Circles, this work, IRSE; squares, Raman scattering measurements, Ref. [43]; dotted line, theory, Ref. [7].

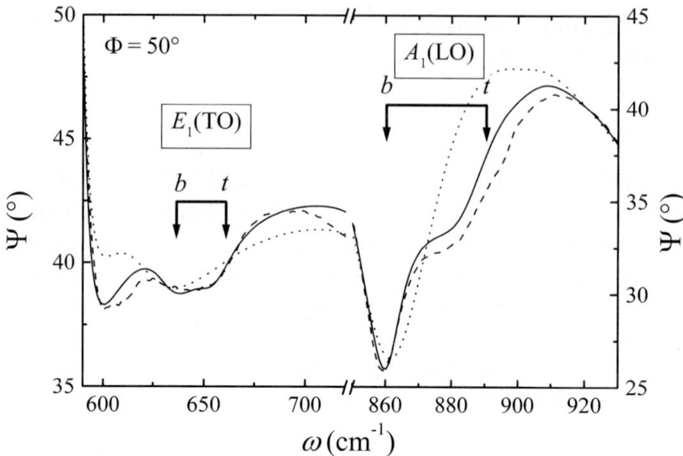

Fig. 9 Ψ spectra of a compositionally inhomogeneous $Al_{1-y}In_yN$ film near the $E_1(TO)$ and $A_1(LO)$ phonon modes. Dashed lines, experimental data; solid lines, calculated best-fit data obtained from a graded $Al_{1-y}In_yN$ layer model; dotted lines, calculated best-fit data obtained from a homogeneous $Al_{1-y}In_yN$ layer model. The arrows mark the positions of the $E_1(TO)$ and $A_1(LO)$ phonon modes, determined at the top (t) and bottom (b) of the ternary layer.

tion of our data from the theoretical frequency dependence reported in Ref. [7] may originate from the partial strain within the $Al_{1-y}In_yN$ films. Our data are in line with results from near-resonance enhanced Raman scattering measurements on $Al_{1-y}In_yN$ films directly deposited on sapphire substrate (squares in Fig. 8) [43].

Elastic recoil detection (ERD) experiments revealed that the alloy composition of a ~580 nm thick $Al_{1-y}In_yN$ film on $GaN/AlN/Al_2O_3$ varies strongly along the growth direction of the film. According to these measurements, the In content decreases from about 0.12 near the interface $Al_{1-y}In_yN/GaN$ to nearly

Table 2 $E_1(TO)$ and $A_1(LO)$ phonon mode frequency ω and broadening values γ derived from the IRSE data analysis of a compositionally inhomogeneous $Al_{1-y}In_yN$ film employing a graded layer model. The uncertainty limits are given in parentheses. ERD measurements provided a nearly vanishing In content at the top and $y = 0.12$ at the bottom of the ternary layer. For comparison, the respective phonon mode frequencies for AlN and $Al_{0.88}In_{0.12}N$ are included. Mode behavior, α_1, and β parameters [according to Eq. (4)] determined for the $E_1(TO)$ and $A_1(LO)$ phonons of the three hexagonal ternary group-III nitride compounds.

	top of graded $Al_{1-y}In_yN$ layer	AlN	bottom of graded $Al_{1-y}In_yN$ layer	$Al_{0.88}In_{0.12}N$
$\omega[E_1(TO)]$ / cm^{-1}	660.7 (0.9)	670.8[a]	637.3 (1.1)	638.9[b]
$\omega[A_1(LO)]$ / cm^{-1}	890.2 (1.3)	890.9[a]	861.0 (0.5)	861.2[c]
$\gamma[E_1(TO)]$ / cm^{-1}	34.1 (1.6)		20.5 (1.9)	
$\gamma[A_1(LO)]$ / cm^{-1}	28.4 (1.8)		8.4 (1.0)	

[a]Raman scattering, Ref. [23].
[b]IRSE, Ref. [41].
[c]MREI calculations, Ref. [7].

zero at the top of the ternary film. X-ray diffraction measurements on the same sample suggested compositional inhomogeneities as well. In the IRSE model analysis, these inhomogeneities were allowed for by employing a graded $Al_{1-y}In_yN$ layer, whose dielectric function parameters were varied linearly with the film depth. The linear grading profiles of the dielectric function parameters were approached by stepwise constant parameter functions employing 20 uniform steps along the film depth. Figure 9 demonstrates the significantly improved match of the simulated spectra to the experimental data in the range of the $E_1(TO)$ and $A_1(LO)$ phonons when employing the graded layer model instead of using a spatially homogeneous dielectric function for the ternary film.

Table 2 shows the best-fit parameters for the $E_1(TO)$ and $A_1(LO)$ phonon modes at the top and bottom of the ternary layer. The best-fit $A_1(LO)$ mode frequency at the top of the layer is close to the respective value reported in Ref. [23] for 5–7 μm thick AlN films. This indicates a small In incorporation into the layer near the sample surface, in accordance with the ERD measurement results. The $E_1(TO)$ mode frequency at the top of the layer determined by IRSE deviates from the respective value for AlN reported in Ref. [23] by about 10 cm^{-1}. In view of the deformation potential constants given in Ref. [44] for AlN the measured $E_1(TO)$ frequency value indicates a considerable tensile strain near the top of the layer. At the bottom of the layer, the $E_1(TO)$ mode frequency determined by IRSE agrees well with the respective value for pseudomorphically strained $Al_{0.88}In_{0.12}N$ derived from the IRSE analysis of Ref. [41]. Moreover, the $A_1(LO)$ phonon mode frequency at the bottom of the layer is highly consistent with theoretical predictions for $Al_{0.88}In_{0.12}N$ [7]. Hence, from the IRSE analysis, the In content of the ternary layer near the interface to the GaN buffer layer is determined as 0.12, in excellent agreement with the results from ERD measurements.

5.6 $Al_xIn_yGa_{1-x-y}N$

IRSE investigations on hexagonal $Al_xIn_yGa_{1-x-y}N$ films with $0 \leq x \leq 0.40$ and $y \approx 0.12$ were published very recently [45] and are thus not repeated here.

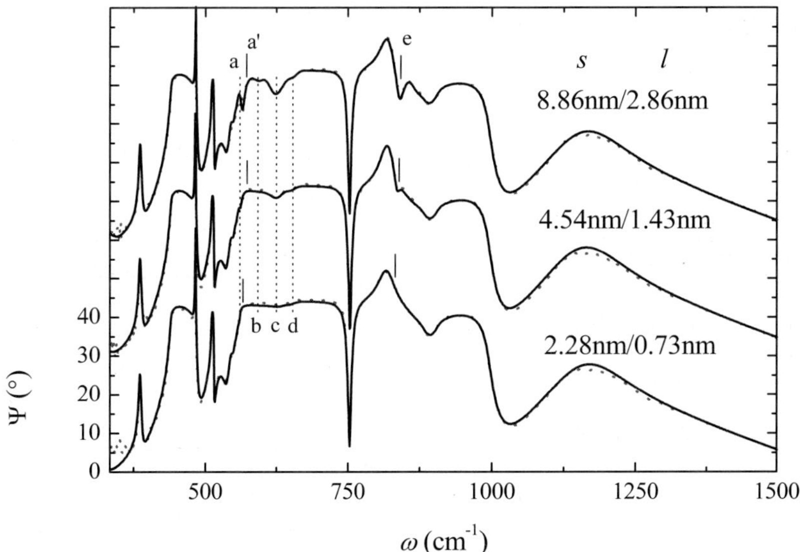

Fig. 10 Ψ spectra of 10-period GaN_sAlN_l-SL's grown on ~2400 nm thick unstrained (label a: $E_1(TO)$ = 558 cm^{-1}) GaN/*c*-plane sapphire with constant Al-fraction. The feature labeled a' indicates the GaN $E_1(TO)$ mode of the SL-sublayer. Features b, d, and c denote SL modes, likely caused by zone-folding (b, d), and the zone-center AlN $E_1(TO)$ mode (c). The position of the AlN-sublayer $A_1(LO)$ mode is indicated by label e. Dotted lines, experimental data; solid lines, calculated best-fit data. The upper two spectra sets are shifted vertically by 30° and 60° for convenience, respectively.

5.7 GaN/Al(Ga)N superlattices

Superlattice (SL) structures with few Ångström periods constitute alternatives to random alloys, and thickness and composition of the SL constituents may be chosen to stabilize strained heterostructures, or to grow strain-compensated SL's. Quantum-confinement and size-affected band-to-band transition energies allow further tailoring of heterostructures constituents properties, such as the direct electronic transition energies, the index of refraction, or the free-carrier mobilities. SL dimension and composition determine the strain in the SL sublayers. Nondestructive and noninvasive IRSE studies of phonon modes and free-carrier properties in MBE and MOCVD grown GaN/AlGaN SL's with different compositions and dimensions were reported in Refs. [46] and [47]. It was found that the GaN sublayers are subject to compressive strain, and that the mobility of the free carriers - the SL's were assumed to be n-type - is favored along the SL boundaries and lowered parallel to the growth direction. Room-temperature data from a (newly investigated) set of undoped high-quality MOCVD grown 10-period GaN_sAlN_l SL's with constant well-to-barrier thickness ratio (1:3) but varying dimensions are depicted in Fig. 10. Defect-induced free carriers (presumably free electrons) are identified upon the IRSE model analysis within the GaN-sublayers. Assuming the GaN effective electron mass values, their concentration and mobility parameters (N_e, $\mu_\perp$, $\mu_\parallel$) are (5×10^{18}cm^{-3}, 173 cm^2/Vs, 144 cm^2/Vs), (5×10^{18}cm^{-3}, 175 cm^2/Vs, 76 cm^2/Vs), and (3×10^{18}cm^{-3}, 195 cm^2/Vs, 65 cm^2/Vs) for $l = 0.73$ nm, 1.43 nm, and 2.86 nm, respectively. Near-IR–UV SE measurements further show a strong blue shift of the lowest direct band-to-band transition energy (3.87 eV, 3.54 eV, 2.95 eV) of the SL's with increasing barrier thickness l. Additionally, the highly tensile strained AlN sublayers begin to relax with increasing l, which is reflected by the increase of the

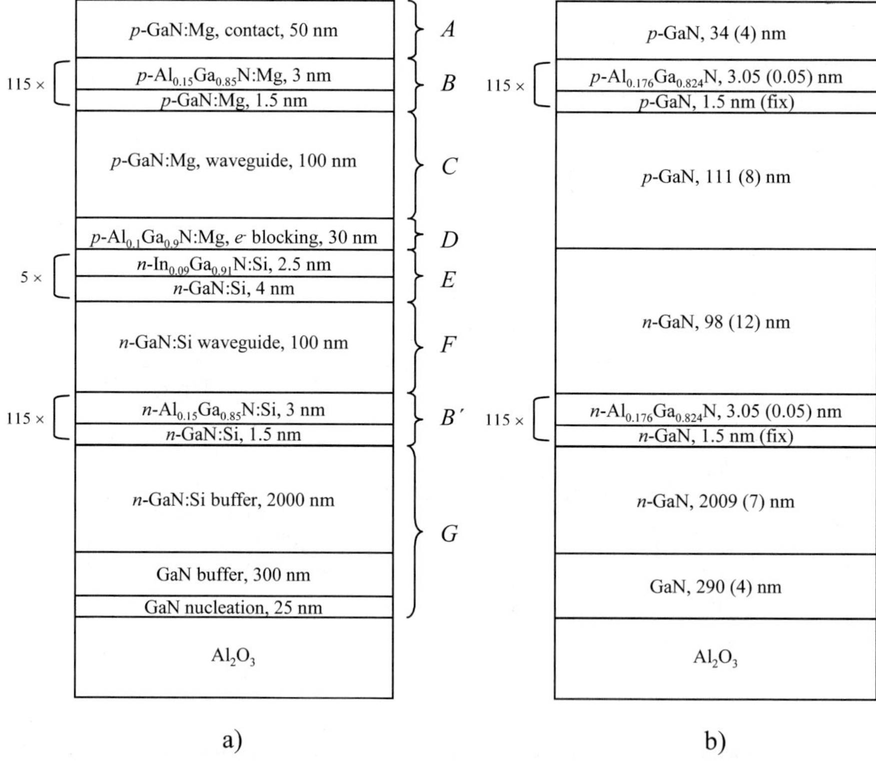

Fig. 11 a) Schematic cross-section of the LD structure studied in this work. The thickness values refer to the intended growth rates and durations. The letters A–G denote the different heterostructure constituents identified in TEM images displayed in Fig. 13. b) Layer model used for the IRSE data analysis with best-fit thickness values. The corresponding best-fit spectra are shown in Fig. 12.

A_1(LO) mode frequency (828 cm^{-1}, 836 cm^{-1}, 841 cm^{-1}) towards the strain-free value. Accordingly, the compressive strain within the GaN sublayers increases, indicated by the blue-shift of the GaN E_1(TO) mode (2 cm^{-1}, 4 cm^{-1}, 9 cm^{-1}) relative to that of unstrained material. Two additional and distinct IR-active modes (~592 cm^{-1}, ~655 cm^{-1}) are identified within the spectra of the SL's and may originate from modes at the zone boundaries activated upon Brillouin-zone folding.

5.8 Group-III nitride device structures

A laser diode (LD) sample, which was grown by MOCVD on c-plane sapphire substrate, is discussed here exemplarily. The layer sequence of this device heterostructure is depicted in detail in Fig. 11a. Two $Al_{0.15}Ga_{0.85}N$/GaN superlattices, which differ only in their doping properties, act as top and bottom optical cladding layers. The top SL is Mg-doped, whereas the bottom SL is Si-doped. The nucleation/buffer layer sequence consists of three different GaN layers. The $In_{0.09}Ga_{0.91}N$/GaN multi quantum well and the $Al_{0.1}Ga_{0.9}N$ electron-blocking layer are embedded from bottom and top by n- and p-type GaN layers, respectively. A highly Mg-doped p-type contact layer terminates the LD structure towards the top. During growth the Mg and Si doping rates were adjusted to obtain a free-hole concentration of ~2×10^{17} cm^{-3} in the p-type contact layer and free-electron concentrations of 1–2×10^{18} cm^{-3} in all n-type conductive layers, respectively. The free-hole concentration in the Mg-doped superlattice was aimed at 1×10^{17} cm^{-3}.

In the case of complex multi-layered samples, obviously not all of the large number of model parameters can be determined simultaneously from a single IRSE measurement, but certain sample parameters need to be known prior to the data regression analysis in order to prevent parameter correlations. Thus, the IR dielectric function database, which has been gradually established in the course of this work, is used for the characterization of the device structure and only physically significant layer parameters with sufficient sensitivity to the IRSE experiment were varied.

The free-carrier response of each sample constituent was treated isotropically, and for n- and p-type

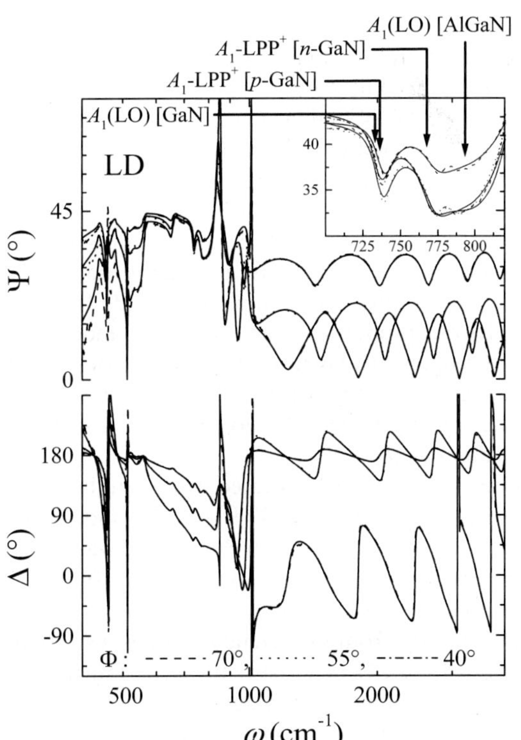

Fig. 12 Measured (broken lines) and calculated best-fit (solid lines) IRSE Ψ and Δ spectra of the LD structure for three different angles of incidence. Note the logarithmic abscissa scale. The best-fit spectra were obtained using the layer model depicted in Fig. 11b. The inset enlarges the spectral range where LO and LPP$^+$ modes of different sample constituents occur.

Table 3   Selected results from IRSE and TEM (see Fig. 13) investigations of the LD sample for the p-type (thickness $d_\mathrm{p} = d_\mathrm{A} + d_\mathrm{B} + d_\mathrm{C} + d_\mathrm{D}$) and n-type (thickness $d_\mathrm{n} = d_{\mathrm{B}'} + d_\mathrm{G} + d_\mathrm{F} + d_\mathrm{E}$) regions, the $Al_xGa_{1-x}N$/GaN SL period $d_{\mathrm{B}'/115}$, and the average Al molar fraction x in the SL barriers. The uncertainty limits are given in parentheses. Further IRSE results concerning thickness values of individual sample constituents are given in Fig. 11b.

	intended growth values	IRSE	TEM
d_p [nm]	698	668 (18)	665 (7)
d_n [nm]	2975	2920 (17)	2879 (29)
$d_{\mathrm{B}'/115}$ [nm]	4.5	4.55 (0.05)	4.54 (0.05)
x	0.15	0.176 (0.005)	[a]
N_e [10^{18} cm^{-3}]	$1-2$	2.2 (0.3)	[a]
μ_e [cm^2/(Vs)]		184 (13)	[a]
N_h [10^{17} cm^{-3}]	2	3.9 (0.5)	[a]
μ_h [cm^2/(Vs)]		22 (11)	[a]

[a]Not available from TEM studies.

GaN the effective carrier masses were assumed to be $m_\mathrm{e}^* = 0.23m_0$ and $m_\mathrm{h}^* = 0.8m_0$, respectively. The free-carrier concentrations in all $Al_xGa_{1-x}N$ sublayers, which act as carrier barriers in the superlattices, were assumed to be below the IRSE detection limits of about 1×10^{17} cm^{-3}. The layer model employed for the IRSE data analysis of the LD sample is delineated in Fig. 11b.

Figure 12 presents experimental (broken lines) and calculated best-fit (solid lines) Ψ and Δ spectra of the LD sample. Below ~1000 cm^{-1} the spectra are dominated by sapphire phonon mode signatures, which are overlaid by features originating from the group-III nitride sample constituents, whereas above the reststrahlen bands of the materials, i.e., for $\omega > 1000$ cm^{-1}, pronounced Fabry-Perot interference oscillations occur. The interference periods change when varying the angle of incidence, which results from different optical path lengths within the layer stack and provides high sensitivity to the thickness values of the device constituents. The spectral positions of the GaN and $Al_xGa_{1-x}N$ modes of particular interest, which were obtained from the SE data analysis, are indicated by arrows in Fig. 12.

The IRSE spectra did not provide sufficient sensitivity to the $Al_xGa_{1-x}N$- nor to the $In_yGa_{1-y}N$-related phonon modes, which are here subsumed by the strong GaN-related phonon and free-carriers signatures. We note however, that these components contribute only ~1.2 % to the overall thickness of the sample.

The SE data analysis distinguishes three differently doped GaN materials: nominally undoped, Si-doped n-type, and Mg-doped p-type. All three GaN materials possess common phonon mode frequencies and TO mode broadening parameters but different free-carrier properties. The nominally undoped GaN nucleation/buffer layer sequence is modeled as a single homogeneous layer without free-carrier contribution. Table 3 summarizes the free-carrier concentration and mobility values obtained for the p- and n-type GaN sample constituents. It was assumed, that the free carriers are homogeneously distributed in each of both n- and p-conductive regions with abrupt changes across interfaces.

Both the free electron and hole concentration values determined by IRSE agree well with the intended values due to the respective doping rates adjusted in the sample growth. The A_1-LPP$^+$ modes of the n- and p-type GaN layers are located at 792 and 736 cm^{-1}, respectively. From the best-fit value for the frequency of the $Al_xGa_{1-x}N$ A_1(LO) phonon (766.8 $\pm$ 0.2 cm^{-1}), an average Al content of 0.176 ($\pm$ 0.005) for the barrier material can be derived employing the results given in Table 1.

The best-fit thickness values given in Fig. 11b deviate slightly from the intended values, especially in the zone near the p-n junction. Cross-section TEM images of the LD heterostructure (see Fig. 13), however, reveal the actual thickness values of the p- and n-type regions, which agree excellently with the IRSE results (see Table 3). Furthermore, the average thickness of the $Al_xGa_{1-x}N$/GaN SL periods obtained by IRSE is – within the error limits – identical with that found in the TEM images. According to

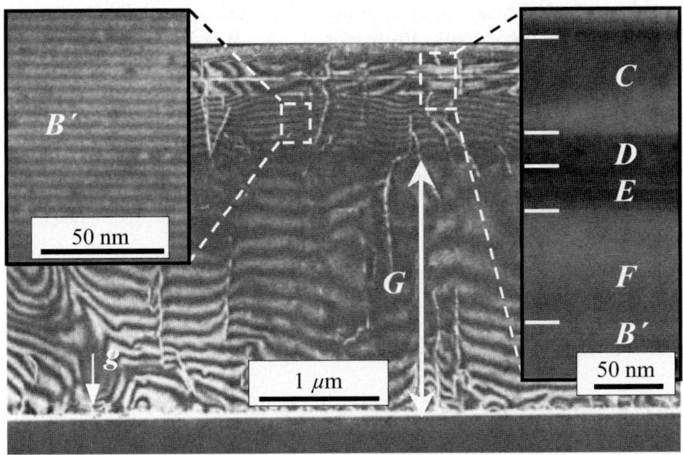

Fig. 13 TEM cross-section of parts of the LD structure studied here. The figure shows a weak beam image $\{g, 3g\}$ with $g = (0002)$. The letters indicate different heterostructure constituents: C and F, GaN waveguide zones; D, $Al_xGa_{1-x}N$ electron blocking layer; E, $In_yGa_{1-y}N$/GaN multi quantum well; B', lower $Al_xGa_{1-x}N$/GaN SL; G, GaN buffer/nucleation layer sequence. The *p-n* junction is located between D and E. The thickness values for individual constituents, derived from the TEM images, are in excellent agreement with the IRSE results (see Table 3).

the IRSE best-fit results, the *p-n* junction within the LD structure is located 668 ($\pm$ 18) nm below the top surface. A more detailed analysis of this LD structure can be found in Ref. [48]. The IRSE investigation of further device heterostructures is given in Refs. [49] and [50].

6 Summary

Spectroscopic ellipsometry in the mid-IR spectral range is presented as a novel characterization method for group-III nitride compounds. The analysis of the derived IR dielectric functions of various binary, ternary, and quaternary group-III nitride films leads to the determination of fundamental material properties, even for layers with some ten nanometer thickness only. Phonon and free-carrier properties of hexagonal GaN, InN as well as of $In_yGa_{1-y}N$, $Al_{1-y}In_yN$, and $Al_xIn_yGa_{1-x-y}N$ films in composition ranges relevant for device applications are studied. In combination with Hall-effect measurements the effective carrier masses in some binary compounds are determined. Otherwise, assuming the effective mass the concentration and mobility of the free carriers can be obtained. For multinary compounds the mode behavior as well as the strain and composition dependences of IR-active phonons are determined. Film inhomogeneities such as composition gradients along the growth direction and thin carrier-depleted surface layers are analyzed. IRSE is further demonstrated to be a powerful tool for the nondestructive IR optical characterization of complex thin-film device heterostructures. Information on the free-carrier concentration and mobility, thickness, alloy composition, average strain state, and crystal quality of individual sample constituents and the *p-n* junction location are derived from the IRSE regression analysis. The application of magnetooptical generalized ellipsometry to group-III nitrides for noninvasive and simultaneous determination of the free-carrier parameters avoiding electrical contacting for Hall-effect measurement is proposed in this review.

Acknowledgements The major part of the infrared ellipsometry investigations was made possible by generous support through the Deutsche Forschungsgemeinschaft under grants Rh 28/3-1 and 3-2 within the funding scheme *Schwerpunktprogramm 1032: Gruppe-III Nitride und ihre Heterostrukturen*. This funding included the purchase of a commercial IR ellipsometer. Part of the research was performed at the Center for Microelectronic and Optical Materials Research at the University of Lincoln-Nebraska, and supported by NSF award DMI-9901510, and by the Nebraska Research Initiative. M. S. and A. K. thank Professor J. A. Woollam, Professor B. Rheinländer, C. M. Herzinger, T. E. Tiwald, and D. W. Thompson for help and support during the initial phase of our work. The Federal Ministry of Education and Research of the Federal Republic of Germany has supported further parts of this work under BMBF project 03WKI09 within the funding scheme *Innovative regionale Wachstumskerne*. M. S. acknowledges a fellowship from the Swedish Foundation for International Cooperation in Research and Higher Education (STINT), and the continued collaboration with Professor H. Arwin. We thank L. Goergens (TU Munich) for performing the ERD measurements on the $Al_{1-y}In_yN$ samples, and U. Teschner (University of Leipzig) for Raman measurements.

References

[1] T. Inushima, V. V. Mamutin, V. A. Vekshin, S. V. Ivanov, T. Sakon, M. Motokawa, and S. Ohoya, J. Cryst. Growth **227**, 481 (2001).

[2] V. Yu. Davydov, A. A. Klochikhin, V. V. Emtsev, S. V. Ivanov, V. V. Vekshin, F. Bechstedt, J. Furthmüller, H. Harima, A. V. Mudryi, A. Hashimoto, A. Yamamoto, J. Aderhold, J. Graul, and E. E. Haller, phys. stat. sol. (b) **230**, R4 (2002).

[3] A. Röseler, Infrared Spectroscopic Ellipsometry (Akademie-Verlag, Berlin, 1990).

[4] T. Hofmann, M. Grundmann, C. M. Herzinger, M. Schubert, and W. Grill, Mat. Res. Soc. Symp. Proc. **744**, M5.32.1-6 (2003).

[5] M. Balkanski, R. Beserman, and J. M. Besson, Solid State Commun. **4**, 201 (1966).

[6] I. F. Chang and S. S. Mitra, Phys. Rev. **172**, 924 (1968).

[7] H. Grille, Ch. Schnittler, and F. Bechstedt, Phys. Rev. B **61**, 6091 (2000).

[8] C. Bungaro and S. de Gironcoli, Appl. Phys. Lett. **76**, 2101 (2000).

[9] V. Yu Davydov, N. S. Averkiev, I. N. Goncharuk, D. K. Nelson, I. P. Nikitina, A. S. Polkovnikov, A. N. Smirnov, M. A. Jacobson, and O. K. Semchinova, J. Appl. Phys. **82**, 5097 (1997).

[10] R. M. A. Azzam and N. M. Bashara, Ellipsometry and Polarized Light (North-Holland, Amsterdam, 1984).

[11] M. Schubert, Phys. Rev. B **53**, 4265 (1996).

[12] M. Schubert, B. Rheinländer, J. A. Woollam, B. Johs, and C. M. Herzinger, J. Opt. Soc. Am. A **13**, 875 (1996).

[13] G. E. Jellison, Thin Solid Films **313-314**, 33 (1998).

[14] A. Kasic, M. Schubert, S. Einfeldt, D. Hommel, and T. E. Tiwald, Phys. Rev. B **62**, 7365 (2000).

[15] A. Kasic, M. Schubert, B. Kuhn, F. Scholz, S. Einfeldt, and D. Hommel, J. Appl. Phys. **87**, 3720 (2001).

[16] A. Kasic, M. Schubert, Y. Saito, Y. Nanishi, and G. Wagner, Phys. Rev. B **65**, 115206 (2002).

[17] A. Kasic, M. Schubert, T. Frey, U. Köhler, D. J. As, and C. M. Herzinger, Phys. Rev. B **65**, 184302 (2002).

[18] M. Schubert, T. E. Tiwald, and C. M. Herzinger, Phys. Rev. B **61**, 8187 (2000).

[19] D. W. Berreman, Phys. Rev. **130**, 2193 (1963).

[20] J. Humlíček, phys. stat. sol. (b) **215**, 155 (1999).

[21] M. Schubert, Phonons, plasmons, and polaritons in III-V semiconductor heterostructures, habilitation thesis, Universität Leipzig, Germany (2002).

[22] M. Schubert, A. Kasic, T. Hofmann, V. Gottschalch, J. Off, F. Scholz, E. Schubert, H. Neumann, I. Hodgkinson, M. Arnold, W. Dollase, and C. M. Herzinger, SPIE Vol. **4806**, 264 (2002).

[23] V. Yu. Davydov, Yu. E. Kitaev, I. N. Goncharuk, A. N. Smirnov, J. Graul, O. Semchinova, D. Uffmann, M. B. Smirnov, A. P. Mirgorodsky, and R. A. Evarestov, Phys. Rev. B **58**, 12899 (1998).

[24] H. Siegle, G. Kaczmarczyk, L. Filippidis, A. P. Litvinchuk, A. Hoffmann, and C. Thomsen, Phys. Rev. B **55**, 7000 (1997).

[25] C. Wetzel, E. E. Haller, H. Amano, and I. Akasaki, Appl. Phys. Lett. **68**, 2547 (1996).

[26] A. Kasic, M. Schubert, B. Rheinländer, V. Riede, S. Einfeldt, D. Hommel, B. Kuhn, J. Off, and F. Scholz, Mat. Sci. & Eng. B **82**, 74 (2001).

[27] A. Polian, M. Grimsditch, and I. Grzegory, J. Appl. Phys. **79**, 3343 (1996).

[28] A. R. Goñi, H. Siegle, K. Syassen, C. Thomsen, and J.-M. Wagner, Phys. Rev. B **64**, 035205 (2001).

[29] P. Perlin, J. Camassel, W. Knap, T. Taliercio, J. C. Chervin, T. Suski, I. Grzegory, and S. Porowski, Appl. Phys. Lett. **67**, 2524 (1995).

[30] D. Kirillov, H. Lee, and J. S. Harris, Jr., J. Appl. Phys. **80**, 4058 (1996).

[31] A. S. Barker Jr. and M. Ilegems, Phys. Rev. B **7**, 743 (1973).

[32] P. Perlin, E. Litwin-Staszewska, B. Suchanek, W. Knap, J. Camassel, T. Suski, R. Piotrzkowski, I. Grzegory, S. Porowski, E. Kaminska, and J. C. Chervin, Appl. Phys. Lett. **68**, 1114 (1996).

[33] V. A. Tyagai, A. M. Evstigneev, A. N. Krasiko, A. F. Andreeva, and V. Ya. Malakhov, Sov. Phys. Semicond. **11**, 1257 (1977).

[34] T. Inushima, T. Shiraishi, and V. Yu. Davydov, Solid State Commun. **110**, 491 (1999).

[35] J. Wu, W. Walukiewicz, W. Shan, K. M. Yu, J. W. Ager III, E. E. Haller, H. Lu, and W. J. Schaff, Phy. Rev. B **66**, 201403 (2002).

[36] K. Osamura, S. Naka, and Y. Murakami, J. Appl. Phys. **46**, 3432 (1975).

[37] N. Wieser, O. Ambacher, H.-P. Felsl, L. Görgens, and M. Stutzmann, Appl. Phys. Lett. **74**, 3981 (1999).

[38] S. Yu, K. W. Kim, L. Bergman, M. Dutta, M. Stroscio, and J. M. Zavada, Phys. Rev. B **58**, 15283 (1998).

[39] F. DemangeoT, J. Frandon, M. A. Renucci, O. Briot, B. Gil and R. L. Aulombard, Solid State Commun. **100**, 207 (1996).

[40] F. Demangeot, J. Frandon, M. A. Renucci, H. Sands, D. Batchelder, S. Ruffenach-Clur, O. Briot, and B. Gil, MRS Internet J. Nitride Semicond. Res. **3**, 52 (1998).

[41] A. Kasic, M. Schubert, J. Off, and F. Scholz, Appl. Phys. Lett. **78**, 1526 (2001).

[42] A. Kasic, M. Schubert, B. Rheinländer, J. Off, F. Scholz, and C. M. Herzinger, Mat. Res. Soc. Symp. **639**, G6.13 (2001).

[43] V. M. Naik, W. H. Weber, D. Uy, D. Haddad, R. Naik, Y. V. Danylyuk, M. J. Lukitsch, G. W. Auner, and L. Rimai, Appl. Phys. Lett. **79**, 2019 (2001).

[44] V. Darakchieva, P. P. Paskov, T. Paskova, J. Birch, S. Tungasmita, and B. Monemar, Appl. Phys. Lett. **80**, 2302 (2002).

[45] A. Kasic, M. Schubert, J. Off, F. Scholz, S. Einfeldt, and D. Hommel, phys. stat. sol. (b) **234**, 970 (2002).

[46] M. Schubert, A. Kasic, J. Šik, S. Einfeldt, D. Hommel, V. Härle, J. Off, F. Scholz, Mat. Sci. & Eng. B **82**, 178 (2001).

[47] M. Schubert, A. Kasic, T. E. Tiwald, J. A. Woollam, V. Härle, F. Scholz, MRS Internet J. Nitride Semicond. Res. **5**, W11.39 (2000).

[48] A. Kasic, M. Schubert, S. Einfeldt, and D. Hommel, Vib. Spectrosc. **29**, 121 (2002).

[49] M. Schubert, A. Kasic, S. Einfeldt, D. Hommel, U. Köhler, D. J. As, J. Off, B. Kuhn, F. Scholz, and J. A. Woollam, phys. stat. sol.(a) **228**, 437 (2001).

[50] M. Schubert, A. Kasic, S. Figge, M. Diesselberg, S. Einfeldt, D. Hommel, U. Köhler, D. J. As, J. Off, B. Kuhn, F. Scholz, J. A. Woollam, and C. M. Herzinger, Proc. SPIE Vol. **4449**, 58 (2001).

phys. stat. sol. (c) **0**, No. 6, 1770–1782 (2003) / **DOI** 10.1002/pssc.200303121

Mg in GaN: the structure of the acceptor and the electrical activity

H. Alves[*, 1], **F. Leiter**[1], **D. Pfisterer**[1], **D. M. Hofmann**[1], **B. K. Meyer**[1], **S. Einfeld**[2], **H. Heinke**[2], and **D. Hommel**[2]

[1] I. Physikalisches Institut, Universität Giessen, Heinrich-Buff-Ring 16, 35392 Giessen, Germany
[2] Institut für Festkörperphysik, Bereich Halbleiterepitaxie, Universität Bremen, Kufsteinerstr. NW1, 28359 Bremen, Germany

Received 4 March 2003, revised 5 May 2003, accepted 12 May 2003
Published online 28 August 2003

PACS 61.18.Fs, 71.55.Eq, 76.70.Hb, 78.55.Cr

We investigated GaN and AlGaN grown by metal organic chemical vapour deposition (MOCVD) and molecular beam epitaxy (MBE) doped with different Mg concentrations by photoluminescence, Hall-effect and SIMS measurements. Optically detected magnetic resonance (ODMR) experiments were used to study the structure of the Mg-acceptors. The ODMR experiments reveal that the spin-density of the hole is equally distributed on the four nearst Nitrogen neighbors. Structural variations observed for Mg in MOCVD material are explained in terms of crystalfield variations caused by potential fluctuations due to the presence of compensating donors. Especially, for highly doped MOCVD samples saturation of the free hole concentration and a strong 2.9 eV recombination are observed. Considering that compensating donors like nitrogen vacancies, hydrogen and complexes of both form during growth, we are able to calculate the experimental determined free hole concentrations. In MOCVD samples codoped with Mg and Si a decrease of the acceptor binding energy, with increasing donor concentration was observed. The effect is caused by Coulomb interaction between the ionized acceptors and the free holes.

1 Introduction

The success of GaN for opto-electronic applications is mostly based on the ability to dope the material p-type conductive by Mg. So far Mg is the only acceptor of choice to achieve free hole concentrations sufficient for p/n junction applications. These unique properties make it worth to investigate its structural and electrical properties in detail. Especially, as the desired linear relation between the free hole concentration and the concentration of incorporated Mg is rarely observed in most of the GaN materials. Passivation and compensation effects complicate the situation. However, the question is raised whether the structure of Mg-acceptors itself is the same in the different materials, or whether structural changes cause variations of the electrical activity.

The first part of this paper deals with the structure analysis of the Mg acceptors in GaN and AlGaN by means of magnetic resonance spectroscopy. Our systematic ODMR investigation on the Mg-acceptors in various GaN samples grown by MBE and MOCVD, in combination with photoluminescence, Hall-effect and SIMS measurements, give strong evidence that the observed variations are caused by potential fluctuations due to the presence of compensating donors in the material.

[*] Corresponding author: e-mail: detlev.m.hofmann@exp1.physik.uni-giessen.de, Phone: +49 641 9933105,
  Fax: +49 641 9933119

In the second part of the paper we use the experimental data on the properties of the acceptors and compensating donors to calculate the free hole concentrations as determined by electrical measurements such as the Hall effect. This gives evidence whether the nature and mechanisms responsible for the compensation are basically understood. Considering that compensating donors like nitrogen vacancies, hydrogen and complexes of both form during growth, we are able to calculate the experimental determined free hole concentrations, for MBE as well as MOCVD grown GaN. In MOCVD samples codoped with Mg and Si a decrease of the acceptor binding energy, with increasing donor concentrations was observed. The effect is caused by Coulomb interaction between the ionized acceptors and the free holes.

2   On the structure of Mg acceptors in GaN and AlGaN

Considering the energy position and lineshape of the 3.27 eV DAP recombination with its LO-phonon replicas and the position of the acceptor bound exciton, the Mg-acceptors show properties rather typical for effective mass acceptors. Effective-mass-theory (EMT) predicts an acceptor binding energy of about 160 meV, which is about 40 meV lower compared to the general accepted value for Mg, the difference is typically attributed to central-cell corrections. Information on the acceptor structure can be obtained by investigating the localisation of the acceptor-hole wavefunction by magnetic resonance- or Zeeman-spectroscopy. The size and symmetry of the Zeeman splitting of the magnetic sublevels, expressed in terms of g-values, gives information on the distribution of the hole wavefunction within the lattice. Taking the GaN valence band structure into account, a simple EMT estimate predicts $g_\parallel \geq 2.00....4.00$ and $g_\perp \equiv 0$ for shallow acceptors [1] (g-values parallel ($\parallel$) and perpendicular ($\perp$) to the crystallographic c-axis). This strong anisotropy of the g-values is caused by the fact that in the EMT model the wavefunction of the hole is localised in one bond direction from the Mg to one of its nearest N-neighbors (for a schematic demonstration the model is shown in Fig. 1a).

In contrast this model the g-values obtained by magnetic resonance and Zeeman-splitting experiments on the Mg bound exciton line are substantially different [2]. Both methods report on g-values rather close to 2.00 without the extreme anisotropy predicted from the effective mass model.

Kunzer and Kaufmann et al. [3, 4] reported, sample dependent, $g_\parallel = 2.067$ to 2.084 and $g_\perp = 1.990$ to 2.022. Koschnik et al. [5] and Glaser et al. [4] observed similar g-values. To date, Glaser and coworkers were able to detect the largest g-anisotropy for Mg acceptor in a lightly doped sample, namely $g_\parallel = 2.113$ and $g_\perp = 1.970$ [7].

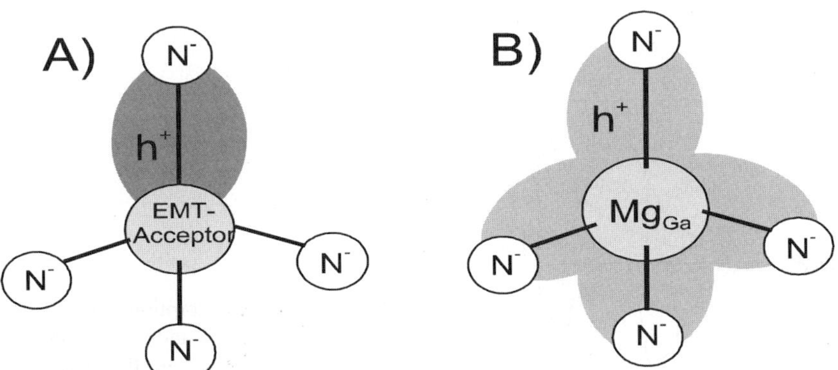

Fig. 1   Schematical illustration of the wavefunction distribution of the hole of Mg acceptors in GaN: a) estimated from a simple effective mass approach; b) model obtained from the magnetic resonance experiments.

The difference between the expected g-values of an EMT acceptor and those observed for Mg in GaN, i.e. $g_\parallel$ and $g_\perp$ close to two and $g_\parallel > g_\perp$, was explained in a model previously developed for self compensating centres in II–VI compounds [8]. For such deep acceptors the hole can be localised in the p-shell of one nearest neighbour-ligands, for the case of Mg in GaN the nitrogen p-shell, this results in $g_\parallel$ and $g_\perp \approx 2$, but $g_\parallel < g_\perp$. The g-factor anisotropy is reversed when the hole is shared among the four N-ligands (Fig. 1b). This model explains the experimental observed order of $g\parallel$ and $g_\perp$ for the Mg-acceptors. However, taking the experimental uncertainties into account, the different groups observe different g-values for the Mg-acceptor in GaN which demands an explanation, because it could be speculated that each set of g-values might reflect a different local structure of the Mg-acceptors, e.g. hydrogen or other species in different positions in the vicinity of Mg, which modifies the distribution of the hole wavefunction.

2.1 Experimental results and discussion

In most of the investigations, the magnetic resonance of the Mg-acceptors was detected on the 3.27 eV DAP recombination or on the broad 2.8 eV–3.2 eV "blue" emission band [3–7]. We were able to show that also the magnetic circular dichroism (MCD) of the acceptor bound exciton absorption:

$$Mg^o + h\nu \rightarrow Mg^oX \tag{1}$$

and the infrared photo-ionisation transition:

$$Mg^o + h\nu \rightarrow Mg^- \tag{2}$$

serve as a sensitive detection channels for the magnetic resonance [9, 10].
Here we give further evidence which settle the assignment of the MCD to the above given two transitions. For transition (1) one expects in $Ga_{1-x}Al_xN$ samples that the Mg bound exciton roughly follow the dependence of the energy gap as a function of the composition x. This can be seen in Fig. 2, with increasing x the bandgap absorption shifts to higher energies (Fig. 2a). Figure 2b shows that the Mg^0 acceptor bound exciton related MCD follows the bandgap absorption to higher energies.

It is worth to note that for higher compositions $x > 0.12$ no Mg^0 related MCD could be detected, neither on the excitonic transition nor on the photo-ionisation transition despite the fact that Mg is present in the samples in concentrations of about $3 \times 10^{19}\,cm^{-3}$ (determined by SIMS).

According to Eq. (2) the intensity of the infrared photo ionisation transition (inset in Fig. 3) should be proportional to the concentration of the number of uncompensated Mg acceptors, thus to the free hole concentration. We show this for a series of GaN:Mg samples grown by MBE (Fig. 3). The MCD intensity shows a linear increase with the the hole concentration determined by Hall measurements. These results further ensure that we detect in the ODMR experiment the species which is responsible for the p-type conductivity of the samples.

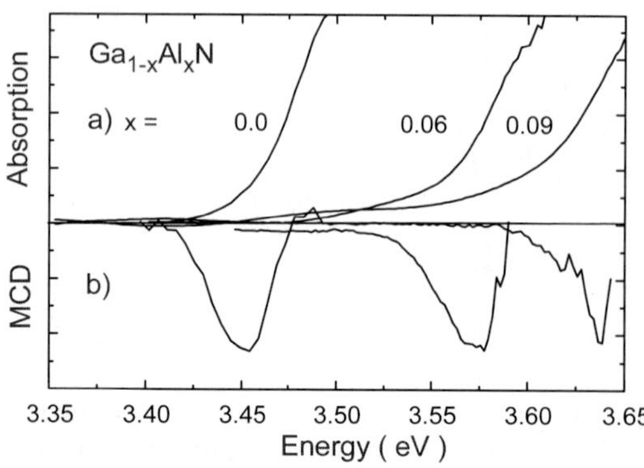

Fig. 2 a) Near band edge absorption of the $Ga_{1-x}Al_xN$ layers for Al contents of 0 %, 6 % and 9 %; b) magnetic circular dichroism (MCD) attributed to the Mg-acceptor bound exciton absorption.

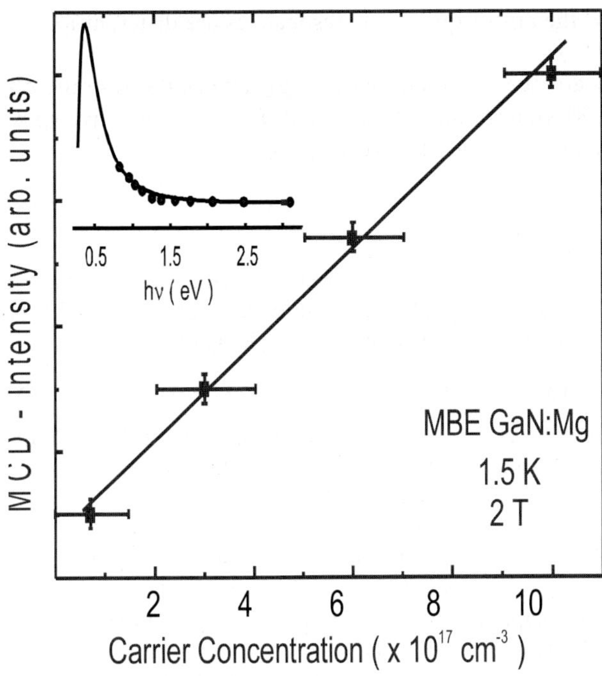

Fig. 3 Intensity dependence of the magnetic circular dichroism (MCD) of the infrared hole ionisation transition of the Mg-acceptors on the free hole concentration (determined by room temperature Hall measurements) for four different MBE grown GaN samples. The inset shows the spectral dependence of the hole ionisation transition.

The magnetic resonances detected on the MCD signals show a typical anisotropy upon rotation of the crystal with respect to the magnetic field (Fig. 4). It is described by

$$g(\theta) = (g_\parallel^2\cos^2(\theta) + g_\perp^2\sin^2(\theta))^{1/2} \tag{3}$$

The $g_\parallel$ and $g_\perp$ are obtained from this plot for the resonance positions at $0°$ and $90°$, respectively. Figure 4 shows the angular dependence of the g-values as a function of the sample orientation θ for the extreme cases. Curve a is observed in a lightly doped MOCVD sample (1×10^{19} cm^{-3}) and has a large anisotropy, while curve b is observed for a higher doping concentration in an MOCVD sample (3×10^{19} cm^{-3}) and is

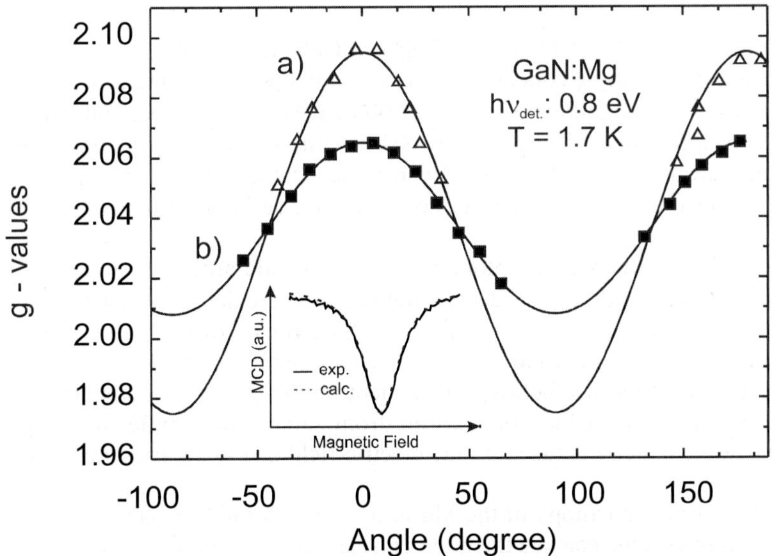

Fig. 4 Variation of the Mg-acceptor g-values in GaN with the crystal orientation ($0°$: $\parallel$ c-axis, $90°$: $\perp$ c-axis) for: (a) a lightly doped MOCVD sample; and (b) a highly doped sample. The inset shows the Lorenzain lineshape (calc.) of the magnetic resonance (exp.).

also typical for most of our MBE samples. Clearly, the anisotropy of the resonances are different for the two cases.

To explain the g-values, i.e. that both $g_{\parallel}$ and $g_{\perp}$ are near two and that $g_{\parallel} > g_{\perp}$, one starts from the p^5-configuration of the hole bound to the acceptor [3, 8]. In a trigonal crystal field δ and for a positive spin-orbit interaction ζ ($\delta/\zeta > 0$), the hole is localised in a Γ_4 state and the g-values are given by

$$g_{\parallel} = k-(k-2)\cos\alpha, \text{ and } g_{\perp} = 1+ (2)^{1/2}k\sin\alpha + \cos\alpha \tag{4}$$

with $\alpha = \arctan((2)^{1/2}/(\delta + 1/2))$, k is a covalency factor.

This gives g-values $g_{\parallel}$ and $g_{\perp} \sim 2$, but $g_{\perp} > g_{\parallel}$. To solve this discrepancy to the experimental observation one has to assume that the hole is not localised in one of the Mg–N bonds, but equally present in all four nearest neighbour bonds, according to the Freiburg group geometrical averaging yields [4]

$$g_{\parallel c} = 1/4\, g_{\parallel} + 1/4\, (g_{\parallel}^2 + 8g_{\perp}^2)^{1/2} \text{ and } g_{\perp c} = 1/4\, g_{\perp} + 1/4\, (g_{\perp}^2 + 8g_{\parallel}^2)^{1/2} \tag{5}$$

Such situation was first observed for A-centres (vacancy–donor complexes) in wide bandgap II–VI compounds which are deep acceptors [8]. Dependent on temperature A-centres were observed in two configurations. At low temperatures the hole was localised in one bond, whereas at temperatures of about 77 K the configuration was observed in which the hole is equally distributed over all bonds to the host lattice neighbours (three neighbours in the case of A-centres, one of the four neighbouring lattice sites is occupied by the donor element). The hopping motion of the A-centre hole in the four bonds to the lattice neighbors was studied by an analysis of the EPR lineshape and hopping frequencies on the order of 10^8 to 10^{10} Hz were obtained for temperatures between 77 K and 120 K [11].

For the Mg-acceptors in GaN the low temperature configuration is not observed, which might be explained by its lower binding energy compared to the A-centres (~ 0.5 eV).

The evidence that this model properly describes the hole properties of Mg in GaN is given by its correct prediction of the $g_{\parallel}$ and $g_{\perp}$-values, further supporting evidence is obtained by the analysis of the lineshape of the MCD detected ODMR signals. In semicondutors with high abundant host-isotopes with nuclear spin, like GaN, the typical lineshape of ground state magnetic resonance signals is of Gaussian shape, due to the unresolved interactions of the paramagnetic species (hole or electron) with the host nuclei (inhomogenous broadening). For systems which are determined by the lifetime of the states, like in excited states or for hopping processes, one can expect Lorenzian lineshapes. The inset in Fig. 4 shows that the ODMR is very well described by a Lorenzian. Thus the halfwidth of the resonance can be taken as a measure of the hopping frequency which is on the order of 10^9 Hz. It should be noted that the Mg-Signals detected by conventional EPR do not show pure Lorentzian lineshapes [13], it demands further explanation.

Table 1 summarizes our results on various MOCVD and MBE grown GaN:Mg samples together with the information on the Mg-content and the free hole concentration at room temperature. One can see that for the MOCVD samples $g_{\parallel}$-values from 2.10 to 2.06 and $g_{\perp}$ from 2.00 to 1.99 are observed, and neither a clear correlation to the total Mg concentration as determined by SIMS (third column) nor a correlation to the free hole concentration (Hall measurement, RT) exists. Even more striking, in all MBE samples very similar $g_{\parallel}$- and $g_{\perp}$-values are obtained although the carrier concentration varies over more than one order of magnitude.

From the above given structure model for the Mg-acceptors in GaN one would attribute variations in the g-values to variations in the crystalfield δ. Additional electric fields can be created in the sample due to the presence of the charged donors and acceptors (compensation). An estimate of this fields can be obtained from energy position of the "blue" emission band, caused by the recombination between compensating donors and the shallow Mg acceptors (see below). Dependent on the degree of compensation its peak energy, measured under low excitation conditions, varies from sample to sample. It gives a measure of the potential fuctuations in the material caused by the electric field due to the presence of the charged donors and acceptors [12].

Figure 5 shows the variation of the g-value anisotopy of the Mg-acceptors in MOCVD GaN versus the peak energy of the blue band in our samples. One can see the good correlation which makes it very likely

Table 1 Free hole-concentrations at room temperature (Hall-effect), total Mg-concentration (secondary ion massspectroscopy), and the g-values of the Mg-acceptors in various MOCVD and MBE grown GaN samples.

sample No.:	hole-concentration $[10^{17}\,\mathrm{cm}^{-3}]$	Mg-concentration $[10^{19}\,\mathrm{cm}^{-3}]$	Mg-acceptor g-values $g_{\parallel}$; $g_{\perp}$
MOCVD			
#1	< 1	1	2.10 ; 1.94
#2	3.6	5	2.10 ; 1.96
#3	4.9	2.6	2.09 ; 1.97
#4	6.9	1.9	2.08 ; 1.99
#5	12.1	2	2.09 ; 1.96
#6	1.,1	5	2.08 ; 1.98
MBE			
#9	10	30	2.06 ; 1.99
#10	6	21	2.05 ; 1.985
#11	3	10	2.06 ; 1.99
#12	0.7	9	2.06 ; 1.99

that the potential fluctuation are the cause of the observed variation of the g-values. In a recent investigation Glaser et al. [13] have shown that different g-values could be measured in one sample dependent on the detection wavelength within the blue band, this results also supports that the potential fluctuations are the cause of the variation of the Mg-acceptor g-values.

In MBE grown GaN the compensation ratio is much smaller, compared to the Mg doped films grown by MOCVD. It explains why we do not observe variations in the Mg-acceptor g-values in MBE material. However, form a first view one would expect then that low compensated MOCVD- and MBE-samples give similar g-values for the Mg-acceptors. But the MOCVD GaN layers are grown on sapphire with buffer layers and are typically compressively strained. This is not the case for MBE grown layers where

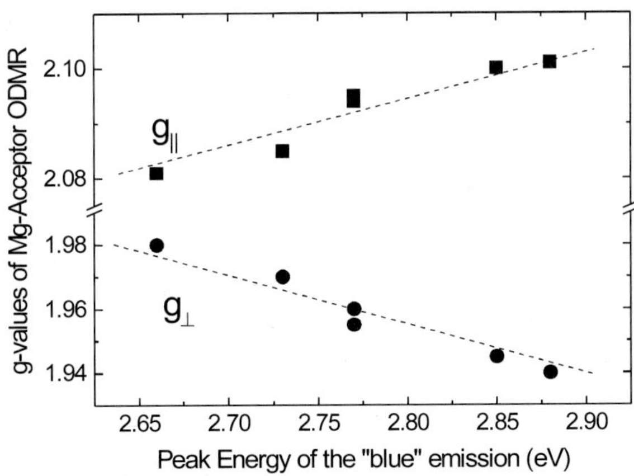

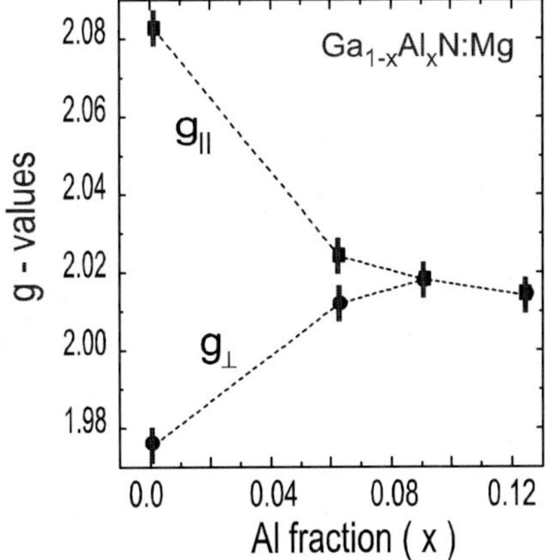

Fig. 5 (left) The variation of the $g_{\parallel}$ and $g_{\perp}$-values of Mg-acceptors in MOCVD grown GaN samples measured by MCD detected ODMR in relation to the peak position of the "blue"-emission caused by compensating donors.

Fig. 6 (right) Variation of the $g_{\parallel}$ and $g_{\perp}$-values of Mg-acceptors in $Ga_{1-x}Al_xN$ for Al fractions up to 12 %.

the concepts of buffer layers, which reduce the total dislocation density, are more difficult due be realized. Thus the strain field conditions for samples grown by the two methods are quite different.

Following the g-values of the Mg acceptors in $Ga_{1-x}Al_xN$ layers with increasing Al fraction x we find that its anisotropy becomes even smaller than in the GaN samples (Fig. 6). For an Al content of 9 % the resonances become isotropic. It can be taken as evidence that Mg-acceptors in $Ga_{1-x}Al_xN$ becomes even more "deep-center-like" than in GaN, this is very likely to be accompanied by an increasing binding energy. Above 12 % Al content the ODMR became undetectable due to the vanishing MCD signal. Obviously, it is a result of increasing compensation effects in the alloy which prevent the observation of the Mg^0-related transitions.

The structural properties of the Mg-acceptors in GaN can meanwhile be regarded as rather well understood and agree resonable well to theoretical calculations [14]. The experimental information on the structure, or even the chemical nature, of the compensating deep donors is less complete [15, 16]. From theoretical as well as experimental considerations most likely Nitrogen vacancies (V_N) or complexes of it with Mg are the dominant compensating centers. However, its clear identification demands further work.

3 Compensation in GaN:Mg

To study the compensation mechanism in GaN:Mg a series of MOCVD and MBE grown samples were investigated. The comparison between these two growth methods is of interest because there are important differences: In MOCVD the growth temperatures are typically around 1300 K and hydrogen is present in high concentrations originating from the precursors. As a result the as-grown samples are H-compensated and post growth treatments are necessary to activate the Mg-acceptors. For the MBE films the growth temperatures are significantly lower, around 1000 K, and they can be regarded as almost hydrogen free, which means that no post growth treatments are needed for the acceptor activation.

In Fig. 7 we see the typical behaviour of the free hole concentration (Hall measurements) versus the Mg concentration (SIMS measurement) for both growth methods. For Mg concentrations up to 1×10^{19} cm^{-3}, the free carrier concentration increases for MOCVD as well as for MBE grown samples. It roughly follows the linear increase expected from Fermi statistics. The scattering of the data in this Mg concentration range is most likely caused by different concentrations of residual shallow donors such as O and Si, which also contribute to reduce the free hole concentrations. The typical concentrations of such species in GaN are about 1×10^{17} cm^{-3}.

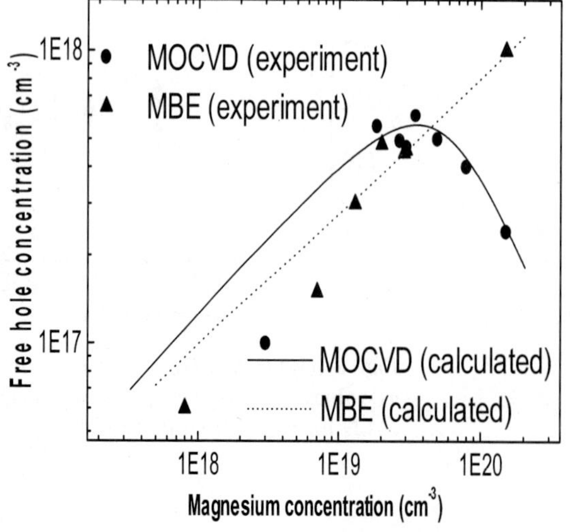

Fig. 7 Triangles and circles corresponds to the free hole concentration versus the magnesium concentration, for MBE and MOCVD respectively. Solid line shows the calculated data for MOCVD and the dotted line the calculated data for MBE.

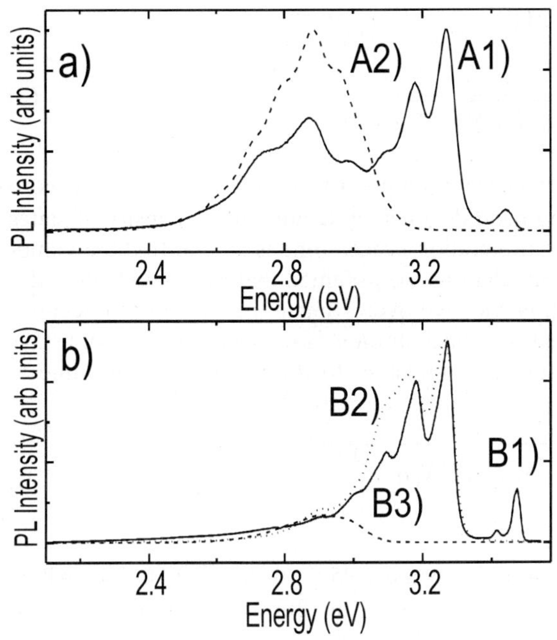

Fig. 8 a) PL of MOCVD grown GaN:Mg: A1) low doped sample, A2) high doped sample. b) PL of MBE grown GaN:Mg: B1) low doped sample, B2) high doped sample and B3) line shape of the 2.9 eV recombination.

For the MOCVD samples with Mg concentrations above 1×10^{19} cm^{-3} first saturation and for even higher concentrations a decrease of the free hole concentration is observed. This is different to the MBE case where the hole concentration continuously increases.

The PL emission of low-doped MOCVD and MBE films is dominated by the shallow donor to Mg acceptor recombination (violet band) positioned at 3.27 eV (Fig. 8).

By increasing the doping concentrations it is observed that the blue band positioned around 2.9 eV develops. For the highly MOCVD doped samples this band, which involves recombinations between the Mg-acceptor and deep compensating donors, dominates the spectrum. In the case of MBE this recombination is also observed but it never dominates. These results indicate that the deep donors necessary for the blue luminescence should be present in higher concentrations in the MOCVD samples.

Theoretical and experimental works gave evidence that the deep donors are caused by the formation of Nitrogen vacancy centers, i.e. self- or intrinsic-compensation happens [17–20].

3.1 Calculation of the free hole concentration vs. Mg concentration

Our goal is know to calculate the free hole concentration as a function of Mg incorporation. Therefore we consider that nitrogen vacancies (V_N), hydrogen (H) and complexes of both (V_NH) are the relevant compensating defects to be formed in p-type GaN. Based on first-principle-calculations the formation of these defects was first proposed by Van der Walle [19]. We also assume that Mg and H form as separated centres. Only during cool down complexes of both are formed. Neugebauer et al. [20] showed that the formation of Mg-H complexes during growth is of minor relevance. Moreover for the statistics there is no difference if Mg and H are separated centres or complexes because for p-type GaN hydrogen acts as a donor and in both configurations it compensates Mg.

The density of compensating defects can be calculated by the following equation:

$$N_D = N_s \exp\left[\frac{-E_{For}}{k_B T}\right] \tag{6}$$

where N_s is the number lattice sites for a defect (4.4×10^{22} cm^{-3} for substitutional defects in GaN), T the growth temperature, k_B the Boltzmann constant and E_{For} the donor formation energy according to Fig. 9. Although these formation energies were basically extracted from [19], we have slightly modified them ($\sim$200meV) to obtain the best agreement with our experimental data. Entropy contributions were shown to be small and are not considered here [20].

The Fermi level (E_F) during growth is calculated by [21]:

$$E_F = -k_B T \times \ln \frac{2\beta^{-1}(N_A - N_D)}{\beta^{-1}N_V + N_D\alpha + \sqrt{(\beta^{-1}N_V + N_D\alpha)^2 + 4\beta^{-1}N_V(N_A - N_D)\alpha}},$$ (7)

where α is $\exp(E_A/k_B T)$ with E_A the acceptor binding energy (for simplicity we have assumed E_A as constant and equal to 160 meV, see below, β is the valence band degeneracy factor, N_V the density of states for the valence band N_A the acceptor concentration. By an iteration process one is able to find the values for N_D and E_F, which satisfy both equations. For the calculations we assumed that residual shallow donors like O and Si are not present. By ignoring their existence we overestimate the free hole concentrations for low Mg concentrations. We further assume that the concentration of compensating donors stays constant during the cool down of the sample from growth temperature to RT. Finally, we are able to calculate the free hole concentration as a function of the Mg incorporation by [21]

$$p = -\frac{N_D + \beta^{-1}N_V\alpha^{-1}}{2} + \sqrt{\beta^{-1}N_V\alpha^{-1}(N_A - N_D) + \left(\frac{N_D + \beta^{-1}N_V\alpha^{-1}}{2}\right)^2}$$ (8)

Typical values for β and N_V are 3.6 and $3.2 \times 10^{19}\,\mathrm{cm}^{-3}$ at RT, 4.6 and $1.95 \times 10^{20}\,\mathrm{cm}^{-3}$ at 1000 K, which is the growth temperature for MBE samples and 5.1, and $2.1 \times 10^{20}\,\mathrm{cm}^{-3}$ at 1300 K, the growth temperature for MOCVD.

In the calculations the information of growth technique enters in two different ways. The first one is the growth temperature and the second one is the presence of hydrogen. For MBE we assume that the hydrogen presence can be neglected thus only the formation of the nitrogen vacancy is considered. For MOCVD and in the presence of hydrogen the three defects, V_N, H and V_NH are considered. The low formation energy of the hydrogen centres (Fig. 9) plays a central rule. The calculations yield that identical amounts of Mg and H centres are present in as-grown MOCVD GaN:Mg. Thus p-type conductivity can be only reached when the hydrogen is removed which is done experimentally by a postgrowth annealing treatment. For the calculations we assumed that all the H^+ centres are removed and that the $V_N H^{2+}$ centres are transformed to the simple nitrogen vacancy defects. The drawn lines in Fig. 7 show the calculated data. A good qualitative agreement between the calculated and the experimental data for both MBE and MOCVD samples is obtained.

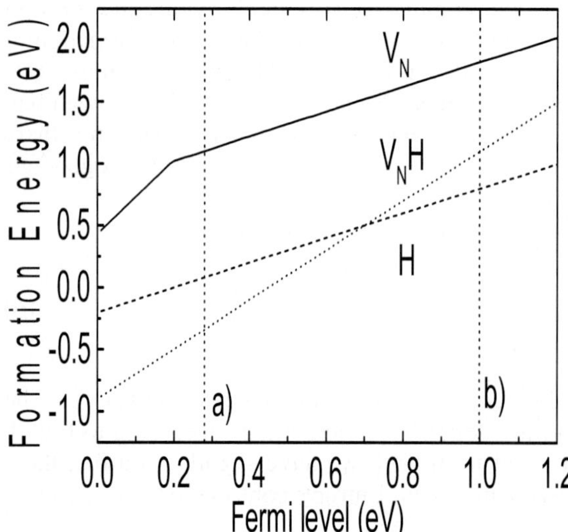

Fig. 9 Calculated formation energies for the three considered compensating donors. Lines a) and b) show the Fermi level position during growth (Mg concentration of $5 \times 10^{19}\,\mathrm{cm}^{-3}$) for MBE and MOCVD respectively

An important point that we can state from these calculations is that the dominant compensating centre for both growth techniques should be the nitrogen vacancy. This is supported by PL. The blue band is observed in MBE and MOCVD material at 2.9 eV. The red shift of the blue band, observed on some highly doped MOCVD grown samples, is related to band fluctuations, which modify the peak position of the emission but do not change the structure of the involved defects in our samples. This situation can be different in other samples where more than one compensating donor is observed.

The calculations show also that the compensation ratio, i.e. the ratio between the concentrations of compensating donors and magnesium acceptors, is almost constant for the MBE samples (<5%). For the highly doped MOCVD samples the compensation can reach up to 20% that is sufficient to account for band fluctuations.

For the two growth methods the formation process of the V_N defect is different. In MBE it is directly formed, and in MOCVD it forms in addition by the intermediate V_NH complex. In fact the H incorporation and V_NH formation are essential to reproduce the strong compensation effect for the highly doped MOCVD samples. Assuming simply a higher growth temperature we could not reproduce this effect.

Several authors [22, 23] reported that more than one compensating centre exits in MOCVD grown GaN:Mg. These centres could be related with remaining V_NH complexes or also, and especially at very high doping levels, with the V_NMg complex suggested by Kaufmann [17, 18].

It also seems that the maximal achievable free hole concentration should not exceed 1×10^{18} cm^{-3}. In the case of MOCVD the formation of V_N limits the free hole concentration and in the case of MBE the Mg solubility limits the concentration of incorporated Mg. To achieve higher values of free hole concentrations higher Mg solubility and lower values of acceptor binding energy would be desirable.

3.2 Codoping of GaN:Mg with Si

Recently it has been suggested that by introducing controlled amounts of donors during the growth of p-type samples (codoping technique), first higher values of solubility and secondly lower acceptor binding energies can be achieved [23]. The increase in solubility can be understood considering that the formation energy (E_{For}) of an electrically charged centre is given by

$$E_{For} = K \pm qE_F \tag{9}$$

where K is a constant, q the electrical charge of the centre and E_F is the Fermi level. In the case of an acceptor q is assumed to be negative. This means that the incorporation of controlled amounts of donors during the growth of a p-type sample raises the Fermi level and decreases the acceptor formation energy. As a result higher amounts of acceptors can be incorporated. The reduction of the acceptor binding energy is not yet well understood. Early models [25, 26] suggest it should be proportional to the concentration of ionised acceptors powered to one third ($Na_i^{1/3}$). More recently Katayama–Yoshida et at [24] suggested the formation of complex-centres involving two acceptors and one donor. Such centre acts effectively as an single acceptor but has a lower binding energy.

To study these effects MOCVD grown samples doped simultaneously with Si and Mg (codoping) were measured by PL and Hall. Fig 10 shows the typical PL spectra of these samples. The spectrum a) shows a sample where $1 \times 10^{19} > N_A > N_D$ cm^{-3}. The spectrum b) shows a sample where the doping concentrations are $N_A = 1.8 \times 10^{20}$ and $N_D = 1.2 \times 10^{20}$ cm^{-3}. The spectrum a) is similar to the simply low/medium Mg doped samples, where the 3.27 eV structure dominates the spectrum. For high codoping concentrations (spectrum b) the PL is still dominated by the UV band but now the luminescence appears to be red shifted (≈ 3.1 eV) and its phonon structure is no longer observed. Its position is also strongly dependent on the excitation power and shifts up to 120 meV can be obtained. This indicates the existence of strong band fluctuations. The blue luminescence at around 2.9 eV is also observed but for these samples it never dominates the PL spectrum, indicating the reduced formation of V_N-type of defects. The Hall effect data of our samples are summarized in Table 2. Figure 11 shows the acceptor binding energy versus the donor concentration N_D. The dependence is explained by

$$E_A = E_{A0} - \alpha N_D^{1/3} \tag{10}$$

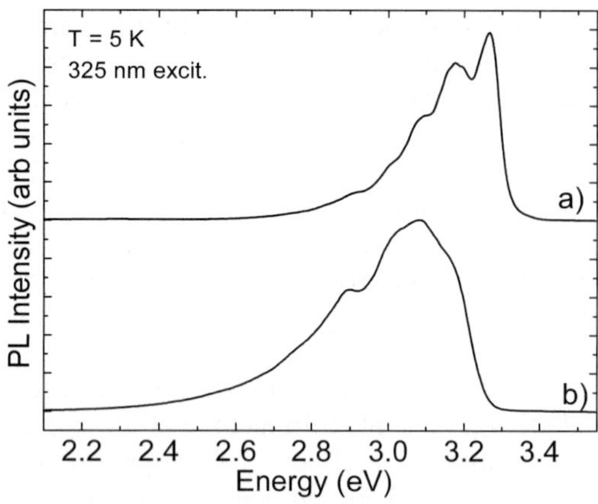

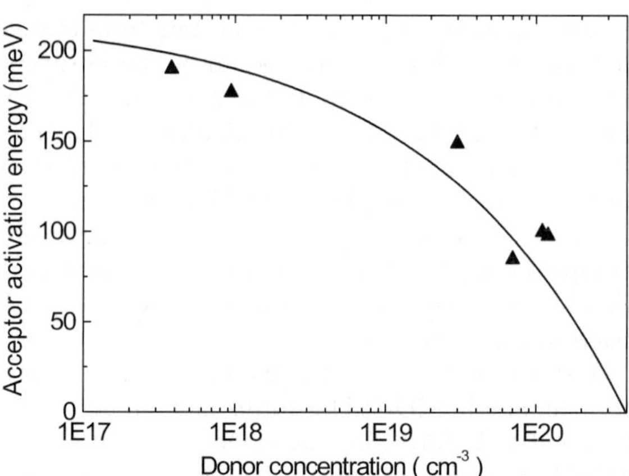

Fig. 10 (left) Photoluminescence spectra of Si and Mg codoped GaN samples: a) „low" doped sample (F in Table 2); and b) shows a sample where the doping concentrations are N_A=1.8 × 10^{20} and N_D = 1.2 × 10^{20} cm^{-3} (B in Table 2).

Fig. 11 (right) Mg-acceptor activation energy as a function of the shallow donor concentration (Si): triangles obtained by Hall effect measurements, drawn line: calculation according to Eq. (10) (for details see text).

where E_A is the Hall measured acceptor binding energy, E_{A0} is the isolated acceptor binding energy and α is proportionality constant. The data was fitted using the values $\alpha = 2.8 \times 10^{-5} \pm 0.2 \times 10^{-5}$ meVcm^{-1} and E_{A0} =220 ± 20 meV. The equation considers that a free hole in the valence band is influenced by the coulomb field produced by the ionised acceptors (N_{Ai}). The average space separation is proportional to $1/N_{Ai}^{1/3}$, so the average increase in energy for one free hole is proportional to $(4\pi\varepsilon)^{-1} \times N_{Ai}^{1/3}$, where ε is the material dielectric constant. This means that the distance between the bottom of the valence band and the acceptor level will be reduced by the factor $\alpha N_{Ai}^{1/3}$, where $N_{Ai} = N_D + p$. For low codoping concentrations generally p is dominant and so $N_{Ai} \approx p$. On the other hand when the codoping concentrations are high enough N_D is dominant and so $N_{Ai} \approx N_D$. This is the approximation used to obtain Eq. (10). A description of the experimental data is obtained (drawn line in Fig. 4) even for the low-codoping conditions (samples E and F in Table 2), where the concentrations for the free holes (p) and shallow donors N_D are similar. For the calculations performed to explain the data in Fig. 7, we have for simplicity assumed E_A to be constant. For the range in which those calculations apply the variations of E_A do not introduce significant changes in the calculations. The value of 160 meV was chose because it is the best average value to fit both MBE and MOCVD data when 10^{19} > [Mg] > 10^{20} cm^{-3} which is the range of interest.

Back to equation (10), one can see that by adding donors a reduction of the acceptor activation energy and probably an increase of the free hole concentration at room temperature (Eq. (8)) can be expected. On the other hand by adding more donors the degree of compensation will increase which produces a decrease of the free hole concentration. The question is to know which of both effects is stronger. To

Table 2 Acceptor-activation energy (E_A), shallow donor concentration (N_D), acceptor concentration (N_A), and free hole (p) concentration in Si, Mg co-doped GaN determined by Hall-effect measurements.

sample	N_A (cm^{-3})	N_D (cm^{-3})	p (cm^{-3})	E_A (meV)
A (codoped)	9.0 × 10^{19}	7 × 10^{19}	3 × 10^{16}	85
B (codoped)	1.8 × 10^{20}	1.2 × 10^{20}	5 × 10^{17}	98
C (codoped)	1.0 × 10^{20}	6 × 10^{19}	8 × 10^{16}	110
D	9.5 × 10^{19}	2.5 × 10^{19}	3 × 10^{17}	155
E	1.8 × 10^{19}	3.8 × 10^{17}	5 × 10^{17}	190
F	3.5 × 10^{19}	9.5 × 10^{17}	4 × 10^{17}	180

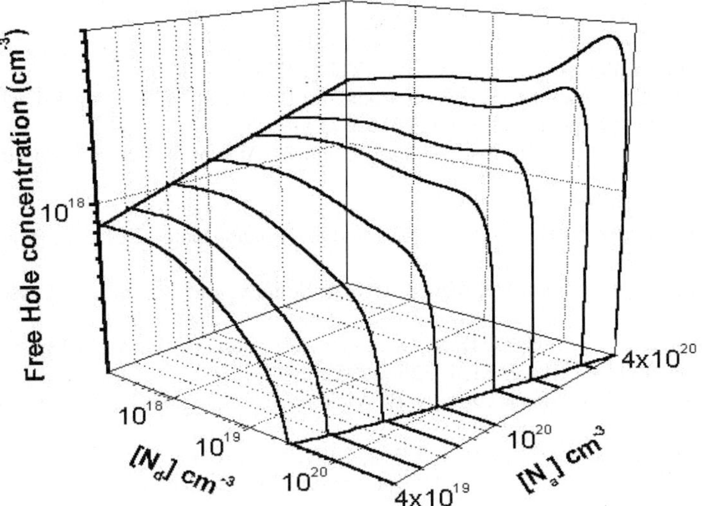

Fig. 12 Calculated dependence of the free hole concentration as a function of Mg and Si concentration. For high doping concentrations it is possible to achieve enhanced free hole concentrations by this „codoping" approach (for details see text).

calculate the free hole concentration at RT we used equation (8) and corrected the value of E_A by equation (10). The results are shown in Fig. 12. For doping concentrations lower than 10^{20} cm^{-3} the corrections given by equation (10) are not strong enough to overcome the compensation effects. This is the range in which the calculations of Fig. 7 apply. Here compensation due to shallow donors (Si or oxygen) or self-compensation (unintentional V_N) is dominant. For doping concentrations higher than 1×10^{20}cm^{-3} and in the range where $N_A > N_D > 1 \times 10^{20}$ cm^{-3} the lowering of the acceptor binding is strong enough to enhance the positive effect of codoping. In this range the free hole concentration goes through a maximum when $N_A \approx 2 \times N_D$ when Mg $= 4 \times 10^{20}$ cm^{-3} and $N_A \approx 2 \times N_D$ a free carrier concentration of 4×10^{18} cm^{-3} is obtained. This means that for high-doped samples there might be an advantage in using the codoping technique. Of course the "positive" effect of codoping on the free hole concentrations, has to be seen in relation to the crystalline quality of the samples that might be influenced by the high doping conditions. For semiconductors with a smaller dielectric constant (higher value of α) the codoping effects can be expected to be more significant.

4 Conclusions

The g-values of Mg acceptors in GaN reflect that the hole is equally distributed in the bonds to the next four N-neighbors. It is a feature typical for medium deep acceptors in wide bandgap semiconductors, but is in contrast to effective mass acceptors in cubic semiconductors. The structural variations of the Mg-acceptors observed in magnetic resonance experiments on MOCVD material are very likely caused by crystalfield variations originating from electric fields introduced by compensating donors in the material. This variations are not observed in MBE material where the compensation ratio is much lower.

The compensation effects observed in GaN:Mg grown by MOCVD and MBE can consistently be explained assuming the formation of V_N, H and V_N–H defects during the growth process. The growth temperature and the presence or absence of hydrogen are sufficient to explain the experimental data observed for the MBE and MOCVD growth samples. The formation of other defects were for simplicity, and in the lack of the formation energies, not considered here. The codoping approach can be expected to increase the free hole concentration, when $N_A \approx 2 \times N_D > 10^{20}$cm^{-3}. Although, one can expect at these high doping and compensation levels problems related to the crystalline quality of the material.

Acknowledgement We would like to thank the Deutsche Forschungsgemeinschaft for support in the frame of the "Schwerpunktsprogramm: III–V Nitride".

References

[1] R. Dingle, D. D. Sell, S. E. Stokowski, and M. Ilegems, Phys. Rev. B **4**, 1211 (1971).

[2] R. Stepniewski, A. Wysmolek, M. Potemski, J. Lusakowski, K. Korona, K. Pakula, J. M. Baranowksi, G. Martinez, P. Wyder, I. Grzegory, and S. Porowski, phys. stat. sol. (b) **210**, 373 (1998).

[3] M. Kunzer, U. Kaufmann, K. Maier, J. Schneider, N. Herres, I. Akasaki, and H. Amano, Mater. Sci. Forum **143-147**, 93 (1994)
M. Kunzer, Dissertation, Freiburg, Germany, 1995.

[4] U. Kaufmann, M. Kunzer, K. Maier, J. Schneider, N. Herres, I. Akasaki, and H. Amano, Mater. Res. Soc. Proc. **395**, 633 (1996).

[5] F. K. Koschnik, K. Michael, J.-M. Spaeth, B. Beaumont, P. Gibart, J. Orf, A. Sohmer, and F. Scholz, J. Cryst. Growth **189-190**, 561 (1998).

[6] E. R. Glaser, T. A. Kennedy, K. Doverspike, L. B. Rowland, D. K. Gaskill, J. A. Freitas, Jr., M. Asif Khan, D. T. Olson, J. N. Kuznia, and D. K. Wickenden, Phys. Rev. B **51**, 13326 (1995).

[7] E. R. Glaser, T. A. Kennedy, J. A. Freitas, Jr., B. V. Shanabrook, A. E. Wickenden, D. D. Koleske, R. L. Henry, and H. Obloh, Physica B **273/274**, 58 (1999).

[8] J. Schneider, W. C. Holton, T. L. Estle, and A. Räuber, Phys. Lett. **5**, 312 (1963).

[9] D. M. Hofmann, B. K. Meyer, F. Leiter, W. v. Förster, H. Alvez, N. Romanov, H. Amano, and I. Akasaki, Jpn. J. Appl. Phys. **38**, L1423 (1999).

[10] D. M. Hofmann, W. Burkhardt, F. Leiter, W. v. Förster, H. Alvez, A. Hofstaetter, B. K. Meyer, N. Romanov, H. Amano, and I. Akasaki, Physica B **273/274**, 43 (1999).

[11] B. Dischler, A. Räuber, and J. Schneider, phys. stat. sol. **6**, 507 (1964).

[12] A more detailed description of this phenomena is given in the article of A. Hoffmann in this volume; phys. stat. sol. (c) **0**, No. 5 (2003).

[13] E. R. Glaser, W. E. Carlos, G. C. Braga, J. A. Freitas, Jr. W. J. Moore, B. V. Shanabrook, R. L. Henry, E. A. Wickenden, and D. D. Koleske, H. Obloh, P. Kozodoy, S. P. DenBaars, and U. K. Mishra, Phys. Rev. B **65**, 085312 (2002).

[14] V. Fiorentini, F. Bernardini, A. Bosin, and D. Vanderbilt, Proc. of the 23rd Int. Conf. on the Physics in Semiconductors, edited by M. Scheffler and R. Zimmermann (World Scientific, Singapore 1996), p. 2877.

[15] M. W. Bayerl, M. S. Brand, O. Ambacher, M. Stutzmann, E. R. Glaser, R. L. Henry, A. E. Wickenden, D. D. Koleske, T. Susuki, I. Grzegory, and S. Porowski, Phys. Rev. B **63**, 125203 (2001).

[16] D. M. Hofmann, B. K. Meyer, H. Alves, F. Leiter, W. Burkhard, N. Romanov, Y. Kim, J. Krüger, and E. R. Weber, phys. stat. sol. (a) **180**, 261 (2000).

[17] U. Kaufmann, M. Kunzer, and H. Obloh, Phys. Rev. B **59**, 5561 (1999).

[18] U. Kaufmann, P. Scholtter, H. Obloh, and K. Kohler, Phys. Rev. B **62**, 10867 (2000).

[19] C. G. Van de Walle, Phys. Rev. B **56**, R10020 (1997).

[20] J. Neugebauer and C. G. Van de Walle, Mater. Res. Soc. **395**, 645 (1996).

[21] J. S. Blackemore, "Semiconductor Statistics" (Pergamon Press, London/Oxford, 1962), p. 134.

[22] L. Eckey and J. Holst, J. Cryst. Growth **189/190**, 523 (1998).

[23] M. A. Reshchikov and B. Wessels, Phys. Rev B **59**, 13176 (1998).

[24] H. Katayama-Yoshida, and N. Orida, J. Phys.: Condens. Matter **13**, 8901-8914 (2001).

[25] G. L. Pearson and J. Bardeen, Phys. Rev. **79**, 1013 (1949).

[26] P. Debye and E. M. Conwell, Phys. Rev. **93**, 693 (1954).

phys. stat. sol. (c) **0**, No. 6, 1783–1794 (2003) / **DOI** 10.1002/pssc.200303120

Local vibrational modes and compensation effects in Mg-doped GaN

A. Hoffmann[*], **A. Kaschner**, and **C. Thomsen**

Institut für Festkörperphysik, TU Berlin, Hardenbergstr. 36, 10623 Berlin, Germany

Received 4 March 2003, revised 9 April 2003, accepted 11 April 2003
Published online 28 August 2003

PACS 63.20.Pw, 71.55.Eq, 78.20.-e, 78.30.Fs

The compensation and self-compensation effects in Mg-doped GaN is studied by low-temperature photoluminescence and Raman spectroscopy using a series of samples with different Mg concentrations. Strongly doped samples are found to be highly compensated in electrical measurement. The compensation mechanism is directly related to the incorporation of Mg leading to the additional formation of three different deep donor levels. Furthermore, hydrogen forms defect complexes with Mg and compensates the acceptor states. These complexes were observed as local vibrational modes in Raman spectra in the range of 2200 cm^{-1}. The direct incorporation of Mg can be controlled by local vibrational modes in the region of GaN host phonons. Investigating the intensity dependence of the different Mg–H complexes and the LVM of activated Mg the Raman spectra give a clear direct evidence of the degree of compensation and p-conductivity.

1   Introduction

One of the major advances in semiconductor technology during the recent years was the realization of p-conductivity in GaN leading to optoelectronic devices in the blue spectral range like high-power light-emitting diodes and laser diodes. The major problem to be solved was the compensation of the incorporated Mg-acceptors which resulted in highly resistive material, even at high dopant concentrations. After the breakthrough by Amano et al. [1] using a low-energy electron beam treatment of their p-doped samples Nakamura et al. [2] showed that hydrogen plays an active role in the compensation process This problem is very similar to the p-doping in II–VI-compound semiconductors with large band gap, like e.g. ZnSe:N [3].

The incorporation of hydrogen is not the only process which leads to a compensation effect. It was found in Mg-doped GaN epilayers that the hole concentration at room temperature is considerable lower than the Mg concentration. The hole concentration as function of the Mg-content in Fig. 1 does not increase continuously, because above a Mg concentration of around 10^{19} cm^{-3} a decrease can be observed [4]. This behavior gives a clear evidence of a strong self-compensation effect. Little is known on the electronic structure of compensating defects and the effects on the optical properties of GaN. We therefore investigated a series of p- and n-type GaN samples with systematically varied concentrations of Mg to gain further insight into the electronically and optical implications of compensation in GaN.

---

[*] Corresponding author: e-mail: hoffmann@physik.tu-berlin.de, Phone: +49 30 314 22001, Fax: +49 30 314 22064

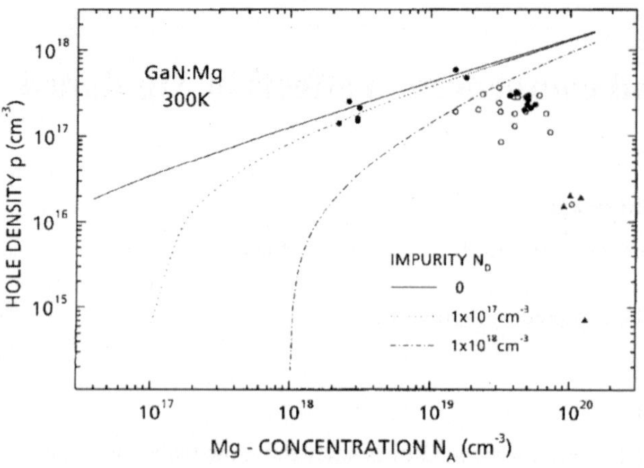

Fig. 1 Hole density in GaN:Mg determined by Hall-measurement at RT as a function of the Mg concentration after [4].

2 Self-compensation effects in GaN:Mg

To examine the self-compensation effects in GaN:Mg a series of samples grown by MOCVD was investigated. The Mg content increases with the Mg/Ga precursor ratio. The samples were annealed at 650°C in a N_2-atmosphere to passivate the H-donors. Below a precursor ratio of 0.3% the samples are p-conductive, at 0.3% they show a high-ohmic behavior which gives a clear hint of an effective compensation mechanism. Figure 2a shows Raman spectra of the samples with lowest and highest Mg-

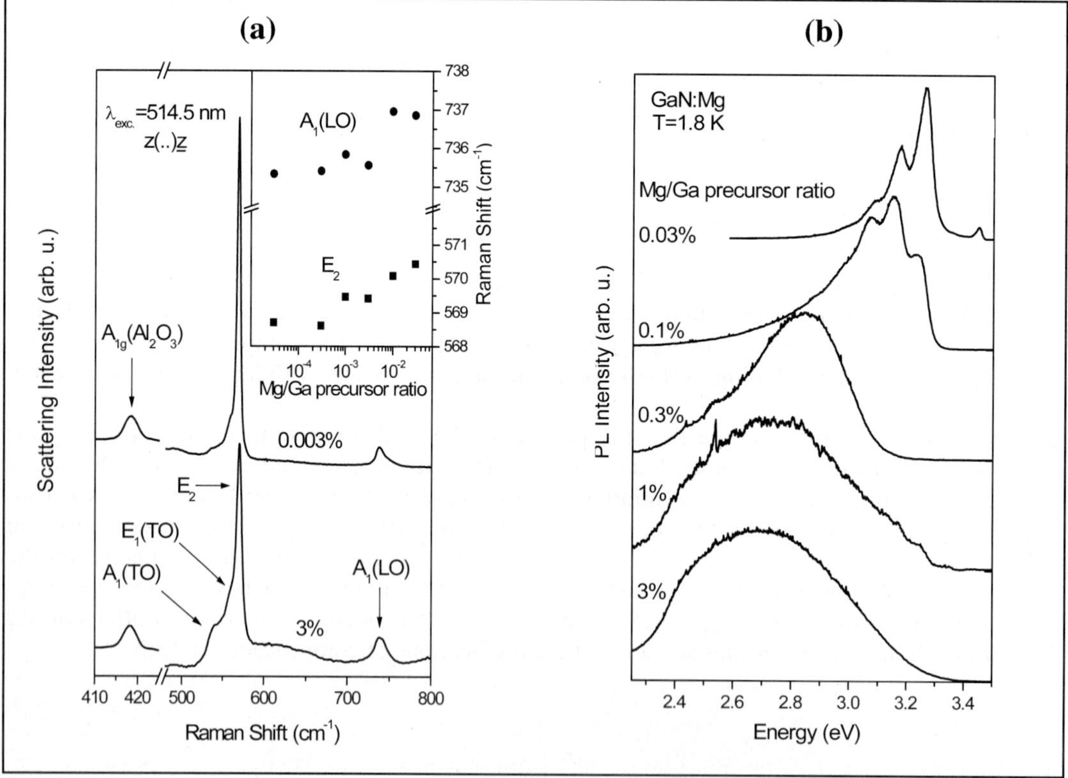

Fig. 2 a) Raman spectra of two GaN:Mg samples having different Mg content. Inset: Line position of the $A_1(LO)$- and the E_2-mode as a function of the Mg concentration. b) Photoluminescence spectra of GaN epilayers as a function of the Mg content at 1.8 K.

concentrations. A strong increase of full widths at half maximum (FWHM) of the E_2-mode from 3.7 to 7.0 cm^{-1} is seen. Additionally, an increase of the intensity of the E_1(TO) mode in the forbidden configuration $z(..)\bar{z}$ is observable. The line broadening and the breakdown of the selection rules indicates that the crystal quality decreases as a function of Mg content through the incorporation of defects. Additionally, the E_2-mode shifts to higher energies which indicates an increase of the compressive strain through the Mg incorporation.

The luminescence spectra in p-conductive GaN is dominated by the donor–acceptor pair (DAP) lines which have typical phonon replicas of the zero-phonon line at 3.27 eV (Fig. 2b). In higher doped samples, only a broad luminescence band appears. The maximum of this band is shifted to lower energies. This behavior is well known from high compensated ZnSe:N [3] and GaAs:Li [5].

For ZnSe:N the intensity behavior of the DAP can be explained in a dynamical model proposed by Shklovskii and Efros [6]. It involves strong fluctuations of the band gap at different positions in the sample caused by electric fields of a high number of compensated and thus, ionized donors and acceptors which are randomly distributed in the sample. At low excitation densities radiative DAP recombination takes place between the energetically lowest neutral donors and acceptors since photo-excited carriers relax quickly to these levels. At high excitation densities, the concentration of photo-excited carriers is high enough to neutralize most donors and acceptors and the band fluctuation should vanish. Thus, at the highest densities the well known structured DAP emission line shape as seen for samples with low compensation should be observed.

To test this model for GaN:Mg we studied the intensity dependence of the PL of the most resistive sample grown with a Mg/Ga precursor ratio of 0.3%. It is shown in Fig. 3 for different excitation conditions. Starting from lowest excitation intensities (left spectra) we observe two broad emission bands around 2.45 eV und 2.8 eV. Assuming a Gaussian line shape for both luminescence bands we fitted the spectra and obtained a very good agreement with the experiments. Raising the excitation density the

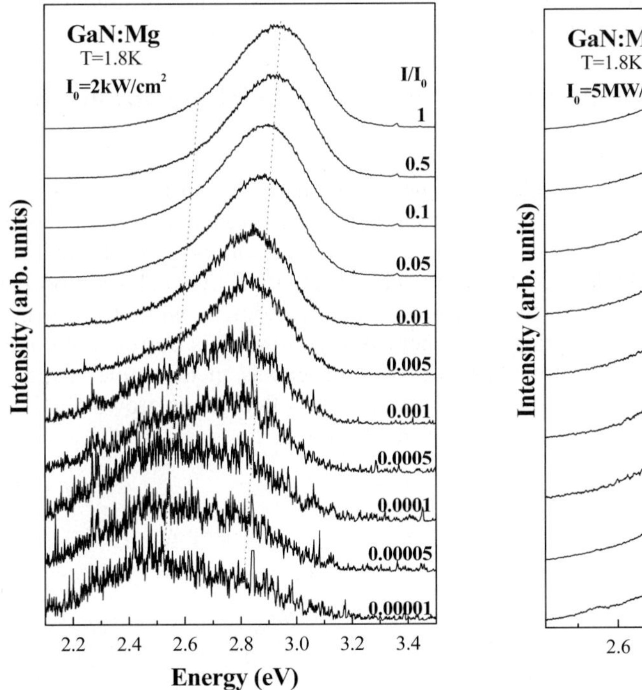
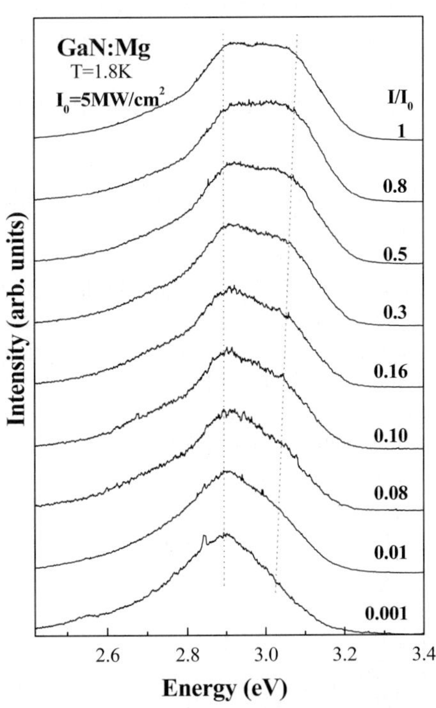

Fig. 3 (online colour at: www.interscience.wiley.com)   Intensity dependence of normalized PL spectra in p-type GaN epilayers doped with 0.3% Mg (left part) measured with a cw HeCd laser up to 2 kW/cm^2 and (right part) with the third harmonic of a pulsed Nd:YAG laser with pulse intensities up to 5 MW/cm^2.

high-energy band exhibits a growth in integrated intensity with a slope of unity, exactly the same as that observed in ZnSe:N. Also the observed logarithmic shift of the peak energy to higher energies with increasing excitation energies is in agreement with other strongly doped and compensated semiconductors. This shift is equal for both emission bands. The total blueshift of the high-energy luminescence between $20\,mW/cm^2$ and $2\,kW/cm^2$ amounts 153 meV. We find that the blueshift saturates at 2.91 eV. On the high-energy shoulder of this luminescence a new emission line grows, again shifting to higher energies with increasing intensity. At $5\,MW/cm^2$ the intensity maximum of this band is found at 3.06 eV, 210 meV below the zero phonon line of the DAP in weakly compensated GaN. Even, at highest excitation densities, no structured DAP band with typical LO-phonon replica was observed. The model of band fluctuations [6] cannot give a satisfying explanation for the behavior in highly compensated GaN:Mg. Therefore, we will now discuss a static model to explain the behavior of the DAP as a function of the excitation density. Certainly, static potential fluctuations here are not responsible for the observed effects but impurity bands of donors. From the energy positions of the saturated luminescence bands at 2.45 eV, 2.91 eV and 3.06 eV we could determine the corresponding donor levels considering the band gap of GaN and the acceptor energy. These levels lie 240 ± 30, 350 ± 30 and 850 ± 30 meV below the conduction band [7, 8]. In the beginning all donors and acceptors are ionized. At low excitation densities a small number of donors and acceptors will be neutralized. The created electrons relax very fast in the deepest donor states. With increasing excitation density the donors will be neutralized one after another from the deepest to the highest energy state. This explains the observed blueshift with increasing excitation intensity. An additional blueshift takes place through the coulomb term between the ionized donors and acceptors. The energies of the deepest donor states are in very good agreement with values of 265 ± 15 meV, ca. 400 meV and 615 ± 20 meV in GaN:Mg determined by Hacke et al. [9] in DLTS investigations.

The shift of the „blue" luminescence between 2.9 eV and 3.1 eV in hydrostatic pressure experiments with 20 meV/GPa is smaller than the shift of the DAP at 3.27 eV (33 meV/GPa) [10]. This is in contradiction to models which take into account shallow donor levels as initial states for this luminescence [11, 12] and supports the above explained model under consideration of deeper localized donor states.

3   Compensation of hydrogen and localized Mg–H vibrations in GaN:Mg (MOCVD)

During the MOCVD growth process of nitrides hydrogen is often present in the growth atmosphere in form of the carrier gases H_2 and NH_3, respectively, or is bound in the organic precursor-molecules. During the dissociation process of these molecules at the surface hydrogen can be incorporated in the crystal. J. Neugebauer and C. G. Van de Walle [13] found in their calculation of the formation energies that under p-type conditions the incorporation of hydrogen is extremely efficient, while for n-GaN the mobility of hydrogen is very slow and so the hydrogen content is very low. This correlation could be found experimentally by Götz et al. [15]. The concentration of hydrogen is in the same range as the Mg-concentration [16]. Since hydrogen forms a shallow donor in GaN, often the as grown GaN epilayers are highly compensated and an activation of the acceptors is necessary.

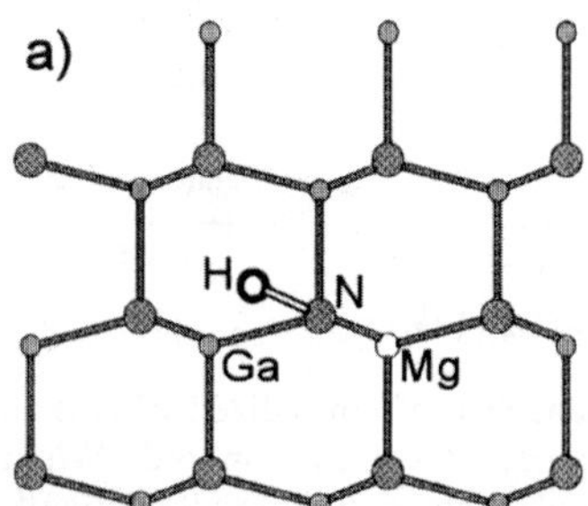

Fig. 4   Structure of the Mg_{Ga}–N–H-complex after [14].

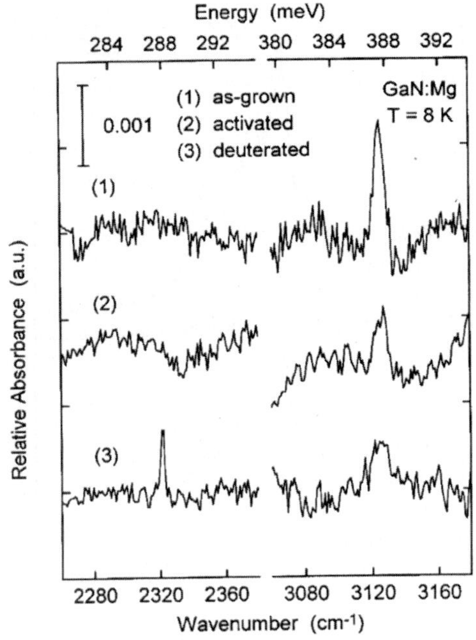

Fig. 5 IR-absorption spectra of GaN:Mg (MOCVD): (1) as-grown, (2) after thermal annealing, and (3) after deuteration [17].

The electrical compensation takes place by the formation of a Mg–H-complex, the corresponding atomic structure (seen in Fig. 4) was proposed by [13] as a Mg_{Ga}–N–H-complex. The exceptional feature is that no direct Mg–H-bonds can be achieved. On the basis of a large ionicity of the Ga–N-bond the hydrogen atom prefers to incorporate on an antibonding site of the nitrogen. The vibronic frequency of the complex is mainly determined by the N–H vibrational mode. The predicted value is 3360 cm^{-1}, nearby the frequency of the NH_3 stretch mode with 3444 cm^{-1}.

In infrared absorption spectra of GaN:Mg (MOCVD) a localized vibrational mode (LVM) was found at 3125 cm^{-1} [17, 18]. After activation of compensated GaN samples through thermal annealing p-conductance and a simultaneously strong decrease of the intensity of the LVM could be observed. Through deuteration a second absorption line having a frequency of 2321 cm^{-1} could be generated (Fig. 5). The frequency ratio between both LVMs is $1.346 \approx \sqrt{2}$. This value would be expected if one take into account a mass change during an isotope substitution from hydrogen through deuterium. This is a strong experimental evidence that the mode at 3125 cm^{-1} can be attributed to a defect complex of Mg_{Ga}–N–H.

4 Mg-doped GaN grown by molecular beam epitaxy – high energy Raman modes

In contrary to GaN:Mg grown by MOCVD the Mg doped MBE samples show p-conductivity in the as grown state [19]. This could be put down to the fact that the MBE chamber has ultrahigh vacuum conditions and so no compensating hydrogen can be incorporated during the growth procedure. One has to take into account that the remaining vapor pressure in the MBE chamber mainly consists of H_2. In addition to that it could be shown by mass spectroscopy that the Mg sources often release hydrogen during the evaporation process [20].

Raman spectra of Mg-doped GaN grown by MBE show at high energies LVMs having at 4 K the following frequencies: 2129, 2148, 2166, 2185, 2204 und 2219 cm^{-1} [21, 22]. Four of these modes (2151, 2168, 2185, 2219 cm^{-1}) has been observed earlier by Brandt et al. [20]. In Fig. 6a the correlation between the Mg concentration and the appearance of the high-energy modes is seen. At doping levels of 8×10^{19} cm^{-3} a fine structure between 2129 and 2219 cm^{-1} occurs. Reducing the Mg concentration to 6×10^{18} cm^{-3} in the chosen spectral range only the stretch vibration of N_2 (2329 cm^{-1} [23]) is seen while the other Raman-active modes are not observable. For concentrations in between the new modes start to grow up.

The high energy defect modes show a distinct A_1-symmetry (Fig. 6b). Only, the modes at 2166 and 2219 cm^{-1} are seen in crossed polarization having a weak intensity in the spectra. However, only these both modes are infrared active [20]. As a result, two different groups of defect complexes can be distinguished, however, the dipole moment of the infrared active mode cannot be parallel to the c-axis. The dipole moment of the other vibrational modes does not change or is oriented parallel to the c-axis.

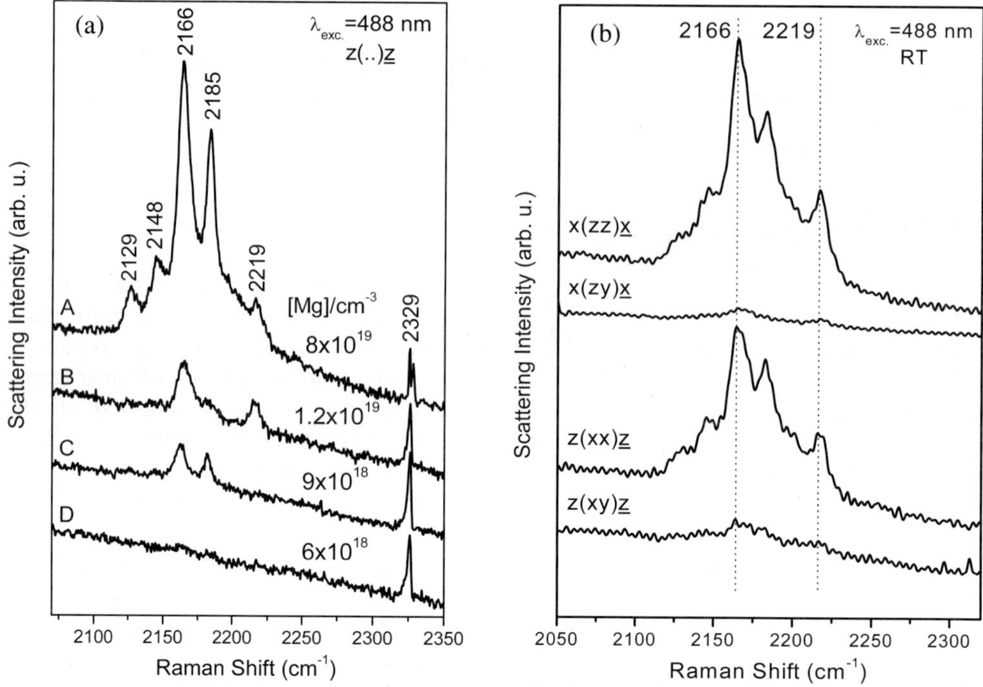

Fig. 6　a) RT Raman spectra of GaN:Mg (MBE) in the high-energy range around 2200 cm^{-1} as a function of the Mg content. b) Polarization dependence of the high-energy modes in GaN:Mg.

The vibrational frequencies of more than 2200 cm^{-1} are typical for defects including hydrogen as complex partner. Similar complexes are wellknown in other semiconductor materials, as i.e. ZnSe:As,H with a frequency of 2165 cm^{-1} or in GaAs:C,H at 2628 cm^{-1} [25]. Similar high-energy vibrations could be expected for N- (see N_2-stretch vibration at 2329 cm^{-1}) or for C with a threefold binding. Nevertheless, it was not clear for a long time, if really the responsible complex for the observed Raman modes includes hydrogen. First of all, the modes were only found in MBE grown material, even though the growth atmosphere includes much less hydrogen compared to the MOCVD growth, if not NH_3 was used as nitrogen source. In the beginning of this chapter was described how hydrogen can contaminate the samples during the MBE growth. Secondly, the predicted Mg_{Ga}–N–H complex could be detected and so no other obvious explanation of a complex in the energy range around 2200 cm^{-1} is possible.
Investigations of Harima et al. [24, 26] could clearly show the correlation of the hydrogen incorporation during their different growth conditions. Through thermal annealing of GaN:Mg (MOCVD) above 600°C the Mg_{Ga}–N–H complex will be destroyed. Simultaneously, the before compensated material will be p-conductive. For temperatures above 700 °C the above mentioned defect modes around 2200 cm^{-1} from the MBE material occur (Fig. 7). If during the MOCVD growth process N_2 was used as carrier gas instead of H_2, neither before nor after thermal annealing the high-energy vibrational modes could be found. These results give a clear evidence for the following interpretation. During the thermal annealing procedure a reorientation of the initial Mg_{Ga}–N–H complex takes place compensating the Mg acceptor through hydrogen. New complexes were formed which were found in GaN:Mg grown by MBE. These defect complexes includes as well hydrogen, however, here the hydrogen is electrical inactive and is not compensating in this configuration. In conclusion we have now an efficient tool to determine the degree of compensation and the p-conductivity with the above mentioned Raman spectroscopy, if one detect the high-energy modes in GaN: Mg.

If one anneal p-type GaN samples grown by MBE in a N_2 or N_2+NH_3 atmosphere [21], it is not possible to increase the p-conductance. The epilayers show at annealing temperatures of 930 °C rather a high-ohmic behavior, while a change of the relative intensities of the different local defect modes is observed

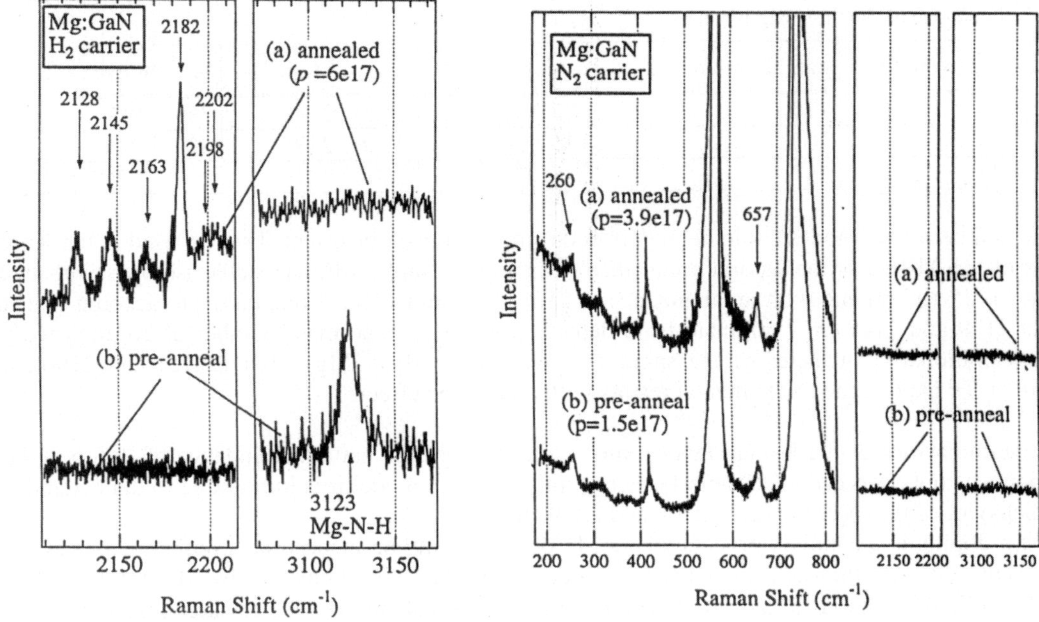

Fig. 7 Raman spectra of GaN:Mg (MOCVD) using H$_2$ (leftpart) and N$_2$ (right part) as transport gas before (b) and after annealing (a) [24].

simultaneously (Fig. 8). At temperatures of 1065 °C, which are typical for the MOCVD growth, all defect modes disappear and the samples are highly compensated. Now, no reorientation to the Mg$_{Ga}$–N–H complex takes place. The worsen conductance can be attributed to the generation of new defects at higher temperatures. A drastic change of the sample quality can be prevented by the N$_2$ partial pressure during the annealing procedure. Since the Mg-correlated vibrational modes disappeared totally, an out-diffusion or reorientation of the Mg atoms is more probable.

The distinct atomic structure of the defect complexes is not understood totally. A comparison with other materials [27] suggests Ga–H- (2171 cm^{-1} in Si) and Mg–H-defects (2144 cm^{-1} in GaAs) as possible candidates. However, V$_{Ga}$–H$_N$-defect complexes [28] are relatively improbable (3000–3100 cm^{-1} in GaN). To explain the multitude of vibrational modes there are different models which have to be considered.

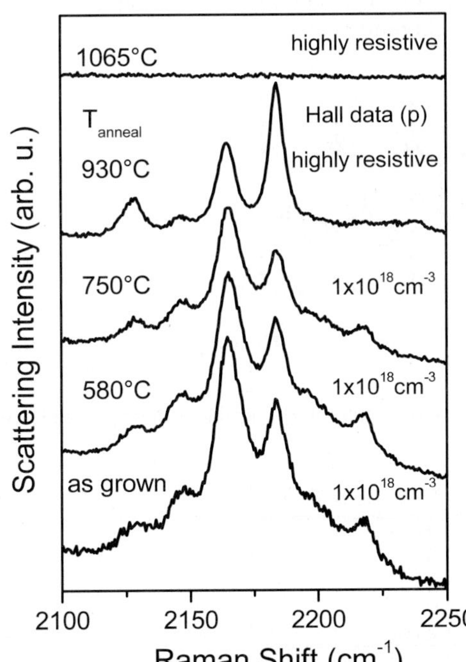

Fig. 8 Raman spectra of the high energy LVMs after annealing at different temperatures

Table 1   Comparison of the relative intensities of the high-energy modes in GaN:Mg with the natural abundance of the three Mg isotopes.

Mg-isotope	nat. abundance	$\omega_{\text{theor.}}$ (cm^{-1})	$\omega_{\text{GaN:Mg}}$ (cm^{-1})	rel. intensity
26	11%	2129 / 2148	2129 / 2148	3% / 10%
25	10%	2171 / 2190	2166 / 2185	40% / 24%
24	79%	2215 / 2235	2204 / 2219	10% / 13%

1. The number of six modes could be understood through an isotope effect. The Mg used in the Knudsen-cells of the MBE has the natural abundance of isotopes. Native Mg has stable isotopes with mass numbers 24, 25 and 26 and with an abundance of 79%, 10% and 11%. If one take into account that the modes at 2129 and 2148 cm^{-1} originate from defects with the isotope mass number of 26, then we receive as vibrational frequencies of the same complex with ^{25}Mg (^{24}Mg) 2171 (2215) and 2190 cm^{-1} (2235 cm^{-1}). To explain the experimental results with an isotope effect

the relative intensities of the Raman modes must correspond to the natural abundance of isotopes. This correlation was not observed (see Table 1). Furthermore, in this model it is impossible to understand the special behavior of the local modes at 2168 and 2219 cm^{-1}.

2. The p-doping is a complex process where Mg does not only be incorporated on lattice places. It is very probable that substitial and interstitial (SI) defect complexes participating by Mg can be created. Reboredo and Pantelides [14] calculated *ab initio* the vibrational frequencies of such defect structures. They proposed SI complexes schematically shown in Fig. 9 with Mg- and Ga-atoms on interstitial sites where hydrogen atoms of different binding length are bound. Since the Ga$_i$–H–bonds are not exactly orientated parallel to the c–axis, it is possible that the defect complex Mg$_{\text{Ga}}$–N–Ga–H$_2$ is responsible for the IR-active modes at 2166 und 2219 cm^{-1}. As seen in Table 2 the presented comparison between experimental and theoretical values are relatively large. An uncertainty of 200 cm^{-1} through the theoretical method is without any additional parameter quite legitimately. For deviations larger than 500 cm^{-1}, the prediction is not acceptable. The authors put forward anharmonic effects for this large deviations. The prediction that the postulated SI–H-defects are reorientated through thermal annealing mechanisms to Mg$_{\text{Ga}}$–N–H-complexes could not be confirmed. Furthermore, our investigations in this paper and the results of Harima et al. [24, 26] differ from the predictions of Reboredo et al. [14].

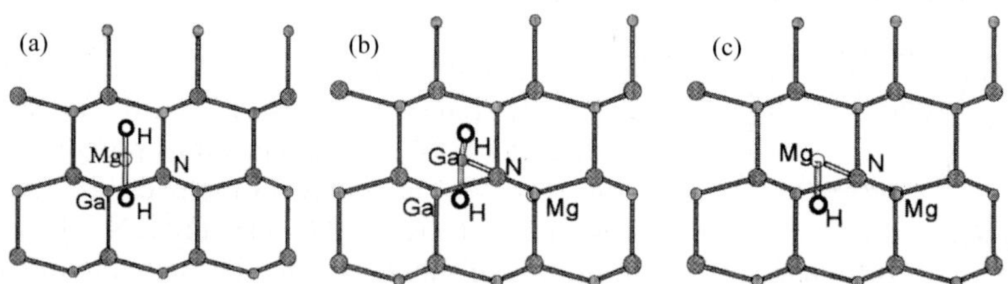

Fig. 9   Postulated Mg–H complexes in GaN [14]: a) Mg$_i$H$_2$, b) Mg$_{\text{Ga}}$–N–Ga–H$_2$, and c) Mg$_{\text{Ga}}$–N–Mg$_i$–H.

Table 2   Comparison of calculated vibrational frequencies of different H-correlated defect complexes with experimental values.

defect complex lt. [14]	$\omega_{\text{theor.}}$ (cm^{-1})	$\omega_{\text{GaN:Mg}}$ (cm^{-1})
Mg$_i$H$_2$ (short bond)	2001	2185
Mg$_i$H$_2$ (long bond)	1570	2148
Mg$_{\text{Ga}}$–N–Ga–H$_2$ (short bond)	2270	2219
Mg$_{\text{Ga}}$–N–Ga–H$_2$ (long bond)	1500	2166
Mg$_{\text{Ga}}$–N–Mg$_i$–H	1320	–

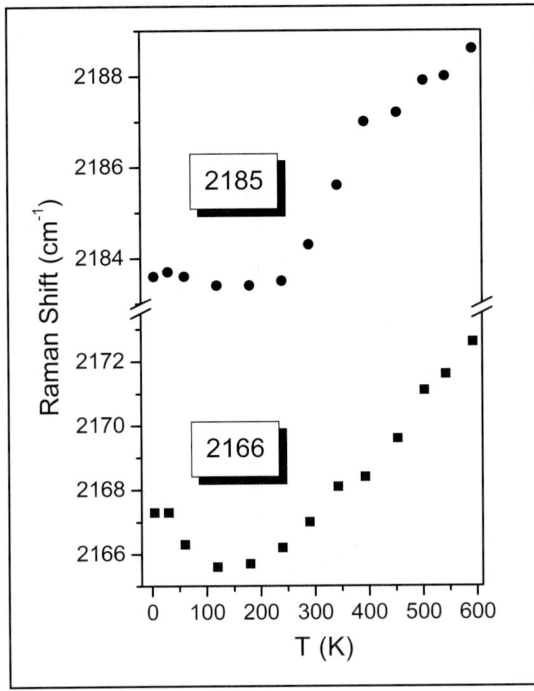

Fig. 10   Raman shift of the two prominent high-energy modes in GaN:Mg as a function of the lattice temperature.

With the discussed interpretations it is not possible to attribute free of doubts the observed Raman modes. Most probable is the attribution to defect complexes which contains as well magnesium and hydrogen as intrinsic defects, like vacancies, atoms on interstitials or dislocations. A possible proof of the correct attribution to the theoretically postulated defects is given by the temperature dependence of the high-energy modes. In Fig. 10 a line shift to higher energies with increasing temperatures is clearly visible. This is remarkable, since through the thermal expansion of the lattice a shift to lower energies should take place [29]. The E_2 mode of the host lattice shifts in the temperature range between 4 K and 590 K around $6\ \mathrm{cm}^{-1}$ to lower energies, while the energies of the defect modes shift in the same temperature range around $5\ \mathrm{cm}^{-1}$ to higher energies. This different temperature dependent behavior of LVMs compared to the host lattice phonons is typical, here the negative lattice expansion coefficient at low temperatures plays an important rule for the binding process at the impurity hydrogen complex.

5   Low-energy modes in Raman investigations

Simultaneously with the appearance of the high-energy modes five local vibrational modes in addition to host modes can be observed in the range of the acoustical and optical phonons in GaN:Mg. The phonon energies at room temperature are 136, 262, 320, 595 und $656\ \mathrm{cm}^{-1}$. The intensity of these modes scales as well with Mg content in the sample. All modes have A_1-symmetry (Fig. 11). Both structures with large FWHM at 320 und $595\ \mathrm{cm}^{-1}$ can be attributed to scattering processes of phonons beyond the Γ–point through a comparison with the phonon density of states (Fig. 12). Through the incorporation of defects the momentum conservation of the Raman process weakens and therefore also the reduction of the observable phonons around $q = 0$. For strongly perturbed crystals the phonon density of states folded with the phonon occupation can be observed in first order Raman spectra. This interpretation is supported through the fact that with increasing Mg concentration the intensity of A_1(TO)-mode in the forbidden $z(..)z$ configuration increases. The energy position of $320\ \mathrm{cm}^{-1}$ is in good agreement with the observed disorder Raman mode at $300\ \mathrm{cm}^{-1}$ which is activated by ion implantation in GaN [30]. The modes at 136, 262 und $656\ \mathrm{cm}^{-1}$ cannot be put down to disorder activated scattering processes, since the phonon dispersion curve in this range have no large density of states. The authors of Ref. [31] report on a mode at $656\ \mathrm{cm}^{-1}$ which occurs in GaN/sapphire after thermal annealing at 1000 °C. Lattice vibrations of disordered sapphire was given as explanation. In contrary the mode we found at the same energy position does not scale with the A_{1g}-mode of sapphire at $418\ \mathrm{cm}^{-1}$ or with the mode of disordered sapphire at $770\ \mathrm{cm}^{-1}$, but with the Mg content. An interpretation of these vibrations as LVMs of Mg can easily be understood. An estimation of LVM frequencies can be done by the effective masses of Ga–N- und Mg–

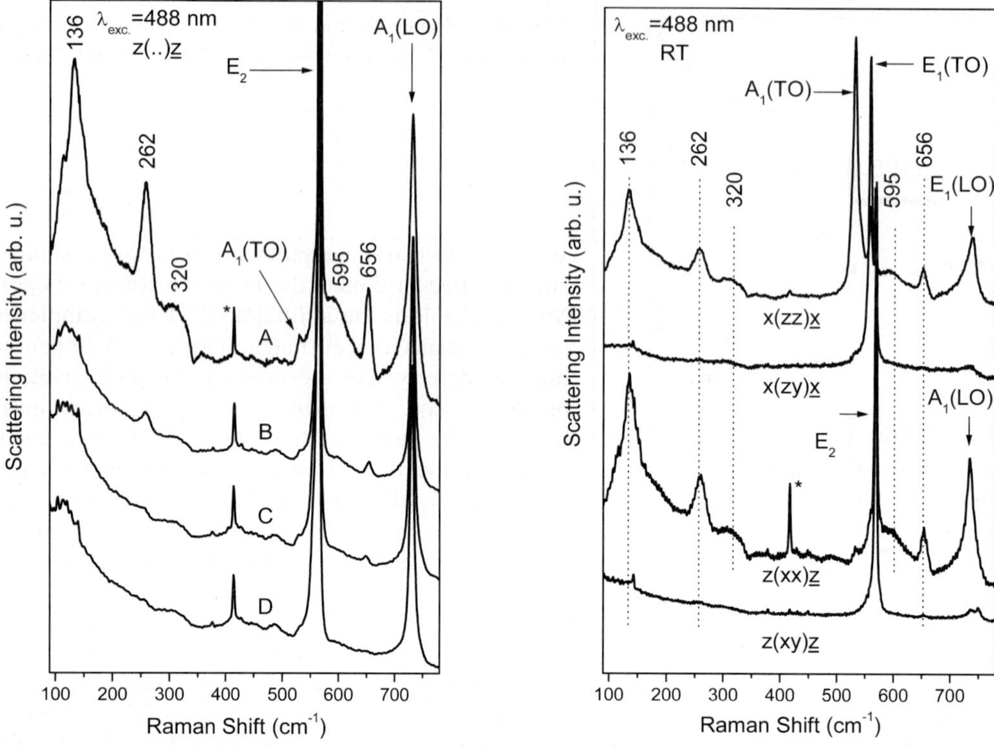

Fig. 11 Normalized Raman spectra in the range of the host phonons of GaN for various Mg doping (left part). The polarization dependence of the low-energy modes is shown in the right figure. The A_{1g}-mode of the sapphire substrate is marked by an asterix.

N-vibrations.

$$\frac{\omega_{GaN}}{\omega_{LVM}} = \sqrt{\frac{\mu_{LVM}}{\mu_{GaN}}} \tag{1}$$

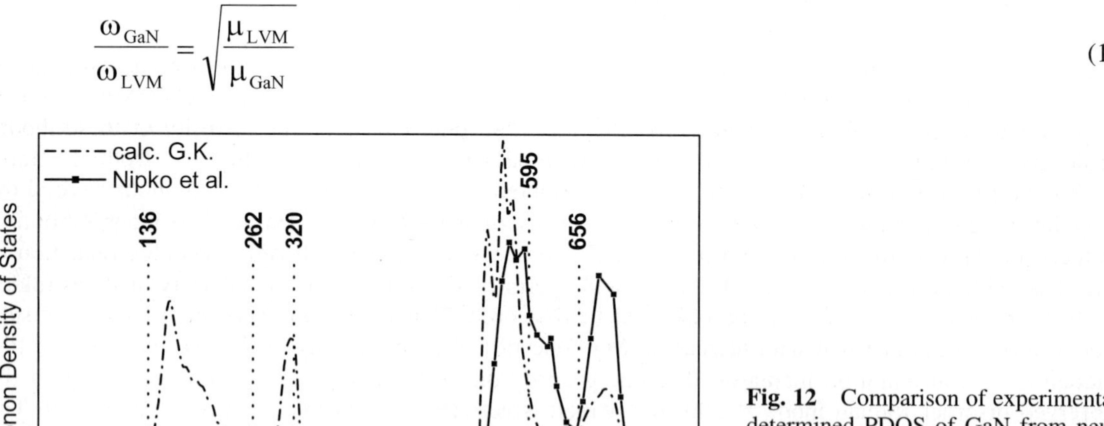

Fig. 12 Comparison of experimental determined PDOS of GaN from neutron scattering investigations [32] and a PDOS calculated by a valence force model [33]. The dashed lines mark the position of the low-energy modes in GaN:Mg.

If one take into account that Mg is incorporated on a Ga lattice site and put in $\omega_{GaN} = \omega[E_l(TO)] = 560$ cm^{-1}, a value of 640 cm^{-1} for Mg–N-vibration can be determined, which is in good agreement with the experinetally found value of 656 cm^{-1}. Calculating the LVMs of Mg$_{Ga}$ in a valence force modell of Kane we received frequencies of 132, 267 und 660 cm^{-1} with a scaling factor of –0.15 [33]. This means that the binding forces in the neighborhood of Mg$_{Ga}$ is around 15% smaller in comparison to the host crystal, which is in agreement with values in other semiconductor materials [34]. The explanation of low-energy modes as LVM of Mg was confirmed by H. Harima et al. [26]. Furthermore, a direct correlation between the hole concentration and the intensity of the LVMs could be found. This shows that the intensities of the modes are a direct measure for the concentration of activated Mg.

6 Conclusion

In GaN:Mg two main compensation mechanisms of Mg acceptors occur. One mechanism is the self-compensation effect. The incorporation of Mg causes a local pertubation of the crystal structure which creates defects acting as donors and compensating acceptors. Through intensity dependent PL-investigations a DAP model for high compensating GaN:Mg could be found and the energy position of compensating donor states could be determined.

In particular for GaN epilayers grown by MOCVD compensation mechanisms with hydrogen play an important role besides the self-compensation effects. Such Mg-doped samples show as grown high-ohmic behavior and an annealing procedure is necessary to activate the acceptors. In MOCVD samples a vibrational mode at 3125 cm^{-1} could be attributed to the Mg$_{Ga}$–N–H defect complex.

Mg-doped MBE samples are p-conductive without any further post growth treatment. Those samples show in Raman spectra a series of LVMs in the range around 2200 cm^{-1} and in the range of the host modes. The high-energy modes originate from defect complexes which are created through a thermal annealing procedure from the Mg$_{Ga}$–N–H-complex. A clear correlation with the incorporation of hydrogen exists. The low-energy modes consist of two groups. The modes at 320 and 595 cm^{-1} are activated through disorder and reflect a range of the phonon density of states. The Raman modes at 136, 262 and 656 cm^{-1} can be attributed to LVMs of Mg. The high- as well as the low-energy modes only appear in p-conductive GaN:Mg. Investigating the intensity dependence of the different Mg–H-complexes and the LVM of activated Mg the Raman spectra give a clear direct evidence of the degree of compensation and p-conductivity.

References

[1]   H. Amano, M. Kito, K. Hiramatsu, and I. Akasaki, Jpn. J. Appl. Phys., Part 2, **28**, L2112 (1989).
[2]   S. Nakamura, N. Iwasa, M. Senoh, and T. Mukai, Jpn. J. Appl. Phys., Part 1, **31**, 1258 (1992).
[3]   R. Heitz, E. Moll, V. Kutzer, D. Wiesmann, B. Lummer, A. Hoffmann, I. Broser, P. Bäume, W. Taudt, J. Söllner, and M. Heuken, J. Cryst. Growth **159**, 307 (1996).
[4]   U. Kaufmann, P. Schlotter, H. Obloh, K. Köhler, and M. Maier, Phys. Rev. B **62**, 10867 (2000).
[5]   H. P. Gislason, B. H. Yang, and M. Linnarson, Phys. Rev. B **47**, 9418 (1993).
[6]   B. I. Shklovskii and A. L. Efros, Electronic Properties of Doped Semiconductors, (Springer-Verlag, Berlin, Heidelberg, New York, 1984), pp. 52.
[7]   L. Eckey, Exzitonen im Galliumnitrid- Struktur und Dynamik, Dissertation, TU Berlin, 1998.
[8]   L. Eckey, U. von Gfug, J. Holst, A. Hoffmann, A. Kaschner, H. Siegle, C. Thomsen, B. Schineller, K. Heime, M. Heuken, O. Schön, and R. Beccard, J. Appl. Phys. **84**, 5828 (1998).
[9]   P. Hacke, H. Nakayama, T. Detchprohm, K. Hiramatsu, and N. Sawaki, Appl. Phys. Lett. **68**, 1362 (1996).
[10] H. Teisseyre, T. Suski, P. Perlin, I. Grzegory, M. Leszczynski, M. Bockowski, S. Porowski, J. A. Freitas, Jr., R. L. Henry, A. E. Wickeden, and D. D. Koleske, Phys. Rev. B **62**, 10151 (2000).
[11] M. Smith, G. D. Chen, H. X. Jiang, A. Salvador, B. N. Sverdlov, A. Botchkarev, H. Morkoç, and B. Goldenberg, Appl. Phys. Lett. **68**, 1883 (1996).
[12] E. Oh, H. Park, and Y. Park, Appl. Phys. Lett. **72**, 70 (1998).
[13] J. Neugebauer and C. G. Van de Walle, Phys. Rev. Lett. **75**, 4452 (1995).

[14] F. A. Reboredo and S. T. Pantelides, Phys. Rev. Lett. **82**, 1887 (1999).

[15] W. Götz, N. M. Johnson, J. Walker, D. P. Bour, H. Amano, and I. Akasaki, Appl. Phys. Lett. **67**, 2666 (1995).

[16] L. Sugiura, M. Suzuki, and J. Nishio, Appl. Phys. Lett. **72**, 1748 (1998).

[17] W. Götz, N. M. Johnson, D. P. Bour, M. D. McCluskey, and E. E. Haller, Appl. Phys. Lett. **69**, 3725 (1996).

[18] B. Clerjaud, D. Côte, A. Lebkiri, C. Naud, J. M. Baranowski, K. Pakula, D. Wasik, T. Suski, Phys. Rev. B **61**, 8238 (2000).

[19] T. D. Moustakas and R. Molnar, Mater. Res. Soc. Symp. Proc. **281**, 753 (1993).

[20] M. S. Brandt, J. W. Ager III, W. Götz, N. M. Johnson, J. S. Harris, Jr., R. J. Molnar, and T. D. Moustakas, Phys. Rev. B **49**, 14758 (1994).

[21] A. Kaschner, G. Kaczmarczyk, A. Hoffmann, C. Thomsen, U. Birkle, S. Einfeldt, and D. Hommel, phys. stat. sol. (b) **216**, 551 (1999).

[22] A. Kaschner, Lokalisierung, Defekte und Verspannung in Gruppe-III Nitriden, Dissertation, TU Berlin, 2001.

[23] F. Rasetti, Phys. Rev. **34**, 367 (1929).

[24] H. Harima, T. Inoue, Y. Sone, S. Nakashima, M. Ishida, and M. Taneya, phys. stat. sol. (b) **216**, 789 (1999).

[25] M. D. McCluskey and N. M. Johnson, J. Vac. Sci. Technol. A **17**, 2188 (1999).

[26] H. Harima, T. Inoue, S. Nakashima, M. Ishida, and M. Taneya, Appl. Phys. Lett. **75**, 1383 (1999).

[27] S. J. Pearton, J. W. Corbett, and M. Stavola, Hydrogen in Crystalline Semiconductors, (Springer-Verlag, Berlin/Heidelberg/New York, 1991), and references therein.

[28] M. G. Weinstein, C. Y. Yong, M. Stavola, S. J. Pearton, R. J. Wilson, R. J. Shul, K. P. Kileen, and M. J. Ludowise, Appl. Phys. Lett. **72**, 2589 (1998).

[29] W. S. Li, Z. X. Shen, Z. C. Feng, and S. J. Chua, J. Appl. Phys. **87**, 3332 (2000).

[30] W. Limmer, W. Ritter, R. Sauer, B. Mensching, C. Liu, and B. Rauschenbach, Appl. Phys. Lett. **72**, 2589 (1998).

[31] M. Kuball, F. Demangeot, J. Frandon, M. A. Renucci, J. Massies, N. Grandjean, R. L. Aulombard, and O. Briot, Appl. Phys. Lett. **73**, 969 (1998).

[32] J. C. Nipko, C.-K. Loong, C. M. Balkas, and R. F. Davis, Appl. Phys. Lett. **73**, 34 (1998).

[33] A. Kaschner, A. Hoffmann, C. Thomsen, T. Böttcher, S. Einfeldt, and D. Hommel, phys. stat. sol. (a) **174**, R4 (2000).

[34] R. C. Newman, E. G. Grosche, M. J. Ashwin, B. R. Davidson, D. A. Robbie, R. S. Leigh, and M. J. L. Sangster, Mater. Sci. Forum **258-263**, 1 (1997).

phys. stat. sol. (c) **0**, No. 6, 1795–1815 (2003) / **DOI** 10.1002/pssc.200303125

Optical micro-characterization of group-III-nitrides: correlation of structural, electronic and optical properties

J. Christen[*][1]**, T. Riemann**[1]**, F. Bertram**[1]**, D. Rudloff**[1]**, P. Fischer**[1]**, A. Kaschner**[2]**, U. Haboeck**[2]**, A. Hoffmann**[2]**, and C. Thomsen**[2]

[1] Institut für Experimentelle Physik, Otto-von-Guericke-Universität Magdeburg, Germany
[2] Institut für Festkörperphysik, Technische Universität Berlin, Germany

Received 4 March 2003, revised 5 May 2003, accepted 7 May 2003
Published online 28 August 2003

PACS 61.72.Ss, 78.30.Fs, 78.60.Hk

For a detailed understanding of complex semiconductor heterostructures and the physics of devices based on them, a systematic determination and correlation of the structural, chemical, electronic, and optical properties on a micro- or nano-scale is essential. Luminescence techniques belong to the most sensitive, non-destructive methods of semiconductor research. The combination of luminescence spectroscopy with the high spatial resolution of a scanning electron microscope, as realized by the technique of cathodoluminescence microscopy, provides a powerful tool for the optical nano-characterization of semiconductors, their heterostructures as well as their interfaces. Additional access to the local electronic and structural properties is provided by micro-Raman spectroscopy, e.g. giving insight into the local free carrier concentration and local stress. In this paper, the properties of group-III-nitrides are investigated by highly spatially and spectrally resolved cathodoluminescence microscopy in conjunction with micro-Raman spectroscopy. Complex phenomena of self-organization and their strong impact on the microscopic and nanoscopic properties of both binary and ternary nitrides are presented. As the ultimate measure of device performance, the microscopic properties of light emitting diodes are assessed under operation. Using micro-electroluminescence mapping in the optical microscope as well as in the near field detection mode of a scanning near field optical microscope, the microscopic origin of the macroscopic spectral red shift in light emitting diodes is identified.

1 Introduction

All ternary or quaternary semiconductor alloys are subject to statistical stoichiometry fluctuations on a sub-microscopic scale. This intrinsic disorder becomes in particular important when lattices mismatch and restricted solubility is involved as given for the case of group-III-nitride alloys. Spatial energy fluctuations extending from atomic scale for perfect alloys up to micrometers for phase separation have a strong impact on both the electronic and optical material properties of the nitrides and the performance of the devices fabricated from them. The situation becomes even more complex for non-homogeneous growth modes of selective local epitaxy. The crystal growth on pre-patterned substrates, as for example used in epitaxial lateral overgrowth GaN (ELOG) [1, 2], results in a complex spatial structure of selforganized micro-domains having dramatic differences in defect density [3, 4] and impurity incorporation [5–9] and, in the case of alloys, stoichiometry [10, 11].

[*] Corresponding author: e-mail: juergen.christen@physik.uni-magdeburg.de,
Phone: +49 391 67 18668, Fax: +49 391 67 11130

InGaN is the material of choice for opto-electronic devices operating in the near ultraviolet (UV), blue, and green spectral range [12, 13]. However, the ternary alloy InGaN is known to be a strongly inhomogeneous system [14] with local variation of the In-concentration [15], and strain, i.e. local band-gap [16]. While remarkable progress has been achieved in the development of InGaN-based opto-electronic devices [17–21] the fundamental emission mechanism of InGaN, i.e. localization in local potential minima [22–25] versus piezoelectric fields [26], still remains a point of controversial discussion. In our measurements we directly evidence the alloy fluctuations in thick InGaN layers (i.e. no quantum wells), resulting in localized electronic states with different localization energies.

AlGaN is an essential component of GaN-based electronic devices [27] as well as for opto-electronic devices operating in the UV region [28, 29]. For efficient performance, high crystalline quality and chemical homogeneity of the AlGaN is mandatory. However, the mismatch of the lattice constants and the thermal expansion coefficients between AlGaN and potential substrates as well as GaN results in high densities of structural defects and in particular in micro-cracks [30–33]. Furthermore, the ternary alloy AlGaN is susceptible to fluctuations of the aluminum content, leading to laterally inhomogeneous strain and bandgap. In this paper, [Al]-fluctuations and local strain distributions in AlGaN layers are quantitatively evaluated. Growth of AlGaN on patterned substrates [34] is demonstrated to be accompanied by the self-organized formation of microscopic and nanoscopic domains with specific aluminum content [10].

Cathodoluminescence (CL) microscopy [35, 36] has proven its immense potential to characterize the evolution of the epitaxial lateral overgrowth of GaN (ELOG) on stripe masks, evidencing the formation of specific growth domains with typical properties for a variety of mask patterns and growth conditions [37–42]. In these investigations, 2D imaging was performed either on cleaved faces perpendicular to the c-plane or on the as-grown sample surface [43]. While this method is fully appropriate for ELOG on stripe masks, it only gives incomplete information for grid-like mask patterns. One approach to perform 3D characterization is to consecutively remove the ELOG layer to clearly defined depths [44, 45]. Here we present such a full 3D characterization of columnar ELOG domains formed during overgrowth of hexagonal masks [46].

The most significant measure for light-emitting diodes (LEDs) eventually fabricated from the nitrides is their electroluminescence (EL) since it probes the relevant properties of the fully processed device under operation [47]. The spectrally and spatially resolved micro-EL and SNOM-EL allow a direct microscopic correlation of the morphological and optical properties of the device.

2 Experimental techniques

Scanning cathodoluminescence (CL) microscopy provides a powerful tool comprising low and variable temperatures, high spatial resolution $\Delta x < 45$ nm and high spectral resolution. In cathodoluminescence the sample under investigation is excited by the highly focused electron beam of a Scanning Electron Microscope (SEM). Our CL-system combines low temperatures (5 K $< T <$ 300 K), an overall spatial resolution of $\Delta x < 45$ nm (under optimum conditions) with high spectral resolution. However, the unique feature of our CL system is the imaging of complete CL spectra [35], [36]. While the focused e-beam is digitally scanned over typically 256×200 pixels a complete CL spectrum $I(\lambda)_{x,y}$ is recorded at each pixel (x,y) and stored. A four-dimensional data set $I_{CL}(x,y,\lambda)$ is obtained and consecutively evaluated. All type of data cross sections through this $I_{CL}(x,y,\lambda)$ tensor can be generated, e.g. sets of monochromatic CL images $I_{CL}(x,y,\lambda_1)$, $I_{CL}(x,y,\lambda_2)$,...; local CL spot spectra $I_{CL}(x_1,y_1,\lambda)$, $I_{CL}(x_2,y_2,\lambda)$, ...; CL spectrum line scans $I_{CL}(s,\lambda)$ (s = arbitrary line scan); as well as CL wavelength images (CLWI) $\lambda_{max}(x,y)$, i.e. mappings of the local CL emission peak wavelength λ_{max} [35, 48].

Our EL measurements are performed using a scanning micro-photoluminescence setup (μ-PL) [49] as well as a combined scanning near field optical microscope (SNOM) / scanning electron microscope (SEM) system. The scanning μ-PL / μ-EL setup [47] is based on an optical microscope using long distance lenses transparent to UV. Here, the sample itself is scanned using computer controlled scanning stages. A DC-motor driven stage enables the scanning with a resolution of 250 nm, piezo scanners are used for higher resolution. When operating in EL-mode, the µ-EL is measured spatially resolved through the microscope in detection mode with an overall spatial resolution of 1 µm. The luminescence is dispersed by a 0.5 m spectrometer with a spectral resolution of 0.5 nm and detected by a liquid nitrogen cooled UV-sensitive Si-CCD camera. The µ-EL measurements reported here were all performed at room temperature. Our unique combined SNOM / SEM setup enables SNOM-EL, SNOM-PL, as well as cathodoluminescence and SEM imaging at the same sample position. The EL is collected via a fiber tip of an atomic force microscope mounted inside a SEM cathodoluminescence setup. The EL is dispersed by a 0.3 m spectrometer, and detected by liquid cooled CCD camera. The data recording procedure for scanning-EL-microscopy is directly adapted from our CL imaging technique. During a single scan over the area of interest a complete EL spectrum is recorded at each pixel and stored together with the morphology image obtained from AFM mode. As for our CL measurements, also for both EL-microscopy techniques described above, all types of data cross sections through the 4-dimensional EL data set $I_{EL}(x,y,\lambda)$ are subsequently generated, such as local EL spot spectra, sets of mono-chromatic EL images, as well as EL wavelength images (ELWI), i.e. mappings of the local EL emission peak wavelength $\lambda_{max}(x,y)$.

Micro-Raman-spectroscopy (µ-Raman) measurements were carried out in backscattering geometry at variable temperature. With the 488 nm line of an Ar^+/Kr^+ gas laser for excitation, frequencies of the Raman signal can be determined with an accuracy of 0.1 cm⁻¹. The lateral spatial resolution is 0.7 µm [50].

3   Compositional fluctuations in thick InGaN layers

To establish the nanoscopic correlation of In-fluctuation with local luminescence spectra, CL microscopy was directly performed on the sample slices prepared for transmission electron microscopy (TEM). The sample was grown in a standard MBE system on a (0001) orientated sapphire substrate, utilizing a radio frequency plasma source for the supply of nitrogen. Details are given elsewhere [51]. The structure consists of 950 nm $In_XGa_{1-X}N$ deposited on a 30 nm GaN epilayer. The average indium content is $x = 0.08$ as determined from high-resolution X-ray diffraction (HRXRD).

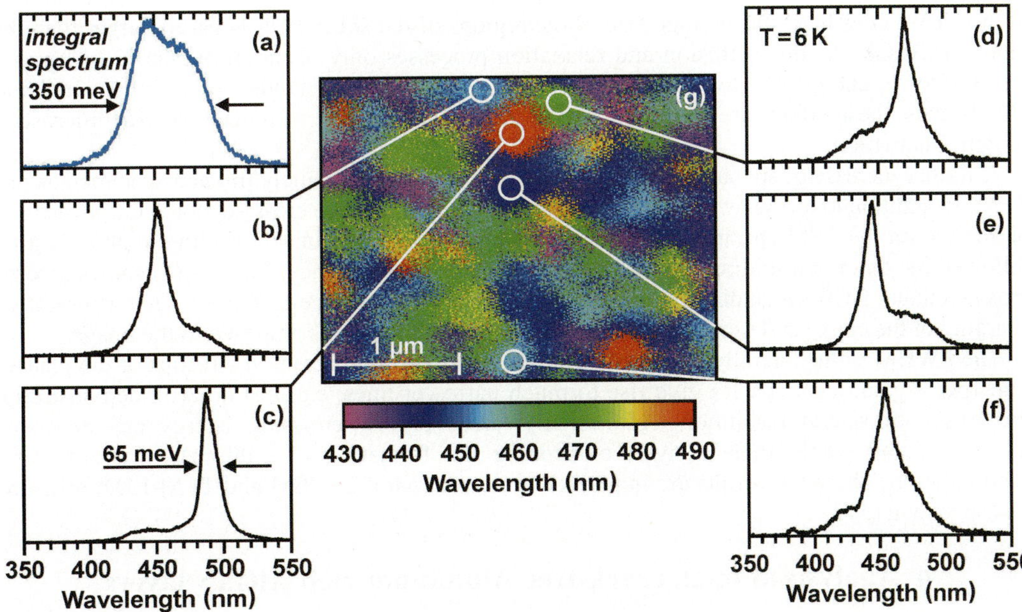

Fig. 1 Compositional fluctuations in a ~ 1 µm thick InGaN layer: a) laterally averaged CL spectrum of the sample surface and visualization of the local bandgap by b)-f) local CL spectra and g) a CL wavelength image.

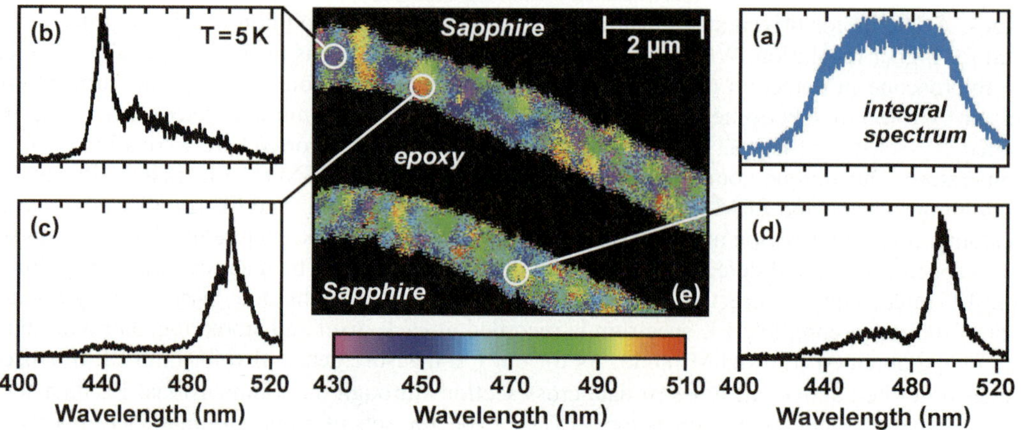

Fig. 2   Compositional fluctuations along the cross-section of a ~ 1 µm thick InGaN layer: a) laterally averaged CL spectrum of a specimen prepared for TEM investigations, b)-d) local CL spectra as well as e) CL wavelength image.

The integral CL spectrum of the sample, spatially averaged over 3 µm × 2 µm, is depicted in Fig. 1(a). A broad emission line centered at about 460 nm with a full width at half maximum (FWHM) of 350 meV is obtained which is also found in spatially integrating photoluminescence spectroscopy. However, a strong lateral fluctuation of the local CL peak wavelength is observed in CL scans, directly visualizing the fluctuations of indium content. Small areas with strongly differing but constant emission wavelength, i.e. specific [In], are clearly visible in the CLWI Fig. 1(g). Local spot-mode CL spectra recorded from the regions indicated in the CLWI are shown in Fig. 1(b)–(f).

As clearly seen, the FWHM of the individual peaks in the local spectra is considerably reduced with respect to the averaged spectrum Fig. 1(a). The size of the areas with constant emission wavelength varies between 50 nm and 500 nm, corresponding to the individual size of the potential minimum.

In principle, capture into the domains of lower energy may deteriorate the spatial resolution. However, in the CLWI (Fig. 1(g)) the average size of the regions of red shifted CL is barely larger than those of the blue shifted CL. As this diffusion and relaxation processes only occur from islands of larger energy to those of lower energy, resulting in a virtual larger size of the red islands and a virtual smaller size of the blue islands, these effects are negligible and the CLWI in Fig. 1(g) visualizes the true microscopic lateral Indium fluctuation.

A further increase of spatial resolution is achieved using an extremely thinned specimen as prepared for TEM. In addition to top-view scans of the as-grown sample surface, cross-sectional CL investigations were performed on the TEM specimens, which were prepared by mechanical grinding to about 5 µm thickness followed by xenon ion milling at 5 kV down to electron transparency. Figure 2(e) shows a corresponding cross-sectional CLWI including two pieces of the sample glued face to face. A drift, caused by a gradual charging of the epoxy and sapphire substrate during scanning, is superimposed to the image.

The FWHM of the laterally averaged spectrum Fig. 2(a) compares to the results of the plan-view scans Fig. 1(a). Again, local spectra give rise to much narrower lines, e.g. in Fig. 2(c), demonstrating the improvement of spatial resolution. Additionally, areas of low emission energy can be seen to reach throughout the whole InGaN layer, evidencing the formation of columnar structures of local In-accumulation. These CL results are in perfect agreement with the TEM and TEM-EDX measurements on this specimen [51].

4 Analysis of thick crack-free Aluminum-rich AlGaN layers

The $Al_xGa_{1-x}N$ layers were grown by MOVPE on sapphire substrates. Prior to $Al_xGa_{1-x}N$ growth, 2 µm GaN were deposited on a 30 nm low-temperature AlN buffer (LT-AlN, [52, 53]), covered again by LT-AlN. The three samples presented here have an average Aluminum molar fraction of $x = 0.44$, $x = 0.61$,

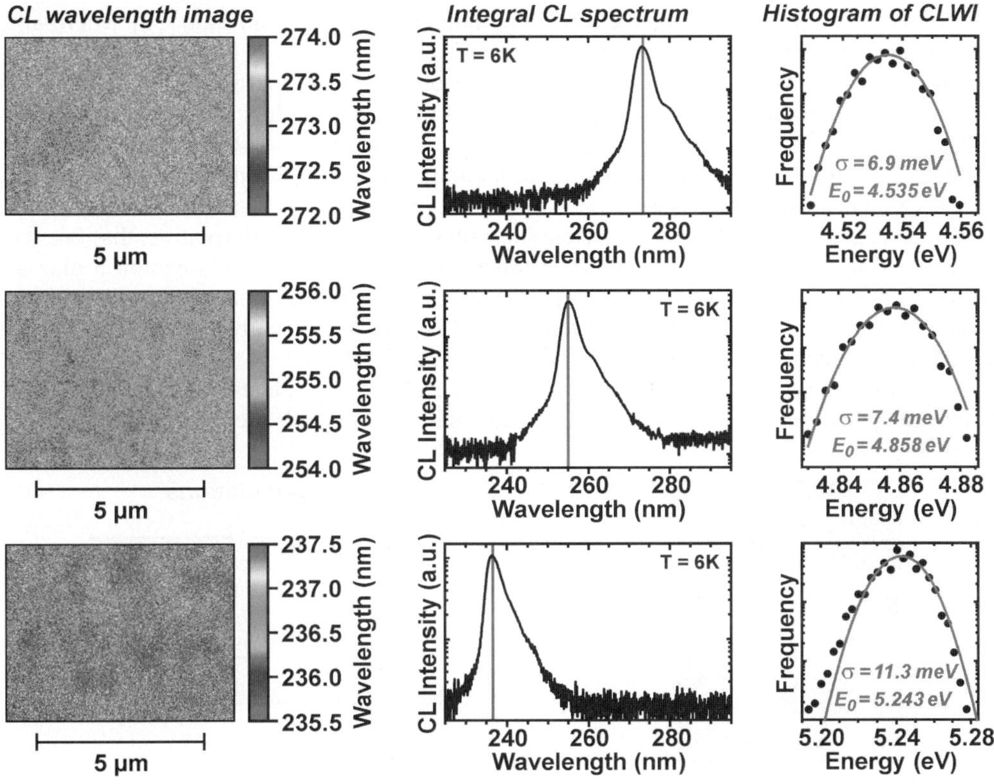

Fig. 3 (online colour at www.interscience.wiley.com)   Compositional fluctuations in AlGaN layers with $x = 0.44$, 0.61, and 0.76 (top to bottom): CL wavelength images, laterally integrated CL spectra, as well as histograms of the CLWIs.

and $x = 0.76$, respectively [54]. Micro-cracks, which often accompany the growth of AlGaN with high aluminum content [30], are completely absent in these $Al_xGa_{1-x}N$ layers even at $x = 0.76$.

The high-energy excitation sources of choice for optical investigations of these samples are either excimer lasers in PL or the electron beam excitation in CL.

In both laterally integrating PL and CL investigations an intense emission of the AlGaN layers is detected, which shows a monotonous dependence on the nominal aluminum content x. As displayed in the laterally integrated CL spectra in Fig. 3, the AlGaN peak position shifts from $\lambda = 273$ nm for $x = 0.44$ to $\lambda = 255$ nm for $x = 0.61$, and finally to $\lambda = 236$ nm for $x = 0.76$, in perfect agreement with the PL results [54].

The statistical fluctuations of the local Al incorporation were evaluated by CL wavelength mapping. The left hand column of Fig. 3 shows CL wavelength images taken at $T = 6$ K. These images qualitatively visualize the extremely low fluctuations of the CL peak position, i.e. the local bandgap. The quantitative analysis of the CLWIs yields the histograms shown in the right hand column of Fig. 3, proving a narrow and perfect random distribution of the peak maximum positions given by a Gaussian function for all samples. No deviation from the purely statistical fluctuation of a perfect random alloy due to local clustering or even large phase separations is observed. The sharpness of the Gaussians evidences the good homogeneity of the ternary alloys. Even at $x = 0.76$ the standard deviation of the peak position of $E_0 = 5.243$ eV is as low as $\sigma = 11.3$ meV over an area of 6 µm × 4 µm, as derived from the histogram of the CLWI in the bottom right corner in Fig. 3.

5 Stress relaxation at micro-cracks in AlGaN

The sample under investigation is a 0.56 µm thick $Al_{0.17}Ga_{0.83}N$ epilayer grown by metal organic vapor phase epitaxy (MOVPE) on a 1.9 µm thick GaN buffer on top of a sapphire substrate. The Al content was determined by X-ray diffraction (XRD) taking into account the strain state [30]. Micro-cracks forming a hexagonal (trigonal) network are observed in the epilayer.

On this $Al_{0.17}Ga_{0.83}N$ epilayer depth resolved CL spectroscopy was performed. A depth dependence of the CL signal was obtained by varying the penetration range R_e which is given by the voltage V_{Beam} accelerating the electron beam inside the SEM. In our experiments, the accelerating voltage V_{Beam} was varied from 3 kV to 20 kV, corresponding to a maximum penetration depth in a range between 0.1 and 2.2 μm, as calculated from the Kanaya-Okayama model [55].

CLWIs corresponding to different accelerating voltages V_{Beam} (i.e. different depths) but the identical sampling position are shown in the left hand column of Fig. 4. The CLWIs exhibit a blue shift of the main emission line from the crack-free areas towards the actual position of the micro-cracks, as also observed elsewhere for cracked AlGaN layers [56], [31]. This blue shift directly visualizes the lateral relaxation of the biaxial tensile stress in the AlGaN layer at the cracks.

In vertical direction, i.e. corresponding to different acceleration voltages, the maximum shift of the CL emission wavelength increases from the AlGaN/GaN interface towards the AlGaN surface.

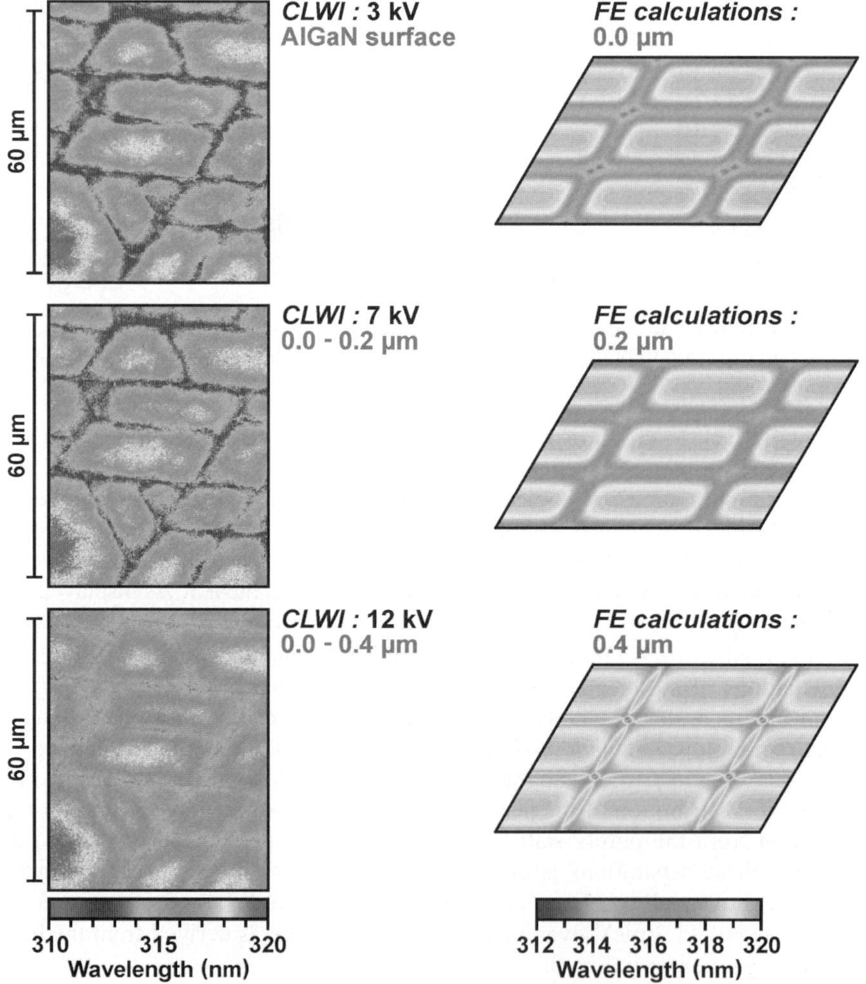

Fig. 4 (online colour at www.interscience.wiley.com)   Local relaxation of tensile stress in AlGaN due to the formation of micro-cracks: CL wavelength images corresponding to different penetration ranges of the electron beam as well as calculations of the emission wavelength derived from a 3D theoretical analysis of the stress distribution by the finite-element method.

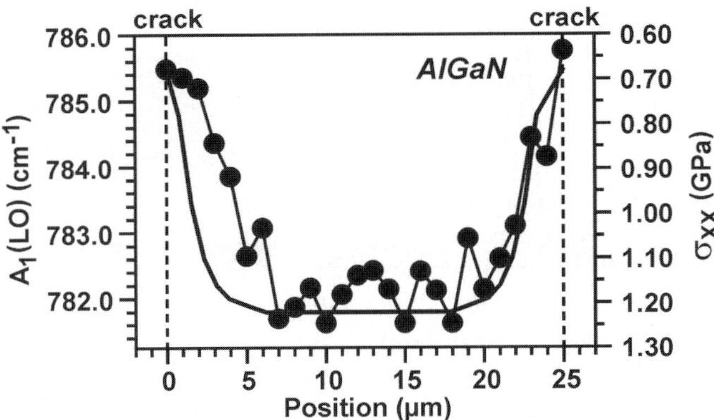

Fig. 5 Evolution of stress between two cracks in the AlGaN layer: Comparison between the experimental stress profile calculated from µ-Raman using the A_1(LO) phonon line (dots) and the theoretical prediction derived from a finite-element analysis (line).

The shift of the local CL emission wavelength directly correlates with the theoretical stress distribution between the cracks calculated from elasticity theory using a 3-dimensional finite-element (FE) approach [57–59]. Details are given elsewhere [60]. As the most frequent and recognizable pattern in the micro-crack network is the parallelogram, such a periodic crack arrangement was used during modeling. The right hand column of Fig. 4 shows the theoretical stress distribution already converted in the distribution of the local band gap, i.e. the local emission wavelength.

At the surface a monotonous decrease of tensile stress towards the cracks, i.e. a monotonous blue shift, is found. This blue shift reaches its maximum at the very crack position, as experimentally observed by CL.

In contrast to the surface case, a stress relaxation towards the cracks followed by a subsequent increase of the tensile stress in the direct vicinity of the cracks is calculated near the AlGaN/GaN interface. Thus, near the cracks a local red shift of the emission wavelength is expected with increasing the maximum penetration depth. In the strained regions far away from the cracks only weak stress dependence is found both in the FE calculations and the CL measurements.

The overall consistent description of the stress distributions by CL microscopy and FE simulation is confirmed by micro-Raman spectroscopy. Due to the spectral overlap of the E_2 mode of the AlGaN layer and the E_2 mode of the GaN buffer, the A1(LO) mode of AlGaN was used to derive the stress distribution, although it is not optimal for the estimation of the stress due to the polar character. As evidenced by Fig. 5, the AlGaN layer is tensile strained between the cracks and continuously relaxes towards the crack position.

6 Lateral seeding epitaxy of AlGaN on trench patterned GaN

For GaN, a strong reduction of dislocation density is possible by applying masking techniques, e.g. epitaxial lateral overgrowth (ELOG) [61, 62] or pendeo epitaxy. However, these methods are not appropriate for AlGaN because the Al-alloys also nucleate on the mask materials [1, 63]. Recently, growth on substrates periodically patterned into terraces and trenches has been proven successful to achieve thick, crack-free AlGaN layers [34, 64, 65],[66]. However this method is liable to the formation of specific micro-domains in the thick AlGaN layers, exhibiting different Al-concentrations [10, 11], as will be directly visualized by cross-sectional scanning CL microscopy in the following (Fig. 6).

The crack-free, 5 µm thick AlGaN layer ([Al] = 0.19) was deposited by MOVPE growth on patterned GaN/LT-AlN/Sapphire substrates. Prior to the final AlGaN growth, a periodic grid of trenches along <1$\underline{1}$00> (5 µm wide / 1 µm deep) was fabricated in the 4 µm GaN layer and subsequently overgrown with a LT-AlN interlayer [52, 53] as schematically illustrated in Fig. 6(c).
Three distinctly separated spectral components **A**, **B**, and **C** appearing in the laterally integrated CL spectra Fig. 6(b) can be unambiguously assigned to three different species of micro-domains.

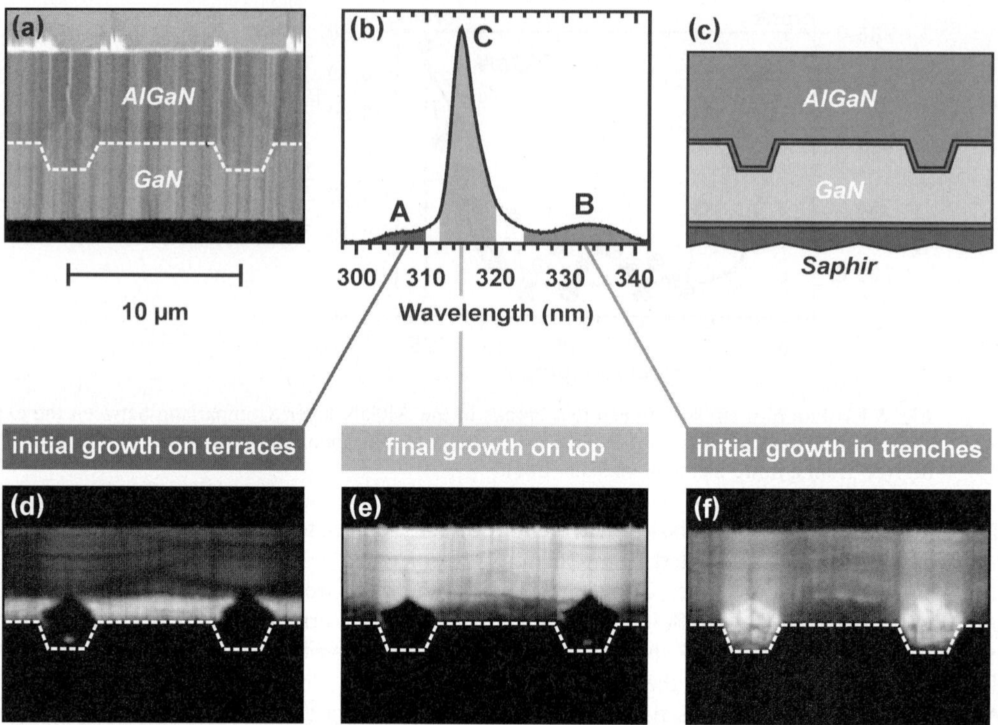

Fig. 6 Self-organized micro-domains with different emission energy, i.e. specific aluminum content, in AlGaN grown on a patterned GaN/Sapphire substrate: a) cross-sectional SEM image, b) CL spectrum averaged over the area of the SEM image, c) schematic sample setup as well as d)-f) CL intensity images in the spectral ranges indicated in b). The profile of the patterned GaN layer is given by the dotted line.

In Fig. 6, the region of initial AlGaN growth on the terraces emits high-energetic luminescence centered at 305 nm (**A**). As seen in Fig. 6(d), with advancing vertical growth this domain also expands laterally over the trenches, finally leading to coalescence above the trench centers.

In strong contrast, the domain of initial AlGaN growth inside the trenches is completely dominated by low-energetic luminescence near 335 nm (**B**), indicating a strong Ga accumulation. This domain **B** gives rise to the bright areas in the intensity image Fig. 6(f). The lateral size of this domain **B** is limited by the trench width. The self-organized lateral expansion of **A** eventually inhibits further vertical growth of **B**, leading to the pentagonal shape in the cross-section Fig. 6(f).

While **A** and **B** clearly reflect the template patterning, the emission wavelength of the final domain **C** is almost independent of the lateral sampling position. Following the self-organized transition from domains **A** and **B** to **C**, we find an average emission wavelength of 315 nm (**C**) up to the sample surface. Thus, domain **C** laterally extends over the whole area of the sample as indicated in Fig. 6(e).

Using µ-Raman to directly determine the local [Al], we find a perfect agreement with the CL results. Here, the Al-concentration is calculated from the A_1(TO) phonon energy [67], [68], [69], [70].

In Fig. 7 we compare µ-Raman line scans along <0001> for two different lateral positions. The line scan Fig. 7(a) crosses the trench center. Above the GaN layer, it initially shows an aluminum concentration of [Al] = 0.09 inside the trenches, exactly where an emission wavelength of 335 nm is measured by CL inside **B**. Following the abrupt transition to **C**, above the trenches we find [Al] = 0.17 consistent with the CL emission wavelength near 315 nm. In comparison, the µ-Raman line scan Fig. 7(b) runs through the terrace center. It proves [Al] = 0.22 directly above the terraces, where CL gives an emission wavelength of 305 nm inside **A**. Underneath the sample surface [Al] = 0.17 is reached, i.e. the identical value as for the upper region above the terraces in (a).

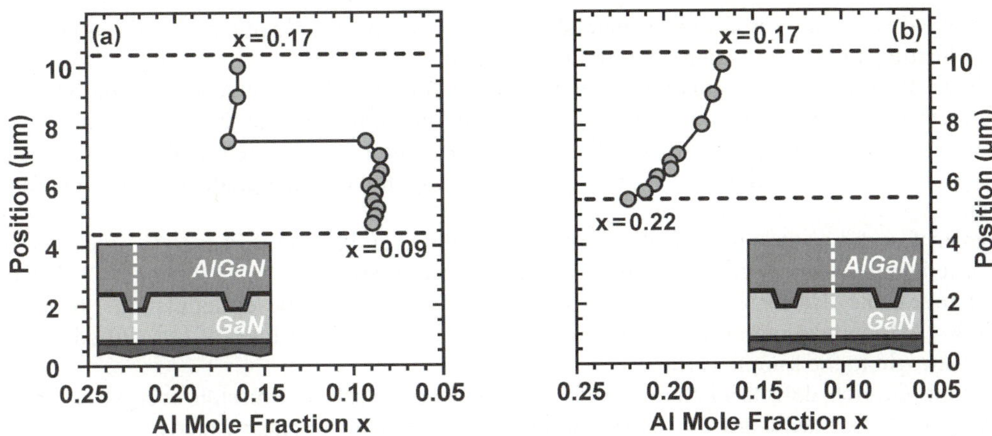

Fig. 7 Evolution of Aluminum content as derived from µ-Raman using the A_1(TO) position: a) line scan along the *c*-axis running through the terrace center, and b) line scan crossing the terrace center.

All three types of micro-domains **A**, **B**, and **C** additionally show specific nanoscopic modulations of their average Al-concentration. For a complete 3D visualization of these complex domain structures, cleavage planes of different crystallographic orientation were prepared and subsequently investigated by CL.

The upper row of Fig. 8 schematically indicates the CL scan direction used for the CLWIs in the lower row of Fig. 8.This way, Fig. 8(a) shows a scan parallel to {11̄20}, where the cleavage plane runs along the trench center. For Fig. 8(c), the sample was cleaved along {112̄0}. In contrast to (a), the cleavage plane now follows the terrace center. Fig. 8(b) shows a cleaved face parallel to {11̄00}, i.e. perpendicular to the trench direction.

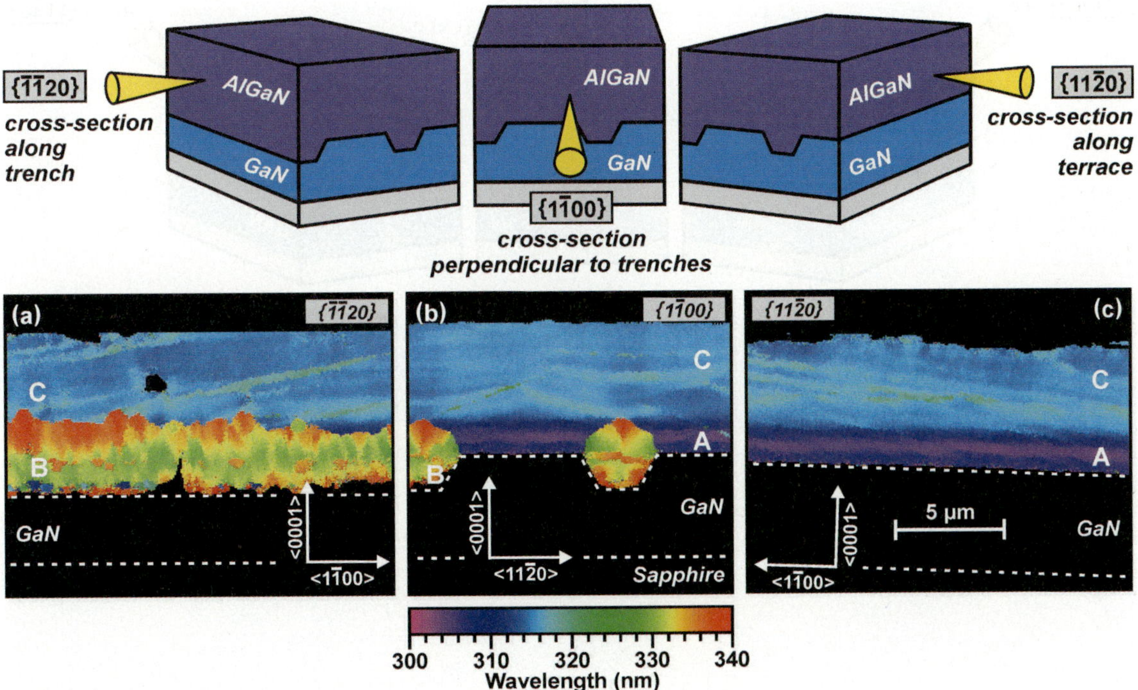

Fig. 8 Three-dimensional characterization of the nanoscopic [Al]-modulations by CL wavelength imaging on differently oriented cross-sections: a) parallel to {11̄20} in trench, b) parallel to {11̄00} as well as c) parallel to {112̄0} on terrace.

In **A** we observe an initial overall blue shift of the peak position compensated by a comparable red shift, indicating a vertical gradient of aluminum content.

In **B** we find statistical fluctuations of the peak position, i.e. the local Al-concentration, for the very first growth stage at the trench bottom. AlGaN growth on the trench side facets is initially accompanied by emission red shifted up to 340 nm. All domains **B** exhibit a strong red shift of the emission wavelength from the outer trench edges towards the domain top, indicating a gradual Ga-accumulation during growth.

Domain **C** exhibits strong vertical modulations of the average emission energy, appearing as curved lines of red shifted luminescence, i.e. increased [Ga], in Fig. 8(b). The cross-sectional CLWIs parallel to {11$\bar{2}$0} reveal the real three-dimensional, marble-like geometry of these modulations. In Fig. 8(a) and (c), these [Al]-modulations inside **C** are visualized by tilted, parallel lines of red shifted luminescence, running straight to the sample surface.

Sporadic Ga-rich defects in the otherwise homogeneous surface emission energy, as visible in the plan-view CLWI of the as-grown (0001) sample surface Fig. 9(b), appear exactly where these planes of the marbled structure penetrate the sample surface. This effect is illustrated by the CLWIs Fig. 9(a) and Fig. 9(c), both showing cross-sectional and surface emission of domain **C** at once.

The positions of these Ga-rich defects always coincide with nanoscopic steps on the planar AlGaN surface. The existence of these steps may give an explanation of the marble-like structure in **C**. As proposed by [71], preferential incorporation of Ga at the step sites caused by the different surface mobility of Ga and Al could lead to a local Ga accumulation. Thus, a lateral movement of the step positions during vertical expansion of the AlGaN layer leaves behind traces of tilted Ga-rich planes, as clearly observed inside domain **C**.

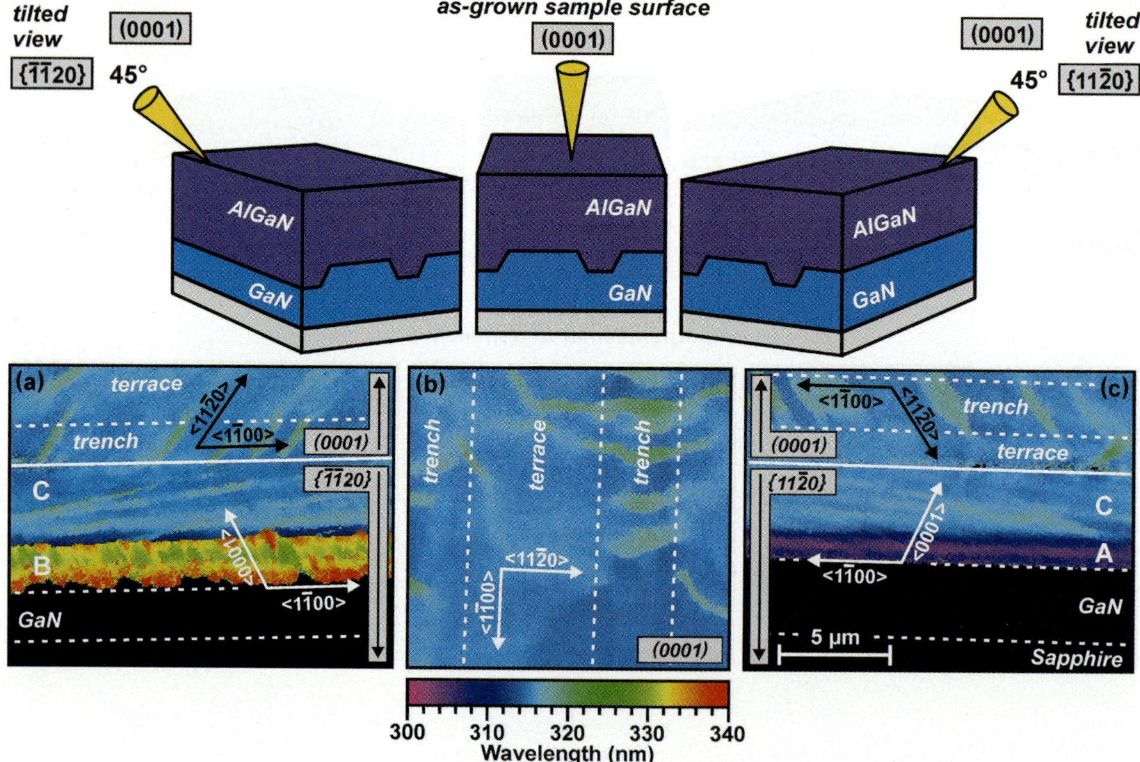

Fig. 9 Marble-like [Al]-modulations in the domain of final AlGaN growth visualized by CL wavelength imaging: a) tilted view of the transition from {11$\underline{2}$0} in trench to surface, b) plan-view image of the as-grown sample surface as well as c) tilted view of the transition from {11$\underline{2}$0} on terrace to surface.

7   Formation of growth domains in ELOG on hexagonal mask patterns

As comprehensively evidenced in our previous studies [5, 6, 7, 38], the selective growth of GaN, especially the epitaxial lateral overgrowth of GaN (ELOG) is always accompanied by the self-organized evolution of characteristic growth domains. The specific optical and electronic properties of the indiviual domains, e.g. caused by specific impurity incorporation, have been shown to directly correlate with the facet structure of the GaN surface during selective growth [36, 40, 41].

While the complete characterization of the domain properties on stripe mask patterns is usually achieved by cross-sectional CL investigations perpendicular to the mask direction, grid-like patterns are experimentally more challenging. Here we present the characterization of columnar growth domains formed in a thick GaN layer during lateral overgrowth of hexagonal masks [46].

The sample growth started with the deposition of a 2 μm MOVPE GaN buffer on (0001) sapphire substrate. On top of the GaN buffer, a SiO2-layer was patterned by photolithography and wet chemical etching into a periodic mask of 7 μm wide SiO2-hexagons set 11 μm apart. Subsequently, these patterned templates were overgrown in an AIXTRON horizontal HVPE system.

In conjunction with our 'standard' cross-sectional CL imaging at cleaved faces perpendicular to the c-plane (Fig. 10), a direct visualization of the 3D domain formation in the HVPE layer was achieved by consecutive vertical series of mappings parallel to the c-plane (Fig. 11). For these depth-resolved CL investigations a 2.4 mm wide spherical pit was fabricated in the ELO-GaN by mechanical grinding and polishing, reaching through the MOVPE buffer layer at its very center. The pit profile was determined using a surface profiler, establishing the correlation between depth and lateral position. CL maps were

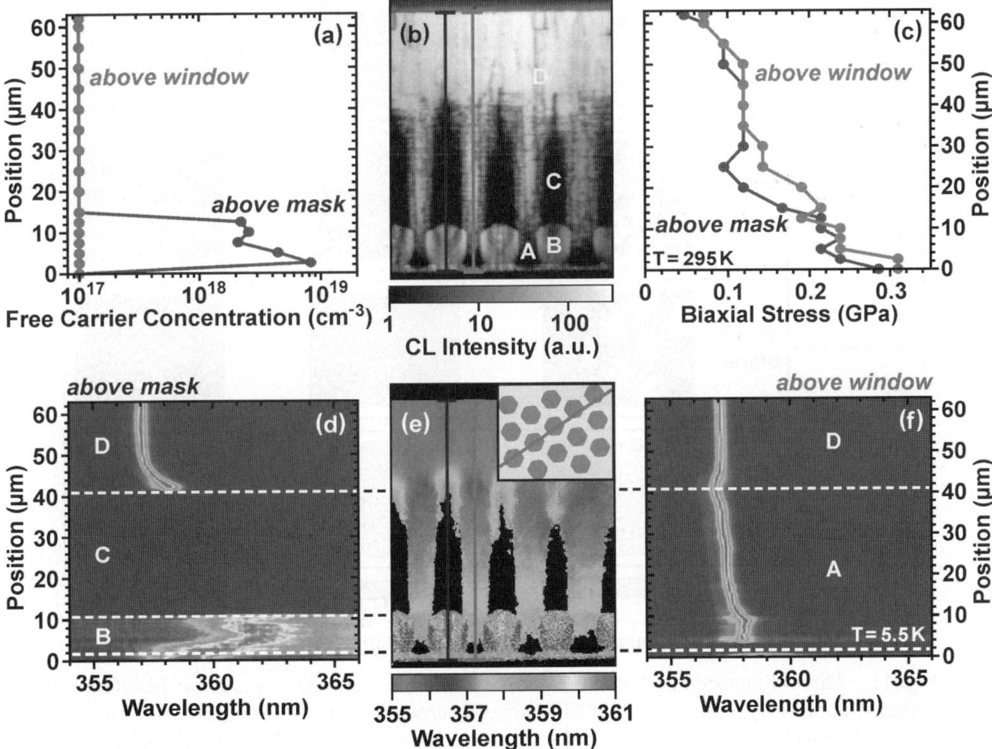

Fig. 10 (online colour at www.interscience.wiley.com)   Columnar growth domains in ELOG on hexagonal mask patterns: a) evolution of local free carrier concentration along the *c*-axis, b) cross-sectional CL intensity image, c) evolution of stress along the *c*-axis, d) and f) evolution of near band edge emission along the c-axis, as well as e) CL wavelength image of the identical area depicted in b).

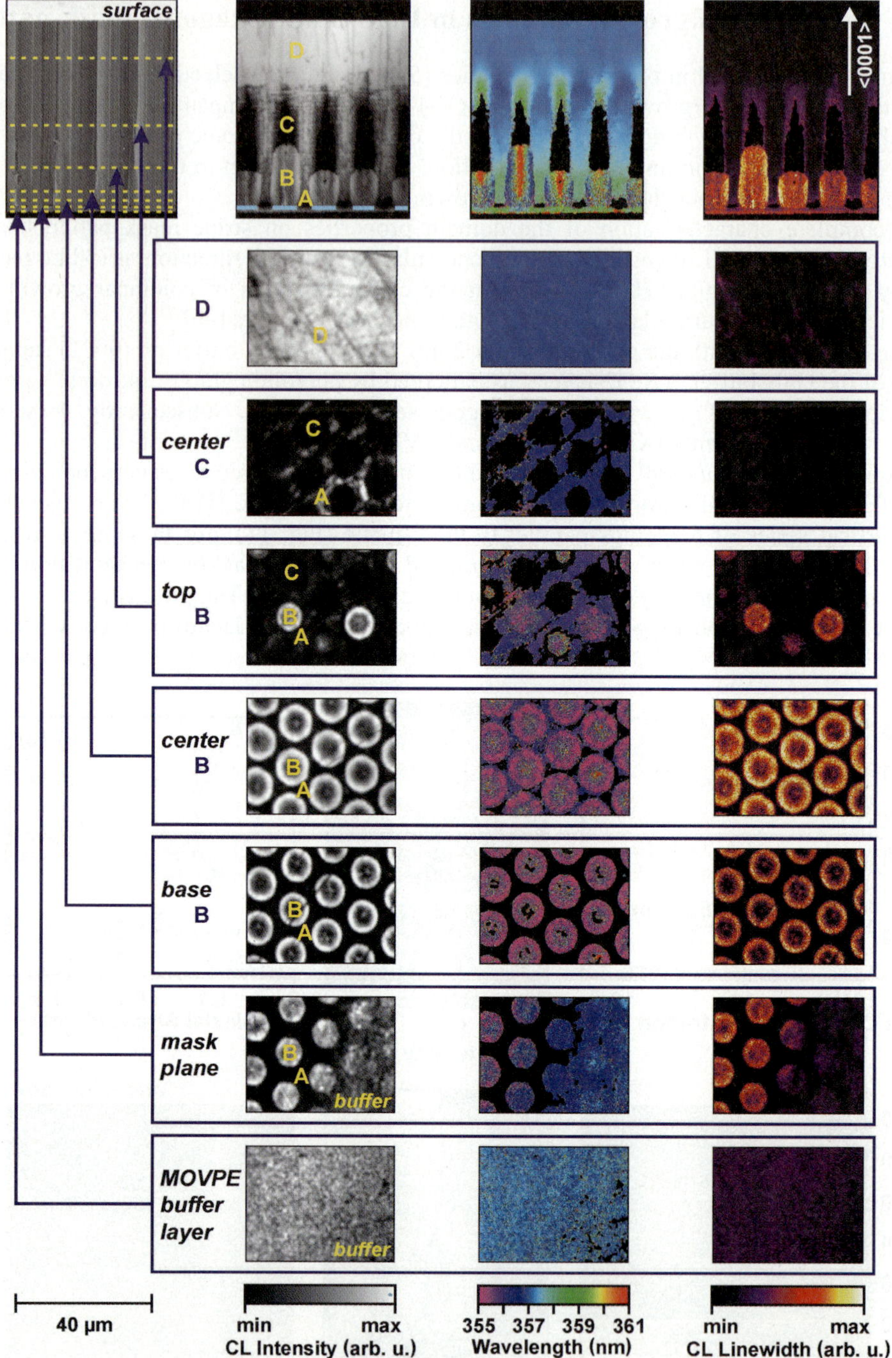

Fig. 11 Three-dimensional characterization of columnar growth domains in ELOG on hexagonal mask patterns: upper row: cross-sectional SEM image perpendicular to the *c*-plane, CL intensity image, CLWI and CL line width map of the identical sampling position; below: series of cross-sectional maps parallel to the *c*-plane at different depths.

taken at different lateral positions with respect to the pit center, leading to a series of cross-sectional images almost parallel to the c-plane but at increasing c-position ranging from the buffer layer up to the ELOG surface. For 50 µm wide maps used here, the spherical inclination of the pit slopes results in a depth difference, which is well below 1 µm (1.1 deg.) in the mask plane and about 5 µm (5.7 deg.) directly underneath the sample surface.

Figures 10(b) and (e) show cross-sectional mappings of the ELOG layer perpendicular to the c-plane. While at the left-hand side of these images the cleaved face almost crosses the centers of the hexagonal masks, at the right-hand side only the mask edges are met. This is illustrated by the magenta line in the schematic mask pattern depicted in the inset of Fig. 10(e).

A variety of distinct features showing characteristic luminescence properties are visible in the integral CL intensity image (CLI) of the near band edge emission Fig. 10(b), and the CL wavelength image (CLWI) Fig. 10(e), both displaying the identical area. These features directly correspond to specific growth domains formed during the evolution of the ELO GaN, marked in Fig. 10(b).

Initial (0001) growth in the unmasked areas (A) is characterized by sharp, however, weak excitonic emission, indicating a low impurity incorporation but high density of non-radiative recombination centers near the MOVPE / HVPE GaN interface. Directly underneath the mask-openings even the luminescence of the buffer is affected, showing an abrupt drop of quantum efficiency. With advancing (0001) growth the excitonic emission intensity above the unmasked areas gradually increases. A high density of dark lines running parallel to the horizontal mask plane is visible in the CL intensity image Fig 10(b) in this area. These lines of low quantum efficiency directly visualize dislocations bent perpendicular to the growth direction. The onset of facetted overgrowth (**B**) at the mask edges coincides with a drastic increase of integral CL intensity. This high integral intensity is exclusively caused by the appearance of broad and blue shifted (e,h)-plasma luminescence, as illustrated in Fig. 10(d). The indication of a high local free carrier concentration n inside **B**, as given by the (e,h)-emission, is confirmed by the µ-Raman line scan Fig. 10(a), as will be discussed below.

The following lateral expansion of the ELO GaN towards the mask centers is accompanied by a gradual red shift of the plasma edge, i.e. a monotonous decrease of n.

In the coalescence region directly above the mask centers we preferentially find strongly red shifted extrinsic luminescence, which is consistent with strong impurity incorporation.

While the (0001) growth in **A** uninhibitedly extends up to the sample surface, the characteristic properties of **B** are restricted to self-limited columns with an average height of 10 µm. Following a totally self-organized transition, further growth above the masks leads to an abrupt drop of the near band edge CL in 30 µm high cone-shaped domains **C** (Fig 10(b)). Exactly here we preferentially find Yellow Luminescence indicating strong local incorporation of deep defect centers. In the final domain of 15 µm (0001)-growth above and between the masks (**D**), no influence of the actual mask positions on the bright excitonic emission is found by CL, evidencing a homogeneous high quality of this upper layer.

Using the µ-Raman line scans to directly determine n, we find a perfect agreement between the optical and electrical properties. In Fig. 10, CL line scans parallel to the c-axis in (d) and (f) (red contrast = high CL intensity) are compared with the corresponding evolution of n in (a). The CL line scan Fig. 10(f), running through a mask opening, is always dominated by the sharp excitonic emission found in **A** and **D**. A spectral blue shift of the excitonic line position along the c-axis is consistent with CL results on thick GaN layers [72], [73], [74]. In Fig. 10(a) the corresponding n never exceeds the detection limit of 10^{17} cm^{-3}. In total contrast, we observe an extremely high n reaching 10^{19} cm^{-3} directly above the masks, exactly where the CL line scan Fig. 10(d) exhibits the broad luminescence band inside **B**. Where the broad CL vanishes at the transition to **C**, n abruptly drops below the detection limit. Region **D** is characterized by sharp excitonic emission and a low n also above the masks.

In these cross-sectional investigations the shape of the columnar ELOG domains **A**, **B** and **C** depends on the actual orientation of the cleaved face. A complete 3D image is obtained by additional depth-resolved mappings parallel to the c-plane. In Fig. 11 a consecutive series of such CLI and CLWI mappings is displayed. The vertical sampling positions of these mappings are indicated in the cross-sectional SEM image in the upper row of Fig. 11. For each cross-section the lateral distribution of the CL intensity, local emission wavelength and CL line width are depicted in one row.

In the lowest row the series shows the granular structure of the MOVPE GaN buffer. The high defect density of the buffer layer is illustrated by a high density of dark spots visualizing the drop of quantum efficiency around dislocations. In the next row the mask plane is met. Due to the slight inclination of the spherical pit surface both the upper part of the buffer and the onset of the HVPE growth are visible. Directly above the MOVPE / HVPE interface, the grid-like pattern of **A** interchanges with the hexagonal shapes of **B**. In the following cross-sections, the high integral CL intensity, blue shifted and broad emission of the (e,h)-plasma luminescence inside **B** is clearly visualized in ring-like structures appearing in the CLI, CLWI and line width map, respectively. Above, at some mask positions the very tops of **B** are still seen, while at other mask positions the transition to **C** has already occurred.

Following the cross-section running through the center of **C**, the homogeneity of **D** is evidenced in the uppermost row of the series. A network of defects appearing in the images corresponds to scratches left from the mechanical treatment of the sample.

8   Microscopic LED device characterization

Scanning electroluminescence microscopy (μ-EL) provides a fast and non-destructive method to analyze the homogeneity of a light-emitting device in its final, completely processed form. It yields a direct image of the spatial and spectral emission characteristics revealing imperfections of epitaxial growth as well as inhomogeneities of contact design and current injection [47, 75].

The light emitting diode (LED) under μ-EL investigation is a commercially available InGaN/GaN single quantum well (SQW) LED of Nichia [76]. The active region of this LED consists of a 2 nm $In_{0.2}Ga_{0.8}N$ SQW embedded in 100 nm $Al_{0.3}Ga_{0.7}N$ and 50 nm $In_{0.02}Ga_{0.98}N$ p- and n-type barriers. The lateral structure of the device is visible in the top-view optical microscope image in Fig. 12(d), showing the position of the p- and n-contacts.

The μ-EL measurements were performed at room temperature using DC injection currents ranging from 0.2 mA up to 60 mA [77].

At 0.2 mA the homogeneous distribution of the EL intensity in Fig. 12(a) proves a laterally uniform current injection over the whole mesa. However, the spot-like appearance of the intensity distribution shows that the radiative recombination is restricted to strongly localized areas. In contrast, at higher injection current a homogeneous distribution of the EL intensity is found on a micro-scale. Nevertheless, at 60 mA the injected current is not longer homogeneously spread over the whole mesa, leading to a drop

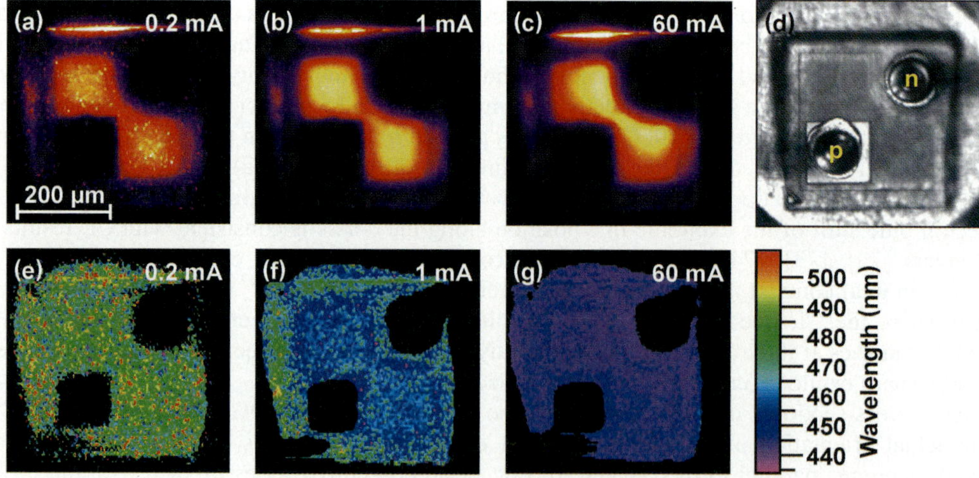

Fig. 12 Micro-electroluminescence maps of the commercial Nichia-LED under different injection currents: a)-c) spectrally integrating μ-EL intensity images, each normalized to the maximum intensity, d) optical microscopy image, as well as e)–g) EL wavelength images.

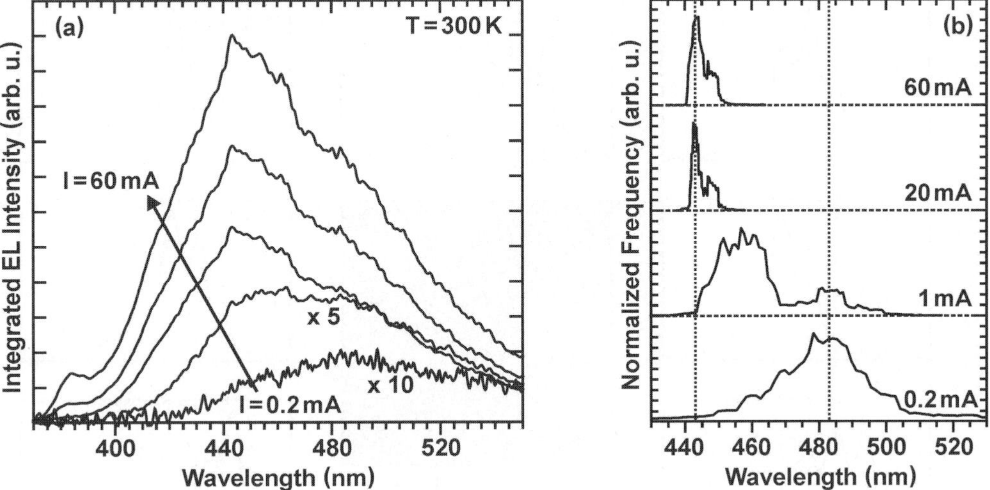

Fig. 13 Spectral blue shift of the emission wavelength with increasing injection current for the commercial Nichia-LED: a) laterally integrated EL spectra, b) histograms of ELWIs (normalized to the maximum frequency and shifted with respect to each other).

of EL intensity from the n-type towards the p-type contact directly visualizing the onset of problems with hole injection at high currents.

The ELWIs Fig. 12(e)–(g) show a drastic overall blue shift of the emission energy with increasing injection current, which is explained by a fundamental transition between the dominant recombination channels.

At a low injection current of 0.2 mA the ELWI Fig. 12(e) shows an emission wavelength around 480 nm. Lateral fluctuations of the emission wavelength are associated with potential fluctuations in the InGaN SQW, exhibiting different localization energies. At low currents, the radiative recombination predominantly takes place after thermalization into these localized states, which are only partly filled.

The blue shift of the emission wavelength with increasing injection current is assigned to a consecutive filling of the local potential minima. Above 20 mA the EL emission wavelength remains constant around 445 nm, proving the complete saturation of localized low-energy states.

Parallel to the spectral blue shift the emission wavelength shows a strong narrowing of its statistical distribution, as quantitatively expressed by the histograms Fig. 13(b). A competing recombination channel appearing at high currents around 380 nm in the EL spectra Fig. 13(a) is attributed to the $In_{0.02}Ga_{0.98}N$ barrier indicating filling of the $In_{0.2}Ga_{0.8}N$ SQW states under strong injection ($I > 60$ mA).

9   Nano-characterization of LEDs: Origin of EL blue shift

In order to gain deeper insight into the microscopic mechanism of the light emission process and the nature of the blue shift always observed with increasing current, we need much higher spatial resolution reaching the nanometer scale. As an example for EL device characterization in our SNOM-EL system, we present blue InGaN/GaN multiple quantum well light emitting diodes under room temperature operation, which were grown on Silicon (111) substrate by MOVPE. Details of the fabrication of these bright blue LEDs on Si are given elsewhere [21].

The SNOM-ELWI in the center of Fig. 14 reveals local emission peak wavelengths between 450 nm and 550 nm. These lateral fluctuations of the EL energy directly image the nanoscopic fluctuations of the stoichiometry in the active InGaN-QWs of the device. The local EL spectra in Fig. 14 exhibit different center positions and intensities, and are superimposed by Fabry-Perot-interferences. The spectral positions of the interference maxima were found to be exactly identical over a wide area, proving a mirror-like surface and constant thickness of the whole diode stack.

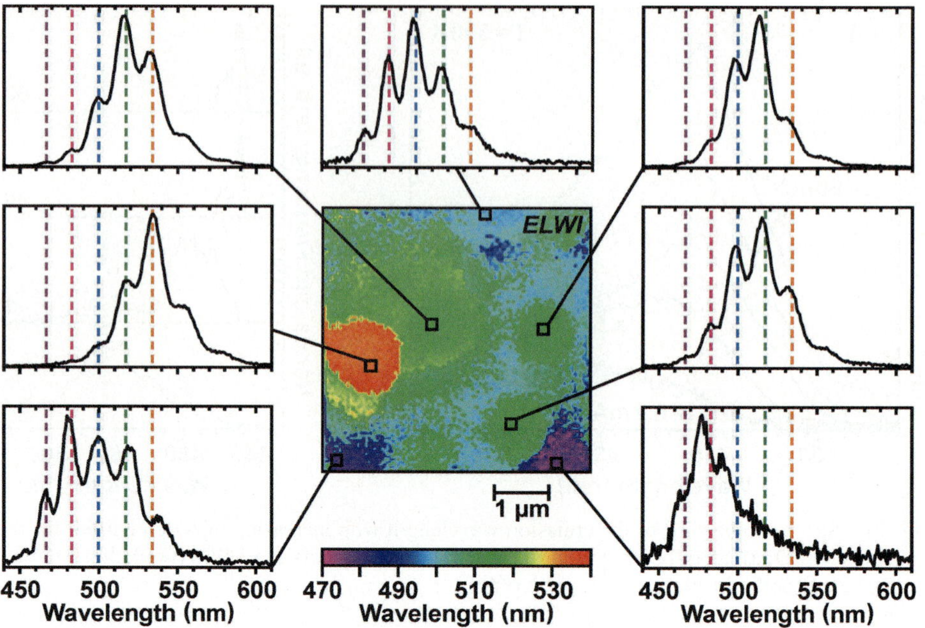

Fig. 14 Electro-Luminescence Wavelength Image (ELWI) and local EL spectra of an InGaN/GaN-LED under operation, taken in SNOM detection mode.

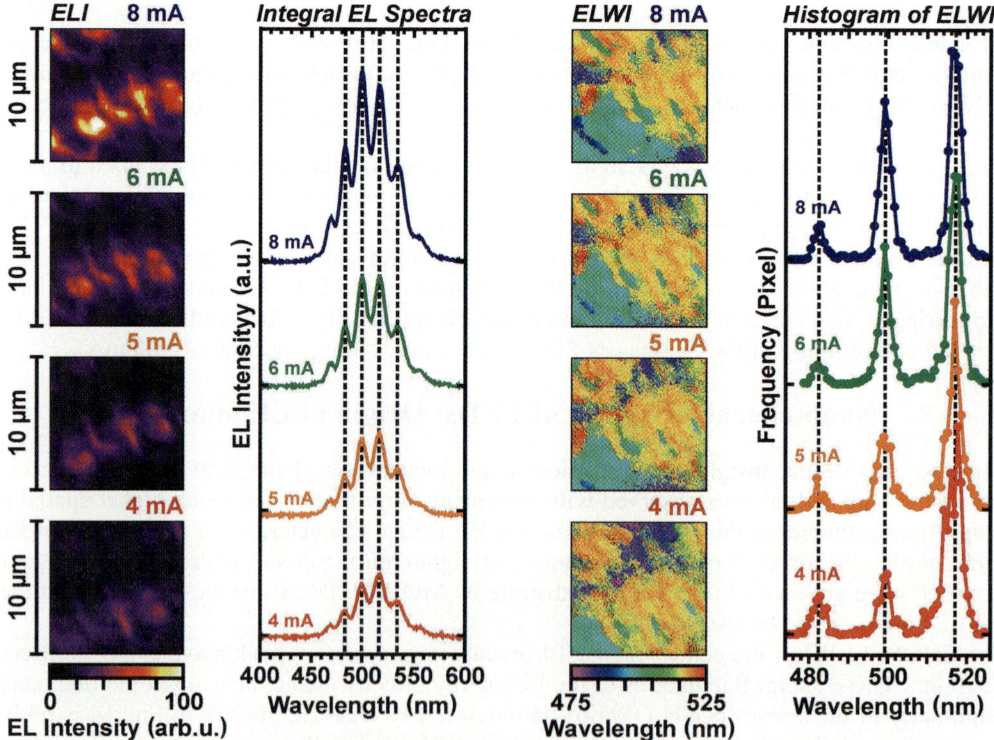

Fig. 15 Electroluminescence intensity images and ELWIs of an InGaN/GaN-LED under different injection currents, taken in the SNOM detection mode.

To understand the nanoscopic nature of the blue shift of EL with increasing current, the SNOM-EL-characterization was performed as a function of injection current. Figure 15 shows EL intensity images (ELIs), i.e. mappings of the spectrally integrated EL intensity (1. column from the left) and corresponding ELWIs of the identical sampling positions (3. column) as a function of increasing injection currents, ranging from $I = 4$ mA to $I = 8$ mA. The integral EL spectra, laterally integrated over the whole area of the ELIs, are depicted in the second column and the histograms of the ELWIs are given in the fourth column of Fig. 15.

A spot like pattern is clearly visible in the ELIs (all four plotted with identical intensity scale) proving that the EL emission only originates from nano-spots under these low injection conditions. The rising light output is accompanied by a spectral blue shift of the total EL, which is clearly visible in the integral EL spectra. This is visualized by the growing lateral size of the 498 nm wavelength domains (blue green contrast) with respect to the decreasing size of the 517 nm domains (yellow contrast) in the ELWIs.

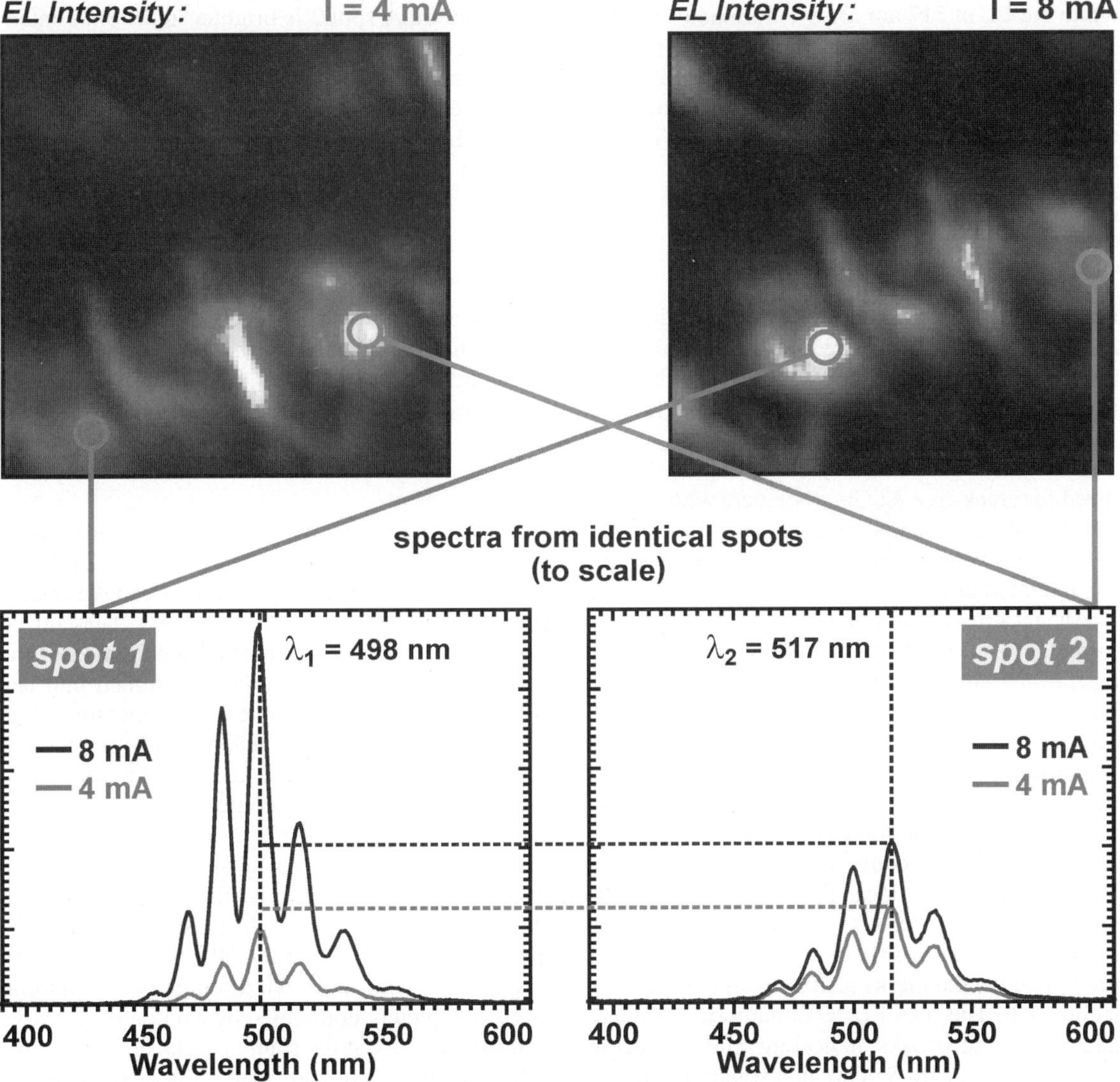

Fig. 16 (online colour at www.interscience.wiley.com) SNOM-EL intensity images (relative scale) and local EL spectra (absolute scale) for injection currents of 4 mA and 8 mA visualizing the growing contribution of high-energy emission due to the saturation of the EL from low-energy potential minima.

The SNOM-EL measurements prove a gradual saturation of the originally dominating emission from low-energy local potential minima with increasing injection current. Thus, in the SNOM-EL maps the lateral size of the short wavelength domains increases, i.e. the blue green regions in the ELWIs in Fig. 15 gain size which is quantitatively expressed by the ELWI histograms in Fig. 15 evidencing the growing contribution of the Fabry-Perot-mode at 498 nm in comparison to that at 517 nm.

Let us now look more detailed into the underlying physics. The two ELI maps of Fig. 15 recorded for 4 mA and for 8 mA injection current are compared in Fig. 16, and two distinct different sub-μm spots are identified in both ELIs. The relative intensity contrast is scaled separately to min/max for each ELI.

At 4 mA, the blue spot 1 is weaker in relative intensity than the red spot 2. In contrast, for 8 mA the situation is vice versa and the red spot 2 is relatively weaker than the blue spot 1. The nano-scale local spot spectra for spot 1 and spot 2 are depicted for 4 mA and for 8 mA in the lower row of Fig. 16.

Here the intensity scale is identical for all four spectra, which thus can be compared in absolute intensity. It is quite obvious that for both injection currents the emission peak at 498 nm dominates spot 1, while the EL at 517 nm always dominates spot 2. However, at 4 mA spot 2 is brighter in *absolute* intensity than spot 1, which gains much stronger in intensity on the current increase. For $I = 8$ mA the ratio of the absolute intensities is inverted and spot 1 becomes dominant. As a result, with higher injection current the short wavelength emitting local micro spots take over in absolute EL intensity as compared to the long wavelength emitting micro spots. This microscopic effect results in an increasing contribution of short wavelength emission to the integral EL and eventually in the macroscopic observed shift of the integral EL spectrum to shorter wavelength.

10   Conclusions

Fluctuations of the indium content in a thick InGaN epilayer are directly imaged by highly spatially resolved CL microscopy. While the spatial resolution of plan-view investigations is still limited by the penetration range of the incident electron beam, a considerable improvement of resolution is evidenced in the TEM-CL mode. Columns of high [In] are found to reach throughout the whole thickness of the 1 μm MBE-InGaN layer.

Extremely low fluctuations of the [Al]-dependent local bandgap are mapped and quantitatively analyzed for crack-free $Al_xGa_{1-x}N$ layers with x up to 0.76.

Combining cathodoluminescence and micro-Raman spectroscopy results with theoretical calculations, a comprehensive analysis of cracked $Al_xGa_{1-x}N$ epilayers is achieved. Here, the shift of the local near bandgap emission line of the AlGaN layer follows the measured and calculated strain profile between the cracks.

AlGaN grown on patterned GaN/Sapphire templates is found to be susceptible to the self-organized formation of Ga-rich AlGaN micro domains, leading to a complex three-dimensional modulation of the optical and structural properties. During initial AlGaN overgrowth of the template (patterned into terraces and trenches), the modulation of the local AlGaN stoichiometry results in periodic structures of Ga accumulation clearly reflecting the pattern periodicity. Following planarization, a homogeneous emission wavelength, i.e. homogeneous aluminum content, is found near the sample surface. All AlGaN microdomains additionally show nanoscopic modulations of their specific aluminum content.

Columnar ELOG domains formed during HVPE overgrowth of hexagonal SiO_2-masks were 3D imaged by CL microscopy. The areas of initial (0001) growth between the masks show sharp excitonic luminescence and a local free carrier concentration n well below 10^{17} cm^{-3}. The first stage of facetted lateral overgrowth leads to 10 μm high self-limited columns exhibiting broad (e,h)-plasma CL above the mask edges, extrinsic luminescence in the coalescence areas, and n reaching 10^{19} cm^{-3}. Subsequent growth above the masks results in an abrupt drop of n below 10^{17} cm^{-3} inside 30 μm high cones emitting Yellow CL. The final ELOG domain (uppermost 15 μm) is completely dominated by excitonic CL above and between the masks, indicating a homogeneous high crystal quality at the sample surface.

As evidenced by SNOM investigations in the detection mode, the overall blue shift of the laterally averaged EL spectrum of the LED is a result of the microscopic fluctuations in local EL emission wavelength due to composition fluctuations and their competition in quantum efficiency as a function of carrier injection density.

Acknowledgements   The authors thank the groups of D. Hommel (University of Bremen, Germany), H. Amano and I. Akasaki (Meijo University, Nagoya, Japan), K. J. Ebeling (University of Ulm, Germany), and A. Krost (Otto-von-Guericke University Magdeburg, Germany) for supplying the excellent samples. The authors are oblidged to Q. K. K. Liu (Hahn-Meitner-Institute, Berlin, Germany) for contributing the finite-element calculations. We are indebted to T. Hempel (Otto-von-Guericke-University Magdeburg, Germany) for the SNOM investigations and to M. Zacharias (Otto-von-Guericke-University Magdeburg, Germany) for help with the µ-EL measurements. The authors gratefully acknowledge financial support by the Deutsche Forschungsgemeinschaft (DFG) in the framework of contracts Nos. CH 87/4-1, CH 87/4-2, THO 662/4-1, and THO 662/4-2.

References

[1]   Y. Kato, S. Kitamura, K. Hiramatsu, and N. Sawaki, J. Cryst. Growth **144**, 133 (1994).

[2]   A. Usui, H. Sunakawa, A. Sakai, and A. Yamaguchi, Jpn. J. Appl. Phys. **36**, L899 (1997).

[3]   P. Vennéguès, B. Beaumont, V. Bousquet, M. Vaille, and P. Gibart, J. Appl. Phys. **87**, 4175 (2000).

[4]   Y. Honda, Y. Iyechika, T. Maeda, H. Miyake, and K. Hiramatsu, Jpn. J. Appl. Phys. Part 4A **40**, L309 (2001).

[5]   F. Bertram, T. Riemann, J. Christen, A. Kaschner, A. Hoffmann, C. Thomsen, K. Hiramatsu, T. Shibata, and N. Sawaki, Appl. Phys. Lett. **74**, 359 (1999).

[6]   A. Kaschner, A. Hoffmann, C. Thomsen, F. Bertram, T. Riemann, J. Christen, K. Hiramatsu, T. Shibata, and N. Sawaki, Appl. Phys. Lett. **74**, 3320 (1999).

[7]   A. Kaschner, A. Hoffmann, C. Thomsen, F. Bertram, T. Riemann, J. Christen, K. Hiramatsu, H. Sone, and N. Sawaki, Appl. Phys. Lett. **76**, 3418 (2000).

[8]   J. W. P. Hsu, M. J. Matthews, D. Abusch-Magder, R. N. Kleiman, D. V. Lang, S. Richter, S. L. Gu, and T. F. Kuech, Appl. Phys. Lett. **79**, 761 (2001).

[9]   M. J. Matthews, J. W. P. Hsu, S. Gu, and T. F. Kuech, Appl. Phys. Lett. **79**, 3086 (2001).

[10]   T. Riemann, J. Christen, A. Kaschner, A. Laades, A. Hoffmann, C. Thomsen, M. Iwaya, S. Kamiyama, H. Amano, and I. Akasaki, Appl. Phys. Lett. **80**, 3093 (2002).

[11]   M. Iwaya, S. Terao, T. Sano, T. Ukai, R. Nakamura, S. Kamiyama, H. Amano, and I. Akasaki, J. Cryst. Growth **237-239**, 951 (2002).

[12]   S. Nakamura, M. Senoh, N. Iwasa, S. Nagahama, T. Yamada, and T. Mukai, Jpn. J. Appl. Phys. Part 2 **10B**, L1332 (1995).

[13]   A. Dadgar, J. Christen, T. Riemann, S. Richter, J. Bläsing, A. Diez, A. Krost, A. Alam und M. Heuken, Appl. Phys. Lett. **78**, 2211 (2001).

[14]   I.-h. Ho and G. B. Stringfellow, Appl. Phys. Lett. **69**, 2701 (1996).

[15]   R. Singh, D. Doppalapudi, T. D. Moustakas, and L. T. Romano, Appl. Phys. Lett. **70**, 1089 (1997).

[16]   F. Bertram, S. Srinivasan, L. Geng, F. A. Ponce, T. Riemann, and J. Christen, Appl. Phys. Lett. **80**, 3524 (2002).

[17]   S. Nakamura, S. Senoh, N. Iwasa, and S. Nagahama, Jpn. J. Appl. Phys. Part 2 **34**, L797 (1995).

[18]   S. Nakamura, M. Senoh, S. Nagahama, N. Iwasa, T. Yamada, T. Matsushita, H. Kiyoku, Y. Sugimoto, T. Kozaki, H. Umemoto, M. Sano, and K. Chocho, Jpn. J. Appl. Phys. Part 2 **37**, L309 (1998).

[19]   A. Dadgar, A. Alam, T. Riemann, J. Bläsing, A. Diez, M. Poschenrieder, M. Straßburg, M. Heuken, J. Christen, and A. Krost, phys. stat. sol. (a) **188**, 155 (2001).

[20]   A. Dadgar, M. Poschenrieder, O. Contreras, J. Christen, K. Fehse, J. Bläsing, A. Diez, F. Schulze, T. Riemann, F. A. Ponce, and A. Krost, phys. stat. sol. (a) **192**, 308 (2002).

[21]   A. Dadgar, M. Poschenrieder, I. Daumiller, M. Kunze, A. Strittmatter, T. Riemann, F. Bertram, J. Bläsing, F. Schulze, A. Reiher, A. Krtschil, O. Contreras, A. Kaluza, A. Modlich, M. Kamp, L. Reißmann, A. Diez, J. Christen, F. A. Ponce, D. Bimberg, E. Kohn, and A. Krost, to appear in phys. stat. sol. (c) 0, No. 5 (2003) (this volume).

[22]   S. Chichibu, T. Azuhata, T. Sota, and S. Nakamura, Appl. Phys. Lett. **70**, 2822 (1997).

[23]   S. Chichibu, K. Wada, and S. Nakamura, Appl. Phys. Lett. **71**, 2346 (1997).

[24]   Y. Narukawa, Y. Kawakami, M. Funato, S. Fujita, S. Fujita, and S. Nakamura, Appl. Phys. Lett. **70**, 981 (1997).

[25]   T. Riemann, D. Rudloff, J. Christen, A. Krost, M. Lünenbürger, H. Protzmann, and M. Heuken, phys. stat. sol. (b) **216**, 301 (1999).

[26]   H. Kollmer, J. S. Im, S. Heppel, J. Off, F. Scholz, and A. Hangleiter, Appl. Phys. Lett. **74**, 82 (1999).

[27]   M. Asif Khan, X. Hu, A. Tarakij, G. Simin, J. Yang, R. Gaska, and M. S. Shur, Appl. Phys. Lett. **77**, 1339 (2000).

[28] S. Nakamura, M. Senoh, S. Nagahama, N. Iwasa, T. Yamada, T. Matsushita, Y. Sugimoto, and H. Kiyoku, Appl. Phys. Lett. **69**, 4056 (1996).

[29] D. J. H. Lambert, M. M. Wong, U. Chowdhury, C. Collins, T. Li, H. K. Kwon, B. S. Shelton, T. G. Zhu, J. C. Campbell, and R. D. Dupuis, Appl. Phys. Lett. **77**, 1900 (2000).

[30] S. Einfeldt, V. Kirchner, H. Heinke, M. Dießelberg, S. Figge, K. Vogeler, and D. Hommel, J. Appl. Phys. **88**, 7029 (2000).

[31] D. Rudloff, T. Riemann, J. Christen, K. Vogeler, S. Einfeldt, D. Hommel, A. Kaschner, A. Hoffmann, and C. Thomsen, Proc. Int. Workshop on Nitride Semicond., IPAP Conf. Ser. **1**, 475 (2000).

[32] M. Iwaya, S. Terao, N. Hayashi, T. Kashima, H. Amano, and I. Akasaki, Appl. Surf. Sci. **159-160**, 405 (2000).

[33] S. Einfeldt, M. Dießelberg, H. Heinke, D. Hommel, D. Rudloff, J. Christen, and R. F. Davis, J. Appl. Phys. **92**, 118 (2002).

[34] M. Iwaya, R. Nakamura, S. Terao, T. Ukai, S. Kamiyama, H. Amano, and I. Akasaki, Proc. Int. Workshop Nitride Semiconductors (IWN2000), IPAP Conf. Series **1**, 833 (2000).

[35] J. Christen, M. Grundmann, and D. Bimberg, J. Vac. Sci. Technol. B **9**, 2358 (1991).

[36] J. Christen and T. Riemann, phys. stat. sol. (b), **419** (2001).

[37] J. A. Freitas, Jr., O.-H. Nam, R. F. Davis, G. V. Saparin, and S. K. Obyden, Appl. Phys. Lett. **72**, 2990 (1998).

[38] T. Riemann, J. Christen, A. Kaschner, A. Hoffmann, C. Thomsen, O. Parillaud, V. Wagner, and M. Ilegems, in: Proc. of the IWN2000, IPAP Conf. Series **1**, 475 (2000).

[39] E. Feltin, B. Beaumont, P. Vennéguès, T. Riemann, J. Christen, J. P. Faurie, and P. Gibart, phys. stat. sol. (a) **188**, 733 (2001).

[40] S. Gradecak, V. Wagner, M. Ilegems, T. Riemann, J. Christen, and P. Stadelmann, Appl. Phys. Lett. **80**, 2866 (2002).

[41] V. Wagner, O. Parillaud, H. J. Bühlmann, M. Ilegems, S. Gradecak, P. Stadelmann, T. Riemann, and J. Christen, J. Appl. Phys. **92**, 1307 (2002).

[42] H. Miyake, S. Bohyama, M. Fukui, K. Hiramatsu, Y. Iyechika, and T. Maeda, J. Cryst. Growth **237-239**, 1055 (2002).

[43] S. J. Rosner, G. Girolami, H. Marchand, P. T. Fini, J. P. Ibbetson, L. Zhao, S. Keller, U. K. Mishra, S. P. DenBaars, and J. S. Speck, Appl. Phys. Lett. **74**, 2035 (1999).

[44] S. Dassonneville, A. Amokrane, B. Sieber, J. L. Farvacque, B. Beaumont, and P. Gibart, J. Appl. Phys. **89**, 3736 (2001).

[45] S. Dassonneville, A. Amokrane, B. Sieber, J.L. Farvacque, B. Beaumont, G. Gibart, J.-D. Ganiere, and K. Leifer, J. Appl. Phys. **89**, 7966 (2001).

[46] T. Riemann, J. Christen, A. Kaschner, A. Hoffmann, C. Thomsen, M. Seyboth, F. Habel, R. Beccard, and M. Heuken, phys. stat. sol. (a) **188**, 751 (2001).

[47] P. Fischer, J. Christen, M. Zacharias, V. Schwegler, C. Kirchner, and M. Kamp, phys. stat. sol. (a) **176**, 119 (1999).

[48] J. Christen, Adv. Solid State Phys. **30**, 239 (1990)

[49] P. Fischer, J. Christen, M. Zacharias, H. Nakshima, and K. Hiramatsu, Solid State Phenom. **63**, 151 (1998).

[50] H. Siegle, L. Eckey, A. Hoffmann, C. Thomsen. B.K. Meyer, D. Schikora, M. Hankeln, K. Lischka, Solid State Commun. **96**, 943 (1995).

[51] H. Selke, M. Amirsawadkouhi, P. L. Ryder, T. Böttcher, S. Einfeldt, D. Hommel, F. Bertram, and J. Christen, Mater. Sci. Eng. B **59**, 279 (1999).

[52] H. Amano, N. Sawaki, I. Akasaki, and Y. Toyoda, Appl. Phys. Lett. **48**, 353 (1986).

[53] Y. Koide, N. Itoh, K. Itoh, N .Sawaki, and I. Akasaki, Jpn. J. Appl. Phys. **27**, 1156 (1988).

[54] B. K. Meyer, G. Steude, A. Göldner, A. Hoffmann, H. Amano, and I. Akasaki, phys. stat. sol. (b) **216**, 187 (1999).

[55] K. Kanaya and S. Okayama, J. Phys. D: Appl. Phys. **5**, 43 (1972).

[56] L. H. Robins and D. K. Wickenden, Appl. Phys. Lett. **71**, 3841 (1997).

[57] MARC, McNeal-Schwendler Corp, USA, User's Guide 2000.

[58] Q. K. K. Liu, A. Hoffmann, H. Siegle, A. Kaschner, C. Thomsen, J. Christen, and F. Bertram, Appl. Phys. Lett. **74**, 3122 (1999).

[59] Q. Liu, A. Hoffmann, A. Kaschner, C. Thomsen, J. Christen, P. Veit, and R. Clos, Jpn. J. Appl. Phys. Part 2 **10A**, L958 (2000).

[60] D. Rudloff, T. Riemann, J. Christen, Q. K. K. Liu, A. Kaschner, A. Hoffmann, C. Thomsen, M. Dießelberg, S. Einfeldt, and D. Hommel, to appear in Appl. Phys. Lett. **81**, (2002).

[61] K. Hiramatsu, J. Phys.: Condens. Matter **13**, 6961 (2001).

[62] B. Beaumont, Ph. Vennéguès, and P. Gibart, phys. stat. sol. (b) **227**, 1 (2001).

[63] T. Detchprohm, S. Sano, S. Mochizuki, S. Kamiyama, H. Amano, and I. Akasaki, phys. stat. sol. (a) **188**, 799 (2001).

[64] M. Yano, T. Detchprohm, R. Nakamura, S. Sano, S. Mochizuki, T. Nakamura, H. Amano, and I. Akasaki, Proc. Int. Workshop Nitride Semiconductors (IWN2000), IPAP Conf. Series **1**, 292 (2000).

[65] T. Detchprohm, M. Yano, S. Sano, R. Nakamura, S. Mochizuki, T. Nakamura, H. Amano, and I. Akasaki, Jpn. J. Appl. Phys. Part 2 **40**, L16 (2001).

[66] S. Mochizuki, T. Detchprohm, S. Sano, T. Nakamura, H. Amano, and I. Akasaki, J. Cryst. Growth **237-239**, 1065 (2002).

[67] H. Siegle, Ph. D. thesis, Technical University Berlin, Germany (1998).

[68] A. Cros, H. Angerer, O. Ambacher, M. Stutzmann, R. Höpler, and T. Metzger, Solid State Commun. **104**, 35 (1997).

[69] L. Bergman, M. D. Bremser, W. G. Perry, R. F. Davis, M. Dutta, and R. J. Nernanich, Appl. Phys. Lett. **71**, 2157 (1997).

[70] F. Demangeot, G. Groenen, J. Frandon, M. A. Renucci, O. Briot, S. Clur, and R. L. Aulornbard, Appl. Phys. Lett. **72**, 2674 (1998).

[71] A. Petersson, A. Gustafsson, L. Samuelson, S. Tanaka, and Y. Aoyagi, MRS Internet J. Nitride Semicond. Res. **7**, 5 (2001).

[72] H. Siegle, A. Hoffmann, L. Eckey, C. Thomsen, J. Christen, F. Bertram, M. Schmidt, D. Rudloff, and K. Hiramatsu, Appl. Phys. Lett. **71**, 2490 (1997).

[73] F. Bertram, S. Srinivasan, F. A. Ponce, T. Riemann, J. Christen, and R. J. Molnar, Appl. Phys. Lett. **78**, 1222 (2001).

[74] W. Zhang, T. Riemann, H. R. Alves, M. Heuken, D. Meister, W. Kriegseis, D. M. Hoffmann, J. Christen, A. Krost, and B. K. Meyer, J. Cryst. Growth **234**, 616 (2002).

[75] P. Fischer, J. Christen, M. Zacharias, V. Schwegler, C. Kirchner, and M. Kamp, Jpn. J. Appl. Phys. Part 1 **39**, 2414 (2000).

[76] S. Nakamura, M. Senoh, N. Iwasa, and S. Nagahama, Appl. Phys. Lett. **67**, 13 (1995).

[77] P. Fischer, J. Christen, and S. Nakamura, Jpn. J. Appl. Phys. Part 2 **39**, L129 (2000).

phys. stat. sol. (c) **0**, No. 6, 1816–1834 (2003) / **DOI** 10.1002/pssc.200303127

Optical properties of nitride heterostructures

Andreas Hangleiter[*]

Institut für Technische Physik, Technische Universität Braunschweig,
Mendelssohnstrasse 2, 38106 Braunschweig, Germany

Received 4 March 2003, revised 16 April 2003, accepted 28 April 2003
Published online 28 August 2003

PACS 77.65.Ly,78.67.De, 85.60.Jb

Low-dimensional III-nitride heterostructures exhibit quite a number of unusual optical properties. We show that while compositional fluctuations do contribute to these phenomena, the dominant contribution comes from the huge internal fields due to piezoelectric and spontaneous polarisation. On a microscopic scale, the unexpectedly low recombination activity of defects in GaInN/GaN quantum wells appears to be related to potential variations correlated to the defect structure itself.

1 Introduction

GaInN/GaN/AlGaN heterostructures and quantum wells are at the heart of high performance blue, green, and amber LED's as well as violet laser diodes [1–3], which are now commercially available. Despite the very large number of defects still present in such structures [4], the efficiency and the lifetime [5] of these devices are extremely good. It appears that for GaInN-based LED's even a significant reduction of the defect density does not improve the efficiency any further [6].

Nevertheless, the optical properties of nitride quantum wells (QW's) are still subject to quite some controversial discussions. Typical nitride quantum wells exhibit a fairly large emission linewidth [7], a huge "Stokes" shift between emission and absorption [8], an energy and intensity dependent decay time of the emission [9], and a large spectral shift between spontaneous and stimulated emission [10]. Initially, it was proposed that carrier localisation due to compositional fluctuations [8] or even phase separation [11] in GaInN quantum wells could explain their unusual optical properties. It was also speculated that localised states could have an impact on the emission properties of violet laser diodes [12]. Moreover, such localisation is also invoked to explain the immunity of GaInN-based quantum wells to nonradiative recombination [6].

More recently, it was recognised that nitride quantum wells are subject to large piezoelectric fields which have a strong impact on their optical properties [13–16]. This is related to the fact that the critical thickness of strained nitride layers is surprisingly large [17–19]. Therefore, all nitride quantum well with thicknesses less than 20 nm are subject to strong biaxial strain leading to internal piezoelectric fields. Moreover, from theoretical considerations [20], there should also be a huge spontaneous polarisation in addition to the piezoelectric one, which might have an impact on the optical properties of nitride heterostructures.

---

[*] e-mail: a.hangleiter@tu-bs.de, Tel.: +49 531 391 8501, Fax: +49 531 391 8511

2 Unusual properties of nitride quantum wells

The most unusual property of nitride heterostructures has already been mentioned: their quantum efficiency is large even though there exists a large number of potentially nonradiative defects. In a more detailed picture, there is a number of peculiar features in the optical properties of low-dimensional nitride heterostructures, which affect device operation and performance.

1. In typical nitride quantum wells, there is a large Stokes-like shift between emission and absorption, i.e. the emission appears at longer wavelength than the absorption edge [8]. This shift is larger for wide quantum wells and for large In mole fractions in GaInN/GaN structures and Al mole fractions in GaN/AlGaN structures, respectively [21]. For GaN/AlGaN structures, it is obvious that the quantum well emission sometimes appears even below the bulk GaN bandgap [22].

2. With increasing excitation level by either optical pumping or electrical injection, the emission is subject to a considerable blue-shift, particularly for wide quantum wells [15, 23].

3. The emission decay times after pulsed excitation become rather large for wide wells and for large In or Al mole fractions. The decay is generally non-exponential and is accompanied by a red-shift of the emission peak [9, 14].

4. The emission linewidth of GaInN/GaN quantum wells is fairly large and increases with increasing In mole fraction [7] or increasing quantum well width [24].

5. Due to the low symmetry of the wurtzite structure, nitride heterostructures exhibit a considerable birefringence [25, 26].

Some of the early results along those lines together with high resolution electron microscopy led to speculations on carrier localisation [8] or even quantum dot formation [11] due to phase separation in GaInN. This picture was also advertised as an explanation for the exceptional quantum efficiency of GaInN QW's, where localisation at potential minima would save carriers from nonradiative recombination at dislocations [27].

However, it was later recognised that polarisation effects such as piezoelectric fields and spontaneous polarisation have a number of consequences for optical properties of nitride quantum wells [13–15]. In particular, strong piezoelectric fields of the order MV/cm [20] arise from the biaxial strain due to the lattice mismatch between well and barrier material.

3 Internal polarisation fields

3.1 Internal fields in polar crystals

Non-centrosymmetric polar crystals, when subject to external stress, may generate so-called piezoelectric fields due to a strain-induced polarisation. The electric polarisation field is given by

$$P_i = d_{ik}\sigma_k \qquad (i = x, y, z; k = xx, yy, zz, yz, zx, xy) \tag{1}$$

where P_i is the electric polarisation field, d_{ik} is the piezoelectric tensor, and σ_k is the *stress* tensor. We note that the polarisation may also be related to the *strain* tensor ϵ_k by another piezoelectric tensor denoted by e_{ik}.

In the wurtzite structure typically occurring for the III-nitrides, biaxial strain in the (0001) plane results in a polarisation field, which is most conveniently expressed by [28]

$$P_z = 2d_{31}(c_{11} + c_{12} - \frac{2c_{13}^2}{c_{33}})\epsilon_{xx}, \tag{2}$$

where d_{31} is the piezoelectric constant, c_{ij} are the elastic constants, and ϵ_{xx} is the in-plane strain. Equivalently, it may be expressed by the e tensor as

$$P_z = -(e_{33} - \frac{c_{33}}{c_{13}}e_{31})\epsilon_{zz}. \tag{3}$$

In fact, we prefer the former form over the latter, since only a single piezoelectric constant d_{31} is needed in the practically important case: Such a coherent in-plane strain is realized in pseudomorphically grown heterostructures with dissimilar lattice constants. The sign of the polarisation of course depends on whether the strain is compressive or tensile.

Both for GaInN quantum wells embedded in GaN and for GaN quantum wells embedded in AlGaN the natural lattice constant of the quantum well material would be larger than that of the barrier material. Therefore, such quantum wells are under biaxially compressive strain and should exhibit large piezoelectric fields.

3.2 Quantum confined Stark effect

The internal piezoelectric field in strained wurtzite heterostructures should give rise to a strong quantum confined Stark effect (QCSE) [29], leading to a red-shift of the quantum well emission with respect to an ideal square potential well. The effect of the electric field within the quantum well is depicted schematically in Fig. 1, which shows that the electron and the hole

wavefunction are localised in opposite corners of the quantum well. In the limit of large fields and wide wells this red-shift is proportional to the field times the well width [29]. Therefore, the most unambiguous test for the influence of piezoelectric fields is the dependence of the emission energy on well width.

3.3 Verification of QCSE for nitride quantum wells

The emission energy as a function of well width has been studied both for GaN/AlGaN and for GaInN/GaN quantum wells [13, 14, 22]. Fig. 2 and 3 shows the

results of these investigations. In both cases, only very narrow quantum wells emit at energies higher than the bulk bandgap of the well material. For well thicknesses larger than about 3 nm, the emission is

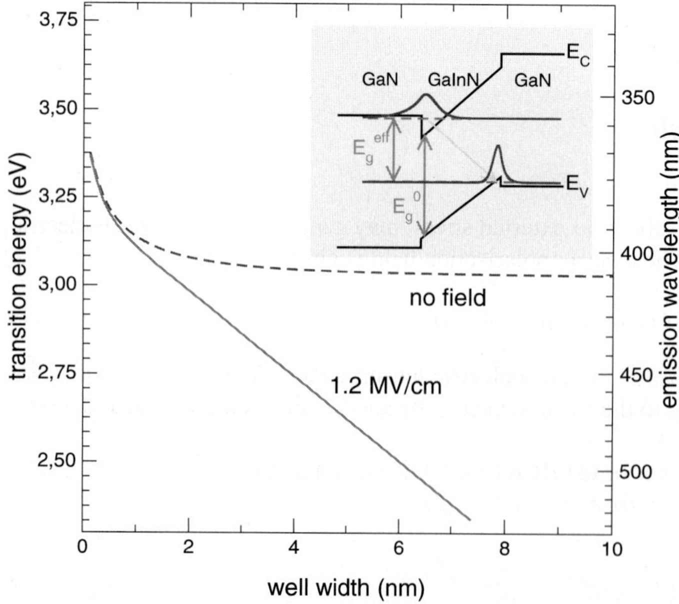

Fig. 1 (online colour at: www.interscience.wiley.com) Effective band gap of quantum wells with and without electric field. A field of 1.2 MV/cm is typical for piezoelectric polarisation in a GaInN/GaN quantum well with 10 % In. Inset: Schematic view of a quantum well with built-in field. Electrons and holes are localised in opposite corners, leading to a red-shift of the lowest transition and to a reduction of the oscillator strength.

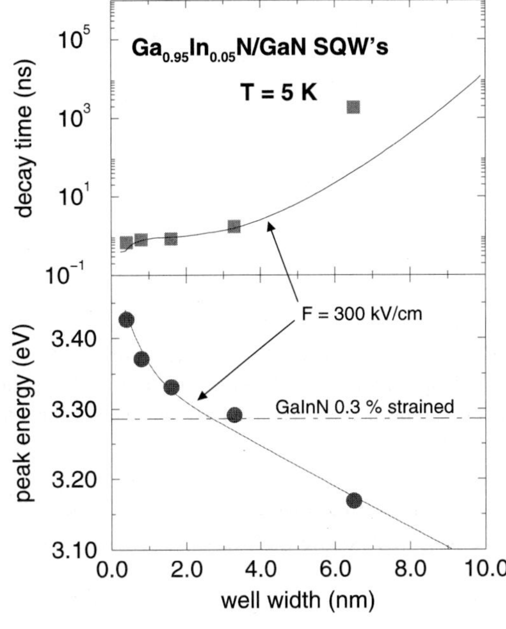

Fig. 2 (online colour at: www.interscience. wiley.com) Measured luminescence emission energy and decay time vs. well width for GaInN/GaN quantum wells (symbols). The full lines represent calculations of the effective band gap and the oscillator strength [22].

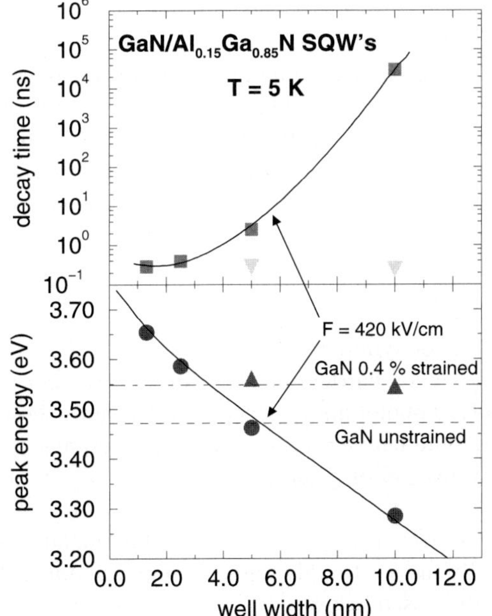

Fig. 3 (online colour at: www.interscience. wiley.com) Measured luminescence emission energy and decay time vs. well width for GaN/AlGaN quantum wells (symbols). The full lines represent calculations of the effective band gap and the oscillator strength [14].

below the bulk gap, shifting linearly to the red with increasing well width. Since the quantisation energy becomes constant for wide wells, the rate of red-shift is then equal to the electric field across the quantum well.

As already observed by Mendez *et al.* [29] for GaAs quantum wells, the oscillator strength of the lowest transition is reduced along with the red-shift. Fig. 2 and 3 also include the measured luminescence decay times vs. well width, which are inversely proportional to the oscillator strength. As can be seen from the figure, the oscillator strength is reduced by several orders of magnitude for wide nitride quantum wells. This effect is much more pronounced for nitride wells than in case of GaAs, since, due to the large effective

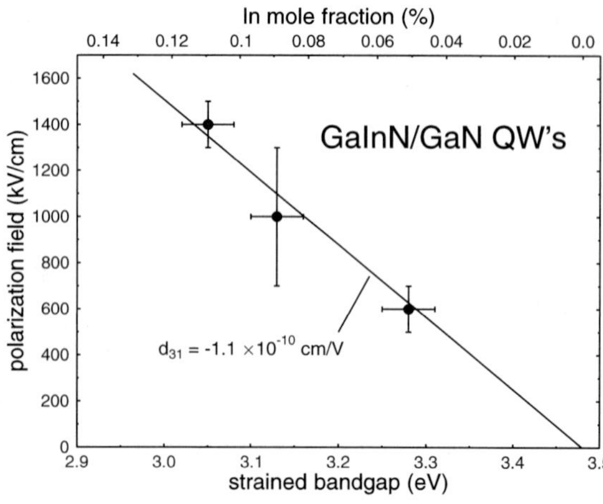

Fig. 4 Piezoelectric field vs. well composition and bandgap for GaInN/GaN quantum wells as determined from the measured well width dependence of the emission energy [30].

masses, quantisation effects set in only for well width of less than 3 nm. Moreover, since the fields are of the order of hundreds of kV/cm, the field induced potential drop across the quantum well easily becomes larger than the heterobarriers.

Even quantitatively, there is a remarkable agreement between the measured data and the predictions of the QCSE model. The full lines drawn in Fig. 2 and 3 are calculated on the basis of a very simple QCSE model, with the field being the only unknown parameter [14]. As clearly shown in the figure, the model does simultaneously account both for the red-shift and for the dramatic change of oscillator strength.

From a variation of the In mole fraction in GaInN/GaN quantum wells one can determine the piezoelectric field as a function of well composition. According to Vegard's law, the biaxial strain in the wells should be proportional to the In mole fraction, resulting in a linear variation of the piezoelectric field (provided the piezoelectric coefficient remains constant). As shown in Fig. 4 this is indeed observed. Fig. 4 also demonstrates that even for moderate In mole fractions of about 10 %, the internal field can be as large as 1400 kV/cm. The piezoelectric coefficient derived from these data is $d_{31} = -1.1 \times 10^{-10}$ cm/V.

3.4   Broken symmetry of QW's with internal field

The existence of internal piezoelectric fields breaks the mirror symmetry of a quantum well. As a consequence, it turns out to make a big difference whether an AlGaN barrier is placed "on top" or "below" of a GaInN/GaN quantum well as in Fig. 5 [31]. Time resolved photoluminescence measurements show that the low-temperature PL decay time, which is dominated by radiative processes, decreases dramatically if an AlGaN barrier is placed on top of a GaInN/GaN QW instead of placing it below the QW (in growth direction). These results are shown in Fig. 6.

Thus carrier confinement is much better if the AlGaN barrier is grown on top of the quantum well, leading to a strongly enhanced oscillator strength. Actually this is strongly related to the question of band discontinuities in nitride heterostructures, which will be discussed below.

Moreover, symmetry is also broken in terms of doping. Quantum well samples with either the lower or the upper barrier layers doped (n-type doping), or none of the barrier layers doped, differ strongly in emission wavelength, as shown in Fig. 7.

While N-type doping of the upper barrier layer of a quantum well does not lead to a significant shift in emission wavelength, doping of the lower barrier layer does lead to a strong blue-shift of the emission. This can be easily explained by the fact that electrons are confined close to the upper interface of the quantum well in the presence of an internal field. Thus electrons from the upper barrier do not lead screening of the field, whereas doping of the lower barrier layer reduces the field by screening [32].

Under high excitation conditions, piezoelectric fields are also screened by injected carriers. This leads to a blue-shift of the emission, the amount of which depends on the width of the quantum well and the quantum well composition [15]. Only due to this screening effect, optical gain and lasing of nitride laser structures becomes possible. If there were no screening, the extremely small oscillator strength would prevent such structures from lasing. On the other hand, under lasing conditions, the internal fields are largely screened and the full oscillator strength is almost recovered [33, 34].

3.5 Optical transitions in asymmetric double quantum wells

So far, all considerations were related to single quantum wells. When multiple quantum wells come into play, one has to face a more complex situation. Similar to the case of a superlattice subject to an electric field, the effect of highly spatially indirect transitions between electron and hole states in nearest neighbour and next nearest neighbour quantum wells has to be considered. The importance of such transitions has been clearly demonstrated by Kollmer *et al.* by comparing the emission spectra of single, double, and triple GaInN/GaN quantum wells [16].

By modifying the structures of Kollmer *et al.* and using two quantum wells of different thickness, three different transitions can be identified spectroscopically. The relevant optical transitions in such an asymmetric double quantum well based on GaN/AlGaN are shown in Fig. 8.

The thicknesses have been chosen such that the thicker well (4 nm) is twice as thick as the thinner well (2 nm), which lets us expect an equal spacing of the emission lines [35].

The properties of such a structure can be most directly seen from time-resolved emission spectra [35], as shown in Fig. 9.

As expected, the three emission lines at 3.57 eV, 3.46 eV, and 3.35 eV are equally spaced with a spacing of 110 meV/2 nm, indicating a field of about 550 kV/cm. The two higher energy emission lines have to be interpreted as the intra-well transitions labelled I and II in Fig. 8. The lowest energy emission corresponds to the inter-well transition labelled III in Fig. 8.

Clearly, the highest energy emission at about 3.57 eV decays most rapidly, whereas the lowest energy transition at about 3.35 eV survives for the longest time. Indeed, we expect a hierarchy of oscillator strengths, since the electron and hole wavefunctions are separated by 2, 4, and 8.5 nm for the respective transitions. This corresponds to a decreasing wavefunction overlap and to decreasing oscillator strength [35].

3.6 Stokes shift and linewidth in nitride quantum wells

Among the observations which are commonly cited as supporting localisation effects are the large "Stokes" shift between emission and absorption [8] and the fairly large photoluminescence (PL) linewidth [7]. Let us first address the relation between emission and absorption spectra. For the usual quantum-confined

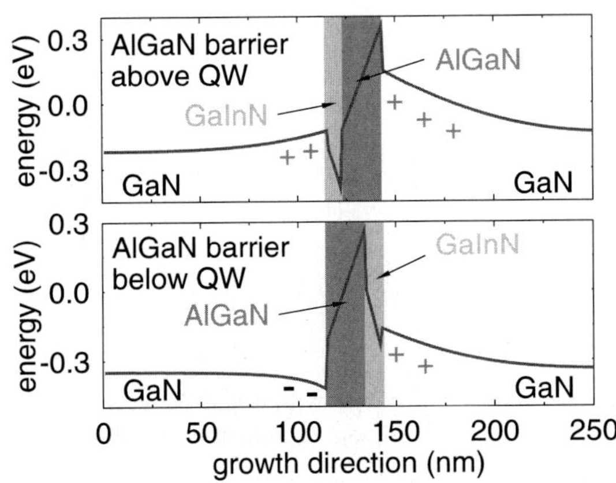

Fig. 5 (online colour at: www.interscience.wiley.com) Schematic conduction band profile for GaInN/GaN QW's with AlGaN barriers below and above the QW. Growth direction is from left to right.

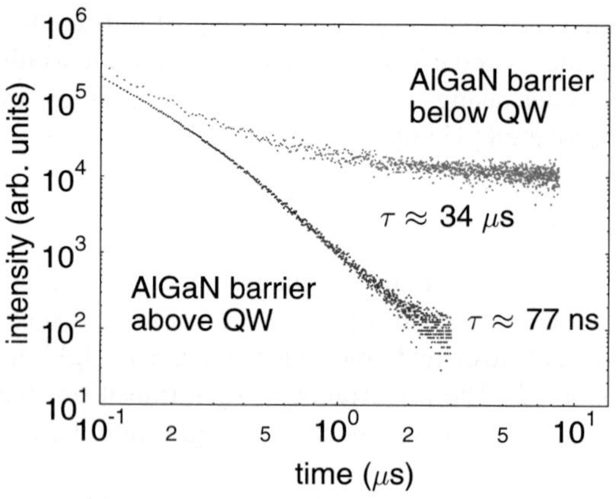

Fig. 6 (online colour at: www.interscience.wiley.com) Photoluminescence decay at low temperature for GaInN/GaN QW's with AlGaN barriers below and above the QW [31].

Stark effect [29] taking place at moderate fields, the absorption and the emission of a quantum well shift in the same way, with a slight loss in oscillator strength. For the huge internal fields present in the nitrides, however, the reduction of oscillator strength may amount to several orders of magnitude [14].

From a calculation of the energies and wavefunctions of both the confined and the unconfined states in piezoelectric quantum wells, we have calculated the absorption spectra of quantum wells with a strong built-in field. In Fig. 10 one can see that even though the lowest electron and hole states have very little spatial overlap, there are higher excited (hole) states, which may provide much higher oscillator strength.

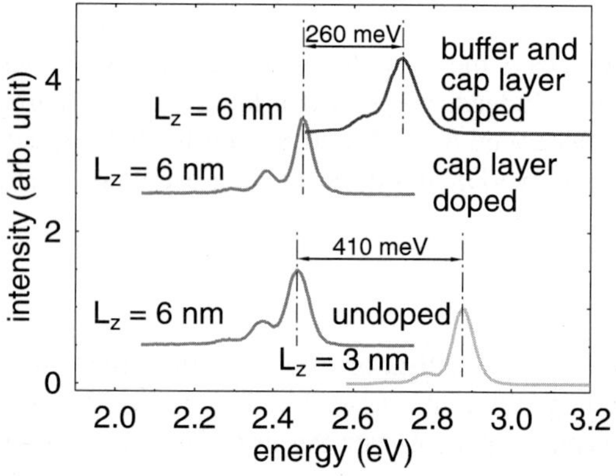

Fig. 7 (online colour at: www.interscience.wiley.com) Low-temperature photoluminescence spectra of GaInN/GaN quantum wells with different doping configurations in the barrier layers. For the 6 nm QW's, a strong blue-shift due to screening of polarisation fields is obtained only if the lower buffer layer is n-doped [32].

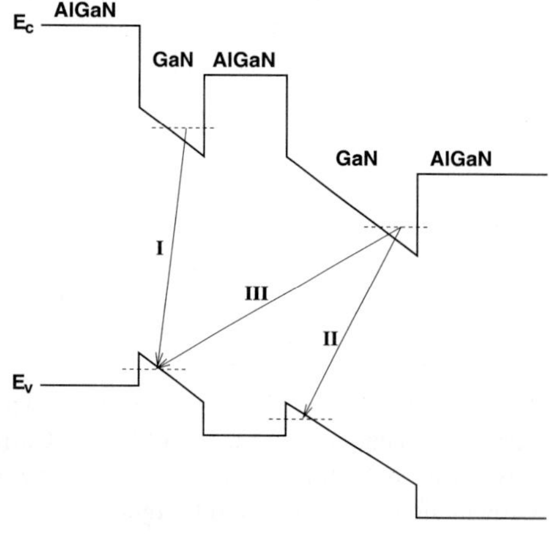

Fig. 8 Optical transitions within and between the quantum wells in an asymmetric double quantum well structure.

This means that significant absorption is possible only through excited states well above the ground state. However, the emission still comes from the ground state, which is the only one occupied at low temperature.

Fig. 11 shows calculated absorption and emission spectra taking into account the whole multitude of excited state transitions. We note that only for well widths of less than 3 nm there is significant absorption at the energy of the emission. for wider wells, there is an apparent Stokes shift of up to 500 meV. This shift is simply due to the fact that the ground state oscillator strength becomes extremely small and that states with sufficient overlap exist only at much higher energy. It is also interesting to note that the effective absorption edge remains at about the bandgap of the bulk well material, independent of the well width.

When analysing PL linewidth data both for GaInN/GaN and for GaN/AlGaN QW's, we noted that for well widths larger than about 2 nm, the linewidth increases linearly with increasing well width. Typical results of this kind are shown in Figs. 12 and 13. In both cases, a straight line has been fitted to the data.

In order to understand these results, we consider the effective gap energy in a piezoelectric quantum well

$$E_g^{eff} = E_g^0 + E_q - e \times F \times L_z \tag{4}$$

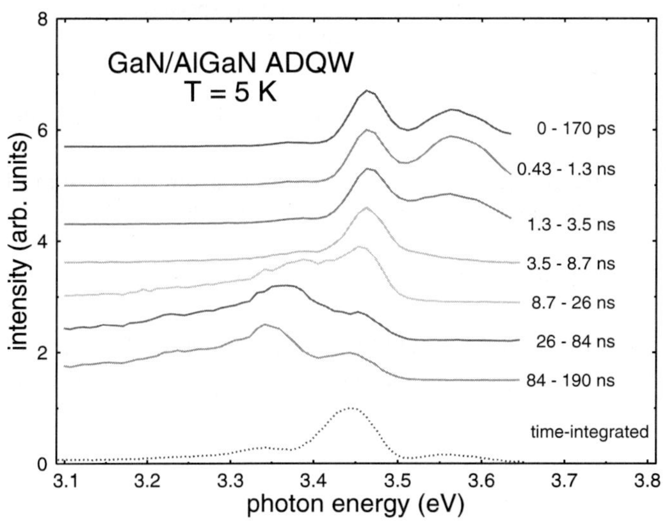

Fig. 9 (online colour at: www.interscience.wiley.com) Time-resolved photoluminescence spectra of a GaN/AlGaN asymmetric double quantum well [35].

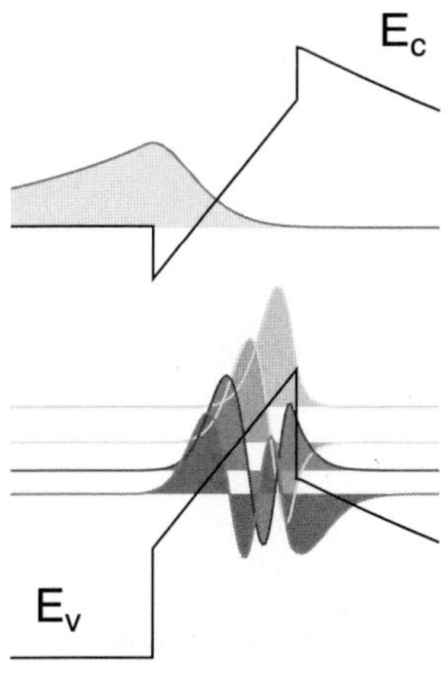

Fig. 10 (online colour at: www.interscience.wiley.com) Schematic view of optical transitions in a piezoelectric quantum well showing the calculated wavefunctions of the individual quantized states.

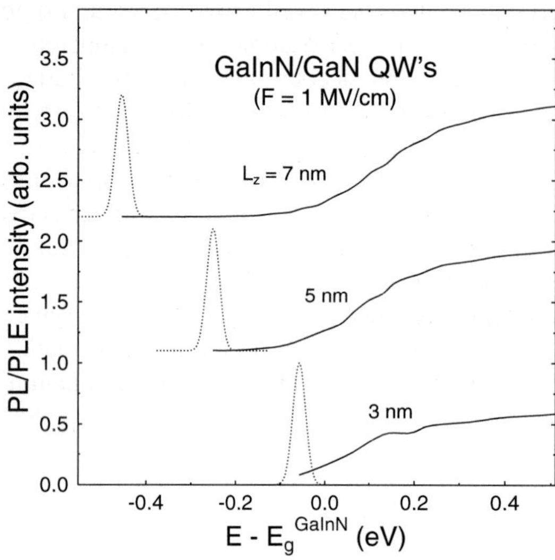

Fig. 11 Calculated emission and absorption spectra (Photoluminescence excitation spectra) of piezoelectric GaInN/GaN quantum wells.

where E_g^0 is the bulk bandgap of the well material, E_q is the quantisation energy, and $e \times F \times L_z$ is the field-induced red-shift of the gap.

In a first approximation, the variation of the effective gap, i.e. the linewidth, is given by the sum of fluctuation terms due to the bulk gap, the quantisation energy, the field, and the well width.

$$\delta E_g^{eff} = \delta E_g^0 + \delta E_q + e \times L_z \times \delta F + e \times F \times \delta L_z \tag{5}$$

If we neglect the variation of the quantisation energy (which is justified for these triangular wells except for extremely narrow ones) and take into account that the piezoelectric field is coupled to the well composition, we find

$$\delta E_g^{eff} = \delta E_g^0 \left(1 + e \times L_z \times \frac{\partial F}{\partial E_g^0}\right) + e \times F \times \delta L_z \tag{6}$$

$$= \left(\delta E_g^0 + e \times F \times \delta L_z\right) + \left(e \times \delta E_g^0 \times \frac{\partial F}{\partial E_g^0}\right) \times L_z \tag{7}$$

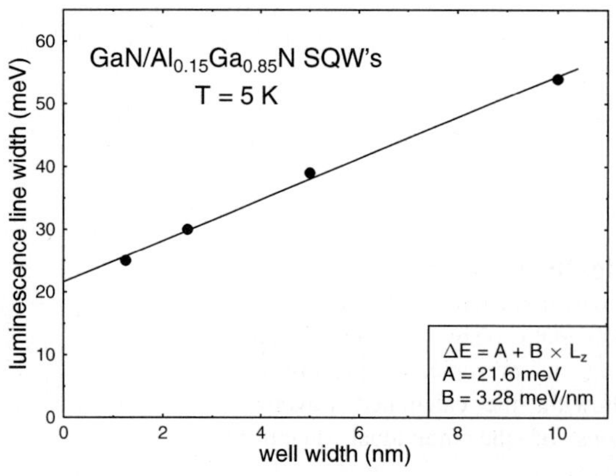

Fig. 12 Measured photoluminescence linewidth of GaN/AlGaN quantum wells of different well width. The data are well described by a straight line.

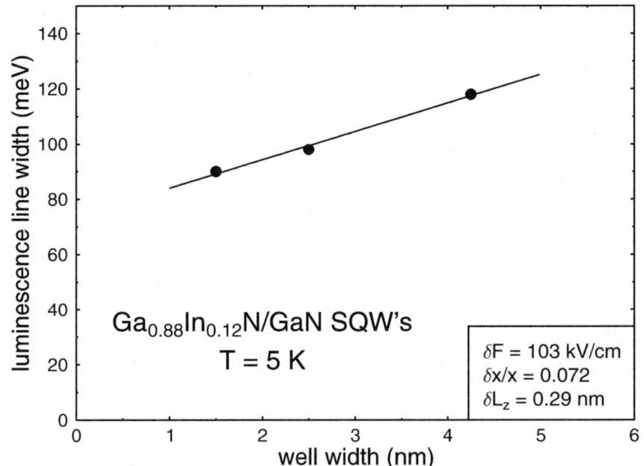

Fig. 13 Measured photoluminescence linewidth of GaInN/GaN quantum wells of different well width. The data are well described by a straight line.

This means that we have one term independent of well width, which is the fluctuation of the bulk gap plus a term proportional to the well width fluctuation. The second contribution is proportional to the well width L_z. In other words, we find a straight line dependence on well width as observed in our experiments.

Interestingly, since $\frac{\partial F}{\partial E_g^0}$ in the second equation is known from the experimentally observed Stark shift, the slope of this straight line directly gives us the fluctuation of the bulk bandgap. Knowing this, we extract the well width fluctuation from the constant term, thereby separating the two contributions quantitatively.

For our typical GaInN/GaN QW's with about 10 % In, we find a relative In fluctuation well below 10 % and a typical thickness fluctuation around 0.3 nm, i.e. 1 monolayer. For our GaN/AlGaN QW's with about 15 % Al in the barriers we find a fluctuation of barrier height and strain of 10 % and a thickness fluctuation of 0.43 nm.

4   Spontaneous polarisation fields

Besides piezoelectric fields, wurtzite nitrides also are subject to so-called spontaneous polarisation [20]. Spontaneous polarisation arises from the fact that the wurtzite unit cell of the III-nitrides has a certain non-ideality, i.e. differences in bond length, which is equivalent to a built-in strain and a resulting "piezo-electric" polarisation even in relaxed crystals. Under "normal" equilibrium conditions, these fields are macroscopically screened by surface charges or background carriers. However, the so-called pyroelectric fields then arise from the temperature dependence of the spontaneous polarisation when temperature is changed.

The general consequence of these considerations is that internal electric fields (piezoelectric and spontaneous ones) tend to mask themselves on a macroscopic "global" scale by attracting charged surface adsorbates, which screen the internal fields. Consequently, any sample under equilibrium conditions will not exhibit a macroscopic potential difference between their polar surfaces, regardless of any internal fields.

When studying GaN-based heterostructures by low-temperature cathodoluminescence (CL) in an electron microscope, we had noted for quite some time that the CL spectra change during observation, except for the lowest excitation power densities. The beam energies ranged from 3 keV to 20 keV. A typical example is shown in Fig. 14, where the CL spectra of an GaInN/GaN quantum well

are shown during the first two minutes of recording [36]. In this particular case, the emission line shifts to the blue by about 25 nm, gaining about one order of magnitude in intensity. By reducing the excitation power again, the original line position can not be recovered. Only by moving the excitation spot to a different place on the sample, the original spectrum can be observed again. Due to the quasi-permanent nature of this change, we denote the original spectrum as representing the "original state" and the blue-shifted spectrum as the "switched state".

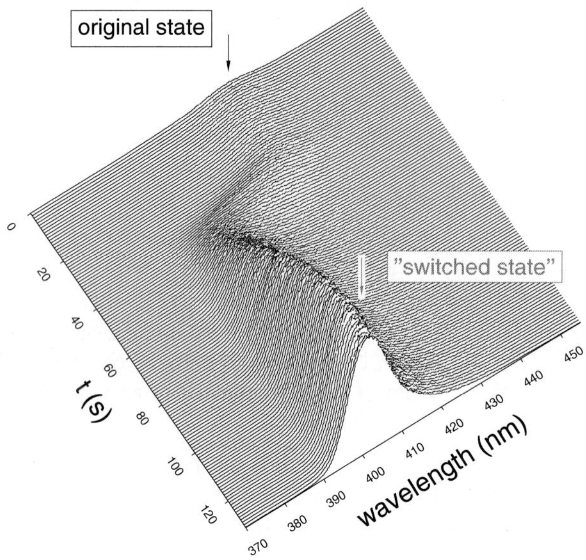

Fig. 14 (online colour at: www.interscience.wiley.com) Temporal series of low-temperature cathodoluminescence spectra of a $GaIn_{0.11}N/GaN$ quantum well. After about 30 s, the emission wavelength jumps to the blue, accompanied by a drastic increase in intensity [36].

To exclude a charging effect of the exciting electron beam (which could also be observed as a bright spot in the electron microscope), we designed an experiment, where we were able to measure photoluminescence (PL) excited by a 325 nm He-Cd laser within the electron microscope. In order to assure that the laser beam and the electron beam were exciting the same area on the sample, a 50 μm pinhole was mounted on the sample. Then the PL was observed while irradiating the pinhole area with the electron beam. Fig. 15 shows that again the emission is shifted to

higher energy by as much as 500 meV. The low-energy PL vanishes after complete irradiating of the pinhole area.

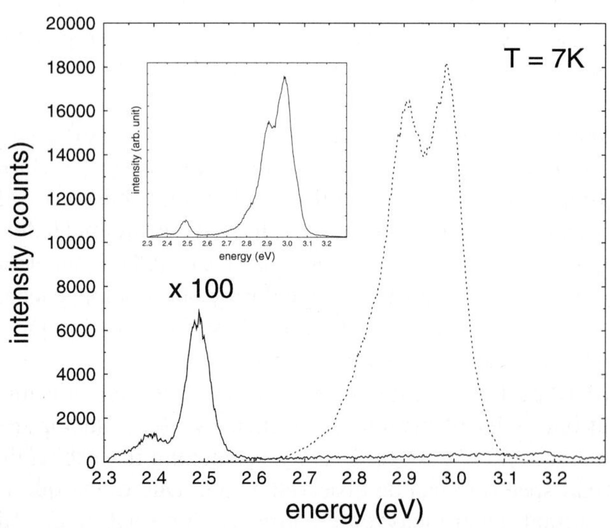

Fig. 15 Low-temperature photoluminescence spectra of a 6 nm $GaN/In_{0.11}GaN/GaN$ quantum well structure before (solid line) and after (dotted line) irradiation with an relatively strong electron beam (10 nA beam current at 10 kV). The inset shows a mixed state where only a fraction of the area was irradiated [36].

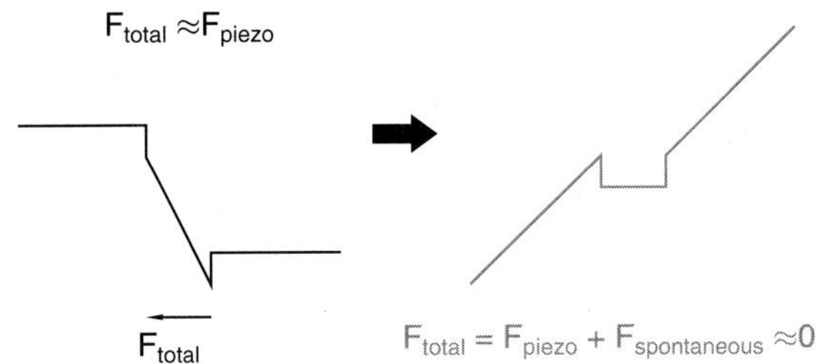

Fig. 16 (online colour at: www.interscience.wiley.com) Schematic view of the change of the internal field arising from changing the screening charge at the surface. Since the spontaneous and the piezoelectric field are of opposite sign and similar magnitude, they may actually cancel in the quantum well.

Using the same setup, we used a QW sample with a 200 nm thick GaN cap layer. At a beam energy of 3 keV, the penetration depth of 80 nm is too small for the electron beam to reach the quantum well. This is verified by no detectable CL signal under these conditions. The PL, however, can be readily detected. As above, we observe a large blue-shift of the emission line. This means that the change of emission wavelength is not a direct effect of the quantum well, but is associated with the topmost layer or the surface.

Within the electron microscope at low-temperature, the "switching" seems to be irreversible. However, when the sample is kept at room temperature under ambient conditions, the original emission wavelength is recovered. At elevated temperature the recovery is accelerated, while in vacuum it takes longer. All these observations indicate that the change of emission wavelength is not associated with defects but rather directly with the surface. It is interesting to note here that Achour *et al.* studied the effect of electron irradiation on CdS, ZnO, and GaAs [37] and found a desorption of one species of molecules covering the surface followed by a occupation with more stable adsorbates.

We propose [36] that the origin of all these observations is related to unmasking the spontaneous polarisation field, due to a change in surface coverage. Consider that the spontaneous polarisation is originally screened by charged adsorbates at the surface. A rough estimate shows that this needs a density of adsorbates of about $2 \cdot 10^{13} \text{cm}^{-2}$. Now the electron beam removes the adsorbates and unmasks the spontaneous polarisation field. Finally, the surface is covered by more stable neutral adsorbates, which allow for the stability of the "switched state". The effect of that on the quantum well is shown schematically in Fig. 16. As long as the spontaneous field is screened,

the electric field in the quantum well is determined by the piezoelectric polarisation. Since the spontaneous polarisation is of the same order of magnitude as the piezoelectric one, with the opposite sign, unmasking the spontaneous field leads to a reduction or even cancellation of the field in the quantum well. This reduction of the field in the quantum well leads to a strong blue-shift of the quantum well emission.

We have verified this picture by studying the change of the luminescence decay time (i.e. the oscillator strength) before and after irradiating the sample [36]. This was possible thanks to the rather stable nature of the "switched state", which allowed for a transfer of samples between different setups. Fig. 17 shows the

measured luminescence decay traces for different energetic positions of the emission line of a 6 nm GaInN/GaN quantum well. For the lowest energy position, the decay time is too long to be resolved. For partly and completely switched states, respectively, the decay times become much shorter being of the order 100 ns. This clearly supports the idea of a reduced field in the QW after irradiating the surface with electrons.

5   Internal fields and LED efficiency

It has been argued that internal polarisation fields are detrimental to the efficiency of light emitting diodes [6] and that only localisation of carriers by compositional fluctuations can provide immunity to nonradiative recombination at threading dislocations. Within this line of reasoning, it is thought that indeed

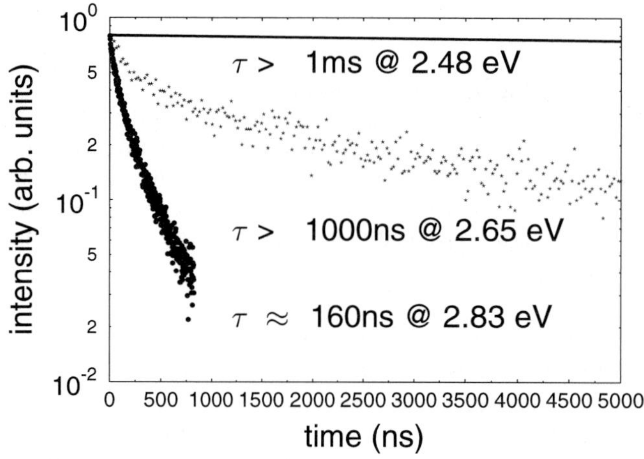

Fig. 17 Carrier lifetimes for different energetic positions of the luminescence. For the original state at 2.487eV the lifetime is in the millisecond-range, for the switched states however it clearly decreases with increasing emission energy [36].

polarisation fields lead to a strongly reduced oscillator strength for radiative transitions and therefore to a reduced quantum efficiency.

However, though it is true that the probability of **radiative recombination** is reduced by internal fields, we have little *a priori* knowledge on the effect of internal fields on **nonradiative recombination**. It is certainly not appropriate to assume that nonradiative transitions are unaffected by strong internal fields.

The physical origin of the reduced radiative oscillator strength is the spatial separation of the electron and hole wavefunctions. In fact, the very same mechanism may also apply to nonradiative transitions. Fig. 18 shows a sketch of the situation

representative for nonradiative recombination in a piezoelectric quantum well. The recombination proceeds through a deep defect level, whose wavefunction is localised on an atomic scale. As with conventional two-step recombination involving deep defect states [38, 39], the capture rate for electrons and holes is proportional to the overlap of their respective wavefunction with that of the defect state. Since the total recombination rate is governed by the smaller of the two capture coefficients, the maximum nonradiative rate will be reached if the defect is located such that the two overlaps are equal. Moreover, since the wavefunctions decay exponentially, the spatial separation of the electron and hole wavefunctions will lead to a dramatic decrease in the nonradiative recombination rate.

Obviously this scenario is very similar to that applicable to radiative recombination as discussed earlier. Both the radiative and the nonradiative recombination rate can be expected to exhibit a dramatic decrease

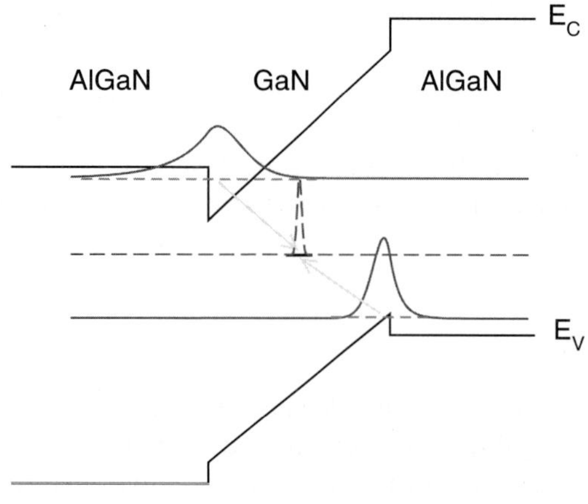

Fig. 18 (online colour at: www.interscience.wiley.com) Schematic view of a nonradiative recombination process through a deep defect level in a piezoelectric quantum well.

for large internal fields and for wide quantum wells. Now, the quantum efficiency is given by the ratio of the radiative to the total recombination rate. It is extremely difficult to predict its behaviour, which will depend on all the details of the specific defect involved. However, it is certainly not justified to assume that internal fields will automatically lead to a reduced quantum efficiency.

6   Band discontinuities at GaInN/GaN heterojunctions

The band lineup at heterojunctions in the nitrides has not been studied to great detail up to now. Experimentally, only a few groups used photoemission type experiments to investigate the band offsets of InN/GaN and GaN/AlN heterojunctions [40, 41]. Deep level emissions were also used as a reference to estimate the band offsets in GaInN/GaN [42] and GaN/AlGaN heterostructures [43, 44]. While the results GaN/AlN are quite similar, yielding a band offset ratio $\Delta E_c : \Delta E_v$ around 75:25, the results for InN/GaN vary widely between 30:70 [40] and 60:40 [42].

On the other hand, theoretical calculations for the unstrained band offsets [45, 46] agree fairly well on a valence band offset of 0.7-0.8 eV for GaN/AlN and 0.3-0.5 eV for InN/GaN. Those calculations also indicate that there is little difference between the wurtzite and zincblende cases. For GaN/AlN the theoretical results are in fair agreement with the experimental data mentioned earlier. In contrast, for GaInN/GaN theory predicts a small valence band offset while at least some of the experimental data indicate a rather large valence band offset.

In quantum well structures, the band offsets govern the degree of carrier confinement. This is particularly important in nitride quantum wells, since pseudomorphic quantum wells in this highly polar material system are subject to large internal polarisation fields [13, 14], due to both piezoelectric and spontaneous polarisation [20]. As a consequence of these internal fields, carrier confinement in such quantum wells effectively occurs in a triangular potential well.

Due to the smaller electron effective mass, quantisation is more pronounced in the conduction band. The degree of wavefunction penetration into the barrier material and therefore the oscillator strength of the band-to-band transition depends sensitively on the band offset in relation to the field strength. For GaN/AlGaN structures at 15 % Al, for example, the conduction band offset is about 240 meV with a polarisation field of about 550 kV/cm. This means that for wide wells there remains a triangular potential with about 4.4 nm width.

In the previous literature, it has been assumed that the band offsets in nitride heterostructures scale linearly with the composition of ternaries. However, it is well known that the bandgap of GaInN is subject to strong bowing [17]. This has also been shown by earlier calculations [47], which demonstrate that bowing takes place almost exclusively in the valence band.

A detailed analysis of calculated results for the band energies as a function of composition shows that the valence and conduction band discontinuities are subject to strong bowing as well [48]. These results (for the unstrained case) are shown in Fig. 19, which demonstrates the strongly nonlinear behaviour.

In particular, the strong bowing of the valence band leads to a quick increase of the valence band discontinuity with In mole fraction. In contrast, the conduction band offset varies almost linearly with composition.

This leads to a situation where the conduction band offset between GaInN and GaN is larger than the valence band offset for In compositions less than about 30 %. For the binaries, on the other hand, the valence band offset is still much smaller than the conduction band offset. In fact, we have to conclude that for the range of In compositions below 30 %, which is the most important in practice, the conduction band offset is only about one half of the valence band offset.

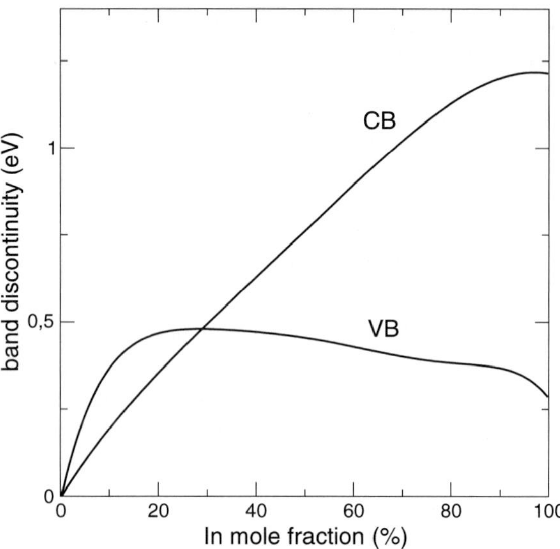

Fig. 19 Calculated valence and conduction band off-sets as a function of In mole fraction. The strong bowing leads to a reversal of the band offset ratio for low In mole fractions [48].

6.1   Experimental confirmation

In order to confirm this unexpected theoretical conclusion, we have performed an experimental study of carrier confinement in nitride quantum wells with different barriers. In particular, we have compared symmetric GaInN/GaN and GaN/AlGaN quantum wells with asymmetric AlGaN/GaInN/GaN and $Al_yGa_{1-y}N/GaN/Al_xGa_{1-x}N$ quantum wells. The In and Al composition in the symmetric structures was chosen to 5 % and 15 %, respectively, in order to obtain very similar magnitudes of the internal field. The well width was set to 7 nm. For the asymmetric structures, the top AlGaN layer contained 15 % Al in the AlGaN/GaInN/GaN case and 22 % Al in the AlGaN/GaN/AlGaN case.

Since the internal electric field points towards the substrate [15], electrons are confined in a triangular potential at the top of the quantum well in such structures. Due to the smaller electron effective mass, the electron wavefunction penetrates much more into the barrier and therefore electrons are much more sensitive to variations in barrier height. A very sensitive measure of electron confinement is therefore given by the interband oscillator strength.

Using an envelope function approach the electron and hole wavefunctions as well as the oscillator strength for asymmetric quantum wells were calculated for various values of the band offsets [48]. The results show that starting from a low conduction band barrier, corresponding to a small conduction band to valence band offset ratio, the oscillator strength increases rapidly when increasing the barrier height. On the other hand, starting from a larger conduction band barrier, corresponding to a large conduction band to valence band offset ratio, there is only a modest increase in oscillator strength [48].

The oscillator strength of the test structures described earlier was investigated using photoluminescence decay measurements at low temperature. The results are summarised in Fig. 20.

While replacing the to GaN barrier in a GaInN/GaN quantum well by an AlGaN barrier reduces the carrier lifetime about threefold (i.e. increases the oscillator strength), the effect of increasing the height of the top AlGaN barrier in a GaN/AlGaN quantum well is very small.

These observations are consistent with the picture that for GaN/AlGaN the conduction band offset is already relatively large, and increasing the barrier height means only a slight improvement of carrier confinement. On the other hand, the oscillator strength in GaInN/GaN quantum wells appears to be very sensitive with respect to the height of the top electron barrier. This would not be consistent with a large conduction band and small valence band offset for GaInN/GaN. It is consistent with a rather small conduction band offset as predicted by Fig. 19.

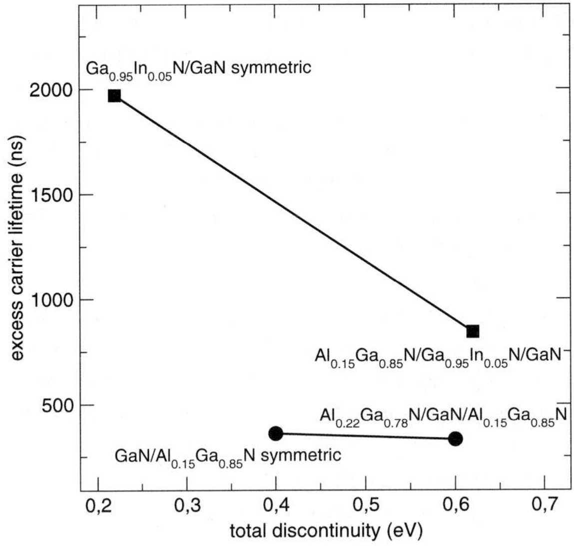

Fig. 20 Measured excess carrier lifetime of various quantum wells vs. total band offset at the top interface [48].

Our experimental results concerning electron confinement in nitride quantum wells confirm the notion of a small conduction band offset in practical GaInN/GaN quantum wells. Our calculated results predict that the conduction band to valence band offset ratio is strongly composition dependent, due to strong and anomalous bowing mostly in the valence band. Simple linear interpolation of the band offsets does not seem to be appropriate. As the strong bowing of the bandgap has been observed experimentally by several groups [17, 18], it is reasonable to expect similar effects for the band offsets.

The small conduction band offset in practical GaInN quantum wells is also highlighted by the need for an AlGaN barrier on top of the quantum wells both in light-emitting diode (LED) and in laser structures. Any lack of carrier confinement in such structures results in a reduced quantum efficiency due to leakage currents.

7 Microscopic origin of quantum well luminescence

As already discussed earlier, the microscopic origin of the luminescence from nitride quantum wells is subject to quite some debate. While many of the presumed macroscopic signatures of carrier localisation are likely to be at least partly due to internal field effects, there is little doubt that potential fluctuations do exist in nitride quantum wells. One possible way to quantify them has been discussed here in the context of emission linewidth. Other possibilities include a determination of localisation energies from the temperature dependence of the radiative lifetime of carriers [49].

On a microscopic scale there still remains the question why defects present in abundance do little harm to the emission efficiency. In order to answer this question, the electronic and optical properties of nitride heterostructures have to be studied on a length scale well below the mean distance between defects, i.e. below about 100 nm.

To this end, we have worked on scanning near field spectroscopy, which should in principle be able to provide a spatial resolution around 1/10 of a wavelength, i.e. about 40 nm. Our near field microscope now allows to aquire a complete spectrum with high sensitivity at every point of the near-field image, both at room temperature and at low temperature. Using home-made fibre tips based on UV fibres, the spatial resolution under optimum conditions is better than 100 nm.

Quite a number of GaInN/Gan quantum well structures have been studied using the near-field microscope. Since it relies on a mechanism similar to an atomic force microscope to keep the fibre tip at a constant distance from the surface, a topographic image is automatically available with each near-field image. This is beneficial to our purpose, since in nitride quantum wells it is quite common that defects are associated with pits visible on the surface [50]. These pits can be observed in the topography image and can be correlated to features in the near-field images.

Fig. 21 shows a room-temperature scan of 4-QW sample on a 2x2 μm scale [51]. The topography image clearly shows partially interconnected pits. Both the intensity image and the peak wavelength image, which have the outlines of the pits superimposed, show a clear correlation with the pits. While it is expected that bright areas are located between defects, it is somewhat unexpected to observe changes in emission wavelength correlated with the defect structure. It appears that in the present case, regions close to defects exhibit longer emission wavelengths.

A low-temperature (50 K) SNOM scan of a similar sample (2-QW) shown in Fig. 22 shows a similar correlation [51]. Even though no useful topography could be obtained in this case, intensity and wavelength images do show a correlation: Bright emission comes from long-wavelength areas, while short-wavelength areas are dim. Assuming that dim areas are close to defects we conclude that there is a blue-shift in the vicinity of defects in this case.

These studies, even though the results do not yet provide really enough resolution to be sure, tend to show that there is a considerable variation of emission wavelength in GaInN/GaN quantum wells on a scale of the distance between defects. The apparent correlation between this wavelength variation and the defect structure suggests that the formation of the former is related to the latter. While any smaller scale variations are beyond our resolution limit, the mere existence of a potential landscape in the quantum well correlated to the defect structure necessarily has an influence on the emission efficiency. Therefore the present results provide an important guideline for future studies.

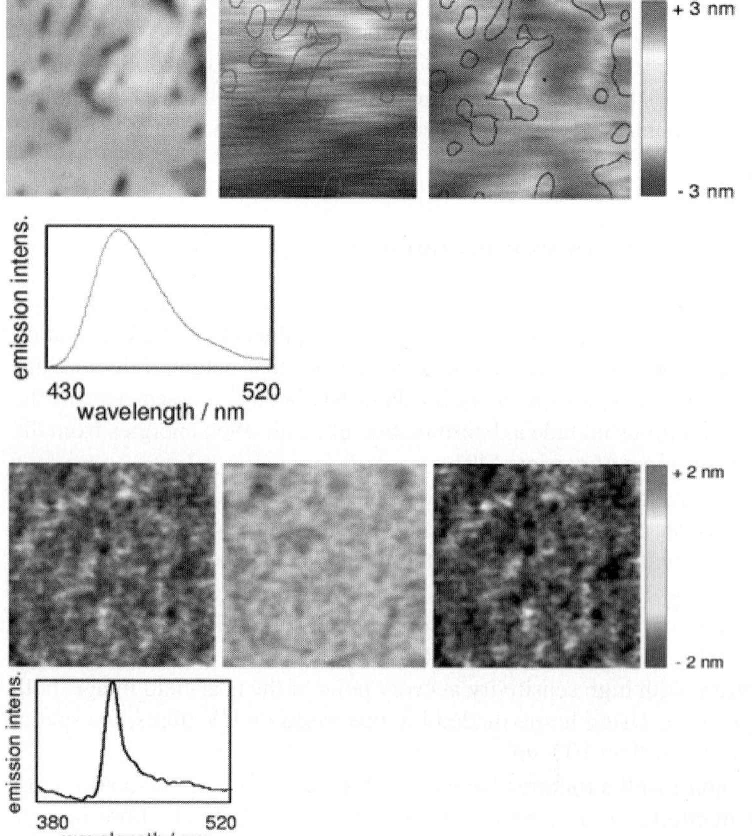

Fig. 21 (online colour at: www.interscience.wiley.com) 2 x 2 micrometer SNOM scan of a 4 QW GaInN/GaN sample at room-temperature [51]. Top row from left to right: topography, integrated emission intensity, peak wavelength map (center wavelength 440 nm). Bottom row: Typical local emission spectrum.

Fig. 22 (online colour at: www.interscience.wiley.com) 4 x 4 micrometer SNOM scan of a 2 QW GaInN/GaN sample at low temperature (50 K) [51]. Top row from left to right: integrated emission intensity, peak wavelength map (center wavelength 440 nm), product of both. Bottom row: Typical local emission spectrum.

8 Conclusions

The optical properties of nitride heterostructures have been investigated to great detail. Most prominently, the huge internal polarisation fields give rise to a large Stark shift of the emission and to a strongly reduced oscillator strength in quantum wells. The field breaks the symmetry of quantum wells and introduces the possibility of inter-well transitions. Even the so-called Stokes shift has a large contribution due to the Stark effect. The unusually large width of emission lines is attributed to fluctuations of composition and well width, but is strongly amplified by the internal field. On a microscopic scale, near-field studies of GaInN/GaN QW's show that local potential variations are correlated with the defect structure of the underlying buffer layer. This may provide further clues to the origin of the unusually large quantum efficiency of GaInN/GaN quantum wells.

Acknowledgements The present review summarises work done in collaboration with a large group of people at the Universities of Stuttgart and Braunschweig. Special thanks are due to O. Gfrörer, F. Hitzel, J.S. Im, S. Lahmann, C. Netzel, U. Rossow, and F. Scholz. The generous financial support by the Deutsche Forschungsgemeinschaft (DFG) in the framework of the programme "Gruppe-III-Nitride und ihre Heterostrukturen" is gratefully appreciated.

References

[1] S. Nakamura, M. Senoh, N. Iwasa, and S.-I. Nagahama, Appl. Phys. Lett. **67**, 1868 (1995).

[2] S. Nakamura, M. Senoh, N. Iwasa, S.-I. Nagahama, T. Yamada, and T. Mukai, Jpn. J. Appl. Phys. **34**, L1332 (1995).

[3] S. Nakamura, M. Senoh, S.-I. Nagahama, N. Iwasa, T. Yamada, T. Matsushita, H. Kiyoku, and Y. Sugimoto, Jpn. J. Appl. Phys. **35**, L74 (1996).

[4] S. D. Lester, F. A. Ponce, M. G. Craford, and D. A. Steigerwald, Appl. Phys. Lett. **66**, 1249 (1995).

[5] S. Nakamura, M. Senoh, S. Nagahama, N. Iwasa, T. Yamada, T. Matsushita, Y. Sugimoto, and H. Kiyoku, Jpn. J. Appl. Phys. **36**, L1059 (1997).

[6] T. Mukai, K. Takekawa, and S. Nakamura, Jpn. J. Appl. Phys. **37**, L839 (1998).

[7] K. P. O'Donnell, T. Breitkopf, H. Kalt, W. van der Stricht, I. Moerman, P. Demeester, and P. G. Middleton, Appl. Phys. Lett. **70**, 1843 (1997).

[8] S. Chichibu, T. Azuhata, T. Sota, and S. Nakamura, Appl. Phys. Lett. **69**, 4188 (1996).

[9] Y. Narukawa, Y. Kawakami, S. Fujita, S. Fujita, and S. Nakamura, Phys. Rev. B **55**, R1938 (1997).

[10] S. Chichibu, D. A. Cohen, M. P. Mack, A. C. Abare, P. Kozodoy, M. Minsky, S. Fleischer, S. Keller, J. E. Bowers, U. K. Mishra, L. A. Coldren, D. R. Clarke, and S. P. DenBaars, Appl. Phys. Lett. **73**, 496 (1998).

[11] Y. Narukawa, Y. Kawakami, M. Funato, S. Fujita, S. Fujita, and S. Nakamura, Appl. Phys. Lett. **70**, 981 (1997).

[12] S. Nakamura, M. Senoh, S. Nagahama, N. Iwasa, T. Yamada, T. Matsushita, Y. Sugimoto, and H. Kiyoku, Appl. Phys. Lett. **70**, 2753 (1997).

[13] T. Takeuchi, S. Sota, M. Katsuragawa, M. Komori, H. Takeuchi, H. Amano, and I. Akasaki, Jpn. J. Appl. Phys. **36**, L382 (1997).

[14] J. S. Im, H. Kollmer, J. Off, A. Sohmer, F. Scholz, and A. Hangleiter, Phys. Rev. B **57**, R9435 (1998).

[15] A. Hangleiter, J. S. Im, H. Kollmer, S. Heppel, J. Off, and F. Scholz, MRS Internet J. Nitride Semicond. Res. **3**, 15 (1998).

[16] H. Kollmer, J. S. Im, J. Off, F. Scholz, and A. Hangleiter, Appl. Phys. Lett. **74**, 82 (1999).

[17] M. D. McCluskey, C. G. van de Walle, C. P. Master, L. T. Romano, and N. M. Johnson, Appl. Phys. Lett. **72**, 2725 (1998).

[18] C. Wetzel, T. Takeuchi, S. Yamaguchi, H. Katoh, H. Amano, and I. Akasaki, Appl. Phys. Lett. **73**, 1994 (1998).

[19] C. A. Parker, J. C. Roberts, S. M. Bedair, M. J. Reed, S. X. Liu, and N. A. El-Masry, Appl. Phys. Lett. **75**, 2776 (1999).

[20] F. Bernardini, V. Fiorentini, and D. Vanderbilt, Phys. Rev. B **56**, R10024 (1997).

[21] J. S. Im, S. Heppel, H. Kollmer, A. Sohmer, J. Off, F. Scholz, and A. Hangleiter, J. Crystal Growth **189-190**, 228 (1998).

[22] J. S. Im, H. Kollmer, J. Off, A. Sohmer, F. Scholz, and A. Hangleiter, in Nitride Semiconductors, Vol. 482 of MRS. Symp. Proc., edited by F. A. Ponce, S. P. DenBaars, B. K. Meyer, S. Nakamura, and S. Strite (Materials Research Society, Pittsburgh, 1998), pp. 513–8.

[23] P. Riblet, H. Hirayama, A. Kinoshita, A. Hirata, T. Sugano, and Y. Aoyagi, Appl. Phys. Lett. **75**, 2241 (1999).

[24] A. Hangleiter, J. S. Im, J. Off, and F. Scholz, phys. stat. sol. (b) **216**, 427 (1999).

[25] S. Heppel, R. Wirth, J. Off, F. Scholz, and A. Hangleiter, phys. stat. sol. (a) **176**, 73 (1999).

[26] D. Ciplys, R. Gaska, M. S. Shur, R. Rimeika, J. W. Yang, and M. A. Khan, Appl. Phys. Lett. **76**, 2232 (2000).

[27] S. Nakamura, Science **281**, 956 (1998).

[28] A. Bykhovski, B. Gelmont, and M. Shur, J. Appl. Phys. **74**, 6734 (1993).

[29] E. Mendez, G. Bastard, L. Chang, L. Esaki, H. Morkoc, and R. Fischer, Phys. Rev. B **26**, 7101 (1982).

[30] A. Hangleiter, Proc. SPIE - Int. Soc. Opt. Eng. **3944**, 58 (2000).

[31] J. S. Im, H. Kollmer, J. Off, F. Scholz, and A. Hangleiter, Mat. Sci. Eng. B. **59B**, 315 (1999).

[32] J. S. Im, H. Kollmer, J. Off, F. Scholz, and A. Hangleiter, MRS Internet J. Nitride Semicond. Res. **4S1**, G6.20 (1998).

[33] W. W. Chow, H. Amano, T. Takeuchi, and J. Han, Appl. Phys. Lett. **75**, 244 (1999).

[34] S.-H. Park, S.-L. Chuang, and D. Ahn, Appl. Phys. Lett. **75**, 1354 (1999).

[35] J. S. Im, A. Hangleiter, J. Off, and F. Scholz, in GaN and related alloys, Vol. 595 of MRS Symp. Proc., edited by R. Feenstra (Materials Research Society, Pittsburgh, 2000), p. W11.28.

[36] O. Gfrörer, J. Off, F. Scholz, and A. Hangleiter, phys. stat. sol. (b) **216**, 405 (1999).

[37] S. Achour, A. Harabi, and N. Tabet, Mat. Sci. Eng. B **42**, 289 (1996).

[38] W. Shockley and W. Read, Phys. Rev. **87**, 835 (1952).

[39] R. Hall, Phys. Rev. **87**, 387 (1952).

[40] G. Martin, A. Botchkarev, A. Rockett, and H. Morkoç, Appl. Phys. Lett. **68**, 2541 (1996).

[41] A. Rizzi, R. Lantier, H. Lth, F. della Sala, A. di Carlo, and P. Lugli, J. Vac. Sci. Technol. B **17**, 1674 (1999).

[42] C. Manz, M. Kunzer, H. Obloh, A. Ramakrishnan, and U. Kaufmann, Appl. Phys. Lett. **74**, 3993 (1999).

[43] J. Baur, K. Maier, M. Kunzer, U. Kaufmann, and J. Schneider, Appl. Phys. Lett. **65**, 2211 (1994).

[44] D. R. Hang, C. H. Chen, Y. F. Chen, H. X. Jiang, and J. Y. Lin, J. Appl. Phys. **90**, 1887 (2001).

[45] S.-H. Wei and A. Zunger, Appl. Phys. Lett. **69**, 2719 (1996).

[46] C. G. van de Walle and J. Neugebauer, Appl. Phys. Lett. **70**, 2577 (1997).

[47] L. Bellaiche, S.-H. Wei, and A. Zunger, Appl. Phys. Lett. **70**, 3558 (1997).

[48] A. Hangleiter, S. Lahmann, C. Netzel, U. Rossow, P. R. C. Kent, and A. Zunger, in GaN and Related Alloys, Vol. 693 of Mat. Res. Soc. Symp. Proc., edited by J. E. Northrup, J. Neugebauer, D. C. Look, S. F. Chichibu, and H. Riechert (Materials Research Society, Boston, MA, 2002), p. I7.2.

[49] C. Netzel, R. Doloca, S. Lahmann, U. Rossow, and A. Hangleiter, phys. stat. sol. (c) **0**, 324 (2002).

[50] I.-H. Kim, H.-S. Park, Y.-J. Park, and T. Kim, Appl. Phys. Lett. **73**, 1634 (1998).

[51] F. Hitzel, S. Lahmann, U. Rossow, and A. Hangleiter, phys. stat. sol. (c) **0**, 537 (2002).

phys. stat. sol. (c) **0**, No. 6, 1835–1845 (2003) / **DOI** 10.1002/pssc.200303137

The origin of the PL photoluminescence Stokes shift in ternary group-III nitrides: field effects and localization

M. Strassburg[1]**, A. Hoffmann**[*, 1]**, J. Holst**[1]**, J. Christen**[2]**, T. Riemann**[2]**, F. Bertram**[2]**, and P. Fischer**[2]

[1] Institut für Festkörperphysik, Technische Universität Berlin, Hardenbergstr. 36, 10623 Berlin, Germany
[2] Institut für Experimentelle Physik, Otto-von-Guericke Universität Magdeburg, Universitätsplatz 2, 39106 Magdeburg, Germany

Received 4 March 2003, revised 4 August 2003, accepted 4 August 2003
Published online 28 August 2003

PACS 78.20.Hp, 78.30.Fs, 78.45.+h, 78.47.+p, 78.67.De, 78.67.Hc

We report on a systematic analysis of the localization and separation mechanisms of carriers and their co-existence in ternary group-III nitrides. AlGaN/GaN and InGaN/GaN multi-quantum-well (MQW) structures grown on sapphire substrate were investigated. Although the localization and separation mechanisms of carriers result in a similar behavior of the luminescence properties (e.g. both lead to blue shift of emission energy with increasing excitation density and vice versa a transient red shift after pulsed excitation), a clear distinction between these mechanisms is given by optical investigations with resonant and non-resonant excitation. Carrier separation due to a localization induced by internal electric fields described by the quantum-confined Stark-effect exist in all MQW structures and was found to be important in particular in the AlGaN/GaN MQWs. In general, the internal electric fields lead to a reduction of the luminescence efficiency. A fundamental localization mechanism, i.e. the localization in nm-scale islands of local minimum energy directly result from the spatial energy fluctuations due to disorder (e.g. alloy or QW thickness fluctuations). Especially at temperatures below 100 K, this mechanism dominates the carrier localization in the InGaN/GaN system. Evidence is given by the unique S-shape temperature dependence of peak energy and the existence of a mobility edge. Due to the localization of carriers in nm-scaled islands the overlap of its wavefunctions increases, which strongly increases the efficiency of the light output. This effect is significant weaker but still detectable in the AlGaN/GaN system.

1 Introduction

The mechanism of the optical recombination in low-dimensional InGaN and AlGaN quantum structures has been under controversial discussion in recent years [1]. The observed red shift in luminescence and absorption experiments was assigned to spontaneous and strain-induced piezo-electric fields [2] as well as localization effects in islands formed by segregation during the growth of ternary alloys [3]. In the present paper we give an overview of the field effects and the localization properties in ternary group-III-nitride structures of reduced dimensionality. We will show that a similar experimental effect (red shift of luminescence) is caused by two principle different mechanisms. On the one hand, strong electric fields amplify the localization of carriers at the opposite interfaces of the quantum well system and thus, lead to a separation of electrons and holes. Otherwise, quantum dot-like structures provide localization of charge carriers and excitons. Here, the overlap of the electron and hole wavefunctions and therefore the PL efficiency may increase. The localization mechanisms in ternary group-III nitrides will be analyzed, distinguished and assigned to material systems and structural particularities.

* Corresponding author: e-mail: hoffmann@physik.tu-berlin.de, Phone: +49 30 314 22001, Fax: +49 30 314 22064

2   Experimental

To study the impact of internal electric fields AlGaN/GaN multiple-quantum wells (MQWs) were inves-tigated. This material system was chosen, because the carriers are localized in the binary compound (GaN). Hence, no alloy broadening is to consider for the localization of carriers as it is typically for ter-nary compounds accept the neglectable weak effect of alloy fluctuations in the AlGaN barriers. A series of AlGaN/GaN MQWs with a variation of the GaN QW width from 2.2 to 4.5 nm were investigated. The samples were grown by MOCVD consisting of a threefold GaN/AlGaN MQW embedded in AlGaN claddings. To achieve a lattice-matched structure, the aluminum content in the MQWs was chosen to 20-30%, whereas the Al content in the ternary claddings is 10 %. The lower cladding was grown on a GaN buffer on sapphire substrate.

A further AlGaN/GaN sample was grown by MOCVD. It differs from the series described above by the MQW system consisting of a tenfold stack of 3 nm thick GaN layers separated by 20 nm thick Al-GaN barriers. This sample was applied for gain measurements and investigations at high-excitation den-sities.

Additionally, for the pump-and-probe experiments a specially designed sample was grown by MBE. This sample consists of a tenfold $Al_{0.11}Ga_{0.89}N$/GaN MWQ. The thickness of the GaN is 1.5 nm and of the AlGaN barriers 5 nm. The MQW system is embedded in AlGaN claddings of several 100 nm thick-nesses.

A series of (InGaN/GaN MQWs) samples was chosen to investigate the impact of potential fluctua-tions. Sample A belongs to a series of MBE-grown samples on sapphire, with an 18 µm GaN (MOVPE) layer and capped by a 30 nm GaN layer. The optical active region of sample A consists of a tenfold 5 nm InGaN MQW with 4 nm GaN barriers. The indium concentration of 6.7 % was determined by XRD. Sample B is also an tenfold InGaN/GaN MQW with a thickness of 4 nm each, grown with different growth temperatures for the QWs and the barriers by low-pressure MOCVD on ~2 µm GaN buffer layer on c-sapphire substrates in a horizontal production-type reactor AIX 2600 G3 in 2000 HT configuration. The In composition was measured by XRD to be 12.5 % with a GaN barrier thickness of 7.7 nm. The GaN buffer layer was grown at 1180 °C. The growth temperatures of the InGaN QWs and the GaN bar-riers were 750 °C and 950 °C, respectively. For a more detailed description see Ref. [4].

Photoluminescence (PL) was excited by the 325 nm line of a HeCd-laser. The samples were mounted in a cryostat allowing the variation of temperatures between 2 K and 300 K. The spectral resolution of the detection system was better than 0.2 meV. The high-excitation density investigations were performed using a dye laser pumped by an Excimer-laser providing pulses with a duration of 15 ns at a repetition rate of 30 to 50 Hz and a total energy of up to 20 µJ. For the pump-and-probe spectroscopy (PPS) the Excimer- laser emission was used for the pump beam while for the probe a dye laser was applied. The gain measurements were performed using the variable stripe-length method [5]. The experimental set up for the spectrally-, spatially-, and time-resolved cathodoluminescence experiments is described else-where [6,7].

3   Results and discussion

Carrier localization and separation in ternary InGaN- and AlGaN-based heterostructures is reported by many authors and was found to determine the optical properties [1-3]. Two different mechanisms were found. Both, the quantum-confined Stark-effect (QCSE) and the localization in quantum dot (QD)-like structures having similar effect on emission wavelength namely causing a pronounced blue shift of PL maximum with increasing excitation density in the active region of UV, blue or green emitting de-vices. The QCSE is caused by internal electric fields and is attributed to govern the optical properties of AlGaN-based devices. Also in the case of InGaN the presence of strong piezoelectric fields is demon-strated [8,9]. Furthermore, some authors explain the recombination behavior solely by the QCSE [10]. In the following we will show, that the localization in QD-like island structures must not be neglected to understand the optical recombination in group-III nitride heterostructures.

3.1 Influence of internal electric fields

To study the influence of the electric fields and to avoid perturbations of the observed effects by localization in potential fluctuations the AlGaN/GaN system was chosen. Due to the Coulomb interaction between electron and holes, excitons are formed. These excitons having a Bohr radius of several nm. Therefore, internal electric fields may separate the electron and the hole of an exciton to each hetero-interfaces of a quantum well. The main advantage of this GaN/AlGaN heterosystem is the confinement of the excitons in the binary GaN QW. Hence the localized 2D excitons only very weakly experience the residual alloy fluctuations in the AlGaN due to a partial tunneling into the MQW barriers. Nevertheless, a pronounced blue shift of the luminescence peak position is usually observed with increasing excitation density [11]. In the following it will be demonstrated that the emission is governed by spontaneous polarization and piezo-electric fields. As it will be shown later, a good agreement between theoretical predictions and experimental results were achieved. The separation of carriers (i.e., electrons and holes of excitons) in QW due to electric fields is described by the quantum-confined Stark-effect (QCSE). The carriers with opposite charge are located at different interfaces of the well and thus the overlap of their wavefunctions reduces. This strongly influences the opto-electronic properties of the heterostructures.

The main parameters determining the emission properties are the exact alignment of the polarization, the thickness of the quantum well and the Al-content of the barriers.

To illustrate the effect of inner electric fields AlGaN/GaN MQW structures differing in the well and barrier thickness were investigated. The polarization in these samples P_{total} can be written as

$$P_{total} = P_{TS} + P_{mis} + P_{spont} = P_{piezo} + P_{spont} , \tag{1}$$

where P_{TS} is the polarization caused by the thermal strain and P_{mis} is the polarization due to the lattice mismatch. In dependence of the barrier (L_B) and the well thickness (L_W) the total field can be obtained after [12] by

$$F_w = \frac{L_B \left(P_{total}^B - P_{total}^W \right)}{\varepsilon_0 \left(L_B \varepsilon_W + L_W \varepsilon_B \right)} . \tag{2}$$

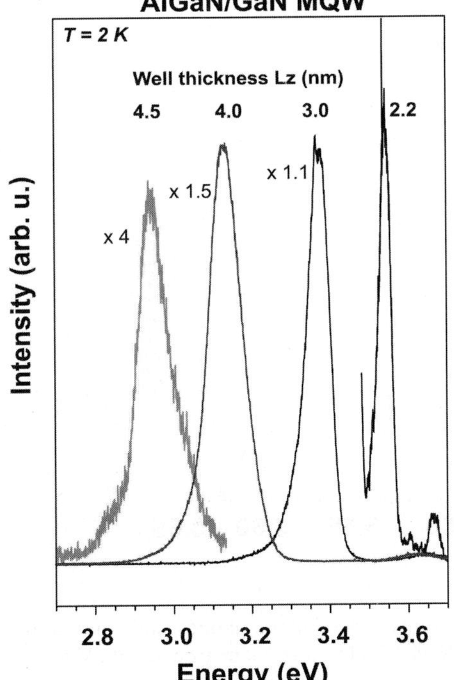

Fig. 1 (online colour at: www.interscience.wiley.com) Photoluminescence (PL) spectra of AlGaN/GaN MQW structures at low temperatures. The varied parameter is the well thickness marked at the respective PL spectra.

The internal electric field strongly influences the emission energy as shown in Fig. 1. The emission energy shifts to lower energies with increasing well thickness. For well thickness above 2.5 nm the emission energy is lower than the GaN bandgap. Obviously, the internal electric fields overcompensates the confinement effect, which causes a blue shift of the emission energy. Furthermore, this QCSE results in reduction of transition probability explaining the observed decrease in PL intensity for thicker wells. The broadening of the PL linewidth is attributed particularly to thickness fluctuations in the AlGaN/GaN MQWs.

Another effect of the QCSE on the emission is the reduction of the transition probability. The internal electric fields limit the optical recombination due to the reduction or break-up of the overlap of electron- and hole-wavefunctions. Thus, the recombination probability is dramatically reduced. Hence, a steep increase of the lasing threshold is observed. Especially in structures with elevated Al content in the barriers, this tendency is pronounced. Their impact on the optical properties are exemplarily illustrated by high-excitation investigations of the MOCVD-grown tenfold AlGaN/GaN MWQ.

In Figure 2 a comparison of low-excitation and high-excitation PL (stimulated emission) is depicted. At low excitation energies the PL is dominated by QW luminescence around 3.4 eV. A second emission band at 3.52 eV is attributed to a spatially direct transition into the GaN well. With increasing excitation energy by a factor of 100 a pronounced blue shift of the QW luminescence is observed with a maximum at 3.51 eV. Above a threshold of $100\,\mathrm{kWcm^{-2}}$ lasing is detected at this wavelength. This behavior is confirmed by gain spectroscopy revealing optical amplification in the same spectral range. Lasing and optical gain is detected 120 meV above the QW luminescence energy demonstrating the strong screening of the QCSE at elevated excitation densities. Theoretical approximations [12] deliver 10 to $20\,\mathrm{kWcm^{-2}}$ per well being in good agreement with the experimentally obtained results presented here. The results underline that the strength of the electric fields is extraordinarily important for the recombination probability and hence for the radiative recombination.

Although the internal electric fields may mainly contribute to the separation of electron and holes, the presence of thickness fluctuations of the GaN QWs can not completely excluded. Since the excitons are localized in the QW areas with larger (thinner) thickness, the separation of carriers increases (decreases) due to their localization at opposite heterointerfaces. Thus the QCSE is to expect larger (smaller) than the

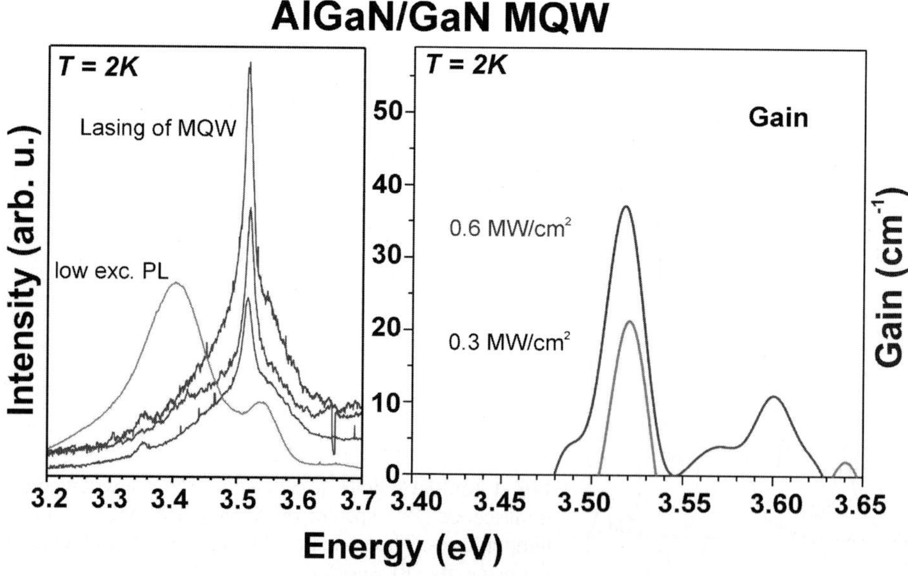

Fig. 2 (online colour at: www.interscience.wiley.com) Comparison of photoluminescence and stimulated emission (left) and gain (right) of a tenfold AlGaN/GaN MQW structure. Gain around 3.6 eV is generated in the 300 nm thick AlGaN cap layer.

average value for a given QW thickness. However, more precise information of the recombination and amplification processes is given by the PPS. The PPS allows a direct observation of the distribution and the recombination channels. Furthermore, the line widths of the emission bands are reduced by the resonant excitation. Figure 3 shows PP spectra of the MBE-grown GaN/AlGaN MQW sample. The excitation densities were varied from 10 kWcm^{-2} to 1 MWcm^{-2}. At the lowest excitation density, a bleaching of absorption is detected around 3.48 eV, while with increasing excitation densities this absorption (gain) structure vanishes. This structure is assigned to the emission of GaN. The observed bleaching is caused by the generation of optical gain. Due to the spectral position and former investigations [13] the gain is attributed to radiative decay of biexcitons.

The most striking feature is a strong induced absorption at 3.65 eV. With increasing excitation density this absorption becomes weaker and its maximum shifts to lower energies. The origin of the absorption is the quantized exciton in the MQW. The decrease of induced absorption (with increasing excitation density) shows the saturation of the excitonic recombination. This behavior is attributed to the rising screening of the Coulomb interaction resulting in a reduction of the excitonic binding energy and thus in a renormalization.

Additionally, an absorption bleaching was observed at 3.56 eV. The maximum of this structure shows a blue shift at elevated excitation densities and is located at 3.59 eV at 1 MWcm^{-2}. This structure is assigned to excitons localized in thickness fluctuations of the MQW [14]. The blue shift is explained by successive filling of localized states with increasing excitation density.

Localized excitons generate optical amplification (gain) meanwhile an induced absorption is observed for the electron and hole states in the MQW. The gain threshold density of excitons localized in fluctuations is by far smaller than that of the excitons localized in the MQW by the internal electric field. The latter amounts to 450 kVcm^{-1} for the active GaN QW layers. The excitons being not localized in fluctua-

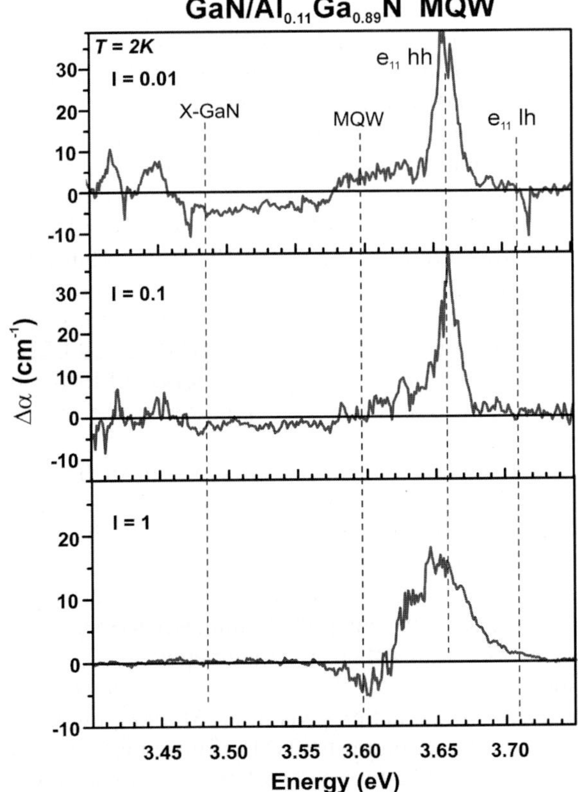

Fig. 3 (online colour at: www.interscience.wiley.com) Derivation of the absorption of the probe signal recorded by pump-and-probe spectroscopy for a 10fold GaN/AlGaN MQW at 2 K. The excitation density of the probe beam was varied from 10 kWcm^{-2} to 1 MWcm^{-2}.

tion potentials contribute to optical gain when the internal electric field is screened. Again, the role of localization and separation of carriers is demonstrated.

As it was mentioned above, to understand the emission properties of GaN/AlGaN structures the quantum-confined Stark-effect has to be included. All luminescence properties are strongly influenced by the presence of internal electric fields. The overlap of electron- and hole-wavefunctions is reduced due to its separation (localization at each side of the heterointerfaces). Therefore, the emission probability decreases lowering e.g. the efficiency of opto-electronic devices. Furthermore, due to the electric fields the exciton binding energy is reduced by a factor of 2-3 [12,15]. This is important so far as the application of excitonic devices is desirable, since high thermal stability and low threshold densities are the main advantages of such devices.

In order to compensate the internal electric fields, an enhanced effort on the sample design (e.g., appropriate quaternary compounds) is necessary to avoid the separation of electrons and holes to each side of the heterointerface. Theoretical investigations predict field free AlInGaN/GaN heterostructures [8,16,17], when the spontaneous polarization is compensated by perfectly matched strain induced piezoelectric fields. Also the reconstruction of the hetero-interface may contribute to the internal electric field as it was theoretically shown by Smith et al. [18].

Exciton localization and carrier separation in group-III nitrides has become a very prosperous field of investigation since it determines the opto-electronic properties of heterostructure devices. It was shown that separation of carriers due to internal electric fields reduces the efficiency of luminescence processes. In the next chapter we report on 3D carrier localization in nm-scaled potential islands.

3.2   Localization in randomly distributed potential fluctuations

The localization mechanism presented in this chapter strongly differs from that described above. Although a blue shift may be detected with increasing excitation densities in PL experiments, there are no further similarities to the behavior of carriers affected by electric fields. Especially, the origin of this behavior differs strongly, because the electrons and holes are localized in islands. Hence, the spatial overlap of their wavefunctions is increased. The InGaN/GaN heterosystem can be seen as a model system for the investigation of carrier and exciton localization in potential fluctuations generated by statistical distribution of local energy. In has a solubility limit in GaN and tends to segregate in the ternary InGaN. Thereby, the growth of homogeneous QWs is prevented and the formation of islands occurs.

More generally speaking, the formation of a ternary compound by epitaxial growth in principle results in chemical disorder leading to more or less pronounced spatial variation of the alloy composition on a microscopic scale. Such material fluctuations are widely studied in different III-V compounds and wide bandgap II-VI compounds [19]. Their size usually ranges from a few nm to several tens of nm. Thus they may act as quantum dots (QDs) providing a 3D localization for excitons and/or single carriers. The localization of carriers and/or excitons in these potential minima strongly influences the optical emission. More over it is found to be responsible for the high efficiency of light generation in InGaN-based LEDs and LDs [20]. Structural investigations by high-resolution TEM indeed revealed dense arrays of InN-rich nano-domains having a size of 3-5 nm and a density of 10^{11} to 10^{12} cm^{-2} [19,21,22]. However, the formation of these QD-like islands strongly depends on the growth conditions [23].

In this chapter evidence of exciton localization in such QD-like structures will be presented. The localization determines either the relaxation dynamics or the optical gain. Degushi et al. [24] have firstly shown that the segregation process in InGaN structures directly influence the lasing processes. Below it will be demonstrated that the localization in QD-like structures has to taken into account to understand the recombination behavior in ternary InGaN heterostructures. Therefore, considerable investigations on InGaN QW structures were performed. Beside time-resolved and time-integrated luminescence spectroscopy, site-selective excitation spectroscopy as well as gain measurements were implemented.

An evidence of the presence of exciton localization was given by the characteristic temperature-dependence of the time-integrated luminescence recorded at low excitation densities [25]. In Figure 4 the evolution of the PL peak position and the full widths at half of maximum (FWHM) of the emission band are shown as a function of the temperature. The GaN luminescence (not shown here) in sample B fol-

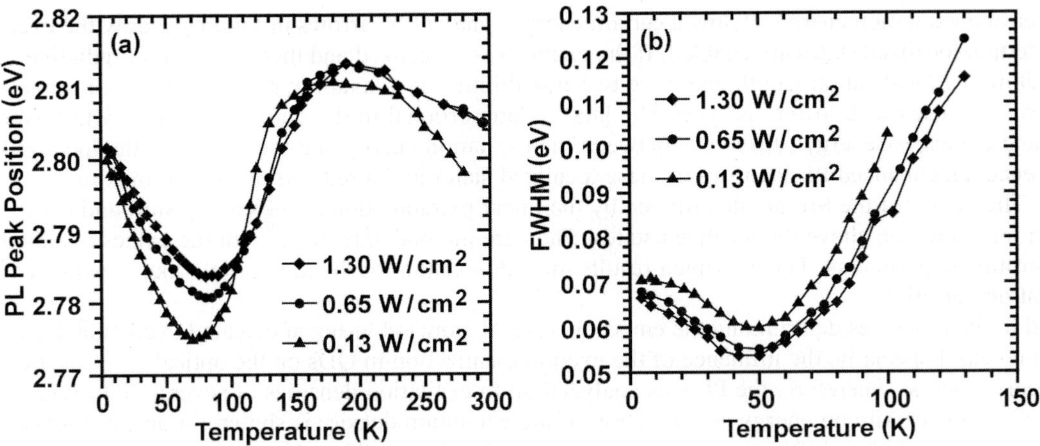

Fig. 4 Temperature dependence of PL peak position (a) and full width at half maximum (FWHM, b) of sample B for different PL excitation power. The increase of peak position energy above 70 K is caused by a carrier freeze out process.

lows Varshni's temperature dependence [26]. In contrast to the luminescence of GaN a strong S-shape behavior is observed for the InGaN-luminescence. Starting at 4.5 K for a PL excitation of 0.13 W/cm^2 an initial red shift of 25 meV leading to a minimum of the emission energy at 70 K. This red shift is compensated by a blue shift of 36 meV up to 170 K. For higher PL excitation densities the minimum becomes less pronounced and vanishes almost completely at an excitation power of 2.7 kW/cm^2. The thermal activation energies of 22 meV and 8 meV for the thermionic emission of the freezed out excitons were determined from an Arrhenius plot of the PL intensity. As it was theoretically shown by Zimmermann and Runge [27] the increase of the PL peak-position energy, namely the S-shape of the temperature dependence is caused by exciton freeze out in a system with randomly distributed potential fluctuations. Further evidence is obtained by absorption and site-selective spectroscopy.

A direct observation of the mobility edge (ME) is enabled by the site-selective spectroscopy. Here, the excitation energy is varied and the PL spectra are detected. In Figure 5 the energy of the PL maximum as a function of the excitation wavelength is depicted. Below a certain wavelength (above a certain energy given by the ME) no change of the PL maximum energy was detected. A further decrease of the excitation energy results in a significant shift of the PL maximum to lower energies. The shift rises linearly

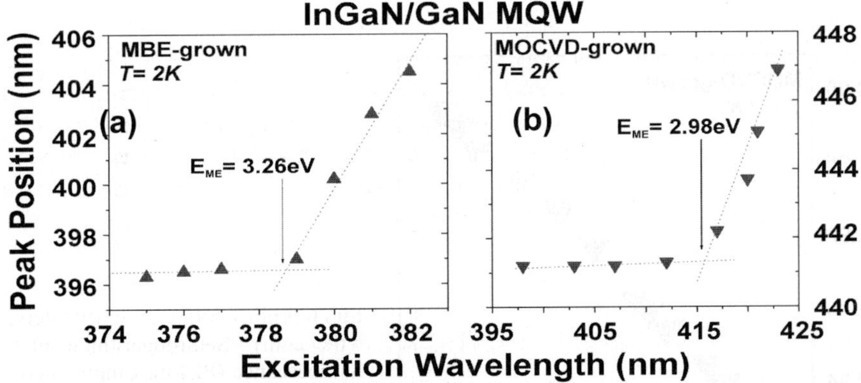

Fig. 5 (online colour at: www.interscience.wiley.com)   Photoluminescence-maximum energy as a function of the excitation wavelength for two InGaN/GaN MQW samples (sample A, (a), and sample B, (b)). The excitation wavelength was varied below and above the mobility edge (ME). The ME (marked by arrows) was determined by extrapolation.

with decreasing excitation energy. Below a certain energy (marked by arrows in Figure 5), only the excitation of certain localized states are enabled. If the excitons are localized and there are no non-radiative centers within the localization length, there are two possibilities of relaxation: radiative decay or tunneling to localized states with lower energies. The latter is proportional to the number of states with lower energies and the excess energy (difference between the excitation energy and the energy of the localized states). Hence, an enhanced PL intensity of these localized states and a red shift of the PL maximum are observed. The values of the ME are determined by the linear extrapolation of the energy shift and the PL maximum for excitation above the localized states. They are in good agreement with the values obtained from absorption experiments. The presented results indicate exciton localization in QD-like structures at low excitation densities.

Beyond it, the processes determining the emission, optical gain and lasing at elevated excitation densities are to reveal. Especially, the influence of the exciton localization in QDs on the optical amplification is of a special interest. Therefore, the PL was analyzed up to excitation densities of several $100 \, \mathrm{kWcm^{-2}}$. In Figure 6 the PL-maximum energy as a function of the excitation density is shown for an InGaN/GaN MQW sample. A total blue shift with increasing excitation density of 190 meV was detected. The excitation density was varied by seven orders of magnitude. The linear dependence in the semi-logarithmic plot proves that the blue shift results from the successive filling of localized states (band-filling effect). Qualitatively it can be described by a Boltzmann distribution of the occupied states and an exponential DOS for excitons. In contrast, a nearly quadratic shift of the PL- maximum would result for piezo-electric fields. Following the estimation of the electric fields after Fiorentini et al. [12], a field of $\sim 1 \, \mathrm{MVcm^{-1}}$ was obtained for the investigated MQW structure. An excitation density of 10 to $20 \, \mathrm{kWcm^{-2}}$ per well is necessary to screen these fields, what is close to the gain threshold density of the MOCVD-grown MQW structure (see also Fig. 8). Although this value is in reasonable agreement with the threshold density for the QCSE, no saturation at elevated excitation densities [12] and no typical nearly quadratic shift of the PL maximum were observed. As it was shown by Miller et al. [28] there are two major effects of an electric field applied perpendicular to the well on confined excitons. First, the exciton energy is reduced due to the bending of the confinement potential. The spatial separation of electron and hole wavefunction increases leading to a rising dipole character of the exciton. This effect leads to a red shift rising quadratically with electric field. Second, decrease of exciton binding energy due to reduced overlap of the wavefunctions. The latter effect amounts to $\sim 10 \, \%$ of the first and has an opposite sign. Thus, it can be concluded that at least at low temperatures the influence of internal fields in InGaN MQW structures is negligible in comparison to the localization of excitons in nm-scaled islands (QD-like structures).

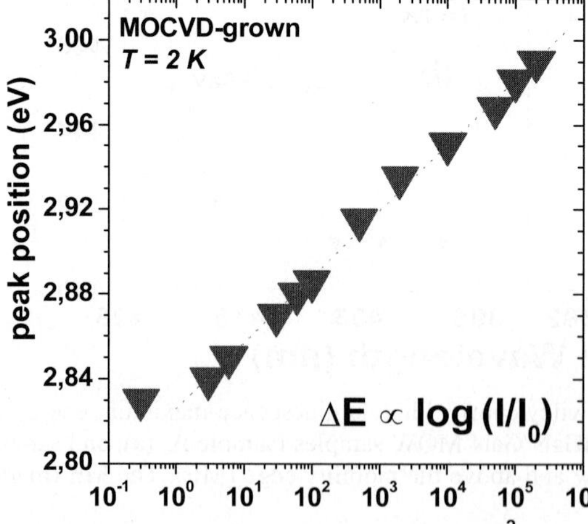

Fig. 6 (online colour at: www.interscience.wiley.com) Semi-logarithmic plot of the photoluminescence (PL)-maximum energy as a function of the excitation density for an InGaN/GaN MQW (sample B). The PL was detected at 6 K.

The results of excitation-density dependent PL are confirmed by time-resolved cathodoluminescence investigations. Here, the peak position of the emission line was detected as a function of the delay time and the excitation density. For the temporal shift of the InGaN peak a function $E_{Peak} = E_0 \pm 25$ meV log (t/t_0) is found (see Fig. 7a). This interpretation as a thermalization process within a statistical distribution of local energy states is strongly supported by excitation dependent cw measurements. Again, a monotonous blue shift of the main emission line according to $E_{Peak} = E_0 + 25$ meV log (P/P_0) is observed (Fig. 7b), directly visualizing the filling of the localized states.

The contribution of localized states to the gain and the lasing processes is revealed in Fig. 8. There, the low-excitation-density PL, the stimulated emission and the gain spectra for two different InGaN/GaN MQW samples are shown. The energy of the mobility edge is marked, too. Again a strong blue shift is observed for the PL maximum of the stimulated emission at elevated excitation densities. Lasing is detected in both samples at low temperatures. The threshold densities are given in Fig. 8. The onset of the lasing was observed at 3.0 eV for 160 kWcm^{-2} and at 3.22 eV for 1.2 MW cm^{-2}, respectively. With increasing excitation density a super linearly growth, line-shape narrowing and the characteristic polarization of the emission were observed. The maximum of the optical gain is blue shifted in comparison to the maximum of the stimulated emission. The energy of the ME is in the energy range of the gain. Thus, the contribution of localized states to the lasing processes is demonstrated. Nevertheless, gain and lasing are generated by de-localized carriers via band-band transitions, too. At room temperature (RT, not shown here) lasing is detected in the MOCVD-grown sample only. The lasing threshold is increased by a factor of four. The blue shift of the stimulated emission (in comparison to the low-excitation-density PL) is decreased, but ranges still between 50 to 100 meV. Because of the large localization energies (100 to 150 meV, note that the $k_B T$ is 24 meV at RT) there is still a contribution of localized states to the emission processes.

Nevertheless, energy-selective spectroscopy did not reveal a shift of the PL maximum. Obviously, the ME behavior is not present at RT and the emission is dominated by de-localized transitions (band-to-band). This can be attributed to the effective temperature of an electron-hole plasma being by far higher than the lattice temperature. Therefore, the most localized states are depopulated by thermionic emission. In that case, the influence of the internal electric field may become more significant. The observed blue shift of the lasing energy with increasing excitation density at RT can be explained by the superposition of the band-filling processes of the remained localization sites and the screening of the electric fields. The lasing threshold matches to the value per QW predicted by Fiorentini et al. [12]. Again, the reduction of the emission efficiency due to the QCSE is shown.

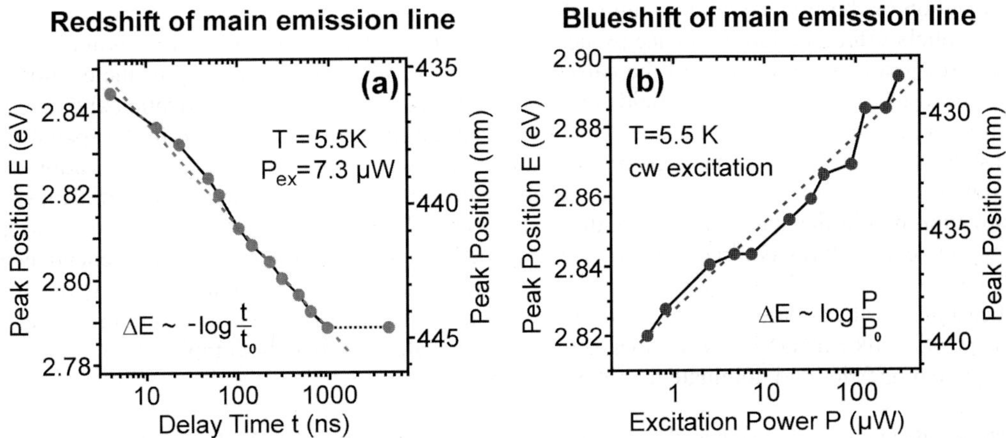

Fig. 7 (online colour at: www.interscience.wiley.com) a) Temporal red shift of the InGaN main emission line (sample B) due to exciton thermalization during decay and b) blue shift of InGaN main emission line (sample B) due to filling of localized states with increasing CL excitation power.

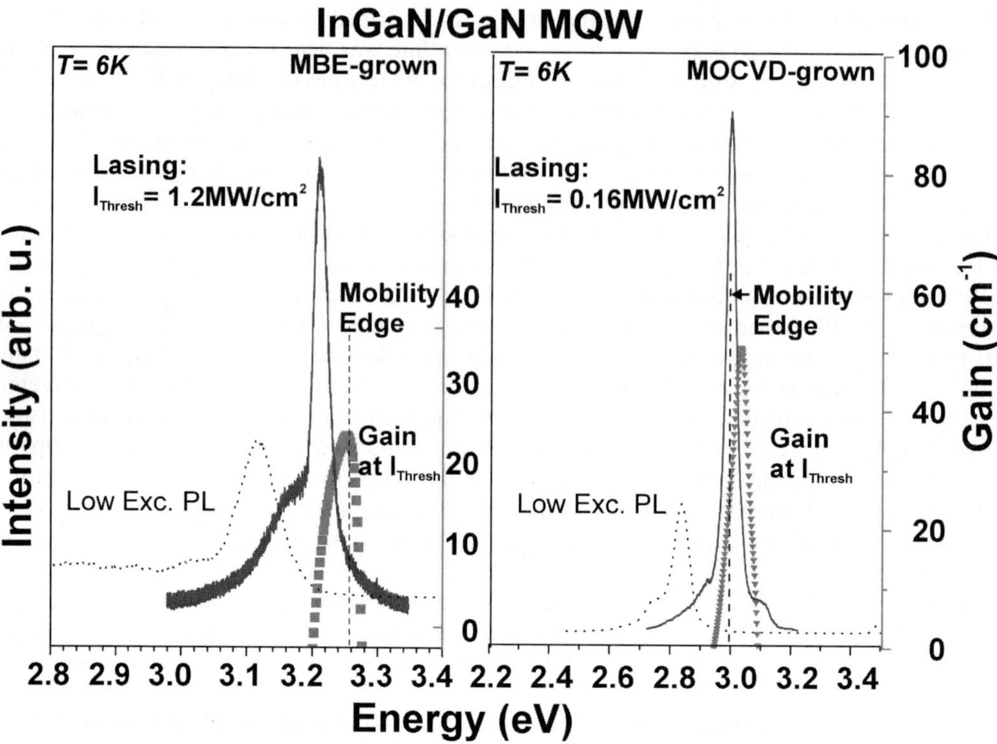

Fig. 8 (online colour at: www.interscience.wiley.com) Low-excitation-density PL (dotted lines), stimulated emission (straight lines) and gain spectra (symbols) for two different InGaN/GaN MQW samples. The energy of the mobility edge is marked by vertical dashed lines.

4 Conclusion

In principle two different localization mechanisms of charge carriers and excitons in ternary group-III nitride structures co-exist: first, the separate localization of electron and holes at the opposite interfaces of the quantum wells (QWs), described by the quantum-confined Stark effect (QCSE), and second, the localization of the free excitons in nm-scaled islands of local energy minima. Both effects superimpose and strongly affect the luminescence properties. The QCSE lead to a drop of the emission efficiency due to the reduction of the overlap of the carrier wavefunctions and the weakening of the excitons. This effect is caused by internal electric fields resulting from both, spontaneous polarization and piezo effect. It is always present in the nitrides. The second localization mechanism results from the spatial inhomogeneity due to nanoscopic fluctuations of the local composition. The strong spatial localization of excitons in nm-scaled islands of minimum energy dramatically improves the luminescence efficiency. The thermalization within the statistical energy distribution and the freeze out of the excitons in local minima at very low temperatures is directly proven by the characteristic S-shape temperature dependence of both, the luminescence peak energy as well as the full width at half of maximum (FWHM) of the luminescence line. Further evidence is given by the presence of a mobility edge for excitons. In particular for ternary QWs like InGaN/GaN the localization in nm-scale islands may completely determine the luminescence properties and dominate the QCSE.

Acknowledgements The authors thank the groups of H. Riechert (Infeneon), M. Heuken (Aixtron AG, Germany), H. Amano and I. Akasaki (Meijo University, Nagoya, Japan), for supplying the excellent samples. The authors gratefully acknowledge financial support by the Deutsche Forschungsgemeinschaft (DFG) in the framework of contracts Nos. Ho 1366/11-1, Ho 1366/11-2, CH 87/4-1, and CH 87/4-2.

References

[1] e.g., S. Chichibu, T. Azuhata, T. Sota, S. Nakamura, Appl. Phys. Lett. **69**, 4188 (1996).

[2] e.g., T. Takeuchi, S. Sota, M. Katsuragawa, M, Komori, H. Takeuchi, H. Amano, and I. Akasaki, Jpn. J. Appl. Phys. **36**, L382 (1997).

[3] e.g., M. Sugawara, Jpn. J. Appl. Phys. **35**, 124 (1996); K.P. O'Donnell, R.W. Martin, P.G. Middleton, Phys. Rev. Lett. **82**, 237 (1999); I.L. Krestnikov, N.N. Ledentsov, A. Hoffmann, D. Bimberg, A.V. Sakharov, W.V. Lundin, A.F. Tsatsul'nikov, A.S. Usikov, Zh.I. Alferov, Yu.G. Musikhin, D. Gerthsen, Phys. Rev. B **66**, 155310 (2002) and references therin.

[4] T. Riemann, D, Rudloff, J. Christen, A. Krost, M. Lünenburger, H. Protzmann, M. Heuken, phys. stat. sol. (b) **216**, 301 (1999).

[5] K.L. Shaklee, R.E. Nahory, and R.F. Leheny, J.Lumin. **7**, 284 (1973).

[6] J. Christen, *Advances in Solid State Physics* **XXXIII**, (Vieweg, Braunschweig, 1990, ed. U. Roessler), p. 239, (1990).

[7] J. Christen, T. Riemann, phys. stat. sol. (b) **228**, 419 (2001).

[8] e.g., F. Bernardini, V. Fiorentini, phys. stat. sol. (b) **216**, 391 (1999).

[9] e.g., T. Takeuchi, S. Sota, M. Katsuragawa, M, Komori, H. Takeuchi, H. Amano, and I. Akasaki, Jpn. J. Appl. Phys. **36**, L382 (1997).

[10] e.g., A. Hangleiter, J. Im, H. Kollmer, S. Heppel, J. Off, F. Scholz, MRS Internet J. Nitride Semiconductor Research **3**, 15 (1998).

[11] T. Takeuchi, S. Sota, M. Katsuragawa, M, Komori, H. Takeuchi, H. Amano, and I. Akasaki, Jpn. J. Appl. Phys. **36**, L382 (1997).

[12] V. Fiorentini, F. Bernardini, F. della Sala, A. di Carlo, P. Lugli, Phys. Rev. B **60**, 8849 (1999).

[13] H. Morkoc, *Nitride Semiconductors and Device* (Springer-Verlag, Berlin, 1999) and references therein.

[14] M. Leroux, N. Grandjean, M. Laugt, J. Massies, B. Gil, P. Lefebvre, P. Bigenwald, Phys. Rev. B **58**, R13371 (1998).

[15] G. Treatta, R. Cingolani, A. di Carlo, F. della Sala, P. Lugli, Appl. Phys. Lett. **76**, 1042 (2000).

[16] e.g., B. Gil, P. Lefebvre, J. Allegre, H. Mathieu, N. Grandjean, M. Leroux, J. Massies, P. Bigenwald, P. Christol, Phys. Rev. B **59**, 10246 (1999).

[17] A.Rizzi; R. Lantier, F. Monti, H. Lüth, F. della Sala, A. di Carlo, P. Lugli, J. Vac. Sci. Technol. B **17**, 1674 (1999).

[18] A. R. Smith, R. M. Feenstra, D. W. Greve, M.-S. Shin, M. Skowronski, J. Neugebauer, and J. E. Northrup, Appl. Phys. Lett. **72**, 2114 (1998).

[19] for a review see: I.L. Krestnikov, N.N. Ledentsov, A. Hoffmann, D. Bimberg, phys. stat. sol. (a) **183**, 207 (2001).

[20] e.g., S. Nakamura, Science **281**, 956 (1998).

[21] N.N. Ledentsov, Zh.I. Alferov, I.L. Krestnikov, W.V. Lundin, A.V. Sakharov, I.P. Soshnikov, A.F. Tsatsul'nikov, D. Bimberg, and A. Hoffmann, Comp. Semicond. **5** (9), 61 (1999).

[22] A.V. Sakharov, W.V. Lundin, I.L. Krestnikov,, V.A. Semenov, A.S. Usikov, A.F. Tsatsul'nikov, Yu.G. Musikhin, M.V. Baidakova, Zh.I. Alferov, N.N. Ledentsov, J. Holst, A. Hoffmann, D. Bimberg, I.P. Soshnikov, and D. Gerthsen, phys. stat. sol. (b) **216**, 435 (1999).

[23] e.g., Yu.G. Musikhin, D. Gerthsen, D.A. Bedarev, N.A. Bert, W.V. Lundin, A.F. Tsatsul'nikov, A.V. Sakharov, A.S. Usikov, I.L. Krestnikov, N.N. Ledentsov, A. Hoffmann, D. Bimberg, Appl. Phys. Lett. **80**, 2099 (2002).

[24] T. Degushi, A. Shikanai, T. Sota, S. Chichibu, S. Nakamura, Proc. ICNS'97, Tokushima; Japan, ed. K. Hiramatsu, 466 (1997).

[25] e.g., J. Holst, A. Kaschner, U. Gfug, A. Hoffmann, C. Thomsen, F. Bertram, T. Riemann, D. Rudloff, P. Fischer, J. Christen, R. Averbeck, H. Riechert, M. Heuken, M. Schwambera, O. Schön, phys. stat. sol. (a) **180**, 327 (2000) and references therein.

[26] Y.P. Varshni, Physica (Netherlands) **34**, 149 (1967).

[27] E. Runge and R. Zimmermann, Adv. Solid State Phys. **38**, 251 Σ1998); Phys. Status Solidi (a) **164**, 511 (1997).

[28] D.A.B. Miller, D.S. Chemal, D.C. Damen, A.C: Gossard, W. Wiegmann, T.H. Wood, C.A. Burrus, Phys. Rev. Lett. **53**, 2173 (1984).; D.A.B. Miller, D.S. Chemal, D.C. Damen, A.C: Gossard, W. Wiegmann, T.H. Wood, C.A. Burrus, Phys. Rev. B. **32**, 1043 (1985).

phys. stat. sol. (c) **0**, No. 6, 1846–1859 (2003) / **DOI** 10.1002/pssc.200303128

GaN based laser diodes - epitaxial growth and device fabrication

T. Böttcher[1], **S. Figge**[1], **S. Einfeldt**[*1], **R. Chierchia**[1], **R. Kröger**[1], **Ch. Petter**[1], **Ch. Zellweger**[2], **H.-J. Bühlmann**[2], **M. Dießelberg**[1], **D. Rudloff**[3], **J. Christen**[3], **H. Heinke**[1], **P. L. Ryder**[1], **M. Ilegems**[2], and **D. Hommel**[1]

[1] Institute of Solid State Physics, University of Bremen, P.O. Box 330440, 28334 Bremen, Germany

[2] Institute for Quantum Electronics and Photonics, Swiss Federal Institute of Technology Lausanne, 1015 Lausanne, Switzerland

[3] Institute of Experimental Physics, University of Magdeburg, P.O. Box 4120, 39016 Magdeburg, Germany

Received 4 March 2003, revised 11 April 2003, accepted 15 April 2003
Published online 28 August 2003

PACS 42.55.Px, 81.05.Ea, 85.60.Jb

Four selected material issues of group-III nitride layer structures grown by metalorganic vapor phase epitaxy and molecular beam epitaxy on basal plane sapphire are reviewed. 1) The decomposition of GaN under the impact of hydrogen gas was measured and described thermodynamically. 2) The nucleation and coalescence of GaN islands on a low-temperature nucleation layer was identified to be the key process for the formation of edge-type threading dislocations and heterogeneous stress. 3) The doping with magnesium was associated with pyramidal defects which formed upon a critical layer thickness and led to self-compensation of the acceptors. 4) The relaxation of plain tensile stress via cracks in bulk layers and superlattices based on AlGaN was investigated. The described studies allowed for a material optimization which finally led to the successful demonstration of a GaN based laser diode under pulsed current injection. The operation of the device is discussed in terms of the perfection of the crystallographically wet etched mirror facets and the lateral current spreading in p-type AlGaN cladding layers.

1 Introduction

Group-III nitride semiconductors with GaN as the key compound have drawn the attention of many physicists and engineers due to their use in electro-optical devices operating in the green, blue, violet and ultraviolet spectral region. Moreover, there is a rising demand for high power and high temperature electronics which can be made with this type of materials. The tremendous efforts in device performance achieved within the last few years were based on the improvement of both the material quality and the processing technology. The first issue requires detailed investigations of the epitaxial growth. Considering a) the complex layer structure of e.g. a laser diode, and b) the large number of parameters which can be varied during the growth with a technique like metalorganic vapor phase epitaxy (MOVPE), there is plenty of room for material optimization. On the other hand, the processing technology is not less of complexity as it involves lateral structuring of chemically inert materials and the formation of low-resistance Ohmic contacts to only moderately doped semidonductors.

It is our belief that a detailed study of the physics involved in epitaxial growth and processing will help to understand the performance of a device and, thus, speed-up the process of device optimization. In this paper we review a number of investigations regarding the epitaxy of group-III nitrides particularly for the

---

* Corresponding author: e-mail: einfeldt@physik.uni-bremen.de, Phone: +49 421 218 7453, Fax: +49 421 218 4581

use of MOVPE. This includes the chemistry of the solid/gas interaction in the reactor (Section 3.1), the growth mode on sapphire substrates involving the nucleation and coalescence of islands and the formation of heterogeneous strain (Section 3.2), the segregation of magnesium and the occurrence of pyramidal defect structures (Section 3.3) and the cracking under tensile stress (Section 3.4). The results obtained during these studies were prerequisites to optimize the material perfection to a level, where sophisticated devices such as laser diodes could not only be fabricated and successfully operated, but their performance was investigated with respect to the structural layout such as the perfection of the mirror facets after wet-chemical etching (Section 4.1) and the design of the p-type cladding layer (Section 4.2).

2 Experimental details

The MOVPE growth was carried out in a 3×2" commercial system from Thomas Swan Inc. It contains a vertical-type reactor with a showerhead in close distance to the susceptor. Trimethylgallium (TMGa), trimethylaluminum, trimethyindium and ammonia were used as precursors in conjunction with hydrogen and nitrogen as carrier gases. Silane and bicyclopentadienylmagnesium served as dopant sources. The growth was always performed on c-plane sapphire substrates. It was monitored in-situ by a reflectometry setup, which utilizes a red laser beam under normal incidence. Few samples were also grown with a molecular beam epitaxy (MBE) system from EPI equipped with an Unibulb radio-frequency plasma source for the activation of the nitrogen gas. Here, the group-III elements were thermally evaporated by conventional effusion cells.

The lattice parameters and the defect structure of epitaxial films were characterized by X-ray diffraction using a Philips X'Pert MRD system equipped with a fourfold Ge(220) monochromator, a threefould Ge(220) analyzer, and an Eulerian cradle. The analysis involved measurements under symmetric, asymmetric and skew symmetric diffraction geometry, respectively. Cathodoluminescence measurements were performed at low temperature in a modified secondary electron microscope (SEM) allowing the aquisition of a complete spectrum at each pixel of the scanned area. The acceleration voltage was properly chosen with respect to the required probing depth.

Light emitting diodes and laser diodes were fabricated from epitaxial layer systems using standard photolithography and metal deposition techniques. The nitrides were etched using reactive ion etching (RIE). SiO_x was used as insulator; Ni/Au and Ti/Al/Au served as p-type and n-type contacts, respectively. The resonator length and the width of the injection stripe were in the range 0.5–2 mm and 5–40 μm, respectively.

3 Metalorganic vapor phase epitaxy

3.1 Thermodynamic

Ammonia, trimethylgallium (TMG) and the carrier gases hydrogen or nitrogen are the mostly used source materials involved in the synthesis of GaN by MOVPE. Various reactions are involved in the growth process both in the gas phase and at the vapor-solid interface. However, the basic net reaction for the formation of GaN at the growth surface is given by

$$Ga(g) + NH_3(g) \rightleftharpoons GaN(s) + \frac{3}{2}H_2. \tag{1}$$

Eq. (1) already points to a difficulty of the growth in a hydrogen atmosphere as an increased hydrogen partial pressure favors the reverse reaction, i.e. the decomposition of GaN. The reaction kinetics is given by the rate constants k_{form} and k_{dec} of the forward (GaN formation) and backward (GaN decomposition) reactions of Eq. (1), respectively. In case of thermodynamic equilibrium, the ratio of both rates is given by the equilibrium constant K through the law of mass actio

$$K = \frac{k_{\mathrm{dec}}}{k_{\mathrm{form}}} = \frac{\hat{p}_{\mathrm{H}_2}^{3/2}}{\hat{p}_{\mathrm{NH}_3}\hat{p}_{\mathrm{Ga}}}. \tag{2}$$

Here, $\hat{p}$ is the equilibrium partial pressure of the corresponding reagent, and an activity of one was assumed for the solid GaN. The state of equilibrium mainly depends on the temperature of the system. Koukitu et al. calculated the equilibrium constants from the free energies of formation of AlN, GaN and InN [1]. AlN exhibits the highest value of K, i.e. it is the most stable material of these three. For GaN, the equilibrium constant is already lower by orders of magnitude, and InN is again one order below GaN.

The concurrence of the backward and forward reaction of Eq. (1) becomes evident by looking at the growth regime which changes from mass transport limited to kinetically limited with increasing temperature through the enhanced desorption of GaN [2]. The significance of the backward reaction of Eq. (1) is also demonstrated by the decomposition temperatures reported for GaN which range from 400 to 1070 °C depending on the specific atmosphere used ([3, 4, 5] and references therein). Experiments performed in an inert atmosphere like nitrogen or helium yielded a higher decomposition temperature compared to experiments performed in hydrogen. Further evidence for the backward reaction was given by annealing GaN in flowing hydrogen and measuring the ammonia concentration in the exhaust gas [6].

Referring to Eq. (2), if the partial pressure of hydrogen is increased, the partial pressure of either ammonia or gallium will also increase to keep the thermodynamic equilibrium. That is, the backward reaction of Eq. (1) will be enhanced and more GaN will decompose [7, 8]. A similar effect can be seen in Fig. 1 which shows the decomposition rate for GaN in a mixed hydrogen/ammonia gas stream. It was measured at a GaN layer in the MOVPE reactor through interference oscillations observed by in-situ reflectometry due to the decrease in the layer thickness. The decomposition is strongest for low ammonia partial pressures. However, even if the ammonia accounts for 30–40% of the total pressure, which is typical for the MOVPE growth of GaN, the decomposition is still significant. The effect can be qualitatively explained using a thermodynamic argument similar to the approaches of Koukitu et al. [9] or Duan et al. [10]. The equilibrium partial pressure of gallium over the GaN surface, which results from Eq. (2) to

$$\hat{p}_{\mathrm{Ga}} = \frac{\hat{p}_{\mathrm{H}_2}^{3/2}}{K\,\hat{p}_{\mathrm{NH}_3}} \tag{3}$$

can be understood as the thermodynamic driving force for the decomposition reaction. The gallium in the atmosphere is continuously removed by the gas flow in the reactor. Therefore, the decomposition reaction is enhanced proportional to the deviation of the actual pressure from the equilibrium value given by Eq. (3).

A quantitative description of the measured GaN decomposition rate is obtained by a rate equation for the gallium partial pressure $p_{\mathrm{Ga}}(t)$ at the GaN surface. If the system vapor/solid is not in thermal equilibrium, Eq. (1) gives

$$\frac{\mathrm{d}}{\mathrm{d}t}p_{\mathrm{Ga}}(t) = k_{\mathrm{dec}}p_{\mathrm{H}_2}^{3/2} - k_{\mathrm{form}}p_{\mathrm{NH}_3}p_{\mathrm{Ga}}(t) - k_{\mathrm{diff}}p_{\mathrm{Ga}}(t). \tag{4}$$

Here, the decomposition (k_{dec}) and formation (k_{form}) of GaN at the surface as well as the diffusion (k_{diff}) of gallium away from the GaN surface are considered. The measured decomposition rate, r_{dec}, should be proportional to $k_{\mathrm{diff}}p_{\mathrm{Ga}}(t)$. Under steady state conditions, i. e. $\mathrm{d}p_{\mathrm{Ga}}(t)/\mathrm{d}t = 0$, Eq. (4) results in

$$r_{\mathrm{dec}} \propto \frac{k_{\mathrm{diff}}k_{\mathrm{dec}}\left(p_{\mathrm{tot}} - p_{\mathrm{NH}_3}\right)^{3/2}}{k_{\mathrm{form}}p_{\mathrm{NH}_3} + k_{\mathrm{diff}}}, \tag{5}$$

where $p_{\mathrm{tot}} = p_{\mathrm{H}_2} + p_{\mathrm{NH}_3}$. If the rates k are assumed to be independent of the pressure p_{NH_3}, Eq. (5) can be fitted to the experimental data shown in Fig. 1. The dashed line in the figure confirms that the model reasonably describes the decomposition of GaN in an MOVPE reactor with hydrogen and ammonia in the atmosphere.

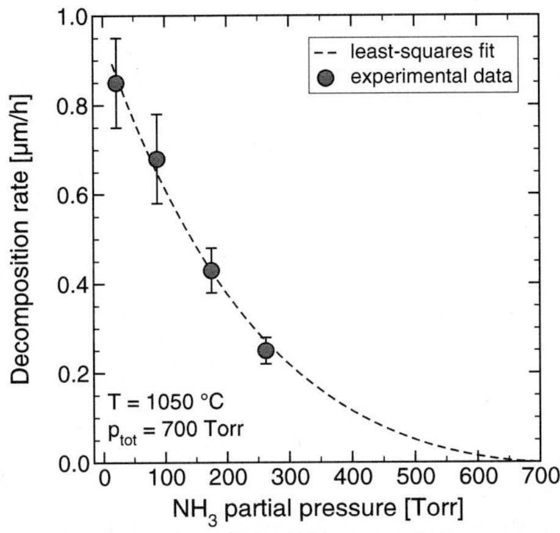

Fig. 1 Decomposition rate of GaN in dependence on the ammonia partial pressure in a mixed H$_2$/HN$_3$ atmosphere. The dashed line is a least squares fit of Eq. (5) to the experimental values.

3.2   Island coalescence and strain formation in GaN

The overgrowth of a recrystallized nucleation layer (NL), which was previously deposited on sapphire (0001) at a relatively low temperature, has been extensively studied [11, 12]. It is marked by an island nucleation, growth and coalescence process starting from the larger hexagonal grains in the recrystallized NL. TEM studies have shown that individual islands grow almost dislocation-free, such that the formation of dislocations can be attributed to the coalescence process of these islands [13, 12, 14, 15]. E.g. grazing incidence X-ray diffraction revealed a crystallographic twist in the annealed NL as large as 1.4° [16]. This results in a misorientation of the islands, which has to be accommodated by threading dislocations with line direction along [0001] at the coalescence front. These threading dislocations will have an edge and screw component for islands with in-plane twist and out-of-plane tilt, respectively. Among the three types of threading dislocations found in GaN (0001) layers grown by MOVPE, pure edge type dislocations with a Burgers vector $1/3[11\bar{2}0]$ are reported to dominate over pure screw and mixed character dislocations with Burgers vectors [0001] and $1/3[11\bar{2}3]$, respectively [17]. The described formation mechanism for threading dislocations was confirmed by Wu et al. [12, 14], who demonstrated the alignment of edge-type threading dislocations along grain boundaries in fully coalesced films having a high dislocation density.

In order to reduce the dislocation density in the layer, it is necessary to control the coalescence process. A prolonged coalescence will generally correspond to a smaller density of islands and, consequently, to a reduced density of grain boundaries and threading dislocations. One possibility to influence the coalescence is to utilize the decomposition of GaN in a hydrogen-containing growth ambient, as was described in Section 3.1. Unless inert gases are used, the growth atmosphere consists mainly of hydrogen and ammonia both of which have opposite effect on the decomposition of GaN. Thus, it is usually not possible to vary the partial pressures of these two gases independently at a constant total pressure, and the ratio of the ammonia and the hydrogen partial pressure is the parameter governing the growth process.

Reflectance transients recorded during the growth of GaN were used in this work to monitor the island growth mode [18, 19]. In Fig. 2 three GaN films are compared, which differ only in the NH$_3$/H$_2$ ratio applied during the recrystallization and the initial overgrowth of the NL. In order to separate the influence of this ratio from the molar V/III ratio, the latter was kept constant by adjusting the TMG flow accordingly. Therefore, the growth rate varied slightly for the layers under consideration. For sample A, both the recrystallization and the overgrowth were performed at equal partial pressures of ammonia and hydrogen. This is a rather stabilizing atmosphere for the GaN growth, as the recrystallization of the NL proceeded slowly, and the island coalescence was completed after a GaN thickness of only ~100 nm. Raising the hydrogen partial pressure during both the recrystallization and the overgrowth (sample B) prolonged the coalescence process up to a film thickness of ~200 nm. This is attributed to a) a faster roughening of the NL during recrystallization, which corresponds to a stronger selection of nucleation sites for the subsequent

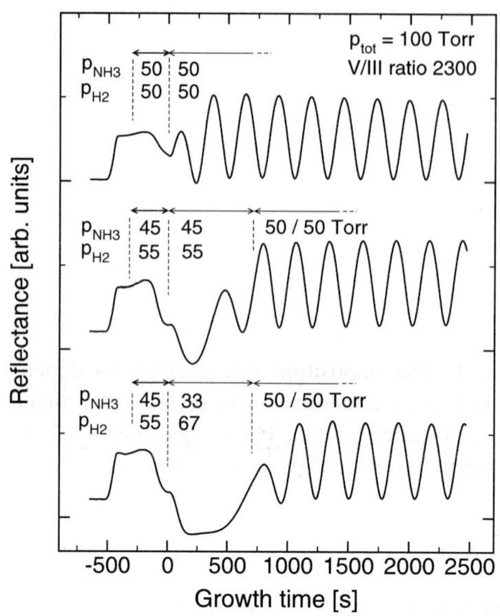

Fig. 2 Reflectance transients of GaN films grown under identical conditions except for the NH_3/H_2 ratio during the recrystallization of the NL and its initial overgrowth. The values for the corresponding partial pressures are indicated.

overgrowth, and b) an enhanced selectivity during the overgrowth itself. To separate the effect of the hydrogen on the recrystallization of the NL from its influence on the overgrowth, sample C was grown under the identical recrystallization conditions like sample B, but the hydrogen flow was increased further during the initial overgrowth. This doubled the film thickness at coalescence to ~400 nm. Therefore, the hydrogen partial pressure during the initial overgrowth seems to mainly determine the coalescence process. It is noted that depending on the chosen set of parameters, the average grain diameter in the grown GaN layers reached values up to 1 µm.

Not only is the distribution and density of threading dislocations in GaN determined by the island coalescence process, but the effective stress is also effected. In spite of its compressive lattice mismatch to sapphire, Hearne et al. [20] found the GaN to be under tension in the c-plane at growth temperature, if a GaN NL was used. Their study also revealed that a compressive extrinsic stress of (-0.66 ± 0.1) GPa was induced in all GaN layers by the cooling from the growth to room temperature due to the mismatch in the thermal expansion coefficients between GaN and sapphire. Therefore, the observed variation in stress at room temperature from sample to sample has to be attributed to variations of an intrinsic stress which is induced during growth. There are two possible origins of the intrinsic stress, namely 1) the pseudomorphic growth on a NL which is under tensile stress and 2) the coalescence of islands. It is the latter which will be detailed in the following.

Following the work of Finegan and Hoffman [21], the coalescence of islands should be associated with the formation of stress as the gap between misoriented islands, which are assumed to be strain-free prior to coalescence, is closed by the elastic deformation of the lattice. Here, the induced strain energy is outbalanced by the reduced surface energy. As the islands are bonded to the substrate, the in-plane intrinsic strain is tensile with the magnitude depending on the grain diameter, d, and the gap width, Δ

$$\epsilon^{\mathrm{intr}}(d) = \frac{\Delta}{d}. \tag{6}$$

From an energetic point of view, the maximum gap at which the system still reduces its free energy amounts to ~1 nm [22]. This would correspond to as much as three a-axis lattice parameters in GaN. Taking the geometry of the islands into account, the corresponding intrinsic stress is obtained as

$$\sigma^{\mathrm{intr}}(d_0) = 0.683 \times \left(c_{11} + c_{12} - 2\frac{c_{13}^2}{c_{33}} \right) \times \epsilon^{\mathrm{intr}}(d_0) = 307.1\mathrm{GPa} \times \frac{\Delta}{d_0}, \tag{7}$$

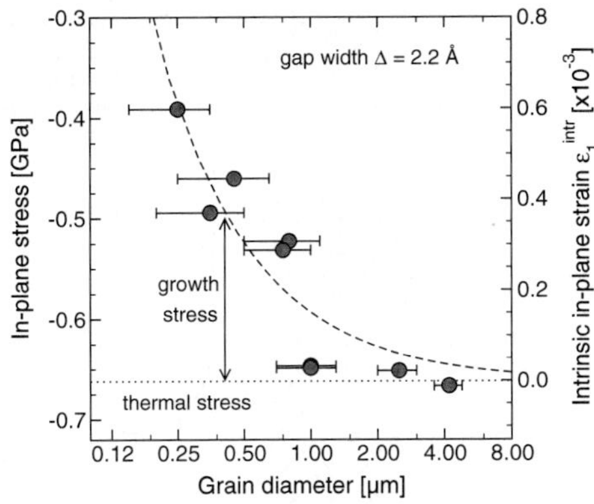

Fig. 3 In-plane stress and intrinsic strain in the GaN layers in dependence on the grain diameter. The dashed line is the theoretically expected sum of intrinsic and extrinsic stress assuming an average gap width of 2.2 Å.

where d_0 is the average grain diameter and c_{ij} are the stiffness constants. Equations (6) and (7) were discussed in more detail in Ref. [23]. The total stress is assumed to be a superposition of this intrinsic stress and an extrinsic stress where the latter is induced after growth by the mismatch of the thermal expansion coefficients between the GaN and the sapphire.

The stresses determined experimentally for all samples by XRD are shown in Fig. 3. They are plotted in dependence on the average grain diameter, which was estimated from the thicknesses of coalescence of the individual layers. The extrinsic compressive stress, which was assumed to be constant for all samples, obviously overcompensates the intrinsic tensile stress, resulting in a total compressive stress, which strongly depends on the grain diameter. For large grains the stress approaches asymptotically the value of the thermally induced stress. The curve in Fig. 3 corresponds to a least-squares fit of Eq. (7) to the experimental data which results in a gap width of $\Delta = (2.2 \pm 1.0)$ Å. This value is below the length of the Burgers vector of the edge type threading dislocations accumulating along the grain boundaries in GaN layers [24]. Furthermore, it is smaller than the lattice parameter a.

3.3 Segregation and defects in magnesium doped layers

One of the crucial problems for the realization of nitride-based laser diodes is to achieve a sufficiently high level of p-type doping. Magnesium is the only dopant known to provide hole concentrations in GaN beyond 10^{18} cm^{-3} at room temperature. However, the high activation energy of this acceptor necessitates magnesium concentrations as large as 5×10^{19} cm^{-3}. At those doping levels the material tends to form self-compensation defects which increase the low Fermi energy [25, 26]. Pyramids with a hexagonal base are the most commonly observed extended defects in heavily magnesium-doped GaN layers grown by MOVPE [27, 28, 29, 30]. The tip of the pyramids points towards the nitrogen face of the matrix material. The exact nature of the defect core is still controversial. On the one hand the high contrast seen in TEM micrographs could indicate an empty core; on the other hand the facets of the defects are rather typical for inversion domains [31, 32]. Nevertheless, it is now commonly accepted that the facets of the pyramids are partially decorated with magnesium [28, 31, 32]. Therefore, the concentration of magnesium in the defects should be higher than that in the surrounding matrix by about two orders of magnitude.

The height and the diameter of the pyramidal defects are typically in the range 10–20 bilayers and 13–27 atoms, respectively. Their density is found to be inhomogeneous along the growth direction. The region of inital growth close to the lower interface of the doped layer is defect-free. Only if the layer exceeds a critical thickness, z_f, pyramidal defects form, whose density strongly oscillates with the depth in the layer, as shown in Fig. 4. This offset in the defect formation is reduced for a higher ratio of the molar flows of magnesium, F_{Mg}, and the group-III elements, F_{III}, as shown in Fig. 5 for samples with a magnesium concentration of 5–6×10^{19} cm^{-3}. These results indicate that the formation of pyramidal defects is connected with the reduction of the Fermi energy and the accumulation of magnesium at the surface due to segregation, respectively. To verify this concept, a model for the surface coverage, N, with

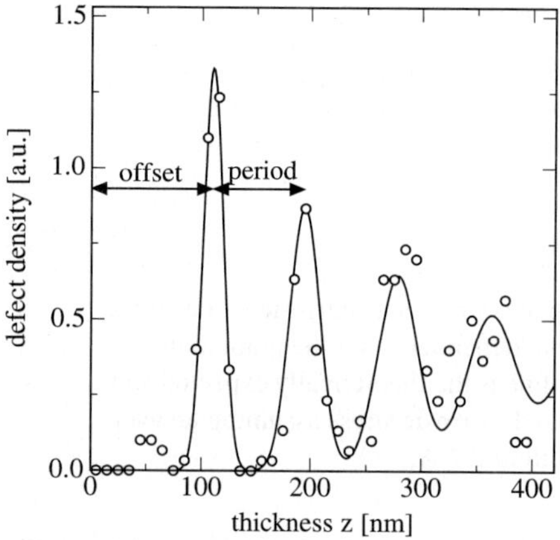

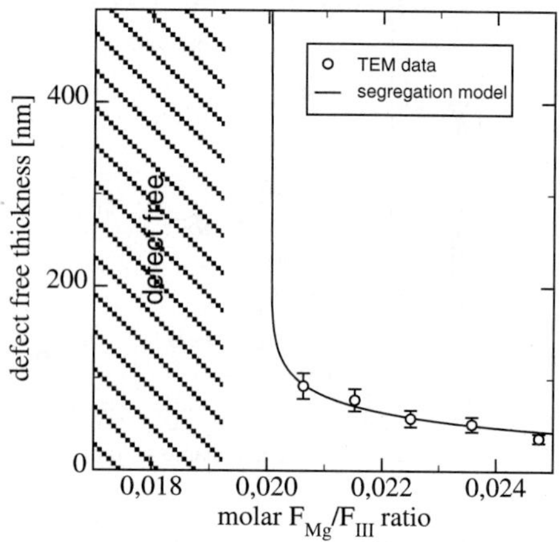

Fig. 4 Local density of pyramidal defects in a magnesium doped layer determined by TEM cross-sectional images in dependence on the distance from the lower interface of the layer.

Fig. 5 Thickness, z_f, of the defect-free region at the lower interace of magnesium doped layers as determined by TEM (bullets) in dependence on the ratio of the molar flows for magnesium and the group-III elements. The solid line correponds to the best fit of Eq. (10) to the experimental data which results in $c_i = 0.038$ nm^{-1} and $c_g/N_c = 1.9$ nm^{-1}.

magnesium was developed. N is assumed to change with the layer thickness, z, due to the supply of new magnesium from the gas phase and the incorporation of magnesium into the crystal. The former process depends on the molar flow of magnesium, F_{Mg}, normalized to the flow of group-III elements, F_{III}; and the latter process is driven by the magnesium coverage of the surface, N, itself, i.e.

$$\frac{dN(z)}{dz} = c_g \frac{F_{Mg}}{F_{III}} - c_i N(z).$$

(8)

The parameters c_i and c_g are constants which scale the two processes. The solution of Eq. (8) is

$$N(z) = \frac{F_{Mg} c_g}{F_{III} c_i} \left(1 - e^{-c_i z}\right).$$

(9)

To derive the defect free thickness, z_f, one can propose a critical magnesium coverage, N_c, of the surface, at which the defect formation starts. Then, Eq. (9) results in

$$z_f = -\frac{1}{c_i} \ln \left(1 - c_i \frac{N_c F_{III}}{c_g F_{Mg}}\right).$$

(10)

Equation (10) suggests that pyramidal defects are only formed for high magnesium flows, i.e. for a positive argument of the logarithm. For those conditions the defect-free layer thickness should be inversely proportional to the magnesium flow. This perfectly fits to the experimental data, as shown by the solid line in Fig. 5. Particularly, layers grown with a flow ratio $F_{Mg}/F_{III} < 0.02$ showed no pyramidal defects. Hall measurements performed on the samples revealed that the hole concentration decreases for flow ratios above 0.02. Therefore, the pyramidal defects are presumably associated with the self-compensation mechanism. It is believed that the defects either form donor-like states or trap high amounts of magnesium [33]. It is noted that with the help of the described defect analysis the epitaxial growth of magnesium-doped layers could be optimized. When the magnesium flow was just below the critical value for the formation of pyramidal defects, hole concentrations as high as 1×10^{18} cm^{-3} corresponding to a mobility of about 5 cm^2/Vs were achieved.

3.4 Cracks in tensile strained AlGaN

Cracking is an inherent problem of $Al_xGa_{1-x}N$ layers on GaN due to large tensile stress associated with coherent growth [34]. It increasingly becomes a problem for high aluminum contents, which, e.g., are beneficial for an efficient optical confinement in laser diodes employing $Al_xGa_{1-x}N$ cladding layers [35]. Short-period $Al_xGa_{1-x}N$/GaN superlattices (SLs) have been used as a replacement for $Al_xGa_{1-x}N$ cladding layers in laser diodes with the objective to suppress cracking. Though the latter is still under investigation, the use of superlattices reduced the threshold current density of the laser diodes [36, 37, 38]. Moreover, the piezoelectric fields in the superlattices were shown to enhance the in-plane carrier transport [39, 40]. The purpose of the studies presented here was to understand the complex process of strain relaxation in $Al_xGa_{1-x}N$ layers and $Al_xGa_{1-x}N$/GaN SLs which particularly includes the formation and propagation of cracks and misfit dislocations, respectively.

$Al_xGa_{1-x}N$ of different composition was grown by either MBE or MOVPE on 1–2 µm thick GaN template layers previously deposited on c-plane sapphire. Two types of samples were investigated: 1) 1 µm thick $Al_xGa_{1-x}N$ layers, which are referred to as bulk layers, and 2) $Al_xGa_{1-x}N$/GaN SLs of different periods whose total thickness was also 1 µm. The onset of cracking occured for Al contents ≥ 0.10–0.11, as shown in Fig. 6. This threshold is independent of whether the samples were grown by MBE or MOVPE and whether they are bulk layers or SLs. Therefore, $Al_xGa_{1-x}N$/GaN SLs are not superior to $Al_xGa_{1-x}N$ bulk layers of the same average aluminum content in view of their resistance to crack formation.

Cracks form in $Al_xGa_{1-x}N$ as soon as the strain energy release rate, G_{ss}, of a channeling crack exceeds a fracture toughness, Γ, i.e. [41]

$$G_{ss} = (1.1215)^2 \sigma_0^2 \pi t (1 - \nu^2)/2/E \geq \Gamma, \tag{11}$$

where σ_0, t, ν and E are the coherency stress, the thickness, the Poisson ratio and Young's modulus of the $Al_xGa_{1-x}N$ layer grown on GaN, respectively. The fit of Eq. (11) to our experimental data and those taken from the literature resulted in $\Gamma = 9$ J/m^2 [42]. Another approach to estimate the fracture toughness is known as Griffith concept, which balances the released strain energy with the induced surface energy of a propagating crack [43, 44]. Using published values for the surface energy of GaN [45], a fracture toughness of 1.89 J/m^2 is obtained. The fact, that this value is considerably smaller than indicated by the experiment, suggests an energetic barrier for the nucleation of cracks.

The number of cracks depends on the built-up strain energy. The strain energy density, which was determined by HRXRD, was found to be nearly constant for all cracked layers [42, 46]. Therefore, an increase of the thickness and/or the aluminum content of the $Al_xGa_{1-x}N$ is counterbalanced by an increase of the crack density in order to keep the strain energy density below a maximum value. The latter was determined to be 4 J/m^2 for bulk layers and 2 J/m^2 for SLs [46]. It is noted that in case of SLs, the saturation is not observed for the total strain energy density, but only for the difference between the total strain energy and the strain energy of a free-standing but coherent SL. This implies that cracks should not destroy the coherency within the SL, i.e. the coherency between the $Al_xGa_{1-x}N$ barriers and the GaN wells; but, they should only terminate the pseudomorphy between the SL and the GaN template. As a result, a heavily cracked SL should be strained as if it was free-standing. Since in a free-standing coherent SL the $Al_xGa_{1-x}N$ barriers and the GaN wells are under opposite stresses, the bulk layers exhibited a higher maximum strain energy density than the SLs, as only the strain energy related to the $Al_xGa_{1-x}N$ barriers but not the GaN wells was considered.

The spatial distribution of the strain in cracked $Al_xGa_{1-x}N$ layers was investigated via mappings of the peak energy in local CL spectra. The CL at cracks was found to be shifted towards higher energies compared to the CL at the center between the cracks, as shown in Fig. 7. This finding agrees with the idea that cracks relieve tensile stress locally, with the result of an increased bandgap. The inset of Fig. 7 summarizes the corresponding CL shifts for cracks of varying spacings. The observed gradual increase was expected, since the amount of elastically relieved stress should gradually decrease when moving away from a crack. The relaxation process was simulated by the twodimensional finite element analysis technique, the results

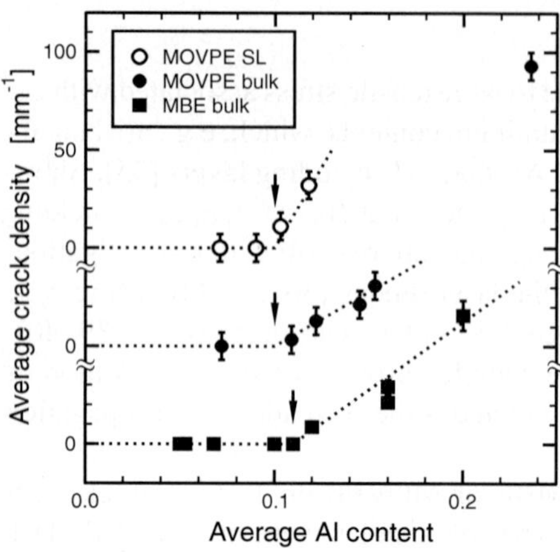

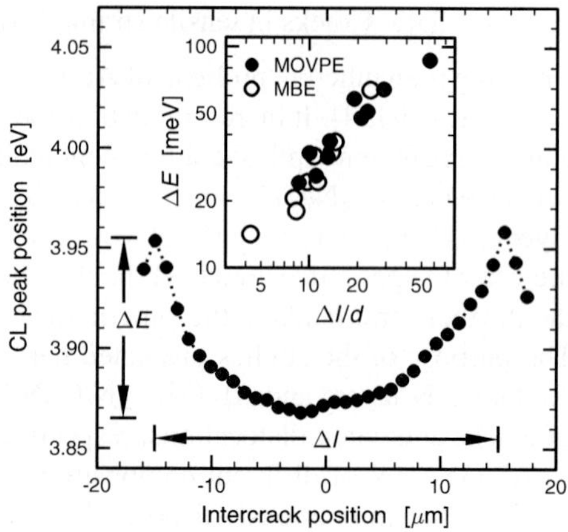

Fig. 6 Average crack density of bulk $Al_xGa_{1-x}N$ layers (filled symbols) and $Al_xGa_{1-x}N$/GaN SLs (open symbols) versus the average Al content. The bulk layers and SLs had an identical thickness of 1 µm. The data points for different sample series are vertically shifted for reasons of clearness. The dotted lines are guides to the eye, and the arrows mark the onset of cracking.

Fig. 7 Linescans of the CL peak energy of an MOVPE sample perpendicular to a pair of cracks of 14.5 µm spacing. Insert: Shift of the CL peak energy, ΔE, from the crack to the center between cracks in dependence on the crack spacing, Δl, normalized by the $Al_xGa_{1-x}N$ layer thickness, d, for an MOVPE sample (open circles) and an MBE sample (filled circles), respectively.

of which are not shown here. For crack spacings less than 20 times the $Al_xGa_{1-x}N$ layer thickness, the experimental data could be reasonably explained by the simulations. For larger crack spacings, however, the experimental data showed significant deviations. Although the discrepancy is not yet fully understood, it was discussed in detail in terms of a) the relief of biaxial or uniaxial stress, b) a variation of the local aluminum content in the layer, c) the occurance of plastic strain relaxation and d) the neglectance of stress components other than the normal stress perpendicular to the crack face [47, 48].

4   Laser diodes

4.1   Device fabrication and operation

The fabrication of GaN based laser diodes from structures grown on basal plane sapphire is hampered by the rotational misfit of the group-III nitride unit cell relative to the substrate as this inhibits the fabrication of smooth mirror facets by cleaving as common for other III-V laser structures [49]. On sapphire the mesa are usually fabricated by RIE. This typically leads to rough sidewalls reducing the device performance. In order to improve the perfection of the dry etched surfaces, one can apply crystallographic wet etching. The procedure exploited here utilizes 50% KOH dissolved by weight in ethylene glycol at a temperature of 140 °C [50]. The planes resulting from etching group-III nitrides in this solution belong to the $\{10\bar{1}0\}$ family, such that it can be used to prepare the mirror facets of laser devices aligned along the $\langle 10\bar{1}0 \rangle$ direction. After patterning the mesa by RIE, the mirror facets are found to exhibit a rough surface and an inclination angle deviating from the normal direction, as depicted in Fig. 8a. The subsequent crystallographic wet etching leads to perpendicular and smooth mirror facets, as shown in Fig. 8b. The final roughness of the facets can hardly be measured but appears to be similar to that of cleaved material [51]. Therefore, the introduced wet etching step allows to fabriacte laser diodes without the usually performed thinning of the sapphire substrate for easier cleavage.

The correct orientation of the mesa with respect to the crystal lattice of the GaN is the crucial parameter to obtain a smooth facet. Therefore, the photolithography masks have to be precisely aligned on the

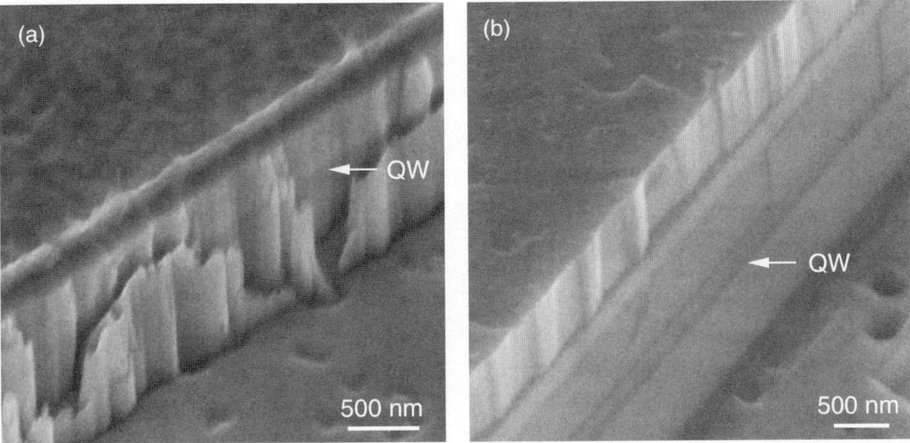

Fig. 8   Secondary electron microscopy images of the mirror facets of a GaN based laser diode a) after RIE and b) after crystallographic wet etching.

wafer. Due to the fixed epitaxial relationship between the GaN and the sapphire, the cleavage plane of sapphire can be used as a reference for the alignment. Here, the laser bars have to be fabricated parallel to this cleaving plane. One of the main advantages of the crystallographic wet etching lies in the excellent reproducibility, since the variability of the surface morphology after the RIE step is of minor importance for the final perfection of the mirror facet. Accordingly, the whole fabrication process of the laser diodes becomes more stable. In addition, the crystallographic wet etching allows to prepare smooth, vertical laser facets at arbitrary positions on the sample, i.e. without the restriction to laser bars.

The described process for the fabrication of mirror facets was applied to realize a gain-guided laser diode with a stripe width and a cavity length of 20 and 500 μm, respectively [52]. One facet was coated with a quarter-wavelength dielectric mirror consisting of four pairs of TiO_2 and SiO_2. The laser structure was grown by metalorganic vapor phase epitaxy and contained a 1.3 μm GaN buffer layer deposited on a GaN NL, a GaN:Si contact layer (1.5 μm), an $Al_{0.07}Ga_{0.93}$N:Si cladding layer (500 nm), a GaN:Si waveguide layer (100 nm), an multiquantum well (MQW) active region, an $Al_{0.2}Ga_{0.8}$N:Mg electron blocking layer (15 nm), a GaN:Mg waveguide layer (85 nm), an $Al_{0.07}Ga_{0.93}$N:Mg cladding layer (500 nm) and a GaN:Mg contact layer (90 nm). The MQW consisted of three 4 nm thick $In_{0.08}Ga_{0.82}$N quantum wells embedded in GaN:Si barrier layers (7 nm). Later structures used $In_{0.02}Ga_{0.98}$N:Si barriers instead. Figure 9a shows a representative L–j curve of the laser diode using pulsed current injection with a pulse length and a duty cycle of 60 ns and 0.06%, respectively. The operation voltage and the threshold current density were 35 V and 16.2 kA/cm^2, respectively. The corresponding spectrum at $1.1 \times I_{th}$ is depicted in Fig. 9b. Under pulsed excitation, maximum output powers of 262 mW per pulse were achieved, as shown in the inset of Fig. 9a.

4.2   Current spreading in cladding layers

To achieve high current densities in the active region of a GaN based laser diode the current path has to be confined from the contacts to the active region. The holes injected from the top contact have to pass the p-type cladding layer, which is often made of an Al_xGa_{1-x}N:Mg/GaN:Mg strained layer superlattice (SLS). Originally it was assumed that these SLSs prevent the tensile strained cladding layers from cracking, as discussed in Section 3.4. Meanwhile, it was shown that the twodimensional hole gas, which forms at the interfaces of the SLS, is a major drawback as it leads to an anisotropy of the conductance [53, 54]. As a consequence, the current spreads laterally in the laser structures and the effective current density in the active region is reduced.

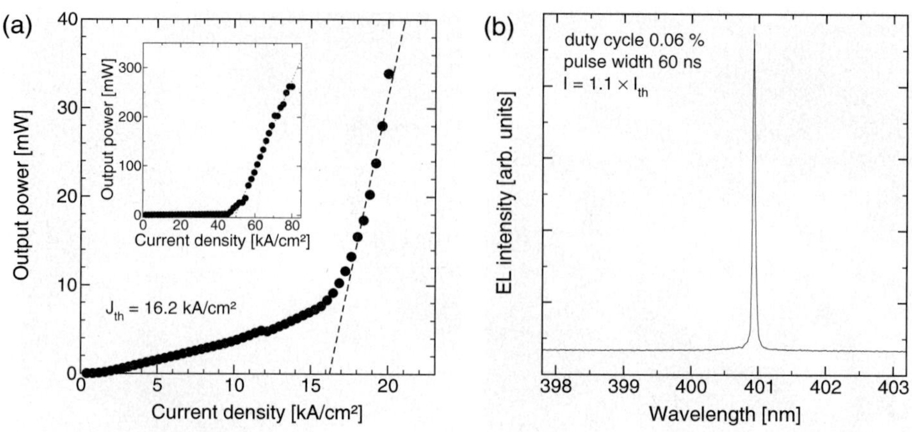

Fig. 9   a) *L-j* characteristic and b) emission spectrum of a laser diode under pulsed current injection. The duty cycle, the pulse width and the temperature were 0.06 %, 60 ns and 10 °C, respectively.

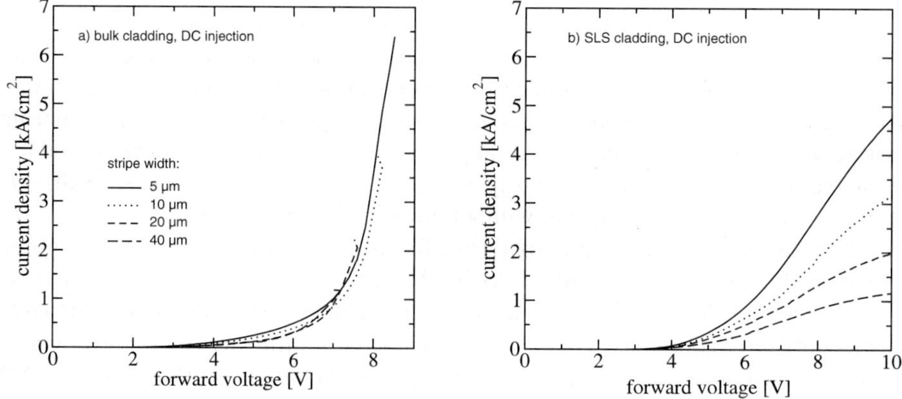

Fig. 10 (online colour at: www.interscience.wiley.com)   Current density versus forward voltage of a gain guided laser diode with a) bulk $Al_xGa_{1-x}N$ cladding layers and b) $Al_xGa_{1-x}N$/GaN SlS cladding layers.

The described effect was investigated by comparing laser diodes containing either a bulk $Al_xGa_{1-x}N$:Mg layer or an $Al_xGa_{1-x}N$:Mg/GaN:Mg SLS with 100 periods as the upper cladding layer. In both cases the cladding layers had an average aluminum mole fraction and an overall thickness of 7 % and 500 nm, respectively. The $Al_xGa_{1-x}N$ electron blocking layer usually grown on top of the MQW was omitted to minimize the current spreading at its interfaces. Gain-guided laser diodes were fabricated which, however, showed no laser emission, since the carrier confinement was low due to the missing electron blocking layer. The results of current-voltage measurements under direct current (DC) injection using different stripe widths are shown in Fig. 10. The serial resistances was as low as 2–8 Ω depending on the stripe width. If the current spreading is low and the serial resistance mainly arises from the p-contact, the current density should only depend on the applied voltage, but not on the stripe width. This is true for the devices with a bulk cladding layer shown in Fig. 10a. In contrast, the current densities increase with decreasing stripe widths for the diodes containing a SLS, as illustrated in Fig. 10b. The current path was obviously broadened by the SLS. Since there is still a change in the current density between stripe widths of 20 and 40 μm, the lateral spreading of the hole current has to be at least 20 μm.

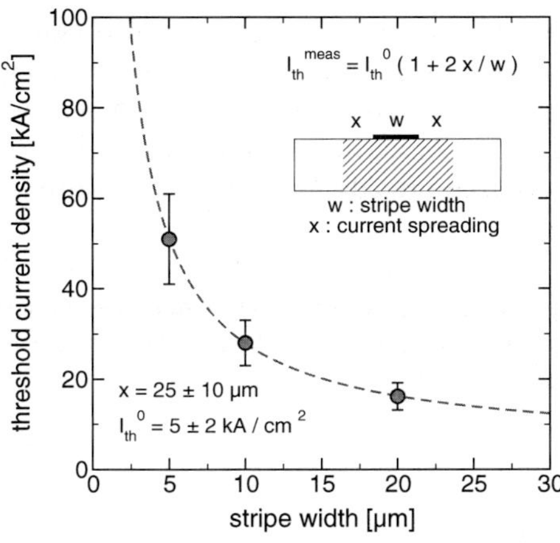

Fig. 11 Threshold current density (circles) of a gain guided laser diode containing a bulk $Al_xGa_{1-x}N$:Mg cladding layer in dependence of the injection stripe width. The dashed line is a fit of a model to the experimental data to derive the lateral current spreading.

The number derived for the lateral current spreading has to be taken as a lower limit, since the resistance arises mostly from the part of the cladding layer in which the current spreading is low. I.e., even the devices with a bulk cladding layer could exhibit current spreading, although it could not be resolved with the measurements described above. Therefore, the threshold current density of gain-guided laser diodes containing a bulk p-type cladding layer and an $Al_xGa_{1-x}N$ electron blocking layer was measured for different widths of the injection stripe. Figure 11 shows that the threshold current density decreases with increasing stripe width. If one assumes the threshold current density in the active region to be identical for all devices under investigation, a simple model can be fitted to the data shown in Fig. 11, indicating a lateral current spreading of 20 µm. It is emphasized that the electron blocking layer also contributes to the current spreading, such that the values obtained from the current–voltage measurements have to be taken as a lower boundary for the widening of the current path.

In conclusion, SLS cladding layers are not feasible for gain-guided laser diodes as the strong current spreading impedes sufficiently high current densities in the active region. However, even with bulk cladding layers the lateral current spreading is larger than typical widths of the injection stripe. This favours the use of ridge waveguide structures to obtain low threshold current densities for laser operation.

5   Summary

A number of issues of the MOVPE growth of group-III nitrides, which are critical for the realization of devices such as laser diodes, were discussed. The thermodynamic of the vapor-solid interaction in the MOVPE reactor was analyzed to describe the enhanced decomposition of GaN in a hydrogen containing ambient. The MOVPE growth of GaN on a low-temperature NL previously deposited on sapphire resulted in an island nucleation and coalescence process which governed the formation of both edge-type threading dislocations at the coalescence front and heterogeneous strain in the film. While the coalescence process was controlled by the composition of the gas phase during growth, the strain was attributed to the elastic closure of the gap between coalescing mismatched islands. Films heavily doped with magnesium suffered from pyramidal defects which formed upon a critical layer thickness and whose density varied periodically with the depth in the layer. A model was developed which assumed the magnesium to accumulate at the surface and induce the pyramidal defects when a critical surface coverage is reached. Bulk $Al_xGa_{1-x}N$ layers and $Al_xGa_{1-x}N$/GaN superlattices, which crack under tensile plain stress, where found to exhibit the same fracture toughnesses for identical average aluminum mole fractions. The cracks resulted in an inhomogeneous stress distribution in the layer with the amount of relieved stress depending on the crack separation.

Laser structures grown by MOVPE were processed to diodes whose laser operation was demonstrated under pulsed current injection at room temperature. The processing technology involved the crystallograhic wet etching of GaN to fabricate mirror facets of superior smoothness. Among other things, the performance of such devices was limited by the lateral spreading of the hole current in the p-type cladding layer particularly when an $Al_xGa_{1-x}N/GaN$ SLS was used. The spreading was estimated to about 20 µm based on measurements of devices with different widths of the injection stripe.

Acknowledgements   This work was supported by the Deutsche Forschungsgemeinschaft under the contract numbers HO 1388/13-1 and HO 1388/13-3. The authors acknowledge technical assistance by K. Vennen-Damm, S. Hesselmann and D. Loske.

References

[1] A. Koukitu, N. Takahashi, and H. Seki, Jpn. J. Appl. Phys. **36**, L1136, (1997).
[2] G. Briot, Group III Nitride Semiconductor Compounds (Clarendon Press, Oxford, 1998).
[3] D. D. Koleske, A. E. Wickenden, R. L. Henry, J. C. Culbertson, and M. E. Twigg, J. Cryst. Growth **223**, 466 (2001).
[4] Z. A. Munir and A. W. Searcy, J. Chem. Phys. **42**, 4223 (1965).
[5] M. Furado and F. Jacob, J. Cryst. Growth **64**, 257 (1983).
[6] R. A. Logan and C. D. Thurmaond, J. Electrochem. Soc. **119**, 1727 (1972).
[7] M. Mayumi, F. Satoh, Y. Kumagai, K. Takemoto, and A. Koukitu, Jpn. J. Appl. Phys. **39**, L707 (2000).
[8] M. Mayumi, F. Satoh, Y. Kumagai, and A. Koukitu, Jpn. J. Appl. Phys. **40**, L654 (2001).
[9] A. Koukitu, T. Taki, N. Takahashi, and H. Seki, J. Cryst. Growth **197**, 99 (1999).
[10] S. Duan and D. Lu, Mater. Res. Soc. Symp. Proc. **468**, 81 (1997).
[11] K. Hiramatsu, S. Itoh, H. Amano, I. Akasaki, N. Kuwano, T. Shiraishi, and K. Oh, J. Cryst. Growth **115**, 628 (1991).
[12] X. H. Wu, P. Fini, E. J. Tarsa, B. Heying, S. Keller, U. K. Mishra, S. P. DenBaars, and J. S. Speck, J. Cryst. Growth **189/190**, 231 (1998).
[13] X. H. Wu, P. Fini, S. Keller, E. J. Tarsa, B. Heying, U. K. Mishra, S. P. DenBaars, and J. S. Speck, Jpn. J. Appl. Phys. (Part 2) **35**, 1648 (1996).
[14] X. H. Wu, L. M. Brown, D. Kapolnek, S. Keller, B. Keller, S. P. DenBaars, and J. S. Speck, J. Appl. Phys. **80**, 3228 (1996).
[15] F. A. Ponce, MRS Bulletin **22**, 51 (1997).
[16] A. Munkholm, C. Thompson, C. M. Foster, J. A. Eastman, O. Auciello, G. B. Stephenson, P. Fini, S. P. DenBaars, and J. S. Speck, Appl. Phys. Lett. **72**, 2972 (1998).
[17] T. Hino, S. Tomiya, T. Miyajima, K. Yanashima, S. Hasimoto, and M. Ikeda, Appl. Phys. Lett. **76**, 3421 (2000).
[18] S. Figge, T. Böttcher, S. Einfeldt, and D. Hommel, J. Cryst. Growth **221**, 262 (2000).
[19] S. Einfeldt, T. Böttcher, S. Figge, and D. Hommel, J. Cryst. Growth **230**, 357 (2000).
[20] S. Hearne, E. Chson, J. Han, J. A. Floro, J. Figiel, J. Hunter, H. Amano, and I. S. T. Tsong, Appl. Phys. Lett. **74**, 356 (1999).
[21] J. D. Fingegan and R. W. Hoffman, J. Appl. Phys. **30**, 587 (1959).
[22] W. D. Nix and B. M. Clemens, J. Mater. Res. **14**, 3467 (1999).
[23] T. Boettcher, S. Einfeldt, S. Figge, R. Chierchia, H. Heinke, D. Hommel, and J. S. Speck, Appl. Phys. Lett. **78**, 1976 (2001).
[24] V. Potin, P. Ruterana, G. Nouet, R. C. Pond, and H. Morkoc, Phys. Rev. B **61**, 5587 (2000).
[25] U. Kaufmann, P. Schlotter, H. Obloh, K. Köhler, and M. Maier, Phys. Rev B **62**, 10867 (2000).
[26] L. T. Romano, M. Kneissl, J. E. Northrup, Chr. G. Van de Walle, and D. W. Treat, Appl. Phys. Lett. **79**, 2734 (2001).
[27] Z. Liliental-Weber, M. Benamara, W. Swider, J. Washburn, I. Grzegory, S. Porowski, D. J. H. Lambert, C. J. Eiting, and R. D. Dupuis, Appl. Phys. Lett. **75**, 4159 (1999).
[28] P. Vennéguès, M. Benaissa, B. Beaumont, E. Feltin, P. De Mierry, S. Dalmasso, M. Leroux, and P. Gilbart, Appl. Phys. Lett. **77**, 880 (2000).
[29] S. Figge, R. Kröger, T. Böttcher, P. L. Ryder, and D. Hommel, phys. stat. sol. (a) **192**, 456 (2002)
[30] R. Kröger, S. Figge, T. Böttcher, and D. Hommel, Mater. Res. Soc. Symp. Proc. **693**, (2002).
[31] L. T. Romano, J. E. Northrup, A. J. Ptak, and T. H. Myers, Appl. Phys. Lett. **77**, 2479 (2000).

[32] V. Ramachandran, R. M. Feenstra, W. L. Sarney, L. Salamanca-Riba, J. E. Northrup, L. T. Romano, and W. Greve, Appl. Phys. Lett. **75**, 808 (1999).

[33] S. Figge, R. Kröger, T. Böttcher, P. L. Ryder, and D. Hommel, to be published in Appl. Phys. Lett. **81**, (2002).

[34] S. J. Hearne, J. Han, S. R. Lee, J. A. Floro, D. M. Follstaedt, E. Chason, and I. S. T. Tsong, Appl. Phys. Lett. **76**, 1534 (2000).

[35] V. E. Bougrov and A. S. Zubrilov, J. Appl. Phys. **81**, 2952 (1997).

[36] S. Nakamura, M. Senoh, S. Nagahama, N. Iwasa, T. Yamada, T. Matsushita, H. Kiyoku, Y. Sugimoto, T. Kozaki, H. Umemoto, M. Sano, and K. Chocho, Jpn. J. Appl. Phys. **36**, L1568 (1997).

[37] S. Nakamura, M. Senoh, S. Nagahama, N. Iwasa, T. Yamada, T. Matsushita, H. Kiyoko, Y. Sugimoto, T. Kozaki, H. Umemoto, M. Sano, and K. Chocho, Appl. Phys. Lett. **72**, 211 (1998).

[38] M. Hansen, A. C. Abare, P. Kozodoy, T. M. Katona, M. D. Craven, J. S. Speck, U. K. Mishra, L. A. Coldren, and S. P. DenBaars, phys. stat. sol. (a) **176**, 59 (1999).

[39] P. Kozodoy, M. Hansen, S. P. DenBaars, and U. K. Mishra, Appl. Phys. Lett. **74**, 3681 (1999).

[40] P. Kozodoy, Y. P. Smorchkova, M. Hansen, H. Xing, S. P. DenBaars, U. K. Mishra, A. W. Saxler, R. Perrin, and W. C. Mitchel, Appl. Phys. Lett. **75**, 2444 (1999).

[41] J. L. Beuth, Jr., Int. J. Solids Struct. **29**, 1657 (1992).

[42] S. Einfeldt, V. Kirchner, H. Heinke, M. Dießelberg, S. Figge, K. Vogeler, and D. Hommel, J. Appl. Phys. **88**, 7029 (2000).

[43] B. R. Lawn and T. R. Wilshaw, Fracture of brittle solids (Cambridge University Press, Cambridge, 1975).

[44] A. P. Parker, The mechanics of fracture and fatigue (E. & F. N. Spon Ltd, London, 1981).

[45] J. E. Northrup and J. Neugebauer, Phys. Rev. B **53**, R10477 (1996).

[46] S. Einfeldt, H. Heinke, V. Kirchner, and D. Hommel, J. Appl. Phys. **89**, 2160 (2001).

[47] S. Einfeldt, M. Dießelberg, H. Heinke, D. Hommel, D. Rudloff, J. Christen, and R. F. Davis, J. Appl. Phys. **92**, 118 (2002).

[48] D. Rudloff, T. Riemann, J. Christen, Q. K. K. Liu, A. Kaschner, A. Hoffmann, Ch. Thomsen, M. Dießelberg, K. Vogeler, S. Einfeldt, and D. Hommel, accepted for publication in Appl. Phys. Lett.

[49] T. Lei, K. F. Ludwig, and T. Moustakas, J. Appl. Phys. **74**, 4430 (1993).

[50] D. A. Stocker, E. F. Schubert, and J. M. Redwing, Appl. Phys. Lett. **73**, 2654 (1998).

[51] T. Miyajima, T. Tojyo, T. Asano, K. Yanashima, S. Kijima, T. Hino, M. Tykeya, S. Uchida, S. Tomiya, K. Funato, T. Asatsuma, T. Kobayashi, and M. Ikeda, J. Phys.: Condens. Matter **13**, 7099 (2001).

[52] T. Böttcher, Ch. Zellweger, S. Figge, R. Kröger, Ch. Petter, H.-J. Bühlmann, M. Ilegems, P. L. Ryder, and D. Hommel, phys. stat. sol. (a) **191**, R3 (2002).

[53] M. S. Shur, A. D. Bykhovski, R. Gaska, J. W. Yang, G. Simin, and M. A. Khan, Appl. Phys. Lett. **76**, 3061 (2000).

[54] S. Figge, T. Böttcher, Ch. Zellweger, M. Illegems, and D. Hommel, Mater. Res. Soc. Symp. Proc. **693**, 683 (2002).

phys. stat. sol. (c) **0**, No. 6, 1860–1877 (2003) / **DOI** 10.1002/pssc.200303124

Optical gain, gain saturation, and waveguiding in group III-nitride heterostructures

M. Röwe* [1], **M. Vehse** [1], **P. Michler** [1], **J. Gutowski** [1], **S. Heppel** [2], and **A. Hangleiter**** [2]

[1] Institut für Festkörperphysik, Universität Bremen, P.O.Box 330440, 28334 Bremen, Germany

[2] Institut für Technische Physik, Technische Universität Braunschweig, Mendelssohnstrasse 2, 38106 Braunschweig, Germany

Received 4 March 2003, accepted 15 April 2003
Published online 28 August 2003

PACS 78.20.Bh, 78.45+h, 78.55.Cr, 78.66.Fd, 78.67.De

The optical gain and waveguiding properties of nitride heterostructures are essential prerequisites for device optimisation. We address the mechanism of optical gain using direct measurements of optical gain spectra and demonstrate the impact of polarisation fields. Limits to the optical gain are shown to be imposed by carrier escape and by gain saturation. Control of guided modes by waveguiding nitride structures differs from other materials due to their birefringence. Careful optimisation is therefore needed in order to obtain good far-field properties.

1 Introduction

Recently, the successful development of blue GaInN/GaN/AlGaN multiple quantum well (MQW) laser diodes operating in continuous wave (CW) mode at room temperature with a lifetime of more than 10000 hours has been reported [1]. Moreover, several other groups have now succeeded in demonstrating GaN-based laser diodes in pulsed and CW operation at room temperature [2–7]. There is a number of technological problems involved in making such laser diodes, such as those related to the hetero-epitaxial growth of nitrides on sapphire or SiC [8], the problem of p-type doping and conductivity [9, 10], and the realization of laser mirrors [11].

Compared to laser diodes based on other material systems like GaAs/AlGaAs, GaInAs/InP, or GaInP/-AlGaInP [12], the threshold current density of nitride-based laser diodes is still extremely large. This has been attributed to the large defect density typically found in hetero-epitaxial nitride layers [8], but may also be due to fundamental limitations imposed by the energy band structure of the nitrides, i.e. the extremely large effective masses [13–16]. Unlike the conventional III–V semiconductors of zinc-blende structure, where the application of built-in strain in thin heterostructures is highly beneficial for achieving low threshold current densities [17], the wurtzite nitrides have little potential for improvement due to built-in strain [18, 19].

Even more fundamentally, the very mechanism of optical gain in GaN-based structures is still subject to some controversy. Whereas for the longer-wavelength III–V semiconductors and quantum well structures the optical gain around room temperature is clearly due to band-to-band transitions, there have been speculations about the role of composition fluctuations in GaInN/GaN laser structures [20, 21]. It has been proposed that localised excitons in random potential wells [20] or even quantum-dot-like structures arising from phase separation of InN and GaN [21] are decisive for the lasing mechanism.

---

* Phone: +49 421 218 4428, Fax: +49 421 218 7318, e-mail: mroewe@physik.uni-bremen.de
** Phone: +49 531 391 8501, Fax: +49 531 391 8511, e-mail: a.hangleiter@tu-bs.de

In this paper, we will review the mechanisms of optical gain relevant for nitride-based lasers as well as the experimental methods used to measure optical gain. Then we will discuss the available experimental data from gain spectroscopy and from other methods, which give information on the gain mechanism. We will restrict the discussion to the hexagonal (wurtzite) phase of the nitrides, which is of technological relevance to date.

2 Theory of optical gain in nitrides

2.1 Band-to-band transitions

In a free-carrier two-band picture of a semiconductor, optical gain is achieved if the separation of the electron and hole quasi Fermi energies exceeds the bandgap energy [22]. Due to the extended nature of the states involved, the spectra of the optical gain tend to be rather broad. Even though a large improvement is expected for low-dimensional structures due to their narrow density of states, inhomogeneous broadening due to size fluctuations diminishes the potential improvement.

From time-dependent perturbation theory one finds for the optical gain of a quantum well heterostructure in the effective-mass approximation [23, 24] (broadening effects are neglected at this point)

$$g_0(E) = \frac{e^2 h}{2\epsilon_0 m_0^2 cnEL_z} \frac{2}{(2\pi)^2} \int |\varepsilon \boldsymbol{M}_{if}|^2 (f_c(k) - f_v(k))\delta(E_i - E_f - E)\mathrm{d}^2 k, \qquad (1)$$

where m_0 is the free-electron mass, n is the refractive index, L_z is the quantum well thickness, and $f_c(k)$ and $f_v(k)$ are the Fermi distribution functions for the conduction and the valence band. The matrix element $\varepsilon \boldsymbol{M}_{if}$ is given by

$$\varepsilon \boldsymbol{M}_{if} = < \varphi_i(k_\parallel)|\varepsilon \boldsymbol{P}|\varphi_j(k_\parallel) >, \qquad (2)$$

where $\varphi_i(k_\parallel)$ are the s-like wavefunctions of the conduction band [25], ε is the polarisation of the emitted light, $\boldsymbol{P}$ is the momentum matrix element, and $\varphi_j(k_\parallel)$ are the eigenfunctions of the valence band.

In fact, when the anisotropy of the valence bands is neglected by using the axial approximation, Eq. (1) can be replaced by a more simple expression, where the angular integration has been carried out and where the matrix element (2) is replaced by the averaged matrix element E_p

$$g_0(E) = \frac{e^2 h}{2\epsilon_0 m_0^2 cnEL_z} \frac{E_p}{3} \sum_{i,j} \frac{1}{\pi} F_{i,j}\xi_{i,j}k_\parallel (f_c(k_\parallel) - f_v(k_\parallel))dk_\parallel, \qquad (3)$$

where $F_{i,j}$ is the overlap integral of the envelope functions and $\xi_{i,j}$ is the polarisation factor [26].

In bulk semiconductors as well as in quantum wells, the optical gain reflects the combined density of states as well as the "combined" distribution function and the energy dependence of the matrix elements.

2.2 Polarisation dependence of the gain

In contrast to cubic semiconductors, the optical gain in wurtzite GaN has a strong polarisation dependence even in bulk material but also in quantum wells. This is a consequence of the nature of the valence band of the wurtzite nitrides, where the spin-orbit splitting is small compared to the crystal-field splitting. The A and B valence band wavefunctions are almost exclusively of "xy" character, whereas the C valence band is only of "z" character close to the Γ point.

By considering Eq. (2) we realize that this means that the A and B valence bands interact only with light polarised in c-plane (usually referred to as TE polarisation), whereas the C valence band has a non-vanishing matrix element only for light polarised perpendicular to the c-plane (TM polarisation). Of course, these strict selection rules are getting weaker if we move out in k-space.

2.3 Broadening mechanisms

Our discussion of the spectral and polarisation dependence of the optical gain in strained quantum well structures so far neglected two major issues: The first is related to the fact that the stimulated optical transition leaves behind two empty states. These empty states are populated again by intraband relaxation processes and therefore have a finite lifetime. This finite lifetime leads to lifetime (Lorentzian) broadening of the optical transition [27].

The second major broadening mechanism is practical in nature and is due to the fact that real quantum wells exhibit some fluctuations in well thickness and (for ternary and quaternary well materials) in material composition and the corresponding fluctuations in transition energy. This leads to an inhomogeneous (Gaussian) broadening of the optical transition.

While the inhomogeneous broadening is straightforward, the homogeneous broadening due to the finite final state lifetime needs density matrix theory for a rigorous treatment [28, 29]. The optical gain as given by Eq. (1) is modified to read

$$g(E) = \frac{1}{2\pi} \int \frac{g_0(E')\Gamma(E')\mathrm{d}E'}{(E - E')^2 + \Gamma(E')^2/4}, \tag{4}$$

where $\Gamma = \hbar/\tau_p$ is the broadening parameter and τ_p is the polarisation relaxation time. We note that in general Γ and τ_p have to be taken as energy dependent [30, 31].

2.4 Many-particle effects

Like for almost all optical properties of semiconductors, the optical gain is subject to the influence of many-particle effects. This is particularly true since optical gain requires large carrier densities. Basically, many-particle effects are a consequence of the Coulomb interaction of electrons and holes. Their influence scales with the effective excitonic Rydberg energy of the material [32]. At low carrier densities, the most prominent many-particle effect is the excitonic enhancement of the optical absorption. It has been shown both theoretically and experimentally that this enhancement decreases when the carrier density is increased, due to the screening of the Coulomb interaction and to phase space blocking [33]. Along with the screening of the excitonic enhancement of the absorption, the fundamental gap is renormalised due to exchange and correlation effects. This leads to a red-shift of the optical spectrum with increasing carrier density.

For most classical III–V semiconductor structures including strained quantum wells, the excitonic Rydberg energy is less than 10 meV. As a consequence, the excitonic enhancement is much reduced for the carrier densities needed to achieve inversion and optical gain. For InGaAs/InP strained quantum wells, the remaining enhancement of the gain is of the order 10 % [34]. Nevertheless, there is a significant amount of bandgap renormalisation [34, 35], making the energetic position of the maximum gain dependent on carrier density. Therefore, many-particle effects have only minor consequences for the static properties of strained quantum well lasers made from arsenide and phosphide-based materials. For the nitrides, due to their larger excitonic binding energy, one might expect a significant excitonic enhancement of the optical gain. Theoretical calculations predict that there is an enhancement of the optical gain by about 50 % [36, 37]. However, in any case, bandgap renormalisation has important consequences for the differential gain which governs the modulation behaviour of semiconductor lasers.

For III–V semiconductors like GaInAs/InP, GaAs/AlGaAs, or GaInP/GaAs and quantum wells from these materials, the optical gain is well understood on the basis of pure band-to-band transitions. Even for strained quantum wells, model calculations of the gain based on 6-band $k \cdot p$ valence band structures yield excellent quantitative agreement between measured and calculated gain spectra [38]. The most important property of the optical gain in our context is the shape of the gain spectra. In case of inversion, optical gain is expected between the (renormalised) band edge and the chemical potential (i.e. the difference of the quasi Fermi energies). At higher energies, a steep absorption edge with up to $\approx 9 \times 10^4 \mathrm{cm}^{-1}$ should be observed.

2.5 Effects of polarisation fields on optical gain

As pointed out earlier, internal polarisation field have a strong impact on the optical properties of nitride quantum wells. In particular, there is a strong Stark shift of the emission towards lower energies, along with a drastic reduction of the oscillator strength [39] and thus the potential optical gain. Qualitatively, one expects some recovery of the gain at high carrier densities, when the internal fields become screened by charge accumulation.

Highly sophisticated calculations of the optical gain in nitride single quantum well structures have already been reported [41, 42], which are based on a $k \cdot p$ model and which are taking into account the internal fields in such structures in a self-consistent manner. However, particularly in the presence of internal fields, multiple quantum wells are likely to behave considerably different from single quantum wells, due to potentially inhomogeneous field distributions. Moreover, it has been shown that in the case of wurtzite nitrides, subband interactions in quantum wells are only of minor importance [43]. Therefore, a somewhat simplified model [40] concentrates on self-consistently calculating the effects of polarisation fields on optical properties for multiple quantum wells by iteratively solving the Schrödinger and Poisson equations. The complications arising from the complex valence band structure are deliberately neglected. On the other hand, the complete multiple quantum well structure, including the commonly used AlGaN electron barrier is taken into account.

As shown in Fig. 1, even under lasing conditions, the quantum well next to the substrate holds considerably more positive charge (i.e. holes) than the others, while there is a broad accumulation of negative charge around the topmost quantum well. Here, a triple quantum well structure was studied, with a GaInN well bandgap of 3 eV, a well width of 2.5 nm, a barrier width of 5 nm, and a AlGaN electron barrier of 10 nm. As can be seen from the conduction band diagram, the overall polarisation across the MQW is largely screened, but there remains a considerable field within the QW's. Accordingly, the oscillator strength is considerably lower than in the field-free case.

Gain spectra calculated using this model will be used in Section 5 to analyse experimental gain spectra.

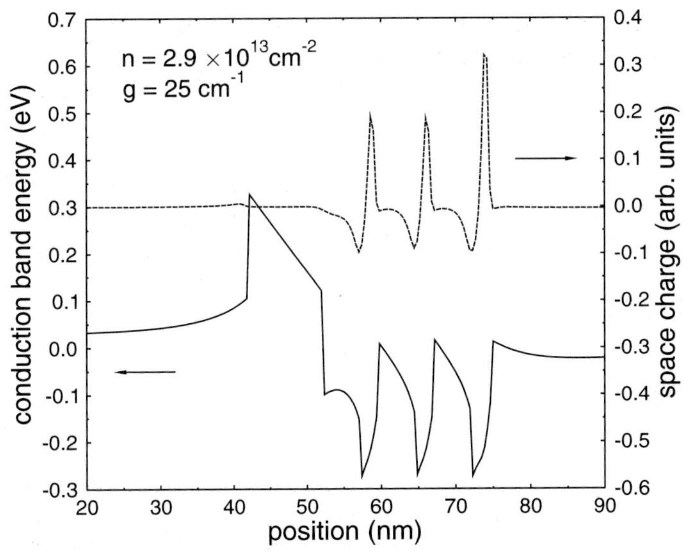

Fig. 1 Conduction band diagram and space charge distribution of a triple quantum well at laser threshold [40]. Clearly, the carrier distribution is strongly inhomogeneous.

3 Experimental methods

The optical gain in semiconductor structures can be determined by a variety of experimental techniques. Basically there are two types of measurements using either plain epitaxial layers without any further processing or complete laser or optical amplifier devices.

3.1 Gain measurements on plain layers

The most basic type of measurement of the optical gain would be to determine the light intensity of a probe beam before and after passing through the inverted medium, just like in the case of optical absorption measurements. For thick bulk-like layers such measurements are sometimes performed under vertical incidence conditions, but for thin quantum well structures the effective amplification of the input signal becomes too small for this method to be practical. On the other hand, such measurements are difficult within the layer plane as well, because of the lack of lateral waveguiding in plain epitaxial layers.

Therefore the method of choice for studying optical gain in plain epitaxial layers is the so-called stripe excitation method as shown schematically in Fig. 2 [44]. Electrons and

holes are generated in the sample by optical excitation with a laser beam, which is focused onto a stripe-like region of the sample. If the excited sample is inverted sufficiently, spontaneously emitted light travelling along the stripe is amplified by stimulated emission. In fact, this situation can also be viewed as an optical amplifier amplifying its own (spontaneous) noise.

The intensity of the light travelling along the stripe is given by

$$\frac{\mathrm{d}I}{\mathrm{d}z} = \Gamma g \cdot I + \beta \Gamma r_{\mathrm{sp}}, \tag{5}$$

where $g(h\nu)$ is the optical gain coefficient of the material, Γ is the mode confinement factor, β is the spontaneous emission factor, and $r_{\mathrm{sp}}(h\nu)$ is the spontaneous emission rate. Solving Eq. (5) for a stripe of length L using $I(0) = 0$ as the boundary condition we find

$$I(L) = \frac{\beta r_{\mathrm{sp}}}{g} \left(e^{\Gamma g L} - 1 \right), \tag{6}$$

which corresponds to an exponential increase of the intensity at the end of the stripe with increasing stripe length in case of positive optical gain.

Experimentally, the sample is optically excited using a pulsed laser (in case of the nitrides usually either a N_2 laser, a XeCl excimer laser, or a dye laser) and the spectrum of the amplified spontaneous emission (ASE) emitted at the end of the stripe is measured for a number of different stripe lengths. Under typical experimental conditions the stripe has a width of 20...30 µm and a maximum length of 200...400 µm on the sample.

Equation (6) is then used to fit these data for every wavelength and to derive the optical gain spectrum. As a matter of fact, this type of analysis is not limited to positive gain. It can also be applied to spectral regions where absorption dominates, provided that the spontaneous emission intensity is sufficiently large.

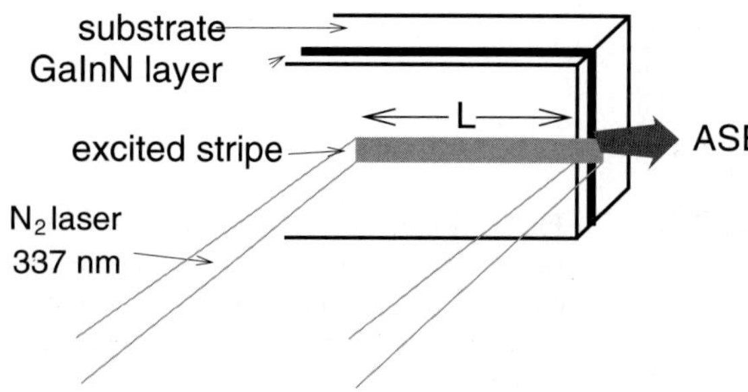

Fig. 2 (online colour at: www.interscience.wiley.com) Schematic view of the experimental arrangement involved in the stripe excitation method for measuring the optical gain.

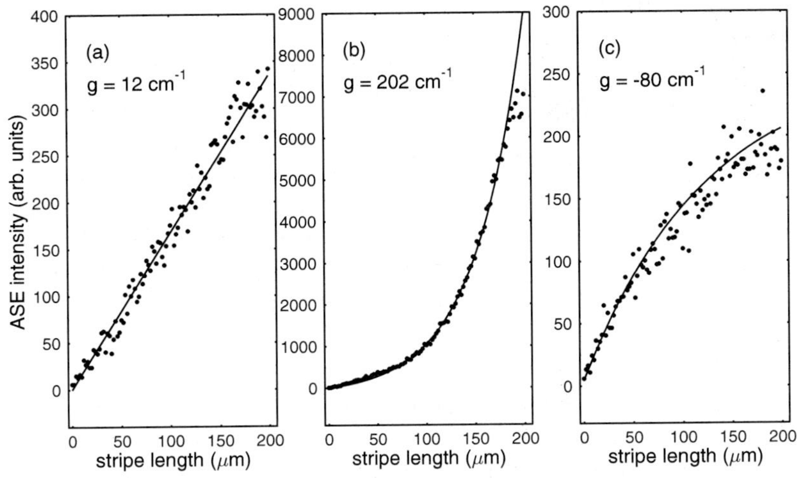

Fig. 3 Intensity of the amplified spontaneous emission as a function of stripe length for three cases: a) very weak gain, low energy side of the gain peak; b) strong gain, on the gain peak; c) moderate absorption, high energy side.

The variation of the ASE intensity with stripe length is illustrated in Fig. 3, where the case of positive gain is shown in the left part of the figure, whereas the case of negative gain (absorption) is shown on the right hand side.

It is important to note that one has to be very careful to avoid saturation of the gain in any spectral region during such measurements. Gain saturation in this context means that the stimulated transition rate becomes comparable with the spontaneous recombination rate leading to a reduction of the inversion. In a semiconductor, the effect of saturation is most pronounced close to the chemical potential, where to gain is most sensitive to changes in carrier density (i.e. the differential gain is largest). Gain saturation manifests itself in a deviation from Eq. (6) towards lower intensities at large stripe length. Since gain saturation is both non-local in frequency (i.e. large intensities at the gain maximum may saturate the gain close to the chemical potential) and space (i.e. due to a partial reflection at the cleaved edge of the sample, gain saturation occurs typically at the "beginning" of the stripe, where the light intensity is highest) it is extremely difficult to model.

In this context we also note that the amplified spontaneous emission (ASE) spectra do not exhibit a strong line narrowing even for significantly positive gain, as long as no gain saturation occurs. We find no significant change in linewidth between very short stripes (almost no stimulated emission) and rather long stripes (strong stimulated emission) [45]. In the literature, it is quite common to associate stimulated emission with a strong line narrowing. This is only true if gain saturation occurs, which is usually most prominent on the high energy side of the spectrum and consequently leads to a narrowing of the line. Therefore it is not appropriate to use strong line narrowing as an indication of the onset of stimulated emission. In fact, stimulated emission occurs already at much lower pumping levels, depending on the experimental geometry.

The optical gain spectra resulting from such experiments are not only useful to analyse the optical gain but also to determine the optical losses of the waveguide. This relies on the fact that the optical gain on the low energy side of the spectrum should approach zero in case of vanishing waveguide losses. Consequently, a negative offset at the gain spectrum can be identified with the optical losses. For long-wavelength lasers, such an analysis has been used to determine the magnitude of the inter valence band absorption in those lasers [46]. For the nitrides, there is a strong variation of the waveguide losses from sample to sample. For good samples the losses can be as low as $10 \, \text{cm}^{-1}$.

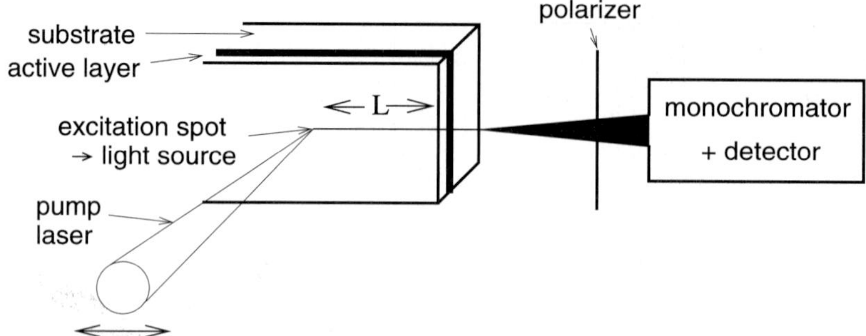

Fig. 4 Schematic view of the experimental arrangement used in the "point excitation method" for measuring the waveguide absorption spectrum.

3.2 Waveguide absorption measurements

As mentioned earlier, it may be rather difficult to experimentally determine the optical absorption spectrum of thin quantum well structures. It is therefore convenient to utilise a waveguiding geometry, where the probe light is travelling along the absorbing medium instead of perpendicular to it. Thus, the effective absorption length can be dramatically increased.

Experimentally, coupling light into a thin quantum well waveguide may prove difficult as well. Rather than using an external light source, the spontaneous emission of the material itself may be used as a light source as indicated in Fig. 4. This is the so-called "point excitation method" [47], where a small highly excited spot (in order to provide a wide spontaneous spectrum) is moved along the sample surface and where the emission from the sample edge is detected. The waveguide is guiding the spontaneous light to the sample edge and at the same time partially absorbing it. By analysing the dependence of the observed intensity on the distance of the excited spot from the sample edge, one can precisely determine the modal absorption spectrum of the waveguide.

4   Thermal activation of carriers and nonradiative recombination

In order to realize an optimal laser design, it is absolutely necessary to understand the physical mechanisms which influence carrier recombination and optical gain. The impact of built-in macroscopic polarisation fields due to piezoelectric and spontaneous polarisation, and that of potential fluctuations in the quantum well area have been examined in detail and will be presented in the following section. The influence of carrier leakage into the barrier is another important aspect which leads to larger threshold densities at room temperature (RT).

In a systematic temperature dependent study of InGaN/GaN laser structures on SiC substrates the profound importance of the thermal activation of carriers and their nonradiative recombination for quantum efficiency and optical gain have been demonstrated.

The temperature dependence of the quantum efficiency recorded via low-density PL spectra indicates the existence of two activation energies, related to different recombination channels. At lower temperatures, carriers are activated with a characteristic activation energy E_{A1} from states localised due to potential fluctuations into higher unbound states and recombine at nonradiative recombination centers [48].

For higher temperatures a stronger quench of the PL intensity is observed which can be attributed to thermal emission of carriers into the barriers. A characteristic correlation between the activation energy E_{A2} obtained from Arrhenius plots and the total confinement energy ΔE of the quantum wells could be demonstrated [49, 50] (see Fig. 5). The found thermal activation energy E_{A2} is about one half the total confinement energy ΔE. This can be explained on the basis of simultaneous thermal emission of electrons and holes into the barriers [51].

The influence of the barrier height is also reflected in the high-density measurements, e.g., in the temperature dependence of transparency power [49] and maximum optical peak gain [50]. In a comparison of samples with different well widths, a serious reduction of the threshold power for increasing quantum well

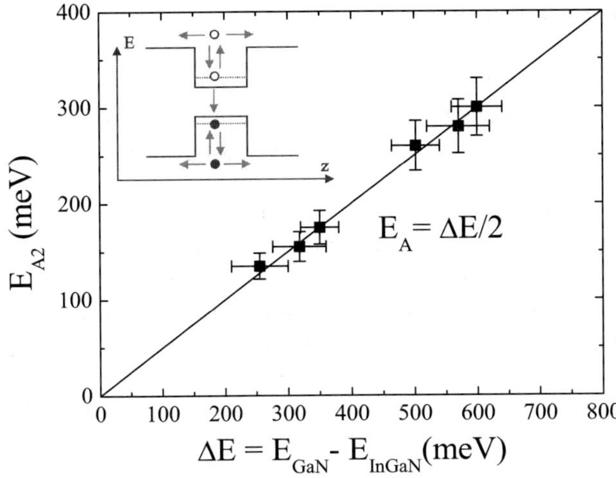

Fig. 5 (online colour at: www.interscience. wiley.com Second, larger thermal activation energy E_{A2} as a function of the total confinement energy ΔE. The solid line corresponds to the funtion $E_{A2} = \Delta E/2$ (from Ref. [49].

width can be recognised (see Fig. 6). This is valid in the whole temperature range investigated (4–300 K). The lower threshold power at larger well width can be well explained by the increasing total confinement of the carriers due to the rising effective barrier height. Just the 4.5 nm thick sample shows a slight increase of the threshold, most probably due to the growing influence of the quantum confined Stark effect with increasing well widths. Additionally, an increasing In content (from 8 to 13 %) in the samples of equal quantum well widths leads to a reduced threshold at RT (see Fig. 6).

In order to avoid drawbacks caused by quantum wells that are too large, the carrier confinement can instead be improved by an enlarged In concentration in the quantum wells. This deepens the wells so that the thermal activation of carriers in the barriers and consequently the threshold densities will be reduced without thicker quantum wells being required.

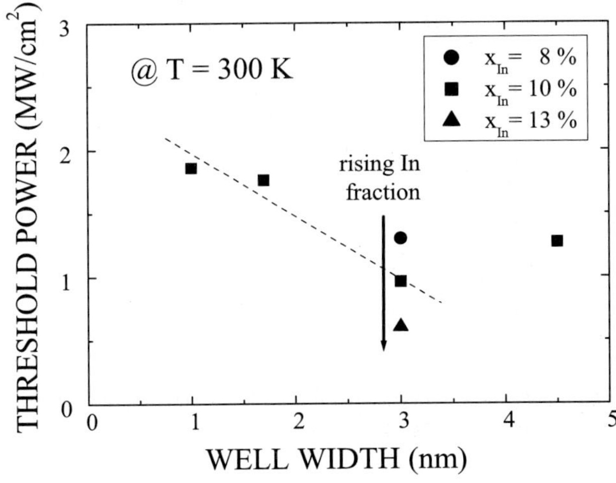

Fig. 6 Comparison of the threshold power at room temperature for all investigated samples. The dashed line is a guide to the eye (from Ref. [49]).

5   Experimental clues to the gain mechanism

5.1   Overview

In general, experimental data regarding optical gain in nitrides have to be divided into two groups, (i) measurements on rather thick epitaxial layers or bulk crystals and (ii) investigations using laser structures incorporating an optical waveguide.

There exists a large number of papers related to nitrides, which belong to the former group of experiments [52–55]. Those results and their interpretation are highly questionable, since only a thin surface layer (≈ 100 nm) is inverted by the pump laser and the propagating mode is subject to strong reabsorption by deeper-lying regions. Since the optical gain spectrum is red-shifted with respect to the bandgap by renormalisation effects, this leads to stimulated emission well below the bandgap. In many cases this is misinterpreted as evidence for "excitonic" optical gain.

The latter group of experiments includes measurements on laser structures grown on sapphire [45, 56–60] and on silicon carbide [40, 61, 62]. Due to the complex waveguiding properties of the nitrides (c.f. Section 7), some of the measurements performed on sapphire-based structures [56, 57] have to be taken with great care [63]: Secondary spectral peaks reported there may as well be due to resonant mode coupling [63].

As shown in Section 7, the waveguiding properties of nitride laser structures grown on silicon carbide are much simpler. A true single-mode waveguide like in other III/V semiconductor laser structures can be easily established. Therefore, such structures are a good choice to analyse for the fundamental gain mechanism with a minimum of complications.

5.2   Optical gain in laser structures grown on SiC

A critical test of any optical gain model for semiconductor structures is the direct comparison of calculated gain spectra with measured ones. This is particularly true for the optical gain, since the gain amplitude is closely related to band filling, which is in turn related to the spectral width via the quasi-Fermi levels.

As shown in Fig. 7, gain calculations performed using the very structural parameters of the real laser structures, nicely reproduce the experimental gain spectra.

For the particular example shown in Fig. 7, the carrier density in the structure is 4.8 and 5.9×10^{13} cm^{-2}, respectively. These values are fairly insensitive to variation of the inhomogeneous broadening parameter.

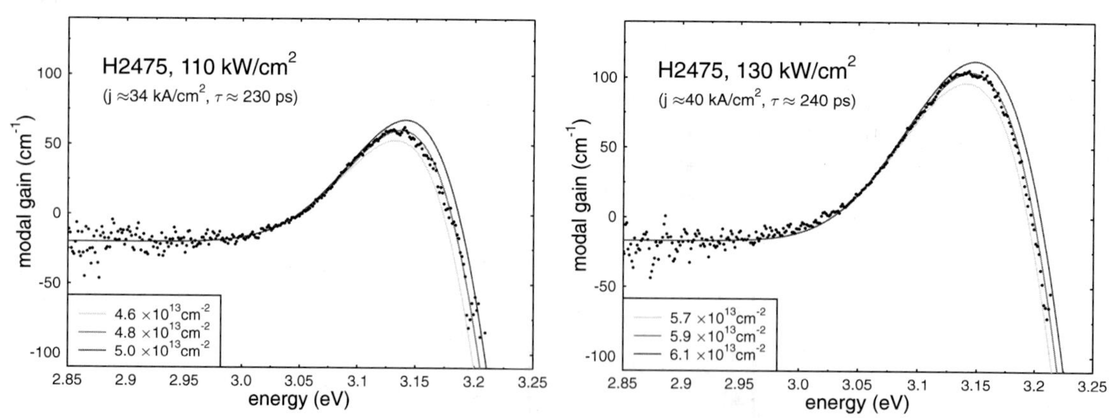

Fig. 7 (online colour at: www.interscience.wiley.com) Comparison of measured and calculated optical gain spectra, allowing the determination of the carrier density, the carrier lifetime, and the optical losses [40].

Together with the (absorbed) pump power densities of 110 and 130 kW/cm^2 used in the experiment, this translates into a carrier lifetime of 240 ps. This has to be compared to a radiative carrier lifetime of 5.8 and 4.8 ns determined from the calculations. Consequently, this laser structure is still dominated by nonradiative recombination losses associated with the high defect density.

Moreover, the optical losses in such structures typically are in the range 10–30 cm^{-1}, which is surprisingly low given the large defect density in heteroepitaxial nitride layers. However, the present result is similar to earlier data obtained for simple double heterostructures [45].

5.3 Comparison of gain and absorption

Using both the stripe-excitation method and the point-excitation method, both the optical gain and the absorption of the un-pumped medium have been compared in a waveguide geometry [62].

Figure 8 shows several measured and modelled gain and absorption spectra for two exemplary structures, which fit all quite satisfactorily. The internal losses for all samples are between 10 and 35 cm^{-1}. To investigate the influence of the number of quantum wells, a series of three samples with two, three, and five (DQW, TQW, QQW) wells was measured and compared with each other. All samples had a nominal quantum well thickness of 3 nm and a barrier thickness of 6 nm. The results for the DQW and the QQW are displayed in Fig. 8. In this figure an absorption spectrum of the unexcited and one of the excited material together with gain spectra for several excitation densities can be seen. With the same set of parameters all spectra could be fitted. For a modal gain (measured gain corrected by the internal losses) of 40 (2 QW) and 31 cm^{-1} (5 QW) carrier densities of 2.4 and 3.2 $\times 10^{13}$ cm^{-2} are necessary. From a comparison of the $g(n)$ of the three different samples one should get the lowest threshold current densities for the DQW, which is in fact the result of electrical measurements. As a measure for the blue shift, which occurs in the presence of a increasingly screened internal field, the energy difference (ΔE) between the energy, where the modal absorption has a value of 0.5 cm^{-1} (arbitrarily chosen), and the energetic position of a (low) gain maximum is taken and plotted in the diagrams. To minimise the potential influence of bandgap renormalisation, gain spectra with low modal gain, i.e. relatively low carrier densities, are considered. Within this set of samples the observed energy shift is quite similar.

On the basis of a second series the influence of the quantum well width L_z is studied. A set of three samples with a quantum well thickness of 1.2, 1.8, and 3.0 nm was measured and compared in the same way as the first series. All samples had five quantum wells and were completely doped. For the one with 1.2 nm and the one with 3.0 nm thick quantum wells the carrier density necessary for a certain gain value and the blue shift ΔE are shown in Table 1. Like for the shown gain value also for higher densities there

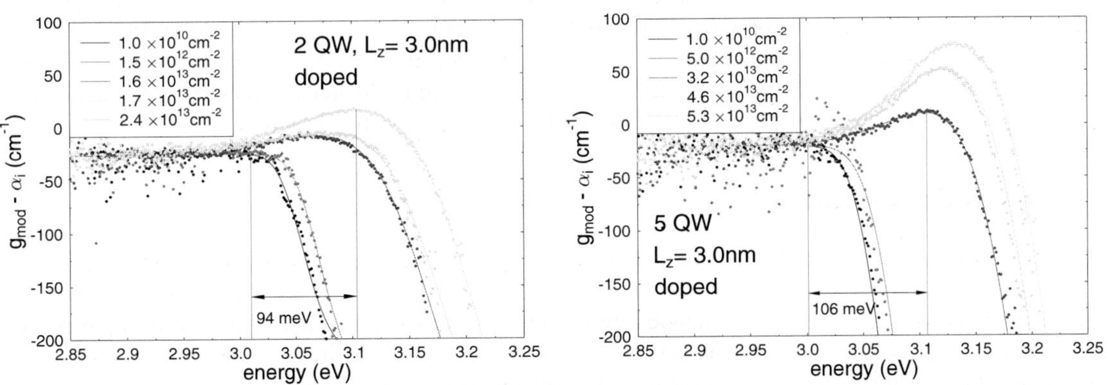

Fig. 8 (online colour at: www.interscience.wiley.com) Comparison of measured and calculated absorption and gain spectra for a sample with two quantum wells (left) and five quantum wells (right) ($L_z = 3$ nm) [62].

sample	Γ (meV)	gain (cm^{-1})	n (10^{13} cm^{-2})	ΔE (meV)
doped; 5 x 1.2 nm	80	35	3.3	62
doped; 5 x 3.0 nm	85	30	3.2	117
undoped; 3 x 2.5 nm	75	50	3.5	191
doped; 3 x 3.0 nm	80	45	2.7	97

Table 1   Typical results for the broadening, the peak gain, the carrier density, and the spectral width for sample series with variation of L_z and doped and undoped samples [62].

is only a small difference in the observed modal gain g for similar carrier densities n, i.e. the dependency $g(n)$ is very similar, despite a more than doubled quantum well thickness, what should result in a larger internal field effect. The cause for this only slight difference, is most likely the not exactly identical In content in the samples. As for the energy shift a pronounced difference is visible with the expected higher value for the thicker quantum wells. Comparing $g(n)$ of all three samples leads to the expectation of the lowest threshold current density for the sample with the quantum well thickness of 1.8 nm, which is confirmed by the results obtained for the processed devices. The lower part of Table 1 shows the results for an undoped and a doped sample with nominally almost the same active region. Assuming a negligible background carrier density for the undoped samples, we calculated the spectra taking the full internal field. In contrast, for all pin laser structures we presumed that the internal field is screened by the doping of cladding and waveguide layers to one third of its unscreened value [64]. The results reflect the higher field to be screened in the undoped sample through a higher carrier density for a similar modal gain and a significantly stronger blue shift in the case of the undoped sample.

5.4   Discussion

Considering the results discussed in the preceeding sections, one notices that experimental gain spectra at room temperature are in fairly good agreement with a model based on band-to-band transitions (i.e. an electron–hole plasma) self-consistently taking into account the internal polarisation fields. While laser structures grown on sapphire tend to exhibit mode-coupling effects [63] and complex gain spectra, laser structures grown on SiC provide well-behaved gain spectra.

The variations of the optical gain with the well width, the number of wells, and the In mole fraction and its relation to the corresponding absorption edge fit well into a polarised quantum well picture.

Inhomogeneities in nitride laser structures arise both from compositional and from well width fluctuations [65]. Such fluctuations lead to an inhomogeneous broadening of the optical gain, which reduces the peak gain and which leads to some small gain below the average renormalised bandgap. In that sense, localised states contribute to the optical gain. However, their density of states is far too small to provide enough optical gain to overcome typical cavity losses in a laser diode. Moreover, there is no evidence for gain due to phase-separated regions forming quantum dots within the active medium.

6   Gain saturation

Knowledge of the saturation behaviour of semiconductor optical amplifiers in particular of the maximum unsaturated modal gain, is essential for device optimisation. Saturation of an optical amplifier can be caused by two reasons. First, if the stimulated emission rate, caused by amplification of an input pulse, becomes comparable to the spontaneous and nonradiative recombination in the amplifier, the electron–hole pair density is reduced and saturation sets in. This mechanism limits the maximum unsaturated output power of the amplifier. Second, saturation can be already caused by internal noise, i.e. spontaneous emission, which is yet present without an input signal. This process limits the unsaturated modal gain of the amplifier device.

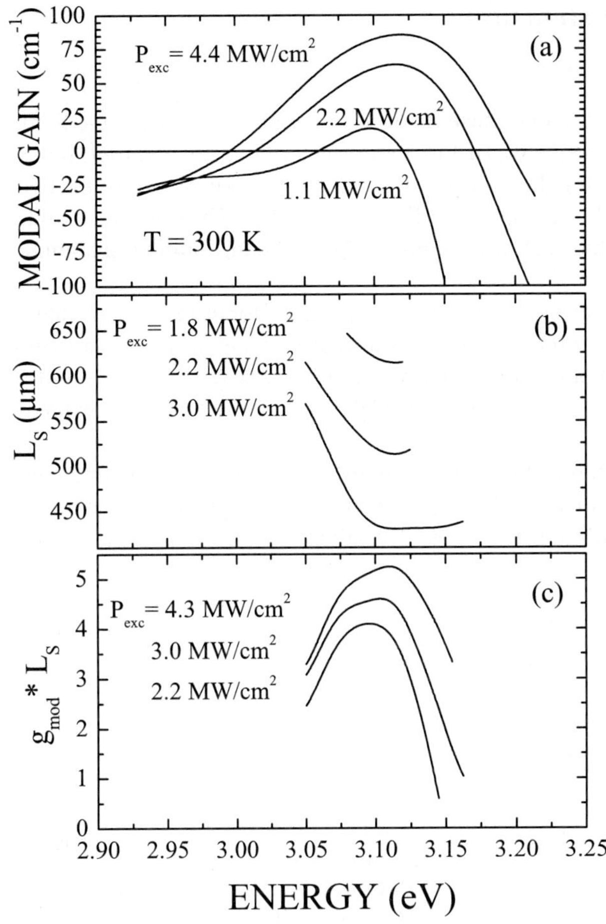

Fig. 9 Energy dependence of a) the measured modal gain, b) saturation length L_s, and c) saturation product $g_{mod}L_s$ for an (In,Ga)N/GaN/(Al,Ga)N laser structure (3 QW sample) at three different excitation densities, respectively (from Ref. [67]).

The saturation of optical gain in (In,Ga)N/GaN/(Al,Ga)N laser structures without application of an external input signal is investigated in detail using the variable stripe-length method [66, 67].

In a comparison of the unsaturated gain factor G all amplifiers show generally the same strong dependence on photon energy and excitation power, whereas there is negligible influence of temperature on gain saturation. The $g_{mod}L_s$ product which expresses the maximum unsaturated gain grows with increasing modal gain (i.e. the highest unsaturated gain factors G can be observed at higher pump power). This can be explained by a weak decrease of the saturation length but a strong increase of optical gain with rising excitation intensity. Spectrally, the unsaturated gain factor reaches a maximum near but not on the energy of the maximum peak gain (see Fig. 9).

Saturation at energies larger than that of the gain maximum occurs earlier since the maximum of the differential gain is reached at energies considerably higher than the latter. Therefore, the gain is much more strongly reduced at higher energies if the carrier density is diminished by ASE.

Very high unsaturated $g_{mod}L_s$ products have been observed for all investigated amplifiers with a maximum saturation product of 9.2 which corresponds to an unsaturated gain factor of 40 dB at 300 K. This shows the considerable application potential of (In,Ga)N/GaN/(Al,Ga)N structures for amplifiers in the blue and ultraviolet spectral region.

7   Waveguiding in nitride laser structures

7.1   Vertical waveguiding

In order to provide a well-defined optical mode and to maximise its overlap with the active gain medium, an optical waveguide is formed by an appropriate sequence of low-index cladding layers and high-index waveguiding layers. Typical layer structures, including their refractive index profile, being used for the two most common substrates, namely sapphire and SiC, are shown in Figs. 10 and 11. The core waveguide structure is formed by the AlGaN cladding layers, the GaN waveguiding layers, and the multiple-quantum-well active gain region.

The most important difference (besides differences in growth and defect density for different substrates) is the refractive index of the substrate, which is lower than that of the waveguide layers in case of sapphire but similar in case of SiC. In addition, in case of sapphire one typically uses a micrometer-thick GaN buffer layer below the actual waveguide, whereas in case of SiC the lower AlGaN cladding layer usually extends down to the substrate.

As a consequence, usual laser structures on sapphire substrates actually represent two competing coupled waveguides, where the coupling is controlled by the thickness of the AlGaN cladding layer in between [68, 69]. Since the thickness of this AlGaN layer is limited to about 0.5 µm due to strain and

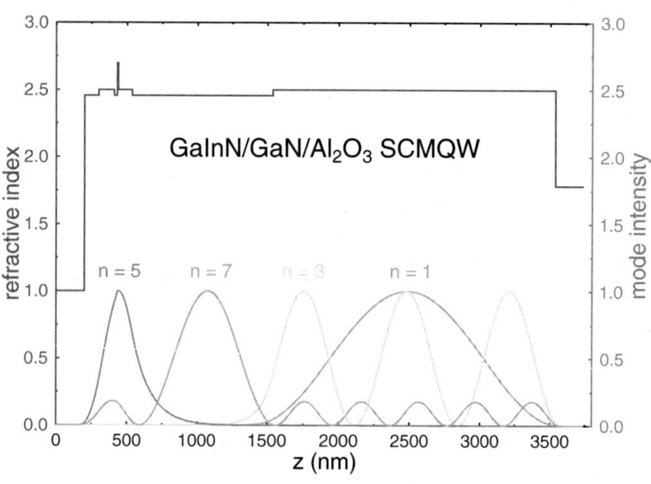

Fig. 10   (online colour at: www.interscience.wiley.com)   Refractive index profile and optical modes in a GaInN/GaN/AlGaN SCMQW laser structure on sapphire. Due to the secondary waveguide formed by the GaN buffer, the lasing mode is not the fundamental mode of the structure [63].

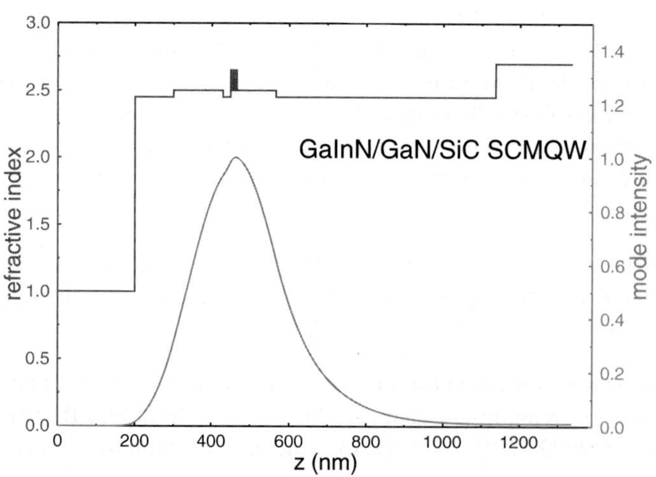

Fig. 11   (online colour at: www.interscience.wiley.com)   Refractive index profile and optical modes in a GaInN/GaN/AlGaN SCMQW laser structure on SiC. Due to the large refractive index of SiC, the structure is single-mode [63].

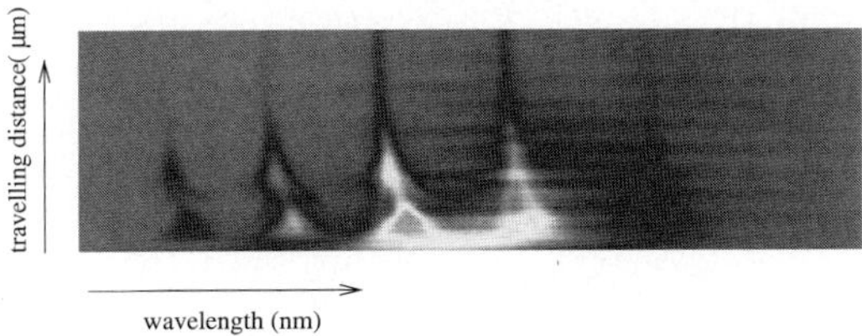
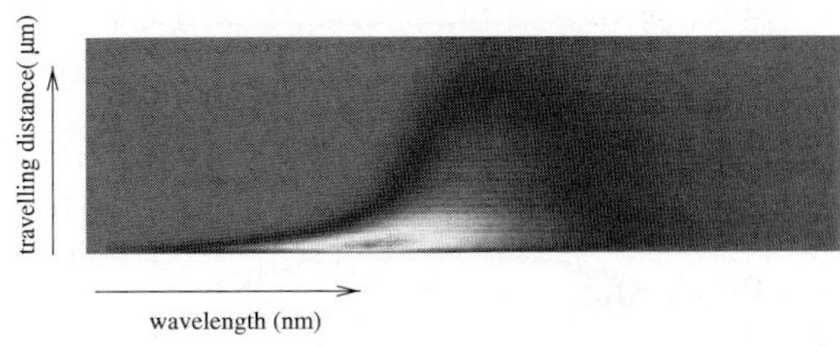

Fig. 12 (online colour at: www.interscience.wiley.com) Two-dimensional representation of the spectral intensity emitted from a nitride waveguide vs. travelling distance. The figure on the left shows the result for a sapphire-based waveguide, on the right a SiC-based laser structure is presented [63].

crack formation [1], this leads to excitation of higher order modes and to multi-lobe vertical far-field patterns [70, 71].

Actually, the detailed mechanism of coupling into higher order modes relies on the birefringence of the wurtzite nitrides [63], which also leads to a spectral resonance behaviour. The most direct proof for this can be obtained from waveguide propagation experiments using the "point excitation method" [63]. Fig. 12 shows two-dimensional intensity maps as a function of wavelength and propagation length. For the sapphire-based structure, both spectral resonances and oscillations with distance are observed, which are clear evidence for resonant coupling of modes. On the other hand, the SiC-based structure exhibits both a smooth spectrum and nice propagation behaviour, which can be related to the fact that only a single waveguide mode is supported in this structure.

While for a conventional waveguide based on isotropic material, phase matching for TE and TM modes is never achieved and therefore there is no effective coupling between those modes, a birefringent waveguide allows coupling of modes. Mode coupling in typical nitride waveguide is mediated by the defects present in large numbers. Improvements to this situation have been found by using modulation-doped superlattices for the cladding layer [1], a thick AlGaN buffer layer instead of the GaN buffer [72], or by using thick HVPE-grown GaN substrates [71, 73].

In contrast to sapphire-based structures, SiC-based laser structures do not suffer from higher order modes, unless the waveguide is not properly designed [63]. Due to the fact that the refractive index of SiC is slightly higher than that of the nitrides *and* that SiC is absorbing wavelengths shorter than about 450 nm, the waveguide structure can be made truly single mode.

8 Far field pattern and lateral waveguiding

The performance of nitride laser diodes (LDs) was much improved within the last years. Further realization of a lower threshold current density and a better beam profile are essential for commercial applications like printing and data storage and require progress of the laser design.

Figure 13 shows the intensity distribution in the far field for a ridge waveguide laser diode with a ridge width of 5 μm. The sample exhibits single-mode activity in vertical as well as in lateral direction. The full width at half maximum (FWHM) in Θ direction (vertical) is $21.6°$ and $5.1°$ in Φ direction (lateral), respectively. The resulting aspect ratio of 4.2 is still too high, but for smaller ridge widths the ratio decreases.

The influence of the layer design on the waveguiding properties of the laser diodes is investigated by calculating the optical field inside and by measuring the angular distribution of the far field intensity behind the structure [74, 75]. Figure 14 shows the profile of the far field pattern in Θ direction for a sample with a 42 nm thick electron blocking layer. The dashed curve is a numerical fit of the calculated optical field to the measured intensity distribution with a FWHM of $19.4°$.

A thicker blocking layer leads to an increasing thickness of the complete waveguide structure, which results in a decrease of the FWHM [75]. Furthermore, the optical confinement factor Γ for the active

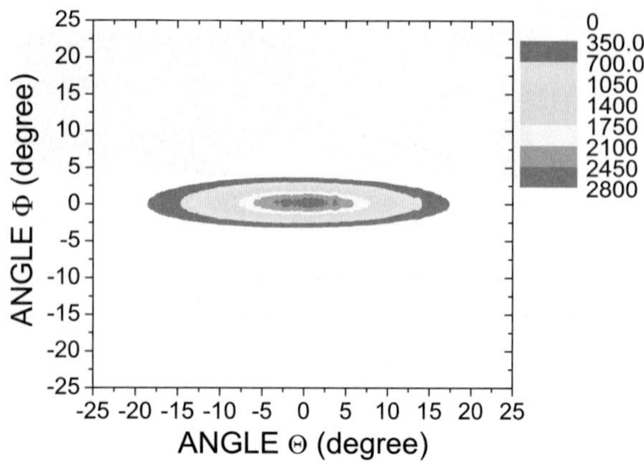

Fig. 13 (online colour at: www.interscience.wiley.com) Intensity distribution in the far field for a ridge waveguide laser diode.

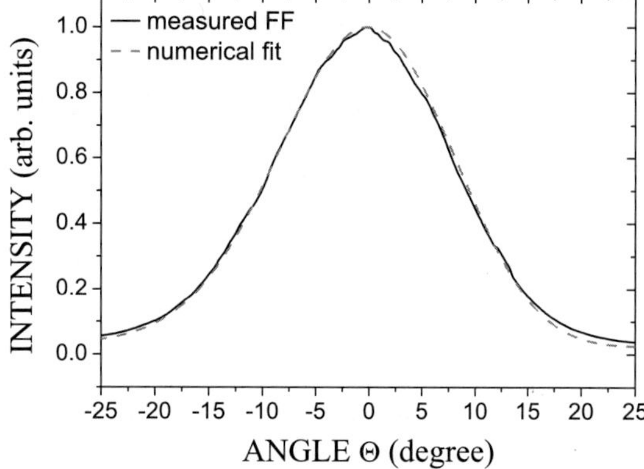

Fig. 14 (online colour at: www.interscience.wiley.com) Intensity distribution in the far field for a sample with a 42 nm thick electron blocking layer. The dashed curve is a numerical fit to the measured far field intensity (solid curve).

region is also strongly influenced by the thickness of the electron blocking layer. A clearly pronounced AlGaN blocking layer causes a strong impact on the mode profile that pushes the field maximum out of the active region [75].

The influence of the cladding layers on the mode profile and also on the far field pattern is more distinct than for every other layer [75]. Figure 15 shows the FWHM of the intensity distribution in the far field as a function of the Al concentration in the n-cladding and p-cladding layer, respectively. The calculated optical confinement factor Γ being normalised to a reference structure is also given.

A better inclusion of the mode into the waveguide due to a higher Al concentration results in an increase of the optical confinement factor by 19 % and an enlargement of the intensity distribution in the far field from 21.6° up to 24.8° (+15 %) in the case of the p-cladding layer. For the n-cladding layer the dependence on the Al concentration is somewhat weaker.

Although the intensity distribution in the far field and the optical confinement factor are nearly independent of the thickness of the p-cladding layer, it should not be too thin to prevent leaking of the zeroth-order mode into the competing waveguide formed by the p-GaN cap layer [75]. In the case of a 261 nm thin p-cladding layer we could show that most of the intensity of the lasing mode was not guided in the waveguide with the active region ïn its center, but in the waveguide formed by the GaN cap layer.

To investigate the lateral waveguiding properties of the laser diodes, we examined the influence of the lateral structure width on the number of modes visible in the far field pattern for electrically pumped oxide stripe laser diodes and for ridge waveguide structures which were electrically pumped or optically probed as passive waveguide structures (see Fig. 16).

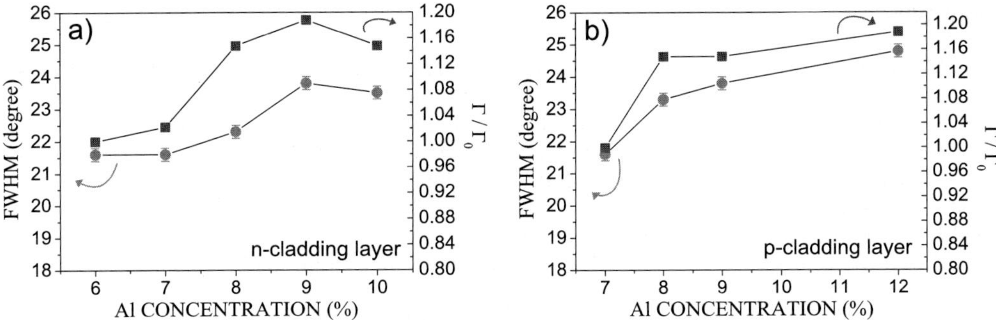

Fig. 15 (online colour at: www.interscience.wiley.com) FWHM of the intensity distribution in the far field and the optical confinement factor as a function of the Al concentration in a) n and b) p-cladding layers (from Ref. [75]).

For structure widths smaller than 4 μm single-mode activity can be observed in lateral direction. The optically probed ridge waveguide laser diodes show compared to the electrically pumped samples 1-2 additional higher-order modes for equal structure widths. The reason for this behaviour is the mode conversion in the case of electrical operation of the diodes. However, the number of lateral modes is comparable for electrically pumped oxide stripe laser and ridge waveguide laser diodes, although the threshold current density is different for these devices.

9   Conclusion

We have analysed the optical gain of nitride heterostructures as well as their waveguiding properties. Fundamentally, the optical gain mechanism is strongly affected by internal polarisation fields. Carrier losses due to thermal escape are limiting the optical gain. Large nonradiative recombination rates lead to a large unsaturated gain factor of 40 dB at 300 K. Waveguiding in nitride laser structures is strongly affected by coupling of guided modes due to the birefringence of wurtzite nitrides. For laser structures on SiC, where true single mode operation is possible, we have discussed the optimisation of the far-field properties.

Acknowledgements   The generous financial support by the Deutsche Forschungsgemeinschaft (DFG) in the framework of the programme "Gruppe-III-Nitride und ihre Heterostrukturen" is gratefully appreciated.

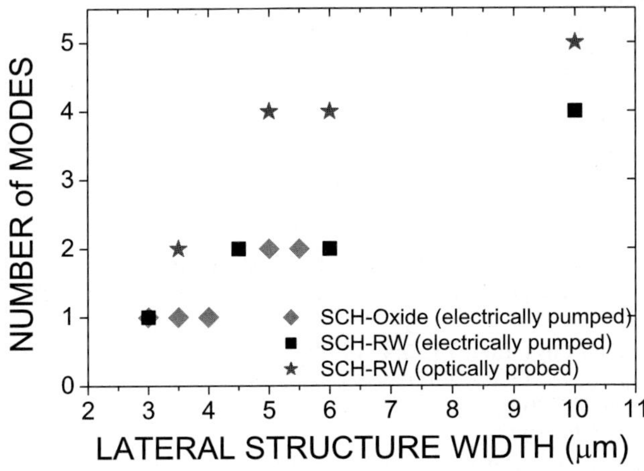

Fig. 16 (online colour at: www.interscience.wiley.com) Number of lateral modes visible in the far field pattern as a function of the lateral structure width for oxide stripe (SCH-Oxide) and ridge waveguide (SCH-RW) laser diodes.

References

[1] S. Nakamura, M. Senoh, S. Nagahama, N. Iwasa, T. Yamada, T. Matsushita, H. Kiyoku, Y. Sugimoto, T. Kozaki, H. Umemoto, M. Sano, and K. Chocho, Jpn. J. Appl. Phys. **36**, L1568 (1997).

[2] I. Akasaki, S. Sota, H. Sakai, T. Tanaka, M. Koike, and H. Amano, Electron. Lett. **32**, 1105 (1996).

[3] K. Itaya, M. Onomura, J. Nishio, L. Sugiura, S. Saito, M. Suzuki, J. Rennie, S. Nunoue, M. Yamamoto, H. Fujimoto, Y. Kokobun, Y. Ohba, G. Hatakoshi, and M. Ishikawa, Jpn. J. Appl. Phys. **35**, L1315 (1996).

[4] G. E. Bulman, K. Doverspike, S. T. Sheppard, T. W. Weeks, H. S. Kong, H. M. Dieringer, J. A. Edmond, J. T. Swindell, and J. F. Schetzina, Electron. Lett. **33**, 1556 (1997).

[5] A. Kuramata, K. Domen, R. Soejima, K. Horino, S.-I. Kubota, and T. Tanahashi, Jpn. J. Appl. Phys. **36**, L1130 (1997).

[6] M. P. Mack, A. Abare, M. Aizcorbe, P. Kozodoy, S. Keller, U. K. Mishra, L. Coldren, and S. DenBaars, MRS Internet J. Nitride Semicond. Res. **2**, 41 (1997).

[7] F. Nakamura, T. Kobayashi, T. Asatsuma, K. Funato, K. Yanashima, S. Hashimoto, K. Naganuma, S. Tomioka, T. Miyajima, E. Morita, H. Kawai, and M. Ikeda, in: Proceedings of The Second International Conference on Nitride Semiconductors (Tokushima, Japan, 1997), p. 460.

[8] S. N. Mohammad and H. Morkoç, Prog. Quantum Electron. **20**, 361 (1996).

[9] H. Amano, M. Kito, K. Hiramatsu, and I. Akasaki, Jpn. J. Appl. Phys. **28**, L2112 (1989).

[10] S. Nakamura, M. Senoh, and T. Mukai, Jpn. J. Appl. Phys. **30**, L1708 (1991).

[11] S. Nakamura, M. Senoh, S.-I. Nagahama, N. Iwasa, T. Yamada, T. Matsushita, H. Kiyoku, and Y. Sugimoto, Jpn. J. Appl. Phys. **35**, L74 (1996).

[12] P. S. Zory, Quantum well lasers (Academic Press, Boston, 1993).

[13] M. Suzuki, T. Uenoyama, and A. Yanase, Phys. Rev. B **52**, 8132 (1995).

[14] J. A. Majewski, M. Städele, and P. Vogl, in: III–V Nitrides, Vol. 449 of MRS Symposium Proceedings, edited by T. Moustakas, B. Monemar, I. Akasaki, and F. Ponce (Materials Research Society, Pittsburgh, 1997), p. 887.

[15] K. Kim, W. R. L. Lambrecht, B. Segall, and M. van Schilfgaarde, Phys. Rev. B **56**, 7363 (1997).

[16] J. S. Im, A. Moritz, F. Steuber, V. Härle, F. Scholz, and A. Hangleiter, Appl. Phys. Lett. **70**, 631 (1997).

[17] A. R. Adams, Electron. Lett. **22**, 249 (1986).

[18] M. Suzuki and T. Uenoyama, Appl. Phys. Lett. **69**, 3378 (1996).

[19] J. B. Jeon, B. C. Lee, Y. M. Sirenko, K. W. Kim, and M. A. Littlejohn, J. Appl. Phys. **82**, 386 (1997).

[20] S. Chichibu, T. Azuhata, T. Sota, and S. Nakamura, Appl. Phys. Lett. **69**, 4188 (1996).

[21] Y. Narukawa, Y. Kawakami, S. Fujita, S. Fujita, and S. Nakamura, Phys. Rev. B **55**, R1938 (1997).

[22] M. G. A. Bernard and G. Duraffourg, phys. stat. sol. **1**, 699 (1961).

[23] G. Lasher and F. Stern, Phys. Rev. **133**, 554 (1964).

[24] H. C. Casey and M. B. Panish, Heterostructure Lasers (Academic Press, New York, 1978).

[25] E. O. Kane, J. Phys. Chem. Solids **1**, 249 (1957).

[26] M. Yamanishi and I. Suemune, Jpn. J. Appl. Phys. **23**, 35 (1984).

[27] P. T. Landsberg, phys. stat. sol. **15**, 623 (1966).

[28] M. Yamada and Y. Suematsu, J. Appl. Phys. **52**, 2653 (1981).

[29] D. Ahn and S.-L. Chuang, IEEE J. Quantum Electron. **QE-24**, 2400 (1988).

[30] P. T. Landsberg and D. J. Robbins, Solid State Electron. **28**, 137 (1985).

[31] M. Asada, IEEE J. Quantum Electron. **25**, 2019 (1989).

[32] S. Schmitt-Rink, C. Ell, H. E. Schmid, and H. Haug, Solid State Commun. **52**, 123 (1984).

[33] H. Schweizer, A. Forchel, A. Hangleiter, S. Schmitt-Rink, J. P. Löwenau, and H. Haug, Phys. Rev. Lett. **51**, 698 (1983).

[34] J. M. F. Pereira, S. W. Koch, and W. W. Chow, Appl. Phys. Lett. **59**, 2941 (1991).

[35] C. Ell and H. Haug, phys. stat. sol. (b) **159**, 117 (1990).

[36] W. W. Chow, A. Knorr, and S. W. Koch, Appl. Phys. Lett. **67**, 754 (1995).

[37] S.-H. Park and D. Ahn, Appl. Phys. Lett. **71**, 398 (1997).

[38] A. Moritz and A. Hangleiter, Appl. Phys. Lett. **66**, 3340 (1995).

[39] J. S. Im, H. Kollmer, J. Off, A. Sohmer, F. Scholz, and A. Hangleiter, Phys. Rev. B **57**, R9435 (1998).

[40] A. Hangleiter, S. Heppel, J. Off, B. Kuhn, F. Scholz, S. Bader, B. Hahn, and V. Härle, J. Crystal Growth **230**, 522 (2001).

[41] W. W. Chow, H. Amano, T. Takeuchi, and J. Han, Appl. Phys. Lett. **75**, 244 (1999).

[42] S.-H. Park, S.-L. Chuang, and D. Ahn, Appl. Phys. Lett. **75**, 1354 (1999).

[43] T. Uenoyama and M. Suzuki, Appl. Phys. Lett. **67**, 2527 (1995).

[44] K. L. Shaklee and R. F. Leheny, Appl. Phys. Lett. **18**, 475 (1971).

[45] G. Frankowsky, V. Härle, F. Scholz, and A. Hangleiter, Appl. Phys. Lett. **68**, 3746 (1996).

[46] G. Fuchs, J. Hörer, A. Hangleiter, V. Härle, F. Scholz, R. W. Glew, and L. Goldstein, Appl. Phys. Lett. **60**, 231 (1992).

[47] A. Moritz, R. Wirth, C. Geng, F. Scholz, and A. Hangleiter, Appl. Phys. Lett. **68**, 1217 (1996).

[48] M. Vehse, P. Michler, J. Gutowski, S. Figge, D. Hommel, H. Selke, S. Keller, and S. DenBaars, Semicond. Sci. Technol. **16**, 406 (2001).

[49] M. Vehse, P. Michler, I. Gösling, M. Röwe, J. Gutowski, S. Bader, A. Lell, G. Brüderl, and V. Härle, Appl. Phys. Lett. **80**, 755 (2002).

[50] M. Vehse, P. Michler, I. Gösling, M. Röwe, J. Gutowski, S. Bader, A. Lell, G. Brüderl, and V. Härle, phys. stat. sol. (a) **188**, 109 (2001).

[51] P. Michler, A. Hangleiter, M. Moser, M. Geiger, and F. Scholz, Phys. Rev. B. **46**, 7280 (1992).

[52] R. Dai, W. Zhuang, K. Bohnert, and C. Klingshirn, Zeitschr. Physik B **46**, 189 (1982).

[53] L. Eckey, J.-C. Holst, A. Hoffmann, I. Broser, T. Detchprohm, and K. Hiramatsu, MRS Internet J. Nitride Semicond. Res. **2**, 1 (1997).

[54] J. Holst, L. Eckey, A. Hoffmann, I. Broser, B. Schöttker, D. J. As, D. Schikora, and K. Lischka, Appl. Phys. Lett. **72**, 1439 (1998).

[55] S. Bidnyk, T. J. Schmidt, B. D. Little, and J. J. Song, Appl. Phys. Lett. **74**, 1 (1999).

[56] T. Deguchi, T. Azuhata, T. Sota, S. Chichibu, M. Arita, H. Nakanishi, and S. Nakamura, Semicond. Sci. Technol. **13**, 97 (1998).

[57] Y. Kawakami, Y. Narukawa, K. Omae, S. Fujita, and S. Nakamura, Appl. Phys. Lett. **77**, 2151 (2000).

[58] A. Ishibashi, I. Kidoguchi, A. Tsujimura, Y. Hasegawa, Y. Ban, T. Ohata, M. Watanabe, and T. Hayashi, J. Lumin. **87/89**, 1271 (2000).

[59] Y. Kimura, A. Ito, M. Miyachi, H. Takahashi, A. Watanabe, H. Ota, N. Ito, T. Tanabe, M. Sonobe, and K. Chikuma, Jpn. J. Appl. Phys. **40**, L1103 (2001).

[60] M. Vehse, P. Michler, J. Gutowski, S. Figge, D. Hommel, H. Selke, S. Keller, and S. P. DenBaars, Semicond. Sci. Technol. **16**, 406 (2001).

[61] M. Vehse, P. Michler, O. Lange, M. Rowe, J. Gutowski, S. Bader, H. J. Lugauer, G. Brüderl, A. Weimar, A. Lell, and V. Härle, Appl. Phys. Lett. **79**, 1763 (2001).

[62] S. Heppel, A. Hangleiter, S. Bader, G. Brüderl, A. Weimar, V. Kümmler, A. Lell, V. Härle, J. Off, B. Kuhn, and F.Scholz, phys. stat. sol. (a) **188**, 59 (2001).

[63] S. Heppel, R. Wirth, J. Off, F. Scholz, and A. Hangleiter, phys. stat. sol. (a) **176**, 73 (1999).

[64] J. S. Im, H. Kollmer, J. Off, F. Scholz, and A. Hangleiter, MRS Internet J. Nitride Semicond. Res. **4S1**, G6.20 (1998).

[65] A. Hangleiter, J. S. Im, J. Off, and F. Scholz, phys. stat. sol. (b) **216**, 427 (1999).

[66] M. Vehse, P. Michler, O. Lange, M. Röwe, J. Gutowski, S. Bader, H.-J. Lugauer, G. Brüderl, A. Weimar, A. Lell, and V. Härle, Appl. Phys. Lett. **79**, 1763 (2001).

[67] P. Michler, O. Lange, M. Vehse, J. Gutowski, S. Bader, B. Hahn, H.-J. Lugauer, and V. Härle, phys. stat. sol. (a) **180**, 391 (2000).

[68] V. E. Bougrov and A. S. Zubrilov, J. Appl. Phys. **81**, 2952 (1997).

[69] M. J. Bergmann and J. H. C. Casey, J. Appl. Phys. **84**, 1196 (1998).

[70] D. Hofstetter, D. P. Bour, R. L. Thornton, and N. M. Johnson, Appl. Phys. Lett. **70**, 1650 (1997).

[71] S. Nakamura, M. Senoh, S. Nagahama, N. Iwasa, T. Yamada, T. Matsushita, H. Kiyoku, Y. Sugimoto, T. Kozaki, H. Umemoto, M. Sano, and K. Chocho, Appl. Phys. Lett. **72**, 2014 (1998).

[72] T. Takeuchi, T. Detchprohm, M. Iwaya, N. Hayashi, K. Isomura, K. Kimura, M. Yamaguchi, H. Amano, I. Akasaki, Y. Kaneko, R. Shioda, S. Watanabe, T. Hidaka, Y. Yamaoka, Y. Kaneko, and N. Yamada, Appl. Phys. Lett. **75**, 2960 (1999).

[73] M. Kuramoto, A. Kimura, A. A. Yamaguchi, H. Sunakawa, N. Kuroda, M. Nido, A. Usui, M. Mizuta, C. Sasaoka, and Y. Hisanaga, Jpn. J. Appl. Phys. **38**, L184 (1999).

[74] M. Röwe, P. Michler, J. Gutowski, S. Bader, G. Brüderl, V. Kümmler, A. Weimar, A. Lell, and V. Härle, phys. stat. sol. (a) **188**, 65 (2001).

[75] M. Röwe, P. Michler, J. Gutowski, S. Bader, G. Brüderl, V. Kümmler, A. Weimar, A. Lell, and V. Härle, phys. stat. sol. (a) , accepted (2002).

phys. stat. sol. (c) **0**, No. 6, 1878–1907 (2003) / **DOI** 10.1002/pssc.200303138

Electronics and sensors based on pyroelectric AlGaN/GaN heterostructures

Part A: Polarization and pyroelectronics

O. Ambacher [*,1], **M. Eickhoff** [2], **A. Link** [2], **M. Hermann** [2], **M. Stutzmann** [2], **F. Bernardini** [3], **V. Fiorentini** [3], **Y. Smorchkova** [4], **J. Speck** [4], **U. Mishra** [4], **W. Schaff** [5], **V. Tilak** [5], and **L. F. Eastman** [5]

[1] Center for Micro- and Nanotechnologies, Technical University Ilmenau, 98688 Ilmenau, Germany
[2] Walter Schottky Institute, TU-Munich, Am Coulombwall, 85748 Garching, Germany
[3] Instituto Nazionale per la Fisica della Materia, Dipartimento di Fisica, Università degli Studi di Cagliari, Italy
[4] Electrical and Computer Engineering Department and Materials Department, College of Engineering, University of California, Santa Barbara, California 93106, USA
[5] Electrical and Computer Engineering, Cornell University, Ithaca, NY 14853-5401, USA

Received 4 March 2003, accepted 20 June 2003
Published online 28 August 2003

PACS: 62.20.Dc, 68.60.Bs, 71.20.Nr, 72.20.Jv, 77.70.+a, 77.84.Bw

Electronic transport in semiconductors that possess high internal spontaneous and piezoelectric polarization opens up a new field of pyroelectronics and pyrosensors. The pyroelectric character of group-III-nitrides with wurtzite crystal structure yields a novel degree of freedom in designing and tailoring devices for modern micro- and nanoelectronic applications. Furthermore, spontaneous and piezoelectric polarization induced surface and interface charges can be used to develop very sensitive but robust sensors for the detection of ions, gases and polar liquids. We present a review of both theoretical and experimental studies of spontaneous and piezoelectric polarization present in AlGaN/GaN heterostructures as well as the electronic transport properties of polarization induced two-dimensional electron gases which are formed at the AlGaN/GaN interface due to the difference in the total polarization of two adjacent III-nitride layers. We demonstrate that the two-dimensional electron gases (2DEGs) achieved without modulation doping are very suitable as channel of high electron mobility transistors optimally suited for high power and high frequency applications (PART A) as well as for various kinds of sensors which can be operated in harsh environments (PART B).

1   Introduction

Solids with a large band gap such as gallium nitride (GaN), aluminum nitride (AlN) and their ternary compounds (AlGaN) are prime candidates for high frequency and high power electronic devices as well as for a variety of sensor applications, particularly operating at high temperatures. On the one hand, the large band gap ensures an increase of device performance due to a low number of optically or thermally generated carriers (e.g. "solar blind" UV detectors [1-3], high temperature gas sensors [4], etc.). On the other hand, the strong chemical bonding between the constituent atoms not only leads to a large band gap, but at the same time gives rise to a quite favorable mechanical, thermal, and chemical stability of this class of materials.

---

[*] Corresponding author: e-mail: oliver.ambacher@tu-ilmenau.de, Phone: +49 3677 693723, Fax: +49 3677 693709

Besides the outstanding mechanical and chemical properties it has been predicted by theory and confirmed by experiment that GaN and AlN with wurtzite crystal structure are pyroelectric materials [5-8]. Characteristic for these crystals, especially for AlN, are large negative values of pyroelectric (spontaneous) polarization up to $P_{AlN}^{SP} = -0.09$ C/m^2, which are present without any external electric field. In contrast to ferroelectric crystals, the orientation of this spontaneous polarization is always along the $\left[000\bar{1} \right]$ direction and cannot be inverted by external electric fields. Although the spontaneous polarization is very strong in group-III nitrides, the thermal pyroelectric coefficients, describing the change of the spontaneous polarization with temperature, are surprisingly small (e.g. $dP_{AlN}^{SP}/dT = -7.5$ µC/Km^2, at room temperature [8, 9]), which is considered to be a great advantage in using these materials for high power and high temperature applications such as high frequency, high power transistors, microwave amplifiers or gas sensors. Pyroelectric crystals like GaN and AlN are always piezoelectric. The piezoelectric constants of group-III nitrides are a factor of 5 to 20 larger than those of InAs, GaAs and AlAs [5]. Using the measured and predicted elastic and piezoelectric constants it has been previously shown that effects induced by spontaneous and piezoelectric polarization can have a substantial influence on the concentration, distribution and recombination of free carriers in strained group-III nitride heterostructures. Observations of Franz-Keldysh-oscillations and Stark-effect in AlGaN/GaN quantum well structures provide direct evidence for the presence of strong electric fields caused by gradients in spontaneous and piezoelectric polarization [10-12]. The importance of polarization induced effects is further emphasized in the case of pseudomorphic, wurtzite AlGaN/GaN based transistor and sensor structures [13-23]. The channel of the transistor can be created by a two dimensional electron gas which is caused by polarization induced interface charges bound at the AlGaN/GaN interface. Furthermore, GaN has important attributes for application to field effect transistors capable of amplifying high power at microwave frequencies. It has a breakdown electric field strength of about 3 MV/cm, which is 7.5 times that of GaAs. Despite of an electron effective mass of 0.22 m$_e$, theoretical drift mobility and saturation velocity values over 2000 cm^2/Vs and 2.5x10^7 cm/s, respectively, have been predicted. GaN has a thermal conductivity that is experimentally more than 1 W/cmK, but theoretically approaches 3 W/cmK. Besides on Al$_2$O$_3$ it grows on SiC substrates, which have a thermal conductivity in the range of 3-5 W/cmK, so that substantially more heat can be removed from power devices than is the case for other semiconductors.

Beside the basic physics of piezoelectric and spontaneous polarization of AlGaN alloys as well as of polarization induced charges at surfaces and interfaces causing the formation of 2DEGs in pyroelectric AlGaN/GaN heterostructures, we will summarize the present status of novel AlGaN/GaN-based high power high frequency devices.

2 First-principles prediction of structural and pyroelectric properties of AlGaN

First models of polarization induced effects in GaN based devices have assumed that polarization in ternary nitride alloys can be calculated by a linear interpolation between the limiting values of the binary compounds. We review theoretical and experimental evidence that the macroscopic polarization in AlGaN is a non-linear function of strain and composition. To facilitate inclusion of the predicted non-linear polarization in future simulations, we give an explicit prescription to calculate polarization induced bound surface and interface charges for arbitrary composition in ternary AlGaN alloys. In addition, the theoretical and experimental results presented here allow a detailed comparison of the predicted electric fields and bound interface charges with the measured sheet carrier concentration of polarization induced 2DEGs. This comparison provides an insight into the reliability of the calculated non-linear piezoelectric and spontaneous polarization of group-III-nitride ternary alloys. In order to predict the pyroelectric polarization of ternary random alloys, the structural, elastic and polarization properties of wurtzite III-V nitrides are calculated from first principles within density-functional theory using the plane-wave ultrasoft pseudopotential method [24, 25]. The pseudopotentials for Ga include the semicore 3d and 4d valence states. A plane wave basis is used to expand the wave functions and a cutoff of 350 eV is found to be sufficient to fully converge all properties of relevance here. For k-space summation a Monkhorst-Pack (888) grid is used for the binaries, while a (444) grid is used in the alloy supercell calulations [5, 6]. Lattice constants and internal parameters are determined using standard total energy calculations, converging forces and stresses to zero within 0.005 eV/Å and 0.01 kbar, respectively. Polarization and related quantities are obtained using the Berry-phase approach as in previous works [5, 6, 26, 27]. Since we are mainly interested in the effects on polarization due to internal strain of ternary alloys related to size mismatch of the two alloyed nitrides GaN and AlN, it is necessary that our calculations reproduce relative mismatches with the highest possible accuracy. Comparing density functional theory calculations we found that the generalized gradient approximation (GGA) [28] has advantages in comparison to local-density-approximations (LDA) in reproducing the relative mismatch between the binary constituents, besides getting closer to the experimental lattice constants and cell-internal parameter u. The microscopic structure of a random alloy is represented by periodic boundary conditions using the special quasi-random structure method [29]. We enforce the periodic boundary conditions needed to predict the macroscopic polarization in the Berry phase approach. As a compromise between computational workload and the description of random structures, a 32-atom 2x2x2 wurtzite supercell is adopted. It is possible to mimic the statistical properties of a random wurtzite alloy for a molar fraction $x = 0.5$ by suitably placing the cations on the 16 sites available in the cell. Since other molar fractions cannot be described as easily and also because non-linear effects are expected to be largest for this concentration, our theoretical approach to random alloys is restricted to $x = 0.5$. The predicted non-linearities of structrual and polarization related properties of $Al_xGa_{1-x}N$ alloys are approximated by quadratic equations of the form:

$$Y_{AlGaN}(x) = Y_{AlN}x + Y_{GaN}(1-x) + bx(1-x) \tag{1}$$

where $b = 4Y_{AlGaN}(x = 0.5) - 2(Y_{AlN} + Y_{GaN})$, $\tag{2}$

is the bowing parameter. In order to determine the quality of the theoretical predictions and the approximation of the non-linearities, the calculated physical properties are compared to experimental results in chapters 3. and 8.

3 Lattice constants, average bond lengths and bond angles in ternary group-III-nitrides

The lattice constants $a(x)$ and $c(x)$ of wurtzite $Al_xGa_{1-x}N$ alloys are predicted to follow the composition weighted average between the binary compounds AlN and GaN (Vegard´s law) [6]:

$$a_{AlGaN}(x) = (3.1986 - 0.0891x) \qquad c_{AlGaN}(x) = (5.2262 - 0.2323x)\,\text{Å}, \tag{3}$$

By a combination of high resolution X-ray diffraction (HRXRD), Rutherford and elastic recoil detection analysis, Vegard´s law can be confirmed for the ternary group-III-nitrides [30, 31]. The agreement between the predicted lattice constants in equation (3) and linear fits of the experimental data:

$$a_{AlGaN}(x) = ((3.189 \pm 0.002) - (0.086 \pm 0.004)x)\,\text{Å}, \tag{4a}$$

$$c_{AlGaN}(x) = ((5.188 \pm 0.003) - (0.208 \pm 0.005)x)\,\text{Å}, \tag{4b}$$

are better than 2% over the whole range of possible compositions (**Fig. 1** and **Fig.2**).

In contrast to the lattice constants, the nearest neighbor bond lengths, $Al\text{-}N(x)$ and $Ga\text{-}N(x)$ change only slightly and in a non-linear way as a function of alloy composition, resembling more their values in the binary constituents rather than an average value corresponding to the virtual crystal limit [32-34].

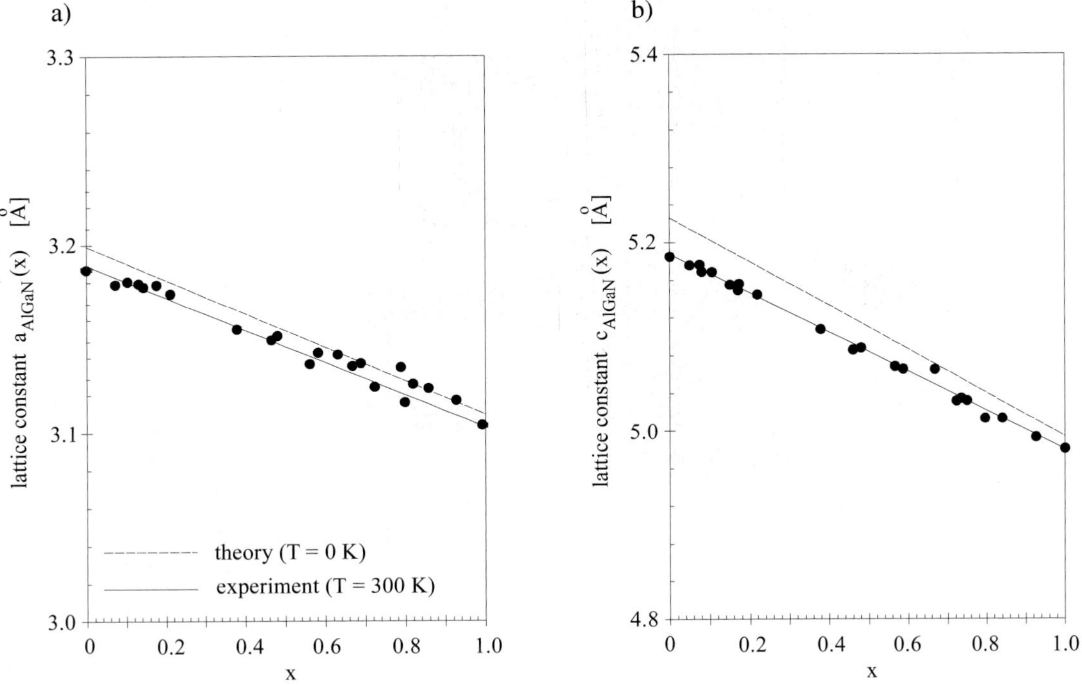

Fig. 1 a) and b) Lattice constants $a_{AlGaN}(x)$ and $c_{AlGaN}(x)$ for AlGaN random alloys measured by high resolution X-ray diffraction (HRXRD) at room temperature (solid lines) and calculated by the method described in chapter 2. for $T = 0$ K (dashed lines).

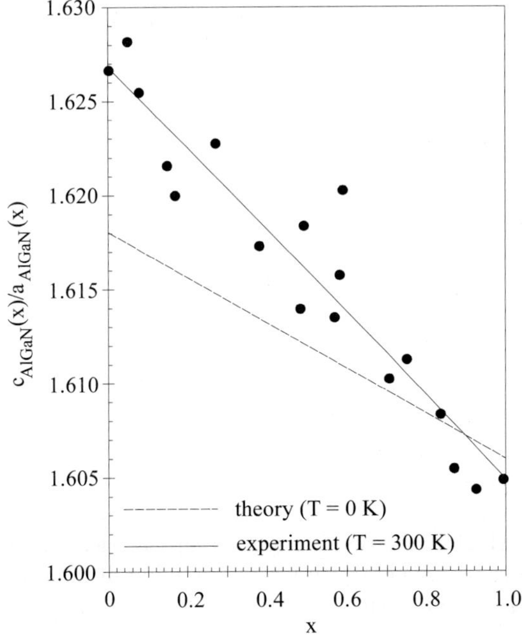

Fig. 2 Measured (solid line) and predicted (dashed line) $c_{AlGaN}(x)/a_{AlGaN}(x)$ ratio for random AlGaN al-

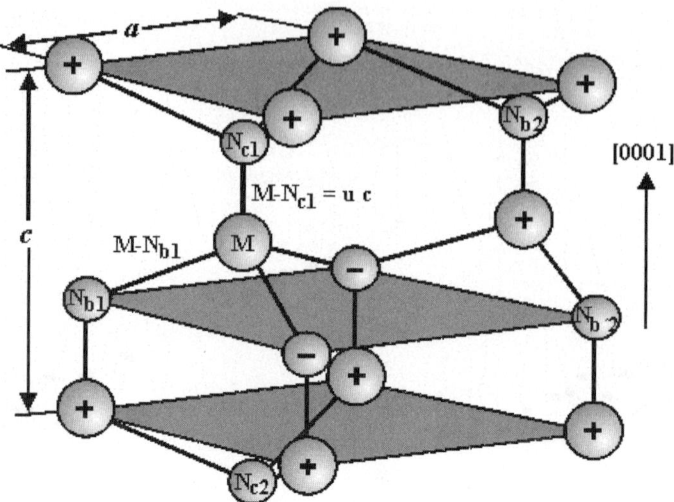

Fig. 3 Schematic drawing of a wurtzite MN crystal (M = Ga or Al) with lattice constants a and c. The signs of the charge of metal and nitrogen atoms are indicated as well as the positions of the nearest and second-nearest neighbors to the metal atom M.

The wurtzite structure of interest here has two types of first neighbor metal-nitrogen bond distances: M-N_c along the c-axis (one bond) and M-N_b in the basal plane (three bonds):

$$M - N_{c1} = uc \, , \tag{5}$$

$$M - N_{b1} = \sqrt{\frac{1}{3}a^2 + \left(\frac{1}{2} - u\right)^2 c^2} \, , \tag{6}$$

and two bond angles $\alpha = \angle\,(M\text{-}N_{c1};\ M\text{-}N_{b1})$, $\beta = \angle(M\text{-}N_{b1};\ M\text{-}N_{b'1})$:

$$\alpha = \frac{\pi}{2} + \arccos\left\{\left(\left(\sqrt{1 + 3\left(\frac{c}{a}\right)^2 \left(\frac{1}{2} - u\right)^2}\right)^{-1}\right)\right\}\, ,$$

$$\beta = 2\arcsin\left\{\left(\left(\sqrt{\frac{4}{3} + 4\left(\frac{c}{a}\right)^2 \left(\frac{1}{2} - u\right)^2}\right)^{-1}\right)\right\}\, , \tag{7}$$

where u denotes the cell-internal parameter, which is the bond length along the c-axis in terms of the lattice constant c (eq. (5)). In addition three types of second neighbor cation-anion distances connecting the cation M to the anions N_{c2}, N_{b2}, and $N_{b'2}$ (**Fig.3**) are present:

$$M - N_{c2} = (1-u)c \text{ (one neighbor along the c-axis)}, \tag{8}$$

$$M - N_{b2} = \sqrt{a^2 + (uc)^2} \quad \text{(six neighbors),} \tag{9}$$

$$M - N_{b'2} = \sqrt{\frac{4}{3}a^2 + \left(\frac{1}{2} - u\right)^2 c^2} \quad \text{(three neighbors).} \tag{10}$$

It should be noticed that in the case of an ideal ratio of lattice constants $\dfrac{c}{a} = \sqrt{\dfrac{8}{3}} = 1.633$ and ideal cell-internal parameter $u = \dfrac{3}{8} = 0.375$ it follows from equations (5-10) that the bond length and the bond angles between the nearest neighbors ($\alpha = \beta = 109.47°$, **Tab.1**) are equal, but the distance to the second nearest neighbor along the c-axis is about 13% shorter than the distance to the other second nearest neighbors (this is not the case in the cubic structure [33]).

Table 1 Calculated cell internal parameters, lattice constants and cation-anion distances between nearest and second nearest neighbors (given in Å) as well as bond angles (given in degrees) of GaN, $Al_{0.5}Ga_{0.5}N$ and AlN with ideal crystal structure or the predicted "real" structure. The structural properties for the random AlGaN alloy are given in the virtual crystal limit.

structural property	GaN		$Al_{0.5}Ga_{0.5}N$		AlN	
	ideal	real	ideal	virtual crystal	ideal	real
u	0.375	0.377	0.375	0.379	0.375	0.382
a	3.199	3.199	3.154	3.154	3.110	3.110
c/a	1.633	1.634	1.633	1.620	1.633	1.606
$M\text{-}N_{c1}$	1.959	1.971	1.930	1.935	1.904	1.907
$M\text{-}N_{b1}$	1.959	1.955	1.930	1.924	1.904	1.890
$M\text{-}N_{c2}$	3.265	3.255	3.220	3.175	3.174	3.087
$M\text{-}N_{b2}$	3.751	3.757	3.698	3.701	3.646	3.648
$M\text{-}N_{b'2}$	3.751	3.749	3.698	3.694	3.646	3.639
α	109.47	109.17	109.47	108.80	109.47	108.19
β	109.47	109.78	109.47	110.14	109.47	110.73

It is known from experiment as well as theoretical predictions, that neither the cell-internal parameter u, nor the c/a-ratio is ideal in GaN and AlN [5, 6, 30, 31]. In order to evaluate the consequences of the non-ideality of the wurtzite structure on polarization and polarization induced effects in group-III-nitrides we have calculated the average bond lengths and angles as well as the second neighbor distances (virtual crystal limit) of AlGaN alloys taking advantage of the measured and calculated lattice constants and average u parameter (**Tab.1**). The average u parameter has been calculated for randomly distributed $Al_{0.5}Ga_{0.5}N$ alloys by the theoretical approach described above (more detailed information can be found in Ref. [35]) and can be approximated by the following quadratic equation:

$$u_{AlGaN}(x) = u_{AlN}\,x + u_{GaN}(1-x) + bx(1-x), \tag{11}$$

$$u_{AlGaN}(x) = 0.3819x + 0.3772(1-x) - 0.0032x(1-x). \tag{12}$$

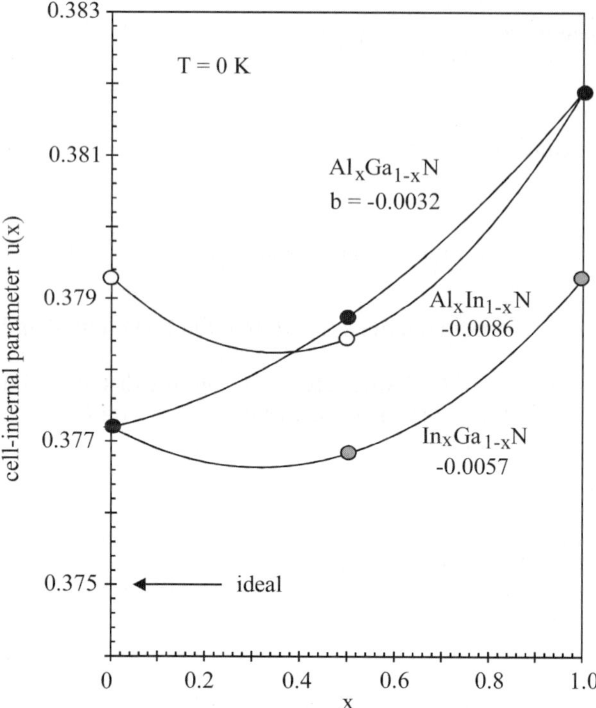

Fig. 4 The cell internal parameter, $u(x)$ (described by eq. (12), of random AlGaN alloys versus Al-concentration, predicted by the theory described in chapter 2 and approximated by a quadratic equation. The nonlinearity of the cell internal parameter can be described by a negative bowing parameter b, given in the figure. In addition the u parameters for random InGaN and AlInN alloys are given for comparison [5, 6].

The cell-internal parameter is increasing from GaN to AlN and the non-linear dependence on alloy composition is described by a negative bowing parameter. It should be pointed out, that although the bowing parameter is negative, the average cell-internal parameter of random alloys is always above the ideal value ($u_{AlGaN} > u_{ideal} = 0.375$).

If the lattice constants scale linearly with the alloy composition but the internal parameter does not, the bond angles and/or the bond lengths of the real and the virtual crystal must depend non-linearly on the alloy composition. The average bond lengths, angles and second nearest neighbor distances calculated by using equations (5)-(10) are shown in **Fig. 5** and listed in **Tab. 1**. The average cation-anion distances to the nearest and second nearest neighbors scale nearly linearly with alloy composition for AlGaN. The average bond length along the c-axis is 0.7 to 0.9% longer than the nearest neighbor bonds in direction of the basal plane. The ratio r, of the distances to the second nearest neighbors along the c-axis (M-N$_{c2}$) and along the basal plane (M-N$_{b2}$, M-N$_{b'2}$) is well above the ratio of the ideal structure ($r = \{1-(M-N_{c2}/M-N_{b2})\}= 13\%$) reaching 15.4% for AlN. The difference between the ideal bond angle $\alpha_{ideal} = 109.47°$ and the average bond angles α and β increase also non-linearly from GaN to AlN, reaching values of $\Delta\alpha = \Delta\beta = 1.25°$ (**Fig. 5**).

Besides the deviation of the AlGaN alloys from the ideal wurtzite structure it should be noticed that the observed non-linearities in the cell-internal parameter, cation-anion distances and bond lengths of the random alloys always tend to decrease the difference between the "real" and ideal structure.

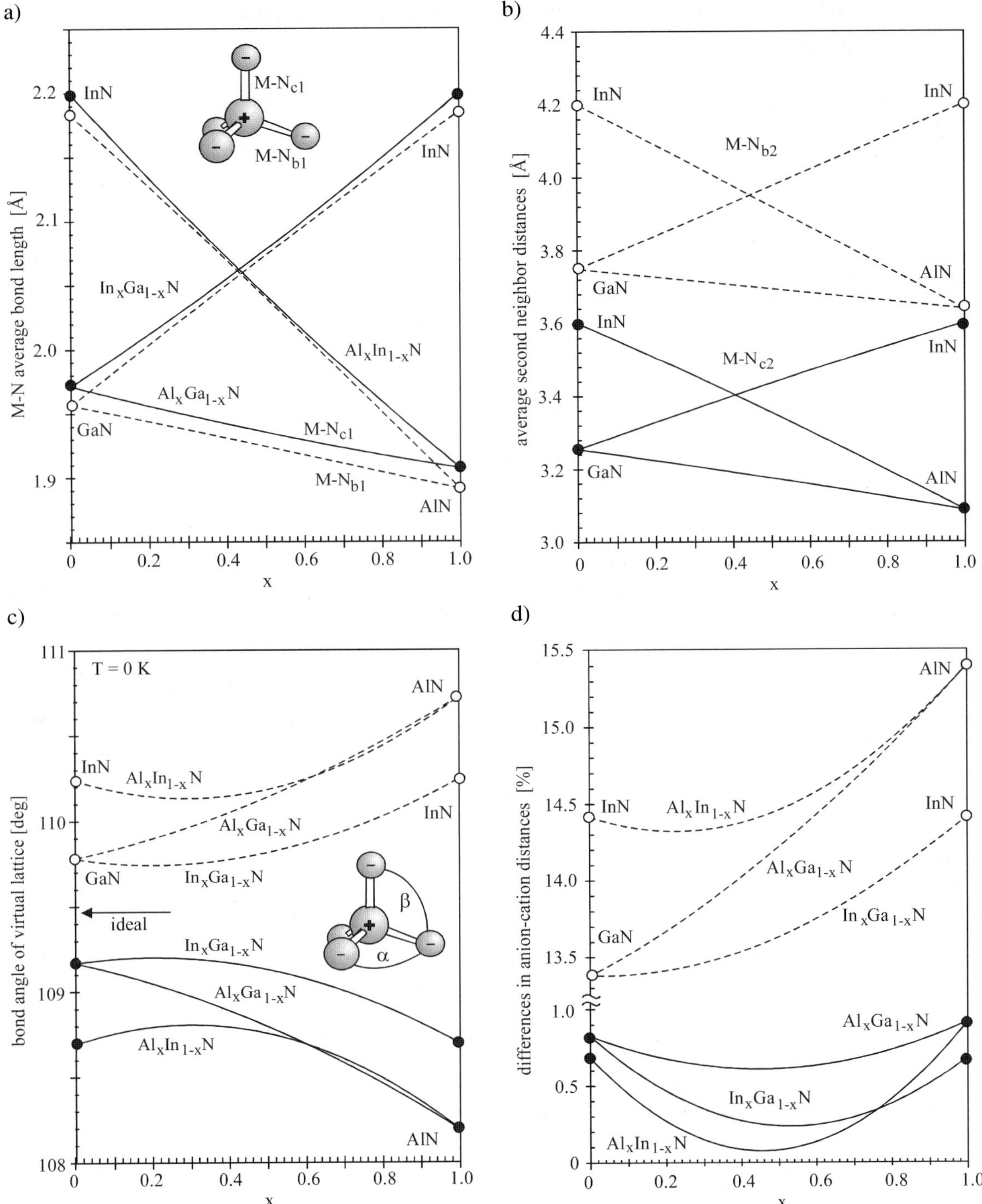

Fig. 5 a) The average bond length (virtual crystal limit) of the metal nitrogen bond along the c-axis (solid line) and in the direction of the basal plane (dashed line) for AlGaN (and for comparison also for InGaN and AlInN [5, 6]) alloys. b) The average second-nearest-neighbor distances of metal and nitrogen atoms in alloys with wurtzite structure versus alloy composition. c) The average bond angles of the hexagonal crystals. d) The ratio r between the nearest- and second nearest-neighbor distances along the c-axis and in the direction of the basal plane.

4 Non-linear spontaneous and piezoelectric polarization in group-III-nitrides

First models of polarization induced effects in GaN based electronic and optoelectronic devices have assumed that polarization in ternary nitride alloys can be interpolated linearly between the limiting values determined by the binary compounds. It is known from earlier work on III-V alloys with zinc blende structure that, although Vegard´s law is found to be valid in predicting the macroscopic structural properties of alloys (a and c lattice constants in this context), the electronic and optical properties like the band gap are non-linear functions of the alloy composition. These non-linearities can be caused by different response to the hydrostatic compression of the binary constituents, charge transfer between cations due to different electronegativity, internal strain effects due to varying cation-anion bond length (bond alternation), and disorder effects due to random distribution of the chemical elements on the cation side.

4.1 Spontaneous polarization

Motivated by the observed non-linearities in the internal structural parameters (u parameter in this context), we have investigated the pyroelectric properties of alloys, looking for possible non-linearities [35]. The results show that the spontaneous polarization of relaxed alloys for a given composition depends linearly on the average u parameter which indicates that spontaneous polarization differences between alloys of the same composition are mainly due to varying cation-anion bond length, whereas disorder has a negligible influence. This idea is supported by the fact that in binaries the polarization depends linearly on the relative displacement of the cation and anion sub-lattices in the [0001] direction [5]. Because of the non-linear dependence of the cell-internal parameter on alloy composition a non-linear behavior of the spontaneous polarization versus x has to be expected.

In addition to the non-linearities which are caused by the structural properties, Bernardini et al. pointed out that the different response to the hydrostatic pressure of the binary constituents of the alloy should contribute significantly to the non-linear behavior of spontaneous polarization in ternary random alloys [5, 6].

Using the theoretical approach described in chapter 2 we have calculated the spontaneous polarization of relaxed $Al_{0.5}Ga_{0.5}N$ (listed in **Tab.1**) in order to determine the bowing parameter of P^{SP}_{AlGaN} as a function of x. The spontaneous polarization of the random AlGaN alloys (in C/m^2) is given to second order in x by

$$P^{SP}_{AlGaN}(x) = -0.090x - 0.034(1-x) + 0.021x(1-x).$$ (13)

The first two terms in the equations are the usual linear interpolation between the binary compounds. The third term is embodying non-linearity to quadratic order. Higher order terms are neglected as their effect was estimated to be less than 10%.

Knowledge of the spontaneous polarization in alloys is not sufficient to describe the pyroelectric properties of AlGaN alloys and GaN-based nanostructures or to predict the values of polarization induced interface charges and electrostatic fields. GaN-based hetero- and nanostructures are usually grown pseudomorphically and strained on substrates and buffer layers with significant mismatch in lattice constants and thermal expansion coefficients. The ensuing symmetry-conserving strain causes a change in the crystal structure and in polarization that amounts to a piezoelectric polarization. Therefore, the polarization of every strained layer is a combination of a spontaneous and a piezoelectric component. Before we focus on the non-linearity of the piezoelectric polarization of AlGaN alloys in dependence of strain, the basics to calculate the piezoelectric polarization in the linear approximation are provided.

4.2 Piezoelectric polarization

By Hooke´s law the deformation of a crystal ε_j, due to external or internal forces or stresses σ_i can be described by:

$$\sigma_i = \sum_j C_{ij}\varepsilon_j \, , \tag{14}$$

where C_{ij} is the elastic tensor (Voigt notation). The 6x6 matrix of the elastic constants C_{ij}, (see **Tab.2**, [36, 37]) for crystals with wurtzite structure is given by:

$$C_{ij} = \begin{pmatrix} C_{11} & C_{12} & C_{13} & 0 & 0 & 0 \\ C_{12} & C_{11} & C_{13} & 0 & 0 & 0 \\ C_{13} & C_{13} & C_{33} & 0 & 0 & 0 \\ 0 & 0 & 0 & C_{44} & 0 & 0 \\ 0 & 0 & 0 & 0 & C_{44} & 0 \\ 0 & 0 & 0 & 0 & 0 & \frac{1}{2}(C_{11}-C_{12}) \end{pmatrix} . \tag{15}$$

At this point we consider uniaxial strain with an orientation parallel to the c-axis and biaxial strain in the basal plane of AlGaN. As long as the C_{6v}^4 space-group symmetry is conserved under the applied forces, the strain tensor is diagonal and possesses the components:

$$\varepsilon_1 = \varepsilon_2 = \frac{a-a_0}{a_0}, \ \varepsilon_3 = \frac{c-c_0}{c_0}, \tag{16}$$

where a_0 and c_0 are the lattice constants of the relaxed wurtzite crystal lattice, respectively.

Hooke´s law then leads to the corresponding diagonal stress tensor with the elements σ_i:

$$\sigma_1 = \sigma_2 = (C_{11}+C_{12})\varepsilon_1 + C_{13}\varepsilon_3 \, , \tag{17}$$

$$\sigma_3 = 2C_{13}\varepsilon_1 + C_{33}\varepsilon_3 \, . \tag{18}$$

In the case of uniaxial stress parallel to the c-axis, the external forces vanish in the plane perpendicular to the stress direction ($\sigma_1 = \sigma_2 = 0$) and there is an elastic relaxation of the lattice in that plane [38]. The ratio of the resulting in-plane strain to the deformation along the stress direction is expressed by the Poisson ratio ν:

$$\varepsilon_1 = -\nu\varepsilon_3 \, , \text{ with } \nu = \frac{C_{13}}{C_{11}+C_{12}} \, . \tag{19}$$

Of main interest here is the fact that the hardness of all binary compounds with wurtzite structure is isotropic in the basal plane. This is important because the strain in epitaxial layers of group-III-nitride heterostructures grown along the c-axis caused by mismatch of the lattice constants a and/or a mismatch of the thermal expansion coefficients of layer and substrate is directed along the basal plane (parallel to the substrate). In this case no force is applied in growth direction and the crystal can relax freely in this direction. The resulting biaxial strain ($\varepsilon_1 = \varepsilon_2$) causes stresses $\sigma_1 = \sigma_2$, whereas σ_3 has to be zero. Using equations (15) and (16), a relation between the strain along the c-axis and along the basal plane can be derived

$$\varepsilon_3 = -v'\varepsilon_1 = -2\frac{C_{13}}{C_{33}}\varepsilon_1 , \tag{20}$$

where $\quad v' = 2\dfrac{C_{13}}{C_{33}} . \tag{21}$

Table 2 Experimental and predicted elastic compliance, elastic constants and piezoelectric constants of wurtzite GaN and AlN (theory 1 [36], theory 2 [5, 6]).

crystal property	unit	GaN			AlN		
		theory 1	theory 2	experiment	theory 1	theory 2	experiment
C_{11}	GPa	367		370[a]	396		410[a]
C_{12}	GPa	135		145[a]	137		140[a]
C_{13}	GPa	103	68	110[a]	108	94	100[a]
C_{33}	GPa	405	354	390[a]	373	377	390[a]
C_{44}	GPa	95		90[a]	116		120[a]
v		0.202		0.214	0.205		0.182
v'		0.509	0.38	0.564	0.579	0.50	0.512
v''		0.980		1.054	1.196		1.207
S_{11}	10^{-12} m^2/N	3.267		3.326	2.993		2.854
S_{12}	10^{-12} m^2/N	-1.043		-1.118	-0.868		-0.849
S_{13}	10^{-12} m^2/N	-0.566		-0.623	-0.615		-0.514
S_{33}	10^{-12} m^2/N	2.757		2.915	3.037		2.828
S_{44}	10^{-12} m^2/N	10.53		11.11	8.621		8.333
e_{31}	C/m^2		-0.34			-0.53	-0.58[d]
e_{33}	C/m^2		0.67			1.50	1.55[d]
e_{15}	C/m^2	-0.22[b]		-0.30[c]			-0.48[d]
d_{31}	10^{-12} C/(m^{2}Pa)		-1.253			-2.298	-2.65
d_{33}	10^{-12} C/(m^{2}Pa)		2.291			5.352	5.53
d_{15}	10^{-12} C/(m^{2}Pa)		-1.579			-2.069	-4.08

[a] Experimental results from [40]   [b] Experimental results from [9]
[c] Experimental results from [37]   [d] Experimental results from [41]

The stresses in the basal plane caused by the mismatch in the lattice constants can be calculated by:

$$\sigma_1 = \varepsilon_1\left(C_{11} + C_{12} - 2\frac{C_{13}^2}{C_{33}}\right) , \tag{22}$$

where $\quad C_{11} + C_{12} - 2\dfrac{C_{13}^2}{C_{33}} > 0 \,.$ (23)

Finally, in the case of hydrostatic pressure the components of the stress tensor are equal ($\sigma_1 = \sigma_2 = \sigma_3$) and from Hooke´s law it follows that:

$$\varepsilon_3 = \frac{C_{11} + C_{12} - 2C_{13}}{C_{33} - C_{13}}\varepsilon_1 \,,$$ (24)

where $\quad v'' = \dfrac{C_{11} + C_{12} - 2C_{13}}{C_{33} - C_{13}} \,.$

The piezoelectric polarization for hexagonal materials [39] is given by:

$$P_i^{pz} = \sum_l d_{il}\sigma_l \,, \; i = 1, 2, 3 \,, l = 1,...,6,$$ (25)

where P_i^{pz} are the components of the piezoelectric polarization and d_{il} are the piezoelectric moduli (**Tab.2**). Using the symmetry relations between the piezoelectric moduli, $d_{31} = d_{32}$, $d_{33} \neq 0$ and $d_{15} = d_{24}$ (all other components: $d_{il} = 0$), equation (25) can be reduced to a set of three equations:

$$P_1^{pz} = \frac{1}{2}d_{15}\sigma_5 \,,$$ (26)

$$P_2^{pz} = \frac{1}{2}d_{15}\sigma_4 \,,$$ (27)

$$P_3^{pz} = d_{31}(\sigma_1 + \sigma_2) + d_{33}\sigma_3 \,.$$ (28)

Keeping in mind that for biaxial stress, as it is of interest here, $\sigma_1 = \sigma_2$, $\sigma_3 = 0$ and sheer stresses are assumed to be negligible ($\sigma_4 = \sigma_5 = 0$), the piezoelectric polarization has only one not vanishing component, which is directed along the growth direction and is given by:

$$P_3^{pz} = 2d_{31}\sigma_1 = 2d_{31}\varepsilon_1(C_{11} + C_{12} - 2\frac{C_{13}^2}{C_{33}}) \,.$$ (29)

More often than the piezoelectric moduli, the piezoelectric constants e_{kl} are used to describe the piezoelectric properties of group-III-nitrides. They can be calculated by:

$$e_{kl} = \sum_j d_{kj}C_{jl} \,, \qquad \text{where } k = 1, 2, 3, \; l = 1,..., 6, \; j = 1,..., 6.$$ (30)

For hexagonal crystals the relations between piezoelectric constants and moduli can be reduced to:

$$e_{31} = e_{32} = C_{11}d_{31} + C_{12}d_{32} + C_{13}d_{33} = (C_{11} + C_{12})d_{31} + C_{13}d_{33}$$ (31)

$$e_{33} = 2C_{13}d_{31} + C_{33}d_{33} \,,$$ (32)

$$e_{15} = e_{24} = C_{44}d_{15} \quad \text{and} \tag{33}$$

$$e_{kl} = 0 \quad \text{for all other components.} \tag{34}$$

The piezoelectric polarization as a function of strain can be written as:

$$P_k^{pz} = \sum_l e_{kl}\varepsilon_l , \quad \text{where } k = 1, 2, 3, \quad l = 1,..., 6. \tag{35}$$

The non-vanishing component of the piezoelectric polarization caused by biaxial strain is:

$$P_3^{pz} = \varepsilon_1 e_{31} + \varepsilon_2 e_{32} + \varepsilon_3 e_{33} , \tag{36}$$

$$P_3^{pz(biaxial)} = 2\varepsilon_1 e_{31} + \varepsilon_3 e_{33} , \quad \text{where} \quad \varepsilon_3 = -2\frac{C_{13}}{C_{33}}\varepsilon_1 , \tag{37}$$

$$= 2\varepsilon_1 \left(e_{31} - e_{33}\frac{C_{13}}{C_{33}} \right). \tag{38}$$

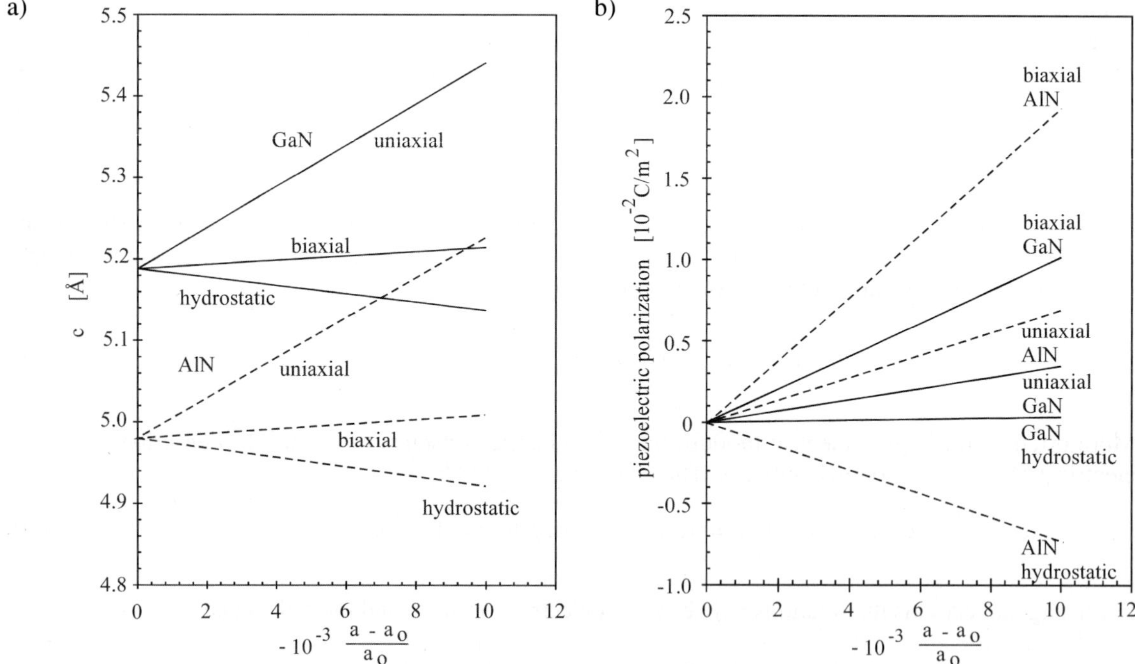

Fig. 6 a) Variation of lattice constant c of AlN (dashed lines) and GaN (solid lines) caused by uniaxial, biaxial or hydrostatic strain, where a_o is the lattice constant of the corresponding relaxed hexagonal crystal. b) The piezoelectric polarization of AlN and GaN under uniaxial, biaxial or hydrostatic strain.

In analogy the piezoelectric polarization along the c-axis caused by uniaxial and hydrostatic pressure can be determined by:

$$P_3^{pz(uniaxial)} = \varepsilon_1 \frac{C_{11} - C_{12}}{C_{11} - \dfrac{C_{13}^2}{C_{33}}}\left(e_{31} - e_{33}\frac{C_{13}}{C_{33}} \right), \tag{39}$$

$$P_3^{pz(hydrostatic)} = \varepsilon_1\left(2e_{31} + \frac{C_{11} + C_{12} - C_{13}}{C_{33} - C_{13}} e_{33} \right). \tag{40}$$

Fig. 6 shows the dependence of the lattice constant c and the piezoelectric polarization P_3^{pz} of hexagonal AlN and GaN crystals in dependence of strain ε_1, caused by uniaxial, biaxial and hydrostatic pressure. It becomes obvious that the deformation of the lattice and therefore the piezoelectric polarization is strongly dependent on the kind of pressure which is applied to the crystal, although the strain along the basal plane can be the same.

Because of its relevance for nitride based devices we are going to determine $P_3^{pz(biaxial)}$ (we will omit the index *3* and *biaxial* in the following discussion) for $Al_xGa_{1-x}N$ pseudomorphically grown on relaxed GaN and buffer layers. Under these assumptions, the strain is defined by:

$$\varepsilon_1 = \frac{a_{GaN} - a(x)}{a(x)}. \tag{41}$$

As a first approach, the lattice, piezoelectric and elastic constants of the alloys are approximated by a linear interpolation between the constants of the relevant binary compounds and implemented into equation (38). For the following calculations we have used the physical properties of the binary compounds predicted by Bernardini et al. (**Tab. 2**) [5, 6]. **Fig. 7** shows the piezoelectric polarization of the alloy pseudomorphically grown on relaxed GaN for different values of x. Very high values of piezoelectric polarization of up to 0.05 C/m^2 are observed. In addition, it should be pointed out that the piezoelectric polarization P^{pz}, calculated with the same piezoelectric constants as before but using the elastic constants predicted by Wright (**Tab. 2**) or measured by Deger et al. [40], deviates from the values shown in **Fig. 7** by only 5% in the worst case.

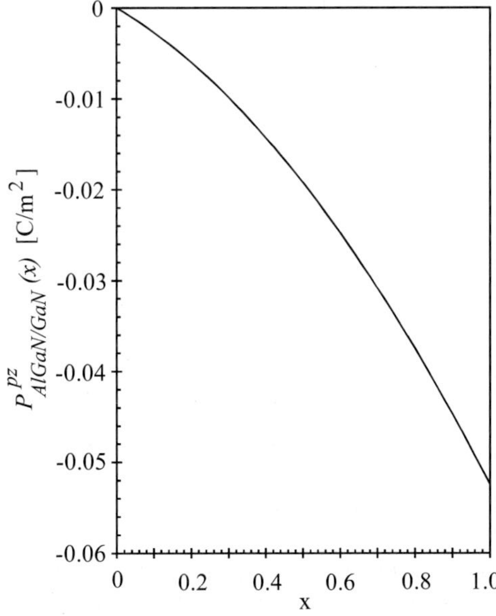

Fig. 7 Piezoelectric polarization of AlGaN barriers with different alloy compositions grown pseudomorphically on relaxed GaN. The piezo-electric polarization caused by biaxial tensile strain was calculated by equation (38).

As a consequence of equation (38) and the linear interpolation of the elastic and piezoelectric constants, the piezoelectric polarization is non-linear in terms of the alloy composition. The non-linear dependence of the piezoelectric polarization on the alloy composition can be approximated with an accuracy of better than 1% by the following quadratic equation:

$$P^{pz}_{AlGaN/GaN}(x) = \left[-0.0525x + 0.0282x(1-x) \right] C/m^2 . \tag{42}$$

For heterostructures with barriers under moderate strain, this equation can be used as an input, either directly as polarization or, as discussed in more detail later, as interface bound sheet charge, depending on the implementation, in a self consistent Schrödinger-Poisson solver based e.g. on effective mass or tight binding theory.

Up to now, we have relied on Hooke´s law assuming that the tensor components C_{ij} and e_{ij} are constant for a given binary or ternary crystal and that the piezoelectric polarization depends linearly on the strain. But it is known from other piezoelectric semiconductors that, if high forces or pressures are applied to the crystals, this relation can become non-linear [42]. In the following, we first demonstrate that the piezoelectricity of GaN and AlN as well as nitride alloys is non-linear in terms of strain, then we suggest how to use this understanding in practice for an improved prediction of the piezoelectric polarization of random (in opposite to ordered) AlGaN alloys caused by high biaxial strain. Again, we consider the technologically most relevant case of a pseudomorphically grown alloy on an unstrained GaN buffer layer. Using the same theoretical approach as described in chapter 2, we calculate the polarization with the constraint $a_{buffer} = a_{GaN}$, reoptimizing all structures. The piezoelectric polarization is then computed as the difference of the total polarization obtained from this calculation and the spontaneous polarization. **Fig. 8** shows the piezoelectric polarization versus strain calculated by equation (38) (linear in strain) and our improved theoretical approach for binary compounds. In both cases the piezoelectric polarization for a given strain increases from GaN to AlN. It is important that calculations by our improved method lead to significantly higher piezoelectric polarizations, especially in cases of high strains.

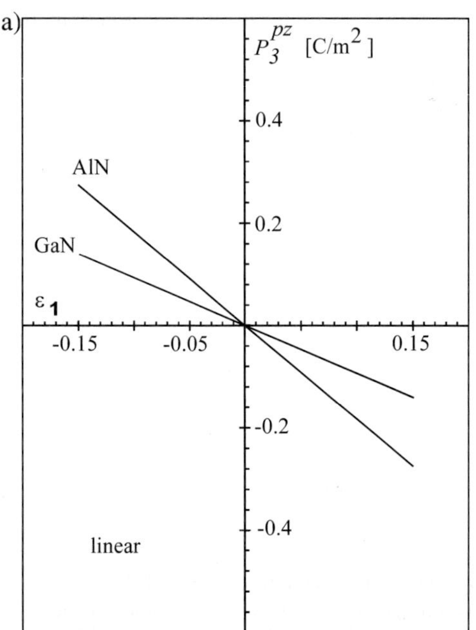

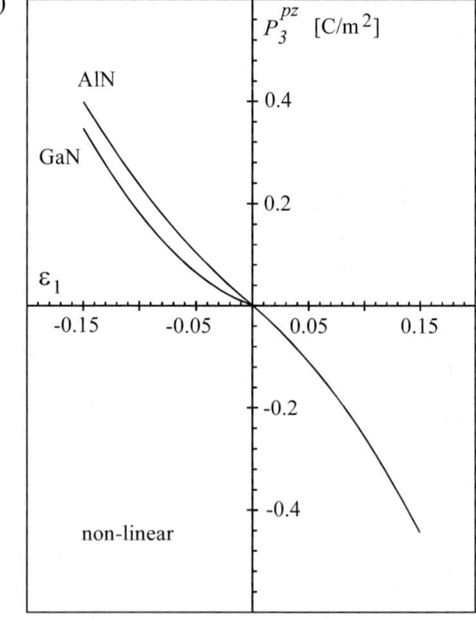

Fig. 8   a) The piezoelectric polarization of GaN and AlN with wurtzite structure under biaxial tensile or compressive strain calculated by equation (41) or b) by equations (43-46) based on the theory described in chapter 2.

The calculated non-linear piezoelectricity of the binary compounds versus biaxial strain ε_l can be described by the relations (in C/m^2):

$$P^{PZ}_{AlN}(\varepsilon_1) = -1.808\varepsilon + 5.624\varepsilon_1^2 \text{, for } \varepsilon_l < 0, \tag{43}$$

$$P^{PZ}_{AlN}(\varepsilon_1) = -1.808\varepsilon - 7.888\varepsilon_1^2 \text{, for } \varepsilon_l > 0, \tag{44}$$

$$P^{PZ}_{GaN}(\varepsilon_1) = -0.918\varepsilon + 9.541\varepsilon_1^2 . \tag{45}$$

The observed non-linearity in the bulk piezoelectricity exceeds any effects related to disorder or bond alternation which were taken into account [35]. Therefore, the calculation of the piezoelectric polarization of an $Al_xGa_{1-x}N$ alloy at any strain becomes straightforward. One can choose a value for x, calculate the strain $\varepsilon_1 = \varepsilon(x)$ from Vegard´s law, and the piezoelectric polarization by

$$P^{PZ}_{AlGaN}(x) = xP^{PZ}_{AlN}(\varepsilon(x)) + (1-x)P^{PZ}_{GaN}(\varepsilon(x)) , \tag{46}$$

where $P^{PZ}_{AlN}(\varepsilon(x)), P^{PZ}_{GaN}(\varepsilon(x))$ are the strain dependent bulk piezoelectric polarizations of the relevant binary compounds given above. A comparison of the calculated values of P^{pz} using a linear interpolation of the elastic and piezoelectric constants (equation (38)) or taking the non-linearity of the piezoelectric polarization into account shows that the linear interpolation leads to an underestimation of the piezoelectric polarization. Therefore, the improved scheme represented by equations (43 - 46) is of interest in modeling hetero- and nanostructures with highly strained AlGaN layers.

5 Polarization induced surface and interface charges

Based on a theoretical understanding of the non-linear dependence of spontaneous and piezoelectric polarization on composition and strain we can provide a more accurate prediction of polarization induced charges bound at surfaces and interfaces of $Al_xGa_{1-x}N/GaN$ heterostructures. As mentioned above, the total polarization P_{AlGaN} is the sum of the piezoelectric and spontaneous polarization,

$$P_{AlGaN} = P^{pz}_{AlGaN} + P^{SP}_{AlGaN} . \tag{47}$$

Associated with a gradient of polarization in space is a polarization induced charge density given by:

$$\rho_P = -\nabla P . \tag{48}$$

As a special case, at the surface of a relaxed or strained $Al_xGa_{1-x}N$ layer as well as at the interface of a $Al_xGa_{1-x}N/GaN$ heterostructure, the total polarization changes abruptly, causing a fixed two dimensional polarization charge density σ, given by

$$\sigma_{AlGaN} = P_{AlGaN} = P^{SP}_{AlGaN} + P^{pz}_{AlGaN} \qquad \text{for surfaces,}$$

$$\sigma_{AlGaN/GaN} = P_{GaN} - P_{AlGaN} , \tag{49}$$

$$= \left(P^{SP}_{GaN} + P^{pz}_{GaN} \right) - \left(P^{SP}_{AlGaN} + P^{pz}_{AlGaN} \right) \qquad \text{for interfaces,}$$

respectively. **Fig. 9** shows the polarization induced surface and interface sheet density σ/e ($e = -1.602 \times 10^{-19}$ C) for pseudomorphic $Al_xGa_{1-x}N/GaN$ heterostructures, grown on dielectric c-Al_2O_3 or 6H-SiC substrates.

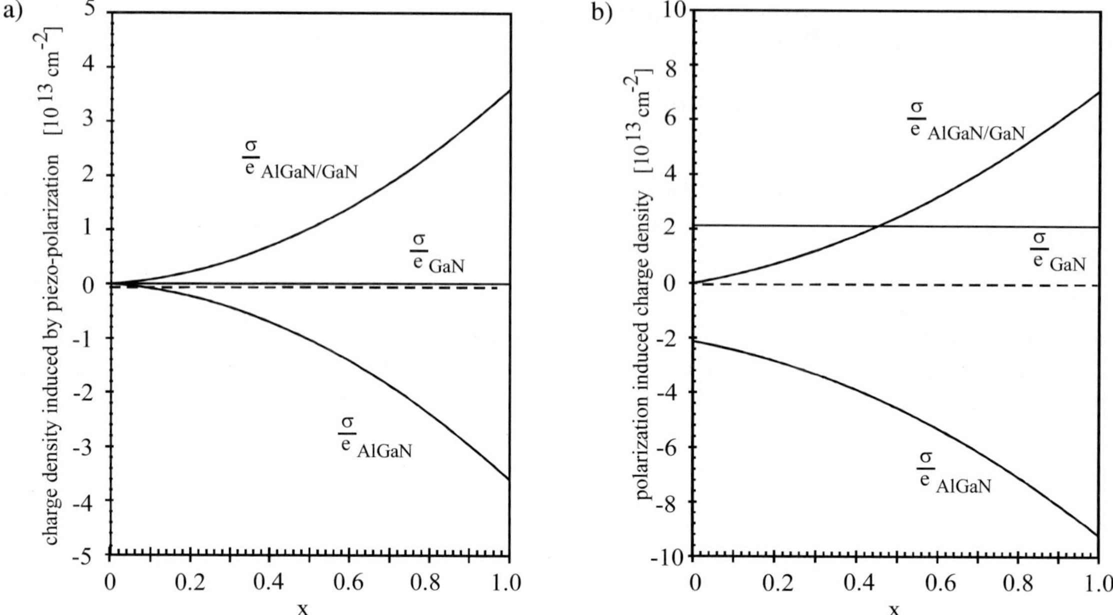

Fig. 9   a) Piezoelectric polarization induced surface (GaN and AlGaN) as well as AlGaN/GaN interface charges of pseudomorphically grown Ga-face heterostructures versus Al-concentration of the barrier. b) Surface and interface charges induced by gradients of piezoelectric and spontaneous polarization at surfaces and interfaces.

Since screening by charges from the ambient at $Al_xGa_{1-x}N/GaN$ interfaces can be excluded, polarization induced effects are much easier to study in heterostructures in comparison to single epitaxial layers. For pseudomorphic Ga-face $Al_xGa_{1-x}N/GaN$ heterostructures, the polarization induced interface charges are predicted to be positive (**Fig. 9**). The bound charge increases non-linearly with x up to 7.06×10^{13} cm^{-2}, estimated for the AlN/GaN heterostructure.

In n-type heterostructures it has to be taken into account that free electrons will accumulate at interfaces with positive bound sheet charges, compensating $+\sigma$. As a consequence, 2DEGs with a sheet carrier concentration close to the concentration of the bound interface density $+\sigma/e$ can be formed. It should be pointed out that these 2DEGs are realized without any need of a modulation doped barrier. This is of special interest for the fabrication of high frequency and high power HEMTs, as the growth and processing of these devices is simplified.

6 Sheet carrier concentrations of polarization induced 2DEGs

With a theoretical understanding of the polarization induced charge we now can predict the sheet carrier concentration of polarization induced 2DEGs and their dependence on alloy composition for pseudomorphic Ga-face AlGaN/GaN heterostructures. For undoped HEMT structures, the sheet electron concentration, $n_s(x)$, can be approximated by taking advantage of the total bound sheet charge $\sigma_{AlGaN/GaN}(x)$ calculated above, and the following equation:

$$n_s(x) = \frac{\sigma_{AlGaN/GaN}(x)}{e} - \frac{\varepsilon_0 E_F}{e^2}\left(\frac{\varepsilon_{AlGaN}(x)}{d_{AlGaN}} + \frac{\varepsilon_{GaN}}{d_{GaN}}\right) - \frac{\varepsilon_0 \varepsilon_{AlGaN}(x)}{e^2 d_{AlGaN}}\left(e\phi_{AlGaN}(x) + \Delta(x) - \Delta E_{AlGaN}^C(x)\right), \tag{50}$$

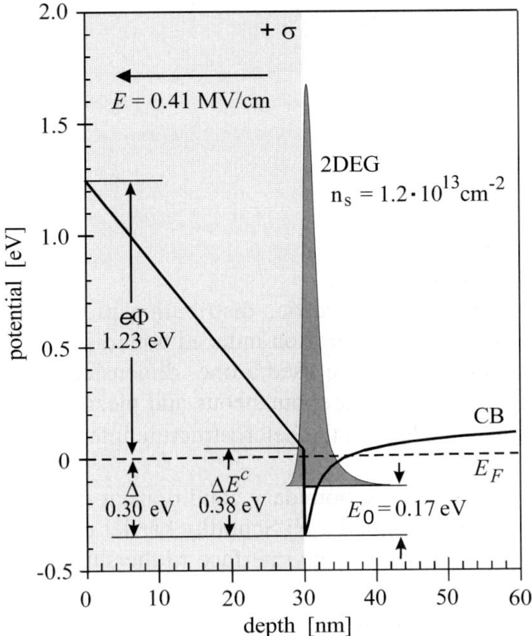

Fig. 10 Self-consistent calculation of the conduction band and electron concentration profile of an undoped Ga-face $Al_{0.3}Ga_{0.7}N/GaN$ (30/2000 nm) heterostructure including the Schottky barrier of the Ni contact.

where ε_0 is the dielectric constant of the vacuum, ε_{GaN} and $\varepsilon_{AlGaN}(x)$ are the relative dielectric constants of the constituent layers, and d_{AlGaN}, d_{GaN} are the thicknesses of the barrier and the buffer layer, respectively. E_F is the position of the Fermi level with respect to the GaN conduction-band-edge close to the GaN/substrate interface, $e\phi_{AlGaN}(x)$ is the Schottky barrier height of the gate contact on top of the barrier (used for CV-profiling measurements), $\Delta E^C_{AlGaN}(x)$ is the conduction band offset between AlGaN and GaN, and $\Delta(x)$ is the penetration of the conduction band edge below the Fermi level at the AlGaN/GaN interface (see **Fig. 10**) [43].

This latter quantity is approximated by [7, 13]:

$$\Delta(x) = E_0(x) + \frac{\pi \hbar^2}{m^*_{GaN}} n_S(x) , \tag{51}$$

where the lowest subband level of the 2DEG is given by

$$E_0(x) = \left\{ \frac{9\pi \hbar e^2}{8\varepsilon_0 \sqrt{8m^*_{GaN}}} \frac{n_S(x)}{\varepsilon_{GaN}} \right\}^{2/3} . \tag{52}$$

We used the following linear interpolations to describe the physical properties of the alloys in our calculations:

dielectric constants [44]: $\varepsilon_{AlGaN}(x) = 0.03x + 10.28 ,$ \tag{53}

Schottky barrier for Ni-contacts [45]: $e\phi_{AlGaN}(x) = (1.3x + 0.84)eV$,   (54)

bandgap [46]: $E^g_{AlGaN}(x) = [6.13x + 3.42(1-x) - 1.0x(1-x)]eV$,   (55)

band offsets [47, 48]: $\Delta E^C_{AlGaN}(x) = 0.63(E^g_{AlGaN}(x) - E^g_{AlGaN}(0))$ (56)

and an electron effective mass of $m_{GaN}* = 0.228\,m_0$ for GaN [49].

In order to determine the sheet carrier concentrations and carrier distribution profiles in undoped HEMT structures including spontaneous and piezoelectric polarization induced bound sheet charges, and to verify the obtained results with the equations above, we have used a one- dimensional Schrödinger-Poisson solver [50]. To incorporate the effects of the non-linear spontaneous and piezoelectric polarization into the program, thin (≈ 6 Å) layers of charge are added to the heterostructure interfaces to simulate the bound sheet density σ/e.

For both kinds of calculations it is necessary to specify the boundary conditions at the surface and at the substrate interfaces. For HEMT structures we have assumed a Ni Schottky barrier contact at the surface, which pins the conduction band according to eq. (55). At the interface to the substrate, the Fermi level was set to mid gap of GaN. Changing this value of the Fermi level has very little impact on the sheet carrier concentration of the 2DEG for GaN buffer thicknesses above 1 µm. For the same boundary conditions, the calculations of the 2DEG sheet carrier concentrations using equation (51) or the Schrödinger-Poisson solver agree within 5% for all HEMT structures investigated.

7 Growth of undoped AlGaN/GaN hetero- and nanostructures

The investigated pseudomorphic Ga-face AlGaN/GaN heterostructures were grown by low pressure metal organic chemical vapour deposition (MOCVD) or plasma induced molecular beam epitaxy (PIMBE) on c-Al_2O_3 substrates for sensor applications or on 6H-SiC substrates for processing of high power transistors. The MOCVD grown undoped AlGaN/GaN heterostructures were deposited at a pressure of 100 mbar, using triethylgallium (TEG), trimethylaluminium (TMA) and ammonia as precursors. A growth rate of about 0.5 µm/h was achieved for a substrate temperature of 1040°C and V/III gas phase ratios of 1800 and about 900 for GaN and AlGaN, respectively [51]. The AlGaN/GaN heterostructures as well as the $Al_{0.1}Ga_{0.9}N$ nucleation layers were grown at a constant substrate temperature and without any growth interruptions to avoid the presence of excessively high free carrier background concentrations. $Al_xGa_{1-x}N$ barriers with thicknesses between 100 and 450 Å and alloy compositions of up to $x = 0.45$ were deposited on GaN buffer layers with thicknesses between 1 and 2.5 µm.

AlGaN/GaN heterostructures were also deposited by PIMBE using conventional effusion cells and a radio frequency plasma source for the generation of nitrogen radicals. The molecular nitrogen flux through the plasma source was fixed at about 1 sccm causing a nitrogen partial pressure in the MBE chamber of 4×10^{-5} mbar during growth. The optimized growth temperature for GaN was determined to be between 780 and 800°C for a Ga flux of 1.5×10^{15} cm^{-2}s^{-1}, resulting in a growth rate of 0.5 µm/h as discussed in more detail in Ref. [20, 52]. AlGaN barriers with thicknesses between 2 and 50 nm were grown on GaN buffer layers using a total flux of Al- and Ga-atoms equal to the Ga-flux optimized for the deposition of high quality GaN. High electron mobility AlGaN/GaN heterostructures were grown by PIMBE on MOCVD-grown GaN/Al_2O_3 templates ($d_{GaN} = 2$ µm). The thickness of the AlGaN barriers were 25 ± 5 nm, the Al-contents were varied between $x = 0.1$ and 0.35 [53]. AlGaN/GaN heterostructures with barriers of different alloy compositions and thicknesses were investigated by electrical characterization methods in order to measure the sheet carrier concentrations of polarization induced 2DEGs and to determine their electronic transport properties related to possible applications for novel transistor and sensor devices described in PART B.

8 2DEGs confined at interfaces of undoped Ga-face AlGaN/GaN heterostructures

In typical undoped AlGaN/GaN heterostructures with insulating GaN layers ($N_d < 10^{16}$ cm^{-3}, thicknesses of about 2 μm), and barrier widths of more than 15 nm, the value of the 2DEG sheet carrier concentration is dominated by the polarization induced sheet charges. These charges are increasing with enhanced Al-concentration of the barrier, because the strain and, therefore, the piezoelectric polarization as well as the spontaneous polarization are increasing. This offers the possibility to control the sheet carrier concentration of the 2DEG by the choice of the Al-concentration of the barrier instead of barrier doping.

In **Fig. 10** the conduction band and electron concentration profiles are shown for an undoped Ga-face $Al_{0.3}Ga_{0.7}N/GaN$ (30/2000 nm) heterostructure with a Ni Schottky contact on top. The polarization induced surface and interface charges are causing an electric field of about 0.4 MV/cm inside the barrier and an accumulation of electrons with a sheet carrier concentration of 1.2×10^{13} cm^{-2} at the AlGaN/GaN interface. In **Fig. 11** the polarization induced bound sheet densities and 2DEG sheet carrier concentrations are shown for heterostructures with the same design but different alloy composition of the barrier. The bound interface sheet density is calculated for barriers grown on relaxed GaN using again (i) a linear interpolation of the physical properties (C_{ij}, e_{ij}, and P^{SP}) of the binary compounds (upper solid curve) and (ii) considering the non-linear behavior of piezoelectric and spontaneous polarization (lower solid curve).

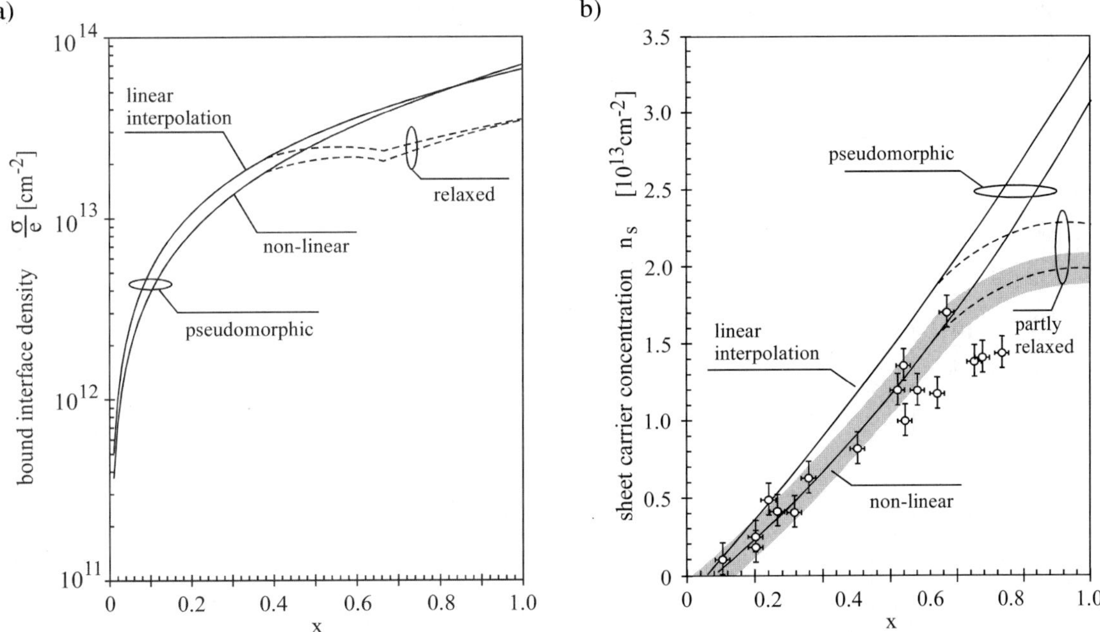

Fig. 11 a) The polarization induced bound interface sheet densities are shown as a function of alloy composition of the barrier for Ga-face AlGaN/GaN heterostructures. The sheet density is calculated for barriers grown on relaxed GaN using (i) a linear interpolation of the physical properties (C_{ij}, e_{ij} and P^{SP}) of the binary compounds (upper solid curve) and (ii) considering the nonlinear behaviour or piezoelectric and spontaneous polarization (lower solid curve). The dashed curves show the change of the sheet density with increasing x taking the measured degree of barrier relaxation into account. b) The 2DEG sheet carrier concentrations are shown as a function of alloy composition of the barrier for Ga-face AlGaN/GaN heterostructures. The sheet carrier concentration is calculated for barriers grown on relaxed GaN using again (i) a linear interpolation (upper solid curve) and (ii) considering the nonlinear behaviour of piezoelectric and spontaneous polarization (lower solid curve). The grey area along this curve indicates the uncertainty in the calculated values of n_s. The measured 2DEG carrier concentrations obtained by CV-profiling are shown as open symbols.

The gray area in **Fig. 11** along one of the curves indicates the uncertainty (which is very similar for the linear and non-linear approach) in the calculated values of n_s, which is mainly caused by the error in the determination of the barrier thickness and the conduction band offsets. To enable a comparison between experimental (open symbols) and theoretical data, the sheet carrier concentration of the 2DEGs is calculated as described above, taking into account the depletion caused by the Ni Schottky contact (**Fig. 11**). Experimentally, the sheet carrier concentrations of 2DEGs confined in AlGaN/GaN heterostructures with alloy compositions of up to $x = 0.5$ are measured by CV-profiling using Ti/Al ohmic and Ni Schottky contacts. The highest measured and calculated sheet carrier concentration for pseudomorphic AlGaN/GaN heterostructures is determined to 2×10^{13} cm^{-2} for $x = 0.37$. For higher Al-concentration, the AlGaN barrier starts to relax, diminishing the piezoelectric component [54]. The dashed extrapolations in **Fig. 11** shows the further variation of the sheet carrier concentration with increasing x, taking the measured degree of relaxation of the barrier into account. For lower Al-concentrations of the (25 ± 5) nm thick barriers, the sheet carrier concentration decreases non-linearly down to $x \approx 0.06$, where the 2DEG disappears.

For pseudomorphic HEMT structures with barrier thicknesses of more than 15 nm, the non-linear dependence of the 2DEG carrier concentration on the alloy composition can be approximated very well by:

$$n_s(x) = \left[-0.169 + 2.61x + 4.50x^2 \right] 10^{13}\, cm^{-2}, \quad x > 0.06. \tag{57}$$

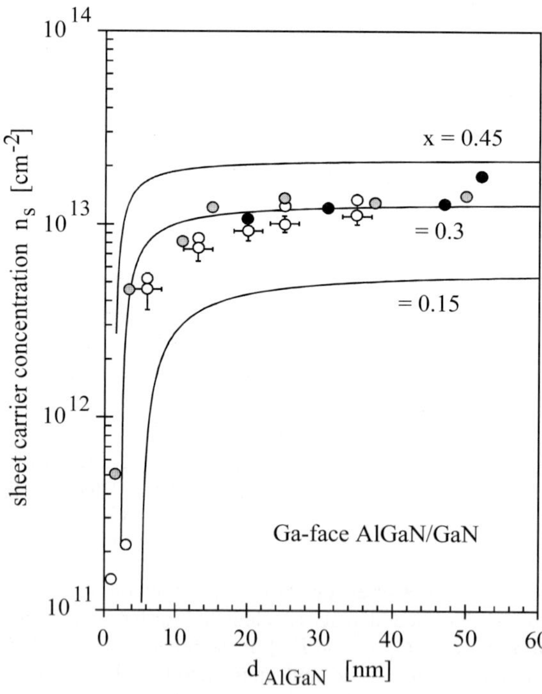

Fig. 12 The dependence of $n_s(x)$ on the barrier thickness calculated using equation (50) for x = 0.15, 0.30, and 0.45 are shown. A drastic decrease in sheet carrier concentration is predicted for thicknesses below 12, 8 and 6 nm, respectively. For x = 0.3 the sheet carrier concentration of AlGaN/GaN heterostructures was measured by CV-profiling for barrier thicknesses between 1 and 50 nm. These data (white open symbols) and experimental results of other groups (black and grey symbols [53, 55]) are in good agreement with the theoretical predictions.

For thinner barriers the depletion by the Schottky contact or the influence of the surface charge of the heterostructure becomes important (for more information see Ref. [43]). In **Fig. 12**, the dependence of $n_s(x)$ on the AlGaN barrier thickness according to equation (50) for $x = 0.15$, 0.30 and 0.45 is shown. A drastic decrease in sheet carrier concentration is predicted for thicknesses below 12, 8 and 6 nm, respectively. For $x = 0.3$ the sheet carrier concentration of AlGaN/GaN HEMTs was measured by CV-profiling for barrier thicknesses between 1 and 50 nm. These data and experimental results of other groups are presented by the data points in **Fig. 12**. A good agreement between our theoretical predictions and the experimental results is observed. More important, the non-linear prediction reproduces the measured data accurately, avoiding almost all of the residual deviations from experiment observed in former approaches [13, 56, 57]. The predicted piezoelectric and spontaneous polarization therefore can be considered as rather reliable for AlGaN/GaN heterostructures.

Besides a detailed understanding of the formation of polarization induced 2DEGs and how to control the sheet carrier concentration, for devices a basic knowledge of the electronic transport properties is important. We have performed Shubnikov-de Haas (SdH) and Hall effect measurements to investigate the electronic transport properties of polarization induced 2DEGs in $Al_xGa_{x-1}N$/GaN heterostructures with alloy compositions between $x = 0.10$ and 0.35 and sheet carrier concentrations between 2×10^{12} and 2×10^{13} cm^{-2}, respectively. To evaluate the transport properties of 2DEGs with different sheet carrier concentrations, we first have performed temperature dependent Hall-measurements (**Fig. 13**).

At room temperature 2DEG mobilities between 1320 and 1610 cm^2/Vs are observed, increasing proportional to $1/T^3$ with decreasing temperature. For temperatures below 100 K, a further decrease in temperature results in no or only a small enhancement of the mobility, indicating significant scattering of electrons due to neutral defects. For a sheet carrier concentration of 2.1×10^{12} cm^{-2}, a maximum mobility of 28000 cm^2/Vs is obtained below 10 K. In contrast to AlGaAs/GaAs heterostructures for which the low

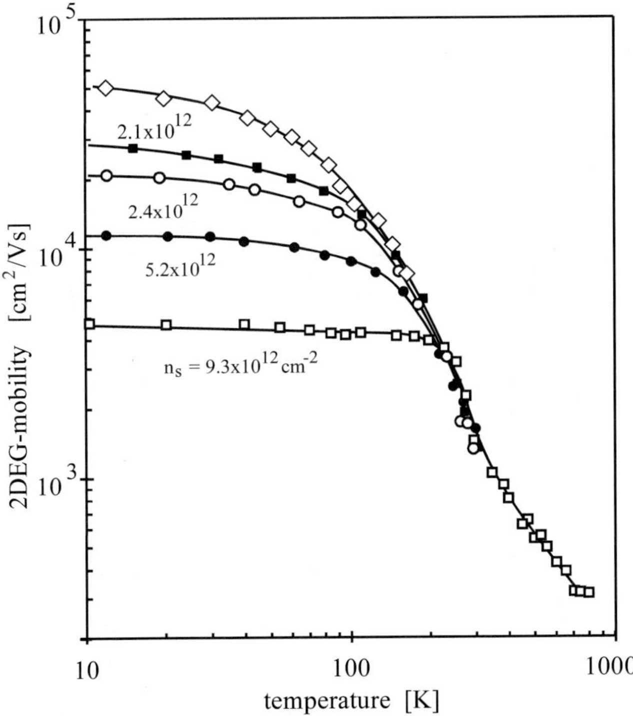

Fig. 13 2DEG Hall-mobility versus temperature for AlGaN/GaN-heterostructures with different sheet carrier concentrations. The low temperature mobility is increasing with decreasing 2DEG density.

temperature mobility, μ_{2DEG}, is increasing with increasing sheet carrier concentration due to enhanced screening of charged defects, a decrease of 2DEG mobility is observed for AlGaN/GaN heterostructures with increasing sheet carrier concentration, n_s ($\mu_{2DEG} \propto n_s^{(-2.1\pm0.2)}$), corresponding to an increase of the Al content in the barrier. These observations were confirmed by Shubnikov-de-Haas (SdH) experiments as discussed in the following.

The effective electron mass, m^*, is a fundamental parameter in electronic transport, as quantum scattering time, effective g-factor and transport scattering time are mass dependent. Applying the theory developed by Ando, Fowler and Stern [58], the oscillatory part of the longitudinal resistance, ΔR_{xx}, can be written as

$$\frac{\Delta R_{xx}}{R_0} = 4\frac{X(T)}{\sinh X(T)}\exp(-\frac{\pi}{\omega_c\tau_q}) , \tag{58}$$

where $X(T) = 2\pi^2 k_B T / \hbar\omega_c$, R_0 is the 2DEG resistance at zero magnetic field B, $\omega_c = eB/m^*$ is the cyclotron frequency, and τ_q is the quantum scattering time. Thus the effective mass can be deduced from the decay of the SdH amplitude A with increasing temperature as follows:

$$\ln\left(\frac{A}{T}\right) \approx C - \frac{2\pi^2 k_B m^*}{e\hbar B}T , \tag{59}$$

where C is a temperature independent constant. From fits of the SdH amplitudes measured at a fixed magnetic field but at different temperatures we obtain $m^* = (0.24 \pm 0.02)m_0$ and $(0.207 \pm 0.013)m_0$ for

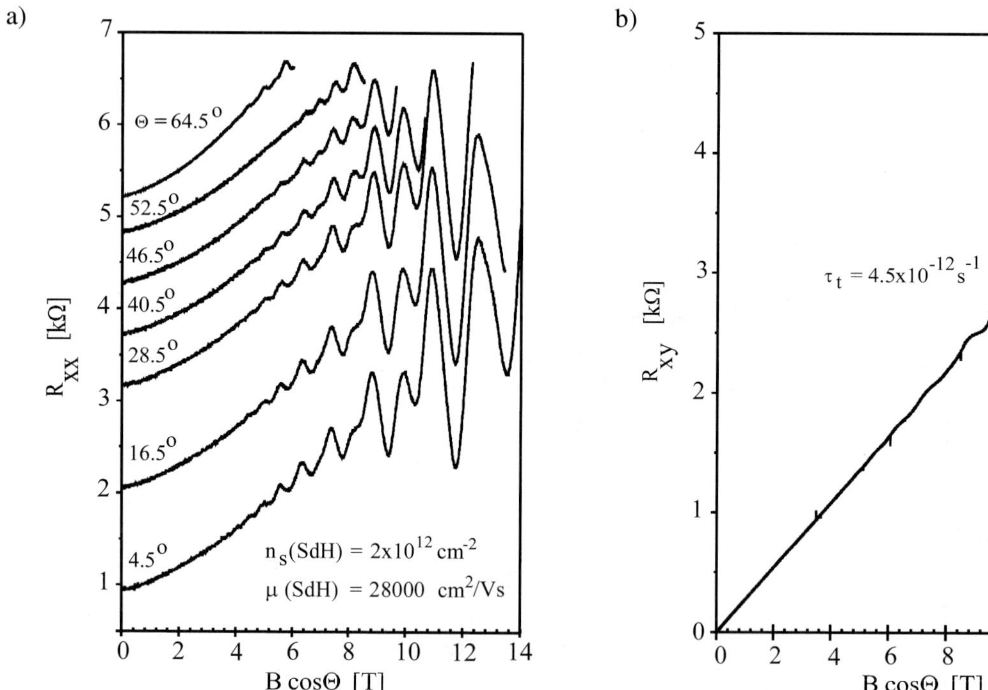

Fig. 14 Shubnikov-de Haas investigations of electronic transport properties of polarization induced 2DEGs in pseudomorphic, Ga-face Al$_{0.1}$Ga$_{0.9}$N/GaN heterostructures. a) Longitudinal resistance for different tilt angles between sample and magnetic field. b) Vertical resistance versus magnetic field, if the direction of the magnetic field is parallel to the surface normal of the sample.

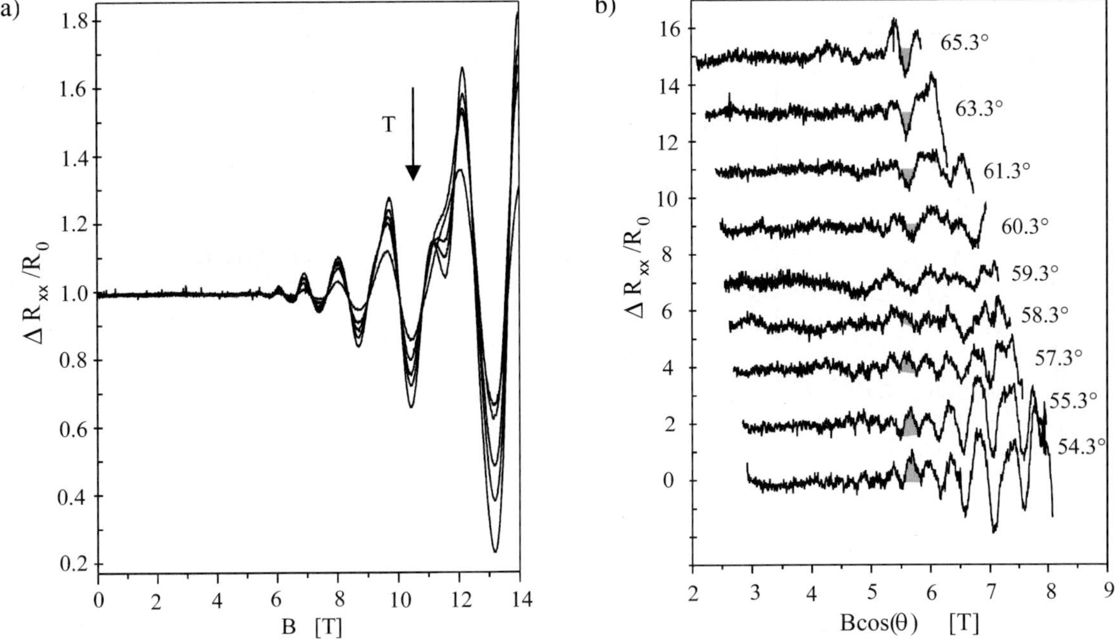

Fig. 15 a) SdH oscillations of R_{xx} measured at T = 0.5 K, 1.3 K, 2.5 K, 4.5 K, and 8.5 K in order to determine the effective electron mass of a 2DEG with a sheet carrier concentration of n_s = 4.6x10^{12} cm^{-2}. b) Normalized longitudinal resistance for different tilt angles between sample and magnetic field, showing phase shift and vanishing amplitude at an angle of about 60°.

2DEGs with n_s = 2.1x10^{12} and 4.6x10^{12} cm^{-2}, respectively. These results are in good agreement with effective masses determined for 2DEGs with similar sheet carrier concentrations determined by other groups [59, 60].

Based on these values for the effective mass, the effective Lande factor g^*, was determined by performing magnetotransport measurements in tilted magnetic fields [61]. SdH measurements of polarization induced 2DEGs with relatively low sheet carrier concentrations (n_s(Hall) = 4.6x10^{12} cm^{-2}) and high electron mobility at low temperature, show clear evidence of spin splitting in the SdH oscillations. By tilting the AlGaN/GaN samples inside the magnetic field we can take advantage of the fact that the Landau level splitting is proportional only to the field perpendicular to the 2DEG plane $B_\perp$, while the spin splitting is dependent on the total field B:

$$E = (n+\frac{1}{2})\frac{\hbar e}{m^*}B_\perp \pm \frac{1}{2}g^*\mu_B B ,$$

(60)

where E is the energy, μ_B the Bohr magneton, , $B_\perp = B\cos(\Theta)$ and Θ is the angle between the direction of the magnetic field and the surface normal of the sample. There exists an angle Θ_0, where the spin splitting is equal to half of the cyclotron energy. At this angle the SdH amplitude goes to zero and the SdH phase changes by π. In **Fig.15** normalized SdH oscillations are plotted for different tilt angles. With a zero angle of about 59.5° we can calculate an effective g*-factor of 2.11 ± 0.23, using the expression

$$g^* = \frac{\hbar e}{2\mu_B m^*}\cos\Theta_0 .$$

(61)

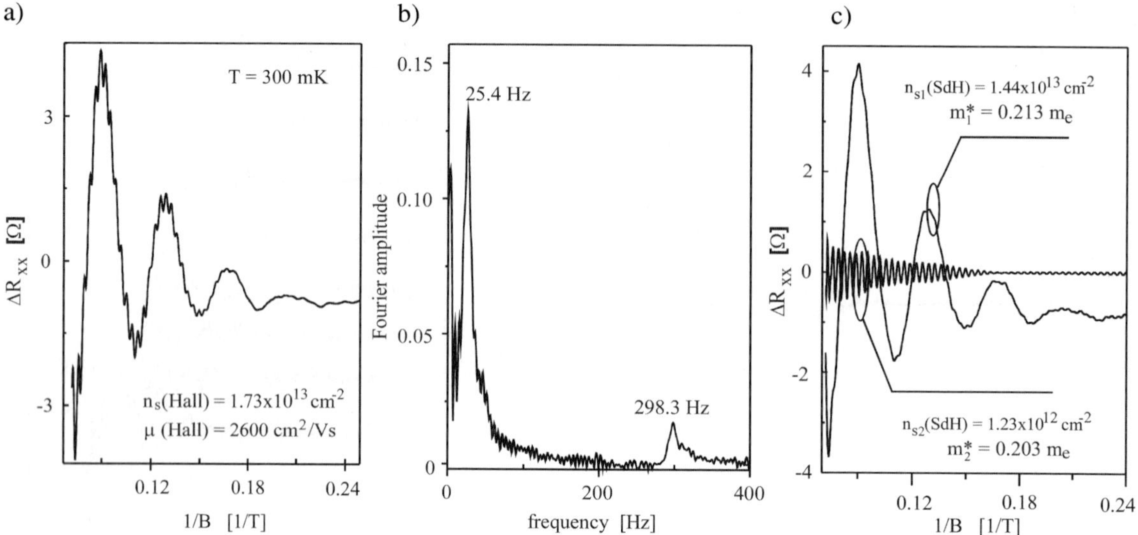

Fig. 16 a) Longitudinal resistance versus reciprocal magnetic field for a pseudomorphic, Ga-face Al$_{0.37}$Ga$_{0.63}$N/GaN heterostructure. b) The Fourier analysis of the SdH oscillations indicates the occupation of the second subbands by electrons. c) Deconvolution of the SdH oscillations allows the determination of effective mass and electron sheet carrier concentrations of two occupied subbands.

In addition, we have observed an exact $\cos(\Theta)$-dependence of the SdH-Oscillations at lower tilt angles, which is a proof for the two-dimensional nature of the polarization induced electron system accumulated at the AlGaN/GaN interface.

SdH measurements of polarization induced 2DEGs with very high sheet carrier concentrations ($n_s(Hall) = 1.73\times10^{13}$ cm^{-2}) show clear signs of the occupation of a second subband. A Fourier analysis of the SdH oscillations reveals two peaks (**Fig. 16**), which correspond to sheet carrier concentrations of 1.44 x 10^{13} cm^{-2} and 1.23 x 10^{12} cm^{-2}. The two frequencies in the Fourier spectrum are well separated, so that frequency filtering allows to investigate separately the contributions n_{s1} and n_{s2} of both peaks to R_{xx}. The result is shown in **Figure 16**. One line represents the low frequency part, the small oscillation around zero the high frequency part of the spectrum. The sum of both are the R_{xx} oscillations as measured (an additional, slowly varying B^2-background was subtracted first). This allows us to calculate the quantum scattering times and effective masses for both subbands. The quantum scattering times, extracted from Dingle plots are 0.143 ps and 0.120 ps for the first and second subband, respectively. The effective masses, which were calculated from the decrease of the SdH amplitude with rising temperature, are also very similar: $m_1^* = 0.213$ m$_0$ and $m_2^* = 0.203$ m$_0$, respectively. The energy separation between the two subbands,

$$\left|E_2 - E_1\right| = \pi\hbar^2 \left(\frac{n_{s2}}{m_2^*} - \frac{n_{s1}}{m_1^*}\right), \tag{62}$$

[62] is ΔE = 147 meV.

Further important physical properties which can be deduced from magnetotransport measurements in dependence of the 2DEG sheet carrier concentration are the electron transport and quantum scattering time, already mentioned above. The quantum scattering time, which takes into account all scattering events, can be extracted from the SdH amplitude dependence on the magnetic field. According to Eq. (58), the logarithm of $\Delta R_{xx} \sinh(X(T))/4R_0 X(T)$ is linearly dependent on the inverse magnetic field. The transport scattering time $\tau_t = \mu m^*/e$, is determined from the electron mobility using the measured electron effective mass.

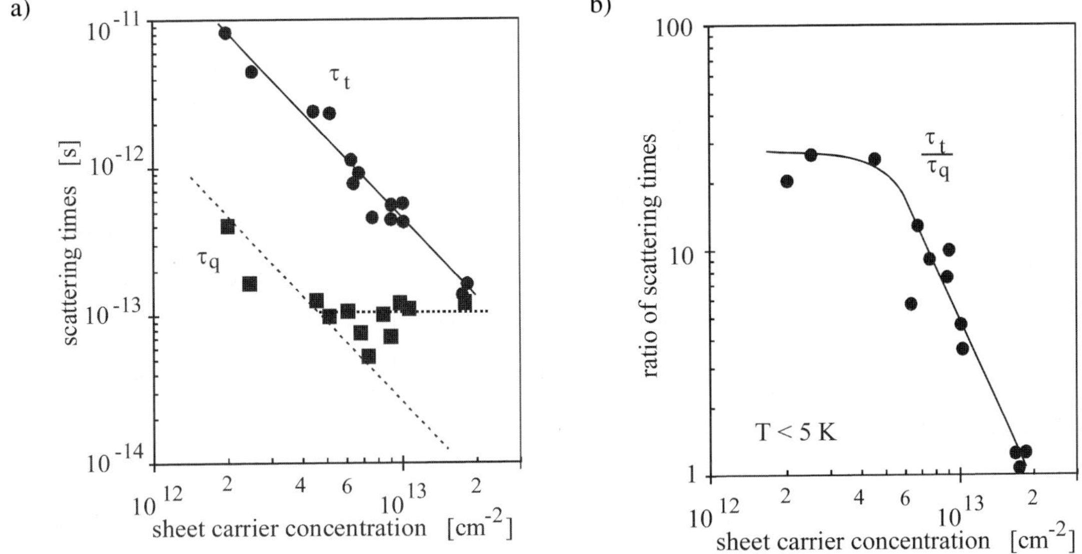

Fig. 17 a) Transport and quantum scattering time of polarization induced 2DEGs versus sheet carrier concentration measured at T < 5 K. The sheet carrier concentration was varied by changing the Al content of the AlGaN barrier. b) Ratio of scattering times versus sheet carrier concentration.

The transport scattering time is decreasing with a slope of about 2 in a semi logarithmic plot versus sheet carrier concentration (shown in **Fig. 17a** (solid line)) from 8×10^{-12}s to 4×10^{-13}s, when the sheet carrier concentration is increased from 2×10^{12} to 2×10^{13} cm^{-2}. The quantum scattering time (dashed line) follows the same trend for sheet carrier concentrations below 5×10^{12} cm^{-2} but is more than one order of magnitude smaller in comparison to the transport scattering time. The ratio between transport and quantum scattering time is about 25 for 2DEGs with $n_s < 5 \times 10^{12}$ cm^{-2} (**Fig. 17a** (dotted line), **Fig. 17b**), indicating that the dominant scattering mechanism is related to long range potential scattering, such as alloy scattering or Coulomb scattering (caused by impurities, charged defects or dislocations). It is significant that for higher sheet carrier concentrations the ratio τ_t/τ_q is reduced. This can be explained by the increasing influence of interface roughness scattering [63]. In undoped AlGaN/GaN heterostructures, n_s is enlarged by increasing the Al concentration of the barrier. The increase of the Al concentration causes an increase in the gradient of polarization at the AlGaN surface as well as at the AlGaN/GaN interface corresponding to an increase in polarization induced surface and interface charges. As a consequence, the electric field inside the GaN channel layer (close to the AlGaN/GaN interface) is enhanced, causing a stronger electron confinement and a reduction of the average distance of the electrons from the interface. At the same time charged point defects, or charged dislocations and extended defects in or close to the GaN channel are screened more efficiently by the higher electron concentration of the 2DEG. In any case it is obvious that the dominant scattering mechanism for polarization induced 2DEGs with low sheet carrier concentrations mainly, gives rise to small angle scattering, whereas for $n_s > 5 \times 10^{12}$ cm^{-2} large angle scattering becomes more significant.

9   AlGaN/GaN heterostructure electronics

AlGaN/GaN heterostructures with high sheet carrier concentrations of about 10^{13} cm^{-2} have been a subject of intense recent investigation and, have emerged as attractive candidates for high electron mobility transistors (HEMTs) operating at high voltage and high-power at microwave frequencies [7, 14, 16-19]. Although the maximum electron drift mobility of two-dimensional electron gases with sheet carrier concentrations of about 5×10^{12} cm^{-2} in GaN is predicted and measured to about 2000 [61] and

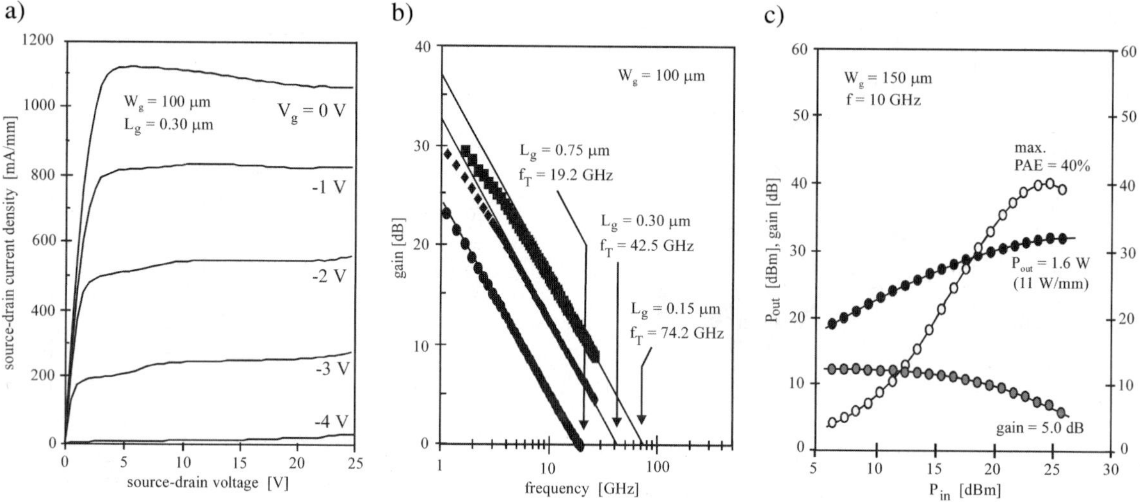

Fig. 18 a) Source-Drain current versus voltage at different gate voltages of an $Al_{0.3}Ga_{0.7}N$/GaN based HEMT with a gate width and length of 100 μm and 0.3 μm, respectively. b) Determination of cut-off frequency for gate length between 0.75 and 0.15 μm. c) Gain, output power and efficiency versus input power at an operation frequency of 10 GHz of a HEMT structure with a gate width and length of 150 μm and 0.3 μm, respectively.

18000 cm^2/Vs [64] at 300 and 77 K, respectively, which is much less than that of GaAs [65], GaN has a larger peak electron velocity, larger saturation velocity, higher breakdown voltage and thermal stability, making this material very suitable for use as channel-material in microwave power devices [66]. To evaluate the potential of pyroelectric heterostructures for electronic devices, we have processed HEMTs using pseudomorphic, Ga-face and undoped AlGaN/GaN heterostructures.

The processing of the HEMT's has been carried out using the following steps. First, the active region mesas are separated using ECR etching with Cl_2. The etch depth (usually 150 nm) is greater than the thickness of the AlGaN barrier layer. Next, the ohmic contacts are made using sequential evaporation of Ti (20 nm), Al (100 nm), Ti (45 nm), and Au (55 nm), which are then annealed at 800°C for 1 minute. The transfer resistance of these contacts is 0.3-0.5 Ωmm, depending on the 2DEG sheet resistance. A tri-level resist set of PMMA is used to form a Ni (20 nm)/Au (370 nm) gate with 0.3 μm footprint, and ≤ 200 Ω/mm DC resistance (per gate width).

The surface potential of the pyroelectric AlGaN/GaN HEMT structures has to be stabilized in order to reduce the number of surface states and to remove mobile surface charge coming from the ambient (compensating the negative polarization induced surface charge of the barrier). At high operation frequencies, mobile surface charge can couple inductively to the 2DEG causing a drastic reduction of the cut-off frequency and the output power. The stabilization of the surface has been accomplished by depositing an insulating film of Si_3N_4 (150 nm) by plasma enhanced chemical vapor deposition (PECVD) at 300°C [14]. As mentioned above, contributing to the low sheet resistance is the ability to achieve polarization induced 2DEG´s with sheet carrier concentrations up to $2x10^{13}$ cm^{-2}, well above those observed in other III-V material systems. These high sheet carrier concentrations enable the fabrication of transistors with gate length between 0.75 and 0.15 μm, channel current densities above 1000 mA/mm, breakdown voltages of up to 35 V and cut off frequencies of $f_t = 74$ GHz and $f_{max} = 140$ GHz (**Fig.18**). HEMT's with 0.3 μm gate length were tested in continuous wave at operation frequencies between 8 and 10 GHz for high normalized saturated power. A single, 150 μm wide channel, with its gate center fed, yielded 11.2 W/mm at 45 V drain-source bias at 32% power-added efficiency. The performance of the device was mainly limited by self heating, indicating that a further improvement of the output power can be achieved by a more efficient heat removal out of the transistor channel. A HEMT with two parallel, 100 μm wide channels separated by 50 μm, yielded 8.8 W/mm at 40 V drain-source bias, at 48% power-added efficiency. Thus polarization induced AlGaN/GaN HEMT's are capable of providing 6-12 W/mm normalized output power, and are predicted to yield up to 50 times as much power into the same load resistance as GaAs HEMT's. The measured performance of these pyroelectronic devices strongly indicates the potential of AlGaN/GaN HEMT´s for microwave amplifiers operating at high frequencies and high power.

10 Summary

Wurtzite GaN and AlGaN are pyroelectric materials with large spontaneous polarization and piezoelectric coefficients. We have presented theoretical and experimental evidence for the non-linearity of spontaneous and piezoelectric polarization in nitride alloys as a function of strain and composition. We have applied our improved theory to model experimental data obtained from a number of AlGaN/GaN transistor structures. We have found that the discrepancies between experiment and ab-initio theory are almost completely eliminated for the AlGaN/GaN based heterostructures when polarization non-linearity is taken into account. To facilitate inclusion of the predicted non-linear polarization in future simulations, we have given an explicit prescription to calculate polarization induced surface and interface charges as well as electric fields for arbitrary compositions of AlGaN random ternary alloys.

In undoped pseudomorphic AlGaN/GaN heterostructures gradients in spontaneous and piezoelectric polarization exist, which cause high sheet concentrations of bound surface and interface charges. The positive polarization induced charge located at the interface of Ga-face AlGaN/GaN heterostructures can be compensated by free electrons leading to a formation of 2DEG´s with high sheet carrier concentrations. We have determined important physical properties relevant to understand the electronic transport properties of polarization induced 2DEGs in AlGaN/GaN heterostructures, like the effective electron mass, effective g-factor, transport and quantum scattering time by a combination of CV-profiling, Hall effect and SdH measurements, indicating different dominant scattering mechanism for 2DEGs with low and high sheet carrier concentrations.

Mainly due to the high electron concentration, the high breakdown field, and an efficient heat removal, AlGaN/GaN HEMT structures on SiC are capable of high microwave power and high efficiency when passivated with Si_3N_4. State-of-the-art microwave power levels of 11 W/mm have been reached at 10 GHz. The power and power-added efficiency levels have approached theoretical levels, with self-heating temperature rise, causing an increase in the knee voltage, as the key remaining limit.

Acknowledgments   The work was partly supported by the German Science Foundation and the Thuringian Ministry of Science, Research, and Art. The calculations performed at Cagliari University were supported in part by MURST, Cofin99 and the INFM Parallel Supercomputing Initiative. The experiments done at Cornell University and the Technical University Ilmenau are supported by the Office of Naval Research under the direction of Dr. C. Wood.

References

[1] M. Razeghi, A. Rogalski, J. Appl. Phys. **79**, 7433 (1996).
[2] H. J. Looi, M. D. Whitfield, R. B. Jackman, Appl. Phys. Lett. **74**, 3332 (1999).
[3] U. Karrer, A. Dobner, O. Ambacher, M. Stutzmann, J. Vac. Sci. Technol. B **18**, 757 (1999).
[4] A. Lloyd-Spetz, A. Baranzahi, P. Tobias, I. Lundström, phys. stat. sol. (a) **162**, 493 (1997).
[5] F. Bernardini, V. Fiorentini, and D. Vanderbilt, Phys. Rev. B **56**, R10024 (1997).
     F. Bernardini, V. Fiorentini, and D. Vanderbilt, Phys. Rev. B **63**, 193201-5 (2001).
[6] A. Zoroddu, F. Bernardini, P. Ruggerone, and V. Fiorentini, Phys. Rev. B. **64**, 45208 (2001).
[7] O. Ambacher, J. Phys. D: Appl. Phys. **31**, 2653 (1998).
[8] M. A. Dubois and P. Muralt, Appl. Phys. Lett. **74**, 3032 (1999).
[9] M. S. Shur, A. D. Bykhovski, and R. Gaska, M. R. S. Inter. J. Nitr. Semic. Res. **4S1**, G1.6 (1999).
[10] C. Wetzel, T. Takeuchi, S. Yamaguchi, H. Katoh, H. Amano, and I. Akasaki, Appl. Phys. Lett. **73**, 1994 (1998).
[11] J. S. Im, V. Härle, F. Scholz, A. Hangleiter, Mat. Res. Soc. Inter. J. Nitr. Semi. Res. **1**, 37 (1996).
[12] N. Grandjean, J. Massies, and M. Leroux, Appl. Phys. Lett. **74**, 2361 (1999).

[13] O. Ambacher, J. Smart, J. R. Shealy, N. G. Weimann, K. Chu, M. Murphy, W. J. Schaff, L. F. Eastman, R. Dimitrov, L. Wittmer, M. Stutzmann, W. Rieger, and J. Hilsenbeck, J. Appl. Phys. **85**, 3222 (1999).

[14] L. F. Eastman, V. Tilak, J. Smart, B. M. Green, E. M. Chumbes, R. Dimitrov, H. Kim, O. Ambacher, N. Weimann, T. Prunty, M. Murphy, W. J. Schaff, and J. R. Shealy, IEEE Trans. Electr. Dev. **48**, 479 (2001).

[15] A. Ozgur, W. Kim, Z. Fan, A. Botchkarev, A. Salvador, S. N. Mohammad, B. Sverdlov, and H. Morkoc, Electron. Lett. **31**, 1389 (1995).

[16] M. A. Khan, Q. Chen, M. S. Shur, B. T. M. Dermott, J .A. Higgins, J. Burm, W. J. Schaff, and L. F. Eastman, IEEE Electron Device Lett. **17**, 584 (1996).

[17] S. C. Binari, J. M. Redwing, G. Kelner, and W. Kruppa, Electron. Lett. **33**, 242 (1997).

[18] R. Gaska, Q. Chen, J. Yang, A. Osinsky, M. A. Khan, and M. S. Shur, IEEE Electron Device Lett. **18**, 492 (1997).

[19] Y. F. Wu, S. Keller, P. Kozodoy, B. P. Keller, P. Parikh, D. Kapolnek, S. P. DenBaars, and U. K. Mishra, IEEE Electron Device Lett. **18**, 290 (1997).

[20] R. Dimitrov, L. Wittmer, H. P. Felsl, A. Mitchell, O. Ambacher, and M. Stutzmann, phys. stat. sol. (a) **168**, R7 (1998).

[21] L. F. Eastman, V. Tilak, J. Smart, B. Green, A. Vertiatchikh, N. Weimann, O. Ambacher, E. Chumbes, H. Kim, T. Prunty, J.H. Hwang, W. J. Schaff, B. K. Ridley, V. Kaper and J. R. Shealy, Proc. of 28th International Symposium on Compound Semiconductors, ISCS 2001, Tokyo, October, 2001.

[22] R. Neuberger, G. Müller, O. Ambacher, and M. Stutzmann, phys. stat. sol. (a) **R1**, 6 (2001).

[23] J. Schalwig, G. Müller, U. Karrer, M. Eickhoff, O. Ambacher, M. Stutzmann, L. Görgens, G. Dollinger, Appl. Phys. Lett. **80**, 1222 (2002).

[24] G. Kresse, and J. Hafner Phys. Rev. B **47**, 558 (1993).

[25] G. Kresse, and J. Furthmuller, Comp. Mater. Sci. **6**,15 (1996).

[26] R. D. King-Smith and D. Vanderbilt, Phys. Rev. B **47**, (1993) 1651.

[27] R. Resta, Rev. Mod. Phys. **66**, (1994) 899.

[28] J. P. Perdew, in Elecronic Structure of Solids '91, edited by P. Ziesche and H. Eschrig, Akademie-Verlag, Berlin, p 11 (1991).

[29] S. H. Wei, L .G. Ferreira, J. E. Bernard, and A. Zunger, Phys. Rev. B **42**, 9622 (1990).

[30] L. Görgens, O. Ambacher, M. Stutzmann, C. Miskys, F. Scholz, and J. Off, Appl. Phys. Lett. **76**, 577 (2000).

[31] H. Angerer, D. Brunner, F. Freudenberg, O. Ambacher, M. Stutzmann, R. Höpler, T. Metzger, E. Born, G. Dollinger, A. Bergmaier, S. Karsch, and H.-J. Körner, Appl. Phys. Lett. **71**, 1504 (1997).

[32] L. Bellaiche, S. H. Wei, and A. Zunger, Phys. Rev. **56**, 13872 (1997).

[33] T. Mattila and A. Zunger, J. Appl. Phys. **85**, 160 (1999).

[34] K. P. O´Donnell, J. F. W. Mosselmans, R. W. Martin, S. Pereira, and M. E. White, J. Phys.: Condensed Matter **13**, 6977 (2001).

[35] F. Bernardini and V. Fiorentini, Phys. Rev. B **64**, 85207 (2001).

[36] A. F. Wright, J. Appl. Phys. **82**, 2833 (1997).

[37] G. D. O´Clock and M. T. Duffy, Appl. Phys. Lett. **23**, 55 (1973).

[38] J.-M. Wagner and F. Bechstedt, Phys. Rev. B. **66**, 115202 (2002).

[39] J. F. Nye, Physical Properties of Crystals; Their Representation by Tensors and Matrices, Clarendon, Oxford, (1985).

[40] C. Deger, E. Born, H. Angerer, O. Ambacher, M. Stutzmann, J. Hornsteiner, E. Riha and G. Fischerauer, Appl. Phys. Lett. **72**, 2400 (1998).

[41] K. Tsubouchi, K. Sugai and N. Mikoshiba, IEEE Ultrason. Symp. **90**, 375 (1981).

[42] M. Rotter, A. Wixforth, A. O. Govorov, W. Ruille, D. Bernklau, and H. Riechert, Appl. Phys. Lett. **75**, 965 (1999).

[43] B.K. Ridley, O. Ambacher, and L. F. Eastman, Semicond. Sci. Technol. **15**, 270 (2000),
J. P. Ibbetson, P. T. Fini, K. D. Ness, S. P. DenBaars, J. S. Speck, and U. K. Mishra, Appl. Phys. Lett. **77**, 250 (2000).

[44] F. Bernardini, V. Fiorentini, Phys. Rev. Lett. **79**, 3958 (1997).

[45] L. S. Yu, D. J. Qiao, Q.J. Xing, S. S. Lau, K. S. Boutros, and J. M. Redwing, Appl. Phys. Lett. **73**, 238 (1998).

[46] D. Brunner, H. Angerer, E. Bustarret, R. Höpler, R. Dimitrov, O. Ambacher, and M. Stutzmann, J. Appl. Phys. **82**, 5090 (1997).

[47] G. Martin, S. Strite, A. Botchkaev, A. Agarwal, A. Rockett, H. Morkoc, W. R .L. Lambrecht, B. Segall, Appl. Phys. Lett. **65**, 610 (1994).

[48] G. Martin, A. Botchkarev, A. Rockett, H. Morkoc, Appl. Phys. Lett. **68**, 2541 (1996).

[49] L. W. Wong, S. J. Cai, R. Li, K. Wang, H. W. Jiang, and M. Chen, Appl. Phys. Lett. **73**, 1391 (1998).

[50] B. E. Foutz, M. J. Murphy, O. Ambacher, V. Tilak, J. Smart, J. R. Shealy, W. J. Schaff, and L. F. Eastman, Mat. Res. Soc. Proc. **572**, 501 (1999).
[51] J. A. Smart, A. T. Schremer, N. G. Weimann, O. Ambacher, L. F. Eastman, and J. R. Shealy, Appl. Phys. Lett. **75**, 388 (1999).
[52] M. J. Murphy, K. Chu, H. Wu, W. Yeo, W. J. Schaff, O. Ambacher, J. Smart, J. R. Shealy, L. F. Eastman, and T. J. Eustis, J. Vac. Sci. Technol. B **17**, 1252 (1999).
[53] I. P. Smorchkova, C. R. Elsass, J. P. Ibbetson, R. Vetury, B. Heying, P. Fini, E. Haus, S. P. DenBaars, J. S. Speck, and U. K. Mishra, J. Appl. Phys. **86**, 4520 (1999).
[54] O. Ambacher, B. Foutz, J. Smart, J. R. Shealy, N. G. Weimann, K. Chu, M. Murphy, A. J. Sierakowski, W. J. Schaff, and L. F. Eastman, R. Dimitrov, A. Mitchell, and M. Stutzmann, J. Appl. Phys. **87**, 334 (2000).
[55] Z. Bougrioua, J.-L. Farvacque, I. Moerman and F. Carosella, phys. stat. sol. (b) (2003) in press.
[56] P. M. Asbeck, E. T. Yu, S. S. Lau, G.J. Sullivan, J. Van Hove, and J. M. Redwing, Electron Lett. **33**, 1230 (1997).
[57] E. T. Yu, G. J. Sullivan, P. M. Asbeck, C. D. Wang, D. Qiao, and S. S. Lau, Appl. Phys. Lett. **71**, 2794 (1997).
[58] T. Ando, A.. B. Fowler, and F. Stern, Rev. Mod. Phys. 54, 437 (1982).
[59] Y. J. Wang, R. Kaplan, H. K. Ng, K. Doverspike, D. K. Gaskill, T. Ikedo, I. Akasaki, and H. Amano, J. Appl. Phys. **79**, 8007 (1996).
[60] W. Knap, S. Contreras, H. Alause, C. Skierbiszewski, J. Camassel, M. Dyakonov, J. L. Robert, J. Yang, Q. Chen, M. A. Khan, M.L. Sadowski, S. Huant, F. H. Yang, M. Goiran, J. Leotin, and M.S. Shur, Appl. Phys. Lett. **70**, 2123 (1997).
[61] W. Knap, E. Frayssinet, M. L. Sadowski, C. Skierbiszewski, D. Maude, V. Falko, M. A. Khan, and M.S. Shur, Appl. Phys. Lett. **75**, 3156 (1999).
[62] M. Ahoujja, S. Elhamri, R. S. Newrock, D. B. Mast, W. C. Mitchel, I. Lo, A. Fathimulla, J. Appl. Phys. **81**, 1609 (1997).
[63] R. Oberhuber, G. Zandler, and P. Vogl, Appl. Phys. Lett.**73**, 818 (1998).
[64] Y. Zhang and J. Singh, J. Appl. Phys. **85**, 587 (1999).
[65] M. Shur, *GaAs Devices and Circuits* (Plenum, New York, 1986).
[66] B. E. Foutz, L. F. Eastman, U. V. Bhapkar, and M. S. Shur, Appl. Phys. Lett. **70**, 2849 (1997).

phys. stat. sol. (c) **0**, No. 6, 1908–1918 (2003) / **DOI** 10.1002/pssc.200303139

Electronics and sensors based on pyroelectric AlGaN/GaN heterostructures

Part B: Sensor applications

M. Eickhoff [1,*], **J. Schalwig** [2], **G. Steinhoff** [1], **O. Weidemann** [1], **L. Görgens** [1], **R. Neuberger** [2], **M. Hermann** [1], **B. Baur** [1], **G. Müller** [2], **O. Ambacher** [3], and **M. Stutzmann** [1]

[1] Walter Schottky Institute, TU-Munich, Am Coulombwall, 85748 Garching, Germany
[2] EADS Germany AG, IRT/LG-MX, P.O. Box 800465, 81663 Munich, Germany
[3] Center for Micro- and Nanotechnologies, Technical University Ilmenau, 98688 Ilmenau, Germany

Received 4 March 2003, accepted 20 June 2003
Published online 28 August 2003

PACS 62.20.Dc, 68,60.Bs, 71.20.Nr, 72.20.Jv, 77.77.+a, 77.84.Bw

In the present article recent results concerning sensor applications of AlGaN layers and AlGaN/GaN heterostructures are summarized. The piezoresistive effect in piezoelectric AlGaN layers is investigated and the dependence of the piezoresistive gauge factor on the Al content is attributed to the influence of strain induced piezoelectric fields. An enhancement of this effect is observed in AlGaN/GaN heterostructures, resulting in high longitudinal gauge factors.

The response of gas sensitive Pt:GaN Schottky diodes to hydrogen and hydrogen containing gases is analyzed up to temperatures of 600°C and employed to realize gas sensitive field effect transistors which are demonstrated to operate up to 400°C. In addition, ion sensitive field effect transistors (ISFETs) have also been fabricated on the basis of AlGaN/GaN heterostructures. The GaN surface shows a high pH sensitivity which is attributed to the presence of a thin native metal oxide layer on the surface.

1   Introduction

In the first part of this review the application of the group III-nitride system for the realization of high frequency and high power electronic devices based on AlGaN/GaN heterostructure high electron mobility transistors (HEMTs) has been outlined. In addition to these applications, wide bandgap semiconductors like silicon carbide, diamond or group III-nitrides exhibit material properties which are very favourable for the realization of different sensors operating at high temperatures or under harsh environment conditions [1-4]. Up to now, the main interest has been focussed on optoelectronic applications [5] of Al(In)GaN compounds as their bandgap can be engineered by adjustment of the Al(In)-content. Recent publications have summarized their applicability in solar blind UV detectors [6, 7].

In addition to the pyroelectric and piezoelectric material properties of AlGaN thin films, which allow the fabrication of HEMT structures with a polarization induced two dimensional electron gas (2DEG), the heteroepitaxial growth of the group III-nitrides on electrically insulating sapphire substrates allows the application of simple planar device structures. Furthermore, the thermal expansion coefficient of the sapphire substrate is close to those of aluminum oxide or aluminum nitride ceramics frequently used as packaging materials for high temperature sensors. Thereby the complicated packaging technologies for high temperature sensors can be simplified. In addition, the low pyroelectric coefficients combined with

---

* Corresponding author: e-mail: Martin.Eickhoff@wsi.tu-muenchen.de, Phone: +49 89 28912889, Fax: +49 89 28912737

the wide bandgap of more than 3.4 eV are advantageous for sensor devices based on AlGaN/GaN heterostucture field-effect transistors, which display high thermal stability.

On the other hand, initial experiments have shown that the channel conductivity in AlGaN/GaN HEMT structures exhibits a high sensitivity towards changes of the surface potential. This effect has been used to detect the presence of polar fluid on the gate surface or to measure contaminations of non polar fluids with polar solvents [8] as well as for the detection of ionized gases [9]. Employing this high sensitivity, the application of the group III-nitride system for the realization of various kinds of surface sensitive sensors [10, 11] has been a subject of research in the last two years.

In the present article, we review recent achievements concerning mechanical and chemical sensor applications of AlGaN layers and undoped AlGaN/GaN heterostructure field effect transistor devices. We summarize experimental results on the piezoresistive effect in AlGaN layers and AlGaN/GaN heterostructures, the sensitivity of GaN Schottky diodes and gas sensitive field effect transistors (GasFETs) for the detection of hydrogen and hydrogen containing gases as well as the performance of GaN based ion sensitive field effect transistors (ISFETs) for highly sensitive pH measurements in electrolyte solutions. Apart from quantitative experimental analysis of the sensitivity we will describe the underlying detection mechanisms.

2 Piezoresistive effect of AlGaN layers and AlGaN/GaN heterostructures

Mechanical sensors such as pressure sensors or accelerometers measure the mechanical strain in a microstructure. The electronic sensor signal can be generated either via the piezoelectric or the piezoresistive effect. Whereas the piezoelectric transduction principle is used for dynamical pressure sensing, static pressure sensors usually employ the piezoresistive transducer principle. Piezoresistive wide bandgap semiconductor pressure sensors for operation at high temperatures have been mainly fabricated on the basis of SiC. Due to heteroepitaxial growth on silicon substrates, cubic silicon carbide sensor elements with a sizeable piezoresistive effect even at temperatures above 300°C [12] can easily be integrated into Si micromaching technologies [13, 14]. Hexagonal SiC layers homoepitaxially grown on bulk substrates also exhibit high temperature stable piezoresistive properties and have been employed for the fabrication of high temperature pressure sensor chips [15, 16].

First investigations of piezoresistivity in GaN layers [17] and AlGaN/GaN heterostructures [18,19] have revealed a piezoresistive effect comparable to that in SiC films, however, the mechanisms of piezoresistivity have not been investigated in detail.

The influence of the piezoelectric properties of AlGaN on the piezoresistive effect has been investigated by variation of the Al-content [20]. The effect of mechanical strain on piezoelectric fields further results in a high strain sensitivity of AlGaN/GaN HEMT structures.

In the case of piezoelectric AlGaN compounds, the lateral current flow can be modulated by perpendicular strain induced piezoelectric fields, allowing dynamic pressure measurements by current monitoring. This procedure is much less sensitive to external disturbances compared to measurements of strain-induced piezoelectric charges. The resulting (longitudinal) piezoresistive effect can be quantified by the piezoresistive gauge factor *GF*, defined by the following relation:

$$GF = \frac{1}{\varepsilon}\frac{\Delta R}{R},$$

where R is the sample resistance, ΔR the strain induced change of the resistance and ε the mechanical strain parallel to the electric field used to measure R. Different values have been reported for the piezoresistive gauge factors of GaN layers and a strong influence of doping type and concentration was observed [17,18,21]. Particularly it has been noticed that the frequency of the strain modulation has a remarkable impact on the resulting piezoresistive gauge factor, indicating that strain-induced piezoelectric polarization charges have a strong influence on the strain dependent conductivity. Measurements at different modulation frequencies, the influence of strain on the electronic characteristics of Schottky contacts [19] as well as different doping types of the underlying GaN layers cause inhomogenities in the reported gauge factors.

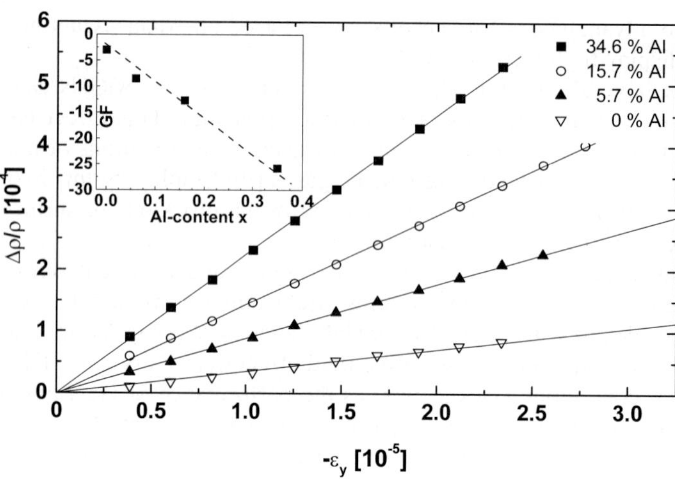

Fig. 1 Strain dependence of the specific resistance in 1 μm Ga(Al)-face AlGaN layers for different Al-contents. The inset shows the resulting longitudinal piezoresistive gauge factor. The measurement frequency was 30 Hz [20].

Fig. 1 shows the relative resistance changes measured for pure, 1 μm thick Ga(Al)-face $Al_xGa_{1-x}N$ layers with different Al concentrations, measured at a frequency of 30 Hz. For a Si-doped GaN layer ($N_e \sim 8 \cdot 10^{18}$ cm^{-3}) a gauge factor of −3 was measured. For higher Al-contents, the absolute value of the negative gauge factor increases, up to 25.8 for an Al mole fraction of 35 % ($N_e \sim 5 \cdot 10^{17}$ cm^{-3} at room temperature), which can be attributed either to the increasing piezoelectric constants or to a reduced density of free carriers [20]. This behaviour is explained in the model of a field effect transistor by considering the strain induced piezoelectric field as an external gate voltage. Within this model, the gauge factor for pieoresistive materials, *GFP*, is defined.

$$GFP(x) = \frac{1}{E_{PE,z}(x)} GF(x)$$

Here $E_{PE,z}(x)$ is the strain-induced piezoelectric field in the growth direction. Within the mentioned model, *GFP* does not depend on the piezoelectric coefficients. The weak dependence on the Al-content x is given via Poisson's ratio. *GFP* strongly depends on the carrier density, as displayed in the inset of **Fig. 2**.

The channel conductivity in AlGaN/GaN HEMT structures is governed by the carrier density of the polarization-induced 2DEG. Mechanical deformation of the sample causes a change of the polarization-induced carrier confinement and changes the channel depletion.

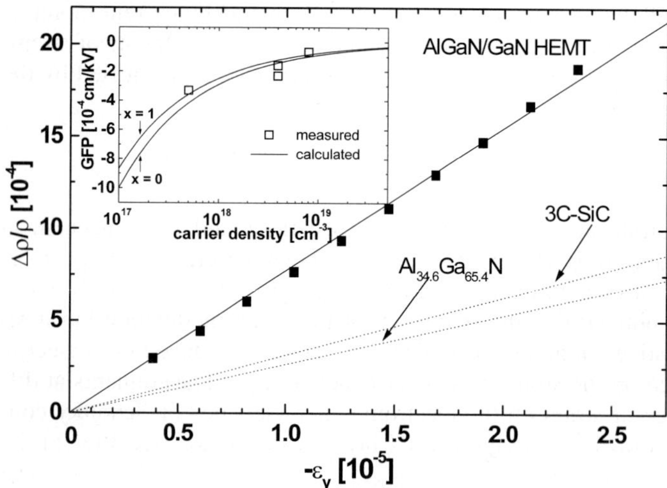

Fig. 2 Strain dependent conductivity of an AlGaN/GaN HEMT structure at modulation frequency of 30 Hz. The observed gauge factor of −85 is approximately three times higher than the highest values reported for 3C-SiC [12]. The inset shows the calculated and measured dependence of the gauge factor for piezoelectric materials, *GFP*, on carrier density [20].

The resulting strain sensitivity of the channel resistance is significantly higher, leading to a gauge factor of *GF = -85* for an inverted HEMT structure, which consists of a heterostructure with N-face polarity and contains a 45 nm thick $Al_{0.39}Ga_{0.61}N$ barrier as well as a 4 nm GaN cap layer ($n_{2DEG} = 9 \cdot 10^{12}\ cm^{-2}$), as shown in **Fig. 2**. The piezoresistive effect in AlGaN/GaN heterostructures has also been investigated in [19], where the measured gauge factor was more than one order of magnitude lower. It should be noted that the piezoresistive gauge factor of AlGaN/GaN hetrostructures is approximately three times higher than the highest gauge factors reported for 3C-SiC [12] or 6H-SiC [16].

From these results it can be concluded that the interaction of piezoelectric and piezoresistive properties in AlGaN layers and particularly in AlGaN/GaN heterostructures leads to high piezoresistive gauge factors. Due to the weak temperature dependence of the piezoelectric constants, it can be expected that the piezoresistive gauge factor of AlGaN/GaN heterostructures exhibits only a weak temperature dependence. However, for the realization of high temperature mechanical sensors based on AlGaN/GaN sensing layers, deposition has to be performed on micromachinable silicon substrates or on pre-structured sapphire, since micromachining technologies for the AlGaN system are currently not available.

3 Chemical sensors for gas detection

Due to the large electronic bandgap and the high chemical stability of AlGaN compounds, electronic devices based on this material system are capable of stable operation at high temperatures. As this is a basic requirement for the application in high temperature gas sensors, the AlGaN system is a promising material for the fabrication of such devices, utilizing either AlGaN Schottky diodes with a catalytic Schottky contact or AlGaN/GaN heterostructure field effect transistors with a catalytic gate metal. The first investigation of Pt:GaN Schottky diodes for the detection of hydrogen and propane up to temperatures of 400°C was reported in 1999 by Luther et al. [22]. In comparison to SiC-based high temperature gas sensors which have been studied in great detail in the last ten years [1,23], single sensor elements or complex sensor arrays based on GaN enable the application of simple planar device designs and lean packaging concepts due to growth on insulating sapphire substrates.

In this paragraph we report on recent achievements in the field of gas sensitive GaN Schottky diodes and AlGaN/GaN HEMT structures. We summarize the measured sensitivities to hydrogen and hydrogen containing gases as a function of device temperature and discuss the detection mechanism of hydrogen-sensitive Pd/GaN or Pt/GaN interfaces.

Schottky diodes were fabricated on 2 μm Si doped GaN layers ($N_e \approx 1 \cdot 10^{18}\ cm^{-3}$) grown by PIMBE as described elsewhere [24]. Ohmic contacts were structured on thermally evaporated Ti/Al/Ti/Au layers. After etching the sample surface by dipping in HCl and exposure to HF vapour (50%), catalytic Pt contacts with a thickness of 15 nm to 30 nm were e-beam evaporated, followed by an annealing for 30 min at 650°C in nitrogen atmosphere. Gas sensitive field effect transistors (GasFETs) were processed on AlGaN/GaN HEMT structures, grown according to the process described in the first part of this article [25].

As shown in **Fig. 3**, exposure of a Pt:GaN Schottky diode to molecular hydrogen causes an increase of both the forward and reverse current which can be attributed to a decrease in the Schottky barrier height [26]. In a constant current operation mode, this reduction of the Schottky barrier height results in a shift to lower bias voltages, which constitutes the sensor signal. This hydrogen induced shift was first observed for Si-based Pd:MOS-structures [27], and later for metal / insulator / silicon carbide (MISiC) diodes [28] with Pt Schottky contacts. It was explained by the formation of a dipole layer of atomic hydrogen, dissociated at the catalytic Schottky contact and adsorbed at the metal/SiO_2 interface [29,30] The hydrogen induced decrease of the barrier height in Pt:GaN diodes can be already observed at room temperature. However, the recovery times can be very long, typically several hours, as illustrated in **Fig. 4**. Shown in the same diagram is the the decrease of both response and recovery time with increasing device temperature. Almost immediate sensor response is observed for temperatures above 200°C (c.f. the curve corresponding to 300°C in **Fig. 4**).

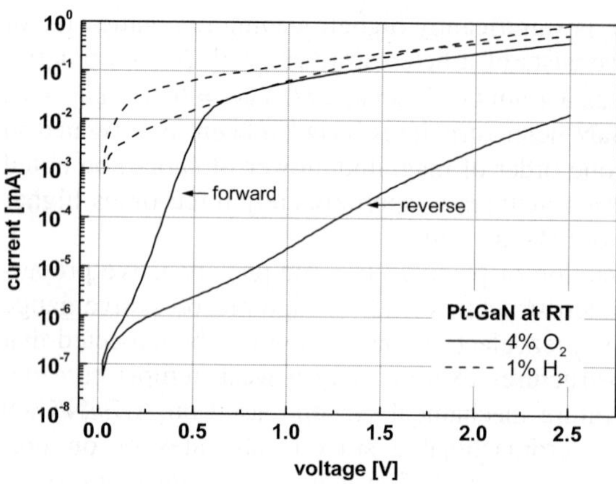

Fig. 3   Response of a Ga-face Pt:GaN Schottky diode to exposure to 1 % H_2 in 4 % O_2 at room temperature. The hydrogen induced decrease of the Schottky barrier height leads to an increase of both forward and reverse current [26]. The diode characteristics in a nitrogen atmosphere with 4% oxygen are shown in the solid curves.

An increase of the device temperature allows the detection of saturated and unsaturated hydrocarbons, as it is known from earlier works on gas sensitive MISiC diodes [23]. These gases have to be thermally cracked on the catalytic metal surface, which is a thermally activated process. In addition, when the activation temperature for desorption of hydrogen from the interface is achieved, a significant decrease of both response and recovery time of the device is observed.

In **Fig. 5**, the sensitivity of a Ga-face Pt:GaN Schottky diode to different constituents of motor vehicle exhaust gases in 4% oxygen atmosphere is plotted as a function of device temperature [31]. The device exhibits a considerable sensitivity towards hydrogen and, at elevated temperatures, towards unsaturated hydrocarbons. Surprisingly, only a weak sensor response to saturated hydrocarbons was observed. Also a sizeable response towards NO_2 (an oxidizing species) could be measured, which leads to a sensor signal of opposite sign. NO, on the other hand, produces a much smaller signal in the same direction as reducing gases. The low sensitivity towards saturated hydrocarbons increases significantly as soon as these reducing gases are offered in excess, i.e. the ratio of reducing and oxidizing gas concentration, α, becomes larger than unity. In 4% oxygen atmosphere this inset of sensitivity was observed at a concentration of 6200 ppm for butane and of 8000 ppm for propane at a device temperature of 500°C respectively [32] (not shown). Pt:GaN Schottky diodes processed on GaN layers with N-face polarity exhibit qualitatively the same characteristics, indicating that piezoelectric surface charges of the bare GaN layer do not affect the sensing mechanism noticeably [31].

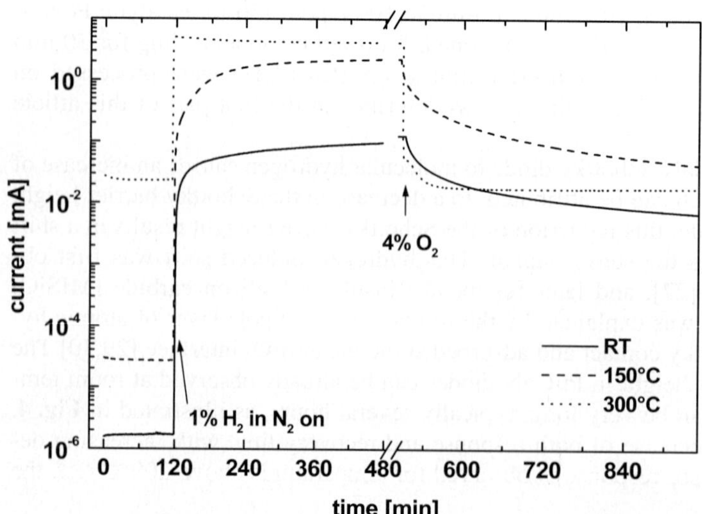

Fig. 4   Transient behaviour of the electrical response of a Pt:GaN Schottky diode to a 1% H_2 in N_2 pulse for different device temperatures. For low tempera-tures, hydrogen is stored at electronically active adsorption sites, leading to a long signal recovery time. Above 200°C, immediate desorption can be observed.

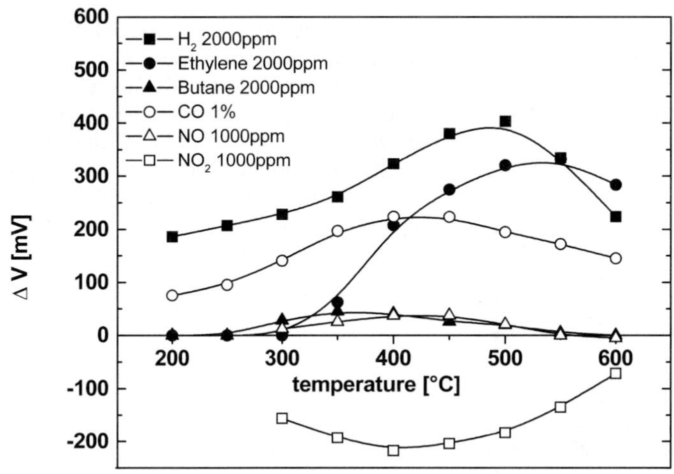

Fig. 5 Response of a Ga-face Pt:GaN Schottky diode to different exhaust gas constituents as a function of device temperature [31].

The gas sensitivity of catalytic Schottky contacts on GaN also can be employed for gas sensitive field effect transistors (GasFETs). In this case, gas induced modulations of the Schottky barrier height are monitored and amplified by the channel current, thus giving rise to sensor signals in the range of milliamperes. The response of GasFETs based on AlGaN/GaN heterostructures to hydrogen was investigated in [33]. The inset in **Fig. 6** shows the hydrogen induced change in the channel current of an AlGaN/GaN HEMT structure with a 3 nm GaN cap layer upon exposure to different hydrogen concentrations at a device temperature of 200°C in 4% oxygen atmosphere. The typical Langmuir behaviour characterized by a high differential sensitivity for low concentrations ($\leq$100 ppm) and a saturating sensitivity for high concentrations [34] was observed. The device response was almost immediate in the whole range of hydrogen concentrations. Similar to Pt:GaN Schottky diodes, AlGaN/GaN GasFETs also respond to other reducing and oxidizing gases, as shown in **Fig. 6** for a device temperature of 400°C. It is noteworthy that no high temperature steps were applied after deposition of the Pt Schottky contact. As a consequence, the Pt contact remained crack-free. Therefore, in contrast to the behaviour of the Schottky diodes shown above, a good selectivity for hydrogen containing gases, reflected by a four times higher sensor signal than for the other test gases, was observed.

As mentioned above, hydrogen adsorption at the metal/SiO₂ interface was demonstrated to be responsible for signal generation in gas sensitive Si-MOS or MISiC devices. In contrast, the origin of hydrogen detection in GaN based sensors is not obvious since the physi- and chemisorption properties of polar III-nitride surfaces and interfaces have not yet been studied in detail. The long recovery time of the electrical characteristics of Pt:GaN Schottky contacts after exposure to hydrogen at room temperature indicates a storage of hydrogen atoms at electronically active adsorption sites, where they are accessible for special analytic techniques. In [26], elastic recoil detection (ERD) was used to identify hydrogen adsorption sites in the Pt:GaN system. The samples were stored in deuterium atmosphere and subsequently analyzed. An accumulation of atomic deuterium and an increased oxygen content, which might originate from a native oxide layer was found at the Pt:GaN interface. Together with the observation that the dependence of the sensor signal on hydrogen partial pressure in oxygen free atmosphere follows a Temkin isothermal behaviour, it was concluded that signal generation is caused by interfacial adsorption of dissociated hydrogen.

To identify the nature of hydrogen adsorption sites at the metal/GaN interface we have recently investigated the hydrogen sensitivity of Pd:GaN Schottky diodes with in-situ deposited Pd contacts (evaporated directly after the PIMBE growth without exposure to atmosphere) [35]. The characterization of the hydrogen sensitivity was carried out in vacuum under application of hydrogen steps at partial pressures of 1 mbar. The hydrogen response of these in-situ fabricated samples was compared to that of ex-situ deposited diodes. The latter exhibit a thin native oxide layer on the surface, formed by natural oxidation due to exposure to atmosphere prior to evaporation of the Schottky contact. As reported in [36], surface treatments with HCl or HF, usually carried out before the deposition of metal contacts, do not remove the native surface oxide from such surfaces [31].

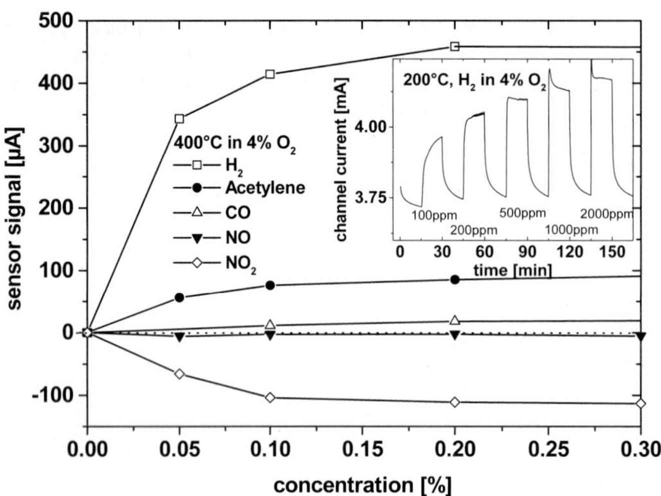

Fig. 6 Sensitivity of an AlGaN/GaN GasFET to different gases at a device temperature of 400°C. Reducing gases cause a positive sensor signal whereas oxidizing gases cause a negative change of the channel current. The inset shows the transient response to hydrogen pulses with increasing H₂-concentration, revealing a Langmuir behaviour [33].

The electric characteristics of the in-situ deposited sample (not shown) indicate a significantly lower Schottky barrier height than that of ex-situ deposited samples, which we attribute to passivation of leakage paths by native oxidation in the latter case [35]. Surprisingly, as shown in **Fig. 7**, the in-situ deposited device shows almost no sensitivity to hydrogen. The hydrogen induced variations of the channel current can hardly be resolved in the background noise. In contrary, the ex-situ deposited sample shows a large hydrogen response. For ex-situ deposited diodes, the hydrogen induced decrease of the Schottky barrier height is 560 mV at a hydrogen partial pressure of 10 mbar, whereas in the case of the in-situ deposited sample it was only 15 mV. The higher background forward current in the case of the ex-situ deposited samples is due to permanent storage of hydrogen (c.f. **Fig. 4**)

These measurements demonstrate that adsorption sites for dissociated hydrogen at the Pd/GaN interface are supplied by the thin intermediate native oxide layer spontaneously formed on the GaN surface when it is exposed to atmosphere. The bare GaN surface does not seem to supply adsorption sites with a sufficient density to give rise to a sizeable sensor signal. Following these results it can be concluded that the fabrication of optimized gas sensitive GaN Schottky diodes or GaN based field effect transistors requires the development of oxidation processes which form a closed, thermally stable oxide layer on the surface and do not lead to degradation of the electronic device properties. Further development of these device processing technologies can make GaN-based devices competitive to silicon carbide high temperature sensors.

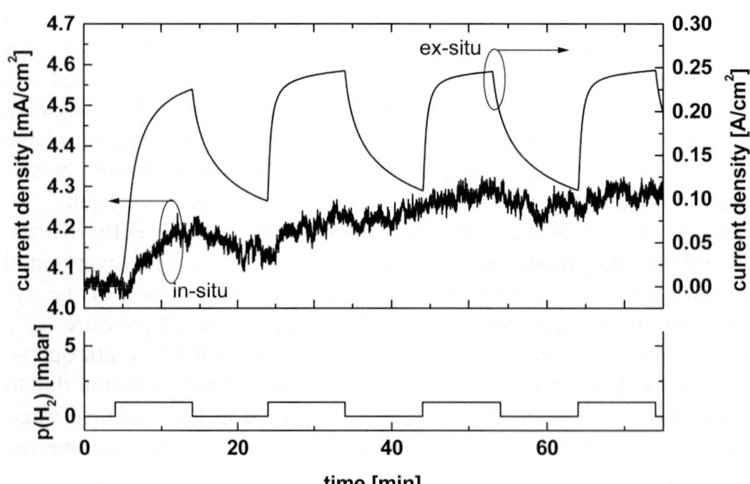

Fig. 7 Electric response of the forward current of in-situ and ex-situ deposited Pd:GaN Schottky diodes upon exposure to hydrogen at a partial pressure of 1 mbar in vacuum at room temperature. The bias voltage is 200 mV. The in-situ deposited device shows almost no sensitivity, whereas the ex-situ deposited device shows a pronounced sensor signal, which corresponds to a decrease in barrier height of 560 mV [35].

4 Ion sensitive field effect transistors (ISFETs) in electrolytes

In addition to the sensitivity of AlGaN/GaN heterostructure field effect transistors to gases and polar fluids, the detection of specific ions in aqueous electrolyte solutions is of great interest. Since the first reports on Si-based ion sensitive field effect transistors (ISFETs), capable of monitoring the pH as well as the concentrations of Na^+ by Bergveld et al. [37,38], the understanding and improvement of Si ISFETs operating as pH sensors [39] or detectors for biochemical processes [38,40,41] has been subject of intense research. The SiO_2 gate surface has been identified as the origin of the relatively low response and poor stability due to degradation in aqueous electrolytes. While SiO_2 surfaces exhibit a poor pH sensitivity of 32– 40 mV/pH, Si_3N_4 or metal oxide layers such as Al_2O_3 (57 mV/pH) or Ta_2O_5 (58.5 mV/pH) exhibit a higher sensitivity which, in the case of the metal oxides, is accompanied by a high chemical stability and consequently by a low, linear drift [40,42,43].

Due to the chemical inertness of $Al_xGa_{1-x}N$ compounds this material system is again a promising candidate for the fabrication of ISFET devices with stable operation in electrolyte solutions. In this paragraph we summarize the performance of GaN based ISFETs as recently reported by Steinhoff et al. [44]. The basic detection mechanism is identified by comparison of different FET devices with untreated and thermally oxidized surfaces. For this purpose, GaN:Si/GaN:Mg field effect transistor and GaN/AlGaN/GaN HEMT structures were grown by plasma induced molecular beam epitaxy [25,45]. The sample structures are shown in the insets of **Figures 8 and 9**. In samples 1 and 2 a *60 nm* thick, Si-doped GaN channel was grown on a *1.5 µm* thick (N-face) GaN buffer layer which was partially compensated with Mg. A carrier density of $3 \cdot 10^{18} cm^{-3}$ at room temperature was measured in the Si-doped top layer by capacitance-voltage (C-V) measurements. The carrier density in the Mg-compensated layer was $5 \cdot 10^{15} cm^{-3}$, so that current flows mainly in the top layer. Sample 2 was thermally oxidized for *2 h* at a temperature of *700 °C*, resulting in a thin Ga_xO_y - layer, as confirmed by XPS measurements [46]. The AlGaN barrier of the investigated HEMT-structure (Sample 3) had an Al-content of *28 %* and a thickness of *35 nm*, resulting in a 2DEG sheet carrier density of $n_{2DEG} = 6.1 \cdot 10^{12} cm^{-2}$, analyzed by capacitance voltage measurements. To obtain a surface comparable to Sample 1, an additional *3 nm* GaN cap layer was deposited on top of the AlGaN barrier. Characterization of the pH-sensitivity was carried out in a *100 mM* NaCl/*10 mM* Hepes solution, titrated with diluted NaOH or HCl for pH adjustment.

The stability of the devices was proven by cyclic I_{DS}-V_G measurements which showed high reproducibility without hysteresis in the investigated voltage range ($|V_G| < 500$ mV). To quantify the chemical response of the gate surface to changes in the electrolyte composition, ion induced changes in the channel current I_{DS} at constant V_{DS} were compensated by adjustment of the gate potential V_G via an Ag/AgCl reference electrode. A linear behavior over the entire investigated range from pH 2 to pH 12 was observed, as shown for Sample 2 and Sample 3 in the insets of **Figs. 8 and 9**.

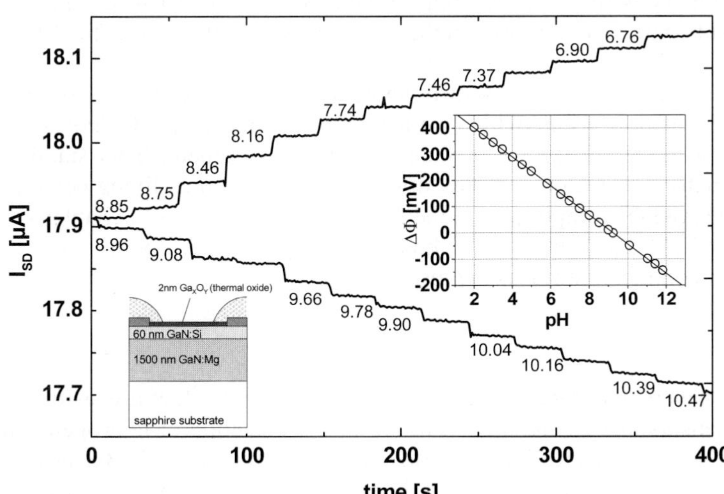

Fig. 8 Transient variation of the channel current in a GaN:Si/GaN:Mg FET structure (Sample 2) with changes in the ambient pH at a source drain voltage of 200 mV. The surface with a thin thermal oxide on the gate area is exposed to the gate electrolyte. The inset shows the measured dependence of the surface potential on the ambient pH. A linear relation with a sensitivity of 56.6 mV/pH is obtained [44].

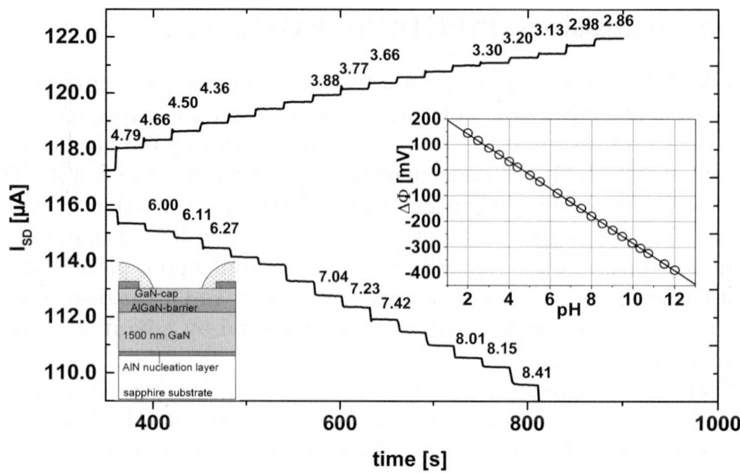

Fig. 9 Transient variation of the channel current in an AlGaN/GaN hetero-structure field effect transistor (Sample 3) with changes in the ambient pH at a source drain voltage of 200 mV. The untreated surface is exposed to the gate electrolyte. The inset shows the measured dependence of the surface potential on the ambient pH. A linear relation with a sensitivity of 56.0 mV/pH is obtained [44].

From the slope of these characteristics an almost Nernstian sensitivity of 57.3 mV/pH, 56.6 mV/pH and 56.0 ± 0.5 mV/pH was determined for Samples 1, 2 and 3, respectively. The similar sensitivity of all investigated structures demonstrates that the pH sensitivity is a pure surface effect and that the sensitivity is governed by a Ga_xO_y layer in all cases, which is formed by natural oxidation in air or in the electrolyte solution.

Figures 8 and 9 show the transient behaviour of the ISFET devices, measured by applying a constant source drain voltage of 200 mV and changing the pH in steps between 0.1 pH and 0.3 pH by titration with diluted NaOH in time intervals of 30 s. An immediate response was observed for all investigated samples. Due to the higher transconduction, the channel current in sample 3 exhibited the highest sensitivity to the pH induced variation of the surface potential. Because of the low number of thermally generated carriers, this wide bandgap semiconductor ISFET exhibited a resolution of better than 0.05 pH. The resolution of Samples 1 and 2 was 0.1 pH.

Based on the metal oxide surface of all investigated devices, the observed pH sensitivities can be explained in terms of the site binding model, which has been suggested by Yates et al. [47] and has been further developed and applied to ISFETs later [48,49]. According to this model, amphoteric hydroxyl groups which, dependent on the ambient pH, can be either neutral, protonized or deprotonized, lead to a pH dependent net surface charge, thereby causing an additional voltage drop at the solid/liquid interface.

These investigations show that GaN based ISFET devices are capable of stable operation in aqueous electrolytes and exhibit linear, almost Nernstian sensitivity and high resolution between pH 2 and pH 12. Further optimization of the device geometry is likely to still improve the device performance. In addition, integration of mixed (AlGaIn)-oxide gate layers allows engineering of the selectivity for the detection of specific ions.

Besides the application as pH-sensors or detectors of specific anorganic ions, these devices offer a variety of possibilities for the application as biosensors, which we have reported recently [50]. For these applications, the chemical stability of AlGaN surface is an indispensable requirement, This stability excludes toxic effects of the semiconductor surface on complex organic systems such as living cells. In addition, the optical transparency of the AlGaN compunds and the fabricated ISFET devices enables the simultaneous combination of optical and electronic analytic techniques for an in depth investigation of interfaces between group III nitride semiconductor surfaces and organic systems.

5 Conclusion

The application of pyroelectric AlGaN layers and AlGaN/GaN heterostructure field effect transistors as piezoresistive material in mechanical sensors and as chemical sensors for the detection of hydrogen and hydrogen containing gases or specific ions in electrolyte solutions has been summarized. Besides an overview over the performance of the specific devices the basic detection mechanisms were analyzed.

Whereas the piezoelectric material properties give rise for high piezoresistive gauge factors especially in HEMT structures, field effect based chemical sensors benefit from the electronic properties of these devices. In both GasFET and pH-sensitive ISFET structures the detection mechanism is governed by a thin metal oxide layer rather than by the clean GaN surface.

A particularly attractive feature of AlGaN/GaN heterostructure sensors is the integrability with electronic or optoelectronic devices or contactless readout schemes based on surface acoustic wave (SAW) devices. However, further development of sensor specific processing technology for III-nitride materials is still necessary before these devices can compete with established sensor designs.

Acknowledgement The authors gratefully acknowledge financial support by the Deutsche Forschungsgemeinschaft (Stu139/2, Stu139/8-1).

References

[1] A. Lloyd Spetz, A. Baranzahi, P.Tobias, and I. Lundström, phys. stat. sol. (a) **162**, 493 (1997).

[2] G. Mueller, G. Krötz, and J. Schalk, phys. stat. sol. (a) **185**, 1 (2001).

[3] E. Kohn, P. Gluche, M. Adamschik, Diam. Relat. Mater. **8**, 934 (1998).

[4] G. Krötz, M. Eickhoff, H. Moeller, Sens. Actuators **74**, 182 (1999).

[5] S. Nakamura, *GaN and Related Materials*, Ed. S. J. Pearton, Gordon and Breach, New York 1997.

[6] Y. A. Goldberg, Semicond. Sci. Technolog. **14**, R41 (1999).

[7] M. Razeghi and A. Rogalski, J. Appl. Phys. **79**, 7433 (1996).

[8] R. Neuberger, G. Müller, O. Ambacher, M. Stutzmann, phys. stat. sol. (a) **185**, 85 (2001).

[9] R. Neuberger, G. Müller, O. Ambacher, and M. Stutzmann, phys. stat. sol. (a) **183**, R10 (2001).

[10] M. Eickhoff, O. Ambacher, G. Steinhoff, J. Schalwig, R. Neuberger, T. Palacios, E. Monroy, F. Calle, G. Müller, and M. Stutzmann, Mat. Res. Soc. Symp. Proc. **693**, 781 (2002).

[11] M. Stutzmann, G. Steinhoff, M. Eickhoff, O. Ambacher, C.E. Nebel, J. Schalwig, R. Neuberger, G. Müller, Diamond and Related Materials **11**, 886 (2002).

[12] J. S. Shor, D. Goldstein, A. D. Kurtz, IEEE Trans. Electron. Dev. **40**, 1093 (1993).

[13] M. Eickhoff, H. Möller, G. Krötz, J. v. Berg, R. Ziermann, Sens. Actuators **74**, 56 (1999).

[14] R. Ziermann, J. v. Berg, E. Obermeier, F. Wischmeyer, E. Niemann, H. Möller, M. Eickhoff, G. Krötz, Mat. Sci. Eng. B **61-62**, 576 (1999).

[15] J. S. Shor, L. Bemis, A. D. Kurtz, IEEE Trans. Electron. Dev. **41**, 661 (1994).

[16] R. S. Okojie, A. A. Ned, A. D. Kurtz, W. N. Carr, IEEE Trans. Electron. Dev. **45**, 785 (1998).

[17] A. D. Bykhovski, V. V. Kamisnkii, M. S. Shur, Q. C. Chen, and M. A. Khan, Appl. Phys. Lett. **68**, 818 (1996).

[18] R. Gaska, J. W. Yang, A. D. Bykhovski, M. S. Shur, V. V. Kaminskii, and S. Soloviov, Appl. Phys. Lett. **71**, 3817 (1997).

[19] R. Gaska, M. S. Shur, A. D. Bykhovski, J. W. Yang, M. A. Khan, V. V. Kaminskii, and S. Soloviov, Appl. Phys. Lett. **76**, 3956 (2000).

[20] M. Eickhoff, O. Ambacher, G. Krötz, M. Stutzmann, J. Appl. Phys. **90**, 3383 (2001).

[21] R. Gaska, J. W. Yang, A. D. Bykhovski, M. S. Shur, V. V. Kaminskii, and S. Soloviov, Appl. Phys. Lett. **72**, 64 (1997).

[22] B. P. Luther, S. D. Wolter, S. E. Mohney, Sensors and Actuators B **56**, 164 (1999).

[23] A. Lloyd Spetz, L. Unéus, H. Svenningstorp, P. Tobias, L.- G. Ekedahl, O. Larsson, A. Göras, S. Savage, C. Harris, P. Martensson, R. Wigren, P. Salomonsson, B. Häggendahl, P. Ljung, M. Mattson, and I. Lundström, phys. stat. sol. (a) **185**, 15 (2001).

[24] O. Ambacher, J. Phys. D: Appl. Phys. **31**, 2653 (1998).

[25] R. Dimitrov, M. Murphy, J. Smart, W. Schaff, J. R. Shealy, L. F. Eastman, O. Ambacher, and M. Stutzmann, J. Appl. Phys. **87**, 3375 (2000).

[26] J. Schalwig, G. Müller, U. Karrer, M. Eickhoff, O. Ambacher, M. Stutzmann, L. Görgens, G. Dollinger, Appl. Phys. Lett. **80**, 1222 (2002).

[27] I. Lundström, M. S. Shivaraman, C. Svensson, and L. Lundkvist, Appl. Phys. Lett. **26**, 55 (1975).

[28] A. Arbab, A. Spetz, and I. Lundström, Sens. Actuators B **15-16**, 19 (1993).

[29] J. Fogelberg, M. Eriksson, H. Dannetun, and L.- G. Petersson, J. Appl. Phys. **78**, 988 (1995).

[30] M. Eriksson and L.-G. Ekedahl, J. Appl. Phys. **83**, 3947 (1998).

[31] J. J. Schalwig, G. Müller, M. Eickhoff, O. Ambacher, M. Stutzmann, Mat. Sci. Eng. B **93**, 207 (2002).

[32] J. Schalwig, G. Müller, O. Ambacher, M. Stutzmann, phys. stat. sol. (a) **185**, 39 (2001).

[33] J. Schalwig, G. Müller, M. Eickhoff, O. Ambacher, M. Stutzmann, Sensors and Actuators B **87**, 425 (2002).
[34] I. Lundström, M. Armgarth, L.- G. Petersson, CRC. Crit. Rev. Solid State Mater. Sci. **15**, 201 (1989).
[35] O. Weidemann, M. Hermann, G. Steinhoff, H. Wingbrandt, A. Lloyd Spetz, M. Stutzmann, and M. Eickhoff, Appl. Phys. Lett. **83**, 773 (2003).
[36] S. W. King, J. P. Barnak, M. D. Bremser, K. M. Tracy, C. Ronning, R. F. Davis, and R. J. Nemanich, J. Appl. Phys. **84**, 5248 (1998).
[37] P. Bergveld, IEEE Trans. Biomed. Eng. **17**, 70 (1970).
[38] P. Bergveld, IEEE Trans. Biomed. Eng. **19**, 342 (1972).
[39] B. H. van der Schoot, P. Bergveld, M. Bos, L. J. Bousse, Sens. Actuators **4**, 267 (1983).
[40] T. Matsuo, K. D. Wise, IEEE Trans. Biomed. Eng. **21**, 485 (1974).
[41] W.H. Baumann, M. Lehmann, A. Schwinde, R. Ehert, M. Britschwein, B. Wolf, Sens. Actuators B **55**, 77 (1999).
[42] H. Abe, M. Esashi, T. Matsuo, IEEE Trans. Electr. Dev. **ED-26**(12), 1939 (1979).
[43] L. Bousse, S. Mostarshed, Sens. Actuators B **17**, 157 (1994).
[44] G. Steinhoff, M. Hermann, W. Schaff, L. F. Eastman, M. Stutzmann, and M. Eickhoff, Appl. Phys. Lett. **83**, 177 (2003).
[45] M. J. Murphy, K. Chu, H. Wu, W. Yeo, W. J. Schaff, O. Ambacher, L. F. Eastman, T. J. Eustis, J. Silcox, R. Dimitrov, M. Stutzmann, Appl. Phys. Lett. **75**, 3653 (1999).
[46] M. Eickhoff, R. Neuberger, G. Steinhoff, O. Ambacher, G. Müller, M. Stutzmann, phys. stat. sol. (b) **228**, 519 (2001).
[47] D. E . Yates, S. Levine, T. W. Healy, J. Chem. Soc. Farady Trans. I **70**, 1807 (1974).
[48] W. M. Siu, R. S. C. Cobbold, IEEE Trans. Electr. Dev. **ED-26**(11), 1805 (1979).
[49] L. Bousse, N. F. De Rooij, P. Bergveld, IEEE Trans. Electr. Dev. **ED-30**(10), 1263 (1983).
[50] G. Steinhoff, O. Purrucker, M. Tanaka, M. Stutzmann, M. Eickhoff, Adv. Funct. Mat. (2003), *in press*.

phys. stat. sol. (c) **0**, No. 6, 1919–1939 (2003) / **DOI** 10.1002/pssc.200303134

Influence of polarization on the properties of GaN based FET structures

M. Neuburger[*1]**, I. Daumiller**[*1]**, M. Kunze**[*1]**,M. Seyboth**[.2]**, T. Jenkins**[3]**, J. Van Nostrand**[4] **and E. Kohn**[1]

[1] Department of Electron Devices & Circuits, University of Ulm, Albert-Einstein-Alle 45, 89081 Ulm, Germany
[2.] Department of Optoelectronics, University of Ulm, Albert-Einstein-Alle 45, 89081 Ulm, Germany
[3.] Air Force Research Laboratory, Sensors Directorate, Wright-Patterson Air Force Base, 202 Avionics Circle, Dayton, Ohio 43240 USA
[4.] Air Force Research Laboratory, Materials & Manufacturing Directorate, Wright-Patterson Air Force Base, 202 Avionics Circle, Ohio 45433 USA

Received 4 March 2003, revised 20 June 2003, accepted 23 June 2003
Published online 28 August 2003

PACS 73.40.Kp, 77.22.Ej, 85.30.Tv

GaN is the first highly polar semiconductor used in field effect transistors. Polarization charge dipoles are an essential part contributing to the device performance. Especially at the surface such charges may influence stability and large signal characteristics. Switching transients in output current of AlGaN/GaN-FETs related to charge storage effects are discussed and related to the polar nature of this material system. Alternative structures not suffering of the surface charge problem are introduced, namely an InGaN-channel FET and a AlGaN/GaN-double barrier structure. For realization doping screening of the polarization field is applied as main tool.

1 Introduction

Presently, the technology of gallium nitride based field-effect transistors, namely AlGaN/GaN HEMTs, has matured to a point, where they have become most attractive as microwave power source surpassing conventional III-V power FET structures by approximately one order of magnitude in power density. Besides their difficult materials growth technology, there are two other important features, which have accompanied the development of these structures. This is the high power loss while operating at large power densities, causing self-heating; and the polar nature of the wurzite material introducing polar dipole charges in an uni-polar device structure. In this study both effects have been analyzed in respect to transients in the time and frequency domain. To circircumvent the polarization induced charge instabilities, novel heterostructure configurations and doping schemes have been developed.

It is generally expected that the transport properties start to degrade with increasing temperature and power will decrease consequently. Indeed this is limiting the power densities of many AlGaN/GaN power device structures especially on sapphire substrate. However, in an early stage of this investigation a case has been encountered, where the saturated power increases with increasing temperature (fig. 1).

---

[*] Corresponding author Martin.Neuburger@e-technik.uni-ulm.de

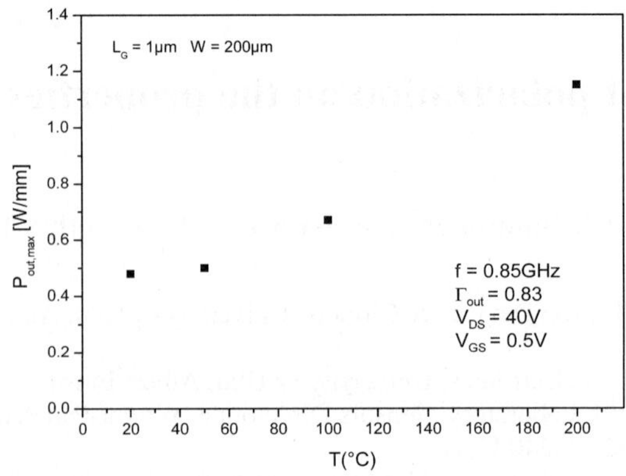

Fig. 1: Saturated RF-power at 850 MHz as a function of temperature of a channel doped AlGaN/GaN-HFET [1].

This could not easily be understood at that time. At first thought this pointed towards a deep trap activation phenomenon in the channel, but a more detailed analysis showed that it is the complex behavior of a surface polarization charge related instability and its temperature behaviour. Another phenomenon, which has been plaguing GaN-based HFET structures for a long time is the RF power slump [2], which could (in many cases) be traced back to a compression in output current as shown later in fig. 12, caused by surface charge storage effects (virtual gate effect) [3]. Both phenomena are specific to GaN based devices because they relate to polarization induced charges at the surface of the FET structure.

The elimination of such polarization related surface charge instabilities was therefore the subject of further investigations. This lead to novel heterostructures namely the InGaN-channel structure sandwiched inbetween two GaN barriers and the double barrier AlGaN/GaN structure. In both arrangements the polarization induced surface charges are removed and placed inside the heterostructure. In addition charge accumulation caused by the polarization field is substituted by ionized doping employing the technique of screening the polarization field by doping [4]. In this case mobile polarization induced interface charges are substituted by immobile donors or acceptors.

2   Polarization induced image charge dipoles

GaN is the first highly polar semiconductor used in unipolar surface devices, namely FETs. The strong spontaneous polarization of the GaN hexagonal wurzite phase implies that fixed bonded charge dipoles are developed between surfaces and interfaces. This dipole charge causes an electrical field, which in turn may attract charges, called induced images charges. Such charges can be highly mobile and are used as channel for conventional AlGaN/GaN-HEMT structures. This is schematically illustrated in fig. 2 for the case of a strained AlGaN barrier layer in between two unstrained GaN layers grown on the substrate with Ga-face. The circles represent image charge densities, which can of course only develop, if the energy bands are tilted enough so that conduction band and valence band states become available.

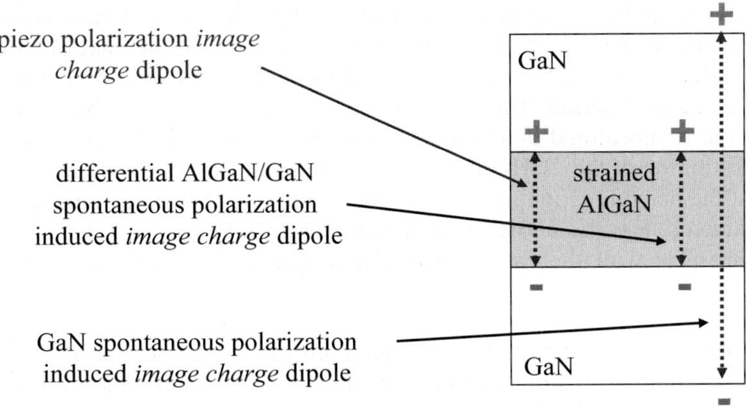

Fig. 2: (online colour at www.interscience.wiley.com) Schematic view of a buried strained AlGaN-layer in between two GaN-layers. Indicated are image charge dipoles which may be induced by the bonded polarization charges, after [5].

Indeed, in the case of AlGaN/GaN HEMT structures the discontinuity in polarization between the Al-GaN and GaN layers (caused by the difference in spontaneous polarization of both materials and the piezo polarization of the strained AlGaN layer) leads to an interfacial 2DEG channel without direct channel doping or modulation doping, a feature called "piezoelectric doping" or "polarization doping" [6]. The 2DEG sheet charge is the image charge induced by the bonded polarization charge at this interface and countercharges of the same density need to be located on the surface of the barrier to keep the system vertically neutral [7]. Assuming, that both image charge densities (the 2DEG at the AlGaN/GaN interface and the image charge on the barrier) are mobile, this will result in two parallel channels of opposite sign. To some extent this may be viewed as an ambipolar FET channel. For the AlGaN/GaN HEMT structure in case the upper barrier charge or surface charge is located in a deep surface state this may cause instabilities, related to the charging and discharging dynamics of the surface trap. Since the domain, in which GaN-FETs are used, is that of high power, deep states may be charged or discharged by high fields or temperatures. The dynamics may be determined by the capture and emission characteristics of the state as well as by a charging path on the surface. The problem has become known to the user as drain current compression at RF-frequencies or RF power slump [2]. In a first attempt the surface charge may be moved into a dielectric passivation layer, if the surface state can be removed at the same time. This has indeed been attempted successfully by passivation with Si_3N_4 [8] and SiO_2 [9], also leading the way to MOSFET structures. However, it is also clear, that this passivation layer needs a bandgap larger than the AlGaN barrier layer and no emission from deep states within the entire field of device operation. This optimization is still an ongoing effort.

Another approach is to remove the counter image charge of the polarization induced channel charge from the surface by novel heterostructure arrangements. Such structures are the topic of this investigation.

3   Doping screening of the polarization fields

Removing the polarization charge in question from the surface and burying the polarization dipole within the heterostructure will however not easily cure the problem of the risk that an ambipolar channel may still develop under certain bias conditions. An ideal situation would be, if the image charge dipole would contain one highly mobile part, representing the 2DEG channel, and one part of strictly localized charge like in the case of shallow ionized impurities. Indeed this can be obtained by introducing a sharp doping spike at the image charge location. Then, the polarization field is screened by the doping profile and the mobile sheet charge is substituted by the ionized impurity density. In the case of n-channel FET device structures the substitution of the 2DHG needs to be realized by a shallow donor doping spike. Of course also deep traps at the heterojunction in question enter fully into the charge balance equation and need to be carefully considered. Has the 2DHG been substituted by a donor sheet charge, this configuration becomes very comparable with the case of modulation doping in other compound semiconductor heterostructures.

The technique to screen polarization fields by doping profiles is well known from the design of optoelectronic GaN-based quantum well structures to avoid internal fields within the wells affecting the emission wavelength [10]. However, in FET devices this has widely been neglected up to now. In the following two cases are discussed where this concept is an essential part.

4   Device technology

In analyzing dispersion, instability and drift related phenomena, it is important to rely on a stable device technology introducing no degradation effects over the entire range of operation.
This is mainly related to a high temperature stability, which may also indicate a high AlGaN/GaN breakdown stability [11]. As a material of wide bandgap nitrides should be able to operate at high temperature, if the chemistry of the heterostructure and its contacts is stable up to high temperatures. The decomposition temperature of GaN in atmosphere is approx. 650 °C (although the material is grown at around 1000 °C by MOCVD) [12]. In a detailed study using a number of devices from various laboratories it was found that operation of devices was possible even above this temperature of up to 750 °C over a short period of time in vacuum as illustrated in fig. 3 [13].

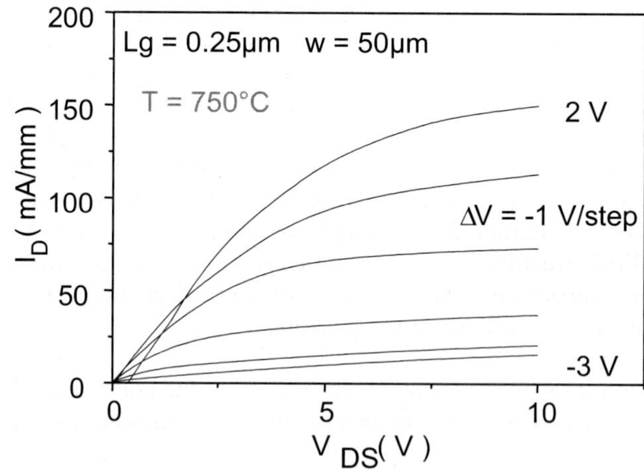

Fig. 3: (online colour at www.interscience.wiley.com) MOVPE grown doped channel $Al_{0.15}GaN_{0.85}$/GaN structure with a 50nm thick doped ($N_D=5*10^{17}cm^{-3}$) channel underneath a 50nm thick AlGaN cap layer. The Hall mobility decreased from 230cm^2/Vs at R.T. to 100cm^2/Vs at 600°C, whereas the sheet charge density remained almost unchanged at $8*10^{12}cm^{-2}$, after [11].

From fig. 4 it can also bee seen, that no degradation occurred after operation at 600 °C over 20 min., comparing the characteristics before and after applying the high temperature. Especially important to note is the high contact stability, when applying refractory metal schemes as used in other semiconductor technologies.

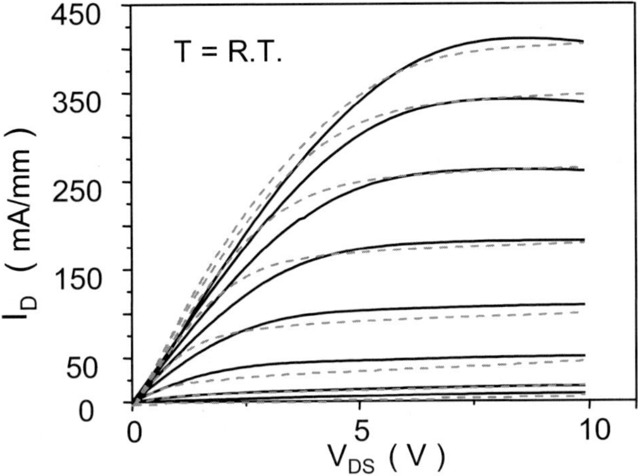

Fig. 4: Overlay of two DC-output characteristics measured at R.T. The dashed line was thermally cycled at 600°C over 20 minutes; The device is a channel doped $Al_{0.15}Ga_{0.85}N$ FET with a sheet charge density $N_S=2.5*10^{12}cm^{-2}$ in a 50 nm thick channel, $L_G=0.25\mu m$ and $W= 50\mu m$, after [11].

In fig. 5, the IV-characteristics a Pt-GaN Schottky diode shows a slightly reduced leakage current after a 600 °C operation in vacuum, indicating that the metal-semiconductor interface is highly stable despite the fact that the leakage current is entirely dominated by defects causing local image force lowering. The high reverse leakage at high temperature is caused by charge injection across defects, but not the chemical/physical degradation of the interface.

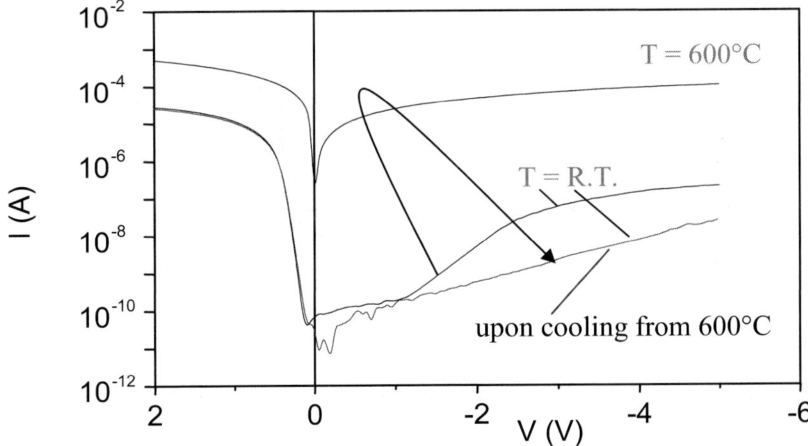

Fig. 5: (online colour at www.interscience.wiley.com) IV-characteristic of a Pt-GaN Schottky diode. The diode was thermal cycled at 600°C for 20 minutes. A light decrease in the reverse current is observed after cooling, after [14].

Ohmic contacts based on Ti/W have shown an essentially unchanged contact resistance up to 700 °C as can be seen from fig. 6. Therefore, it seems that instabilities observed in many devices are indeed of electronic nature and not masked by physical/chemical degradation of the device structure.

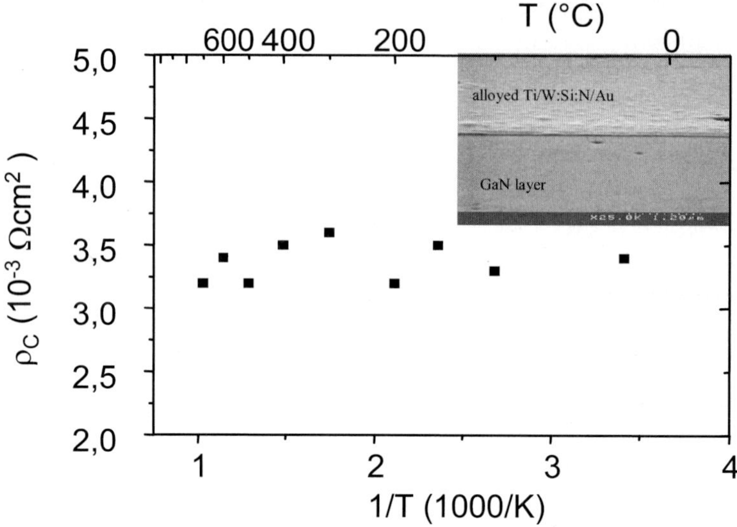

Fig 6: Specific contact resistance of a Ti/W:Si:N/Au contact on n-GaN up to temperatures of 700°C. No major degradation is seen. The insert shows a SEM picture of the contact after thermal cycling. No surface degradation after annealing is observed, after [11].

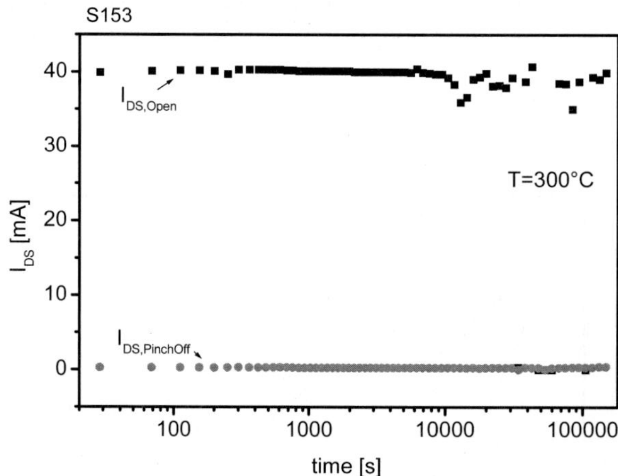

Fig. 7: (online colour at www.interscience.wiley.com) GaN MESFET hold at T=300°C. The drain current for Pinched Off condition and Open Channel mode are monitored. No degradation over 1000 hours is determined.

The above short term analysis indicates that GaN FET structure may be reliable operated at elevated temperatures. In fig. 7, the drain current of a GaN MESFET is monitored, which is hold at a temperature of 300°C for more than 1000 hours. During this time the device was biased in an open channel condition.

It is only pinched off at the measurement event. A comparison of the drain currents in the open channel mode and in pinched off condition shows no degradation. The scatter towards the end of the measurement can be explained by the heavily stressed contact pads because of extensive sampling.

5 Thermal transients

Self-heating of GaN-based devices may be quite severe. This will cause transients in the device characteristics. It is important to separate this effect from electronic transients. It has therefore been investigated separately. Simulation shows that this effect dependents strongly on the buffer and substrate configuration. Channel temperatures of more than 400°C may be reached on sapphire already with an RF power loss of 5 W/mm [15]. Self-heating develops within the channel between gate and drain, where the output power loss is generated. Fig. 8 illustrates the generation of heat in the drift region of a doped-channel HFET. Obviously, the heat flux to the heat sink will determine the steady state temperature for CW operation. This heat transfer is at first across the GaN buffer layer and the substrate, which is either sapphire or SiC. However, a large portion may also be removed from the surface.

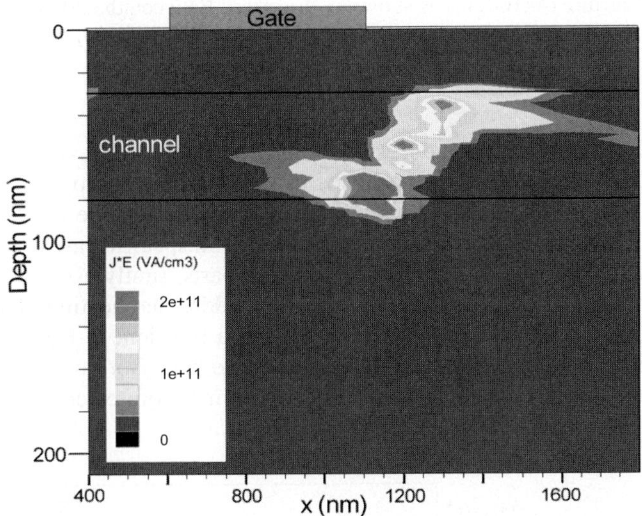

Fig. 8: (online colour at www.interscience.wiley.com) Simulation of the power dissipation of a doped-channel HFET with a power loss of 5W/mm using drift diffusion model (no overshoot). The heat is assumed to be removed through the substrate only, after [15].

Fig. 9 shows the heat removal capability of different heat sink configurations. A diamond surface overlayer is especially attractive, because it will represent a dielectric passivation layer directly extracting the heat. However, the nucleation and growth of thin CVD diamond films with high thermal conductivity on foreign substrates like GaN is still a challenging problem. Nevertheless, such a configuration has already been realized, however, without reporting on the heat extraction properties [16].

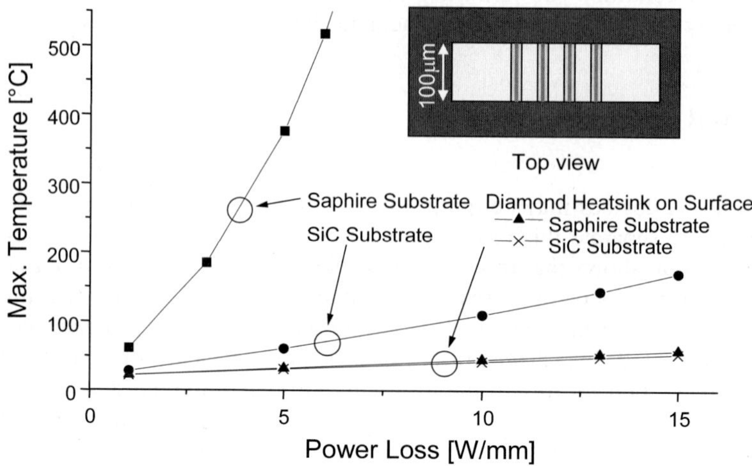

Fig. 9: (online colour at www.interscience.wiley.com) Max. temperature vs. power loss for different heat sink concepts for a 4 finger FET structure (4x100µm) as shown in the insert. Between the 20nm thick diamond heat sink and the device, a 100nm thick Si_3N_4 layer is introduced. The substrate thickness is 500µm, buffer thickness is 2µm, after [15].

Still, in the case of an effective heat spreader, switching to a high power level (like in the case of class A operation) will cause transients related to the heat flow through the setup until equilibrium is reached. The thermal transients will overlap with electronic instabilities during the time where both effects may show opposite tendencies and may even cancel out each other in specific cases.

Simulation reveals that the thermal transient will have two parts, firstly a steep transient due to the fast spreading of the heat through the buffer followed by a slow additional heating phase to reach equilibrium through the substrate (see fig. 10). This self-heating causes a transient in FET output current through a change in channel mobility, series resistance and intrinsic gate bias. Both transient regions can indeed be identified experimentally in a device configuration not suffering from electronic instabilities, namely an InGaN-channel device (which is described below).

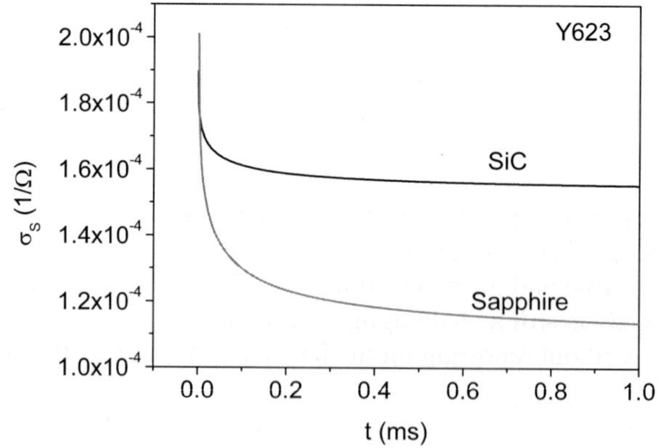

Fig. 10: (online colour at www.interscience.wiley.com) Calculated thermal transient in channel sheet resistance for a 4 finger InGaN-FET structure using experimental temperature dependent hall data. Geometry like insert in fig. 8, after [17].

6 AlGaN-HFETs, analysis of electronic transients

GaN based HFETs suffer in particular from large signal instabilities which can be traced back to a compression in output current illustrated in fig. 12, but show only little small signal dispersion. This points towards charge storage related phenomena in the buffer layer, the Schottky barrier layer or the surface.

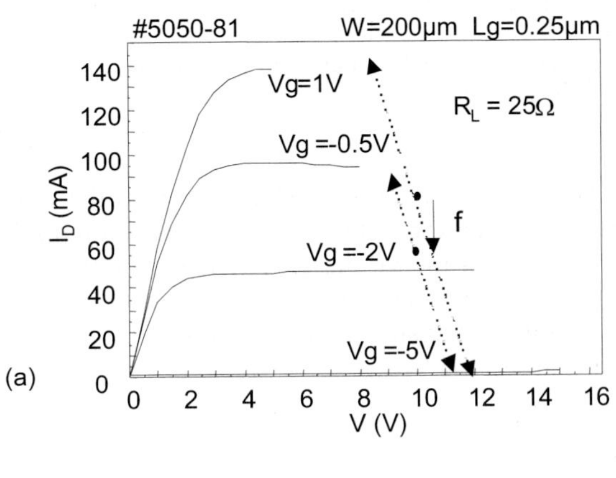

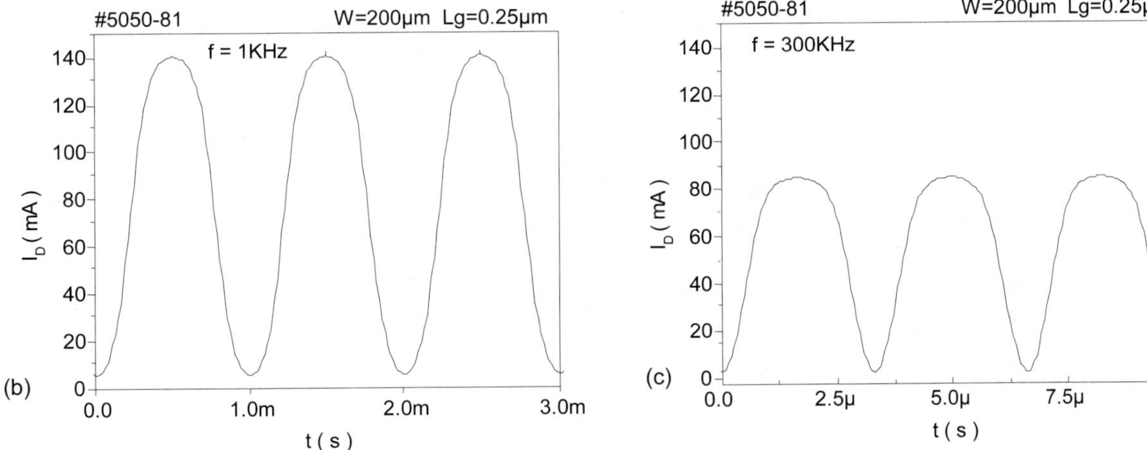

Fig. 12: Current compression in a channel doped $Al_{0.3}Ga_{0.7}N$/GaN device with 200µm gate width and a gate length of 0.25µm after [18].

    (a)   maximum output current amplitude at 1kHz and 300kHz respectively across a 25 Ω load line and output charactersitic.

    (b)   uncompressed waveform at 1kHz

    (c)   compressed waveform at 300 kHz

All these cases have been discussed based on experimental results. However the majority of experimental results point towards a surface related effect, widely refered to as "virtual gate effect" [3, 19]. It resembles the configuration of a dual gate cascode circuit with the second gate acting like a temporary current limiter onto the main gate, indicated in fig. 13.

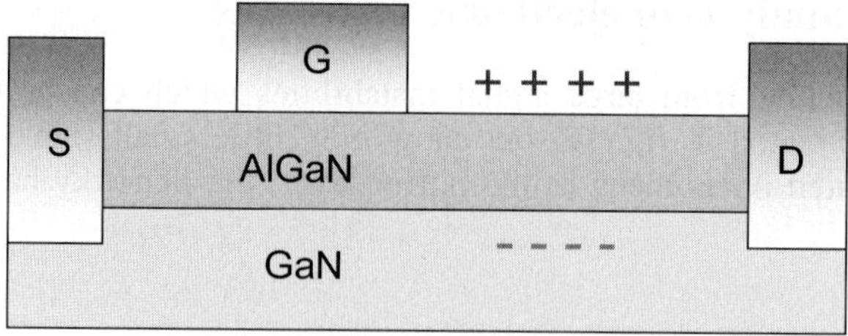

Fig. 13: (online colour at www.interscience.wiley.com) Virtual gate as current limitter is indicated in for of accumulated charge between the gate and the drain.

The FET output current will be limited to a current, which can pass the virtual gate as long as residual injected charge remains on this gate. Therefore this effect will result a transient and current compression at high frequency, where the charge on the virtual gate remains frozen, called RF power slump. In a rather complex study involving a large number of devices from various laboratories the current compression phenomenon has been investigated and it was found to span across a frequency window of nearly 12 orders of magnitude (see fig. 14).

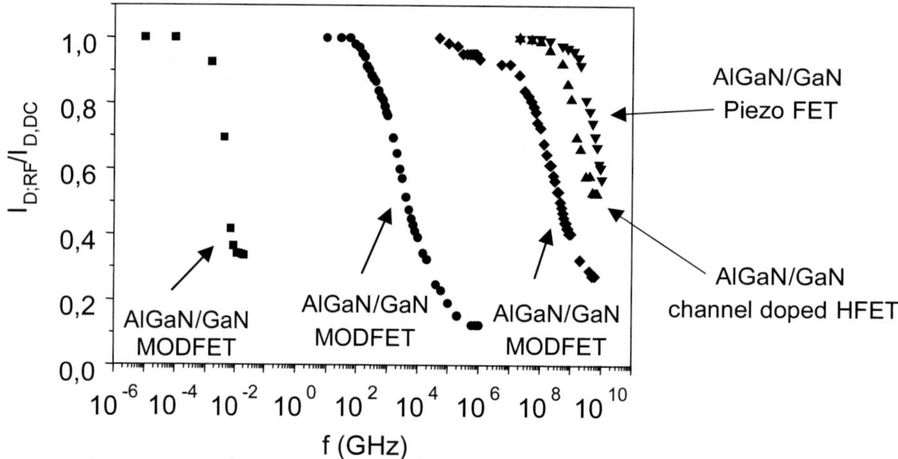

Fig. 14: Current Amplitudes monitored versus frequency as shown in fig. 2 for 1kHz and 500kHz..

Several things were puzzling: it was not reasonable to assume that surface trap charges could respond with GHz frequencies, and why would there only be one major transition seen in each device? It was therefore concluded that the virtual gate would be charged by lateral charge injection from the main gate. This might therefore result in a technology dependent time constant. One possibility for the delay could be the dielectric cut-off of the surface charging path. The different values of the time constant would reflect different surface residual conductivity levels (in the MΩ/ and GΩ/ -range). Thus, this effect would be sensitive to surface treatments and passivation, which is indeed observed. Fig. 15 shows the change of transition frequency in the current compression effect related to surface annealing of an unpassivated AlGaN/GaN HEMT structure after [20]. Changes are seen at temperatures, where only surface conditions are effected.

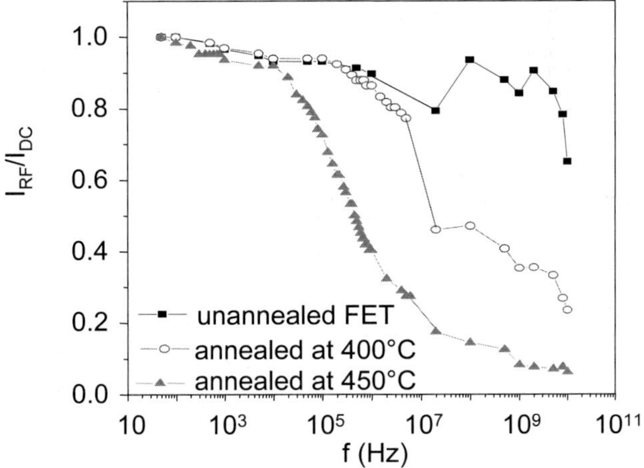

Fig. 15: (online colour at www.interscience.wiley.com) Normalized change of transition frequency with annealing up to 450°C of an AlGaN/GaN HEMT device, after [20].

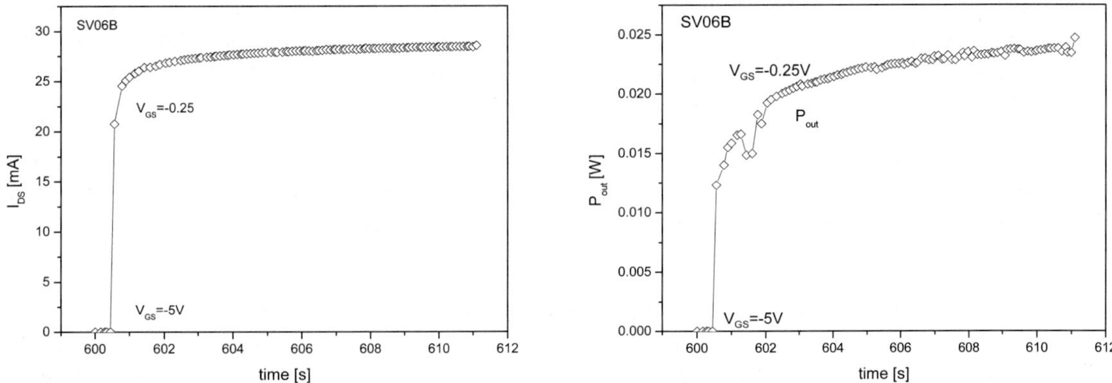

Fig. 16:

Left side: A current transient measurement for an unpassivated AlGaN/GaN HEMT structure grwon by RF-MBE on sapphire.

Right side: The corresponding 4 GHz saturated power transient measurement (50 ohm load line). The device data are: $L_g = 1.0$ μm, $w = 100$ μm, I max = 300 mA/mm. The shown transients are switched from pinch off ($V_{GS} = -5$ V) to class A ($V_{GS} = -0.25$ V). Drain voltage is 10 V.

Recently the current compression effect could be related to the RF power slump characteristic [21]. Fig. 16 shows drain current transient and the corresponding transient of a saturated power measurement at 4 GHz.

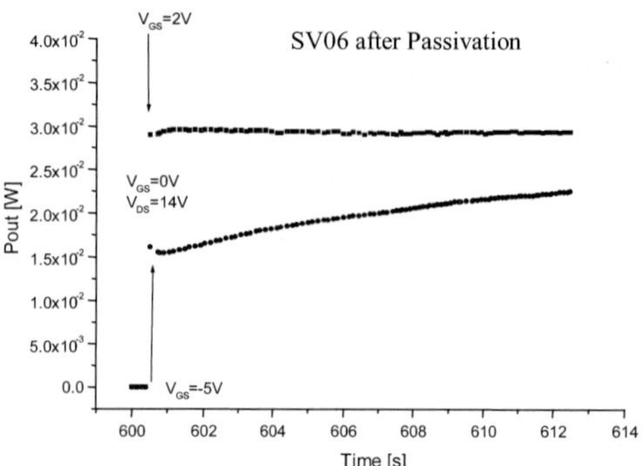

Fig. 17: Power switching measurement as described in the text. The transient is clearly seen for switching from pinch-off to class A bias point. Switching from open channel (at V_{DS}=0V, V_{GS}=2V) to class A bias point reveals the thermal transient.

In these experiments thermal transients and electronic transients became visible. Fig. 17 shows an experiment, where a device is switched into a class A bias point, simultaneously with switching-on of the saturated power level. The device was switched on from two different soaking levels. Firstly, the device was switched on from the pinch-off condition, revealing the electronic transient (the surface charging effect). Initially only 50 % of expected output power is available. Secondly, the device was switched into the class A bias point after a forward gate bias soaking (open channel condition), however with zero drain bias to prevent self-heating. Switching into the class A bias point reveals self-heating, and in the case shown still a small charging effect. These characteristics will be specific for each device, depending on the surface preparation as well as the bias conditions. Charge injection from the gate may be determined by an injection threshold and/or an activation energy of the charging path.

Finally, let us return to the early experiment described in the introduction (fig. 1). Here the saturated power level increased with increasing temperature. In similar devices a transition frequency for the current slump effect was found in the upper MHz-range just below the frequency of the large signal measurement (850MHz). When heating these devices, it was seen that the transition frequency moved to higher values like in the device used in fig. 18. In the case shown an activation energy of 0.3 eV was extracted and related to the activation of the charging current path. Assuming, that the transition frequency will cross the measurement frequency while the device is heated the current and the power will increase. This is what most likely has happened.

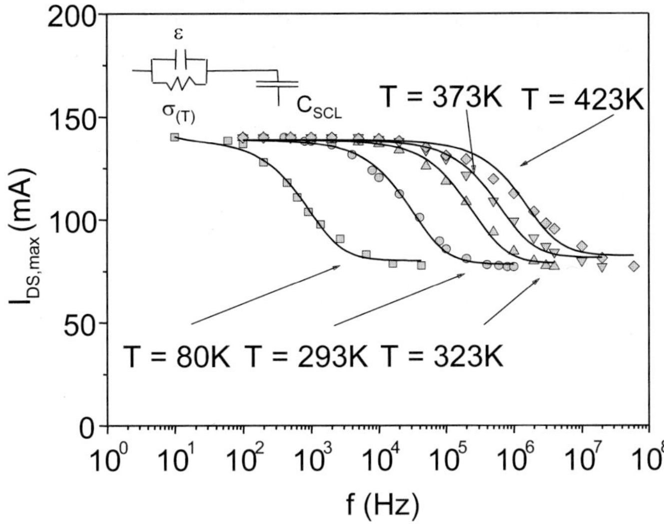

Fig. 18: Thermal activation of current slump transition, steadily moving to higher frequencies with increasing temperature. Fitting these curves with an equivalent circuit as shown in the insert an activation energy of 0.3eV is extracted which is related to the charging resistance.

7 InGaN-channel FET structure

In the case considered, a thin strained InGaN layer may be sandwiched in between two GaN barrier layers. At first this case shall be discussed without introducing additional doping profiles. Fig. 19 shows the schematic vertical distribution of the image charges (similar to what is seen in fig.1 for the case of an AlGaN barrier in between two GaN layers), the main contribution of about 95% being the piezo polarization image charge dipole in the InGaN well.

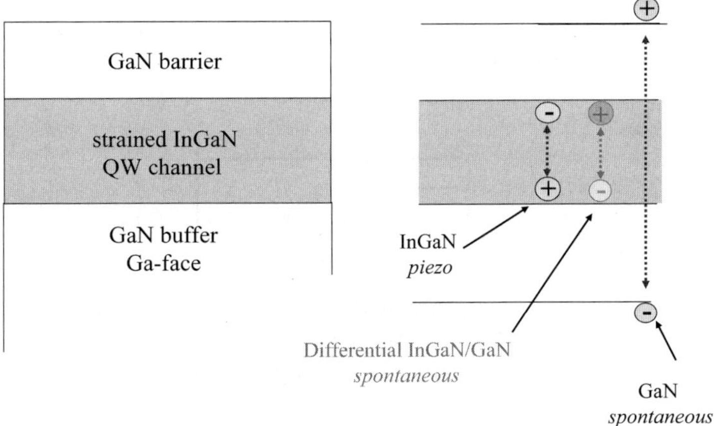

Fig. 19: (online colour at www.interscience.wiley.com) Schematic cross-section of an InGaN-layer placed in between two GaN-layers and sketch of dipole image charge distribution. In case the piezo-polarization is dominating. The 2DEG is located at the top interface due to reversal of the stress situation as compared to the AlGaN barrier structure of fig. 1.

The energy banddiagram of such a structure with an In-content of 10 % in the well is shown in fig. 20. The polarization charges are calculated after [22], assuming that only the InGaN is strained and that there is no clustering and relaxation of the layer.

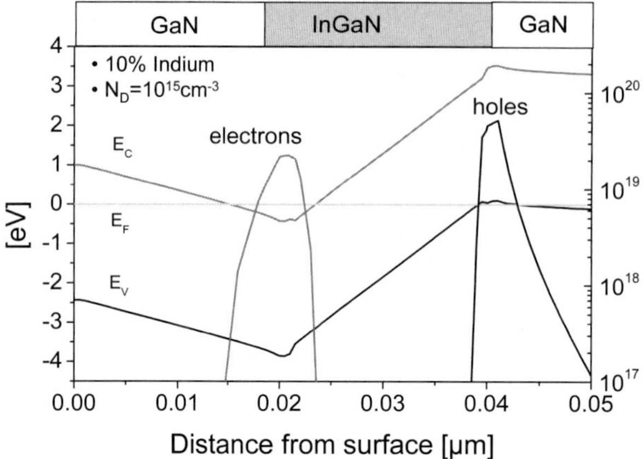

Fig. 20: (online colour at www.interscience.wiley.com) Shown is a schematic cross section of the simulated GaN/InGaN/GaN band structure with 10%-Indium in the InGaN-layer and the electron/hole distribution. A background doping of $N_D=10^{15}cm^{-3}$ is assumed the entire structure. In the simulation the donor doping spike is introduced as a 0.1nm thin spike of corresponding sheet charge. The electron and hole distribution represents the polarization image charge dipoles in the InGaN-quantum well.

The polarization induced 2DEG/2DHG dipole at the InGaN/GaN interfaces is indeed observed, in the case of a strained InGaN QW the polarization field within the well being dominated by the piezo polarization. The slight difference in density of electrons and holes stems from the difference in boundary conditions, namely a potential of 1 V at the surface and a Fermi-level at midgap in the buffer. Clearly this will lead to an ambipolar channel in the FET device. Assuming, ohmic contacts can be realized to both type of charges, what will the FET characteristics be? This is sketched in fig. 21 for a bias point in saturation, where most devices will operate.

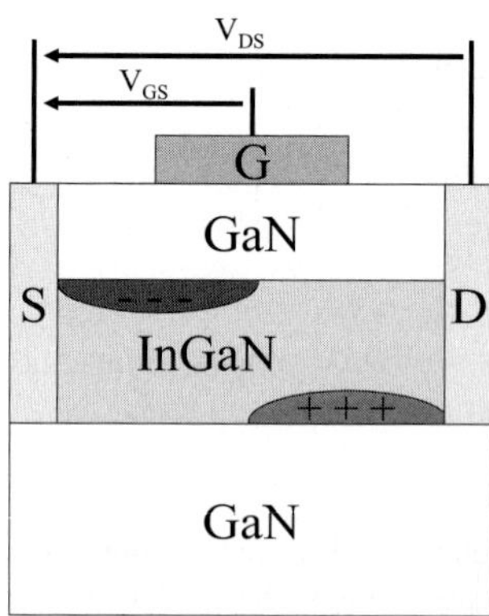

Fig. 21: (online colour at www.interscience.wiley.com) Charge accumulation in an ambipolar channel for operation in saturation, after [22].

As can be seen, charge accumulation will appear at both sides of the gate, resembling a PN-junction. The channel will be widely blocked, if both carriers have high mobility, with the effect strongly depending on the dynamics of both carriers. To obtain a unipolar channel the 2DHG has to be substituted by a donor doping spike. This can be placed within the InGaN quantum well or in the rear GaN barrier layer close to the interface. Both cases have been evaluated [23]. To maintain a high electron mobility in the channel, the placement of the doping spike in the rear GaN barrier layer is preferable, making the configuration very similar to an inverted HEMT structure. This configuration is shown in fig. 22.

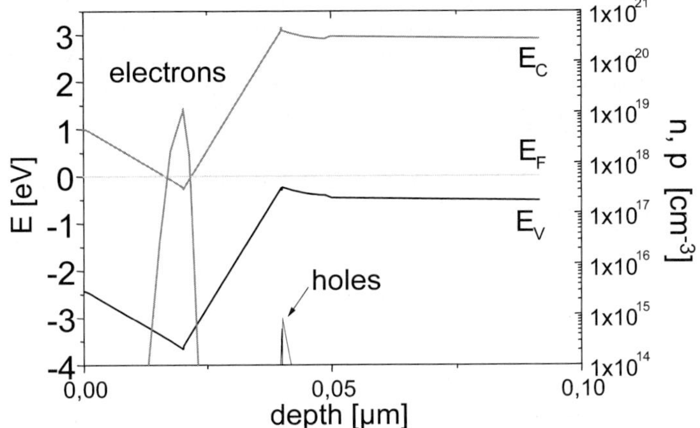

Fig. 22: (online colour at www.interscience.wiley.com) Simulated band structure and electron/hole distribution of an inverted GaN/InGaN/GaN HEMT structure and doping in the rear GaN barrier as shown in fig. 6 to screen the 2DHG sheet charge after [21].

The concept has been experimentally verified by CV-profiling as shown in fig. 23.

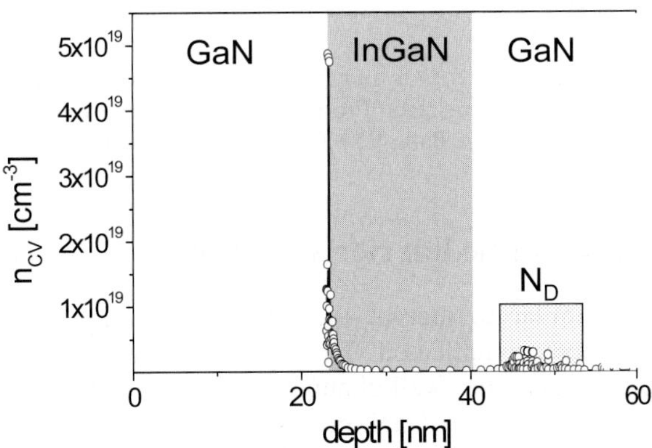

Fig. 23: GaN/InGaN/GaN heterostructure configuration and donor doping as discussed in fig. 5 and a CV-doping profile of an experimental structure with 9% Indium content in the InGaN well. It can be seen that the rear donor doping profile has overcompensated the 2DHG density slightly, after [21].

It can indeed be seen that a 2DEG develops at the upper interface and that the 2DHG has been fully substituted. In the case shown the donor profile has overcompensated the 2DHG and a residual electron profile has remained within the rear GaN barrier.

FET devices fabricated on such "inverted HEMT"-like heterostructures have yielded in DC-output current levels of up to 700 mA/mm for 0.5 µm gate length generated by the polarization doping of the InGaN strain field [24]. The important question at this point is, whether the RF-power slump instability could be avoided by this approach. This has been evaluated carefully and compared with standard AlGaN/GaN HFETs [21]. Large signal characterization shows indeed no dispersion in maximum output current up to 20 GHz [25]. However, the most critical experiment is a switching experiment from a pinch-off bias point into a class A bias point, switching on the saturated power level simultaneously. Such an experiment is shown in fig. 24. No delay in output signal is observed within the resolution limit of the power measurement of a few µs the power level being very close to the steady state (CW) saturated power. It may also be noted that the device was stable within the entire bias and temperature range, only the effect of self-heating was present [23].

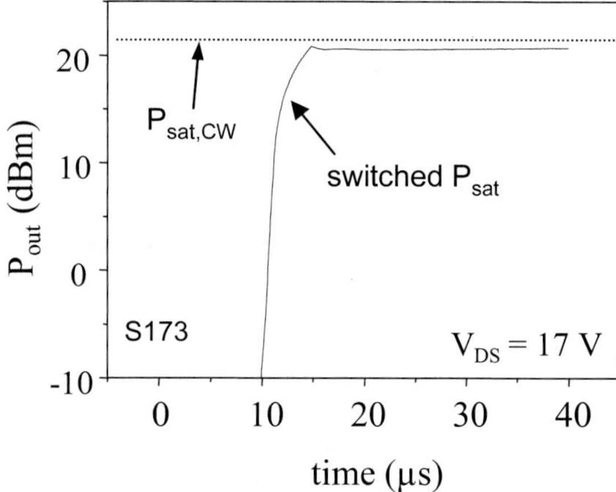

Fig. 24: 2 GHz power switching transient for an InGaN-channel FET with 9% Indium in the InGaN-quantum well. P_{sat} is the steady state power level. Switching data: From pinch-off (V_{GS}=-3V) to a class A (V_{GS}=0V) simultaneously turning on the saturated power level. Device data: L_g = 0.25 µm, W = 100 µm, I_{max} = 420 mA/mm, after [21].

8   AlGaN/GaN double barrier structure

To obtain a configuration similar to that of the above discussed InGaN-channel device in the AlGaN/GaN heterosystem is more complicated. The reason is, that here the AlGaN barrier materials are strained in respect to the GaN quantum well channel. To remove the positive image charge contribution from the surface and still use the AlGaN/GaN interface to generate the 2DEG-channel, a second AlGaN barrier layer has to be introduced as shown in fig. 25. As seen from the bandstructure lineup this will result in an ambipolar 2DEG/2DHG channel in the unstrained GaN quantum well and a parallel 2DEG channel at the rear interface.

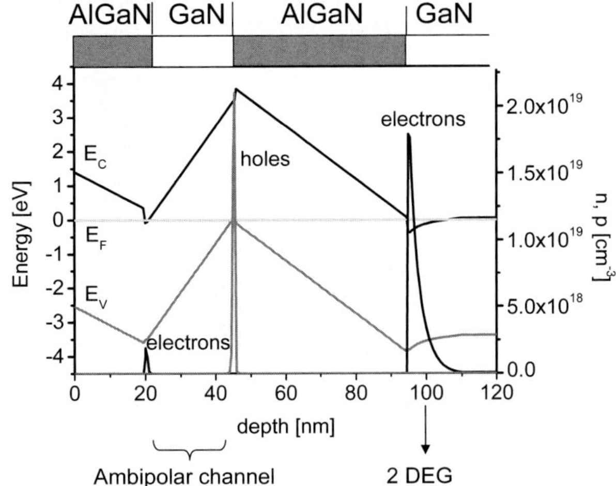

Fig. 25: (online colour at www.interscience.wiley.com) Simulated band structure and the electron/hole distribution of an AlGaN/GaN double barrier structure with 20% Aluminium. Three image charge accumulations can be seen.

This time the 2DHG in the ambipolar GaN channel and the rear 2DEG have to be substituted by a spike of donor doping and an acceptor doping spike respectively. The result of such a configuration is shown in fig. 26. Here the energy banddiagram of a heterostructure has been simulated for two cases: (a) where the FET channel is fully open and (b) where the channel is pinched off. Within the entire field of operation no ambipolar or parallel 2DEG channel is developed. This can also be seen from the simulated output characteristic (c), which shows no anomalities.

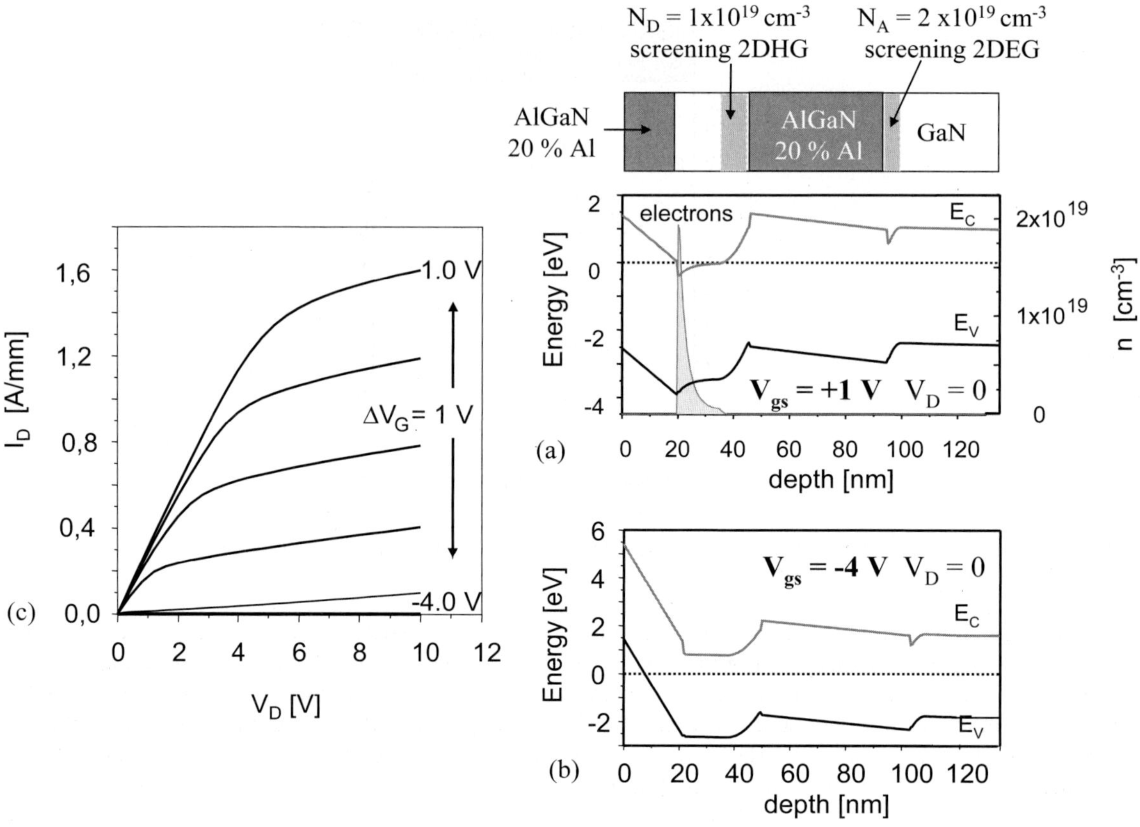

Fig. 26: (online colour at www.interscience.wiley.com) Heterostructure configuration and doping profile of double barrier AlGaN/GaN heterostructure. Hole-charge accumulation in the quantum well channel and the lower 2DEG-channel shown in fig. 8 are substituted by doping profiles.

    a)   Simulated band structure and charge distribution for a fully open channel (V_{GS}=1V) for the pinch-off case (V_{GS}=-4V). No p-channel is developed.

    b)   Simulated band structure for the pinch-off case (V_{GS}=-4V). No p-channel is developed.

    c)   Simulated output characteristic from open channel (V_{GS}=1V) to pinch-off (V_{GS}=-4V), after [19, 26].

Again the concept has been verified in a first experiment [21]. The CV-doping profile is shown in fig. 27.

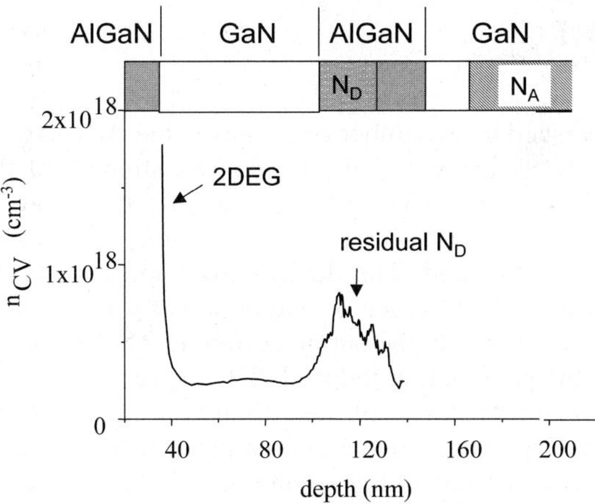

Fig. 27: Heterostructure configuration of the double barrier HFET and comparison with experimental CV doping profile. It can be seen that in the experimental structure the 2DEG at the rear AlGaN/GaN interface was undercompensated, a residual conducting path has remained, after [21].

The rear DEG channel has not been substituted in total and a residual parallel channel is observed, however the concept is verified. The substitution of the 2DEG sheet charge was attempted by a narrow acceptor profile of high peak concentration. This is difficult to realize and certainly not optimized at this time.

Again the device was tested under RF power conditions. This time shown is an experiment where the saturated RF-power level was applied after switching from both the pinch-off condition and the open channel condition into the identical class A bias point (fig. 28). No transients are observed and both levels coincide within a few percent. The drain bias point was close to the onset of saturation and the load line 50 ohm, taking into account the leakage currents and low breakdown voltages in the device. Self-heating was therefore only marginal.

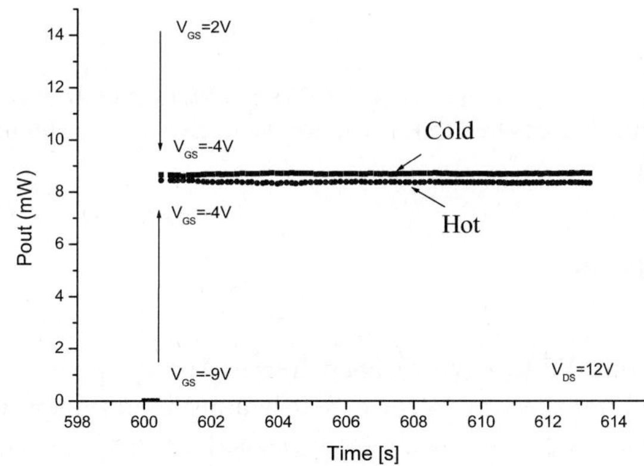

Fig. 28: 4 GHz saturated power switching experiment of an experimental double barrier AlGaN-HFET across a 50 ohm load line from the pinch-off condition and open channel condition as indicated. Device data: L_g = 0.5 µm, W = 100 µm, after [21]

9   AlN/GaN FET

One of the heterostructures persued by a number of groups is the AlN-barrier HFET. Here the polarization induced 2DEG-channel sheet charge and the breakdown strength of the gate barrier layer should reach their maximum value for this heterosystem. However, also the positive surface image charge density would be high and the device possibly highly unstable. Such structures have been grown and FET devices with 3 μm gate length fabricated. The devices exhibited essentially stable characteristics. No dispersion in the small signal transconductance or output power was observed (see fig. 29). What is the reason? Firstly it should be noted that both the output current level (200 mA/mm) and output power level (0.5 W/mm) were low. Thus, the polarization induced 2DEG sheet charge density was low. Two reasons contributed to this fact. Firstly, the thickness of the AlN barrier layer was beyond the critical thickness and may have started to relax. This would also relax the piezo field. At the same time this would introduce a high density of defects. This leads to the second point. The high defect density had introduced a high leakage current across the barrier, the conductivity shortciruciting the polarization field.
The stability was obtained at the expense of a widely short-circuited polarization field and thus low channel sheet charge. Thus, care has to be taken in the analysis of data, if the stress distribution and leakage across the polar dielectric layer are not known.

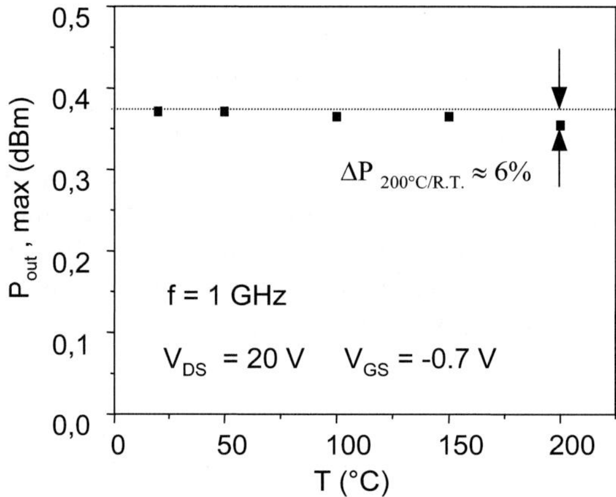

Fig. 29: Measured output power of a channel doped AlN/GaN FET. Only 6% reduction in output power at 1GHz is obtained by heating up to 200°C compared to RT. Device data: L_g=3μm, w=200μm, after [27].

10   Conclusion

AlGaN/GaN heterostructure FETs have reached impressive RF power densities, already reflecting the materials breakdown and transport properties. Self-heating is imminent but this problem has been well addressed and the thermal limit has already been pushed out not to mask the transport limit. However, this heterostructure system is also the first polar materials system used in a uni-polar device configuration. As a consequence of this dipole nature image charges need to be taken into account and the charging dynamics of these charge states need to be carefully considered. Indeed they seem to be the dominant source for the RF power slump phenomenon, which even now prevents to use the full power density available in these devices in the circuit environment. In contrast to many studies where this problem was addressed by improved surface passivation, here alternatives in the heterostructure arrangemen have been investigated. This has raised the problem of amibpolar charge distribution. To avoid ambipolar channel behavior the concept to screen the polarization field by doping profiles has been applied to GaN-FET structures for the first time. The structures investigated were the InGaN-channel HFET and the double barrier AlGAN HEMT. Indeed surface related charge instability have been essentially diminished.

Acknowledgements   We thank our supporting team namely Y. Men and U. Spitzberg for their work in electron beam lithography and RF characterization as well as F. Baloon and T. Zimmermann for supporting the processing. We would like to thank U.K. Mishra (UCSB) as well as A. Vescan and H. Leier (DaimlerChrysler) for stimulating discussions and N. Ngyuen (HRL) as well as D. Theron (IEMN) for some of the measurements undertaken in this comprehensive study.

References

[1]   E. Kohn, GAAS ´99, Munich, Germany, Proceedings, 1999, pp. 240-245.

[2]   C. Nguyen, Electronics Letters, vol 35, 1999, pp. 1380-1382.

[3]   I. Daumiller, IEEE Electron Device Letters, vol 22, 2001, pp. 62-64.

[4]   F. Bernadini, Physical Review B, vol 56, no. 16, 1997.

[5]   M. Neuburger, 26th Workshop on Compound Semiconductor Devices and Integrated Circuits held in Europe, Abstracts, Chernogolovka, Moscow, Russia, May 2002.

[6]   A. D. Bykhovski, Applied Physics Letters, vol 73, 1998, pp. 3577-3579.

[7]   J. P. Ibbetson, Applied Physics Letters, vol. 77, 2000, pp. 250-252

[8]   B. M. Green, IEEE Electron Device Letters, vol. 21, 2000, pp. 268-270

[9]   G. Simin, IEEE Electron Device Letters, vol. 23, 2002, pp. 458-460.

[10]   F. D. Sala, Applied Physics Letters, vol 74, 1999, pp. 2002-2004.

[11]   I. Daumiller, IEEE Electron Device Letters, vol. 20, 1999, pp. 448-450.

[12]   A. Pisch, Journ. Cryst. Growth, vol 187, 1998, pp. 329-332.

[13]   E. Kohn, GAAS '99, Munich (Germany), Oct. 1999, Proceedings 240-245.

[14]   I. Daumiller, 1997 IEEE/Cornell Conference on Advanced Concepts in High Speed Semiconductor Devices and Circuits, Cornell, USA, Proceedings, 1997, pp. 227-235.

[15]   P. Schmid, 9th European Heterostructure Technology Workshop (HETECH 99), IENM, University of Lille, Lille (France), Sept. 27-28 (1999), paper 2 in session E.

[16]   M. Seelmann-Eggbert, Diamond and Related Materials, vol. 10, 2001, pp. 744-749.

[17]   E. Kohn, Electronic Letters, vol 38, 2002, pp. 603-605.

[18]   E. Kohn, Electronics Letters, vol. 35, 1999, pp. 1022-1024.

[19]   R. Vetury, IEEE Tran. Electron Devices, vol. 48, 2001, pp. 560-566.

[20]   A. Vescan, Inst. Of Phys. (IoP), Series 166, 2000, pp. 503-506.

[21]   E. Kohn, IEEE MTTS , vol. 51, 2003, pp. 634-642.

[22]   O. Ambacher, Jorunal of Applied Physics, vol 87, 2000, 334-344

[23]   M. Neuburger, 7th International Workshop on Wide Bandgap III-Nitrides, Richmond, Virginia, March 2002.

[24]   E. Kohn, MRS 2001 Spring Meeting Proceedings, vol. 680E, no. E3.1, 2001.

[25]   E. Kohn, IWPSD, International Workshop on the Physics of Semiconductor Devices, Dehli (India), 2001.

[26]   M. Neuburger, phys. Stat. sol.(C), vol. 0, no. 1, 2002, pp-86-89

[27]   I. Daumiller, Electronics Letters, vol 35, 1999, pp. 1588-1590.

phys. stat. sol. (c) **0**, No. 6, 1940–1949 (2003) / **DOI** 10.1002/pssc.200303123

Gallium-nitride-based devices on silicon

A. Dadgar[*1]**, M. Poschenrieder**[1]**, I. Daumiller**[2]**, M. Kunze**[2]**, A. Strittmatter**[3]**, T. Riemann**[1]**,
F. Bertram**[1]**, J. Bläsing**[1]**, F. Schulze**[1]**, A. Reiher**[1]**, A. Krtschil**[1]**, O. Contreras**[4]**, A. Kaluza**[5]**,
A. Modlich**[5]**, M. Kamp**[5]**, L. Reißmann**[3]**, A. Diez**[1]**, J. Christen**[1]**, F.A. Ponce**[4]**, D. Bimberg**[3]**,
E. Kohn**[2]**, and A. Krost**[1]

[1] Otto-von Guericke Universität Magdeburg, Institut für Experimentelle Physik,
   Fakultät für Naturwissenschaften, Postfach 4120, 39016 Magdeburg, Germany
[2] University of Ulm, Department of Electron Devices and Circuits, 89069 Ulm, Germany
[3] Technische Universität Berlin, Institut für Festkörperphysik, Hardenbergstr. 36, 10623 Berlin, Germany
[4] Department of Physics and Astronomy, Arizona State University, Tempe, AZ 85287, USA
[5] Global Light Industries GmbH, Carl-Friedrich-Gauß-Str. 1, 47475 Kamp-Lintfort, Germany

Received 4 March 2003, revised 22 April 2003, accepted 22 April 2003
Published online 28 August 2003

PACS 68.37.Lp, 78.55.Cr, 78.60.Hk, 78.60.Fi, 81.15.Gh, 85.30.Tv, 85.60.Jb

GaN devices on Si are interesting for low-cost, high-power devices as LEDs and FETs. Until recently, most LED and FET devices suffered from cracking and low output power and additionally, from high series resistances for vertically contacted LEDs. Here, we give a brief overview on state of the art crack-free, bright LEDs with an output power up to 0.42 mW and AlGaN/GaN FETs with an output power of 2.5 W/mm at 2 GHz.

1 Introduction

The integration of high-speed and high-power III-V based devices with low-cost and well established Si electronics has been the motivation of a tremendous research effort during the last decades. The main problem for such devices is usually the high dislocation density generated at the Si/III-V layer interface. Several attempts as special buffer layer sequences [1, 2] and structured substrates [3] were partly successful and led to demonstrators of transistor and photodetector devices [2, 4] which are usually less sensitive to high dislocation densites. However, light emitters, especially laser devices on Si, show only lifetimes of more than a few days [5]. Group-III nitrides are much less sensitive to high dislocation densities and commercially available LEDs have a dislocation density in the range of 10^9 cm^{-2}. Thus, compared to the conventional III-Vs, GaN-based devices on Si have a much higher potential to be reliable. Here, another problem, which is also present for the other III-Vs, is of even bigger importance for device growth. Cracking due to a strong thermal mismatch is detrimental for all devices and a strong tension of the layer might lead to cracks at high power device operation.

But not only the integration with Si electronics is motivating GaN device growth on Si. At present, the growth of GaN involves the use of heterosubstrates as sapphire and SiC. Here the usage of low-cost, large size, and thermally and electrically well conducting Si is a very attractive alternative to these substrates which are either insulating or expensive.

[*] Corresponding author: e-mail: armin.dadgar@physik.uni-magdeburg.de, Phone: +49 391 67 11384, Fax: +49 391 67 11130

An overview on suited growth methods to achieve crack-free, high-quality GaN layers on Si by MOVPE is given in this volume of physica status solidi. Here, we give a brief overview on the state of the art of GaN devices on Si.

2 MOVPE growth of InGaN

The growth of high-quality ternary InGaN layers is prerequisite for devices like bright LEDs. In early growth studies the layers suffered from cracking which can already occur when cooling to InGaN growth temperature or from thin relatively rough buffer layers. With the growth of thicker, crack-free buffer layers the roughness and dislocation density can be reduced and the properties of ternary layers improved. We have performed a study on the growth of InGaN layers which shows that the InGaN quality on Si is as good as on sapphire [6]. The high interface quality obtained enables the measurement of superlattice fringes up to sixth order in X-ray diffraction measurements (XRD) (Fig.1) and also in X-ray reflectometry measurements interference from the InGaN/GaN superlattice can be well resolved (Fig. 2).

In a detailed photoluminescence (PL) study we demonstrated the high potential of InGaN QWs on Si for light emitters from blue to orange [6]. For a variation of the InGaN growth temperature or In/Ga ratio high intensities in the longer wavelength region are achieved. We observe an exponentially decreasing PL intensity for longer wavelengths and an intensity ratio for the blue to red-orange emission around 12:1, similar to samples grown on sapphire [7].

We also observe that an increase in QW thickness which is sometimes used to achieve long wavelength emission [8] does not yield as high luminescence intensities in the green region as the other two methods. This can be expected because of the strong decrease of the electron and hole wavefunction overlap for wider QWs due to the strong band bending in nitride semiconductors.

When comparing electroluminescence (EL) and PL spectra of vertically contacted LED structures we observe a redshift of the EL emission and no or only a very small blueshift at higher injection currents (Fig. 3) [9]. The missing blueshift is likely due to a homogenous current distribution in contrast to top contacted LEDs.

In a sample series where the QW deposition temperature was varied we also observe a decrease of the luminescence intensity which is stronger for the EL than for the PL with increasing wavelength (Fig. 3). The increase in wavelength corresponds to a decreased deposition temperature and with it an increased indium content of the QWs. The I–V spectra of the diodes show an increased scattering in the forward current characteristics which we attribute to leaks at the p-AlGaN cladding layer junction. With decreasing deposition temperature and increasing In-content these leakage paths and with them the loss of the LEDs increase. Thus, the growth of smooth high quality InGaN with high In content is very important for LED performance in the longer wavelength region.

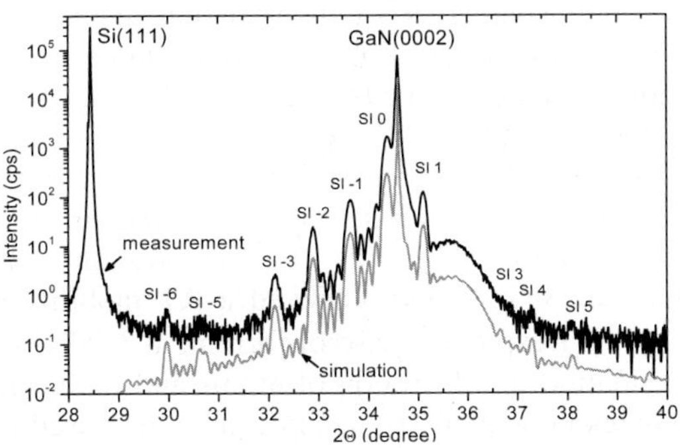

Fig. 1 XRD of a fivefold InGaN/GaN multi-quantum well with, $d_{Barrier} = 8.58$ nm, $d_{Well} = 3.87$ nm, and In = 13.4%.

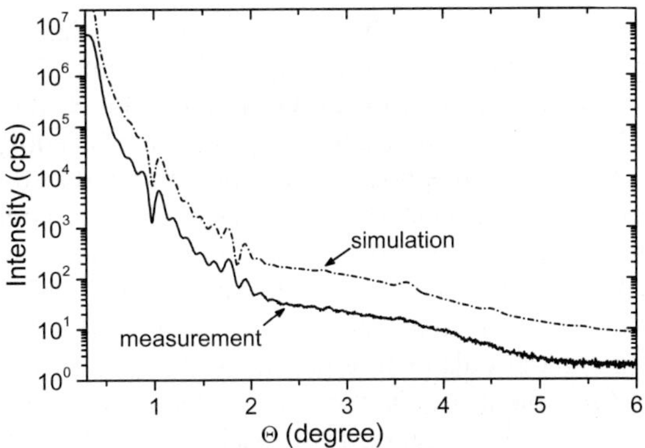

Fig. 2 X-ray reflectometry measurement of a five-fold InGaN/GaN MQW. The multilayer interference signal can be only observed for high quality interfaces

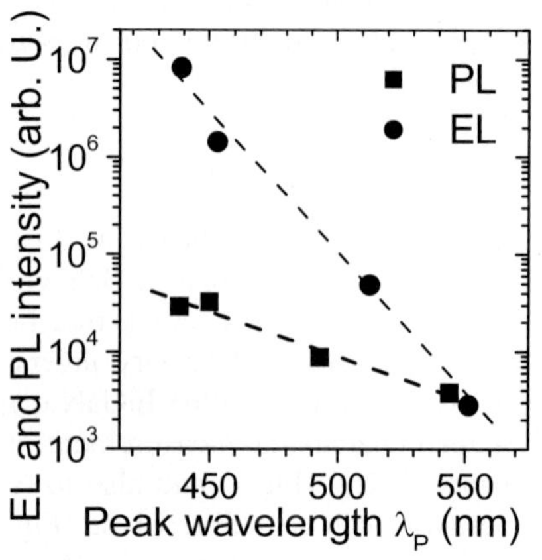

Fig. 3 Comparison of PL and EL intensities and wavelengths for different LEDs. EL decreases stronger with increasing wavelength likely due to a higher non-radiative carrier recombination at the p/n junction.

3 Devices

Optoelectronic devices on Si were the driving force of research efforts on III-Vs on Si in the last century 80s and 90s. In the case of the nitrides, first light emitting devices on Si showed very poor luminescence [10] and a lot of research efforts concentrated on electronic devices which are preferred on substrates with a good thermal conductivity as Si or SiC and which do not necessarily require thick layers with low dislocation densities.

3.1 Transistors

Transistor devices as FETs are usually relatively simple to grow and well suited for Al-rich buffer layers since a good electrical insulation from the Si substrate is mandatory for good device performance. For such devices the use of GaN is interesting for two reasons: A high electron saturation velocity and a high breakdown voltage enable high speed devices in the region up to 100 GHz at high power densities. Therefore, it is very interesting for applications as, e.g. next generation mobile phone base stations or compact microwave generators which would benefit from the usage of low-cost, large-sized Si substrates instead of expensive semiinsulating SiC. For high-power devices the thermal conductivity of the substrate is a key point. Out of the three substrates for GaN growth sapphire, SiC, and Si, SiC is unbeatable with a thermal conductivity above 3 W/cm K. Sapphire on the other hand has a very poor thermal conductivity of only 0.5 W/cm K. The thermal conductivity of Si is about the same as for GaN and yields ~1.5 W/cm K. Thus Si is the preferred substrated for low-cost devices.

Undoped GaN on Si often has a carrier concentration below 10^{15} cm^{-3} which is much lower than for layers grown on sapphire. One possible explanation for this difference is a contamination of the GaN layer by the oxygen in the sapphire substrate. For Si it is usually believed that the substrate causes an unwanted Si doping of the GaN layer. But Si outdiffusion is likely blocked by the rapid formation of Si_xN_y thus only a very thin layer at the interface might be doped with Si.

By spontaneous polarization and strain induced piezo-fields a 2-D electron or hole gas forms at interfaces as AlGaN/GaN or InGaN/GaN. Without modulation doping sheet carrier densities around 10^{13} cm^{-2} and mobilities well above 1000 cm^2/Vs at room temperature can be achieved with such structures. By MBE, Semond et al. grew an AlGaN/GaN FET structure on Si(111) and achieved a RT mobility of 1620 cm^2/Vs with a sheet carrier density of 4×10^{12} cm^{-2} [11]. Despite the ease of forming a 2-DEG the piezo charges also cause problems at the surface since oxidation or adsorption of atoms and molecules modifies the charge at the polar surface and with it the channel conductivity which can be also used for gas sensing applications[12]. Additionally, for the application as high-temperature devices the contact stability is a problem.

Growth of most FETs on Si starts with an AlN seed layer followed by a thick GaN layer while in some cases the buffer layer starts with an AlGaN layer or graded AlGaN followed by a thick GaN layer.

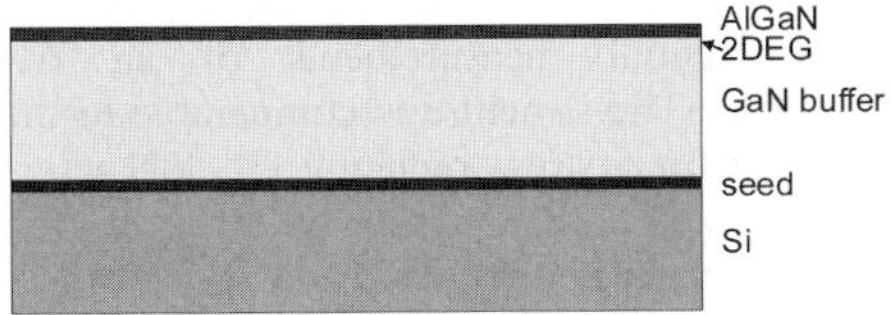

Fig. 4 Layer sequence of an FET on Si. Often, directly on the seed layer an Al-rich buffer layer is grown to improve insulation from the substrate and reduce stress. The 2-DEG is formed by the spontaneous polarisation and piezoelectric fields at the GaN/AlGaN interface and depends on AlGaN thickness and Al-content.

To obtain the two-dimensional electron gas a thin (10–30 nm) AlGaN layer is grown on top of the GaN layer (Fig. 4).

Such a simple FET structure with a 600nm GaN buffer and 35 nm of $Al_{0.32}Ga_{0.68}N$ was grown by Kaiser and coworkers. A sheet carrier concentration of 4×10^{12} cm^{-2} and a mobility of 820 cm^2/Vs were obtained but no device was presented [13].

A similar AlGaN/GaN heterostructure was presented by Schremer and co-workers [14]. They obtained sheet carrier densities in the range of $(0.9–1.3) \times 10^{13}$ cm^{-2} and mobilities ranging from 600 cm^2/Vs to 900 cm^2/Vs. Here, leakage currents to the Si substrate were observed which are a commonly reported problem.

Devices were presented by Egawa and coworkers who showed a MESFET using an AlGaN/AlN intermediate layer [15]. The transconductance and the drain–source current were relatively low with 25 mS/mm and 169 mA/mm, respectively.

Better values were obtained by Javorka et al. [16] who obtained a saturation current of 0.82 A/mm and a transconductance of 110 mS/mm. The devices with 0.3 µm gate length could dissipate 16 W/mm of static heat. In the meantime te same group presented a slightly higher transconductance and a saturation current of 0.91 A/mm and 122 mS/mm and a cutoff frequency of 12.5 GHz [17]. The electron mobilities were rather low (~700 cm^2/Vs) at moderate sheet carrier densities (6×10^{12} cm^{-2}) as determined by Hall effect measurements.

A graded AlGaN buffer layer for strain engineering and better isolation from the substrate was used by Marchand et al. to fabricate a FET [18]. Applying a graded AlGaN layer they could avoid parasitic current paths which were present at the Si interface and degrading device performance. A carrier mobility of 1428 cm^2/Vs and a sheet carrier density of 8.5×10^{12} cm^{-2} was achieved leading to a maximum transconductance of 100 mS/mm and a maximum saturated current of 525 mA/mm.

First power FETs on Si were presented by NITRONEX Inc. who obtained an output power of 3.3 W/mm at 2 GHz [19]. At the same time an output power close to that value was obtained by us with a simple 1.3 µm thick structure as in Fig. 4 with one additional LT-AlN interlayer 400nm above the AlN seed layer [20]. We measured 2.5 W/mm output power at 2 GHz (Fig. 5) and 1 W/mm at 10 GHz [21].

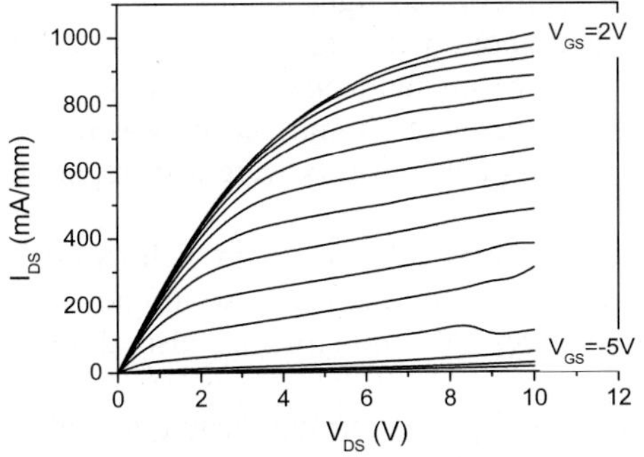
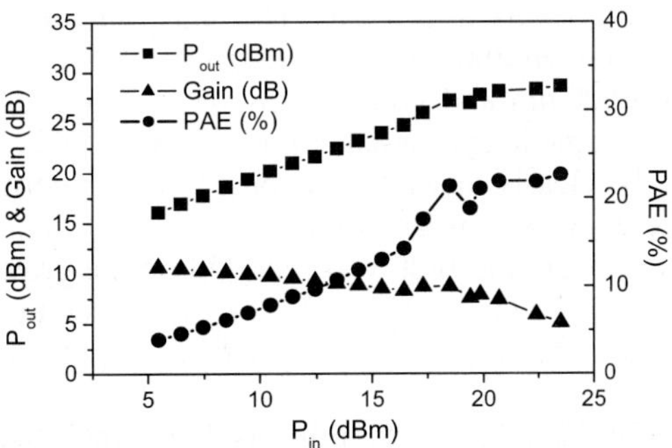

Fig. 5 DC (left) and HF (right) characteristics of an AlGaN/GaN FET on Si. The device structure consists of a 25nm AlN seed layer, 400 nm GaN, 12 nm LT-AlN , 850 nm GaN and 25 nm $Al_{0.25}Ga_{0.75}N$ on a highly resistive p-Si substrate. Device parameters: L_{Gate} = 0.25 µm, W = 200 µm (left); L_{Gate} = 0.5 µm, W = 300 µm, P_{out} = 2.5 W/mm (right)

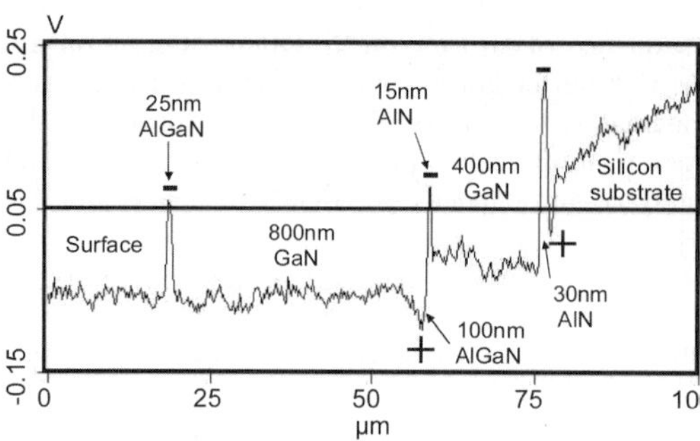

Fig. 6 Surface potential measurements of an Al-GaN/GaN FET on Si. The structure is comparable to the one in Fig. 4 except for a stress reducing LT-AlN inter-layer followed by ~100nm AlGaN in the buffer layer. The first negative charge peak below the 25 nm AlGaN layer is due to the 2-D electron gas of the top AlGaN/GaN channel. All other charges are unwanted and detrimental for device performance. These charges and with it leakage currents in the buffer layer limit I_{DS} to around 100 mA/mm for this device. For the surface potential measurements a bevel of ~1° was fabricated by polishing and ion etching. By this the cross section of the sample can be investigated in higher resolution.

The devices showed a maximum transconductance close to 300 mS/mm and a maximum I_{DS} of ~1 A/mm. Best room temperature mobilities achieved for AlGaN/GaN FETs were 1590 cm^2/Vs at a sheet carrier density of 6.7×10^{12} cm^{-2} [20, 21].

A common problem for GaN-based FETs on Si is a too thin buffer layer possibly leading to a poor electrical insulation from the Si substrate and for thick layers the buffer layer scheme used to avoid cracking. Because strain engineering involves Al-rich buffer layers piezocharges are generated there leading to parasitic currents. Surface potential measurements of a FET structure show the potential problems of such Al-rich layers (Fig. 6). When a graded AlGaN layer is applied as, e.g. described by Marchand et al. [18], this can lead to a positive charge by the piezofields which some authors use as a method of doping [20].

One solution of this problem might be doping with Fe which is known to be a deep acceptor in GaN and was recently reported to be easily applicable to GaN growth [22].

3.2 Light emitting devices

Guha and Bojarczuk [10] reported the first GaN-based light emitting diode on Si which was grown by MBE on n-type Si(111). The diodes started light emission at 4.5–6.5 V. At 12 V the forward currents varied from 14 to 65 mA. These rather high voltages as compared to MOVPE grown devices on sapphire or SiC were attributed to a low p-type doping and non-optimal p-type contacts. A device with a thin Si doped GaN layer showed a near band edge electroluminescence at 360 nm with a FWHM of 17 nm, and a broad long wavelength tail that extended out into the visible spectral range, while a heterostructure with an undoped GaN layer showed a broad emission band centered at 420 nm most probably due to deep radiative levels in the gap. The same authors reported on multicolored light emitters on silicon substrates using similar violet MBE-grown GaN LEDs as described above with higher Al-content [23]. Orange at ~600 nm and green-yellow at ~530 nm electroluminescence on the same Si wafer was obtained using organic dye-based color converters. The „visible part of the electroluminescence was bright enough to be clearly observed by the eye under normal room illumination". Cracks were also reported despite these layers were grown by MBE.

Tran et al. were the first who reported on the growth of InGaN/GaN multiple quantum well (MQW) blue LEDs on Si(111) by MOVPE [24]. The structure showed blue electroluminescence at 465 nm with light emission starting at 4 V. An optical output power was not given and the structure also showed cracks.

Yang et al. [25] fabricated a thin (<1 µm) InGaN/GaN MQW LED by a patterning technique with a combined MBE/MOVPE growth procedure in 300 × 300 µm^2 fields defined by openings in a SiO$_2$ mask. The density of cracks was comparable to similar structures on flat SiC substrates. For the LED a forward turn-on voltage of 3.2 V was measured. The forward differential resistance was a factor of four higher than for comparable LEDs on sapphire substrate. At room temperature the device emitted at 465 nm.

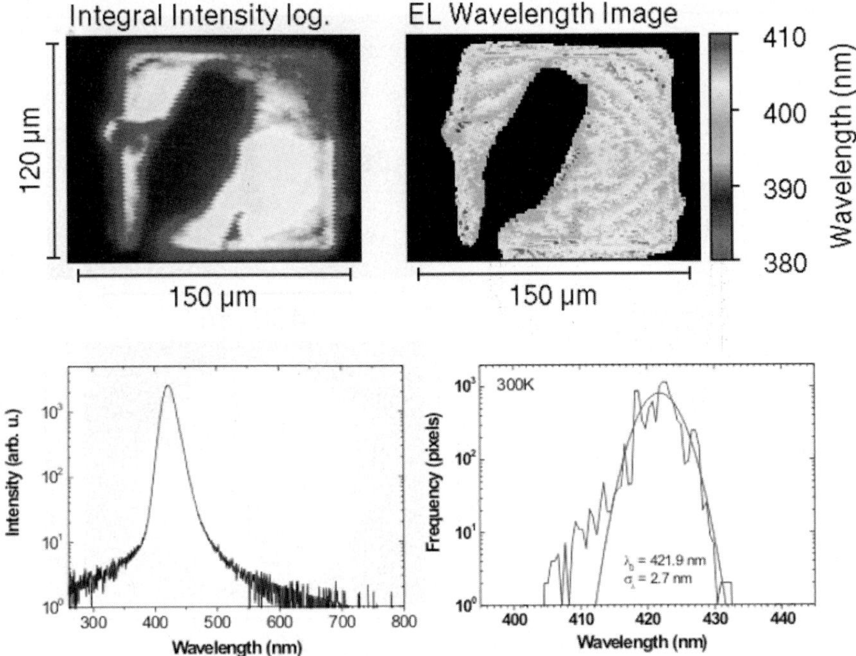

Fig. 7 Scanning µ-electroluminescence measurement of a 100×100 µm^2 LED structure on patterned Si(111) [32] showing the intensity (top left) and peak emission wavelength (top right). The sample was contacted with a bond wire covering about 25% of the top surface (dark spot). The emission wavelength is rather homogenous mostly modulated by interference fringes from thickness modulations. The peak emission wavelength is 421.9 nm with $\sigma = 2.7$ nm as determined from the histogram of the peak intensities (bottom right).

Problems with high series resistance can be usually attributed to an Al-rich buffer layer which is commonly applied for strain engineering [26–30]. However, even when Al-rich buffer layers are used a significant reduction of the series resistance from values well above 200 Ω to values even below 100 Ω can be obtained. For example an LED with a relatively high output power of 20 µW at 20 mA and a series resistance around 100 Ω was reported by Egawa et al. [31]. Lifetime testing of these LEDs over 500 hours showed no degradation. For vertical contacts they further improved the series resistance and obtained a value of only 30 Ω but, at a lower optical output power of only 18 µW at 20 mA and 478 nm [27]. The improvement in series resistance was attributed to a possibly leaky Si/AlN interface region.

With the patterning technique we fabricated a crack-free 3.6 µm thick 100×100 µm^2 LED structure (Fig. 7) [32]. The diode showed bright electroluminescence (~100 µW) with an onset at 3.2 V and a series resistance of 350 Ω.

Using a nitrided AlAs buffer layer on n-type Si(111) a simple LED structure with a single InGaN/GaN quantum well was realized with (Al/Au) n-contacts on the back side and 10 nm Pt p-contacts on the front side of the sample [33]. Despite the formation of cracks and pits, which act as potential short circuits paths, and a high series resistance (20 mA at 12.5 V), the sample emitted at 439 nm, with the luminescence being well visible at daylight. A further improvement could be achieved by the introduction of a AlGaN/GaN multilayer in the buffer which reduces stress and can be also used as a bragg-reflector minimizing absorption losses Fig. 9 [34].

The technique of contacting cracked LEDs on Si was further improved with a two step metallisation process [35]. Here a photoresist is spinned onto the cracked surface and removed to the surface level. In this way the cracks are filled with photoresist. After this a transparent top metallisation is applied followed by a second photolithography step defining a net for a thick Au metallisation which bridges the

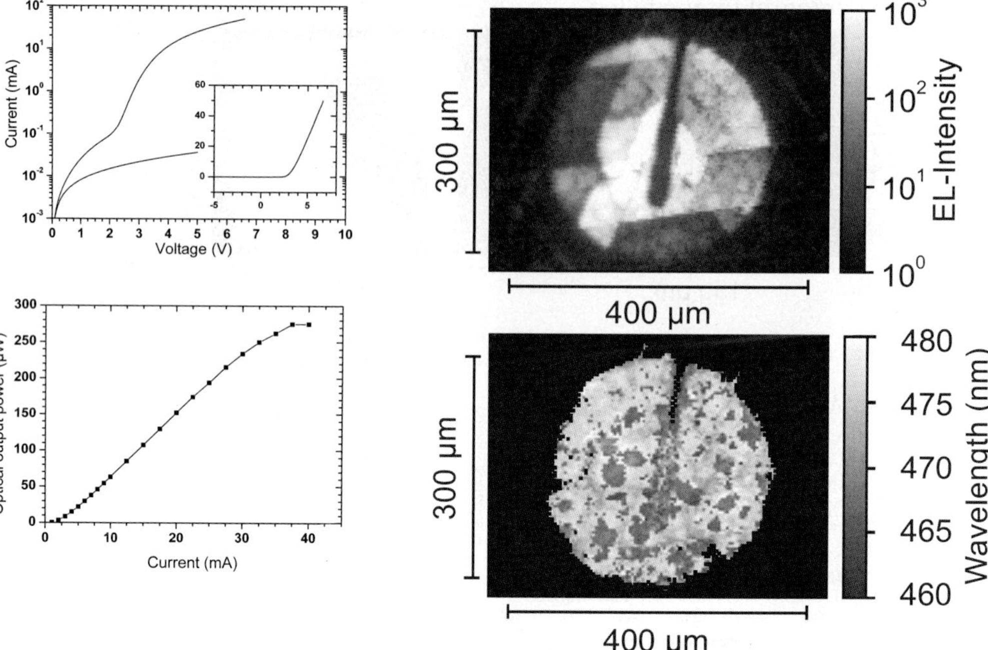

Fig. 8 *I–V* characteristics and optical output power of a vertically contacted LED on Si(111) in an epoxy package using two LT-AlN and one Si_xN_y in-situ mask for stress and dislocation reduction [36, 37].

Fig. 9 Panchromatic (top) and wavelength (bottom) μ-El images of a 2.3 μm thick LED with a 15-fold Al-GaN/GaN buffer layer. The diode was contacted using a transparent circular Pt top contact and a backside contact on the Si substrate. Due to cracks the GaN segments are not homogenously illuminated.

cracks and distributes the current homogenously. In the first experiments already 30–50% of the LEDs were functioning, however, with a low output power around 1 μW at 20 mA (Fig. 10).

Using LT-AlN interlayers for strain engineering and a Si_xN_y in-situ mask we succeeded in the growth of crack-free and thick (<2.8 μm) LED structures on 2" Si(111) wafers [37] (Fig. 10). Best output powers were 155 μW at 20 mA and 455 nm, which is already sufficient for simple indicator light applications. The resistivity could be reduced to 45 Ω [20]. Top contacted LEDs with an active region grown in a production type reactor show an output power of up to 0.42 mW at 20mA at 498 nm and low series resistance.

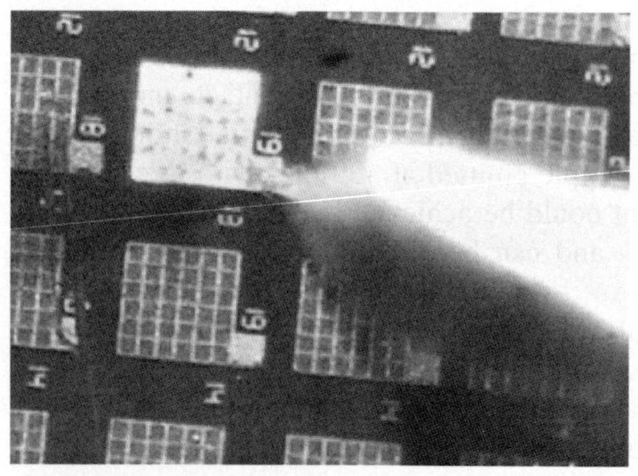

Fig. 10 (online colour at: www.interscience.wiley.com) Blue LED on cracked GaN on Si with transparent top metallisation and a thicker current distributing Au net on top.

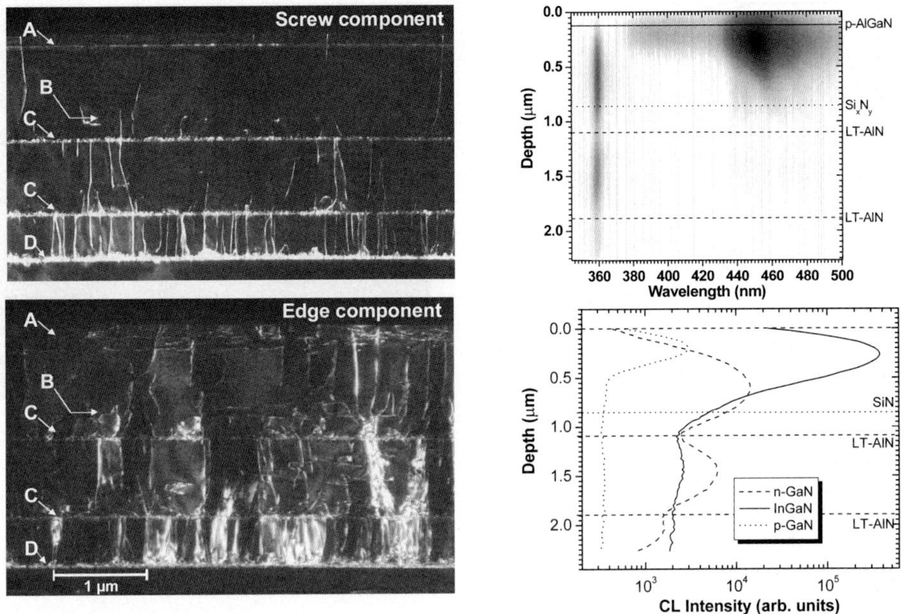

Fig. 11 Weak-beam TEM images using (0001) and $(1\bar{1}00)$ reflection to view screw type (top left) and edge type (bottom left) dislocations, respectively [36]. In the TEM image the reduction of dislocation density due to the LT-AlN interlayers (*C*) and the Si$_x$N$_y$ in-situ mask (*B*) can be clearly seen. *A* denotes the AlGaN:Mg/5× (InGaN/GaN:Si) active region and *D* the AlN seed layer. CL linescans taken at 5 K for the same cross section are plotted on the right side on the same scale. On the top-right, a spectrum linescan is plotted depicting the vertical evolution through the layer sandwich. On the bottom-right, intensity profiles are plotted for the n-GaN, p-GaN and InGaN luminescence, respectively.

Cross section TEM and CL measurements of such a diode structure clearly demonstrate the positive effect of the LT-AlN and Si$_x$N$_y$ in-situ masking (Fig. 11, 12) [37]. About an order of magnitude reduction in threading dislocation density to 10^8 cm^{-2} for screw-type and 10^9 cm^{-2} for edge-type dislocations is achieved. A comparison of the GaN (D^0,X) and the InGaN MQW luminescence of a sample with and without Si$_x$N$_y$ in-situ mask showed a strongly enhanced luminescence intensity of the GaN (D^0,X) and the InGaN luminescence for the sample with the Si$_x$N$_y$ mask [36, 37]. Further improvements of the device structure resulted in a maximum output power of 0.42 mW at 20 mA and 498 nm for top contacted LEDs [20] (Fig. 13). Lifetime testing over 100 hours at 80 °C and 20 mA current shows even a slight improvement in output power without degradation in *I–V* characteristics.

In the last four years since the demonstration of the first GaN based LED on Si device performance improved significantly. Early LEDs were rather dim, showed high series resistance and in most cases cracks. Most of these problems are solved today, especially, cracking is not a major problem any more since several concepts to avoid cracks exist [30, 32, 38]. Remaining problems are still higher series resistances for vertically contacted as compared to top contacted diodes and – a principal problem – a high optical loss. The latter is due to the angle of total reflection of around 24° and 38° for blue to green emission for the bare chip and when an epoxy housing is used, respectively. So, only between ~4% and ~10% of light are refracted out of the diode structure for an ideally smooth surface. Consequently, most of the light is absorbed in the nitride layer or the Si-substrate. Possible ways to reduce absorption are the usage of antireflection coatings, structuring of the surface, etching of facets or faceted holes to couple out the light [39], and the removal of the substrate after the device has been attached with the p-type contact, e.g. to a metal carrier. The latter is likely the ultimate way to fabricate high-brightness LEDs on Si since substrate removal is very easy, e.g. by simple wet chemical etching. Thus, the presently best values of 0.42 mW for LEDs on Si have to be looked at in a different way. First of all even for these LEDs no

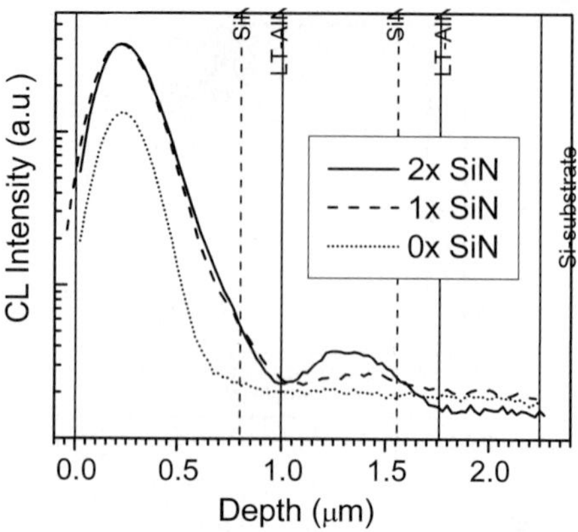

Fig. 12 Cathodoluminescence cross-section measurements of the InGaN luminescence of the three LED samples without, with one, and with two SiN masks. The samples with the SiN mask show a significantly enhanced InGaN luminescence and carrier diffusion length.

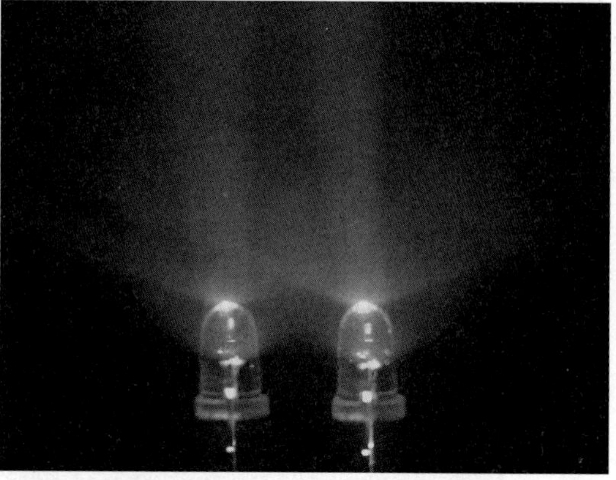

Fig. 13 (online colour at: www.interscience.wiley.com) Packaged bright, blue LEDs on Si.

optimized growth parameters were used. From a comparison with LEDs on sapphire we know that the growth on Si is slightly different especially the In content in the active layers seems to differ resulting in a redshift by ~20 nm. Also most results reported were grown in university research labs which do ususally not get higher output powers even on sapphire. Thus, we believe that the potential of GaN devices on Si is the same as on sapphire or SiC.

4 Summary

GaN based devices on Si have undergone dramatic improvements in recent years. FET devices are to go into market and even for LEDs soon a commercial release can be expected in the low cost market segment. Electrically pumped laser devices on Si have not been reported yet and will be a question of a significant reduction in dislocation density for uncracked films. Therefore, methods as ELOG, pendeo and cantilever epitaxy must be combined with stress reducing layers to achieve uncracked films with a low dislocation density. If no large homoepitaxial GaN substrates are available in the near future, Si substrates will certainly play a significant role in production of FETs and LEDs.

Acknowledgements   Part of this work has been funded by the Deutsche Forschungsgenmeinschaft under contract #Kr1239/10-1, #Ch87/4-1, and #Bi284/25-2. We acknowledge the support of A. Alam and M. Heuken from AIXTRON AG in the beginning of the project. O. Contreras permanent address is Centro de Ciencias de la Materia Condensada, UNAM, Ensenada, Baja California, CP 22860, Mexico.

References

[1] M. Grundmann, A. Krost, and D. Bimberg, Appl. Phys. Lett. **58**, 284 (1991).
[2] K. Eisenbeiser, R. Emrick, R. Droopad, Z. Yu, J. Finder, S. Rockwell, J. Holmes, C. Overgaard, and W. Ooms, IEEE Electr. Dev. Lett. **23**, 300 (2002).
[3] A. Krost, R.F. Schnabel, F. Heinrichsdorff, U. Rossow, D. Bimberg, H. Cerva, J. Cryst. Growth **145**, 314 (1994).
[4] E. Dröge, R.F. Schnabel, E.H. Böttcher, M. Grundmann, A. Krost, and D. Bimberg, Electron. Lett. **30**, 1348 (1994).
[5] see e.g.: Z.I. Kazi ,P. Thilakan, T. Egawa, M. Umeno, and T. Jimbo, Jpn. J. Appl. Phys. **40**, 4903 (2001).

[6] M. Poschenrieder, F. Schulze, J. Bläsing, A. Dadgar, A. Diez, J. Christen, and A. Krost, Appl. Phys. Lett. **81**, 1591 (2002).

[7] R.W. Martin, P.R. Edwards, R. Pecharroman-Gallego, C. Liu, C.J. Deatcher, I.M. Watson, and K.P. O'Donnell, J. Phys. D **35**, 604 (2002).

[8] B. Damilano, N. Grandjean, J. Massies, L. Siozade, and J. Leymarie, Appl. Phys. Lett. **77**, 1268 (2000)

[9] M. Poschenrieder, K. Fehse, F. Schulz, J. Bläsing, H. Witte, A. Krtschil, A. Dadgar, A. Diez, J. Christen, and A. Kros, phys. stat. sol. (c) 0, No.1, 267 (2002).

[10] S. Guha and N.A. Bojarczuk, Appl. Phys. Lett. **72**, 415 (1998).

[11] F. Semond, P. Lorenzini, N. Grandjean, and J. Massies, Appl. Phys. Lett. **78**, 335 (2001).

[12] J. Schalwiga, G. Müller, U. Karrer, M. Eickhoff, O. Ambacher, M. Stutzmann, L. Görgens, and G. Dollinger, Appl. Phys. Lett. **80**, 1222 (2002).

[13] S. Kaiser, M. Jakob, J. Zweck, W. Gebhardt, O. Ambacher, R. Dimitrov, A.T. Schremer, J.A. Smart, and J.R. Shealy, J. Vac. Sci. Technol. B **18**, 733 (2000).

[14] A.T. Schremer, J.A. Smart, Y. Wang, O. Ambacher, N.C. Mac Donald, and J.R. Shealy, Appl. Phys. Lett. **76**, 736 (2000).

[15] T. Egawa, N. Nakada, H. Ishikawa, and M. Umeno, Electr. Lett. **36**, 1816 (2000).

[16] P. Javorka, A. Alam, M. Nastase, M. Marso, H. Hardtdegen, M. Heuken, H. Lüth, and P. Kordoš, Electr. Lett. **37**, 1364 (2002).

[17] P. Javorka, A. Alam, M. Wolter, A. Fox, M. Marso, M. Heuken, H. Lüth, and P. Kordoš, IEEE Electr. Dev. Lett. **23**, 4 (2002).

[18] H. Marchand, L. Zhao, N. Zhang, B. Moran, R. Coffie, U.K. Mishra, J.S. Speck, S.P. DenBaars, and J.A. Freitas, J. Appl. Phys. **89**, 7846 (2001).

[19] A. Vescan, J.D. Brown, J.W. Johnson, R. Therrien, T. Gehrke, P. Rajagopal, J.C. Roberts, S. Singhal, A. Nagy, R. Borges, E. Piner, and K. Linthicum, to be published in phys. stat. sol.

[20] A. Dadgar, M. Poschenrieder, J. Bläsing, O. Contreras, F. Bertram, T. Riemann, A. Reiher, M. Kunze, I. Daumiller, A. Krtschil, A. Diez, A. Kaluza, A. Modlich, M. Kamp, J. Christen, F.A. Ponce, E. Kohn, and A. Krost, J. Cryst. Growth **248**, 556 (2003).

[21] A. Dadgar, M. Poschenrieder, J. Bläsing, O. Contreras, F. Bertram, T. Riemann, A. Reiher, M. Kunze, I. Daumiller, A. Krtschil, A. Diez, A. Kaluza, A. Modlich, M. Kamp, J. Christen, F.A. Ponce, E. Kohn, and A. Krost, Proc. of the State-of-the-Art Program on Compound Semiconductors XXXVII/Narrow Bandgap Opto-electronic Materials Devices, 2002; edited by P. C. Chang, W. K. Chan, D. N. Buckley, and A. G. Bac.

[22] S. Heikman, S. Keller, S.P. DenBaars, and U.K. Mishra, Appl. Phys. Lett. **81**, 439 (2002).

[23] S. Guha and N.A. Bojarczuk, Appl. Phys. Lett. **73**, 1487 (1998).

[24] C.A. Tran, A. Osinski, R.F. Karlicek, and I. Berishev, Appl. Phys. Lett. **75**, 1494 (1999).

[25] J.W. Yang, A. Lunev, G. Simin, A. Chitnis, M. Shatalov, M.A. Kahn, J.E. Van Nostrand, and R. Gaska, Appl. Phys. Lett. **76**, 273 (2000).

[26] M. Adachi, N. Nishikawa, H. Ishikawa, T. Egawa, T. Jimbo, and M. Umeno, IPAP Conf. Series **1**, 868 (2000).

[27] B.J. Zhang, T. Egawa, H. Ishikawa, N. Nishikawa, T. Jimbo, and M. Umeno, phys. stat. sol. (a) **188**, 151 (2001).

[28] H. Ishigawa, G.Y. Zhao, N. Nakada, T. Egawa, T. Soga, T. Jimbo, and M. Umeno, phys. stat. sol. (a) **176**, 599 (1999).

[29] H. Ishigawa, G.-Y. Zhao, N. Nakada, T. Egawa, T. Jimbo, and M. Umeno, Jpn. J. Appl. Phys. **38**, L492 (1999).

[30] E. Feltin, S. Dalmasso, P. de Mierry, B. Beaumont, H. Lahrèche, A. Bouillé, H. Haas, M. Leroux, and P. Gibart, Jpn. J. Appl. Phys. **40**, L738 (2001).

[31] T. Egawa, B. Zhang, N. Nishikawa, H. Ishikawa, T. Jimbo, and M. Umeno, J. Appl. Phys. **91**, 528 (2002).

[32] A. Dadgar, A. Alam, T. Riemann, J. Bläsing, A. Diez, M. Poschenrieder, M. Straßburg, M. Heuken, J. Christen, and A. Krost, phys. stat. sol. (a) **188**, 155 (2001).

[33] A. Dadgar, J. Christen, S. Richter, F. Bertram, A. Diez, J. Bläsing, A. Krost, A. Strittmatter, D. Bimberg, A. Alam, and M. Heuken, IPAP Conf. Ser. **1**, 845 (2000).

[34] A. Dadgar, A. Alam, J. Christen, T. Riemann, S. Richter, J. Bläsing, A. Diez, M. Heuken, and A. Krost, Appl. Phys. Lett. **78**, 2211 (2001).

[35] A. Strittmatter, L. Reißmann, and D. Bimberg, ISBLLED 2002, Cordoba (Spain).

[36] A. Dadgar, M. Poschenrieder, J. Bläsing, K. Fehse, A. Diez, and A. Krost, Appl. Phys. Lett. **80**, 3670 (2002).

[37] A. Dadgar, M. Poschenrieder, O. Contreras, J. Christen, K. Fehse, J. Bläsing, A. Diez, F. Schulze, T. Riemann, F.A. Ponce, and A.Krost , phys. stat. sol. (a) **192**, 308 (2002).

[38] A. Dadgar, J. Bläsing, A. Diez, A. Alam, M. Heuken, and A. Krost, Jpn. J. Appl. Phys. **39**, L1183 (2000).

[39] S.X. Jin, J. Li, J.Y. Lin, and H.X. Jiang, Appl. Phys. Lett. **77**, 3236 (2000).

Author Index

Akasaki, I. . . . . (c) 1750
Alves, H. (c) 1770
Amano, H. . . . . (c) 1750
Ambacher, O. . . . . . . (c) 1627, (c) 1878, (c) 1908
As, D. J. . . . . (c) 1607, (c) 1750

Baur, B. (c) 1908
Bechstedt, F. . . . . (c) 1732
Bernardini, F. . . . . (c) 1878
Bertram, F. . . . (c) 1583, (c) 1795, (c) 1835, (c) 1940
Bimberg, D. . . . . (c) 1583, (c) 1940
Bläsing, J. . . . . (c) 1583, (c) 1940
Blumenau, A. T. . . . . (c) 1684
Böttcher, T. . . . . (c) 1750, (c) 1846
Bühlmann, H.-J. . . . . (c) 1846

Chierchia, R. . . . . (c) 1846
Christen, J. . . . (c) 1583, (c) 1795, (c) 1835, (c) 1846
. . . . (c) 1940
Contreras, O. . . . . (c) 1583, (c) 1940
Correia, M. R. . . . . (c) 1750

Dadgar, A. . . . . (c) 1583, (c) 1750, (c) 1940
Darakchieva, V. . . . . (c) 1750
Daumiller, I. . . . . (c) 1919, (c) 1940
Dießelberg, M. . . . . (c) 1846
Diez, A. . . . . (c) 1583, (c) 1940

Eastman, L. F. . . . . (c) 1878
Eickhoff, M. . . . . (c) 1878, (c) 1908
Einfeldt, S. . . . . . . (c) 1750, (c) 1770, (c) 1846
Elsner, J. . . . . (c) 1684

Fall, C. J. . . . . (c) 1684
Figge, S. . . . . (c) 1846
Finger, T. . . . . (c) 1583
Fiorentini, V. . . . . (c) 1878
Fischer, P. . . . . (c) 1795, (c) 1835
Frauenheim, T. . . . . (c) 1684
Furthmüller, J. . . . . (c) 1732

Gerthsen, D. . . . . (c) 1668
Görgens, L. . . . . (c) 1908
Gutowski, J. . . . . (c) 1860

Haboeck, U. . . . . (c) 1710, (c) 1795
Hahn, E. . . . . (c) 1668
Hangleiter, A. . . . . (c) 1816, (c) 1860
Heggie, M. I. . . . . (c) 1684
Heinke, H. . . . . (c) 1770, (c) 1846

Hempel, T. . . . . (c) 1583
Heppel, S. . . . . (c) 1860
Hermann, M. . . . . (c) 1878, (c) 1908
Hoffmann, A. . . (c) 1710, (c) 1783, (c) 1795, (c) 1835
Hofmann, D. M. . . . . (c) 1770
Holst, J. . . . . (c) 1835
Hommel, D. . . . . . . (c) 1750, (c) 1770, (c) 1846

Ilegems, M. . . . . (c) 1846

Jenkins, T. . . . . (c) 1919
Jones, R. . . . . (c) 1684

Kaluza, A. . . . . (c) 1940
Kamp, M. . . . . (c) 1940
Kaschner, A. . . . . (c) 1783, (c) 1795
Kasic, A. . . . . (c) 1583, (c) 1750
Kelly, Michael K. . . . . (c) 1627
Köhler, U. . . . . (c) 1750
Kohn, E. . . . . (c) 1919, (c) 1940
Kröger, R. . . . . (c) 1846
Krost, A. . . . . (c) 1583, (c) 1750, (c) 1940
Krtschil, A. . . . . (c) 1583, (c) 1940
Kuhn, B. . . . . (c) 1750
Kunze, M. . . . . (c) 1919, (c) 1940

Leiter, F. . . . . (c) 1770
Link, A. . . . . (c) 1878
Lischka, K. . . . . (c) 1607

Meyer, B. K. . . . . (c) 1571, (c) 1770
Michler, P. . . . . (c) 1860
Mishra, U. . . . . (c) 1878
Miskys, Claudio R. . . . . (c) 1627
Modlich, A. . . . . (c) 1940
Monemar, B. . . . . (c) 1750
Müller, G. . . . . (c) 1908

Nanishi, Y. . . . . (c) 1750
Neubauer, B. . . . . (c) 1668
Neuberger, R. . . . . (c) 1908
Neuburger, M. . . . . (c) 1919
Neugebauer, J. . . . . (c) 1651

Off, J. (c) 1750

Pereira, S. . . . . (c) 1750
Petter, Ch. . . . . (c) 1846
Pfisterer, D. . . . . (c) 1770
Ponce, F. A. . . . . (c) 1583, (c) 1940

Poschenrieder, M. . . . . (c) 1583, (c) 1940
Potin, V. . . . . (c) 1668

Reiher, A.. . . . . (c) 1583, (c) 1940
Reißmann, L. . . . . (c) 1940
Riemann, T.. . . . . . . (c) 1583, (c) 1795, (c) 1835
. . . . (c) 1940
Rosenauer, A. . . . . (c) 1668
Röwe, M.. . . . . (c) 1860
Rudloff, D. . . . . (c) 1795, (c) 1846
Ryder, P. L.. . . . . (c) 1846

Saito, Y. . . . . (c) 1750
Schaff, W. . . . . (c) 1878
Schalwig, J.. . . . . (c) 1908
Schikora, D.. . . . . (c) 1607
Scholz, F.. . . . . (c) 1750
Schowalter, M. . . . . (c) 1668
Schubert, M. . . . . (c) 1583, (c) 1750
Schulze, F.. . . . . (c) 1940
Seyboth, M.. . . . . (c) 1919

Siegle, H. . . . . (c) 1710
Smorchkova, Y.. . . . . (c) 1878
Speck, J. . . . . (c) 1878
Steinhoff, G. . . . . (c) 1908
Strassburg, M. . . . . (c) 1835
Strittmatter, A. . . . . (c) 1583, (c) 1940
Stutzmann, M. . . . . . (c) 1627, (c) 1878, (c) 1908

Thomsen, C. . . . . . . (c) 1710, (c) 1783, (c) 1795
Tilak, V.. . . . . (c) 1878

Van Nostrand, J. . . . . (c) 1919
Vehse, M. . . . . (c) 1860
Veit, P.. . . . . (c) 1583

Wagner, G.. . . . . (c) 1750
Wagner, J.-M.. . . . . (c) 1732
Weidemann, O. . . . . (c) 1908

Zellweger, Ch. . . . . (c) 1846
Zhang, Wei. . . . . (c) 1571

Information for conference organizers and guest editors

The third journal section *physica status solidi (c) – conferences and critical reviews* is devoted to the publication of proceedings, ranging from large international meetings to specialized workshops, as well as collections of topical reviews on various areas of current solid state physics research. The new series has been launched in December 2002 and will appear with several issues in its starting volume **0** (2002/03). It is available both as an online journal and in hardcover print volumes, to be delivered to conference contributors and participants (upon arrangement with the organizers). Single copies of pss (c) may be ordered as a book using its ISBN number. Regular subscriptions to pss (c) will be offered from 2004. In 2003, a free trial access is available online through Wiley InterScience (www.interscience.wiley.com). Subscribers to pss (a) or (b) will receive free pss (c) sample copies.

Essential details concerning layout and organization of the new journal series are:

- pss (c) is published as a full hardcover-bound series, carrying a standard green-coloured cover design, individually adapted according to the organizers' request which includes conference designation, logo, names of Guest Editors etc.

- Proceedings issues contain all conference contributions which have been peer-reviewed and accepted by the Guest Editors. Upon special agreement between the pss journal editors and the Guest Editors, part of the conference papers may also be published simultaneously in an issue of pss (a) or (b). For all papers, strict criteria for journal publications, i.e. positive peer-review by independent referees, are obligatory. All papers are unambiguously citable as phys. stat. sol. (a), (b), or (c) journal articles and will be covered by standard reference databases.

- All articles are published online in PDF format at Wiley InterScience. Access for registered users (e. g. conference participants with special password) may be installed. The online version contains colour figures at no additional cost, regardless of their colour or black/white representation in print.

- The Editorial Office provides document templates and style files for Word and LaTeX, respectively, to be used by all authors, allowing an easy manuscript preparation and length estimate of their paper with respect to the page limits given by the organizers.

- The issue is completed by a table of contents in topical order, an author index, a preface, listings of conference committee members, organizers and sponsors, and any additional material, if desired.

- The usual service of the Editorial Office is available and includes support in the refereeing process, acceptance messages, PDF proofs (for typesetted papers), free PDF reprints (hardcopy reprints may be ordered) as well as individual communication with authors and organizers. The use of a Web-based software system for online submission and refereeing of papers is offered to Guest Editors.

- The editors of pss (c) aim at a timely, professional, and high-quality print and online publication of proceedings, typically within only four to six months after a conference.

- Various service packages for production are available, including either full typesetting of papers using electronic manuscript data or publication-ready delivery of manuscript files (prepared using the template/style files) by the organizers.

For further details as well as an individual offer for the publication of the proceedings of your forthcoming conference or of a special issue containing topical reviews, please contact the Editorial Office at pss@wiley-vch.de (for other contact information see the title page).